HANDBOOK
OF PROBIOTICS
AND PREBIOTICS

HANDBOOK OF PROBIOTICS AND PREBIOTICS

Second Edition

Edited by

YUAN KUN LEE
National University of Singapore, Singapore

SEPPO SALMINEN
University of Turku, Turku, Finland

A John Wiley & Sons, Inc., Publication

Copyright © 2009 by John Wiley & Sons, Inc. All rights reserved
Published by John Wiley & Sons, Inc., Hoboken, New Jersey
Published simultaneously in Canada

No part of this publication may be reproduced, stored in a retrieval system, or transmitted in any form or by any means, electronic, mechanical, photocopying, recording, scanning, or otherwise, except as permitted under Section 107 or 108 of the 1976 United States Copyright Act, without either the prior written permission of the Publisher, or authorization through payment of the appropriate per-copy fee to the Copyright Clearance Center, Inc., 222 Rosewood Drive, Danvers, MA 01923, (978) 750-8400, fax (978) 750-4470, or on the web at www.copyright.com. Requests to the Publisher for permission should be addressed to the Permissions Department, John Wiley & Sons, Inc., 111 River Street, Hoboken, NJ 07030, (201) 748-6011, fax (201) 748-6008, or online at http://www.wiley.com/go/permission.

Limit of Liability/Disclaimer of Warranty: While the publisher and author have used their best efforts in preparing this book, they make no representations or warranties with respect to the accuracy or completeness of the contents of this book and specifically disclaim any implied warranties of merchantability or fitness for a particular purpose. No warranty may be created or extended by sales representatives or written sales materials. The advice and strategies contained herein may not be suitable for your situation. You should consult with a professional where appropriate. Neither the publisher nor author shall be liable for any loss of profit or any other commercial damages, including but not limited to special, incidental, consequential, or other damages.

For general information on our other products and services or for technical support, please contact our Customer Care Department within the United States at (800) 762-2974, outside the United States at (317) 572-3993 or fax (317) 572-4002.

Wiley also publishes its books in a variety of electronic formats. Some content that appears in print may not be available in electronic formats. For more information about Wiley products, visit our web site at www.wiley.com.

Library of Congress Cataloging-in-Publication Data:
Lee, Y. K. (Yuan Kun)
  Handbook of probiotics and prebiotics / Yuan Kun Lee, Seppo Salminen. – 2nd ed.
     p. ; cm.
  Rev. ed. of: Handbook of probiotics / Yuan-Kun Lee ... [et al.].
  Includes bibliographical references and index.
  ISBN 978-0-470-13544-0 (cloth)
 1. Probiotics–Handbooks, manuals, etc. 2. Intestines–Microbiology–Handbooks, manuals, etc. 3. Food–Microbiology–Handbooks, manuals, etc. 4. Microorganisms–Therapeutic use–Handbooks, manuals, etc. I. Salminen, Seppo. II. Handbook of probiotics. III. Title.
  [DNLM: 1. Bacterial Physiology. 2. Food, Formulated. 3. Probiotics–pharmacology. 4. Probiotics–therapeutic use. QW 52 L482 2009]
  QR171.I6H36 2009
  664.001′579–dc22

                                                                         2008033278

Printed in the United States of America

10 9 8 7 6 5 4 3 2 1

CONTENTS

PREFACE   xv

CONTRIBUTORS   xvii

PART I   PROBIOTICS   1

1   Probiotic Microorganisms   3

   1.1   Definitions, 3
   1.2   Screening, Identification, and Characterization of *Lactobacillus*
       and *Bifidobacterium* Strains, 4
       1.2.1   Sources of Screening for Probiotic Strains, 5
       1.2.2   Identification, Classification, and Typing of
             Bifidobacterium Strains, 7
             1.2.2.1   Taxonomy, 7
             1.2.2.2   Identification and Typing, 8
       1.2.3   Identification, Classification, and Typing of
             Lactobacillus Strains, 14
             1.2.3.1   Taxonomy, 14
             1.2.3.2   Identification and Typing, 15
       1.2.4   Characterization of Probiotic Properties
             in *Bifidobacterium* and *Lactobacillus* Strains, 19
             1.2.4.1   Survival to GIT Stressing Conditions, 19
             1.2.4.2   Adhesion/Colonization to/of GIT, 23
             1.2.4.3   Antimicrobial Activity, 24
             1.2.4.4   Other Probiotic Properties, 24

1.2.5 Conclusion, 24
1.3 Detection and Enumeration of Gastrointestinal Microorganisms, 25
    1.3.1 Methods for Intestinal Microbiota Assessment, 25
        1.3.1.1 Culture-Dependent Methods, 25
        1.3.1.2 Culture-Independent Methods, 30
    1.3.2 Detection and Enumeration in Dairy Products, 37
    1.3.3 Detection and Enumeration of Specific Probiotics in the Gut, 38
    1.3.4 The Problem of the Viability and Physiological State of Intestinal Bacteria, 41
    1.3.5 Conclusions, 42
1.4 Enteric Microbial Community Profiling in Gastrointestinal Tract by Terminal-Restriction Fragment Length Polymorphism (T-RFLP), 43
    1.4.1 T-RFLP, 43
    1.4.2 Universal and Group-Specific Primers, 44
    1.4.3 Fluorescent Dyes, 44
    1.4.4 DNA Extraction, 46
    1.4.5 PCR Amplification, 46
    1.4.6 Generation of Terminal Restriction Fragments (TRF) by Digestion of Amplicons with Restriction Enzymes, 46
    1.4.7 Software and Data Processing, 47
    1.4.8 Microbial Diversity in Different Intestinal Compartments of Pigs, 47
    1.4.9 Tracking the Fate of Orally Delivered Probiotics in Feces, 48
    1.4.10 Conclusion, 51
1.5 Effective Dosage for Probiotic Effects, 52
    1.5.1 Acute (Rotavirus) Diarrhea in Children, 53
    1.5.2 Antibiotic-Associated Diarrhea, 54
        1.5.2.1 Combination of *L. acidophilus* + bifidobacteria or *Streptococcus thermophilus*, 54
        1.5.2.2 *L. rhamnosus* GG or *Saccharomyces boulardii* Applied Singly, 57
    1.5.3 *Helicobacter pyroli*, 58
1.6 Incorporating Probiotics into Foods, 58
    1.6.1 Probiotic Ingredients, 60
    1.6.2 Factors Affecting the Viability of Probiotics in Foods, 60
        1.6.2.1 Choice of Probiotic Organism/Food Combinations, 61
        1.6.2.2 Physiologic State of the Probiotic, 63
        1.6.2.3 Temperature, 63
        1.6.2.4 pH, 64
        1.6.2.5 Water Activity, 64
        1.6.2.6 Oxygen, 65
        1.6.2.7 Toxicity of Ingredients, 66

                1.6.2.8   Growth Factors, Protective, and
                          Synergistic Ingredients, 67
                1.6.2.9   Freeze–Thawing, 67
                1.6.2.10 Sheer Forces, 67
        1.6.3   Synbiotics, 67
        1.6.4   Delivery Systems, 68
                1.6.4.1   Microencapsulation, 68
                1.6.4.2   Delivery Devices, 69
        1.6.5   Probiotic Foods, 69
        1.6.6   Conclusions, 69
1.7 Safety of Probiotic Organisms, 75
        1.7.1   Current Proposals for Probiotic Safety, 77
        1.7.2   Taxonomic Identification, 79
        1.7.3   Pathogenicity, 81
        1.7.4   Antibiotic Resistance and Susceptibility, 83
        1.7.5   Immune Modulation, 87
        1.7.6   Clinical Studies, 90
        1.7.7   Postmarket Surveillance, 92
        1.7.8   GMO Probiotics, 93
        1.7.9   Conclusion, 94
1.8 Legal Status and Regulatory Issues, 95
        1.8.1   Human Probiotics, 95
                1.8.1.1   Asia, 95
                1.8.1.2   Europe, 106
                1.8.1.3   The United States of America, 111
        1.8.2   Animal Probiotics, 123
                1.8.2.1   United States, 123
                1.8.2.2   European Union, 123
                1.8.2.3   China, 125
                1.8.2.4   Japan, 125
                1.8.2.5   Korea, 125
                1.8.2.6   Thailand, 125
                1.8.2.7   Australia, 125
                1.8.2.8   New Zealand, 135
                1.8.2.9   Indonesia, Malaysia, Philippines, and Vietnam, 139
References, 139

2   Selection and Maintenance of Probiotic Microorganisms   177

2.1 Isolation of Probiotic Microorganisms, 177
2.2 Selection of Probiotic Microorganisms, 178
        2.2.1   Manufacturing Criteria (General Criteria), 179
        2.2.2   Shelf Life and Gut Transit (General Criteria), 179
                2.2.2.1   Shelf Life of Viable Probiotics Under Different
                          Storage Conditions, 179

    2.2.2.2 Tolerance to Digestive Juices, 180
    2.2.2.3 Adhesion and Colonization onto Specific Site of Body Surface, 181
  2.2.3 Health Properties (Specific Criteria), 181
  2.2.4 Safety, 182
  2.2.5 Identification, 182
 2.3 Maintenance of Probiotic Microorganisms, 184
References, 187

3 Genetic Modification of Probiotic Microorganisms   189

 3.1 Mutants Obtained from Probiotic Microorganisms by Random Mutagenesis, 189
 3.2 Plasmids, 202
 3.3 Vectors for Lactobacilli and Bifidobacteria, 212
 3.4 Genetic Recombination, 222
References, 229

4 Role of Probiotics in Health and Diseases   257

 4.1 Cell Line Models in Research, 259
 4.2 Laboratory Animal Models in Research, 263
 4.3 Effects on Human Health and Diseases, 267
  4.3.1 Nutritional Effects, 267
    4.3.1.1 Lactose Maldigestion, 268
    4.3.1.2 β-Galactosidase in Fermented Milk Products, 269
  4.3.2 Prevention and Treatment of Oral Infection and Dental Caries, 270
  4.3.3 Prevention and Treatment of Diarrhea, 272
    4.3.3.1 Acute (Rotavirus) Diarrhea in Children, 272
    4.3.3.2 Antibiotic-Associated Diarrhea, 276
    4.3.3.3 *Clostridium difficile* Associated Diarrhea, 279
    4.3.3.4 Radiation-Induced Diarrhea, 279
    4.3.3.5 Traveler's Diarrhea, 280
    4.3.3.6 Diarrhea in Tube-Fed Patients, 281
  4.3.4 Treatment of Irritable Bowel Syndrome, 282
  4.3.5 Prevention and Treatment of Inflammatory Bowel Diseases, 287
  4.3.6 Treatment of *H. pylori* Infection, 292
  4.3.7 Prevention of Postoperative Infections, 295
  4.3.8 Prevention and Treatment of Respiratory Tract Infections, 299
  4.3.9 Prevention and Treatment of Allergic Diseases, 302
  4.3.10 Antitumor Effects, 310
  4.3.11 Reduction of Serum Cholesterol, 313
  4.3.12 Enhancement of Vaccine Responses, 318

4.4 Effects on Farm Animals, 321
    4.4.1 Poultry, 322
    4.4.2 Swine, 323
    4.4.3 Ruminants, 331
    4.4.4 Rabbits, 339
    4.4.5 Pets, 339
References, 350

5 Mechanisms of Probiotics   377

5.1 Adhesion to Intestinal Mucus and Epithelium by Probiotics, 377
    5.1.1 Adhesion to Gastrointestinal Epithelial Cell Lines, 378
    5.1.2 Adhesion to Intestinal Mucus, 378
    5.1.3 Colonization of Probiotics in Human Intestine as Assessed by Biopsies, 379
    5.1.4 Comparisons Between *In Vitro* and *In Vivo* Results, 379
    5.1.5 Adhesins, 379
    5.1.6 Factors Affecting the Adhesion Properties of Probiotics, 379
    5.1.7 Adhesive and Inhibitory Properties of Nonviable Probiotics, 380
    5.1.8 Role of Age and Diseases on Adhesion, 383
5.2 Combined Probiotics and Pathogen Adhesion and Aggregation, 384
    5.2.1 Aggregation, 385
    5.2.2 Adhesion, 385
    5.2.3 Assay for Adhesion, 386
    5.2.4 Assay for Aggregation, 386
    5.2.5 Factors that Determine Adhesion, 389
    5.2.6 *In Vitro* Models, 389
    5.2.7 Probiotics in Combination, 390
    5.2.8 Conclusion, 391
5.3 Production of Antimicrobial Substances, 391
    5.3.1 Organic Acids, 392
    5.3.2 Hydrogen Peroxide, 392
    5.3.3 Carbon Dioxide, 393
    5.3.4 Bacteriocins, 393
    5.3.5 Low Molecular Weight Antimicrobial Compounds, 394
    5.3.6 Other Antimicrobial Agents, 394
5.4 Immune Effects of Probiotic Bacteria, 395
    5.4.1 The Neonatal Intestinal Microbiota, 395
    5.4.2 The Importance of the Intestinal Microbiota in Immune Development, 395
    5.4.3 Interaction of Commensal and Pathogenic Bacteria with the Intestinal Immune System, 396
    5.4.4 Probiotic Effects on Immune Responses, 396

- 5.4.5 Probiotic Effects on Epithelial Cells, 397
- 5.4.6 Probiotic Effects on DCs, 397
- 5.4.7 Probiotic Effects on Adaptive Immune Responses: T Helper Cells and T Regulatory Cells, 397
- 5.4.8 Delivery of Probiotic Bacteria, 398
- 5.4.9 The Specificity of Probiotic Effects, 399
- 5.4.10 Summary, 399
- 5.5 Alteration of Microecology in Human Intestine, 399
  - 5.5.1 Impact on Human Health: in Infants and the Elderly, 399
    - 5.5.1.1 Stepwise Establishment of Microbiota, 400
    - 5.5.1.2 Methodological Improvements in Microbiota Assessment, 401
    - 5.5.1.3 Microbiota After Infancy, 403
    - 5.5.1.4 Host–Microbe Cross Talk, 403
    - 5.5.1.5 Microbiota in the Elderly, 404
    - 5.5.1.6 Maintenance of Healthy Microbiota, 405
    - 5.5.1.7 Conclusion, 405
  - 5.5.2 Impact on Animal Health: Designer Probiotics for the Management of Intestinal Health and Colibacillosis in Weaner Pigs, 406
    - 5.5.2.1 The Farrowing Environment, 406
    - 5.5.2.2 The Weaning Environment, 406
    - 5.5.2.3 Colibacillosis in Pigs, 407
    - 5.5.2.4 Control of Colibacillosis, 408
    - 5.5.2.5 Mechanism of Action, 408
    - 5.5.2.6 Pathogenic and Commensal *E. coli*—the Concept of Gene Signatures, 409
    - 5.5.2.7 Mosaicism and Genome Plasticity in Porcine *E. coli* (Clone Gene Signatures), 410
    - 5.5.2.8 Population Gene Signatures in Epidemiological Study, 412
    - 5.5.2.9 Designer Lactic Acid Bacteria as Probiotics, 415
    - 5.5.2.10 Population Gene Signatures as a Measure of Probiotic Bioefficacy, 417
    - 5.5.2.11 Creation of Enteric Microbial Communities for Sustainable Intestinal Health (Probiosis), 419
- References, 421

6 Commercially Available Human Probiotic Microorganisms 441

- 6.1 *Lactobacillus acidophilus*, LA-5®, 441
  - 6.1.1 Gastrointestinal Effects, 441
    - 6.1.1.1 Intestinal Microbial Balance, 441
    - 6.1.1.2 Diarrhea, 442
    - 6.1.1.3 Other Gastrointestinal Effects, 442

        6.1.2   Immunomodulatory Effects, 443
                6.1.2.1   Nonspecific Immune Responses, 443
                6.1.2.2   Specific Immune Responses, 443
        6.1.3   Other Health Effects, 443
        6.1.4   Safety, 444
6.2   *Lactobacillus acidophilus* NCDO 1748, 444
        6.2.1   Origin and Safety, 445
        6.2.2   *In Vitro* and Animal Studies, 445
        6.2.3   Human Studies, 446
6.3   *Lactobacillus acidophilus* NCFM®, 447
        6.3.1   *L. acidophilus* NFCM Basic Properties, 447
        6.3.2   Survival of Intestinal Transit and Change in Intestinal Microbiota Composition and Activity, 447
        6.3.3   Lactose Intolerance, 448
        6.3.4   Relief of Intestinal Pain, 448
        6.3.5   Prevention of Common Respiratory Infections and Effects on Immunity, 449
        6.3.6   Application, 449
        6.3.7   Conclusion, 449
6.4   *Lactobacillus casei* Shirota, 449
        6.4.1   Effects on Intestinal Environment, 450
        6.4.2   Adhesive Property, 450
        6.4.3   Intestinal Physiology, 451
        6.4.4   Immunomodulation, 452
        6.4.5   Effects on Cancer, 453
        6.4.6   Prevention of Infectious Diseases, 454
        6.4.7   Prevention of Life Style Diseases, 454
        6.4.8   Clinical Application, 455
        6.4.9   Safety Assessment, 456
6.5   *Lactobacillus gasseri* OLL2716 (LG21), 457
        6.5.1   *Helicobacter pylori*, 458
        6.5.2   Selection of a Probiotic for *H. pylori* Infection, 458
        6.5.3   Effects of LG21 on *H. pylori* Infection in Humans, 458
        6.5.4   Mechanisms of Therapeutic Effects of LG21 on *H. pylori* Infection, 461
        6.5.5   Conclusion, 462
6.6   *Lactobacillus paracasei* ssp. *paracasei*, F19®, 462
        6.6.1   Identification and Safety, 462
        6.6.2   *In Vitro* Studies, 463
        6.6.3   Global Gene Expression, 463
        6.6.4   Human Studies, 464
6.7   *Lactobacillus paracasei* ssp *paracasei*, *L. casei* 431®, 466
        6.7.1   Adhesion and Survival Through the GI Tract, 466
        6.7.2   Gastrointestinal Effects, 466

    6.7.2.1 Intestinal Microbial Balance, 466
    6.7.2.2 Diarrhea, 466
  6.7.3 Immunomodulatory Effects, 468
  6.7.4 Other Health Effects, 468
  6.7.5 Safety, 468
 6.8 *Lactobacillus rhamnosus* GG, LGG®, 469
  6.8.1 Storage Stability, 469
  6.8.2 Gastrointestinal Persistence and Colonization, 469
  6.8.3 Health Benefits, 469
  6.8.4 Source of LGG®, 470
 6.9 *Lactobacillus rhamnosus*, GR-1® and *Lactobacillus reuteri* RC-14®, 470
  6.9.1 The Strains, 471
  6.9.2 *In Vitro* Properties, 471
  6.9.3 Animal Safety, Toxicity, and Effectiveness Studies, 471
  6.9.4 Clinical Evidence, 472
    6.9.4.1 Safety, Effectiveness, and Efficacy, 472
  6.9.5 Summary, 473
 6.10 *Lactobacillus rhamnosus* HN001 and *Bifidobacterium lactis* HN019, 473
  6.10.1 Basic Properties of *L. rhamnosus* HN001 and *B. lactis* HN019, 473
  6.10.2 Survival During the Intestinal Transit and Modulation of the Intestinal Microbiota, 474
  6.10.3 Modulation of the Immune System, 474
  6.10.4 Reduction of Disease Risk, 477
  6.10.5 Application, 477
  6.10.6 Conclusions, 477
 6.11 LGG®Extra, A Multispecies Probiotic Combination, 477
  6.11.1 Strain Selection for the Combination, 477
  6.11.2 Adhesion and Gastrointestinal Survival, 478
  6.11.3 Health Benefits, 478
  6.11.4 Technological Characteristics, 479
  6.11.5 Source of LGG®Extra, 480
 6.12 *Bifidobacterium animalis* ssp. *lactis*, BB-12®, 480
  6.12.1 Adhesion and Survival Through the GI Tract, 480
  6.12.2 Gastrointestinal Effects, 480
    6.12.2.1 Intestinal Microbial Balance, 480
    6.12.2.2 Diarrheas, 481
    6.12.2.3 Gastrointestinal Health of Infants, 482
    6.12.2.4 Other Gastrointestinal Effects, 482
  6.12.3 Immunomodulatory Effects, 483
    6.12.3.1 Nonspecific Immune Responses, 483
    6.12.3.2 Specific Immune Responses, 483
    6.12.3.3 Other Immunomodulatory Effects, 484

        6.12.4 Other Health Effects, 484
        6.12.5 Safety, 485
    6.13 *Bifidobacterium breve* Strain Yakult, 485
        6.13.1 Effects on Intestinal Environment, 485
        6.13.2 Intestinal Physiology, 485
        6.13.3 Effects on Cancer, 486
        6.13.4 Prevention of Infectious Diseases, 486
        6.13.5 Prevention of Life Style Diseases, 486
        6.13.6 Clinical Application, 487
    6.14 *Bifidobacterium longum* BB536, 488
        6.14.1 Evaluation of Safety of BB536, 488
        6.14.2 Physiological Effects of BB536, 489
            6.14.2.1 Improvement of Intestinal Environment, 489
            6.14.2.2 Effects on Immunity and Cancer, 490
            6.14.2.3 Antiallergic Activity, 490
        6.14.3 Technologies in BB536 Applications, 491
    6.15 *Bifidobacterium longum* Strains BL46 and BL2C—Probiotics for Adults and Ageing Consumers, 492
        6.15.1 Safety of BL2C and BL46, 492
        6.15.2 The Health Effects of BL2C and BL46, 493
            6.15.2.1 BL2C and BL46 Stabilize the Gut Function in the Elderly, 493
            6.15.2.2 Modulation of Gut Microbiota by BL2C and BL46, 493
            6.15.2.3 BL46 is Effective Against Harmful Bacteria, 493
            6.15.2.4 Effects of BL2C and BL46 on the Immune System and Infections, 493
            6.15.2.5 BL2C and BL46 Can Bind Toxic Compounds, 493
        6.15.3 Technical Properties and Sensory Qualities of BL2C and BL46, 494
        6.15.4 Conclusions, 494
    References, 494

PART II PREBIOTICS   533

7 Prebiotics   535

    7.1 The Prebiotic Concept, 535
    7.2 A Brief History of Prebiotics, 536
    7.3 Advantages and Disadvantages of the Prebiotic Strategy, 536
    7.4 Types of Prebiotics, 537
    7.5 Production of Prebiotics, 540
    7.6 Prebiotic Mechanisms, 546

7.7 Modulating the Intestinal Microbiota in Infants, 546
    7.7.1 Breast Milk, 546
    7.7.2 Infant Milk Formulas, 547
7.8 Modulating the Intestinal Microbiota in Adults, 548
    7.8.1 Effects at the Genus Level, 548
    7.8.2 Effects at the Species Level, 548
    7.8.3 Altering the Physiology of the Microbiota, 549
7.9 Modifying the Intestinal Microbiota in the Elderly, 549
7.10 Health Effects and Applications of Prebiotics, 549
    7.10.1 Laxatives, 550
    7.10.2 Hepatic Encephalopathy, 550
    7.10.3 Primary Prevention of Allergy in Infants, 551
    7.10.4 Amelioration of Inflammatory Bowel Disease, 551
    7.10.5 Prevention of Infections, 555
    7.10.6 Mineral Absorption, 556
    7.10.7 Prevention of Colorectal Cancer, 556
    7.10.8 Reduction in Serum Lipid Concentrations, 559
    7.10.9 Use in Weight Management and Improving Insulin Sensitivity, 559
7.11 Functional Foods for Animals, 559
7.12 Safety of Prebiotics, 560
7.13 Regulation of Prebiotics, 560
7.14 Conclusion, 561
References, 562

AUTHOR INDEX   **583**

SUBJECT INDEX   **585**

PREFACE

The first edition of the *Handbook of Probiotics* was published in 1999 when probiotics was still a relatively new scientific discipline. The idea of compiling a handbook came from our review article, "The coming of age of probiotics," published in the *Trends in Food Science and Technology* (61: 241–245, 1995), confirming probiotics to be a scientific discipline. The handbook was meant to serve as a source book for aspiring scientists, and it was the first handbook of its kind.

Probiotics have since developed into a major research focus area. Product applications include several commercially successful functional foods, health supplements, and therapeutic components and preparations. Cutting-edge methodologies, such as molecular approaches for the identification and quantification of intestinal probiotics, viability of probiotics under processing and storage conditions, and markers for host immune modulation, have been developed. Therefore, it is timely to update the scientific research and clinical trial data and to review and compile advances in methodology for easy reference.

At the time of publication of the first edition of the handbook, prebiotics were only at a concept level. Substantial research and clinical interventions on specific prebiotics have since been published to provide scientific basis for their reported effects. It is timely to include prebiotics in this updated handbook.

The aim of this updated handbook is to put together information and technology required in the development of a successful probiotic and prebiotic product from the laboratory to the marketplace. The book would continue to serve as a resource material for students, researchers, and company product development technologists.

This second edition of the *Handbook of Probiotics and Prebiotics* includes the following changes:

1. New chapters on methods for the analysis (enumeration, identification) of gastrointestinal microbiota.
2. The safety issue in novel probiotic bacteria is expanded, in view of the new regulation requirements for novel food products in Asia, Europe, and North America.
3. Understanding on probiotic mechanisms is incorporated in a new chapter.
4. A new chapter on commercially available human probiotic microorganisms covers in detail most of the early and new strains and preparations as well as the scientific information.
5. The chapter on "Enhancement of Indigenous Probiotic Organisms" is renamed as "Prebiotics" and expanded to accommodate the most recent findings.

YUAN KUN LEE

SEPPO SALMINEN

CONTRIBUTORS

Andrew W. Bruce, Canadian Research and Development Centre for Probiotics, Lawson Health Research Institute, Canada

Toni Chapman, Immunology and Molecular Diagnostic Research Unit (IMDRU), Elizabeth Macarthur Agricultural Institute, Australia

James J.C. Chin, Immunology and Molecular Diagnostic Research Unit (IMDRU), Elizabeth Macarthur Agricultural Institute, NSW Department of Primary Industries, Australia.
Immunology, John Curtin School of Medical Research, Australian National University, Canberra, Australia.
School of Veterinary Science, University of Queensland, Brisbane, Australia

M. Carmen Collado, Functional Foods Forum, University of Turku, Finland

Ross Crittenden, Food Science Australia, Australia

Fred H. Degnan, King & Spalding LLP, Washington, DC, USA

Clara G. de los Reyes-Gavilán, Instituto de Productos Lácteos de Asturias (CSIC), Spain

Diana Donohue, Toxicology Centre, School of Medical Sciences, RMIT University, Australia

Dorte Eskesen, Chr. Hansen A/S, Health & Nutrition Division, Denmark

Rangne Fondén, Finnboda Kajväg 15, Sweden.

Rafael Frias, Central Animal Laboratory, University of Turku, Finland

Miguel Gueimonde, Instituto de Productos Lácteos de Asturias (CSIC), Spain

Camilla Hoppe, Chr. Hansen A/S, Health & Nutrition Division, Denmark

Kajsa Kajander, Valio Ltd, R&D, Helsinki, Finland

Katsunori Kimura, Food Science Institute, Division of Research and Development, Meiji Dairies Corporation, Japan

Mayumi Kiwaki, Yakult Central Institute for Microbiology Research, Tokyo, Japan

Riitta Korpela, Valio Ltd, R&D, Helsinki, Finland

Sampo Lahtinen, Health & Nutrition, Danisco, Finland. Functional Foods Forum, University of Turku, Finland

J.M. Laparra, Department of Food Science, Cornell University, Ithaca, NY, USA

Charlotte Nexmann Larsen, Chr. Hansen A/S, Health & Nutrition Division, Business Unit, Denmark

Yuan Kun Lee, Department of Microbiology, National University of Singapore, Singapore

Allan Lim, Kemin Industires (Asia) Pte Ltd, Singapore

Abelardo Margolles, Instituto de Productos Lácteos de Asturias (CSIC), Spain

Baltasar Mayo, Instituto de Productos Lácteos de Asturias (CSIC), Spain

Koji Nomoto, Yakult Central Institute for Microbiology Research, Tokyo, Japan

Päivi Nurminen, Health & Nutrition, Danisco, Finland

Arthur Ouwehand, Health & Nutrition, Danisco, Finland

Martin J. Playne, Melbourne Biotechnology, Australia

Gregor Reid, Canadian Research and Development Centre for Probiotics, Lawson Health Research Institute, Canada

Patricia Ruas-Madiedo, Instituto de Productos Lácteos de Asturias (CSIC), Spain

Jose M. Saavedra, Johns Hopkins University School of Medicine, Baltimore, MD, USA

Seppo Salminen, Functional Foods Forum, University of Turku, Finland

Reetta Satokari, Functional Foods Forum, University of Turku, Finland

Maija Saxelin, Valio Ltd, R&D, Helsinki, Finland

Ulla Svensson, Arla Foods, Sweden

Hai-Meng Tan, Kemin Industires (Asia) Pte Ltd, Singapore

Mimi Tang, Department of Allergy and Immunology, Royal Children's Hospital, Australia

William Hung Chang Tien, Lytone Enterprise Inc., Taiwan

Hirokazu Tsuji, Yakult Central Institute for Microbiology Research, Japan

Satu Vesterlund, Functional Foods Forum, University of Turku, Finland

Jin-Zhong Xiao, Food Science and Technology Institute, Morinaga Milk Industry Co., Ltd, Japan

PART I

PROBIOTICS

1

PROBIOTIC MICROORGANISMS

1.1 DEFINITIONS

YUAN KUN LEE

Department of Microbiology, National University of Singapore, Singapore

"Probiotics" is derived from Greek and means "prolife." It has been redefined throughout the years as more scientific knowledge and better understanding on its relationship between intestinal health and general well-being has been gained. The following are definitions of "probiotics" derived through times.

Lilly and Stillwell in 1965 (5) defined probiotics as "Growth promoting factors produced by microorganisms."

Parker in 1974 (7) suggested an interaction between microorganisms with the host: "Organisms and substances with beneficial effects for animals by influencing the intestinal microflora."

Fuller in 1989 (3) defined it as "A live microbial feed supplement which beneficially affects the host animal by improving its intestinal microbial balance."

Havenaar and Huis Int Veld in 1992 (4) said probiotics are "A mono- or mixed culture of live microorganisms which, applied to animal or man, affect beneficially the host by improving the properties of the indigenous microflora."

ILSI (International Life Sciences Institute) Europe Working Group (1998) (9): "A viable microbial food supplement which beneficially influences the health of the host."

Handbook of Probiotics and Prebiotics, Second Edition Edited by Yuan Kun Lee and Seppo Salminen
Copyright © 2009 John Wiley & Sons, Inc.

Diplock et al. in 1999 (1) puts it as
"Probiotic food is functional if they have been satisfactorily demonstrated to beneficially affect one or more target functions in the body beyond adequate nutritional effects, in a way that is relevant to either an improved state of health and well-being and/or reduction in the risk of diseases."

Naidu et al. in 1999 (6) said "A microbial dietary adjuvant that beneficially affects the host physiology by modulating mucosal and systemic immunity, as well as improving nutritional and microbial balance in the intestinal tract."

Tannock in 2000 (11) observed that long-term consumption of probiotics was not associated with any drastic change in the intestinal microbiota composition, and thus proposed an alternative definition: "Microbial cells which transit the GI tract and which, in doing so, benefit the health of consumer."

Schrezenmeir and de Vrese in 2001 (10) defined probiotics as "A preparation of a product containing viable, defined microorganisms in sufficient numbers, which alter the microflora (by implantation or colonization) in a compartment of the host and by that exert beneficial health effects in this host."

FAO/WHO (Food and Agriculture Organization and World Health Organization) (2001) (2) and Reid et al. (2003) (8) concentrated exclusively on its health purpose: "Live microorganisms which when administered in adequate amounts confer a health benefit on the host."

1.2 SCREENING, IDENTIFICATION, AND CHARACTERIZATION OF *Lactobacillus* AND *Bifidobacterium* STRAINS

ABELARDO MARGOLLES, BALTASAR MAYO, AND PATRICIA RUAS-MADIEDO

Instituto de Productos Lácteos de Asturias (CSIC), Villaviciosa, Asturias, Spain

Several genera of bacteria (and yeast) have been proposed as probiotic cultures, the most commonly used are *Lactobacillus* and *Bifidobacterium* species. However, the selection of a strain to be used as an effective probiotic is a complex process (Fig. 1.1). The work begins with the source of screening of strains, the most suitable approach being the natural intestinal environment.

According to FAO/WHO guidelines it is necessary to identify the microorganism to species/strain level given that the evidence suggests that the probiotic effects are strain specific (60). It is recommended to employ a combination of phenotypic and genetic techniques to accomplish the identification, classification, and typing. For the nomenclature of bacteria, scientifically recognized names must be employed and it is recommended to deposit the strains in an internationally recognized culture collection. Further characterization of strains must be undertaken taking into account the "functional" or probiotic aspects and safety assessment. *In vitro* tests, some of them summarized in Fig. 1.1, are useful to gain knowledge of both strains and mechanisms of the probiotic effect. In addition, even if these genera have a long history of safe consumption in traditionally fermented products and several species have been awarded a

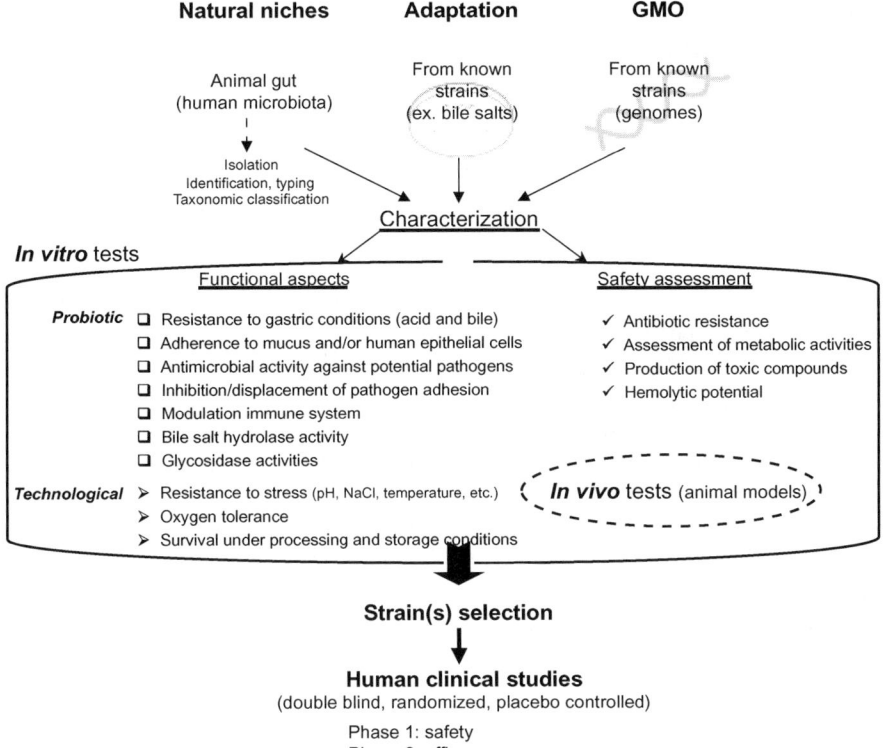

FIGURE 1.1 Procedure for the isolation and characterization of novel strains with putative probiotic status.

"General Recognised As Safe" (GRAS) status by the American Food and Drug Association (63) or a qualified presumption of safety (QPS) consideration by the European Food Safety Authority (EFSA) (59), some characteristics (Fig. 1.1) must be studied to ensure the safety of the novel lactobacilli and bifidobacteria strains. Several of the *in vitro* tests can be correlated with *in vivo* studies with animal models, but probiotics for human use must be validated with human studies covering both safety (phase 1 trials) and efficacy (phase 2 trials) aspects. Phase 2 studies should be designed as double-blind, randomized, and placebo-controlled to measure the efficacy of the probiotic strain compared with a placebo and also to determine possible adverse effects (60).

This chapter focuses on the current techniques for bacterial identification, taxonomic classification, and typing of *Lactobacillus* and *Bifidobacterium* strains, and also reviews the *in vitro* probiotic characterization of strains based on their functional aspects.

1.2.1 Sources of Screening for Probiotic Strains

Even though essentially all animals contain strains of both *Lactobacillus* and *Bifidobacterium* genera, it is well accepted that an effective human probiotic should be of human origin. The underlying reason for this is that human intestines are sufficiently different from those of animals, such that the isolates suited to those environments would not necessarily be suited to the human intestine (121). The human gastrointestinal tract (GIT) is a very complex ecological niche and its bacteria inhabitants can achieve the highest cell densities recorded for any ecosystem. Nonetheless, diversity at a division level is among the lowest (19) and the lactobacilli and bifidobacteria comprise less than 5% of the total microbiota (92). A number of articles have been published in the last few years studying the diversity of the GIT ecosystem employing several culture-independent genetic tools. But, for the isolation of novel strains, classical cultivation techniques must be employed. Enrichment, selective media, and specific culture conditions are employed for the isolation of strains from human samples that are initially identified by morphological characterization under the microscope. Molecular tools, mainly based on the sequencing of the 16S rRNA gene, allow identification down to the species level. Using this basic scheme several collections of strains have been isolated from human (and other animal) samples. Commonly, fecal samples are donated by healthy adult or infant volunteers (49, 156). But other GIT sections obtained from healthy individuals and patients submitted to biopsies such as the terminal ileum (56) or colonic mucosa (49) can be screened. Also the oral cavity seems to be the origin of some allochthonous lactobacilli of the intestine (44). Recently, it has been indicated that the infant fecal microbiota reflects the bacterial composition of the breast milk (79, 101). Therefore, the natural microbiota of human milk could be proposed as a source for the isolation of novel probiotic bacteria.

Another approach to search for improved probiotic strains (Fig. 1.1) is the adaptation of wild types to the intestinal stressful conditions. After ingestion, the probiotic bacteria must survive the passage through the GIT and reach the colon in order to exert their beneficial effect. The low pH in the stomach and the high concentration of bile salts in the small intestine, which act as biological detergents disrupting the cell membrane, are the principal challenges that probitics must overcome (21). Margolles and coworkers (100) obtained sodium-cholate-resistant *Bifidobacterium* derivatives by exposure to gradually increasing concentrations of this compound. The resistant phenotype remained stable and promoted some physiological changes that improved the survival of the adapted bacteria into the colon environment (52). Similarly, Collado and Sanz (39) developed a method for direct selection of acid-resistant *Bifidobacterium* strains by prolonged exposure of human feces to stressful conditions. The recovered strains were intrinsically resistant to acid gastric conditions (pH 2.0) and also showed good tolerance to high concentrations of bile salts and NaCl. This cross-resistance between low pH and bile salts was previously described in bile-adapted strains (118). Several strains with improved tolerance to these and other stressful factors have been described in literature (34, 111, 130, 146) as a method of selecting lactobacilli and bifidobacteria strains with improved viability to GIT and technological conditions.

Finally, taking advantage of the genome sequences, novel strains with improved or "designed" probiotic characteristics can be constructed toward specific therapies (157, 165). However, the use of recombinant strains is still far from being applied in functional foods, at least in the European legal frame. Some *Bifidobacterium* strains have been genetically engineered for therapy against tumors after oral administration (74) and to fight against intestinal pathogens (114, 168).

Recombinant *Lactobacillus* strains are currently under study for the enhancement of the immune system (77, 78), treatment against *Helicobacter pylori* (41) and improvement of inflammatory colitis (76). Although the species *Lactococcus lactis* is generally not considered as a probiotic, recombinant strains have been constructed for the oral delivery of therapeutic molecules (87) for the treatment or alleviation of diverse diseases such as allergies (12) and colitis (164).

1.2.2 Identification, Classification, and Typing of *Bifidobacterium* Strains

1.2.2.1 Taxonomy
Microorganisms of the genus *Bifidobacterium* are nonsporeforming, nonmotile, and nonfilamentous rods, which can display various shapes, with slight bends or with a large variety of branchings, from which the most typical ones are slightly bifurcated club-shaped or spatulated extremities. They can be found singularly, in chains, in aggregates, in "V," or palisade arrangements when grown under laboratory conditions. They are strictly anaerobic, although some species can tolerate low oxygen concentrations, and they have a fermentative metabolism (151). Tissier described these bacteria at the beginning of the twentieth century (173). They were first included among the family *Lactobacillaceae*, but in 1924 Orla-Jensen proposed the reclassification of the species *Lactobacillus bifidum* into the new genus *Bifidobacterium* (151).

The species of the genus *Bifidobacterium* form a coherent phylogenetic group and show over 93% similarity to the 16S rRNA sequences among them (150). This genus is clustered in the subdivision of high G + C Gram-positive bacteria, and it is included in the phylum *Actinobacteria*, class *Actinobacteria*, subclass *Actinobacteridae*, order *Bifidobacteriales*, and family *Bifidobacteriaceae*. According to the DSMZ Bacterial Nomenclature database (http://www.dsmz.de/microorganisms/bacterial_nomenclature), the species included in the genus *Bifidobacterium* are 29: *B. adolescentis, B. angulatum, B. animalis, B. asteroides, B. bifidum, B. boum, B. breve, B. catenulatum, B. choerinum, B. coryneforme, B. cuniculi, B. dentium, B. gallicum, B. gallinarum, B. indicum, B. longum, B. magnum, B. merycicum, B. minimum, B. pseudocatenulatum, B. pseudolongum, B. psychraerophilum, B. pullorum, B. ruminantium, B. saeculare, B. scardovii, B. subtile, B. thermacidophilum,* and *B. thermophilum*. In turn two subspecies constitute the species *B. animalis* (subsp. *animalis* and *lactis*), *B. pseudolongum* (subsp. *globosum* and *pseudolongum*), and *B. thermacidophilum* (subsp. *thermoacidophilum* and *porcinum*), and the species *B. longum* is subdivided in three different biotypes (longum, infantis, and suis).

All the currently known *Bifidobacterium* isolates are from a very limited number of habitats, that is human and animal GITs, food, insect intestine, and sewage (65, 196). Among the strains most commonly found in human intestines and feces are those belonging to the species *catenulatum, pseudocatenolatum, adolescentis, longum, breve, angulatum, bifidum,* and *dentium*, and the typical species isolated from functional foods is *B. animalis* subsp. *lactis* (104); therefore, strains belonging to these species are the first target for health-promoting studies.

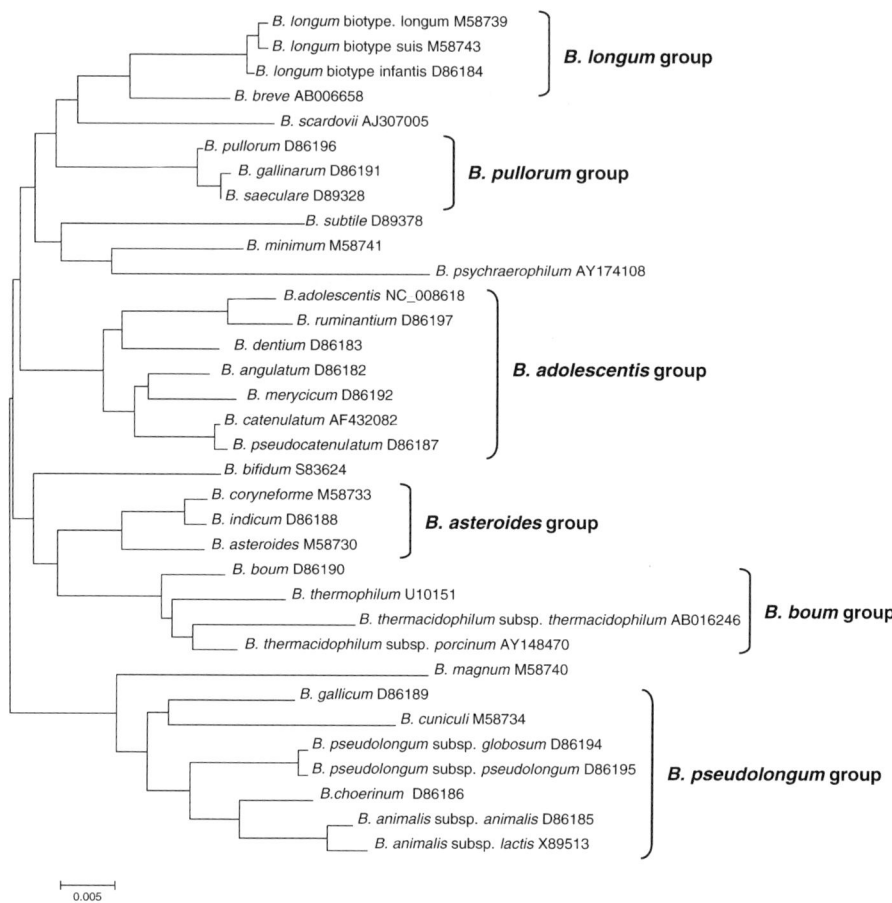

FIGURE 1.2 Evolutionary relationships of *Bifidobacterium* strains obtained using 16S rDNA sequences. The evolutionary distances were inferred using the neighbor-joining method and were computed using the maximum composite likelihood method. Units indicate the number of base substitutions per site. All positions containing gaps and missing data were eliminated from the dataset.

A number of phylogenetic studies carried out during the last few years (108, 148, 196, 200), mainly based on sequence comparison of total or partial sequences of the 16S rRNA genes and other housekeeping genes, have grouped the bifidobacterial species in six groups, *B. boum* group, *B. asteroides* group, *B. adolescentis* group, *B. pullarum* group, *B. longum* group, and *B. pseudologum* group (Fig. 1.2).

1.2.2.2 Identification and Typing Currently, there is great concern that the correct identification of a probiotic strain is the first prerequisite to be able to state its microbiological safety. Many studies have revealed deep deficiencies in the microbiological quality and labeling of currently marketed probiotic products for human and animal use. The incorporation of incorrectly identified probiotic bacteria in functional food products clearly has public health implications, by undermining the efficiency of probiotics and by affecting public confidence in functional foods (83). Thus, the use of adequate tools to provide proper strain identification for legal and good manufacturing practices, and to track probiotics during food production, as well as during their intestinal transit, are strictly necessary.

Traditionally, bifidobacteria have been identified on the basis of phenotype investigations. The host from which the bifidobacteria was isolated (e.g. *animalis, adolescentis, pullorum, dentium*, etc.) often represented the first identification criteria for many of these bacteria. Cell morphology, determination of metabolites, enzyme activities, and the ability to utilize sugars are the most commonly analyzed phenotypic characteristics for this genus, and until the 1960s the only identification criteria used. Specifically, the association of a branched shape with the presence of fructose-6-phosphate phosphoketolase (F6PPK) activity in a strain indicates that it belongs to the genus *Bifidobacterium* (20, 170, 196). However, several problems become apparent when the identification is carried out at species level, and the classical phenotyping, such as sugar fermentation profiles, transaldolase serotyping, cell-wall composition, and the study of the F6PPK isoforms, is clearly not discriminative enough to reach species, subspecies, and biotype level identification with confidence. Furthermore, these phenotypic methods suffer from a certain lack of reproducibility due to the culture conditions, metabolic status of the cells, and sometimes the lack of stability of the genetic determinants responsible for such phenotypes. As a matter of fact, most cases of probiotic misidentifications stem from the use of inappropriate phenotypic methods (83).

Mainly in the last decade molecular tools have been developed for identifying probiotics, based on the analysis of nucleic acids and other macromolecules because of the high potential provided by using polymerase chain reaction (PCR) amplification and hybridization with DNA and RNA (22). A summary of the molecular techniques used for identification and typing of potential probiotic bacteria is presented in Table 1.1.

The study of ribosomal rRNA genes (rDNA) is the most common methodology for bifidobacteria identification up to date. Bacterial ribosomes are formed from proteins and three ribonucleic acids: 5S RNA, 16S RNA, and 23S RNA. The rRNA genes are organized in *rrn* operons, bifidobacteria harboring from two to six depending on the species (152, 196). The 16S rDNA has nine variable regions (V1 to V9), and the three genes are separated by variable spacer regions. The detailed analysis of the 16S rDNA, as well as the 16S–23S spacer region (intergenic transcribed sequence, ITS), showed nucleotide fingerprints with different discriminatory levels. The 16S sequencing is being employed to discriminate all bifidobacterial species and their respective subspecies and biotypes (67, 106–108, 200), whereas the 16S-23S rRNA ITS sequence is much more variable than the 16S rRNA structural gene both in size and sequence, even within closely related taxonomic groups, which makes it a suitable target for both identification and typing by using species-specific primers (72); its analysis has a higher discriminatory capacity and allows the differentiation of different *Bifidobacterium* strains among the same species (94).

TABLE 1.1 Main Methods for Identification and Typing of *Bifidobacterium* and *Lactobacillus* Species

Methods[a]	Target	Outcome	Advantages	Limitations
Phenotyping	Cell metabolism	Identification and biotyping	Special equipment not required	Phenotypic variability, standardization of culturing and reading conditions
Protein profiles of whole cells	Whole-cell proteins	Identification	Easy to perform, high repeatability	Laborious, need of reference strains
FAME profiles	Fatty acids	Identification	Cheap	Special equipment and technical skill needed, mathematical treatment of the results
Species-specific PCR/ multiplex PCR	rRNA genes, end intergenic regions, other genes	Identification of single/ multiple strains, detection of microbial species/strains	Simple, highly reproducible, fast and sensitive, availability of primer sequences for most species, possibility of sequencing	Only few species can be detected at a time; unknown members are not identified; requires sequence data for specific primer design
ARDRA	rRNA operons	Identification, typing	Simple, digestion with many restriction enzymes	Availability of rRNA gene sequences
rRNA 16S sequencing	rRNA genes	Identification, phylogenetic relationships	Simple, sequencing of PCR fragments and sequence comparison, availability of many "universal" and group-specific primers	Absence of a clear definition of genetic species

Method	Target	Application	Advantages	Disadvantages
PCR amplification and sequencing of housekeeping genes	recA, groESL, dnaK, hsp60, tuf, etc.	Identification and typing	More resolution for closely related bacteria than the 16S rRNA sequences	Availability of sequences
Ribotyping	rRNA genes	Typing and identification	Automated (Riboprinter, Qualicon, DuPont)	Laborious except automated, high conservation of rRNA operons, expensive equipment
DGGE, TGGE	16S rRNA gene, other genes	Identification of bacteria, typing of microbial communities	Large numbers of samples can be analyzed simultaneously; superior taxonomic information regarding unknown bands	PCR biases; co-migration of different species; dominant species; heterogeneous copies of rDNA operons; lysis and extraction efficiency
PFGE	Whole genome	Typing	High resolution and reproducibility, use of many restriction enzymes	Laborious, time-consuming, analysis of small number of samples
AFLP	Whole genome	Typing	High reproducibility	Laborious, time-consuming, technical skill and high standardization needed, expensive
AP-PCR, RAPD	Whole genome	Typing and identification	Simple, easy to perform, many primers, cheap	Low repeatability, small number of bands
ERIC-PCR	Whole genome	Typing	Sensitivity and specificity, simple, simultaneous comparison of many samples	PCR biases, careful standardization of the technique
REP-PCR	Whole genome	Typing	Simple, PCR-based technique	PCR biases

(continued)

TABLE 1.1 (*Continued*)

Methods[a]	Target	Outcome	Advantages	Limitations
TAP-PCR	Whole genome	Typing	RADP variant, 18 mer oligonucleotide with a degenerated position at the 3′ end, more reproducibility	PCR biases
RISA, ARISA	rRNA intergenic spacer region	Typing of bacteria, fingerprinting of microbial community structure	Highly reproducible; automated	PCR biases; requires large quantities of DNA
Microarray hybridisation	16S rRNA genes, whole genome, specific genes	Bacterial detection and community analysis in complex environments	High power of identification; thousands of genes can be analyzed, automated analysis	Laborious; expensive; in early stage of development

[a] FAME: fatty acids methyl esther; ARDRA: amplified ribosomal DNA restriction analysis; DGGE, TGGE: denaturing gradient gel electrophoresis, termal gradient gel electrophoresis; PFGE: pulsed field gel electrophoresis; AFLP: amplified fragment length polymorphism; AP-PCR, RAPD: arbitrarily primed PCR, also know as random amplification of polymorphic DNA; ERIC: enterobacterial repetitive intergenic consencus sequence; REP-PCR: repetitive extragenic palindromic elements-PCR; TAP-PCR: triple arbitrary primed-PCR; RISA, ARISA: ribosomal intergenic spacer analysis, automated ribosomal intergenic spacer analysis.

Recently, more robust and powerful typing methods have been applied to *Bifidobacterium* species and strains, such as the multilocus sequence typing (MLST) scheme. The MLST method was first utilized for bacteria in 1998, and it made use of an automated DNA sequencing procedure to characterize the alleles present at different housekeeping gene loci (95). As it is based on nucleotide sequences, it is highly discriminatory and provides unambiguous results that are directly compared between laboratories. Several authors (103, 191, 192, 194, 195, 197–200, 208, 211), analyzed several gene sequences for detailed identification and classification purposes (*tuf, recA, xfp, atpD, groEL, groES, dnaK, hsp60, clpC, dnaB, dnaG, dnaJ1, purF, rpoC*). Other gene sequences (*pyk, tal*) have also been studied proving to be valuable for species and subspecies identification (137, 183).

Methods based on the PCR are widely used and allow the differentiation between strains of the same species and to some extent, also between species. By examining fingerprint patterns generated by amplification of DNA fragments these methods offer considerable potential for probiotic strain typing. The random amplification of polymorphic DNA (RAPD) technique uses short random sequence primers that are able to bind under low stringency to partially or perfectly complementary sequences of unknown location along the genome. Fingerprint patterns generated with this technique were useful to differentiate *Bifidobacterium* strains from human and food origin (50, 110, 201). Amplified ribosomal DNA restriction analysis (ARDRA) consists of the amplification of rDNA genes (totally or partially) and subsequent digestion with restriction enzymes, thus the choice of the enzyme(s) is critical for the discriminatory power. Species-specific identification is usually achieved with this technique (89, 143, 185, 188, 190), although *B. animalis* subsp. *lactis* and *B. animalis* subsp. *animalis* can also be distinguished (191). ERIC (enterobacterial repetitive intergenic consensus sequence)-PCR, and REP (repetitive extrogenic palindromic)-PCR examine specific patterns of repetitive DNA elements. ERIC sequences are 126-bp inverted repeats and REP sequences are short DNA fragments (between 21 and 65 bases) detected in the extragenic space; both are dispersed throughout the bacterial genomes (193, 174). The application of ERIC-PCR for bifidobacterial identification at species and subspecies level has been reported (158, 189), and REP-PCR can be considered as a promising genotypic tool for the identification of bifidobacteria potentially up to strain level (102, 103).

Although the aforementioned PCR techniques are the most common methods for identification and typing *Bifidobacterium* strains, other PCR-based approaches include TGGE/DGGE (temperature gradient gel electrophoresis/denaturing gradient gel electrophoresis) (62, 99, 172), Amplified fragment length polymorphism (AFLP), PCR coupled to enzyme-linked immunosorbent assay (ELISA), triplicate arbitrarily primer (TAP)-PCR, restriction fragment length polymorphism (RFLP)-PCR (48, 147), and multiplex PCR (26, 55, 91, 116, 187) to some extent have been utilized to type bifidobacteria to species, subspecies or strain level (Table 1.1).

Some methods for bifidobacteria identification and typing using total (or partial) DNA profiles, including plasmid analysis and RFLP of total DNA, have been used (14, 27, 142). However, pulsed-field gel electrophoresis (PFGE), which involves the digestion of genomic DNA with rare-cutting restriction enzymes and the subsequent separation of the macrofragments through a continuously reorienting electric field, is often considered by microbiologists the best technique for strain-specific typification. PFGE protocols have been established for different *Bifidobacterium* species (29, 110, 159, 207) and have shown a high discriminatory power to differentiate, for example, *B. animalis* subsp. *lactis* strains, which are often not discriminated using other methodologies due to the close genetic background among strains (66).

The ribotyping is the most popular and widespread hybridization method for bacterial typing. It combines southern hybridization of genomic DNA restriction patterns with rDNA probes. Furthermore, the availability of commercial systems allows the analysis of a wide range of bacteria in an automated manner. Although it is generally believed that ribotyping has a lower discriminatory power than the PFGE analysis (150), it has been extensively used for bifidobacterial typing (88, 97, 98, 110, 148).

Among these techniques, southern blot and microplate blot have also been used to type *Bifidobacterium* strains (97, 205), and microarray hybridization has arisen during the recent years as a valid alternative to discriminate between *Bifidobacterium* species (203), although the need for specific equipment and specialized personnel for the analyses severely limits its current applicability as an ordinary method of probiotic identification.

Finally, it is worthwhile pointing out that other methods have also been applied to the identification and typing of *Bifidobacterium*. The chromatographic analysis of organic acids (93) appears to be a useful tool for rapid identification of *Bifidobacterium* spp. at the genus level. Also, the analysis of the intrinsic fluorescence of aromatic amino acids (16) was shown to be an inexpensive and convenient means of rapidly identifying intestinal bifidobacteria, which could be of help for large probiotic surveys.

1.2.3 Identification, Classification, and Typing of *Lactobacillus* Strains

1.2.3.1 Taxonomy The genus *Lactobacillus* is the largest group among the lactic acid bacteria (LAB) containing, at present, more than 120 species and 20 subspecies (http://www.dsmz.de/microorganisms/bacterial_nomenclature_info.php?genus= LACTOBACILLUS (65, 150)); though its number increases every year (13 new species have been proposed in 2005, 9 in 2006, and 7 in 2007 up to the time of writing; http://www.ncbi.nlm.nih.go/sites/entrez). The lactobacilli are a broad, morphologically defined group of Gram-positives, nonspore-forming rods or coccobacilli with a G+C content usually below 50 mol% (86). Lactobacilli are clustered in the subdivision of low G+C Gram-positive bacteria, and are included in the phylum *Firmicutes*, class *Bacilli*, order *Lactobacillales*, and family *Lactobacillaceae*. They are strictly fermentative (either homo- or heterofermenters), aerotolerant or anaerobic, aciduric or acidophilic having complex nutritional requirements (carbohydrates, amino acids, peptides, fatty acid esters, salts, nucleic acid derivatives, vitamins) (86). They are naturally associated with a large variety of nutritive-rich plant- and animal-derived environments, and many species are involved in the manufacture and preservation of fermented foods and feed from raw agricultural materials (such as milk, meat, vegetables, and cereals) in which they are present as contaminants (166). Moreover, some species and strains are broadly used as starters and adjunct cultures to drive food and feed fermentations; notably dairy products (yogurt and cheese), fermented vegetables (olives, pickles, and sauerkraut), fermented meats (salami, sausages), and sourdough bread and other cereal-based food commodities. Although less numerous than bifidobacteria, lactobacilli are natural inhabitants of the GIT and genitourinary (GUrT) tracts of animals and humans, where they are thought to play pivotal roles in the maintenance and recovery of a healthy state (136, 182). Not surprisingly, a number of strains have been used as probiotics for more than 70 years (138). Beneficial effects attributed to indigenous and probiotic lactobacilli include colonization of intestinal and genital mucosa (85), inhibition of pathogens (36, 81), immunomodulation (88), and cholesterol assimilation (132).

1.2.3.2 Identification and Typing   Reliable identification of bacterial species and strains and correct naming are primary aims of taxonomic studies, but it also has important consequences for industrial application of bacteria. Morphology, Gram staining, and biochemical tests (fermentation of carbohydrates, growth at different temperatures, salt concentration, etc.) have traditionally been used as the primary methods for classifying *Lactobacillus* species; these methods are still in use. Based on phenotypic and biochemical characteristics, lactobacilli were divided into three groups according to the type of sugar fermentation (86). Obligate homofermentative lactobacilli ferment hexose sugars by glycolysis and produce mainly lactic acid, while obligatory heterofermentative species use the 6-phospho-gluconate/phosphoketolase (6PG/PK) pathway and produce other end products (CO_2, ethanol) in addition to lactic acid (18). A third group includes the facultative heterofermentative lactobacilli that ferment hexoses via the glycolysis and pentoses via the 6PG/PK pathways, respectively. Phenotypic analyses are time consuming and require technical skill and standardized assays and reading conditions, in order to avoid subjective results. Furthermore, it has been widely recognized that *Lactobacillus* species and strains display an inherent high level of phenotypic variability (86). Thus, phenotypic heterogeneity makes classical microbiological methods ambiguous and unreliable. In fact, many studies emphasize that the phenotypic classification of lactobacilli is unsatisfactory (13, 40, 113, 133, 162). As a recent example, Boyd et al. (28) have reported that the API 50 CH identification system failed to identify the seven *Lactobacillus* reference strains utilized in their study, and 86 out of 90 vaginal isolates, as compared to the identification obtained by hybridization using whole-chromosomal DNA probes. Of particular complexity are the phenotypically and genetically closely related species belonging, among others, to the *Lactobacillus casei* group (115) or to the *Lactobacillus acidophilus* complex (131, 144). Moreover, the phenotypic identification does not reflect the phylogenetic relation of the different species (51).

The taxonomy of the *Lactobacillus* species has changed considerably with the increasing knowledge of their genomic structure and phylogenetic relationships gathered with molecular methods (65). The DNA–DNA hybridization technique is still of reference, although this technique is labor intensive and time consuming. Fatty acid methyl ester (FAME) analysis has also been applied to the identification of lactobacilli from dairy and probiotic sources (68, 206). This is an inexpensive procedure that is also of help to study diversity, composition and dynamics of microbial communities, but FAME profiles are rather difficult to interpret and have to be subjected to mathematical treatments. Identification and classification of *Lactobacillus* species has also been accomplished by analysis of whole-cell protein patterns (133, 204). Highly standardized SDS-PAGE conditions allow a rapid and precise identification of a large number of strains. Profiles of unknown strains are compared to a pattern database of known species. In spite of all these techniques, at present, a majority of the molecular identification methods of *Lactobacillus* strains rely on the analysis of rRNA genes, mostly after their partial or complete amplification by the PCR technique (Table 1.1).

rRNA genes have been generally accepted as the potential target for identification and phylogenetic analysis of bacteria (15). Consequently, PCR amplification and sequencing of 16S rDNA- or 23S rDNA-targeted primers have successfully been used for the detection and identification of *Lactobacillus* species (115, 171, 186). Amplicons are usually digested with restriction enzymes for some techniques (such as ARDRA), and, more frequently, subjected to double-stranded sequencing. It has been experimentally determined that species having 70% or greater DNA similarity (at the DNA–DNA hybridization or re-association level) share, in fact, more than 97% of 16S rDNA sequence identity (127, 161). Isolates having such a percentage of identity belong to what has been called an operational taxonomic unit (OTU). Comparison of the rRNA gene sequences (mainly 16S rRNA) allows a precise identification and, at the same time, tracking of the evolutionary relationships among the distinct species. The analysis of 16S rDNA sequences has shown that the division of lactobacilli species in three groups is not in accordance with their natural relationships (18). In fact, *Lactobacillus* species branch into several groups and do not form a coherent phylogenetic unit (65, 150) (Fig. 1.3). At present, specific primers are available for targeting most *Lactobacillus* species (24, 90, 150, 154). Besides genes of both rRNA molecules, the analysis of ITS has also been utilized for identification purposes (24, 75, 119). Based on either the genes or the ITS regions, some authors have developed multiplex PCR of species-specific primer pairs for the detection of up to eleven different LAB species (90, 154, 160). In the same way as oligonucleotide primers, oligonucleotide probes can also be used in hybridization experiments for specific detection, identification, and quantification of *Lactobacillus* species (80, 128, 133, 155). Nucleotide differences in the 16S rRNA genes can also be exploited for the electrophoresis separation of PCR-derived amplicons by DGGE technique or its relative temporal temperature gradient electrophoresis (TTGE). These techniques can either be used for the identification of individual strains (61, 184) or for the analysis of the diversity and evolution of whole populations in complex bacterial mixtures (80, 120, 135).

Coding genes of highly conserved proteins such as RecA (64, 175), GroESL (202), and the elongation factor (EF) Tu (33, 192) have all been used to identify lactobacilli species and to determine their phylogenetic relationships. These gene sequences provide phylogenetic resolutions comparable to that of the 16S rRNA gene at all

SCREENING, IDENTIFICATION, AND CHARACTERIZATION   17

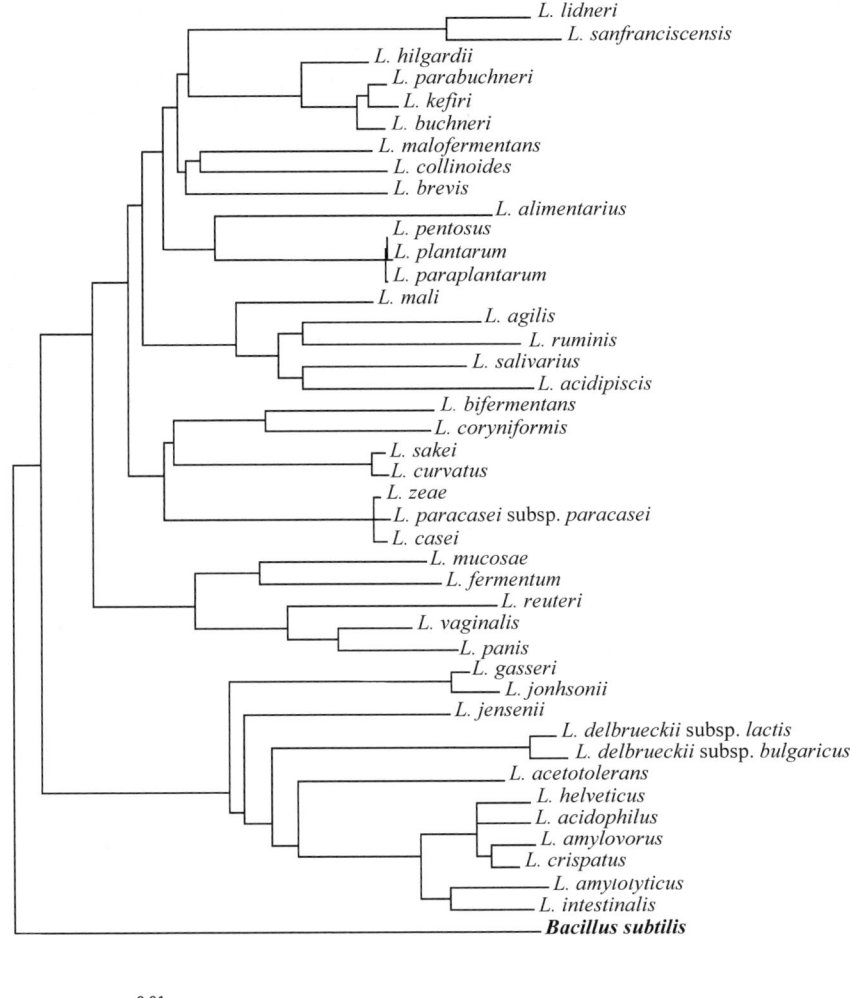

FIGURE 1.3   Phylogenetic tree showing the relationships of the 16S rDNA sequences of type strains of selected *Lactobacillus* species. Sequences were obtained from the Ribosomal Database Project (http://rdp.cme.msu.edu/) and the phylogenetic tree was constructed by an online tree builder resource that uses the Weighbor-weighted neighbor-joining algorithm. The 16S rDNA sequence from the *Bacillus subtilis* type strain was selected as an outgroup.

taxonomic levels, and better resolution between closely related organisms, as rates of evolutionary substitution in protein-coding genes are one order of magnitude higher than those for 16S rRNA genes. The use of protein-coding sequences further avoids the biases due to multicopy and intragenomic heterogeneity associated to rRNA sequences The comparison of the sequences of the fructose-1,6-bifosphatase (*fbp*) gene has been recently used for identifying and typing food borne and clinical strains of *Lactobacillus rhamnosus* (145). Beyond sequence data, the polyphasic approach, which integrates phenotypic, genotypic, and phylogenetic information, has been recognized by the International Committee on Systematic Bacteriology as a new tool for the description of species and for the revision of the present nomenclature of some bacterial groups (181).

Intraspecific differentiation of bacteria is highly relevant for the selection of starter and probiotic cultures, because technological, sensorial, antimicrobial, and probiotic attributes are strain specific. Typing methods are very helpful in distinguishing patent protected strains, as well as the distinction of starter, adjunct, and probiotic cultures (strain tracking) from natural isolates. As for the safety aspects, it is crucial to be able to compare clinical (pathogenic) isolates with biotechnological strains in use. Besides phenotypic methods, many PCR-based typing methods have been used for the typing of lactobacilli strains, such as ribotyping (141, 210), RAPD (46, 58, 117, 153, 175), PFGE (144, 179, 190), TAP-PCR (43), AFLP (176), REP elements PRC amplification (REP-PCR), ERIC-PCR (190), etc. PCR-RFLP of intragenic DNA fragments of protein-coding genes involved in primary metabolism (β-galactosidase, lactose permease, and proline dipeptidase) has also been used as a typing method for dairy *Lactobacillus delbrueckii* strains (69). Chromosome typing (restriction endonuclease fingerprinting of chromosomal DNA, chromotyping) has been applied to the discrimination of strains of lactobacilli and found to be specific and highly reproducible (163, 210); although the large number of bands requires careful standardized electrophoretic conditions.

The powerful MLST technique has recently been applied for species identification and phylogenetic studies of *Lactobacillus* strains. A MSLT method based on the analysis of six loci (*pgm, ddl, gyrB, purK1, gdh,* and *mutS*) has been developed for the analysis of *L. plantarum* strains (47). Even more recently, Diancourt et al. (54) developed and applied a MLST variant, called multilocus variable-number tandem repeats (VNTR) analysis (MLVA), for the fine subtyping of *L. casei/L. paracasei* strains. A high concordance between the profiles obtained by MLVA and those obtained by AFLP and MLST was observed.

Whole-genome sequencing and comparative genomics providing insights on bacterial evolution will surely influence bacterial taxonomy in the near future. In fact, gene and genomic sequence information has recently been proposed as a tool for defining a new genomic–phylogenetic species concept for prokaryotes (163). Comparative genomics has further strengthened the idea that the lactobacilli as a whole do not form a coherent phylogenetic group, supporting the recognition of new subgeneric divisions (32, 96). In fact, it seems that some species (*L. salivarius, L. plantarum*) are more closely related to *Enterococcus faecalis* than they are to other lactobacilli (32), and other non-lactobacilli such as *Pediococcus pentosaceus* will likely cluster within or close to some species of the genus (32, 96). Genome techniques such as comparative genome hybridization (CGH) can quickly be used to determine the genome content of a bacterial strain whose genome sequence is not known (112). CGH has also been found to be valuable for clarifying controversial taxonomical issues. It has already been used for comparison of members within the *L. acidophilus* group, addressing both intra- and interspecies diversity (23). Microarrays based on the *L. johnsonii* NCC533 genome were hybridized with total DNA from strains of this and other species of the *L. acidophilus* complex. A clear stepwise decrease in similarity between members of the complex was found, suggesting that these species belong to a natural phylogenetic unit. Exhaustive phylogenetic analyses based on genome data will be performed when more genome sequences are completed and analyzed. To date, 10 *Lactobacillus* genomes have been published, and at least 11 more sequencing projects are ongoing (35, 96).

1.2.4 Characterization of Probiotic Properties in *Bifidobacterium* and *Lactobacillus* Strains

Several criteria have been used for the selection of probiotic strains (Fig. 1.1), the most commonly employed being the survival of the stressful GIT conditions (low pH and high bile salts concentrations), the ability to transitory colonize the GIT, which is related with the adhesion to mucus and/or intestinal epithelium and the antimicrobial activity through the production of antimicrobial molecules or the ability to inhibit/displace the adhesion of pathogens. Several *in vitro* and *in vivo* tests are employed for the screening of these characteristics (45, 57, 178, 209), although there is a lack of standardized or unified methodology for the assessment of probiotic functionality. Table 1.2 summarizes some works that report the screening of the most common probiotic characteristics within collections of *Lactobacillus*, *Bifidobacterium*, and other LAB strains, mainly isolated from human samples but also from other sources.

1.2.4.1 Survival to GIT Stressing Conditions The transit of probiotics included in foods through different sections of the GIT takes variable times and is submitted to different stressful conditions. After mastication, the first barrier that bacteria must overcome is the low pH values of the stomach with values ranging from 1 to 3 and mean exposure times of 90 min. Into the duodenum the pH value rises to 6–6.5, but bile salts are poured from the gallbladder to reach concentrations ranging from 1.5 to 2% during the first hour of digestion and decreasing afterwards to 0.3% w/v or lower (118). The residence period in the small intestine until 50% emptying oscillate between 2.5 and 3 h and the transit through the colon could take up to 40 h (31). In this location pH values are close to neutral (from 5.5 to 7) and the physiological concentration of bile salts is lower. For the screening of putative probiotic bacteria most works (Table 1.2) simulate *in vitro* these GIT conditions. Several pH values and bile concentrations are tested for variable times in order to determine the survival of the strain(s) under test. Bacteria are enumerated by culture dependent and/or independent techniques, such as those employing fluorescent probes that allow knowledge of the population of dead and live bacteria. The results of viability obtained are strain dependent and, in general, bifidobacteria strains are less tolerant to acidic conditions than lactobacilli, whereas the first seems to be more tolerant to bile challenge. Few studies have been carried out employing human samples of gastric juice and bile and interestingly, the source of bile (bovine, porcine, or human) modifies the tolerance pattern (56). Therefore, it would be

TABLE 1.2 Screening of Probiotic Characteristics within Collections of *Lactobacillus* and *Bifidobacterium* Isolated from Different Sources

Strains Tested (number)	Strain Origin	Probiotic Character Studied	Observations	References
Bifidobacterium (25)	Infant feces	Survival GIT conditions Adhesion/colonization GIT Antimicrobial activity	Not all desirable characteristics were present in a single strain; most adherent and inhibitory were less resistant	(209)
Bifidobacterium (40)	Human, dairy	Survival GIT conditions	One strain (*B. lactis*), survive the simulated GIT conditions	(42)
Bifidobacterium (280)	Human feces	Antimicrobial activity	Inhibition ability varied according to pathogen tested	(25)
Bifidobacterium (11)	Human, animals	Adhesion/colonization GIT Antimicrobial activity	Adhesion and antimicrobial ability varied according to strain and its origin	(38)
Bifidobacterium (15)	Human, culture collection (70)	Adhesion/colonization GIT	The acquisition of bile resistance modify the adhesion ability of strains	(70)
Bifidobacterium (8)	Human, culture collection	Adhesion/colonization GIT	Adhesion is strain-dependent and is modified by pH and pathogens presence	(139)
Lactobacillus (12)	Culture collections	Adhesion/colonization GIT	Adhesion was strain-dependent	(177)
Lactobacillus (47)	Human, fermented foods, unknown	Survival GIT conditions Adhesion/colonization GIT Antimicrobial activity	The survival of three strains seemed to be linked to adhesion ability and tolerance to pH 2.5.	(84)

Species (n)	Source	Properties studied	Main findings	Ref.
L. acidophilus group (35)	Human, animals, dairy	Adhesion/colonization GIT	The adhesion to rat colonic mucin ability was strain dependent	(109)
Lactobacillus (7)	Culture collection	Adhesion/colonization GIT Antimicrobial activity	The growth media and the food matrix affect de adhesion properties	(125)
Lactobacillus (35)	Infant feces	Antimicrobial activity	More effective against Gram- and inhibitory activity was strain-dependent	(17)
Lactobacillus (88)	Dairy	Survival GIT conditions Adhesion/colonization GIT Antimicrobial activity In vitro and in vivo safety assessment	One strain (L. plantarum) showed probiotic characteristic and survive incorporated in cheese	(37)
Lactobacillus (65)	Dairy	Antigenotoxic properties	High number of strains showed antigenotoxic and antimutagenic properties	(30)
Lactobacillus (4)	Culture collection	Adhesion/colonization GIT Antimicrobial activity	High specificity in the inhibition/displacement of pathogens by lactobacilli	(71)
Lactobacillus (22)	Dairy	Survival GIT conditions Antimicrobial activity Bile salt deconjugation	Most strains were resistant to biological barriers. Only 3 strains were able to deconjugate bile salts.	(180)
Lactobacillus (26)	Human oral cavity	Survival GIT conditions Antimicrobial activity	Two strains (L. salivarius and L. gasseri) as putative probiotics	(167)
L. plantarum (31)	Animals, dairy, fermented foods	Adhesion/colonization GIT	Adhesion depends on both the adhesion model employed and the strain tested	(169)

(*continued*)

TABLE 1.2 (*Continued*)

Strains Tested (number)	Strain Origin	Probiotic Character Studied	Observations	References
Lactobacillus and *Bifidobacterium* (200)	Human, dairy	Survival GIT conditions	Three lactobacilli and 1 bifidobacteria strains survive to GIT conditions	(134)
Lactobacillus and *Bifidobacterium* (7)	Culture collections	Antimicrobial activity	Antagonistic activity variable according to strain tested	(82)
LAB (120)	Human feces	Survival GIT conditions Antimicrobial activity Bile salt hydrolase activity (BSH) *In vivo* cholesterol reduction	One strain (*L. plantarum*) showed high *in vitro* BSH activity and after oral administration was able to reduce cholesterol levels	(73)

recommendable that probiotic strains intended for human consumption are tested in the presence of human intestinal fluids.

1.2.4.2 Adhesion/Colonization to/of GIT Some of the health effects attributed to probiotics are related to their capability to adhere to the intestinal mucosa. Adhesion is a prerequisite for intestinal colonization, stimulation of the immune system, and for antagonistic activity against enteropathogens through competitive exclusion (57). The intestinal mucosa is covered by a layer of different types of epithelial cells, which are distinctly different in the different regions of GIT, and is in contact with the lumen, the outside of the body. In addition to secretory and absorption cells, an important part of the immune system is placed in this location and it is collectively referee to the GALT (gut-associated lymphoid tissue). The intestinal epithelium is almost completely covered by a protective mucus gel composed predominantly of mucin, glycoproteins acting as the anatomical GIT site in which the host first encounters gut bacteria (53). Genomic information of some probiotic strains revealed the presence of several molecules able to adhere to different components of the intestinal mucosa and to exchange signals with the intestinal immune system (149), which indicate a good adaptation of probiotics to the gut environment.

Several models have been employed to study the ability of putative probiotic strains to adhere to the intestinal epithelium. Studies have often been carried out with cellular lines obtained from human colon adenocarcinomas such as Caco-2 (ATCC HTB-37) and HT-29 (ATCC HTB-38) the last one being able to produce mucin (37, 84, 139, 169, 177). Frequently, the adhesion ability of putative probiotics from different collections has been extensively tested against mucus obtained from human (38, 70, 123, 125) or animal origin (109). Interestingly, some strains of *Bifidobacterium* adhere better to human mucus than to porcine mucus indicating that adhesion is property strain dependent (124), because mucus from different origins (human, canine, possum, bird, and fish) did not modify the adhesion of probiotic strains (140). In addition, bacterial adhesion to human mucus decreased with the age of the donor of the mucus sample, which could be one of the reasons for low bifidobacteria colonization in elderly subjects (122). A good correlation between the human mucus model and the adhesion to Caco-2 has been demonstrated by Gueimonde and collaborators (71) employing three *Lactobacillus* strains. Both methods are adequate for *in vitro* adhesion studies but some *ex vivo* models employing resected tissue of the intestinal mucosa from human or animals have also been shown to be useful (105, 126). In the human intestinal mucus model proposed by Ouwehand and coworkers (126) the material is obtained from patients with colon cancer submitted to surgery. The healthy sections of resected tissue obtained from different sites of the colon are employed in these studies. In general, the strains tested showed higher adhesion to mucus than to colonic tissue and, depending on the strain, the location of the colonic tissue but not that of mucus, also influenced the adhesion properties of the probiotics tested. This is a good model for the assessment of the adhesion of LAB to GIT epithelium and to mucus.

1.2.4.3 Antimicrobial Activity   In the complex GIT ecosystem probiotics have developed mechanisms to survive in competition with other microorganisms. Essentially, the antagonism is exerted by competition for nutrients and for physical location, but also through the production of antimicrobial substances. In connection with the previous paragraph, the ability of probiotics to produce antimicrobials is one mechanism to inhibit, exclude or compete with adherent enteropathogens for the ecological niche. Several works (Table 1.2) have been carried out to test *in vitro* the interference on adhesion between probiotics and pathogens such as *Salmonella enterica* serovar Typhimurium, *Escherichia coli, Clostridium difficile, Enterobacter sakazakii,* and *Listeria monocytogenes*. Using human intestinal mucus it has been demonstrated that the adhesion antagonism is clearly both, probiotic- and pathogen-strain dependent (38, 71). This specific interaction indicates the need for a case-by-case assessment in order to select probiotics with the ability to inhibit or displace certain pathogens. Most often, cocultures probiotic/enteropathogen are carried out to test the antimicrobial ability of probiotic strains (17, 25, 37, 82, 84, 167). Viability of both types of bacteria is determined and in some cases the antimicrobial activity is tentatively assigned to the production of substances such as organic acids, ethanol, H_2O_2, or proteinaceus components bacteriocin-like. The general conclusion that arises from these *in vitro* studies suggests again that the inhibition ability is strain- and culture-condition- dependent and that several molecules and mechanisms are involved in the interrelationship between probiotics and pathogens.

1.2.4.4 Other Probiotic Properties   In addition to the previously reviewed properties, other characteristics could be tested to consider a strain as putative probiotic. From these screenings it has been reported that some strains are able to modulate the immune system (129), to produce antigenotoxic compounds (30), to deconjugate bile salts (73, 180), and to decrease cholesterol levels (73).

1.2.5 Conclusion

The selection of a strain to be used as an effective probiotic is a complex process. For human consumption, it is widely accepted that an effective human probiotic should be of human origin the most suitable source being the human GIT. A vast array of specific and reproducible molecular techniques is now available for identification and typing of lactobacilli and bifidobacteria. Molecular techniques have allowed the precise and rapid identification and typing of novel probiotic stains, providing new ways to check for their presence and monitor their development. Nevertheless, for microbial characterization, a polyphasic combination of phenotypic assays and molecular techniques is preferred, since these approaches may provide complementary results. On the other hand, several *in vitro* and *in vivo* tests have been found to be useful for the screening of novel strains with putative probiotic properties. In general, the probiotic characteristics are strain-dependent and properties are not all simultaneously present in a single strain. Of note is the realization that the efficacy and safety of a probiotic should be validated in phase 2 clinical trials, a pending subject for most current-in-use probiotic strains.

1.3 DETECTION AND ENUMERATION OF GASTROINTESTINAL MICROORGANISMS

MIGUEL GUEIMONDE AND CLARA DE LOS REYES-GAVILÁN

Instituto de Productos Lácteos de Asturias (CSIC), Villaviciosa, Asturias, Spain

1.3.1 Methods for Intestinal Microbiota Assessment

Understanding the cross talk that occurs between intestinal microbiota and its host promises to expand our views about the relationship between intestinal microbiota and well-being. Unfortunately, we are still far from knowing the qualitative and quantitative composition of the intestinal microbiota and the factors governing its composition in an individual. Several different methodologies, culture dependent and culture independent, have been used for intestinal microbiota assessment (Table 1.3). The aim of this chapter is to review these methodologies, which are divided into two main groups: the culture-dependent and culture-independent methods.

1.3.1.1 Culture-Dependent Methods

1. Nonselective and selective culture media for intestinal microorganisms. The study of intestinal microbiota composition, both qualitatively and quantitatively, has been traditionally carried out by cultivation of feces. In some cases it has also been considered that there are mucosa-associated intestinal microbiota in biopsies of healthy individuals (229, 293) or patients (225, 252, 347). The classical method has been culturing fecal samples on suitable growth media, the sample generally being handled in anaerobic cabinets and processed immediately or within a few hours after collection. The bacterial counts of a given microbial group are determined after incubation in the appropriate conditions. Both nonselective- and selective-differential media have been used for growth and counting and the choice for one or the other was dependent on the microbial group being screened and on the method used for subsequent identification. Some of the more widely used media in recent years includes Wilkins-Chalgren (257, 330, 331, 341) and Columbia blood agar (225, 296) as general media for total anaerobic bacteria. The same media were also employed as selective and/or differential after the addition of the appropriate antibiotics and selective agents for the enumeration of Gram-positives including *Clostridium,* as well as *Bacteroides, Prevotella,* and other Gram-negatives (257, 341). MRS, Rogosa, and trypticase phytone yeast extract (TPY) agar were frequently used as base media for counting *Bifidobacterium* with or without the addition of selective agents (231, 257, 322, 323, 330, 331, 341), and MRS and Rogosa agar were also employed for the enumeration of *Lactobacillus* (212, 231, 269, 293, 322, 323, 330, 331). Bile–Esculin agar and derived supplemented media were among the most frequently employed for the isolation and enumeration of enterococci (225, 269, 323) and *Bacteroides* (229, 231, 296, 323). Enterococci have been also counted and isolated

TABLE 1.3 Advantages and Disadvantages of Some Techniques Widely Used for Intestinal Microbiota Research

Techniques	Advantages	Disadvantages
Culture-Dependent Techniques		
All	Widely used	Provides information only on culturable microorganisms
	Quantitative	Time consuming
	Possibility to isolate strains for further study and as source of potential probiotics	Complicated sample manipulation (e.g. anaerobes)
		Sublethally injured or dormant cells not detected
Culture in selective or differential media	Direct quantification of the microorganisms of interest	The presence of selective and/or inhibitory agents may inhibit the growth of part of the population (e.g. injured cells)
Culture in nonselective media followed by specific counting	Avoids the use of selective and/or inhibitory agents in the media	Requires an extra step for specific counting
Culture-Independent Techniques		
All	Provides information on both culturable and non-culturable microorganisms	Bias due to the different methodologies
	High throughput, rapid, and sensitive	Difficult standardization
	Possibility to use frozen samples	
Detection by PCR	Fast and easy to perform	Not quantitative
	Versatile (primers with different specificities can be used)	Previous knowledge required on the target microorganisms (to design specific primers instead of using universal primers)
	Different degrees of specificity possible (from group to strain specific)	

Sequence analyses of randomly amplified 16S rRNA genes	Provides information on previously unknown bacteria	Not quantitative
TGGE/DGGE	No previous knowledge of the microbiota present is needed (when using universal primers) Provides important information at population level (e.g. diversity)	Bias due to different cell lysis/PCR/Cloning efficiencies Methodologically difficult
T-RFLP	Versatile (primers with different specificities can be used) No previous knowledge of the microbiota present is needed (when using universal primers) Provides information at population level (if using appropriate primers)	Not quantitative Bias due to different cell lysis/PCR efficiencies Methodologically difficult.
	Versatile (primers with different specificities can be used) Provides information at population level (if using appropriate primers)	Not quantitative Bias due to different cell lysis/PCR efficiencies Methodologically difficult and expensive
DNA-Arrays	Very high throughput. Allows detection of thousands of sequences in a single assay	Not quantitative Bias due to different cell lysis/PCR efficiencies Methodologically difficult and expensive
FISH	Quantitative Possibility to observe bacteria *in situ* (e.g. position in the mucosa) Possibility to couple with automatic image analysis or flow cytometry	Bias due to different cell permeability to probe Difficult visual counting (e.g. cell aggregates) Previous knowledge of the target microorganisms required for probe design (if not universal)

(*continued*)

TABLE 1.3 (*Continued*)

Techniques	Advantages	Disadvantages
	Fast and easy to perform	Previous knowledge of the target microorganisms required for primer design (if not universal)
RT quantitative PCR	Allows to determine *in situ* metabolic activity or specific gene expression	Previous knowledge of the target genes required for primer design Does not provide information on cell numbers
Metagenomics	No previous knowledge on the microbiota present is needed Allows identification of novel genetic features in the ecosystem Provides important information at population level (e.g. diversity)	Not quantitative Time consuming Bias due to different cell lysis/PCR/cloning efficiencies Methodologically difficult and expensive
Metaproteomics	No previous knowledge on the microbiota present is needed Provides information on protein coding genes that are being expressed in the ecosystem	Time consuming Methodologically difficult

frequently in Slanetz-Bartley (230, 330) and KF media (296). It is also worthy to note the extensive use of the selective and differential McConkey agar for counting coliforms and enterobacteria (225, 293, 231, 232). Finally, for heterogeneous and complex groups of intestinal bacteria such as clostridia, a great diversity of culture media has been used provided that a unique medium is not appropriate for cultivation of all microbial clusters (257, 269, 296, 322, 331).

2. Identification and typing of intestinal strains. Traditionally, intestinal strains isolated from solid media were identified by means of some general phenotypic characteristics such as carbohydrate fermentation profiles, enzymatic tests, cell morphology, and colony appearance, which have lead to numerous misidentifications. The development of new and different phenotypic methods contributed to the improvement of the accuracy of identification and it is currently possible to identify not only colonies isolated from solid culture media but also nonisolated bacteria present in mucosa and feces. The genus *Bifidobacterium* is the only intestinal Gram-positive displaying fructose-6-phospate phosphoketolase (F6PPK) activity and its determination constitutes a reliable test for this genus among bacteria from the intestinal environment (255, 330). Vlková and coworkers (330) used F6PPK, α-galactosidase, and α-glucosidase as enzymatic methods for detection of the abundance of bifidobacteria directly in infant feces. Determination of catabolic end products was also useful for the identification of some particular and characteristic groups of microorganisms (257). One of the most accurate phenotypic methods currently used for identification at the species level is based on the determination of the cellular fatty acid composition (234) with the help of the chromatographic MIDI system (http://www.midi-inc.com/pages/literature.html) both directly in feces (269, 324) and in previously isolated cultures (263, 341).

The development of genetic methods for the identification and typification of the bacterial isolates greatly contributed to the improvement our knowledge of intestinal microbiota, although its level of sensitivity is obviously limited by the accuracy and sensitivity of the previous culture media on which microorganisms were isolated. Partial amplification and sequencing of the 16S rRNA gene from previously isolated colonies has been extensively used for identification at the genus and species level (212, 228, 323, 330, 331). Monoclonal antibodies have been used for selective enumeration of *Bacteroides vulgatus* and *Bacteroides distasonis* on fecal samples after dilution and plating in a nonselective medium, avoiding the use of selective agents to which part of the *Bacteroides* population could be sensitive (226). Oligonucleotide probes have also been used for the identification and the quantification of intestinal microbiota by means of colony hybridization (232, 271). Typing of the strains from a given species of intestinal microorganism has often been carried out by pulsed field gel electrophoresis (PFGE) (322, 323) or RAPD by PCR (212, 228) among other genotypic techniques.

Several different culture-dependent techniques have been used for quantitative and qualitative characterization of human intestinal microbiota although large differences in species composition and quantitative contents can be found among the results obtained by different authors. These differences could be attributed to the different culture media employed, different methodologies used for the subsequent identification, etc. The high variability of the intestinal microbiota among individuals and the analysis of a low number of individuals in some studies, could also have contributed to the different results observed.

1.3.1.2 Culture-Independent Methods It is estimated that less than 25% of the intestinal bacteria have been cultivated so far, suggesting that many bacteria in the human gut have not been cultured yet and that classical culture based methods have not provided an accurate representation of this community (233, 259, 319). Consequently, the study of intestinal microorganisms has been restricted to the cultivable species and from these, only to the cultivable fraction of each population. These facts led to the overestimation of some species and the underestimation of others, limiting our understanding of intestinal microbiota composition and function. By using different bifidobacterial selective culture media it has been shown that they differ in their selectivity and some media even fail in the recovery of certain species, which could lead to a biased representation of the population (213). These results clearly show the limitations of the culture-based approaches for the study of complex communities. Therefore, more rapid, accurate and specific methods of detection and quantification have been developed.

During the last few years, developments in molecular biology have led to alternative culture-independent methods in addition to the traditional culture. One of the most widely applied approaches deals with the use of 16S rRNA and its encoding genes as target molecules. The 16S rDNA gene contains highly conserved regions, present in all bacteria, and highly variable ones that are specific for certain microbes. Specific PCR primers and probes can thus be designed based on these variable regions to detect certain species or groups of bacteria. These culture-independent approaches include 16S rRNA measurements, PCR amplification with specific primers of 16S rDNA extracted from fecal or mucosal samples, universal or group 16S rDNA PCR amplification followed by cloning and sequencing, TGGE, DGGE, terminal restriction fragment length polymorphism (T-RFLP) analysis, fluorescence *in situ* hybridization (FISH), real-time quantitative PCR, and oligonucleotide-microarrays. In more recent years metagenomic and metaproteomic approaches have also been applied to the intestinal microbiota assessment.

1. *Design of PCR Primers for DNA Amplification.* Several authors have developed species or group-specific primers for the detection of different microorganisms in the GIT including members of the genera *Bacteroides, Clostridium, Fusobacterium, Peptostreptococcus, Eubacterium, Bacteroides, Prevotella, Lactobacillus*, and *Bifidobacterium* (251, 254, 289, 291, 335).

    Nowadays, rDNA-targeted PCR primers enable a rapid and specific detection of a wide range of bacterial species. Therefore, procedures in which these primers are used have a widespread use in intestinal microbiota assessment. By means of PCR amplification with species-specific primers Wang and collaborators (335) analyzed the fecal microbiota, showing that *Fusobacterium prausnitzii, Peptostreptococcus productus*, and *Clostridium clostridiforme* had the highest PCR titters followed by *Bacteroides thetaiotaomicron, Bac. vulgatus* and *Eubacterium limosum*. Matsuki et al. (292) studied the bifidobacterial microbiota in adults' fecal samples and found that *Bifidobacterium catenulatum* group *Bifidobacterium longum* and *Bifidobacterium adolescentis* were the most common species whereas *Bifidobacterium breve, Bifidobacterium infantis*, and *B. longum* were the predominant species in infants. Other DNA sequences, such as ERIC sequences, have also been used as targets of PCR primers to fingerprint the microbial community of the human gut (337).

2. *Design of Hybridization Probes.* Several probes have been developed for the assessment of intestinal microbiota. There are probes for specific detection of *Bifidobacterium* (271, 280), some *Clostridium* groups (241, 338), *Bacteroides/Porphyromonas/Prevotella* group (232), *Bacteroides fragilis* group, *Bac. distasonis* or *Streptococcus/Lactococcus* group (241). Also probes for some species of *Bifidobacterium* (343) and *Eubacterium* (313) have been developed. Some other probes for specific intestinal groups such as *Phascolarctobacterium* group, *Veillonella, Eubacterium hallii* and relatives, *Lachnospira* group, *Eubacterium cylindroides* and relatives and *Ruminococcus* and relatives are also available (258).

These probes have been used for specific culture-independent detection and quantification of different intestinal microorganisms by means of FISH (241, 258, 280, 313, 338) or dot blot hybridization (264, 287, 314).

Combinations of PCR amplification and hybridization have also been reported. In a study by Wei and coworkers (337) ERIC-PCR amplicons from a sample were labeled and used to hybridized against other samples in order to identify those amplicons common to different fecal samples. This approach may be helpful for the identification of specific microbiota aberrancies related to different diseases by comparing healthy and ill individuals.

3. *Polymerase Chain Reaction – Enzyme-Linked Immunosorbent Assay (PCR-ELISA).* This technique combines PCR amplification of DNA and ELISA. The amplified DNA is labeled, commonly with digoxigenin, and hybridized with the specific detection probe that is immobilized in microtiter plate wells. The presence of hybridized DNA is determined by using digoxigenin-targeted antibodies. This methodology has not been extensively used, but it has been applied to the analysis of *Bifidobacterium* species composition in human feces during a feeding trial (285).

4. *Sequence Analysis of Randomly Amplified 16S RNA Genes.* Another procedure that has been used in intestinal microbiota research is the PCR amplification of 16S rRNA genes in a sample, using universal or group-specific primers followed by cloning and sequencing of the amplified DNAs.

By using universal primers different studies (259, 261, 319) have shown that the predominantly cloned sequences from fecal samples belonged to

Clostridium coccoides group (*Clostridium* rRNA cluster XIVa), *Clostridium leptum* group (*Clostridium* rRNA cluster IV), and *Bacteroides*, which is in agreement with the results recently reported by Eckburg and coworkers (233). By using this approach it was found that, according to the results obtained by culture, even though clostridia tends to increase with age, the *Clostridium* rRNA cluster XIVa tends to decrease in elderly individuals. This is probably due to a decrease in the number of *Ruminococcus obeum* and related phylotypes, indicating a possible relation between *R. obeum* and aging (261). Regarding the species composition, it was found that the predominant fecal species in the *Bacteroides* group are *Bacteroides uniformis* and *Bac. vulgatus* (319). In the *Clostridium coccoides* group, *Eubacterium eligens, Eubacterium rectale*, and *Eubacterium hadrum* were the predominant species. *Ruminococcus bromii, Eubacterium siraeum*, and *F. praustnizii* were the unique species detected in the *C. leptum* group, being *F. praustnizii* one of the most frequent and numerous species detected by 16S rDNA analysis of human fecal samples (259, 261, 319). Sequencing of amplified 16S rRNA genes has also been used to characterize the differences between the fecal and mucosal microbiota. Eckburg and coworkers (233) analyzed over 11,800 bacterial 16S rDNA sequences from different intestinal locations of three subjects and found that 62% of the phylotypes were novel (244 out of 395) and 80% represented noncultivable species. On the other hand, the same authors indicated that different phylotypes were present in fecal and mucosal samples.

Surprisingly, in some studies carried out using this methodology (261, 319) no sequences belonging to the genus *Bifidobacterium* were detected, probably indicating some problems during the amplification. In this regard, the number of PCR cycles can significantly distort the representation of some organisms in the ecosystem due to preferential amplification of some rDNAs (339).

5. *Temperature Gradient Gel Electrophoresis (TGGE) or Denaturing Gradient Gel Electrophoresis (DGGE) Analyses of 16S rRNA Genes.* This technique has been one of the most widely used for intestinal microbiota assessment. It consists of the PCR amplification of the 16S rRNA genes with universal or group-specific primer pairs, one of which has a GC clamp attached to the 5' end in order to avoid a complete dissociation of the two DNA strands of the amplified product. Then, amplification products are separated by denaturing gel electrophoresis, through a gradient of temperature (TGGE) or denaturant agent (DGGE), in which the double-stranded DNA will migrate until it reaches its denaturing conditions in the gradient. This method has been shown to be a powerful tool for monitoring bacterial succession phenomena. In addition, the predominant bands obtained can be sequenced in order to know the identity of the most abundant microorganisms. Using PCR-TGGE, Zoetendal et al. (348) studied the diversity of predominant bacteria in fecal samples from adults. A remarkable stability of the profiles over time was observed and *E. hallii,*

R. obeum, and *F. prausnitzii* were the most commonly encountered species. In spite of the stability of the predominant fecal microbiota, it was possible to detect variations in some subpopulations over time by using these techniques (327).

Satokari and collaborators (311) studied the bifidobacteria microbiota in fecal samples from adults by means of PCR amplification with genus-specific primers and DGGE. Their results highlighted *B. adolescentis* as the most common species in feces from adults. Other species also found were *B. catenulatum*, *Bifidobacterium pseudocatenulatum*, *Bifidobacterium dentium*, and *Bifidobacterium ruminatum*. By using these techniques *Bifidobacterium* and *Ruminococcus* were reported to be the dominant groups in the intestinal microbiota of babies (236). Favier and coworkers (237) studied the establishment and development of gut microbiota in babies during the first 4 months of life, finding that *Escherichia coli* and *Clostridium* spp. were the initial colonizers followed by the appearance of other microorganisms, such as *Bifidobacterium* and *Bacteroides* after 2–5 days. On the other hand, comparison of the babies PCR-DGGE profiles with those of their parents suggested a vertical transmission of some microorganisms.

With regard to the lactobacilli population, PCR-DGGE results showed that there is a relatively stable *Lactobacillus* population in each individual (262). *Lactobacillus ruminis* and *Lactobacillus salivarius* have been reported to be the true autochthonous lactobacilli whereas other species frequently used in food manufacture can be also detected in feces of individuals (332). In addition, by using PCR-DGGE Zoetendal and coworkers (350) showed that the mucosa-associated bacteria in the colon differ from those recovered from feces, and found host-specific profiles of the mucosa-associated microorganisms. This suggests that the intestinal microbiota composition is influenced by some host factors.

Most of the studies carried out using these techniques are aimed at the assessment of microbiota composition by targeting the rDNA genes. However, the rRNA (RT-PCR DGGE/TGGE) has also been used instead of rDNA allowing the identification of the metabolically active microorganisms in the gastrointestinal ecosystem (294, 348).

6. *Denaturing High-Performance Liquid Chromatography.* This recently developed technique (342) has also been applied to the study of the intestinal microbiota (246). It consists of the PCR amplification of the 16S rRNA genes followed by the separation of the amplification products by means of denaturing high-performance liquid chromatography. Separated PCR products are fluorescent dyed and detected using a fluorescence detector.

7. *Terminal-Restriction Fragment Length Polymorphism (T-RFLP) Analysis.* The 16S rDNA T-RFLP analysis consists of the amplification of the 16S rDNA with a primer fluorescent labeled and an unlabeled primer so that the PCR product is labeled at only one end. After digestion of the PCR

products with one or more endonucleases the length of the labeled terminal restriction fragments is determined by capillary electrophoresis. A rapid assessment and identification of predominant human intestinal bacteria can be accomplished with this method using the appropriate restriction enzymes (160, 260, 272, 289).

Sakamoto and coworkers (289) assessed the fecal microbiota in adults using T-RFLP analysis showing that the patterns are host specific. These results are in agreement with previous results obtained by DGGE. This technique has also been used for assessment of fecal microbiota in elderly people (261) and to study the effect of a vegetarian diet on intestinal microbiota (50). Interestingly, differences have been found between the results obtained by T-RFLP and cloned 16S rDNA analysis, indicating a possible bias related with a large number of cycles in PCR amplification (260, 261).

8. *Oligonucleotide Arrays.* Wang and coworkers (333, 334) developed an oligonucleotide-microarray using species-specific probes for the detection of the predominant human intestinal bacteria in fecal samples. Microarray technology can be used for simultaneous detection of thousands of target DNA sequences at one time. Thus, its use could permit the detection of many bacterial species in a sample in a rapid and accurate manner. In order to avoid the use of the expensive microarray equipment needed for this technique, a membrane-array procedure has also been reported (336). Recently, Palmer and coworkers (303) developed a microarray containing over 10,000 16S rDNA probes and applied it to the assessment of the colonic mucosa, which allowed the detection and the determination of the relative abundance of species present at levels of 0.03% or greater.

In the next few years, with the increasing availability of genome sequences from intestinal bacteria, microarrays analysis will become a powerful and valuable tool to assess microbial composition of the human intestinal tract and to study how different members of the intestinal microbiota modulate the expression of genes from both intestinal cells and other intestinal bacteria.

The culture-independent approaches discussed earlier has led to a better understanding of the qualitative content and the predominant species of the intestinal microbiota. Unfortunately, they have failed to provide reliable data on its quantitative content or on the less abundant groups or species that are also present in the GIT. Thus, some quantitative culture-independent methods have been recently developed and are discussed next.

9. *Relative Amount of Group or Specie-rRNA.* One approach for the quantitative study of the intestinal microbiota is the quantification of the relative amounts of 16S rRNA of each group or species with regard to the total amount of 16S rRNA in the sample by using specific probes and, for example, dot blot hybridization. The amount of 16S rRNA provides not so much a measure of cell numbers as a measure of the metabolic status of each microbial group. By using this procedure it was shown that six bacterial groups represented up to 70% of the total fecal rRNA, *Bacteroides-Prevotella* being the dominant

group with 37% of the total 16S rRNA (314). In addition, by means of this procedure Marteau et al. (287) showed that the human cecal and fecal microbiota differs quantitatively and qualitatively.

10. *Fluorescence In Situ Hybridization (FISH)*. Using FISH with different group-specific probes around 90% of the total fecal bacteria can be detected; *Bacteriodes/Prevotella* and the *Clostridium coccoides/E. rectale* groups being the microorganisms present at higher numbers (10^{10}) (241, 258), followed by *Eubacterium* low G+C group (258). Other bacterial groups present at high levels (over 10^9 cells/g feces) included *Ruminococcus* (258) and *Bifidobacterium* (241, 258, 280). *Enterobacteriaceae, Veillonella*, and the group *Lactobacillus/Enterococcus* showed counts under 10^8 cells/g feces.

FISH has also been used for the assessment of changes in levels of the predominant groups of intestinal bacteria as a result of the consumption of prebiotics or probiotics (273, 325) or to assess the influence of the mode of delivery on intestinal microbiota (309). Also the effect of breast-feeding was studied by means of FISH, and it has been shown to be related to the predominance of bifidobacteria, whereas formula-fed infants showed similar amounts of *Bacteroides* and *Bifidobacterium*. In addition, changes in the minor components of the fecal microbiota were also observed (256). By means of FISH it has also been shown that there are differences in the gut microbiota between infants who later do or do not develop atopy (269).

Although FISH has been widely used for intestinal microbiota assessment this technique is laborious, there is some difficulties for the visual counting of the samples and is extremely time consuming, thus limiting its further applicability. Multi-color FISH would allow the detection of a few microorganisms by a single hybridization reaction. This approach has been applied to the analysis of seven bifidobacterial species in human feces (320), but still the visual counting is very laborious. Because of that, alternative methods have been developed in order to solve difficulties in manual-visual counting, such as automated image analysis (267) or flow cytometry (349).

In FISH results can be influenced by differences in the availability of the target region, cell permeability or by the ribosome content of the cells. Low fluorescence levels in positively hybridized cells can also significantly overlap signals of the negative controls (280). Coaggregation of bacteria rests of broken cells or contaminating compounds make the counting difficult. Therefore, more rapid and accurate procedures have been developed and are commented on next.

11. *Quantitative Real-Time PCR*. Quantitative real-time PCR is a promising tool to study the composition of complex communities such as the GIT. This procedure has attracted the attention of researchers in recent years as a consequence of the need for new rapid and accurate quantitative culture-independent techniques for intestinal microbiota analyses.

Different real-time quantitative PCR assays have been developed. By using the SYBR Green dye both total fecal bifidobacteria and specific bifidobacterial species or groups have been quantified (290, 305). In addition, 5′nuclease assays have also been developed for *Bifidobacterium* and *Lactobacillus* quantification by using TAQMAN probes (253, 254, 305) or probes labeled with fluorescent lanthanide quelates (248, 251).

Real-time quantitative PCR has also been applied to quantification of other intestinal microorganisms, such as *Clostridium difficile* in feces by using molecular beacons (217), *Escherichia coli* and *Bac. vulgatus* in gastrointestinal mucosa by means of the 5′ nuclease assay with TAQMAN probes (265) or *Desulfovibrio* in feces and mucosa by using the SYBR Green assay (239). Real-time PCR has been employed to characterize and compare the fecal microbiota between healthy and hospitalized elderly subjects (216), proving to be a useful tool for quantitative microbiota monitoring. This procedure has also been used for the quantification of total bacteria and some characteristic species of dental plaque and caries dentine (284, 299).

Similarly to other techniques based on the PCR, the use of 16S rRNA instead of DNA (reverse-transcriptase quantitative PCR) would provide data on the activity/viability of the microorganisms rather than on cellular levels. This methodology may also be applied to monitor *in situ* the expression of specific genes by targeting the corresponding mRNA, as has been demonstrated by Fitzsimons and coworkers (240) using the gene *slp*A of *Lactobacillus acidophilus*. In this regard reverse-transcriptase quantitative PCR provides a very useful tool for monitoring bacterial activity and gene expression in gastrointestinal conditions.

Nowadays, the 16S rRNA genes are being used as target molecules, but as more bacterial sequences are becoming available, new specific primers and probes targeting other genes will also be available in the near future to be used in cases in which the 16S rDNA is not an adequate target. In this regard, it must be taken into account that the bacterial quantification by real-time PCR can be influenced by differences in the number of rRNA operons among the quantified species or groups, sequence heterogeneity among different operons within the same species or by differential amplification of different DNA molecules (312, 340).

12. "*Omics.*" During recent years the so-called "omics" revolution (genomics, proteomics, metabolomics) has provided an impressive amount of new information allowing the development of new very powerful molecular techniques. The genome information about some gut microbiota members has increased our understanding on the adaptation of these microorganisms to the intestinal environment (310). Metagenomic and metaproteomic approaches have been applied to the study of the intestinal microbiota. These approaches consist of the procurement and study of a genetic library containing all the genetic material present in a sample (metagenomics) or the study of all the proteins present (metaproteomics). Metagenomic analyses have been used to study the microbiota of the large intestine (245) or to assess the diversity of fecal microbiota in Crohn's disease (286). Klaassens and

coworkers (274) applied metaproteomics for the first time to the study of the intestinal microbiota in infants.

When using omics, as with any other techniques, the possible bias due to the methodologies used must be considered. In this regard, in metagenomic studies, possible biases due to differences in bacterial lysis or cloning efficiencies among different bacteria or DNA sequences should be taken into account (244).

13. *Other Methods.* There are also other methods that have been applied to the assessment of intestinal microbiota without the need for cell culture. The analysis of cellular fatty acids profiles in fecal samples (264) has been frequently used. Metabolic activities (bile acids deconjugation or dehydroxilation, vitamin K production, some enzymatic activities, etc.) can be also used as a crude signature of the microbiota and compositional changes may be tracked by noting changes in these metabolic activities (295). Flow cytometry coupled with fluorescent labeling of live–dead bacteria has been applied to the identification of the viable and active populations in the gut (219). Another interesting approach is the rRNA-stable isotope probing to identify the specific microorganisms responsible for the utilization of a substrate among those present in the complex intestinal ecosystem (235). Two-dimensional polyacrylamide gel electrophoresis of the amplified rRNA genes from a population has also been found to be a high performing technique for the study of complex microbial populations (268) although it has not yet been used to assess gut microbiota composition.

1.3.2 Detection and Enumeration in Dairy Products

Fermented dairy products are considered as one of the most suitable vehicles for the administration of probiotic bacteria. *Lactobacillus* and *Bifidobacterium* species are the most commonly used probiotics, which are often implemented in dairy products in combination with other LAB. In spite of the availability of culture-independent molecular tools for quantification of probiotics in commercial products, most manufacturers still use conventional culture techniques for enumeration purposes. In addition, culture-dependent methods are crucial to determine possible physiological or biochemical changes in the population of probiotic bacteria during the refrigerated storage of the product (329).

From a practical point of view, differential enumeration of probiotic and starter bacteria in food products is rather difficult due to the presence of several closely related species of LAB. The majority of media currently available for the selective enumeration of probiotics and LAB included in dairy products are based on differentiation by colony appearance (223, 300, 307, 321, 328). However, this is not always a stable phenotypic feature and, in addition, it is largely dependent on the subjectivity of each one. Therefore, for a more conclusive identification and enumeration of probiotic bacteria, some selective media for each targeted species have also been developed (326). The disadvantage of these media is that they can underestimate counts of the microbial group selected.

Among the great variety of general, modified, selective, and differential media, only a few of them have proven in comparative studies to be suitable for quantification of a given probiotic species on the basis of their high recovery, and clear differentiation from or inhibition to other LAB also present. In spite of that, authors in general agree that no unique, selective or differential medium provides reliable counts of probiotic bacteria in all dairy products available and the most representative for fermented milks are indicated in Table 1.4.

It is worthy to mention an enzyme-based most probable number (MPN) method for the enumeration of *Bifidobacterium* in dairy products developed by Bibiloni et al. (220). It is based on the selectivity for bifidobacteria of MRS broth containing 0.3% bile and subsequent analysis of the F6PPK activity in grown tubes.

Several studies have been performed using selective and differential media for the correct identification of bacterial species claimed in the product label and for following the viability of probiotics and starter cultures during the refrigerated storage of fermented milks. Probiotics often show poor viability in market preparations (217, 249, 304, 316). Several factors could be involved in affecting the viability of probiotic cultures in fermented milks such as fat content (329), temperature, oxygen content, acidity, pH, and the presence of other LAB, among others (315).

1.3.3 Detection and Enumeration of Specific Probiotics in the Gut

To detect or enumerate a specific probiotic strain among the vast array of microorganisms present in the intestinal environment is often a challenging issue. However, it is essential in order to study the survival in the gut or the colonization ability of probiotic strains. Several different methods, both culture-dependent and culture-independent, have been used to this end (Table 1.5).

Traditionally, culture followed by morphological colony characteristics or strain isolation for genotypic or phenotypic characterization has been used. Nevertheless, this approach shows all the limitations of culture-dependent techniques and if the probiotic strain is outnumbered by similar microorganisms present in the gut, the proper isolation and further identification of the specific strain is difficult to achieve. In some cases antibiotics are used as selective agents in the media. After culture in appropriate media the identity of the isolated strains is confirmed by a highly discriminatory technique such as RAPD (242, 247, 288), ARDRA (224, 297), or REA-PFGE (275, 283). Fluorescent hybridization has also been used for this purpose (283). Antibiotic-resistant variants of the probiotic strains may also be used to allow specific enumeration by using media supplemented with the appropriate antibiotic (242, 302). The combination of selective culture media with monoclonal antibodies has also been applied (346).

PCR primers have been developed for some probiotic strains (222, 243, 318) and then used to confirm colony identity (243, 250, 318, 344) or for direct detection in the samples (250). Molecular biology offers also the possibility to label the strain by transforming it with a plasmid containing a gene marker (238). However, it must be taken into consideration that the use of GMOs may imply certain limitations, especially in the setting of clinical studies.

TABLE 1.4 Media for Viable Cell Counts of Probiotic *Lactobacillus* and *Bifidobacterium* in Fermented Milks Containing the Yogurt Starters *S. thermophilus* and *L. delbrueckii* subsp. *bulgaricus*

Agar Medium	Basic Medium	Microorganisms Counted	Supplements Added	Type of Medium	Differential Count Based on	Oxygen Conditions for Incubation	References
G-MRS	MRS	S. thermophilus L. bulgaricus L. acidophilus B. bifidum	Galactose (carbon source)	Modified	Colony appearance	Anaerobiosis	(317)
T-MRS	MRS	L. acidophilus	Trehalose (carbon source)	Modified	No growth of other LAB	Anaerobiosis	(266)
Bile-MRS	MRS	L. acidophilus	Bile	Selective	Inhibition of other LAB	Aerobiosis	(266)
MRS-clindamycin	MRS	L. acidophilus	Clindamycin	Selective	Inhibition of other LAB	Anaerobiosis	
LC medium	Basic medium	L. casei L. paracasei L. rhamnosus	HCl until pH 5.1 Bromocresol green Ribose	Selective and modified	Inhibition/no growth of other LAB	Anaerobiosis	(304)
MRS-AC	MRS	L. rhamnosus L. paracasei	Acetic acid until pH 5.2	Selective	Inhibition of other LAB	Anaerobiosis	
NA-salicin	Nutrient agar	L. acidophilus	Salicin	Selective/ differential	Inhibition/ colony appearance	Anaerobiosis	(281)
LP-MRS	MRS	B. bifidum	Lithium chloride Sodium propionate	Selective	Inhibition of other LAB	Anaerobiosis	(282)

(*continued*)

TABLE 1.4 (*Continued*)

Agar Medium	Basic Medium	Microorganisms Counted	Supplements Added	Type of Medium	Differential Count Based on	Oxygen Conditions for Incubation	References
MRS-NPLN	MRS	*Bifidobacterium* sp.	Neomycin sulfate Paromomycin sulfate Nalidixic acid Lithium chloride	Selective	Inhibition of other LAB	Anaerobiosis	(227)
AMC	Reinforced clostridial medium	*Bifidobacterium* sp.	Nalidixic acid Polymyxin B Iodoacetate 2,3,5-triphenyltetrazolim chloride Lithium propionate	Selective	Inhibition of other LAB	Anaerobiosis	(214)
DP	Columbia agar base	*Bifidobacterium* sp.	Dicloxacillin Propionic acid	Selective	Inhibition of other LAB	Anaerobiosis	(220)
BFM	—	*Bifidobacterium* sp.	Lactulose (carbon source) Propionic acid Methylene blue Lithium chloride	Selective	Inhibition of other LAB	Anaerobiosis	(300)

TABLE 1.5 Methods Used for Detection of Some Probiotic Strains in Human Fecal/Intestinal Samples

Strains	Culture Step	Identification Techniques	References
B. animalis DN173010	Yes	ARDRA	(224)
B. animalis BB-12	Yes	PFGE or fluorescent hybridization	(283)
L. johnsonii La1	Yes	PCR	(243)
L. rhamnosus GG	Yes	PCR	(250)
L. plantarum 299v	Yes	RAPD	(247)
L. gasseri SBT2055SR	Yes	Use of streptomycin-rifampicin resistant mutant and RAPD	(242)
L. casei DN114001	Yes	Use of rifampicin-resistant spontaneous mutant	(302)
L. casei Shirota	Yes	Monoclonal antibodies	(346)
L. paracasei B21060	Yes	ARDRA	(297)
L. paracasei CRL-341	Yes	PFGE	(283)
L. rhamnosus GG	No	PCR	(250)

1.3.4 The Problem of the Viability and Physiological State of Intestinal Bacteria

Microorganisms in different ecological niches, including the GIT and acidic food products, may exist in several physiological states of viability. Traditionally, microorganisms were considered viable if they were capable of multiplying in an appropriate medium, being culture-based methods such as plate counts and MPN counts largely used to enumerate viable cells. However, certain microorganisms, which are readily cultivable can also exist in other states where the cell fails to replicate, but retains some metabolic activities typical of viable cells and may return to be cultivable under certain conditions. This is the case of the so-called viable but uncultivable cells (306), starved cells (301), dormant cells (270), or sublethally injured cells (345). While in pathogenic and environmental bacteria the phenomenon of the different states of viability has received quite a lot of attention, these studies are considerably less developed in probiotics and intestinal bacteria.

Bacteria under stressful conditions may modify their viability maintaining equilibrium between multiplication and survival activities (301). Recently, strains of *Bifidobacterium* used as health-promoting probiotic bacteria, have been shown to become dormant during storage of fermented products (277) or sublethally injured following stress treatment (218). Moreover, Ben-Amor et al. (219) demonstrated a great physiological heterogeneity within separated populations of viable, injured, and dead fecal bacteria.

Methods measuring multiplication as the sole criterion of viability have been extensively used although temporally uncultivable cells fail to be detected. Therefore, other viability assays apart from those based on multiplication in culture media have been developed. For example, using the antibiotic ciprofloxacin as an inhibitor of cellular division, Barcina et al. (215) used changes occurring in cell morphology and elongation of cells to determine microscopically "direct viable counts." Nevertheless, most currently available methods for measuring the viability are based on the employment of fluorescence techniques that generally use two fluorochromes with different emission wavelengths in combination to discriminate between intact or viable cells, injured or damaged cells, and dead cells. Membrane integrity has been employed as a criterion of viability for intestinal bacteria and probiotics (278–280, 299). The commercial LIVE/DEAD® BacLight™ kit contains two nucleic acid stains: the green fluorochrome SYTO 9 is a small molecule that can penetrate all membranes whereas the larger red fluorochrome propidium iodide can penetrate only compromised membranes, thus rendering cells green when they are viable or red when they are dead. Other criteria that have been used as markers for viability and different cell states of probiotic and intestinal microbiota include the assessment of intracellular esterase activity (279), the maintenance of intracellular pH (279), and quantification of the 16S rRNA (276).

1.3.5 Conclusions

The intestinal microbiota is a complex ecosystem showing great variations among individuals and which is influenced by environmental and physiological factors of the host, making its study difficult. Culture-dependent methodologies have been traditionally used for intestinal microbiota assessment, allowing the isolation of some cultivable intestinal microorganisms for their further characterization. The development of culture-independent methods provides more rapid and accurate tools for the study of complex microbial intestinal populations, which has lead in recent years to a significant increase in our understanding of intestinal microbiota composition and its interaction with the host. Most culture-independent techniques target the rDNA, although the rRNA has also been used instead of rDNA, allowing the identification of the metabolically active microorganisms. Techniques such as the PCR-DGGE/PCR-TGGE and especially DNA microarrays that use the information of genome sequences available, greatly contributed to the study of the qualitative content and the predominant species of the intestinal microbiota. However, they failed to provide reliable data on its quantitative content or on the less abundant groups. Among the quantitative culture-independent methods, FISH has been widely used for the assessment of changes in the levels of predominant and minor components of the intestinal microbiota. However, it is laborious and time consuming, which is limiting its further applicability. Real-time quantitative PCR is becoming very promising for studies of intestinal microbiota composition and when it is targeted to the rRNA also provides a very useful tool for monitoring bacterial activity and gene expression in gastrointestinal conditions. The recent development of metagenomics and metaproteomics allows the study, at the same time, of all the genetic material or the proteins present in a sample. In spite of all that is indicated here, since the GIT is a very stressful environment the possibility that microorganisms may exist in several physiological states that could condition their metabolic activity should be taken into consideration and investigated further. All the classical and molecular techniques currently available enhance our understanding of microbial ecology in the gut but at the same time have evidenced that our current knowledge of intestinal microbiota composition and interactions with the host is still limited.

1.4 ENTERIC MICROBIAL COMMUNITY PROFILING IN GASTROINTESTINAL TRACT BY TERMINAL-RESTRICTION FRAGMENT LENGTH POLYMORPHISM (T-RFLP)

TONI A CHAPMAN AND JAMES JC CHIN

Immunology and Molecular Diagnostic Research Unit (IMDRU), Elizabeth Macarthur Agricultural Institute, NSW Department of Primary Industries, Menangle, New South Wales, Australia

Traditional culture-dependent methods of analyzing complex microbial communities such as those found in the GIT of all life forms have been limited because of cultural bias when selective culture medium is used for bacteria isolations. This is almost always associated with the tedium of having to conduct colony enumerations followed by characterization and identification based on metabolic chemistries. Since the primary objective in many community studies is aimed at understanding the diversity and richness of bacteria species that have colonized various niche compartments in the GIT, culture-independent methods of enumeration with minimal cultural bias would be desirable, especially if these methods also enabled identification of OTU or phylotypes. With a rapidly burgeoning database of 16S rDNA sequences as well as a suite of software tools to query alignment homologies and primer design, molecular PCR-based techniques such as denaturing/thermal gradient gel electrophoresis (DGGE/TGGE) (369)), single-stranded site conformational polymorphism (SSCP) (357), and terminal-restriction fragment length polymorphism (T-RFLP) (366) are becoming viable alternative tools for dissecting and analyzing complex microbial communities.

1.4.1 T-RFLP

T-RFLP is a quantitative molecular technique for the analysis of microbial communities and is based on the use of common or universal primers where one of them (usually the forward or *f*-primer) has been fluorescent labeled at the 5′ end (364) with a DNA dye. PCR products amplified in this way from source DNA are then subjected to carefully selected restriction enzyme (usually 4-base cutters) digestion. DNA fragments or digestion products from generated amplicons representing various OTUs are identified by variations in the length of the fluorescent and terminally labeled restriction fragments (*T*RFs). The entire mix of *T*RFs is analyzed by sequencing capillary electrophoresis. Only fluorescent peaks are visualized and profiled based on the length of the nucleotide sequence. Restriction fragments (RF) that are not terminally labeled by the fluorescent primer remain as invisible debris.

1.4.2 Universal and Group-Specific Primers

A review of recent publications on T-RFLP describes more than 95% of citations base microbial diversity analysis on polymorphisms around the small subunit-rRNA gene (SSU 16S rDNA) sequence. The ribosomal database project (RDP) (365) has available an unaligned SSU rDNA sequences of 14,870 nucleotides from which primer sequences can be designed depending upon degeneracy or conservation. The first key requirement of T-RFLP is the selection of "universal" primers to amplify the targeted region of 16S rRNA that is representative of the domain *Bacteria*. The most commonly used universal primer is the 8f-926r domain primer pair proposed by the original developers of T-RFLP (364). Depending upon the needs of the investigator, primer pairs can be universal but relatively specific such as the detection of *Bacteroides/ Prevotella* group in feces (352) or pathogens in prosthetic joints (372). Table 1.6 provides a list of some of the more commonly used primer/probes in T-RFLP. It is important that various applications require more rigorous scrutiny of the universality of universal primers as different combinations of primer pairs can provide quite different levels of diversity coverage for Gram-positive and -negative lineages to division level (358). With time, as the microbial genome database is built from the gold standard approach of assessing sequence polymorphisms and phylogenetic diversity via clone libraries and high-throughput sequencing, it can be anticipated that intergenic spacer (IGS) (371) or internal transcribed spacer (ITS) (356) regions between or within ribosomal operons will be deployed in community profiling analyses. Intending users of T-RFLP are advised to consult the T-RFLP analysis program (TAP) located at the RDP website (367) for guidance in primer designing.

1.4.3 Fluorescent Dyes

The most common fluorophore used in terminal labeling of the forward primer is the blue dye FAM (http://docs.appliedbiosystems.com/pebiodocs/00115046.pdf). It is theoretically possible to use a number of other colored dyes such as HEX, VIC, JOE, and TET (green); TAMRA, NED (yellow); ROX, PET (red), and LIZ (orange). It is possible to increase the interrogative potential of T-RFLP by combining various universal primer pairs in a multiplex PCR (MPX). In this case, different dyes have been tagged to the forward primer of each universal primer pair or different color combinations for different forward primer pairs. We have observed an effect of dyes in shifting the *T*RF size of OTUs when the same universal primer is being evaluated in different dye configurations. The reason for testing different dyes is the potential to combine one-colored universal primer pair with a differently colored group specific primer pair or various other combinations thereof in a MPX reaction. The resolution is dependent on certain dye combinations for each MPX and great care must be taken to optimize the PCR reaction for each specific application to ensure that artifacts are not generated in the TRF profiles. Despite this limitation, the added versatility of different colored TRFs generated from the use of dual-labeled forward and reverse primers increases the complexity of analysis as there will now be two sets of terminally labeled restriction fragments. In this case, one color can provide matching confirmation or not

TABLE 1.6 List of Universal and Specific Primer/Probes Used in T-RFLP[a]

Application (Reference)	16S rDNA Position	Primer or Probe	5′–3′ Sequence
Domain primers bacteria (364)		8f	f-AGAGTTTGATCCTGGCTCAG
		926r	r-CCGTCAATTCCTTTRAGTTT
Domain primers archaea (355)		Ar109F	f- ACKGCTCAGTAACACGT
		Ar912R	r- CTCCCCCGCCAATTCCTTTA
Environmental bacteria (379)	349–365	Bac349F	f-AGGCAGCAGTDRGGAAT
	787–806	Bac806R	r-GGACTACYVGGGTATCTAAT
	516–540	Bac516F	p-TGCCAGCAGCCGCGGTAATACRDAG
Intestinal microflora (375)	46–65	ANA1F	f-GCCTAACACATGCAAGTCGA
	518–536	K2R	r-GTATTACCGCGGCTGCTGG
Human feces (377)	516–532	516f	f-TGCCAGCAGCCGCGGTA
	1510–1492	1510r	r-GGTTACCTTGTTACGACTT
Dental bacteria (376)	331–349		f-TCCTACGGAGGCAGCAGT
	772–797		r-GGACTACCAGGGTATCTAATCCTGTT
	506–528		p-CGTATTACCGCGGCTGCTGGCAC
Eubacteria (378)		27F	f-AGAGTTTGATCCTGGCTCAG
		1492R	r-GGTTACCTTGTTACGACTT
This article universal primer		7f	f- AGAGTTTGAT(C/T)(A/C)TGGCTCAG
		1510r	r- ACGG(C/T)TACCTTGTTACGACTT
This article LAB group specific r-primer		7f	f- AGAGTTTGAT(C/T)(A/C)TGGCTCAG
		LbLMA1r	r- CTCAAAACTAAACAAAGTTTC

[a] Modified from Horz (358).

of community diversity and richness by the other. By using different restriction enzymes, an entire series of profiles can be generated from a simple reaction. Even more complex diversity analysis will become feasible if current T-RFLP protocols can be overlaid with MPX-enabled fluorescent primer pairs for IGS or ITS. These strategies will offset the limitation that more than one OTU may be associated with each *T*RF (368).

1.4.4 DNA Extraction

Various protocols have been used to extract total community DNA. Intestinal washings and fecal suspensions are frequently particulate in texture and can be further disrupted by homogenization in stomacher bags in tryptone–salt solution. Aliquots can be removed at this stage for culture-dependent enumeration. Otherwise, the uniformly dispersed suspension is pelleted by centrifugation, resuspended, and washed in buffer or water, and the bacteria lysed mechanically in the presence of glass or ceramic beads with a bead beater (374). Mechanical disruption of DNA should be limited to 30–60 s to avoid excessive DNA shearing (363). Final selection for DNA extraction methods should be decided upon by trialing different protocols dependent upon the samples undergoing evaluation (363). Washed bacteria cell pellets can also be ground in a mortar/pestle in the presence of 4 M guanidine thiocyanate–150 mM Tris-HCl (pH 7.5)–1% *N*-lauroyl-sarcosine; de-proteinized in phenol–chloroform and precipitated with isopropanol. Occasionally, substances present in feces can be inhibitory to PCR polymerase. Further purification of DNA is then required and an excellent protocol involves the use of benzyl chloride – sodium dodecyl sulfate (373). Genomic DNA is then captured on a membrane in spin column format, and eluted and rehydrated for PCR analysis. Alternatively a number of commercial kits such as UltraClean soil DNA isolation kit by MO BIO Laboratories (354, 359, 360) and Fast DNA kit by QBiogene (363) can be used. The concentration (µg/mL) of purified genomic DNA can be determined by spectrometry as follows: $[-36(A_{280}-A_{320})] + [62.9(A_{260}-A_{320})]$ and its integrity evaluated by agarose gel electrophoresis.

1.4.5 PCR Amplification

Conditions of PCR amplification vary with different applications. In particular, the number of amplification cycles must be optimized to reduce artifacts such as the formation of chimeric amplicons. This bias can be minimized by limiting the number of PCR cycles to 20–35 (370). The pooling of multiple PCR reactions from a single sample can also ensure the minimization of random artifacts (361). With some samples, the presence of "interfering" agents that inhibit PCR reactions can be eliminated by the addition of aluminum ammonium bisulfate (353).

1.4.6 Generation of Terminal Restriction Fragments (TRF) by Digestion of Amplicons with Restriction Enzymes

The selection of restriction enzyme is very important to generate fluorescent-labeled *T*RFs following community DNA amplification with universal primers. Most T-RFLP applications utilize 4-base cutters and depending upon the GC content of the community, restriction enzymes like CfoI (GCG/C); HaeIII (GG/CC); HpaII and MspI (C/CGG) are commonly used for GC-rich communities while others such as AfaI or RsaI (GT/AC) AluI (AG/CT) and MseI (T/TAA) can provide phylotype discrimination for less GC-rich communities. A number of websites are available without subscription fees to assess what restriction enzyme to use against different universal primers under evaluation. One of these, microbial community analysis III or MiCA3 can be accessed at http://mica.ibest.uidaho.edu/trflp.php. *In silico* modeling of *T*RFs can be quite useful in trying to establish actual and theoretical phylotype diversity. On a precautionary note, secondary *T*RF can be a complicating artifact, generating "pseudo" T-RFLP peaks. These can be eliminated by digestion of amplicons with single-strand-specific mung bean nuclease prior to analysis of *T*RF (355).

1.4.7 Software and Data Processing

*T*RFs can be resolved by capillary electrophoresis using systems such as the Applied Biosystems 3730 DNA Analyzer. The 48-capillary analyzer is fitted with argon–ion multi-line, single-mode laser with primary excitation lines of 488 and 514.5 nm. The machine utilizes in-capillary detection by dual-side illumination. Multiple filter sets allow for the reading of five dyes in a single run. Community profiles can be further processed using the STRand software developed by the Davis' Veterinary Genetics Lab at University of California. This software is freeware available at http://www.vgl.ucdavis.edu/informatics/STRand/. Sequencer data files generated by ABI 3730; ABI 377, ABI 373, and MJ GeneSys Base Station are supported by this software. Readers interested in statistical methods for processing and comparing *T*RF data sets including binning, clustering, and statistical analysis should refer to Abdo (351).

1.4.8 Microbial Diversity in Different Intestinal Compartments of Pigs

To illustrate the enteric microbial community profiling capabilities of T-RFLP, we show in Fig. 1.4 the distribution of phylotypes based on *T*RFs in different gastrointestinal compartments of subclinical pigs (see Section 5.4.3). Amplicons were generated using the universal primers 7f and 1510r (Table 1.6). Two restriction enzymes—Hha1 and MspI—were used on different aliquots of the same PCR reaction. In profiles (Fig. 1.4, Table 1.7) representing the duodenal contents, HhaI resolved a very strong *E. coli* peak (*T*RF 370, >2000 FI) while this is not visible in Msp1 digests. Clostridium and Corynebacterium complex are seen as smaller peaks at *T*RF 231 and 358 respectively in HhaI digests while the second most prominent peak in MspI digests is represented by Campylobacter species. Both restriction enzymes generate increased richness of phylotypes in the small intestine (ileum) but the relative abundance in terms of fluorescence intensity (FI) is decreased by about fourfold compared to the duodenum or colon (500 FI vs. 2000 FI). In general, MspI produced a richer OTU profile in the cecum and colon compared to HhaI. These results highlight the importance of using different restriction enzymes to generate more information about community diversity. If this were combined with the use of universal primers and group-specific primers, then it would be possible to mine a plethora of information on

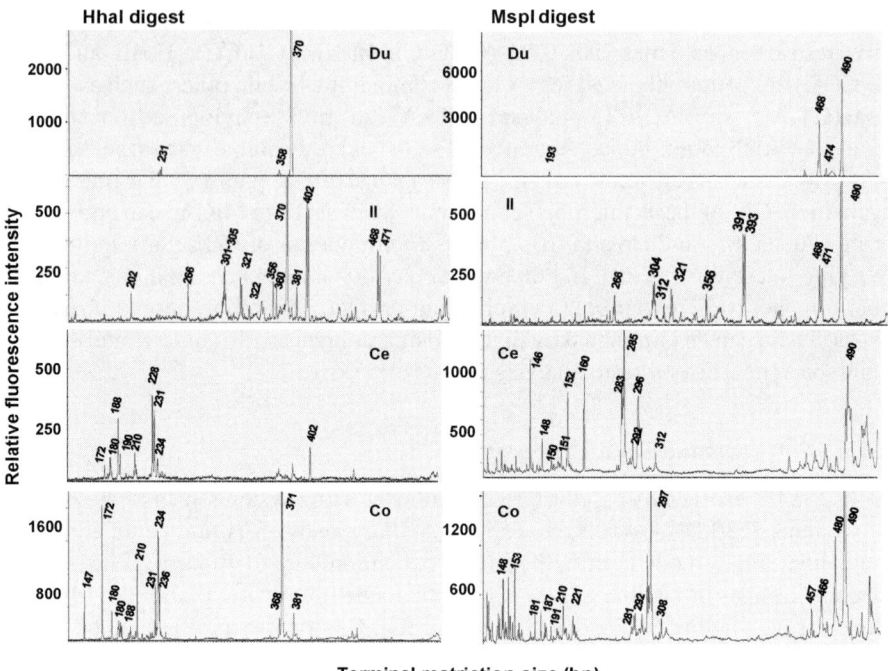

FIGURE 1.4 Enteric microbial community profiles of genomic DNA extracted from different intestinal compartments of pig. HhaI and MspI digests are profiled on the left and right columns respectively, showing profiles from duodenum (Du), ileum (Il), caecum (Ce), and colon (Co) from top to bottom.

the microbial community. We have found the EMCoP profiles very reproducible within assays and between animals in any one treatment group. This enables temporal comparisons to be conducted with treatment protocols such as growth promotants, dietary changes, and therapeutic antibiotics.

1.4.9 Tracking the Fate of Orally Delivered Probiotics in Feces

In Chapter 5, Section 5.4.3, we described a LAB formulation consisting of *L. salivarius, L. casei, L. plantarum,* and *L. acidophilus* that had been developed specifically to target enterotoxigenic *Escherichia coli* (ETECs) strains responsible for neonatal and postweaning diarrhea in pigs. In one on-farm trial, this LAB formulation (ColiGuard) was administered (10^9 cfu/g weaner mash) to newly weaned pigs for 10 days. At this time, rectal swabs or feces can be taken from control and probiotic-supplemented pigs, DNA extracted, and T-RFLP analysis carried out with VIC-LAB-specific group primer pairs 7*f* and LbLMA1*r* (see Table 1.6). Under normalized PCR conditions, the profiles (Fig. 1.5) of pigs in the control group show the presence of only *L. acidophilus* at a comparatively low intensity (738 FI). After 10 days of ColiGuard,

TABLE 1.7 List of Operational Taxonomic Unit Identities Based on Terminally Labeled Restriction Fragments (*T*RF) Generated Following HhaI (A) or MspI (B) Restriction Digestion of Amplicons

Fragment Size (bp)	Organism Name
2A HhaI Digest	
147	*Fibrobacter intestinalis*
172	*Enterococcus saccharolyticus*
180	*Bifidobacterium thermophilum*
188	*Eubacterium hallii*
190	*Clostridium symbiosum, Clostridium clostridiiforme*
202	*Eubacterium ruminantium*
210	*Fusobacterium simiae*
231	*Clostridium botulinum, Clostridium scatologenes, Clostridium tetani, Clostridium sporogenes, Clostridium collagenovoransm, Clostridium ljungdahlii, Clostridium algidicarnis*
234	*Lactobacillus* sp.
236	*Clostridium subterminal, Clostridium* sp.
266	*Lactobacillus mucosae*
301–305	*Flexibacter filiformis*
321–322	Unidentified
356	*Corynebacterium variabilis*
358	*Corynebacterium genitalium, Corynebacterium pseudogenitalium*
360	*Clostridium tetanomorphum*
368	*Desulfovibrio desulfuricans*
370	*E. coli*
381	*Desulfotomaculum thermosapovorans*
402	*Lactobacillus fermentum*
	Lactobacillus reuteri
468	*Streptomyces tendae, Streptomyces diastatochromogenes, Streptomyces bottropensis, Streptomyces scabiei, Streptomyces coelicolor, Streptomyces ambofaciens, Streptomyces ornatus, Streptomyces nodosus, Streptomyces caelestis*
471	*Streptomyces bluensis, Streptomyces mashuensis, Streptomyces vellosus*
2B MspI Digest	
132	*Fibrobacter intestinalis*
134	*Desulfovibrio gigas*
141	*Eubacterium dolichum*
148	*Bacillus badius,* Bacillus *firmus*
150	*Fusobacterium russii*
151	*Fusobacterium varium*
152	*Fusobacterium moriferum, Fusobacterium gonidiaformans*

(*continued*)

TABLE 1.7 (*Continued*)

Fragment Size (bp)	Organism Name
153	*Bacillus macroides, Bacillus benzoevorans, Bacillus sphaericus, Bacillus fusiformis*
160	*Desulfotoaculum halophilum*
181	*Lactobacillus gallinarum*
187	*Leptotrichia* sp.
191	*Lactobacillus gasseri, Lactobacillus crispatus*
193	*Clostridium irregularis, Clostridium bifermentans*
210	*Desulfotomaculum thermobenzoicum*
221	*Eubacterium hallii, Eubacterium barkeri, Eubacterium limosum*
266	*Clostridium filamentosum*
281	*Fusobacterium simiae*
282	*Ruminococcus flavefaciens*
283	*R. flavefaciens*
285	*Bifidobacterium inopinatum*
292	*Clostridium ramosum*
296	*Eubacterium desmolans, Eubacterium yurii* subsp.
304	*Eubacterium* sp.
308	*Eubacterium* sp.
312	*Butryvibrio fibrisolvens*
321	*B. fibrisolvens*
356, 391, and 393	Unidentified
468	*Camplyobacter* sp., *Camplyobacter showae, Camplyobacter concisus, Camplyobacter rectus*
471	*Camplyobacter* sp.
474	*Peptostreptococcus anaerobius*
480	*Flexibacter flexilis*
490	*E. coli*

the probiotic-treated group shows significant increases in the *L. acidophilus*, *L. casei*, and *L. plantarum* peaks (note log increase in FI scale). *L. acidophilus* is located at position TRF247 and has an increased FI of 29.3-fold relative to nonprobiotic supplemented pigs. Due to redundancy in the 16S rDNA sequence for the LAB primer pair, *L. plantarum* and *L. casei* are phylotyped at positions TRF 327 and 328 respectively and therefore appear as two very close proximity peaks with fluorescent intensities of 8061 and 32,088 respectively. Since each animal would consume on the average about 300 g of feed per day, one would expect delivery of about 7.5×10^{11} cfu of each of the four LAB strains per day. This would contribute to the significant increases observed in FI of all LAB phylotypes with the possible exception of *L. salivarius* (projected location at TRF 279 and 281). Pending further analysis, the fate of *L. salivarius* in the GIT of probiotic-treated pigs can only be speculated at this point. It may be that *L. salivarius* is a very efficient colonizer with most community members domiciled in as yet undefined intestinal compartments. Alternatively,

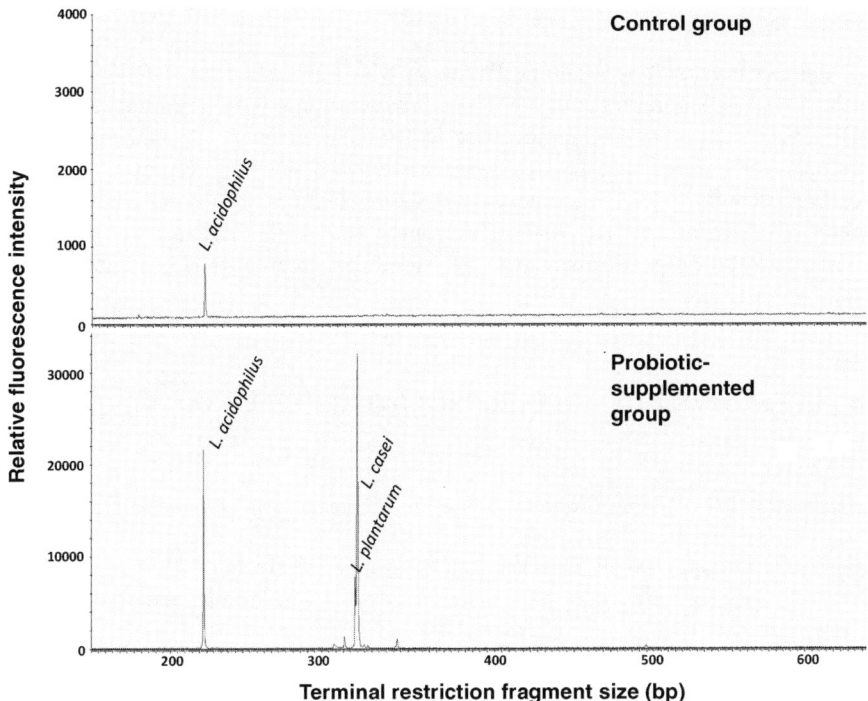

FIGURE 1.5 EMCoP profile of LAB in DNA extracted from the feces of pigs treated with ColiGuard probiotic and nonprobiotic-supplemented animals (control).

L. salivarius may have limited viability and may not be as competitive as other strains, dying off rapidly after it has delivered its impact in altering the gene signatures of ETECs in the GIT of subclinical pigs (Section 5.5.2). In any event, this example clearly documents the usefulness of T-RFLP in tracking the LAB community and if combined with time course sampling, will provide a temporal estimate of persistence of orally delivered probiotics.

1.4.10 Conclusion

Even though TRFLP was first developed for the analysis of community diversity in soil and environmental samples, its adaptation to gut microflora has provided a strong impetus in its use as a tool to unmask the impact of pre- and probiotics in the GIT. It is a far simpler procedure to use than DGGE, highly reproducible, amenable to fine-tuning at the level of primer design, and restriction fragment generation. T-RFLP also does not suffer from the disadvantage of longer primer sequences needed to design GC clamps for the DGGE procedure that can in turn cause artifacts during the annealing step; as well as the production of heteroduplexes that are innately unstable under the denaturing conditions of a DGGE run (362). We anticipate that T-RFLP will become a very important analytical protocol in enteric microbial community profiling.

Acknowledgments

The authors acknowledge Jannine Patterson who established the T-RFLP protocol in the IMDRU laboratory and Kent Wu for contributions of group-specific LAB T-RFLP. The expertise and input of Bernadette Turner merits special mention for technical work well done. This program was funded substantively by a Commonwealth Research and Development Start Grant awarded to International Animal Health (C. Lawlor and K. Healey) in conjunction with CSIRO and NSW DPI. ColiGuard® probiotic strains were provided by International Animal Health, Sydney, Australia.

1.5 EFFECTIVE DOSAGE FOR PROBIOTIC EFFECTS

YUAN KUN LEE

Department of Microbiology, National University of Singapore, Singapore

Among the human clinical trials cited in Section 4.3 (Effects on human health and diseases), the probiotics were administered in great variation in accordance to the following:

- Type of probiotics (lactobacilli, bifidobacteria, yeasts, enterococci);
- Daily dose (10^7–10^{10} cfu);
- Daily frequency of administration (1–4 times);
- Timing of administration (before, during, and after meal);
- Duration of administration (1 day to several months);
- Method of delivery (fermented food, beverage, capsule, tablet, or powder);
- Viability.

To achieve probiotic effects, the probiotics and their products need to be delivered to the desired gastrointestinal site in sufficient quantity. The importance of viability depends on the mechanism of the probiotic effect and each probiotic bacteria needs to be evaluated respectively. The method of delivery appears to have minimal effect on probiotic efficacy, as different preparations of the same dose were reported to achieve the same preventive or therapeutic efficacy. An example is the treatment of diarrhea (Section 4.3.3). The duration of administration would depend largely on the needs and nature of the diseases; for example treatment for diarrhea is short term whereas cancer prevention is of longer term.

There is no information as to when is the best time to administer probiotic preparation. It is logical to assume that probiotics administered orally before meal should have the capability to tolerate the extreme pH condition and digestive enzymes and bile present in the intestinal tract. Probiotics taken together with meal would be diluted by food materials, which could reduce the chances and frequency of physical encounter between the probiotic organisms and the mucosal receptors. Moreover, food matrix may compete with mucosal receptors for probiotic and product binding. Hence it is reasonable to assume that the best period for the administration of probiotics is between meals, and be carried in liquid media.

It is not so clear if the frequency of administration has any effect on probiotic efficacy. Microbiologically, 1×10^{10} cfu administered four times daily has little different from 4×10^{10} cfu administered once a day. It could nevertheless be assumed that a probiotic strain, which does not adhere well on the mucosal receptors and is unable to colonize temporarily would need to be administered more frequently. And probiotic strains that are denatured readily by the gastrointestinal conditions should be administered in larger dosage to counter the wastage.

This leaves us with the remaining factors of dosage and strain of probiotics. How much a probiotic needs to be consumed to achieve the probiotic effects is crucial information in the formulation of probiotic functional foods as well as the therapeutic products. So far, there is no systematic study on the effective dosage of the respective probiotics for specific applications. We could nevertheless obtain a glimpse from the human clinical studies available. The dose effect of probiotics on specific disease also shed light on the mechanisms of probiotic effects and the interaction between probiotic organisms and the host.

1.5.1 Acute (Rotavirus) Diarrhea in Children

Probiotics used as adjunctive therapy appeared to improve the treatment of acute diarrhea in 18 out of the 23 clinical trials conducted among children (Section 4.3.3.1). The trials involved more than 1800 children. When the mean days of reduction in diarrhea in the probiotic group in comparison to that of the placebo group were plotted against the dosage of the respective probiotic administered per day, no direct correlation between the treatment efficacy and the dosage could be recognized, as shown in Fig. 1.6. No direct correlation was observed even among the same probiotic of different dosage.

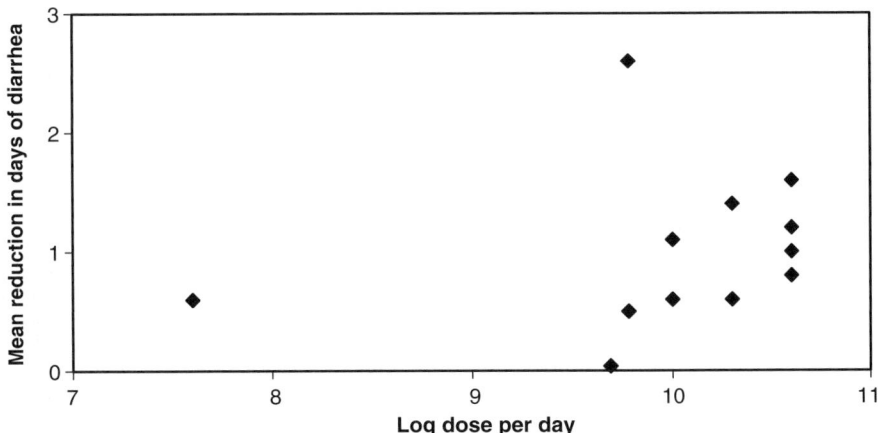

FIGURE 1.6 The efficacy of daily probiotic dose on the recovery from acute diarrhea among children. The data were extracted from Section 4.3.3.1.

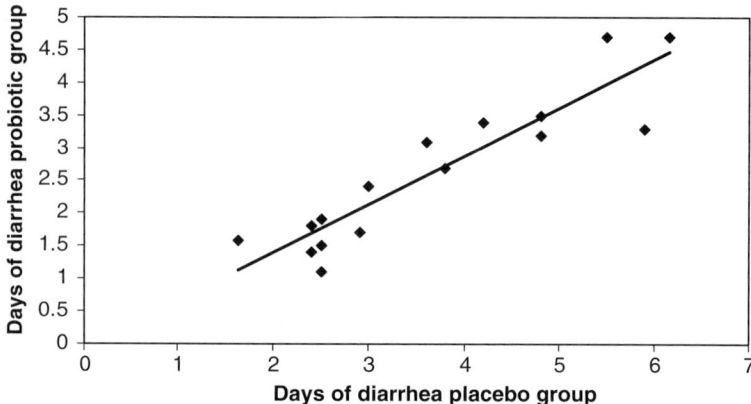

FIGURE 1.7 Correlation between days of diarrhea in probiotic group and corresponding placebo group among children suffering from acute diarrhea. The data were extracted from Section 4.3.3.1.

When the days of diarrhea in the probiotic group was plotted against the corresponding placebo group, a linear correlation was obtained (Fig. 1.7) with a slope of 0.78, which intercepts the origin. This implies that a constant reduction of 22% in the number of days of diarrhea was achieved through the consumption of all the probiotics tested, namely *L. acidophilus, L. bulgaricus, L. reuteri, L. rhamnosus, Streptococcus thermophilus, B. infantis*, and *Saccharomyces boulardii*, applied singly or in various combination. The efficacy was the same with daily dose of 4×10^7 to 6×10^{10} cfu, suggesting that the probiotics have a common mechanism in relieving acute diarrhea, which contributes to 22% of the cure of diarrhea. It can be concluded that probiotics could assist in the speedy recovery of acute diarrhea among children but are not able to prevent and cure diarrhea.

1.5.2 Antibiotic-Associated Diarrhea

Probiotics were widely reported to reduce the incidence of antibiotic-associated diarrhea (Section 4.3.3.2). The probiotics were used singly (*L. rhamnosus GG, Saccharomyces boulardii*) or in combination (*L. acidophilus* + bifidobacteria or *Streptococcus thermophilus*).

1.5.2.1 Combination of L. acidophilus + bifidobacteria or Streptococcus thermophilus When the probiotics were used in combination involving 194 subjects, a 0.5 risk ratio was achieved at a daily dose of 6×10^8 cfu (Fig. 1.8A). The risk factor of the combined probiotics decreased with the daily dose in a hyperbolic manner to reach a value of 0.3 in antibiotic-associated diarrhea at a daily dose of 1×10^{11} cfu. The interaction could be described as the competition for specific receptor binding between the probiotics, pathogens and the host surface (380).

Let us assume that (a) diarrhea is caused by the adhesion of diarrheic microbes on the intestinal surface and the biochemical reactions initiated, (b) the protective effect

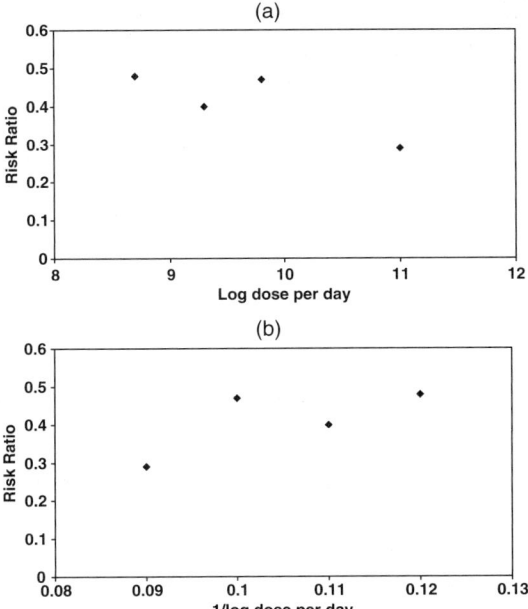

FIGURE 1.8 Dose-dependent efficacy of combination of *L. acidophilus* + bifidobacteria or *Streptococcus thermophilus* in the prevention of antibiotic-associated diarrhea. The data were extracted from Section 4.3.3.2.

of probiotic bacteria is due to the competition for binding onto intestinal surface and it is a simple dissociation process:

$$\text{Microbial cell} + \text{Intestinal cell} \underset{K-1}{\overset{K+1}{\rightleftarrows}} \text{Microbe} - \text{Intestinal Cell Complex}$$

Where $K + 1$ and $K - 1$ represents the association and dissociation constant of the reaction, respectively. The process is similar to the interaction between a substrate and the receptor on an enzyme that forms a substrate–enzyme complex, but without the subsequent formation of products.

There are three assumptions in the relationship:

1. The interaction between the microbial cells and the intestinal cell surface receptor remains in equilibrium. This condition should be achieved if the microbial cells do not penetrate the intestinal cells.
2. Microbial concentration remained essentially unchanged throughout the clinical studies, so that the concentrations of the microbial cells can be considered equal to the initial microbial concentrations. This condition could be achieved when the total number of microbial cells present is much greater than the number of microbial cells adhering to the intestinal surface. This is probably the case, where the concentration of the probiotic bacterial cells consumed is

usually in the range of 10^5–10^8 per mL, whereas the number of bacterial cells adhering to the intestinal cells is fewer than 10 per cell.

3. In the simple dissociation equation described above, if X is the concentration of the microbial cell suspension, e is the epithelial cell or mucus concentration, and e_x is the concentration of the microbe–intestinal cell–mucus complex, then the concentration of free epithelial cells or mucus will be $(e - e_x)$.

Since the process is in equilibrium, the dissociation constant for the process (K_x) can be defined as

$$K_x = \frac{K-1}{K+1} = (e-e_x)\frac{X}{e_x}.$$

This equation can be rearranged to give an expression for the concentration of the microbe–intestinal cell–mucus complex,

$$e_x = \frac{eX}{K_x+X}.$$

When X is very much larger than K_x, the intestinal cells or mucus is saturated with microbial cells (i.e., e_x approaches e), and the maximum value of e_x, e_m is obtained. As it is technically easier to estimate the maximum concentration of adhered microbial cells (e_m) than the epithelial cells/mucus concentration (e), the equation could thus be re-written as

$$e_x = \frac{e_m X}{K_x+X}. \tag{1.1}$$

The equation could be further re-arranged to give a linear relationship,

$$\frac{1}{e_x} = \frac{1}{e_m} + \frac{K_x}{e_m X}. \tag{1.2}$$

The values of e_x and K_x are independent of each other. That is, a microbe that adheres on intestinal surface in large number could have a low affinity for the intestinal surface receptors and vice versa.

In the case where the probiotic bacteria and the diarrheic microbes are present at the same time and compete for the same receptors on the intestinal surface, the competition for adhesion of each of the microorganisms is determined by the affinity of the competing organisms to the intestinal surface (K_x) and the concentration of the microbial cells (X). Thus, the ratio of e_x for probiotic bacteria (p) and diarrheic microbes (d) in the mixed microbial system can be described as

$$\frac{e_{xp}}{e_{xd}} = \frac{e_{mp}}{e_{md}} \frac{X_p}{X_d} \frac{K_{xd}+X_d}{K_{xp}+X_p}. \tag{1.3}$$

The relationship in Equation 1.3 above suggests that the outcome of competition between two microorganisms for adhesion on the same receptors on intestinal surface is determined by the ratio of the respective microbial concentrations around the receptors and the affinity of the respective microbes for the receptors.

If the concentrations of the diarrheic microbes were of comparable magnitude in the clinical trials reported, the values of e_{xd}, (e_{mp}/e_{md}), X_d, K_{xd}, K_{xp} were constant values.

$$\frac{e_{xd}}{e_{xp}} = \frac{1}{e_{mp}/e_{md}} \frac{1}{X_d} (K_{xd}+X_d) \frac{K_{xp}+X_p}{X_p} \frac{e_{xd}}{e_{xp}} = \frac{K' K_{xp}}{X_p+K'}. \quad (1.4)$$

Since e_{xd}/e_{xp} determines the risk ratio in the clinical studies, the plot of the risk ratio versus the 1/(daily dose of probiotic, X_p) should yield a linear relationship as shown in Fig. 1.8B. The data points in Fig. 1.8B are too few to obtain a statistically meaningful linear plot. Nevertheless the dosage where risk ratio is 1 (total prevention of diarrhea) is estimated at 10^{13} probiotic bacteria per day.

1.5.2.2 *L. rhamnosus GG* or *Saccharomyces boulardii* Applied Singly

In the cases of *L. rhamnosus* GG involving 281 subjects or *S. boulardii* involving 888 subjects, applied singly, near 90% of preventive efficacy was achieved at a daily dose of about 1×10^{10} cfu (Fig. 1.9). At above a daily dose of 1×10^{10} cfu, the diarrhea-preventing efficacy dropped precipitously. Negative corporative effect, where the binding of a probiotic on the mucosal surface receptor resulted in a reduction in affinity for the subsequent bacterial binding could have occurred (380). It is a demonstration that maximal probiotic efficacy is achieved at the optimal probiotic dose, which needs to be determined for respective probiotic strains.

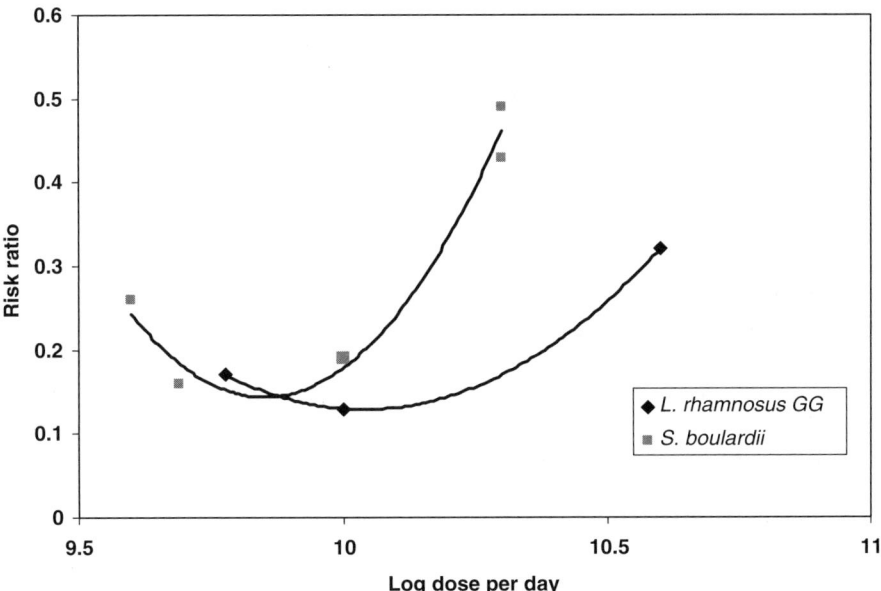

FIGURE 1.9 Dose-dependent efficacy of *L. rhamnosus* GG and *S. boulardii* in the prevention of antibiotic-associated diarrhea. The data were extracted from Section 4.3.3.2.

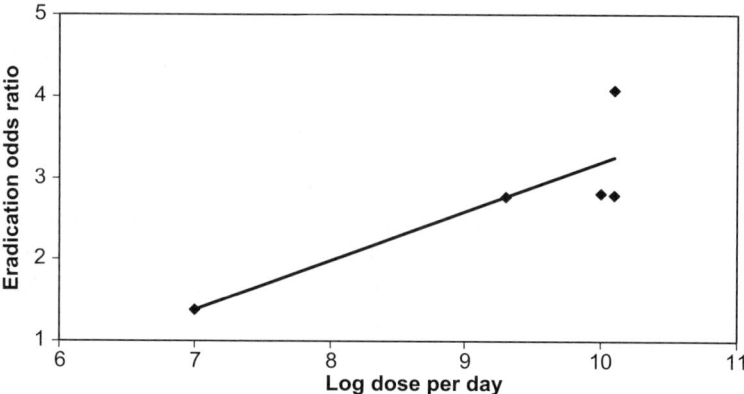

FIGURE 1.10 Dose-dependent efficacy of probiotics (*L. acidophilus, L. casei, L. rhamnosus, Propionibacterium freudenreichii, B. breve, B. animalis* singly or in combination) in enhancing the eradication of *H. pyroli* during antibiotic treatment. The data were extracted from Section 4.3.3.2.

1.5.3 *Helicobacter pyroli*

Some probiotics have been demonstrated to enhance the eradication of *H. pyroli* during antibiotic treatment and reduce the occurrence of side effects. The eradication efficacy involving 478 subjects was dose dependent (Fig. 1.10), and the minimal effective dose is estimated to be about 5×10^6 cfu per day, when the graph in Fig. 1.10 intercepts the eradication odds ratio of 1.

1.6 INCORPORATING PROBIOTICS INTO FOODS

Ross Crittenden
Food Science Australia, Australia

Probiotics have for decades been used in fermented dairy products such as yogurts and fermented milks. The techniques and technologies to incorporate these organisms into fresh, refrigerated dairy products are now relatively mature. The continuing emergence of clinical evidence for benefits to consumers and the subsequent marketing power these ingredients bring have now seen probiotics become the fastest growing category of functional food ingredients (458). Food companies worldwide are seeking ways to incorporate these ingredients into a much broader range of foods and beverages. However, incorporating live probiotic microorganisms into foods and then keeping them alive throughout shelf life is a significant challenge for food technologists. Indeed, it is an anathema to usual food-processing methods and matrices that have always been designed to minimize the survival of microorganisms with food safety considerations foremost in mind.

Although the need for probiotic viability for some health impacts, such as immunomodulation (390, 402), may not require the bacteria to be alive, viability may still be an essential property of probiotics for some health effects (421). Indeed, probiotics remain defined as *live* microorganisms that when administered in adequate amounts confer a health benefit on the host (407). Viability (in fact, more accurately defined as cultivability) is in reality only a convenient surrogate marker of probiotic activity. However, since the ability of probiotic organisms to impart benefits to the health of the host are usually only quantifiable through animal or clinical studies, viability remains the only really practical quality assurance measure for probiotics. Health benefits have usually been attributed in clinical studies to doses of probiotics in excess of 10^8–0^9 viable cells per day (427). Therefore, food regulatory/advisory bodies generally stipulate that foods containing probiotic organisms need to have $>10^6$–10^7 cfu/g at the time of consumption (e.g., International Standard of Fédération Internationale de Laiterie/International Dairy Federation (403)).

The viable count of probiotic organisms generally declines during product storage (10–100-fold or more) (433). An acceptable viable count can sometimes be achieved by introducing higher numbers of probiotics during manufacture (called overage). The consumption of probiotic organisms at high doses is safe (440, 452, 463), and so oversupplying consumers does not appear to pose a health risk. However, in practice the addition of considerable overage can be an expensive proposition given the relatively high cost per weight of probiotic cultures as ingredients. There may also be organoleptic limitations to the amount of a probiotic that can be acceptably added to foods. Therefore, there is a strong imperative to maintain the viability of probiotics in foods during production and shelf storage.

Overall, there are five main points to address when incorporating probiotics into foods:

1. Select a compatible probiotic strain/food type combination.
2. Use food-processing conditions that are compatible with probiotic survival.
3. If fermentation is required, ensure that the food matrix will support probiotic growth.
4. Select a product matrix, packaging, and environmental conditions to ensure adequate probiotic survival over the product's supply chain and during shelf storage.
5. Ensure that addition of the probiotic does not adversely impact on the taste and texture of the product.

This chapter summarizes the main parameters that affect probiotic survival during manufacture and storage of foods and provides examples of successful incorporation of probiotics into a range of shelf-stable foods. It aims to provide food technologists with the knowledge to select and incorporate suitable probiotics into foods beyond the traditional fermented dairy food sector, and to maximize their survival over extended shelf lives.

1.6.1 Probiotic Ingredients

Probiotic organisms are predominantly bacteria selected from the genera *Lactobacillus* and *Bifidobacterium*, which are normal constituents of the human intestinal microbiota. A range of different species within each genus is commonly used and within each species there are particular strains that have been shown to have probiotic attributes (Chapter 6). Considerable strain-to-strain differences have been observed within species and probiotics are generally defined down to the strain level (for example, *Lactobacillus rhamnosus* GG and *Lactobacillus rhamnosus* LC-705) (406). Probiotic organisms have typically been selected via screening regimes to perform well technologically, to survive intestinal transit, and to impart health benefits on consumers. Good probiotic strains have demonstrated health and safety data from randomized, controlled clinical trials.

Probiotics organisms are usually supplied by manufacturers of these ingredients as either dry powders (freeze-dried or spray-dried) at 10^{10}–10^{12} cfu/g or as frozen "direct vat set" concentrates at 10^9–10^{10} cfu/g (430). When received as ingredients, it is important that the probiotic be correctly stored as per the manufacturer's instructions in order to avoid rapid losses in probiotic viable counts. For dried powders, this means storing the probiotics cold and avoiding moisture or humidity, while for frozen cultures is it important to maintain constant temperatures and to avoid repeated freeze–thawing.

Probiotics can be incorporated into foods and beverages in a variety of ways.

- Dry blended into foods and powders such as infant formulas.
- Dispersed into liquid or semiliquid products such as juice or ice-cream.
- Inoculated into fermented products such as yogurts and fermented milks.

In the first two cases, the probiotics do not multiply in the product and are generally added at doses in the order of 10^7–10^8 cfu/g. For a standard probiotic freeze-dried powder at 10^{11} cfu/g, this represents addition of the probiotic at 0.01–0.1% (w/w) of the final product. In fermented products there may be some growth and increase in probiotic numbers during fermentation, allowing a lower number of organisms to be initially added (for example 10^6 cfu/g). The number of viable probiotic organisms then usually declines during product storage, with the rate of decline dependent on a range of factors as discussed in the following sections. Ensuring losses in probiotic viability are minimized is the one of the main goals for food technologists developing foods containing probiotics.

1.6.2 Factors Affecting the Viability of Probiotics in Foods

A number of intrinsic and extrinsic factors influence the survival of probiotics in foods. It is important to consider these factors at all stages between addition of the probiotic to the food and delivery of the probiotic to the gut of the consumer. These include manufacturing processes, food formulations and matrices, packaging materials, and environmental conditions in the supply chain and during self-storage. The main factors to be considered that may influence the ability of the probiotics to survive in food products include:

1. the physiological state of the added probiotic;
2. the physical and chemical conditions of food processing;
3. the physical conditions of product storage (e.g. temperature);
4. the chemical composition of the product (acidity, nutrients, moisture, oxygen);
5. interactions with other product components (inhibitory or protective).

The first stage of product development is to align a compatible combination of probiotic strain(s) and food product(s).

1.6.2.1 Choice of Probiotic Organism/Food Combinations Probiotic organisms are generally selected from constituent intestinal lactobacilli and bifidobacteria, which have evolved to grow and survive in environmental conditions within the human intestinal tract. In the small intestine and colon, the pH is generally close to neutral, the temperature is constant (37–39°C), a complex nutrient supply is constantly available and there is little oxygen. These conditions are of course very different to those found in food processes and food matrices. Nonetheless, bacteria show a remarkable ability to survive in adverse environments and probiotics can survive in food environments, to a point.

Probiotic ingredients are not all the same. Differences extend from the genus to the species and even strain level, and apply both to their physiological impacts on the consumer and to their technological attributes in foods (406, 410). The closer probiotic organisms are related, usually the more similarly they will perform. However, considerable strain-to-strain differences are still apparent that can significantly impact on the performance of probiotics in foods (Fig. 1.11).

The differences in the technological characteristics of different probiotic species and strains means that care must be taken in selecting the most appropriate strain for a particular food application. Indeed, the first step in incorporating a probiotic into a food is identifying compatibilities between the attributes of the selected strains and the food production steps, food matrix and storage conditions. This may involve a compromise between the desired health attributes and technological capabilities of particular strains for particular food applications. When developing new products some research may be required to ensure that the selected strain is able to survive well in the food, provide the appropriate technological properties (e.g. acidification during fermentation, if required) and importantly, that the added probiotic does not adversely affect the taste, smell, and texture of the food or beverage.

While emphasizing the importance of strain specificity of technological attributes of probiotics, some generalizations can still be made on the robustness of probiotic organisms. Generally, lactobacilli are more robust than bifidobacteria (406, 431, 449). There is a wider range of probiotic *Lactobacillus* species that are technologically suitable for food applications than bifidobacteria. Common examples include *L. acidophilus, L. johnsonii, L. rhamnosus, L. casei, L. paracasei, L. fermentum, L. reuterii* and *L. plantarum*. Often, the *L. acidophilus* group of organisms, while resistant to low pH, prove less robust than other lactobacillus

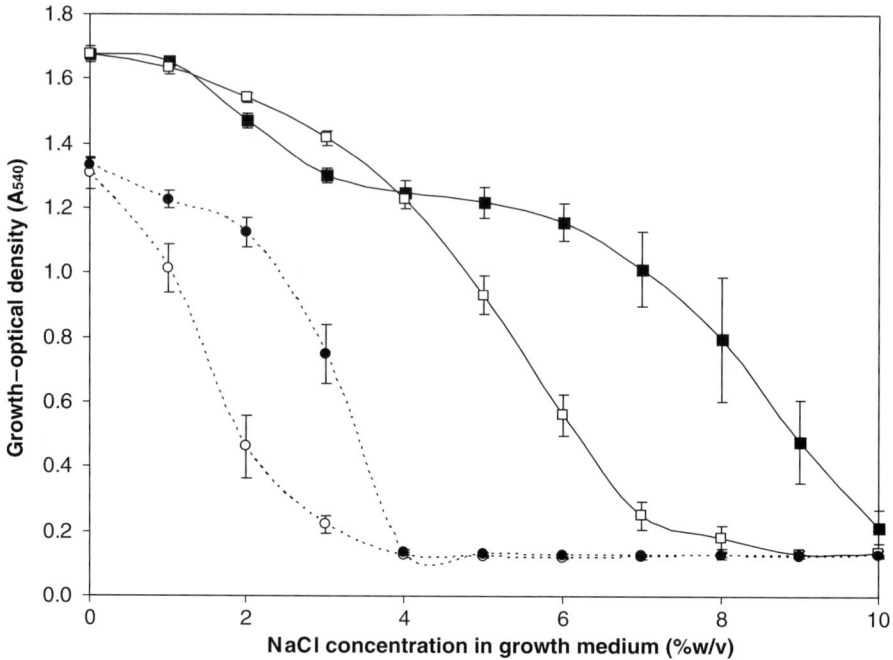

FIGURE 1.11 Inter- and intraspecies variation in the ability of probiotic lactobacilli to grow in the presence of salt. This characteristic is important, for example, in meat fermentation. The figure demonstrates the importance of selecting probiotic species and strains that are compatible with a particular food. The lactobacilli were grown in MRS broth containing various concentrations of NaCl, at 37°C, for 24 h. Error bars represent 1 standard deviation from the mean of three biological replicates. *Lactobacillus rhamnosus* GG (■); *Lactobacillus rhamnosus* CSCC 5277 (□); *Lactobacillus acidophilus* MJLA1 (●) *Lactobacillus acidophilus* CSCC 2401 (○). From Crittenden R, Morris L, and Playne MJ. Unpublished data, Food Science Australia.

species in non-traditional probiotic food applications (391, 413, 415, 445, 453, 461). The *Bifidobacterium* species most commonly used in foods is *B. animalis* subsp. *lactis* (398). This species is significantly more robust than human intestinal species such as *B. longum* (*infantis*), *B. breve*, and *B. bifidum*, although certain strains of these species are able to survive well in some foods (395, 398, 409). *B. adolescentis* is a common species in the intestinal tract of adult humans, but tends to be sensitive to environmental conditions in foods (395, 457, 462) and is rarely used commercially as a probiotic.

The metabolism of the probiotic organism is an important consideration in fermented probiotic foods, not only for probiotic growth and survival, but also for food quality. For example, heterofermentative lactobacilli that produce CO_2 as a metabolic end product are not suitable where gas formation adversely impacts on food quality (384). Bifidobacteria produce acetate and lactate as end products of carbohydrate fermentation, and a more vinegar-like taste profile if they are actively fermenting in food products (395, 404).

The ability to utilize the available carbon and nitrogen substrates in a product may be required for probiotic growth and acidification (455). Lactobacilli and bifidobacteria can generally utilize a wide range of carbon substrates, with differences in the carbon substrate profiles occurring between species and strains. Probiotic strains may also be selected on the basis that they metabolize desirable bioconversions, such as deconjugation of isoflavones in soy (375), or indeed, because they do not metabolize other ingredients in the food. For some fermented foods where LAB form part of the native microbiota (e.g. fermented meats), an approach has been taken where the probiotic attributes of cultures isolated from these foods (and therefore known to survive well) have been examined to select new probiotic strains (422, 426, 443).

1.6.2.2 Physiologic State of the Probiotic An important factor in probiotic survival is the physiological state of the bacteria when prepared, and the physiological state of the bacteria in the product itself. If the food product is dry (e.g., a powdered infant formula) the probiotic will also be dried and in a quiescent state during storage. However, when included in a wet product such as a yogurt, the bacteria will be in a vegetative state and potentially metabolically active (albeit slowly at refrigeration temperatures). The state of the bacteria will have a large bearing on the possible shelf life of the bacteria, with long-term survival of vegetative cells only possible at low temperatures (403). In comparison, dried, quiescent cells may have longer shelf lives at ambient temperatures, though they too will be more stable at lower temperatures.

Bacteria are able to respond to stressful environments through the induction of various stress tolerance mechanisms. The induction of stress proteins by exposure of the cells to sublethal stresses such as heat, cold, starvation, low pH, and osmotic tension can condition probiotics to better tolerate environmental stresses in food production, storage, and gastrointestinal transit (449, 456). Cross-protection has often been observed, where exposure to one stress provides protection against other stresses (403, 449, 456). The main point to emphasize is that while different probiotic strains have their own intrinsic tolerances to environmental conditions, tolerance can also be influenced by how the culture is prepared. Stress responses can be exploited to make probiotic strains more resilient and likely to survive in food matrices.

1.6.2.3 Temperature The temperature at which probiotic organisms grow is important in food applications where fermentation is required. The optimum temperature for growth of most probiotics is between 37°C and 43°C (395, 403). Species of bifidobacteria isolated from the human intestinal tract such as *B. longum (infantis), B. breve, B. bifidum,* and *B. adolescentis* have optimum growth temperatures in the range of 36–38°C, whereas *B. animalis* subsp. *lactis* can grow at higher temperatures of 41–43°C (398, 403). Usually no growth is observed for bifidobacteria at temperatures below 20°C or above 46°C (403). Probiotic lactobacilli can grow well over a similar temperature range though some can grow at up to 44°C and at mesophilic temperatures down to 15°C (454).

Temperature is also a critical factor influencing probiotic survival during manufacture and storage. In practical terms, the lower the temperature the more stable probiotic viability in the food product will be. During processing, temperatures above 45–50°C will be detrimental to probiotic survival. The higher the temperature, the shorter the time period of exposure required to severely decrease the numbers of viable bacteria, ranging from hours or minutes at 45–55°C to seconds at higher temperatures. It is obvious that probiotics should be added downstream of heating/cooking/pasteurization processes in food manufacture.

Elevated temperature also has a detrimental effect on stability during product shipping and storage. Again, the cooler a product can be maintained, the better probiotic survival will be. For vegetative probiotic cells in liquid products, refrigerated storage is usually essential (403). In dried products containing quiescent bacterial cells, acceptable probiotic viability can be maintained in products stored at ambient temperatures for 12 months or more. As discussed later, there is a substantial interaction between temperature and water activity. Therefore, producing and maintaining low water activities in the foods is the key to maintaining probiotic viability during nonrefrigerated storage.

1.6.2.4 pH Lactobacilli and bifidobacteria produce organic acid end products from carbohydrate metabolism. Hence, these genera can tolerate lower pH levels than many bacteria. Indeed, numerous *in vitro* and *in vivo* studies have demonstrated that probiotic organisms can survive gastric transit where the cells are exposed to pH values as low as 2.0, though the time of exposure (1–2 h) is relatively short (398, 452). Adapted vegetative cells are usually able to survive better in acidic environments compared to quiescent cells (448).

In food products, lactobacilli are able to grow and survive in fermented milks and yogurts with pH values between 3.7 and 4.3 (395). Bifidobacteria tend to be less acid tolerant, with most species surviving poorly in fermented products at pH levels below 4.6 (395, 449). Again, *B. animalis* subsp. *lactis* is more acid tolerant than human intestinal species and, hence, is the species of *Bifidobacterium* most commonly used in acidic foods (398). A recently described phenotypic group, *B. thermoacidophilum* is even more tolerant to low pH (and heat) (398, 403), but has not been characterized thoroughly for probiotic traits and is not used commercially, at least so far.

Survival in low pH beverages such as fruit juices (pH 3.5–4.5) posses a significant challenge to probiotic survival, but commercially successful products have been produced, such as Gefilus (Valio Ltd, Finland), which contains *Lactobacillus rhamnosus* GG. Carriers such as dietary fibers have been shown to improve viability at low pH (451). Survival of lactobacilli in acidic environments has also been enhanced in the presence of metabolizable sugars that allow cell membrane proton pumps to operate and prevent lowering of intracellular pH (397). This can improve survival during gastric transit, but may not be applicable to improving probiotic survival over the time frames of shelf-storage.

1.6.2.5 Water Activity For quiescent, dried, probiotic bacteria water activity is a crucial determinant of survival in food products during storage (430, 434, 436, 444). As

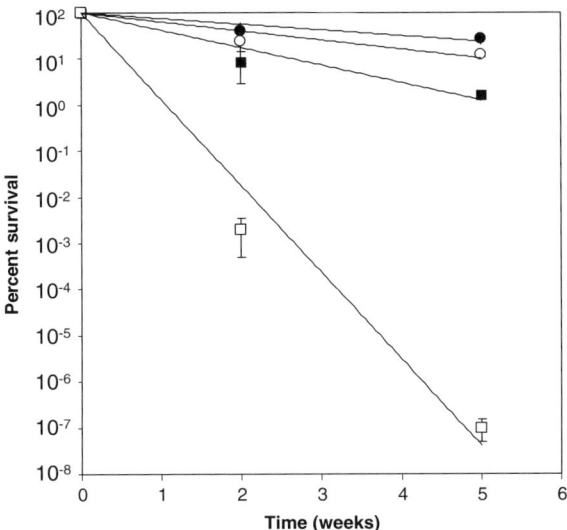

FIGURE 1.12 The impact of water activity on the survival of a freeze-dried probotic *Lactobacillus rhamnosus* strain stored at 38–40°C. The probiotic was stored in the same type of food matrix, but the water activity controlled at various levels: $a_w = 0.1$ (●); $a_w = 0.15$ (○); $a_w = 0.22$ (■); $a_w = 0.25$ (□). Error bars represent the standard deviation from two to three biological replicates. From Crittenden R, Weerakkody R, and Sanguansri L. Unpublished data, Food Science Australia.

moisture levels and water activity are increased the survival of probiotics is substantially decreased (Fig. 1.12). Probiotics can survive well over long shelf lives (12 months or more) at ambient temperatures in dried products as long as the low moisture levels in the products can be maintained (at least below a_w 0.2–0.3). In general, the lower the water activity, the better the bacterial survival will be (434). There is a substantial interaction between water activity and temperature with respect to their impact on the survival of quiescent probiotics. As the storage temperature is increased the detrimental impact of moisture is magnified (Fig. 1.13). Although the precise mechanisms of cell death remain unclear, osmotic stresses appear to play a role, with the presence of smaller molecules resulting in poorer bacterial survival (434).

Despite the clear evidence that very low water activities improve probiotic survival there may be technological limitations to reducing water activity to very low levels. These include the energy costs of drying, adverse impacts on the palatability of foods and difficulties in wetting and dispersing powders. Moisture barrier packaging may be applied to prevent the egress of moisture from the environment during storage. Maintaining probiotic viability in moderate water activity foods (0.4–0.7) is a major challenge and solutions such as microencapsulation or incorporation of probiotics into fat phases of products can provide improved survival.

1.6.2.6 Oxygen Both bifidobacteria and lactobacilli are considered strict anaerobes and oxygen can be detrimental to probiotic growth and survival (419). However,

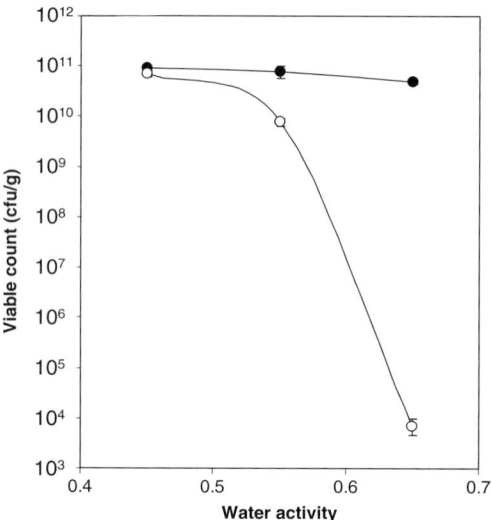

FIGURE 1.13 The viable count of a freeze-dried probiotic *Lactobacillus acidophilus* strain after 3 weeks of storage at moderate water activities at either 20°C (●) or 30°C (○). Increased temperature magnified the detrimental impact of water activity on the survival of the dried probiotic. Error bars represent the standard deviation from triplicate biological replicates. From Crittenden R, Weerakkody R, and Sanguansri L. Unpublished data, Food Science Australia.

the degree of oxygen sensitivity varies considerably between different species and strains (419). In general, lactobacilli, which are mostly microaerophilic, are more tolerant of oxygen than bifidobacteria, to the point where oxygen levels are rarely an important consideration in maintaining the survival of lactobacilli. Most probiotic bifidobacteria do not grow well in the presence of oxygen (398). However, many bifidobacteria have enzymatic mechanisms (via NADH-oxidase and NADH-peroxidase) to limit, oxygen toxicity (457). *B. animalis* subsp. *lactis* is relatively resistant to oxygen stress, and *B. longum* (*infantis*) and *B. breve* are more resistant to oxygen than *B. adolescentis* (395, 432, 457).

For oxygen sensitive strains, some strategies are available to prevent oxygen toxicity in food products. Antioxidant ingredients such as ascorbic acid or cysteine have been shown to improve probiotic survival (394, 401), as well as the use of oxygen barrier or modified atmosphere packaging (459). Since oxygen toxicity can sometimes influence probiotic survival in foods, it is advisable to minimize processes that are highly aerating, particularly when using bifidobacteria.

1.6.2.7 Toxicity of Ingredients The compatibility of probiotics with other ingredients within food formulations can have a significant impact on bacterial survival. Interactions between probiotics and other ingredients can be protective, neutral, or detrimental to probiotic stability (430). Obviously, the inclusion of antimicrobial preservatives can inhibit probiotic survival. Elevated levels of ingredients such as salt, organic acids, and nitrates can inhibit probiotics during storage (387, 395, 423), while starter cultures can sometimes inhibit the growth of probiotics during fermentation through the production of specific bacteriocins (384, 395).

1.6.2.8 Growth Factors, Protective, and Synergistic Ingredients Probiotic lactobacilli and, in particular, bifidobacteria are only weakly proteolytic and grow relatively slowly or poorly in milk (395, 423, 455). The growth of bifidobacteria can be enhanced by the presence of suitable companion cultures, including starter cultures, which can aid in protein hydrolysis and through the production of growth factors (430, 455). Alternatively, growth substrates such as carbon sources (e.g. glucose), nitrogen sources, and growth factors (e.g. yeast extract or protein hydrolysates) or antioxidants, minerals, and vitamins can be added to improve growth (430, 455). Other ingredients can protect the viability of probiotics in foods by acting as carriers (449–451). Finally, the food matrix itself can be protective. An example is cheese, where the anaerobic environment, high fat content and buffering capacity of the matrix helps to protect the probiotic cells both in the product and during intestinal transit (395).

1.6.2.9 Freeze–Thawing Freezing probiotic cells damages cell membranes and is detrimental to survival (388, 410). Protectants are usually added to cultures to be frozen or dried in order to prevent, or at least mitigate, cell injury. The most common protectants used at industrial scale are lactose or sucrose, monosodium glutamate, milk powders, and ascorbate (430). Once frozen, probiotics can survive well over long shelf lives in products such as frozen yogurts and ice-cream. Use of slow cooling rates, or conditioning cells with prefreezing stress, can significantly improve cell survival (388). Repeated freeze–thawing cycles are highly detrimental to cell survival and should be avoided. The cell membrane damage caused by freezing can also render probiotic cells more vulnerable to environmental stresses (410). In the example shown in Fig. 1.14, the survival of a probiotic *Lactobacillus paracasei* in a low pH fermented whey drink was studied during shelf storage. In one sample the culture was frozen during supply chain transport and then thawed. A parallel sample was only refrigerated during transport. It was evident that freeze–thawing increased the sensitivity of the cells to the acidic environment.

1.6.2.10 Sheer Forces Probiotic lactobacilli and bifidobacteria are Gram-positive bacteria with thick cell walls that are able to tolerate the sheer forces generated in most standard food production processes. Some high-sheer processes such as high-speed blending or homogenization may result in cell disruption and losses in viability.

1.6.3 Synbiotics

Probiotics are not the only functional food ingredients developed to improve human health by modulating the intestinal microbiota. Prebiotic ingredients represent an alternative and potentially synergistic approach. These nondigestible carbohydrates pass through to the colon where they selectively stimulate the proliferation and/or activity of beneficial microorganisms within the intestinal microbiota (412). Ingredients and foods that contain both prebiotics and probiotic are called synbiotics. A

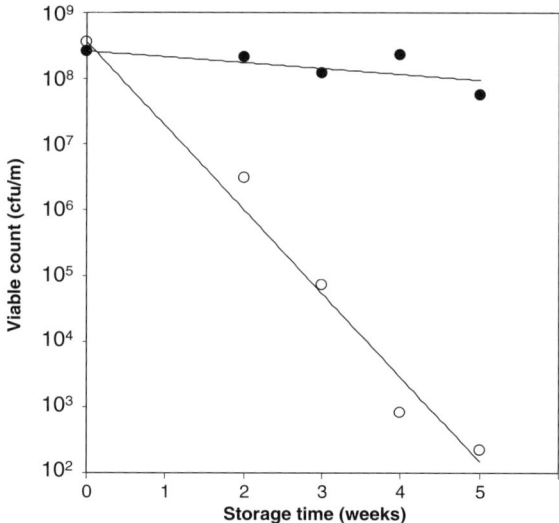

FIGURE 1.14 The impact of freeze–thawing on the subsequent refrigerated storage stability of a probiotic strain in a low-pH (3.8) whey drink. A culture of *Lactobacillus paracasei* was divided into two aliquots. One was frozen at −20°C for 72 h (○) while the other was stored refrigerated at 4°C for 72 h (●). The frozen culture was then thawed and the viability of the probiotic in both aliquots was monitored during subsequent storage at 4°C for 5 weeks. A single freeze–thaw cycle proved highly detrimental to the storage stability of the probiotic. From Crittenden R. Unpublished data, Food Science Australia.

range of nondigestible sugars, oligosaccharides, and polysaccharides can act as prebiotics, and possess a range of physiological and physicochemical properties that make them attractive food ingredients. Prebiotics are discussed in more detail in Chapter 7.

1.6.4 Delivery Systems

1.6.4.1 Microencapsulation Providing probiotics with a physical barrier to environmental conditions by microencapsulating the bacteria is an approach that has been trialed using a range of materials and techniques. Microencapsulation of probiotics, though, is not a simple undertaking. There are many demands for a successful probiotic microencapsulant:

- The materials have to be food grade, inexpensive, and compatible with the food into which the probiotic will be encapsulated.
- The microencapsulation process must be simple, inexpensive and must not reduce probiotic viability.
- The encapsulation efficiency must be high (i.e. close to 100% of the bacteria in a suspension should be encapsulated).

- The microcapsules must contain a high loading (%v/v) of probiotics.
- The microcapsules must not adversely impact on the taste and texture of food and beverages (small capsules <30 μm).
- The microcapsules must protect the probiotics against a range of environmental stresses during manufacture and storage. Protection against moisture and low pH are two of the most common stresses to protect against.
- It is also an advantage for the microcapsules to protect the probiotics during gastrointestinal transit.
- The microcapsule must be able to release the probiotic bacteria in the gut at the required site of action.

A range of experimental microencapsulation technologies have been reported, including entrapment in polymers such as alginate, carrageenan, and starch; coating in emulsions or fat; or dry impacting of prebiotics and enteric coats (385, 418, 424). Commercial microencapsulation systems for probiotics available currently include Priobiocap™, a fat-coating system developed by Institut Rosell, Canada; Micro-MAX™, an emulsion-based, synbiotic-coating system developed by Food Science Australia (399, 400), and EnCoate™, a biopolymer system produced by EnCoate Ltd, New Zealand.

1.6.4.2 Delivery Devices Another approach to maintaining the viability of probiotics for long periods at ambient temperatures is to physically separate the probiotics from the food and atmosphere. This can most simply be achieved by keeping dried probiotics in sealed sachets with the food or beverage to be mixed immediately prior to consumption. More innovative packaging and delivery systems have been developed to deliver dried probiotic ingredients into beverages. These include drinking straws that contain dried probiotics, which are released into the beverage as it passes through the straw (438) and a drink cap that contains the dried probiotics, which are released into the drink as the cap is opened (439).

1.6.5 Probiotic Foods

Probiotics have been successfully incorporated in a number of foods including the traditional vehicles of fermented milks and yogurts, and increasingly in other fermented and nonfermented foods beyond the dairy sector. Examples of the major product classes that have successfully incorporated probiotics are listed in Tables 1.8–1.10 along with the main technological points involved in incorporating probiotics and maintaining their viability.

1.6.6 Conclusions

Industry demand to include probiotics as functional ingredients in foods will no doubt continue to grow as the clinical evidence of health benefits builds. Indeed, the number of randomized, controlled trials reported that involve probiotic interventions has

TABLE 1.8 Examples of Fermented Dairy Products Containing Probiotics and the Major Technological Considerations Affecting Probiotic Survival

Products	Probiotic Species	Main Technological Points	References
Yogurt/ fermented milk	Wide range of lactobacilli and bifidobacteria B. animalis subsp. lactis is the most robust and commonly used Bifidobacterium species used	Probiotics can be inoculated together with traditional starters or, in the case of stirred yogurts, added after the fermentation. The combination of starter culture and probiotic must work technologically. Probiotic lactobacilli and especially bifidobacteria are not highly proteolytic and do not grow well in milk alone. The addition of growth factors can improve probiotic growth. Starter cultures can both improve growth or inhibit probiotic cultures. Relative proportions of starter and probiotic inoculation and/or timing of inoculations can be manipulated to maximize probiotic survival. Mild (pH 4.3–4.6) yogurts improve survival of bifidobacteia compared to traditional yogurts (pH 3.7–4.3). Mild yogurts can be manufactured with the substitution of *L. delbrueckii* subsp. *bulgaricus* with the probiotic cultures, though the inclusion of this traditional starter is mandatory for yogurts in some jurisdictions.	(395, 403, 417, 430, 455)
Matured cheeses	Strains of *L. casei*, *L. paracasei*, *L. rhamnosus* and *B. animalis* subsp. *lactis* usually survive well	Probiotics can be inoculated with the traditional starters to ferment and grow or to the curd after scalding.	(391, 392, 395, 396, 411, 414, 417, 441, 442, 445, 464)

	L. acidophilus survival is strain variable B. adolescentis, B. breve, and B. longum (infantis) usually survive poorly	The action of rennet releases peptides, which aid the growth of the typically weakly proteolytic probiotics. The probiotics are usually added at approximately the same dose as the starter cultures and then increase 10-fold in numbers during fermentation. Probiotic viability then declines slightly during ripening. The pH value of cheeses (4.8–5.6) is markedly higher than the pH of yogurts and fermented milks, which aids long-term survival of probiotics. High salt levels can inhibit the survival of bifidobacteria. The metabolism of starter and nonstarter bacteria within the cheese results in an almost anaerobic environment within a few weeks of ripening. Probiotics can sometimes cause flavor defects such as bitterness and acidity. Relative to the starter organisms, the probiotic inoculum size, time of addition, and growth conditions must be balanced to promote probiotic viability without excessive acid production. The inclusion of probiotics can increase secondary proteolysis and accelerate ripening.	
Fresh cheeses	L. rhamnosus, L. casei survive well Bifidobacteria including B. infantis and B. bifidum. B. animalis subsp. lactis survive poorly	Probiotics can be added with traditional starters or added with the cream. Addition with the starters can increase probiotic numbers but may have an adverse impact on taste by increasing acidity Synergies with or antagonism by the starter cultures	(395, 396, 417, 461)

TABLE 1.9 Examples of Nonfermented Dairy Products Containing Probiotics and the Major Technological Considerations Affecting Probiotic Survival

Products	Probiotic Species	Main Technological Points	References
Ice-cream	*L. johnsonii, L. casei, L. rhamnosus.*	Ice-cream is an excellent matrix for the long-term stability of probiotics	(381, 382, 389, 409, 416, 429)
	B. longum, B. bifidum, B. animalis subsp. *lactis*	Even relatively sensitive species of probiotics can remain stable for many months	
	L. acidophilus survives, but not as well as other species.	Can be made by direct addition of probiotics or with addition of fermented milk	
		Losses in viability occur mainly in ice-cream preparation during addition of overrun (aeration) and cell damage during freezing.	
		The added sugar acts as a cryoprotectant.	
		The higher pH of ice-cream compared to fermented products probably aids survival of probiotics.	
		Occasional impacts on sensory properties, but very good sensory properties achievable	
Dairy desserts	*L. paracasei* generally survives well	Neutral pH and refrigerated storage of these products is conducive to probiotic survival	(383)
	Survival of *L. acidophilus* is variable	No fermentation by probiotics desirable since acid production will lead to undesirable flavors	
		Absence of preservatives, neutral pH, and high water activity means spoilage organisms are a risk and may limit shelf life	
		Important to find compatible strains and products	
Dry products such as infant formula and nutritional powders	*L. rhamnosus, L. acidophilus,*	Probiotics usually dry blended into infant formula powder	(430, 444)
		Mixing probiotic with formula blend prior to spray drying may improve survival if the strain is robust enough to survive spray drying	
	B. animalis subsp. *lactis*	Water activity is a major determinant of survival and should be as low as possible	

TABLE 1.10 Examples of Nondairy Products Containing Probiotics and the Major Technological Considerations Affecting Probiotic Survival

Products	Probiotic Species	Main Technological Points	References
Fermented soymilk	L. acidophilus, L. casei, L. rhamnosus, L. johnsonii. B. animalis subsp. lactis.	Bifidobacteria and lactobacilli often grow better in soymilk than in cows' milk Able to use the soy oligosaccharide substrates raffinose and stachyose Approx. 10-fold increase in probiotic numbers during fermentation which then decline over shelf storage Deconjugation of soy isoflavone phytoestrogens may increase bioavailability Some manipulation of inoculation timing required to achieve correct sensory balance Minimize acetate formation which has undesirable sensory attributes	(404, 4408, 460)
Fermented cereal products	L. acidophilus, L. plantarum, L. paracasei B. animalis subsp. lactis	Fermented-oat products such as YOSA® (Bioferme Oy, Finland) Growth nutrients such as simple sugars and proteins often needed for fermentation No requirement for additional starter bacteria	(386, 413)
Juice	L. rhamnosus B. animalis subsp. lactis	Low pH value of juice (3.5–4.5) means only acid-tolerant strains survive well Probiotic survival is achievable (e.g. Gefilus produced by Valio Ltd, Finland) Probiotics added after pasteurization/UHT of juice Probiotics are particulate and tend to settle, so not applicable for clarified juices Can slightly reduce palatability, but can be masked Carriers such as dietary fibers can improve stability	(428, 450, 451)

(continued)

TABLE 1.10 (Continued)

Products	Probiotic Species	Main Technological Points	References
Fermented meats	L. rhamnosus, L. casei, L. paracasei, L. plantarum. B. animalis subsp. lactis Survival of L. acidophilus sometimes poor Heterofementative lactobacilli are not suitable due to gas formation	Probiotics inoculated along with starter cultures Numbers increase during fermentation and then decline slowly during ripening and storage Some inhibit pathogens such as Listeria monocytogenes and enterotoxigenic E. coli or spoilage organisms Should not inhibit rapid acidification by starter bacteria and should not produce excessive levels of lactic acid which can affect sensory quality Need to tolerate high salt and nitrite/nitrate levels	(384, 387, 405, 420, 422, 435, 443, 446, 453)
Confectionary	L. casei, L. paracasei B. longum	Probiotics mixed with confectionary (e.g. chocolate) at approximately 40°C No growth of probiotics in confectionary Only dried probiotics can be added Can be added to chocolate without impact on tempering conditions, texture, hardness, and flavor Crucial factors affecting the viability of probiotics in confectionary are water activity, osmotic tension, and temperature Entrapment in fat provides protection against oxygen, moisture and gastric transit	(425, 436, 437)
Food service products	L. acidophilus, B. animalis subsp. lactis	Probiotics can be added to fresh service products such as sandwiches, sushi, and smoothies stored refrigerated Short shelf life required Frozen cultures survive better than freeze-dried cultures	(447, 448)

grown rapidly in recent years. Not only is the evidence of specific health benefits emerging, but some potential mechanisms of action are also beginning to be elucidated (393). Such knowledge may allow the development of specific quality assurance tests for probiotic potency in foods that go beyond simply viability. However, today, viability remains the only practical measure of the quality of probiotic ingredients in foods and the measure employed by regulatory bodies.

It is already possible to include probiotics into a variety of foods beyond traditional fermented milks and yogurts. Numerous examples exist where probiotics have been successfully incorporated without adversely impacting sensory quality of the food and stability maintained above 10^6–10^7 cfu/g over long shelf lives. However, the maintenance of adequate viability in many food types remains a major challenge. Moderate water activity foods (0.4–0.8), in particular, are difficult matrices in which to maintain the viability of probiotics in shelf-stable products. For many foods, adjunct technologies such as microencapsulation are likely to be required in order to stabilize probiotics. As "off-the-shelf" probiotic microencapsulation technologies that work in all food matrices are not yet available, these generally have to be adapted to suit the particular food matrix.

The five critical points to address when incorporating probiotics into foods remain the following:

1. Select a compatible probiotic strain/food type combination
2. Use food-processing conditions that are compatible with probiotic survival.
3. If fermentation is required, ensure that the food matrix will support probiotic growth.
4. Select the product formulation, packaging and environmental conditions to ensure adequate probiotic survival over the product's supply chain and during shelf storage.
5. Ensure that addition of the probiotic does not adversely impact on the taste and texture of the product.

1.7 SAFETY OF PROBIOTIC ORGANISMS

DIANA DONOHUE

RMIT University, Melbourne, Victoria, Australia

Consumers searching on the Internet can readily find a multitude of probiotic products. Among them is one stated to be a prize-winning preparation constituted by a celebrated microbiologist. It contains multiple strains of natural live LAB, including an award-winning strain proven to be an order of magnitude stronger than any other probiotic. This mixture, consisting of *Enterococcus faecalis*, *Streptococcus thermophilus*, three species of *Bifidobacterium*, and seven *Lactobacillus* species, is guaranteed to colonize the colon with good microorganisms essential for a healthy immune system and general good health.

We are informed that scientific studies conclude that this preparation strongly enhances immunity and is effective against deadly bacteria such as methicillin- and vancomycin-resistant *Staphylococcus aureus*, *Helicobacter pylori*, food poisoning *Escherichia coli*, and *Bacillus cereus*. It is suggested that in the age of bioterrorism the product may have the potential to protect against anthrax as it is effective against *Bacillus cereus*, a member of the same genus.

The recommended dose is two to four capsules daily, at a cost of US$52 for 60 capsules. Regular consumption is strongly recommended for a plethora of conditions including irritable bowel syndrome, leaky gut, peptic ulcers, bloating, heartburn, ulcerative colitis, Crohn's disease, constipation, or diarrhea. Consumers of natural products such as psyllium, which are purported to remove good intestinal bacteria, are similarly advised. It is suggested that the product may be useful in treating diseases as diverse as asthma, cystic fibrosis, diabetes, Epstein-Barr virus, acne, psoriasis, eczema, arthritis, multiple sclerosis, human immunodeficiency virus (HIV), acquired immune deficiency syndrome (AIDS), and hepatitis B and C.

A search of five scientific and medical databases however was unable to identify a single scientific publication for the product mentioned above.

In a community, which is often suspicious of chemicals, probiotics are marketed, and seen by many, as a natural and appealing alternative to maintain and promote good health. In contrast to pharmaceuticals, traditional probiotics appear safe and serious adverse effects appear rare. Each year over 20 billion doses of probiotics are used by both healthy people and those with medical conditions. New species and more specific strains of organisms are constantly being sought and their probiotic attributes characterized to generate an expanded range of probiotic foods and targeted therapies (466–470). The demonstration of efficacy in probiotics offers untold opportunities for the development and marketing of human and veterinary products. Probiotics consumed in foods and dietary supplements do not have to comply with more rigorous guidelines for probiotics that claim amelioration or prevention of disease in clinical use. Probiotics that claim specific nutritional, functional, or therapeutic characteristics blur the boundaries between food, dietary supplement, or medicine, posing challenges for regulators. Their safety cannot be assumed. There is currently a lack of standard safety requirements for new probiotic organisms (471–474).

In November 2006, a Google search on the Internet for probiotics returned 2,810,000 hits. In contrast five scientific databases located 4447 journal papers, of which 345 related to aspects of probiotic safety. It is apparent that there are significant discrepancies between health claims made for what are being promoted as probiotics and verifiable scientific evidence to substantiate these claims. The probiotic concept has been appropriated for commercial exploitation, to the detriment of those probiotics tested to be safe and beneficial by rigorous science. When confronted with inflated and dubious claims the consumer cannot be expected to judge what a proven probiotic is and what the twenty-first-century equivalent of snake oil is. As efficacy is inextricably linked to safety, any claims of health benefits for a probiotic require substantiation by scientific evidence. What is a "probiotic" and who is entitled to use the term? What standards should be attained to ensure a product labeled "probiotic" is safe and effective?

The use of "history of safe use" as a criterion for the safety of food organisms is an arbitrary classification. Lactic acid bacteria and yeasts intrinsic to the production of traditional foods have been accepted as safe without any real scientific criteria, partly because they are normal commensal flora, and because of their consumption through centuries presumably without adverse effect. Evidence for the safety and efficacy of probiotic organisms has until recently been largely anecdotal or based on relatively little, and often poorly designed research.

The initial concept of probiotics has shifted from one of traditional dairy bacteria that ferment milk to "promote gut health" to that of a complex range of bacteria potentially capable of colonizing any human mucosal surface, not only the gut (477, 478). Probiotics have expanded to include nontraditional bacteria, with applications targeted to clinical conditions and not limited to oral therapy (477). The change from traditional bacteria in food to designer probiotic has led to considerably increased numbers of organisms per dose to achieve health effects (472). Where previously a limited number of food products existed, now there are single or multiple species of organisms in a range of quasi or proven therapeutic products.

The drift from dairy foods to complementary or prescribed medicines with therapeutic claims has elevated probiotics to a class that was once the sole province of pharmaceuticals. How to assess the safety of new probiotic products needs to be re-evaluated in these altered circumstances. Regulators must judge whether a probiotic is a food, a supplement or a clinical therapy and develop enforceable safety standards accordingly. If probiotics are intended for therapeutic use they must be evaluated for quality, safety, and efficacy in the same manner as any other therapeutics—with documented and verifiable characterization of the active ingredient, dose, efficacy, safety, and adverse effects (479). This raises a quality control issue for food and therapeutic manufacturers, who have the responsibility to market accurately characterized, stable, and viable organisms, in appropriate doses and formulations, with clearly defined applications and attendant health benefits. Consumers need to have accurate information from a credible source to ascertain whether a product is safe and the purported health benefits are genuine.

1.7.1 Current Proposals for Probiotic Safety

Conventional toxicology and safety evaluation has limitations for the safety assessment of probiotic bacteria. Vigorous debate continues on what constitutes appropriate safety testing for novel probiotic strains proposed for human use. In recent years several organizations have formulated approaches to assess the safety of probiotics. For the most part these have been predicated on oral applications but it should be remembered that probiotics are not limited to oral products, but may be applied vaginally, topically, and intranasally (478).

The Joint FAO/WHO Working Group on Drafting Guidelines for the Evaluation of Probiotics in Food (475) proposed a framework consisting of strain identification and functional characterization, followed by safety assessment and Phase 1, 2, and 3 human trials. It recommended that probiotic foods be properly labeled with the strain designation, minimum numbers of viable bacteria at the end of shelf life, storage conditions, and manufacturer's contact details. Importantly, the Working Group further proposed that the use and adoption of the guidelines should be a prerequisite for calling a bacterial strain "probiotic."

The Working Group considered the minimum tests required to characterize safety are

- determination of antibiotic resistance patterns;
- assessment of metabolic activities (e.g. D-lactate production, bile salt deconjugation);
- assessment of side effects during human studies;
- postmarket epidemiological surveillance of adverse incidents in consumers.

If the strain being evaluated belongs to a species known to produce a mammalian toxin or to have hemolytic potential, it must be tested for these characteristics.

The EFSA has proposed a scheme based on the concept of QPS, defined as "an assumption based on reasonable evidence" and qualified to allow certain restrictions to apply (480). The scheme aims to have consistent generic safety assessment of microorganisms through the food chain without compromising safety standards. Individual evaluations would be limited to aspects particular to the organism, such as acquired antibiotic resistance determinants in LAB. QPS status would not apply to a microorganism that commonly causes pathogenicity. A microorganism would not necessarily be considered a potential pathogen where there are infrequent reports of clinical isolates from severely ill people.

Broadly the characteristics to be evaluated for QPS approval are

- unambiguous identification at the claimed taxonomic level;
- relationship of taxonomic identity to existing or historic nomenclature;
- degree of familiarity with organism, based on weight of evidence;
- potential for pathogenicity to humans and animals;
- the end use of the microorganism.

The latter would influence any qualifications imposed, depending on whether the organism is to be directly consumed; is a component of a food product not intended to enter the food chain, but which may adventitiously; or is used as a production strain in a product intended to be free of live organisms.

Bernardeau et al. (481) consider this generic approach to safety assessment of microorganisms is not relevant to the *Lactobacillus* genus and have proposed modifications. They contend that LAB are not a homogeneous group as some species are pathogens, and that the rarely pathogenic *Lactobacillus* genus should undergo its own limited safety assessment. The genus should be accorded the status of Long Standing Presumption of Safety based on its long history of safe use in fermented foods. Individual species could then be assessed for safety based on one, two, or a full suite of tests, depending on the intended use. The first safety test would be to demonstrate an absence of antibiotic resistance and its ability for transference. The second, a high dose tolerance test in animals, would be required if the organism was not resistant to antibiotics and was a known lactobacillus for which a new application was being proposed.

A full safety assessment would be required if the body of knowledge was insufficient. In the event of risks being identified an experimental presumption of safety (EPS) status would be given for 10 years as the body of knowledge accumulated. If specified risks were identified the organism would be granted restricted EPS status (EPSr) with conditions placed on its use for 10 years. If the body of knowledge had increased after 10 years a restricted QPSr status could be granted, with the restrictions on use standing. Bernardeau et al. (481) consider this approach would be preferable to granting non-QPS status, as it allows the community at large to benefit while simultaneously protecting the small proportion of the population at risk.

A more stringent perspective is evident in the regulations of the Canadian National Health Product Directorate, where products can be considered "traditional" or "nontraditional." To be considered traditional a product must have at least 50 consecutive years of use, two independent references to traditional use in support of health claims, and safety reports on adverse reactions and interactions (482).

Gueimonde et al. (483) propose a sequence to determine safety of new bacteria of nondairy origin sought specifically for their probiotic effect and frequently isolated from the human or animal intestinal tract. While the scheme has many features in common with that of the Working Group, its emphasis is on the absolute necessity for correct strain identification to allow comparisons of potential risk with taxonomically related organisms, to avoid use of potential pathogens, and for continuous quality control in postmarker surveillance of bacteremia. Gueimonde et al. (483) suggest that the increasing availability of probiotic genome sequences will facilitate identification of potential risk factors.

While guidelines for probiotics have been proposed and refinements suggested, as yet agreement has not been arrived at on a universal or enforceable standard. There is general consensus on the need for standards, enforced regulations, and improved quality control over products (483, 472, 477, 478, 482).

1.7.2 Taxonomic Identification

The safety of a putative novel probiotic is contingent on its unequivocal identification at the genus, species, and strain level, as probiotic effects are strain specific. Sophisticated phenotypic and molecular techniques enable species identification and discriminate between closely related strains. Reliable taxonomic identification of both species and strain is a safety issue for quality control of the product, consumer or prescriber information, diagnosis, and appropriate treatment of suspected clinical cases and epidemiological surveillance of the exposed population.

The taxonomy of lactic acid and other bacteria has changed significantly with the advent of genetic methods of classification. Strains previously thought to be dissimilar have merged, while other strains have been added or reassigned to different genera. The persistent use of incorrect or nonexistent species names on product labels despite taxonomic re-assignment is a significant issue for the credibility and safety of probiotics.

Yeung et al. (484) used partial 16S rDNA sequencing to identify named commercial strains obtained directly from the manufacturer and found discrepancies in 14 of 29 species designations. Lourens-Hattingh and Viljoen (485) concluded that probiotic cultures in South African yogurt were little more than a marketing tool upon finding the initial counts of *Bifidobacterium bifidum* in three different sources of commercial yogurts were lower than the therapeutic minimum. Weese (486) identified isolates from eight veterinary and five human probiotics to find accurate descriptions of organisms and concentrations for only 2 of the 13 products.

Temmerman et al. (487) found that of isolates from 55 European probiotic products, 47% of food supplements and 40% of dairy products were mislabeled. The food supplements yielded either no viable bacteria (37%) or significantly lower counts than the dairy products, contradicting the concept that health benefits derive from the presence of a minimum concentration of live probiotic bacteria.

In six products, all species isolated conformed to the label description; in 19 products they differed from those listed. *Enterococcus faecium* was isolated in such high numbers that contamination was unlikely to be the source. Only 2 of the 22 food supplements purporting to contain *Lactobacillus acidophilus* did. Bifidobacteria were isolated from 5 of 27 products claiming to contain them, despite the use of different selective media. The organism most frequently claimed to be in, and isolated from dairy products was *L. acidophilus*, though it was not necessarily found where claimed.

Huys et al. (488) used a suite of validated and standardized molecular methods to taxonomically re-identify 213 cultures of LAB and propionibacteria obtained for the PROSAFE project, a European Commission project into safety of probiotic LAB for human use. Probiotic strains, candidate probiotic strains, and nutritional strains, with their identity, were submitted from international culture collections, commercial manufacturers, and a research institute. The genus was confirmed correct for 194 strains (91%), with the genus of 87% of probiotic strains being confirmed as accurate. Of the 186 cultures submitted with species identity, that identity was confirmed for 159 (86%) and for 83% of the probiotic strains.

More probiotic strains (28.1%) were misidentified than nutritional (11.4%) or research strains (14.0%). The 34 misidentified probiotic strains were submitted by 10 commercial companies, with 18 of the 34 being provided by two companies. The finding that 28% of probiotic strains were misidentified at either genus or species level corroborates the reports from these and other studies (489, 490) that inaccurate identification and mislabeling of probiotic products continue. The evidence from this study suggests mislabeling originates at the start of production with incorrect identification of strains. The authors suggest misidentification may in many instances result from use of methods that are technically inadequate for reliable taxonomic characterization of a bacterial species (488).

A new probiotic culture must be at least as safe as its conventional counterparts. Inaccurate nomenclature provides no scientific or regulatory validity, misinforms or confuses the consumer, and compromises quality and safety of the product. Consumers are entitled to expect that the label on a probiotic product accurately reflects its contents: the organism is what it purports to be, it is present alive in a specified concentration range for a stated period, and the suggested serving size contains sufficient organisms to achieve the proven benefit.

1.7.3 Pathogenicity

It is an obvious requirement that a probiotic should not cause infection. This is a significant issue where the intestinal barrier is immature as in infants; where its integrity is impaired from radiotherapy, antibiotic treatment or disease; and in immunocompromized states, such as HIV infection. With advances in medical care, an increasing proportion of the community may at some time be immunocompromized, or at risk of opportunistic infection.

Lactobacillus species are commonly used probiotics and considered nonpathogenic in most situations. Vesterlund et al. (491) tested 44 fecal, 52 blood, and 15 probiotic isolates to compare the presence of properties that are known virulence factors in recognized pathogens. No significant differences in adhesion to collagen, fibrinogen, or mucus were observed between blood, fecal and probiotic isolates, although blood isolates had a higher tendency to adhere to mucus than probiotic strains. No lactobacilli tested positive for α- or β-hemolysis. Probiotic strains induced a respiratory burst activity lower than, but close to, that of the blood isolates. Although probiotic strains showed a higher resistance to the bactericidal effect of complement-activated serum than did fecal strains, no significant differences were seen in serum resistance between fecal, clinical, or probiotic isolates. In summary, the tested properties varied greatly between strains and no unequivocal virulence factors for lactobacilli were identified. Of the three groups of isolates the probiotic strains induced the lowest respiratory burst in polymorphonucleocytes and showed the highest serum resistance, an observation that warrants further examination (491).

Lactobacillus species in general are thought to have low pathogenicity or be opportunistic pathogens in immunocompromised individuals or those with serious underlying disease. It has been suggested that *Lactobacillus rhamnosus* in particular warrants surveillance because it is associated with more cases of bacteremia than other lactobacilli. *L. rhamnosus* is among the most common Lactobacillus species in the human intestine so the incidence of bacteremia may be relative to its extensive presence in the intestine (492).

Two clinical cases have been reported in which a lactobacillus indistinguishable from an ingested probiotic strain has been identified in association with infection. A 74-year-old woman with hypertension and diabetes mellitus developed a liver abscess in association with pneumonia and pleural empyema. She had a history of drinking probiotic milk containing *L. rhamnosus* GG and a strain indistinguishable from that was isolated from the abscess (493). A 67-year-old man with mild mitral regurgitation developed endocarditis after dental extractions. His blood cultures were positive for a strain of *L. rhamnosus* indistinguishable from that in probiotic capsules he chewed (494).

Wolf et al. (495) assessed the safety of probiotic *Lactobacillus reuteri* in a double-blind, placebo-controlled study in HIV adults, and found the organism to be well tolerated with no significant safety problems. A review of probiotic safety found no published evidence that immunocompromised patients had an increased risk of opportunistic infection from probiotic lactobacilli or bifidobacteria (496).

Probiotic *L. acidophilus* has been identified as the cause of persistent bacteremia complicated by recurring pulmonary emboli associated with a catheter infection in an AIDS patient, post chemotherapy for Hodgkin's disease (497). The patient had undergone probiotic treatment three times daily for 3 weeks. Isolation of the probiotic from the catheter site suggested that, rather than gastrointestinal translocation, to be the origin of infection. The bacteremia resolved with clindamycin and gentamicin, the only antibiotics to which *L. acidophilus* was sensitive.

Land et al. (498) reported two instances of bacteremia attributable to therapy *with L. rhamnosus* GG in young children. A 6-week-old baby was treated enterally with *L. rhamnosus* GG to ameliorate diarrhea following prolonged postoperative complications of cardiac surgery and broad spectrum antibiotic therapy. Despite improvement in the diarrhea the baby remained ill. A thrombus was found adherent to the right atrial wall and cultures from blood and a central venous catheter yielded penicillin-sensitive isolates of *Lactobacilli*.

Another child aged 6 years received enteral *L. rhamnosus* GG therapy for antibiotic-associated diarrhea, following treatment for infections with *E. coli*, *S. aureus*, and enterococcal sepsis. Blood cultures taken after the child developed fever yielded a penicillin-sensitive *Lactobacillus* species.

The DNA fingerprint of the lactobacillus isolated from blood cultures of the two children was indistinguishable from that of the probiotic *L. rhamnosus* GG. The most likely mechanism propounded for the bacteremia was bacterial translocation following enteral administration, rather than contamination of the central venous catheter (498).

These first reports of probiotic bacteremia in children testify that this outcome is not limited to rare cases in severely compromised adults, but needs to be borne in mind in severely compromised patients regardless of their age.

Srinavasan et al. (499) studied the safety of *Lactobacillus casei* Shirota in a randomized controlled trial of pediatric patients admitted to intensive care with severe conditions such as meningococcal septicemia and respiratory failure. Known immunodeficiency and intolerance to cow's milk or lactose were exclusion criteria. *L. casei* Shirota (10^7 cfu/day in three divided doses) was given enterally to 28 children for up to 5 days. *L. casei* Shirota was cultured from five of the six stool samples produced, but not from any normal sterile body fluid or surface. Although the probiotic appeared to be tolerated without adverse effects, the small group size, short dosing period, and exclusion of vulnerable immunodeficient subjects limit the conclusions from the study, while its efficacy remains to be established.

It would appear that the general population is not at risk from exposure to probiotics. The rare cases of infection associated with probiotics or very similar organisms have occurred in groups of patients whose conditions predispose them to opportunistic infection. In contrast other patients with very serious underlying diseases have benefited from probiotics (471). Elucidation of the mechanisms underlying rare cases of probiotic bacteremia in immunocompromised or seriously ill patients will assist clinicians in identifying those patients for whom probiotics may be contraindicated. Until then a cautious use of probiotics in this group is suggested until those individuals in whom probiotics are contraindicated can be identified with a greater degree of certainty (471, 477, 500).

Clearly the potential for infection in this group of patients must be kept in mind, but it should not prevent the use of probiotics in the general population. Case studies are isolated events particular to an individual's condition and need careful evaluation relative to the safety of the population as a whole. Rare adverse effects need to be interpreted in the context of relative risk, in a manner similar to pharmaceuticals. For the ill patient, the risk of forgoing the benefits of treatment needs to be assessed against the lesser risk of an adverse effect (482).

While hospitalized patients are monitored, those in the wider community who self-medicate with probiotics are without the benefit of clinical oversight. It has been proposed that individuals with serious gastrointestinal or blood conditions should inform their doctors if they are consuming probiotics and report symptoms such as fevers or chills (471).

Administration of a living probiotic differs from that of a pharmaceutical drug where a specific characterized chemical is given. The risk of infection from a probiotic could be eliminated if its active constituent was identified and isolated to design an inanimate equivalent. The likelihood of this will depend on elucidation of the mechanisms by which a probiotic modulates immune effects, and the constitutive functions of the probiotic as a carrier mechanism to intestinal sites. Probiotic candidates could be selected based on their active constituents and pharmacokinetic characteristics to enable delivery of the active constituent to a target site in the intestine (501).

1.7.4 Antibiotic Resistance and Susceptibility

Lactic acid bacteria are naturally resistant to many antibiotics by virtue of their structure or physiology. In most cases the resistance is not transferable and the species are also sensitive to antibiotics in clinical use. However it is possible for plasmid-associated antibiotic resistance to spread to other species and genera. The transmissible resistance of enterococci to glycopeptide antibiotics such as vancomycin and teicoplanin is of particular concern, as vancomycin is one of the remaining antibiotics effective in the treatment of multidrug-resistant pathogens (492).

Antibiotic resistance mechanisms, their genetic nature, and transfer characteristics of resistance determinants have been studied comparatively recently in anaerobic bacteria. It has been shown that the plasmid that encodes for macrolide resistance can be transferred from *L. reuteri* to *E. faecium* and from *E. faecium* to *E. faecalis* in the mouse GIT (502). The properties of enterococci render them not suitable both as probiotics and in the production of fermented foods. They are not infrequently associated with hospital-acquired infections such as endocarditis and bacteremia. Their superior survival properties coupled with innate and acquired antibiotic resistance make them difficult to eliminate once they have become pathogenic. Little is known of the mechanisms by which enterococci become pathogenic.

A study of virulence genes in 13 *E. faecalis* strains isolated from clinical, food, and animal sources found 8–13 virulence genes in all isolates (503). Study of two of the clinical isolates found patterns of virulence gene expression were dependent on the growth phase, environmental conditions and the bacterial isolate, rather than the source of the isolate or the combination of virulence genes. The observation that gene expression of virulence in enterococci is modulated by environmental conditions such as temperature and pH has implications for manufacturers of foods and probiotics.

A study by Temmerman et al. (487) found 68.4% of probiotic isolates were resistant to two or more antibiotics. Strains of lactobacilli were found resistant to kanamycin (81%), tetracycline (29.5%), erythromycin (12%) and chloramphenicol (8.5%). The disc diffusion method showed 38% of *E. faecium* isolates were resistant to vancomycin, while the PCR-based van gene detection assay showed they were susceptible.

Salminen et al. (504) characterized 86 clinical lactobacillus blood isolates at species level and tested them for antimicrobial sensitivity. Of the eleven species identified, 46 isolates were *L. rhamnosus* ($n = 22$ *L. rhamnosus* GG type), *Lactobacillus fermentum* ($n = 12$) and *Lactobacillus casei* ($n = 12$). All lactobacillus isolates showed low minimum inhibitory concentrations (MICs) of imipenem, piperacillin-tazobactam, erythromycin and clindamycin. The range of MICs of cephalosporin varied widely with species while MICs of vancomycin were high except for *Lactobacillus gasseri* and *Lactobacillus jensenii*. The antimicrobial susceptibility pattern for probiotic *L. rhamnosus* GG was similar to those of *L. rhamnosus* GG type and other *L. rhamnosus* clinical isolates. This study of a large number of blood culture isolates of lactobacillus indicates their antimicrobial sensitivity to be species dependent.

Sullivan and Nord (505) characterized the Lactobacillus blood isolates from bacteremic patients in Stockholm, Sweden, between January 1998 and March 2004 to identify the possible presence of three probiotic strains of lactobacillus consumed in Sweden. The majority of the 59 isolates were *L. rhamnosus* ($n = 17$), *L. paracasei ssp. paracasei* ($n = 8$), and *L. plantarum* ($n = 8$). No isolates were identical to the probiotic strains. All isolates of *L. rhamnosus, L. paracasei ssp. Paracasei*, and *L. plantarum* were resistant to vancomycin and teicoplanin, while the majority of isolates were susceptible to clindamycin.

Opinions differ on the clinical significance of lactobacillus isolated in infections. In their retrospective review of 241 cases of lactobacillus-associated infections reported between 1950 and 2003; Cannon et al. (506) found the species most commonly isolated were *L. casei, rhamnosus, plantarum*, and *acidophilus*, at 35.7, 22.9, 10, and 10% respectively.

Bacteremia was identified in 129 cases, in association with conditions such as cancer, diabetes, broad-spectrum antibiotic treatment, transplantation, or abscesses. *L. rhamnosus* (32.1%) and *L. casei* (28.3%) were the most common species isolated. Over 90% of isolates were sensitive to clindamycin, erythromycin, and gentamicin, while 26.7% were sensitive to vancomycin.

Seventy-three cases of endocarditis were identified, 75% of whom had either existing structural heart disease or a previous episode of endocarditis. *L. casei* (40.6%), *L. rhamnosus* (17.2%), and *L. plantarum* (17.2%) were the most commonly identified species. Isolates were most sensitive to ciprofloxacin, erythromycin, and ampicillin (84.6–100% of isolates) with 26.7% sensitive to vancomycin.

Assessment of the clinical significance of lactobacillus isolated from blood is often confounded by the concomitant presence of other organisms. While Cannon et al. (506) found 38.8% of lactobacillus bacteremia cases were polymicrobial, it is interesting to note that in 61.25% of cases other organisms were not isolated. It is noteworthy that in 95.9% of endocarditis cases *Lactobacillus* was the only species isolated.

These data lend further support to recommendations that probiotics should be taken cautiously or not at all by people with specified serious conditions, in-dwelling devices, and prosthetic or abnormal heart valves (471, 472).

Danielson et al. (507) used molecular methods to characterize 23 *Lactobacillus* strains isolated from blood cultures of Danish patients between 1997 and 2004; prior to testing individual susceptibility to antibiotics. Their findings corroborated the high prevalence of *L. rhamnosus* (43%), *L. paracasei* (22%), and *L. plantarum* (17%) previously observed (506). Although lactobacilli are often regarded clinically as one group, Danielson et al. (507) found distinct variations in susceptibility patterns for the same antibiotic between species, with susceptibility varying widely between species.

Lactobacilli are ubiquitous commensals in humans and whether or not their presence is indicative of infection or contamination is a topic of contention. The source of lactobacilli is frequently unknown and they are present in supplements and dairy foods that may be consumed in high volume. It can be seen that rather than relying on a general recommendation for the genus, the selection of antibiotic treatment for lactobacillus infections should be based on sensitivity data for a species (504, 506, 507).

Organisms have intrinsic resistance to antibiotics or can acquire resistance, either naturally or deliberately. It is a significant reason to select strains lacking the potential to transfer genetic determinants of antibiotic resistance. There is little basis for scientific regulation of strains with intrinsic resistance, as little is known about the levels of intrinsic resistance in current probiotic and food strains.

Mathur and Singh (508) noted a lack of studies examining acquired antibiotic resistance in food LAB, and proposed that LAB used in probiotic or starter cultures may be a source of antibiotic-resistant genes, which if transferable have potential to be transferred to endogenous and pathogenic bacteria. The antibiotic resistance of *Enterococcus* species in particular has been studied because some strains cause serious infections in humans (509). In contrast less is known of the antibiotic resistance of other LAB that consists of numerous genera and species, with varied susceptibility to antibiotics.

Kastner et al. (510) questioned the extent of antibiotic resistance among desirable food bacteria so surveyed the antibiotic resistance of starter and probiotic cultures in Swiss foods by several molecular methods. For the first time the *lnu*(A) gene, which confers lincomycin and clindamycin resistance was found in an *L. reuteri* SD2112 isolate. Tetracycline resistance gene *tet*(W) was also detected in probiotic *B. lactis* DSM10140 and *L. reuteri* SD2112.

Hummel et al. (511) determined and verified the antibiotic resistance of 40 starter cultures and 5 probiotic cultures at the genetic level. Probiotic and starter strains of *Lactobacillus, Pediococcus, Leuconostoc*, and *Streptococcus* were found to be relatively sensitive to penicillin, ampicillin, tetracycline, erythromycin, and chloramphenicol, while more than 70% were resistant to gentamicin, streptomycin, and ciprofloxacin.

In the process of this study Hummel et al. (511) identified factors thought likely to hinder the implementation of the safety evaluation scheme proposed in EFSA's QPS system. The factors are listed below.

- There are no approved standards for phenotype and genotype evaluation of antibiotic resistance in food isolates.
- There is no optimal growth medium capable of growing the majority of *Lactobacillus* species.
- There are no approved standard MICs at which an organism is considered resistant or susceptible to an antibiotic, except for *Enterococcus* species.
- When the genetic basis for resistance is unknown, whether resistance is intrinsic or transferable is unable to be ascertained for many antibiotics.

Antibiotic treatment is known to modify the diversity of normal intestinal microbiota and cause other unwanted side effects. Probiotics are often given concomitantly with antibiotic therapy in attempts to reduce the side effects caused by the antibiotics. In these instances the probiotic should carry only the antibiotic resistance specific for that antibiotic.

Studies have examined the effect of antibiotic treatment combined with probiotic therapy on the fecal microflora. Conversely Saarela et al. (512) investigated the effect in patients of oral doxycycline on the gastrointestinal survival and tetracycline susceptibility of simultaneously administered probiotics *L. acidophilus* LaCH-5 and *Bifidobacterium animalis subsp. lactis* Bb-12. The gastrointestinal survival of both probiotics was similar in the control and probiotic groups. Doxycycline therapy increased tetracycline resistance in fecal anaerobic bacteria and the ingested probiotic *B. animalis subsp. lactis*. Consumption of the two probiotics with doxycycline did not increase transference of resistance genes in these two strains after 10 days of doxycycline therapy. *L. acidophilus* remained susceptible to tetracycline and no resistance genes were detected in *B. animalis* additional to preexisting *tet*(W).

Systematic screening for antibiotic resistance in probiotic strains is not undertaken at present. It is essential that probiotic organisms be sensitive to broad spectrum and commonly used antibiotics. The inability to transfer antibiotic resistance cannot be assumed for all members of a species and like many other probiotic properties this must be assessed on a strain-by-strain basis.

A decision strategy has been proposed (513) using molecular techniques to assess the risk of antibiotic resistance in bacterial strains. The steps are to

- identify the resistance gene;
- attempt to transfer resistance to normal gastrointestinal flora;
- characterize the biochemical mechanism of resistance;
- elucidate the genetic basis for resistance.

If after following this protocol it were shown that a resistance gene was not associated with a mobile genetic element then the risk of transfer of resistance would be assessed as low.

1.7.5 Immune Modulation

The relationships between host immune system and gut microflora and the many mechanisms underlying the beneficial effects of probiotics have yet to be elucidated.

It was originally thought that gut health was achieved by probiotic organisms binding to sites on the epithelial cells of the gut, thus excluding pathogens (477). Subsequent mechanistic studies have shown several possible mechanisms may be involved: stimulation of cell-mediated immune effects (514); altered immunity at mucosae distant from ingested probiotics mediated by Peyer's patches (515); and suppressing of IgE-mediated allergic hypersensitivity by oral probiotics (516).

Modulation of the immune system has a potential to ameliorate allergic, inflammatory, and autoimmune disorders. While an enhanced immune response is desirable in conditions such as infection and cancer, it may not be in allergic disorders where the response needs to be attenuated (473). Current evidence suggests regulation of effects on the immune system may differ between healthy and ill subjects (465, 517). It is thought that immunomodulation may depend not only on the dose of probiotics but also on the immune status of the host (518) and the probiotic strain.

In a murine study of the effect of oral probiotics on lymphocyte proliferation, responses were seen to vary from suppression of lymphocyte proliferation to enhanced T and B cell mitogenesis, depending on the strain (519).

It has been shown that probiotic bacteria can colonize and persist in the GIT of germ-free athymic probiotic-treated mice and their untreated progeny (520). Mortality associated with two of the tested strains (*L. reuteri* and *L. casei* GG) was observed in the immunodeficient-colonized pups. Athymic mice colonized with *Candida albicans* were also treated with the probiotics to test their ability to protect the immunodeficient mice from infection. Survival of the mice increased and dissemination of *Candida* decreased in athymic and control mice, but varied with probiotic strain (521). This paper is reportedly the first to describe an enhanced inflammatory infiltration by probiotics (*L. acidophilus* and *B. animalis*) in response to infected mucosal tissue.

The observation of mortality in probiotic-colonized immunodeficient pups appears to be the first report of neonate mortality associated with colonization by probiotics. It suggests that the safety of probiotic bacteria should be assessed cautiously in the immunodeficient, particularly in neonates (520).

The gut microflora are the major source of microbial stimulus in infancy. The initial colonization of the gut by microflora and their composition are pivotal to the development of immune responses and normal gut barrier function. Kalliomäki et al. (516) demonstrated that the composition of gut microflora differs between healthy and allergic infants. In a standardized double-blind placebo-controlled trial *L. rhamnosus* GG was given to mothers prenatally for 2 weeks before delivery and 6 months postnatally if breast feeding or to the infant if not. No adverse effects were observed in the mothers, and in infants the incidence of atopic eczema in the first 2 years of life was halved compared to that in infants given placebo.

The finding that a specific strain of probiotic bacteria strongly influences immune regulation in infants brings into question the use of probiotics in infancy. Several commercially available infant formulae contain probiotics. Long-term ingestion of probiotics while the gut microbiota is being established in theory has the potential for the gut to be colonized with the probiotic organisms. The long-term effects of probiotics on the composition of the gut flora and gut immunity during maturation are unknown. It has been questioned whether probiotic safety can be assessed solely by an absence of adverse effects, and longer term endpoints have been proposed to determine whether there is increased risk of incurring diseases such as diabetes and inflammatory disorders (471).

The properties and effects of a probiotic are specific to genus, species, and strain, thus a single probiotic is unlikely or unable to elicit universal health benefits across a spectrum of diseases. In theory this could be overcome by combining several probiotics in one product, the notion being that an individual probiotic would either compensate for any inadequacies of another or neutralize its adverse effects. Synergistic or additive effects could potentially be gained by combining multiple species into a single probiotic product (522). Multispecies probiotics are available but at present no evidence base or criteria exist for selection of the optimal number of strains and their most desirable properties.

Timmerman et al. (522) attempted to resolve this by systematically testing 69 culture collection strains to design a multiplespecies probiotic for preventing infection in critically ill patients. Probiotic strains selected for high survival in the environment of the digestive tract were tested *in vitro* for their capacity to inhibit growth of gut pathogens and modulate immune responses. Candidate strains were tested and ranked on their ability to inhibit clinical isolates and induce high concentrations of anti-inflammatory cytokines or low concentrations of pro-inflammatory cytokines. Strains with negative selection criteria were then excluded, notably *L. rhamnosus* W71 and *Lactobacillus plantarum* W59 (from a species lacking a long history of safe use, and showing resistance to a range of antibiotics respectively). The resultant multi-species disease-specific probiotic consisted of *Bifidobacterium bifidum, Bifidobacterium infantis, L. acidophilus, L. casei, Lactobacillus salivarius*, and *Lactococcus lactis*. In *in vitro* tests the probiotic combination demonstrated a wider antimicrobial spectrum, greater induction of anti-inflammatory cytokines and suppression of pro-inflammatory cytokines than its constituent strains. Proof of efficacy in human clinical conditions remains to be demonstrated in randomized, double-blind, placebo-controlled trials.

Baken et al. (523) assessed the effects of oral doses of *L. casei* Shirota on T helper 1 (Th1) responses and development of autoimmunity in a panel of four assays and found varied effects between assays. In a modified local lymph node assay in BALB/c mice lymphocyte proliferation was significantly reduced only at the highest concentration of topical exposure to the Th1 cell-dependent antigen dinitrochlorobenzene. Treatment with *L. casei* Shirota exacerbated experimental autoimmune encephalomyelitis, a rat model of multiple sclerosis, in treated animals compared to controls, with increased incidence of disease, earlier appearance of neurological symptoms, longer duration of the disease and higher cumulative clinical scores. Neither inhibitory nor stimulatory effects on modulation were found in mitogen-induced cell proliferation and cytokine release assays in mesenteric lymph nodes of rats from treated and control groups. Gene expression profiling by microarray analysis in rat spleen, liver, thymus and mesenteric lymph nodes found no clear changes in gene expression induced by *L. casei* Shirota. Inhibition or stimulation of Th1 mediated immune responses were found depending on the assay performed, an indication that probiotics can affect the Th1/Th2 balance in either direction. The observation of varying effects on Th1 responses indicates that probiotic consumption may have beneficial as well as harmful effects in immune-related conditions (523).

In conditions where acute immunosuppressive intervention is desirable, as in individuals suffering from conditions such as inflammatory bowel disorder (IBD), it has been proposed that treatment could be targeted through probiotics rendered more powerful through genetic engineering (524). The probiotic design could be customized for differing mechanisms of action and the specific end functions required. Examples already exist of organisms modified for functions as diverse as sequestration of toxins, competitive replacement of harmful bacteria, antibody production, correction of enzyme deficiencies, *in situ* production of detoxification enzymes, and *in situ* synthesis of cytokines (524, 525).

Additional criteria for safety characterization of a probiotic have been proposed to address concerns about immunomodulation (477).

- An infectivity test using high doses of viable organisms in immunocompromised animals.
- Measurement of changes in cytokine balance.
- Mixtures of probiotics to be assessed *in vitro*, to exclude an isolate that can inhibit the cytokine stimulation of another.

It is apparent that an evaluation of immunomodulatory effects should form part of the safety and efficacy assessment for a probiotic. Modifications to previous schemes have been proposed to include immunomodulatory effects for the evaluation of new products (473). As immune effects vary with the probiotic organism and the experimental model (523), a suite of tests is considered essential. Ezendam et al. (473) have initially proposed *in vitro* assays in monocytes and macrophages to determine cytokine profiles, and studies of dendritic cell maturation and activation. Subsequent experimental animal studies would include models for cellular immunity, allergy, autoimmunity, and contact hypersensitivity. Safety of candidate probiotics should then be evaluated in well-conducted clinical trials as evidence for safety and efficacy in humans is lacking. The comprehensive safety and efficacy data should finally be evaluated by an expert to consider the intended use, plausibility of health claims, possible adverse effects, and likelihood of high-risk groups in order to make a risk-benefit assessment (473). The effects of long-term consumption would be monitored by postmarketing surveillance.

The response of normal gut microflora to probiotic intervention varies with age and clinical status of the subject, so immunological effects need to be assessed in specific at-risk populations. Safety evaluation of long-term health effects will be important in the selection of, and characterization studies for a probiotic. The molecular factors modulating immunoregulation need to be elucidated. There are currently no agreed guidelines for the safe use of probiotics in immunocompromised patients. Immunocompromized people undertaking treatment with probiotics should be observed clinically over a long period to assess the effects and safety of immunomodulation (471, 472).

Mass marketing of probiotics is directed at healthy individuals despite an absence of long-term studies to verify claims that long-term use of probiotics helps to maintain good health in this population. There are no studies measuring the effect of probiotics on the immune system of healthy individuals, or on their innate resistance to disease (465).

1.7.6 Clinical Studies

Clinical studies in humans have investigated the effect of oral administration of probiotics on the balance of intestinal microbiota and in a variety of disorders. Until recently many studies were of inadequate design and produced unreliable data. Inadequate studies have had an absence of a patient control group; small treatment groups; undefined treatment groups; a wide age range within a treatment group; a diversity of antibiotic treatments; an absence of dosing criteria such as dose and duration; or subjects with symptoms of concurrent disease with the potential to confound an observation of adverse effects. The gold standard remains a controlled study with randomized, blind assignment to treatment, placebo, and untreated groups.

Immunosuppressive therapy is considered a risk factor in bacteremia from opportunistic pathogens. Salminen et al. (526) evaluated the efficacy and safety of *L. rhamnosus* GG (LGG) in moderating gastrointestinal symptoms of HIV-positive patients on antiretroviral therapy, in a placebo-controlled double-blinded crossover study. Subjects with HIV infection and persistent noninfectious diarrhea taking highly active antiretroviral therapy were standardized to receive twice daily LGG (viable LGG $1-5 \times 10^{10}$ cfu/dose) for 2 weeks and 2 weeks placebo in standardized order. No probiotic products were permitted during the washout periods before and after each treatment, to reduce the likelihood of a carryover effect from persistent probiotic. Although the LGG preparation was well tolerated it gave no significant reduction in gastrointestinal symptoms. No adverse events or clinical infections were observed in the subjects during the study or in the 6-month follow-up period. The evidence from this study suggests that LGG is unlikely to be a health risk in HIV patients.

Weizman et al. (527) conducted a 12-week double-blind, placebo-controlled, randomized trial of infant formula supplemented with either *Bifidobacterium lactis* (BB-12) or *L. reuteri* (ATCC 55730) and no probiotics, in healthy infants in child-care centers. The rate and duration of respiratory illness was unaffected by probiotic supplementation. In contrast, children supplemented with *B. lactis* and *L. reuteri* had fewer and shorter episodes of diarrhea compared with the placebo group, with *L. reuteri* showing a significant decrease. These probiotics are safe for infants.

Elderly hospitalized patients were treated with a commercial probiotic in a randomized double-blind placebo-controlled trial to determine the efficacy of the probiotic to prevent diarrhea associated with antibiotic use or caused by *Clostridium difficile* (528). One-hundred grams of probiotic (*L. casei* DN—114001; 1.0×10^8 cfu/mL; *S. thermophilus*, 1.0×10^8 cfu/mL; *L. bulgaricus*, 1.0×10^7 cfu/mL) was consumed twice daily during and for 1 week after antibiotic treatment. The reduction observed in the incidence of antibiotic-associated diarrhea and *C. difficile* associated diarrhea in the treated group was statistically significant, and the probiotic was well tolerated with no adverse effects. Criticisms of the study included highly selective inclusion and exclusion criteria such that subjects were unrepresentative of elderly hospital patients, giving rise to results with low generalizability to the wider hospital population (529).

It is thought that infants younger than 3 months may be at risk of acidosis from ingestion of high concentrations of D(−)-lactate-producing probiotic organisms (472). In a study reported by Connolly et al. (530) infants with a family history of allergy were supplemented daily from birth with *L. reuteri* ATCC55730 (SD2112) (10^8 cfu/day) in a double–blind, placebo-controlled clinical trial. *L. reuteri* ATCC55730 produces D(−)–lactic acid which it was thought may abnormally elevate levels in infants. Blood levels of D(−)-lactic acid measured in the infants after 6 and 12 months showed no differences between placebo and treated infants and no adverse effects from long-term supplementation, attesting to the safety of *L. reuteri* ATCC55730 in infants (530).

Severe acute pancreatitis is a serious illness associated with a significant mortality rate for a proportion of those patients who contract infections and necrozing pancreatitis. As antibiotic treatment has not proven effective in reducing infection it has been thought probiotic treatment may be beneficial.

Besselink et al. (531) recently addressed the deficiencies of earlier clinical studies in an elegant multicenter randomized, double-blind, placebo-controlled trial of a multispecies probiotic (522) in 298 patients with predicted severe acute pancreatitis, with unanticipated results.

The probiotic (total daily dose 10^{10} bacteria) or placebo was administered twice daily in enteral nutrition for a maximum of 28 days. The primary endpoint was a composite of several specified infectious complications, including infected pancreatic necrosis, bacteremia, and pneumonia.

Contrary to expectations the incidence of infectious complications was not significantly reduced in the probiotic group, being 30% compared to 28% in the placebo group. None of the infections were shown to be due to the probiotic strains.

Mortality was significantly higher ($p = 0.01$) in the probiotic group (16%) compared with the placebo group (6%). Most of the deaths resulted from multiorgan failure. Nine patients in the probiotic group developed bowel ischemia (eight with a fatal outcome) while no patient did in the placebo-treated group.

The focus on pathogenicity of probiotics has centered largely on the potential for bacteremia and immunomodulation. *In vitro* and animal studies were not predictive of the serious adverse effects seen in this study (531). The probiotic mixture consisted of *Lactobacilli* and *Bifidobacteria* strains that are common ingredients in probiotic supplements or dairy foods individually enjoy European Union QPS status and have not shown adverse effects in previous small clinical studies.

Whether the adverse effects resulted solely from probiotic administration or from this and other factors is unclear. The authors propose putative mechanisms for further investigation.

The evidence from this study is that therapy with this multispecies probiotic is contraindicated in patients with predicted severe acute pancreatitis. Importantly, contrary to a hope for amelioration of the disease, the high mortality seen in the probiotic group raises doubts about the safety and efficacy of probiotics in such critically ill patients.

1.7.7 Postmarket Surveillance

Two Finnish studies have investigated the incidence of infections associated with LAB. In the first study 16S rRNA methods were used to characterize and identify LAB isolated from blood cultures of bacteremic patients in Southern Finland (532). The number of infections caused by lactobacilli was extremely low and the infections were not associated with the probiotic strain newly introduced in fermented milks.

In a subsequent study, lactobacilli isolated from bacteremic patients between 1989 and 1994 were compared to common dairy or pharmaceutical strains (533). From a total of 5192 blood cultures 12 were positive for lactobacilli, an incidence of 0.23%. None of the clinical cases could be related to lactobacilli strains used by the dairy industry. In both studies, patients with LAB bacterium had other severe underlying illnesses.

Salminen et al. (504) examined the incidence of lactobacilli bacteremia in the Finnish population for the period corresponding to a rapid increase in consumption of the probiotic strain *L. rhamnosus* GG (ATCC 53103). This strain was isolated from human intestinal flora and introduced into dairy products in 1990. By 1999 the annual per capita consumption was estimated at 6 L (3×10^{11} cfu) per person/year.

The Helsinki University Central Hospital collected all lactobacillus isolates from blood cultures and cerebrospinal fluid in its catchment area from 1990 to 2000. Blood culture isolates were also collected for all cases of lactobacillus bacteremia reported (and unreported) by mandatory notification to the National Infectious Disease Register, from its inception in 1995 to 2000. Species were characterized and compared to *L. rhamnosus* GG strain by molecular epidemiological methods.

Ninety cases of lactobacillus bacteremia were identified between 1995 and 2000, when the population in Finland was 5.2 million. Of the 66 isolates available for species-level identification 48 were lactobacillus isolates, with the most common species being *L. rhamnosus* (26, 54%), *L. fermentum* (9, 19%) and *L. casei* (7, 15%) respectively. In 35 cases more than one additional bacterial species other then *Lactobacillus* was also identified. Eighteen of the 66 isolates (27%) were organisms other than lactobacillus. Eleven of the 26 *L. rhamnosus* strains were indistinguishable by PFGE from the probiotic *L. rhamnosus* GG.

No increase in the incidence or proportion of lactobacillus bacteremia was observed, despite a clear increase in the number of cases of bacteremia over the period. Lactobacillus isolates as a proportion of all blood culture isolates was 0.24%, consistent with previous Finnish reports (533). The average annual national incidence of lactobacillus bacteremia was estimated to be 0.29 cases per 100,000 people per year. The study provides evidence that increased consumption of *L. rhamnosus* GG had not led to a corresponding increase in lactobacillus bacteremia.

Borriello et al. (496) was unable to find published medical literature regarding the consumption of viable probiotics by hospital patients, some of whom may be predisposed to infection by probiotic bacteria. They suggested that because of the low incidence of probiotic bacteremia and the sophisticated methods and experience needed to confirm it, identification and confirmation of species and strain characteristics of suspect clinical isolates should be referred to national reference centers. National clinical and epidemiological databases could include identity of organism, status of the patient's underlying conditions, coexisting infections and outcomes, and data on the patient's use of probiotics.

1.7.8 GMO Probiotics

While the search continues for nonpathogenic organisms with therapeutic potential, genetic engineering of an organism to produce an identified desirable bioactive molecule may represent a technically more efficient and attractive approach (501). Administration of therapies by the systemic route is recognized to cause unwanted side effects at sites other than those of interest. The concept of localizing delivery to a region of the intestinal mucosa where synthesis of a bioactive molecule *in situ* may bring about the desired effect without the disadvantages of systemic side effects has appeal.

Genetic manipulation offers the potential to enhance the existing probiotic properties of an organism or to imbue an organism with probiotic properties (524). Elucidation of mechanisms of activity of a probiotic will enable manipulation of organisms to create specific and targeted probiotics. Consumer resistance to genetically modified organisms (GMO) is such that GMO probiotic foods are unlikely in the near future, but clinical applications to ameliorate or prevent chronic intractable diseases may be more readily accepted.

Steidler et al. (534) and Kaur et al. (535) treated mice models with GM bacteria to prevent colitis and enhance the efficacy of antitumor therapy respectively, demonstrating in principle that probiotics can be designed to produce potent bioactive chemicals. Having engineered *L. lactis* to deliver mouse cytokine IL-10 at the intestinal mucosa Steidler (524) then constructed a biologically contained *L. lactis* to produce human IL-10. In the first clinical trial of its kind Braat et al. [539] treated Crohn's disease patients with this GM *L. lactis* in a phase 1 placebo-uncontrolled trial. A decrease in disease activity was observed with minor adverse effects, and containment of the organism was achieved through its dependency on thymidine for growth and IL-10 production.

The incorporation of GMO bacterial strains into therapeutic products will necessitate stringent procedures for safety assessment. To treat an individual with a living recombinant microorganism is to release a GMO into the environment and in such instances safety is of paramount importance (476, 524, 537). Of no lesser importance introducing and exposing an individual to foreign protein in this manner has potential to provoke an immune response that may preclude clinical applications (524).

The organism will need to be "biologically contained" to prevent its undesirable release and accumulation into the environment, and to prevent transmission of the genetic modification to other bacteria (524). Methods for biological containment have been demonstrated previously. Control of the organism may be active through, for example, production of a bacterial toxin that is regulated through genetic expression controlled by an environmental response. Passive control may be employed where growth of an organism depends on either the presence of an essential substrate or the gene facilitating its production (524).

1.7.9 Conclusion

Consumers are increasingly managing their health by self-medication, generating expanding market opportunities for the food and pharmaceutical industry. Food and drink manufacturers are adopting genomic and proteomic technologies previously the domain of the pharmaceutical industry to design more sophisticated and novel products.

Because exploitation of the probiotic concept is still associated with unsubstantiated claims reliable and proven products need to be readily distinguished from those of dubious quality. A significant proportion of consumers mistrust manufacturers' claims. Forty-five percent of Americans claim to largely or entirely disbelieve food and drink manufacturers' health claims, a figure similar to France and greater than in the Netherlands (538). The evolution of probiotic products thus necessitates changes in the regulations related to labeling, safety, and health claims.

The credibility of health claims for healthy individuals remains to be established. Viable probiotic bacteria have to be consumed in large quantities for a prolonged period to achieve a health benefit. The long-term effects of probiotic consumption on a healthy population are unknown, and yet the general population is being encouraged to consume probiotics regularly, to promote good health and well-being. Studies have yet to be undertaken to demonstrate what effect, if any, there is on the well-being of healthy individuals.

While probiotic cultures are incorporated into foods or dietary supplements without making specific health claims they avoid the need to conform to the more rigorous regulatory procedures for therapeutic products, which require demonstrated quality, safety and efficacy. There is evidence from well-conducted clinical trials of beneficial health effects from probiotics in a range of clinical conditions. The probiotic effects have been shown to be strain specific so health effects cannot be generalized between strains.

Standardized, verifiable clinical studies are needed to demonstrate the safety, efficacy, and limitations of a putative probiotic, and whether it is superior to existing therapies. Additional studies are needed to determine effects on the immune system in healthy and diseased individuals and effects of long-term consumption. More rigorous quality control, standards, and regulations have been called for (481, 482, 477, 483). The prospect of GM probiotics targeted for clinical conditions demands a rigorous safety strategy to prevent spread into the environment and dissemination of the genetic modification.

Permission to label a product probiotic should remain contingent on its compliance with the FAO/WHO definition of probiotic (475). Labels should include consumer or prescriber information about the identity of the organism(s); its GMO status; viability count and shelf life; dosing and duration; conditions for which its use is and is not appropriate; proven benefits; side effects, particularly symptoms that require clinical assessment; and a recommendation to advise health practitioners of probiotic use (471). Where an adverse reaction is suspected the facility to report it along with the product details to a national database should be available, as it is for adverse effects of other therapies.

Guidelines have been proposed to assess the efficacy and safety of probiotics but international agreement on these has yet to be arrived at. Consensus on uniform regulations is desirable to ensure unequivocal identity, quality manufacturing processes, accurate labeling, proven safety and efficacy for a product that will then merit the label "probiotic."

1.8 LEGAL STATUS AND REGULATORY ISSUES

1.8.1 Human Probiotics

1.8.1.1 Asia

WILLIAM TIEN HUNG CHANG

Lytone Enterprise, Inc., Taipei, Taiwan

Most of the early probiotic studies were related to fermented dairy products enriched in LAB. Items such as yogurt, sour milk, cottage cheese, etc. were in themselves regular foods with thousands of years of tradition. Beneficial microbes contained within these foods were identified and developed into specific products with health augmentation purpose. The market has been growing rapidly from 7% in the United States to 15% in China. (Business Communications Co. Ltd. (BCC) July 2005). Total value may reach US$1.1 billion 2010 (539). This does not include food items that already have live microorganisms as a part of their original production process, such as yogurt, Kefir, cheese, sauerkraut, Kimchi, etc. In the interest of public health protection, national governments tend to view probiotics as food rather than drug, unless specific claim on therapeutic effects were attempted. Regulatory attitudes also gradually change as more scientific information become available. This report intends to review the current status of regulatory policies regarding probiotics in Asian countries, to the extent that such information is publicly available.

The various definitions of probiotics are listed in Section 1.1.

Summarizing the common features of the definitions listed, probiotics is a term referring to products that

- are living organisms;
- requires sufficient dosage to be effective;
- confer health benefits.

Lactic acid bacteria are the most commonly mentioned example of probiotics. However, other microorganisms such as *Bacillus* sp., yeast, and algae (blue-green algae, *Spirulina*, *Chlorella*, etc.) were also grouped into the definition of probiotics, since the propagation and management of these organisms as human food seems to require similar technologies. Therefore, regulatory considerations usually include these products into the same category. On the contrary, larger organisms such as mushrooms, and fungi, such as *Gonoderma* sp., *Cordyceps* sp., are usually not included in probiotics but are considered as regular foods. Therefore, the definition suggested by FAO/WHO (540) seems to have been accepted by most official organizations in Asia: "Probiotics are live microorganisms which when administered in adequate amounts confer a health benefit on the host."

Probiotics has been recognized as an important product due to numerous scientific publications that have become available in the past 20 years. The beneficial functions of probiotics are described in Section 4.3.

Probiotics has been demonstrated to be effective in relieving certain symptom of illness that has not been successfully addressed by conventional medical practice. As more data on the safety and successful performance of probiotics become available, traditional western health professionals have changed their suspicious attitude to welcome food supplement products that could help conventional health-care practice. The market of probiotics thus has grown significantly in the past 10 years. Products with clear label and proven functional properties, as well as stable specifications are more likely to succeed. Market survey of probiotic-related products varies depending on the scope of definition. Dairy products containing LAB was estimated to be worth over US$30 billion in 2005, with Japan alone accounting for US$5 billion (541). The market size for pure probiotics is smaller. Probiotics sold as food supplement in the USA was slightly over US$240 million (542). However, the annual growth rate was 14%, which is highly remarkable for any food item. The sale of probiotics in Europe was 12 million. The annual growth rate was also 15%, indicating an optimistic trend for the near future.

However, a series of legal incidences in recent years also caused concern for business development (543). Consumer advocates and insurance companies have also started demanding more guidelines when it comes to using probiotics as a part of regular health care portfolio. Some would insist that so long as probiotic products are used for therapeutic purposes, they should be regulated as any other conventional drugs, notwithstanding the fact that the product may have been consumed as a food item safely for many years. The others would argue that since the safety record of a probiotic product has been acceptable as a food item, the relevant requirement on safety tests should be rationally reduced, while emphasizing on the efficacy demonstration. The result of these tests, and the regulatory position pertaining such "therapeutic" products should be fully disclosed and inform the public aggressively to reduce confusion. These differences in attitude has been subject for debate in many forums, and resulted in legal action in some cases. Companies with major stakes in the matter have therefore stepped forward and tried to establish general consensus on product concepts based on strong scientific foundation that would encourage healthy market growth. Stabilization technology for microorganisms has been developed as a result. Food products with special functional purposes were also developed using probiotics. Items include baby formula, fruit juice, breakfast cereals, yogurts, even chocolates with over 1 year shelf life. It was felt that only with sound scientific data could a positive guideline be established for healthy development of probiotic products that could benefit human health. This report intends to review the phenomenon in Asian countries, especially on the regulatory part.

Study of the Asian country's position regarding probiotics should start from the position of FAO/WHO of the United Nations (UN). In our survey, it was noticed that with a few exceptions, most of the Asian countries started their own official regulatory policy after the UN has established a relevant guideline. In fact a conference on functional food was organized by FAO/WHO in 2004 in Bangkok for the very reason of increasing awareness for the need of regulating such products. Probiotic was only part of the discussion during this conference. The two UN organizations have noticed the increased consumption of probiotic products in the world since 1990. Discrepancy among member countries, and even among regulatory authorities within the same country, has led to confusion in the general public. Inferior products competed under false pretense, and indeed may cause health risk to the consumer due to lack of accurate information. An expert panel on the subject of probiotic therefore was convened in 2001 at Cordoba, Argentina. Consensus was reached during the meeting regarding the importance to have a common guideline for probiotics regulation. A recommendation was drafted as a result and was approved in May 2002 during a follow-up meeting in Canada (544). This guideline recommends how a probiotic food product could be approved for marketing by the national regulatory authorities based on scientific principle. Member countries could establish their own national regulation based on this recommendation. The expert panel consciously excluded probiotic products designed for medical use, animal use, and microorganisms that have been modified through genetic modification, so as to limit the scope of discussion.

The principle involved in the guideline is that the probiotic product must first be qualified as a food; it can then be further regulated as microorganisms that confer beneficial effect to human health.

The leading issue in the guideline seems to be the identity of the microorganism within the product in question. It would be necessary to identify the culture at subspecies or strain level for most probiotic products, except some of the traditional LAB that are used for conventional fermentation foods. The microbiology associated with the safety, functional property and physiological effects cannot be studied until the identity of the bacteria has been confirmed.

The technology that was preferred in the FAO/WHO guideline for culture identification was DNA–DNA hybridization. However, the procedure involves complicated process and expensive equipment that may not be readily available. Therefore, 16sRNA sequence comparison was also recognized as a reliable technique to identify microbial taxonomy, in addition to biochemical reactions and phenotypic observations. Such data package was recommended to be the first step for probiotic product registration. The other review process and data required for safety, *in vitro* and *in vivo* trials, toxicity data, etc. are usually more lenient than those required for pharmaceuticals, more so if the product has been classified as GRAS, in lieu of Food and Drug Administration of USA. The recommended registration process for probiotic product is as shown in Fig. 1.15.

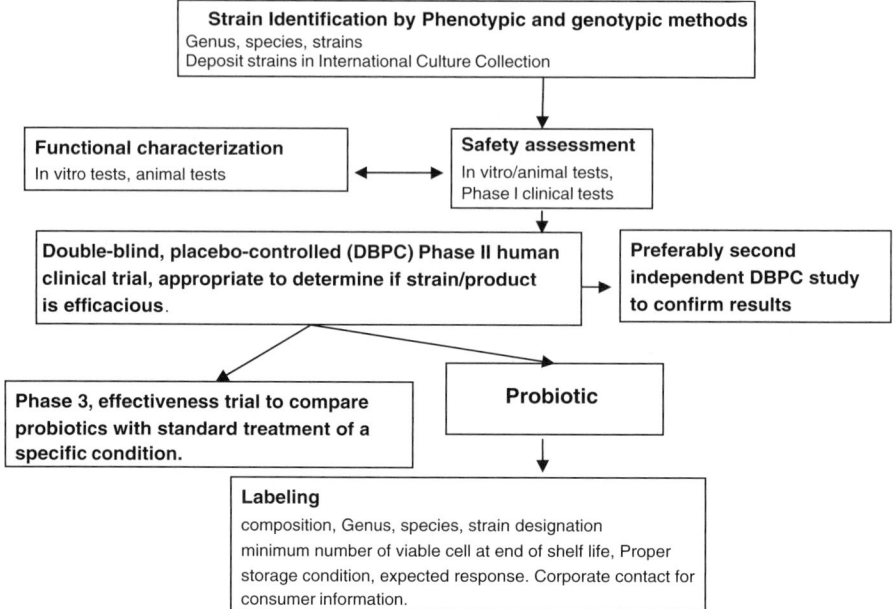

FIGURE 1.15 FAO/WHO recommended registration process for probiotic product (544).

United Nations has recommended a number of product specifications that should be considered when a probiotic product is to be approved. These include stability against pH, ability to colonize in the GIT, taxonomical identification, and proper dosage.

1. *Viability*: Probiotic should survive passage through the GIT. Since the most severe threat comes from acidity in the stomach, *in vitro* simulation in dilute HCl could be used. Bile acid and other digestive enzymes may also be tested *in vitro*.
2. *Colonization*: Good probiotic culture should be able to colonize and/or become attached in the GIT of the host, and multiply in the presence of bile acid. The growth should benefit the host by reproducible evidence. Different microbes may colonize at different locations within the GIT of different host.
3. *Probiotic Culture*: The culture should be able to compete successfully against pathogenic bacteria for the attachment on intestinal surface.
4. *Inhibition*: Probiotic may show inhibition of pathogenic bacteria by *in vitro* tests.
5. *Resistance*: Probiotics used for vaginal protection should show resistance against other sanitary agents or disinfectants.
6. *Identity*: Probiotic products should indicate on its label the genus and species name according to international nomenclature and expected viability during the shelf life of the product.

7. *Dosing*: The label should indicate recommended dosage and expiration date.
8. *Data*: It will be desirable if there are data indicating the product will not infect immunocompromized animal.

The FAO/WHO guideline constituted the background for each member nations to establish their own regulation. However, such is not always the case for every member nation. In Asian countries this is especially so. The following are some of the examples.

1. *Japan*. Japan is one of the few countries that actually have an endogenous regulation regarding probiotics before the UN guideline was published. One of the main reasons was that Japan had established its own microbiology industry almost simultaneously with those in western countries. Japanese scholars published widely in the field of pathogenic bacteriology in early 1900s and a robust fermentation industry had been established during early and mid-twentieth century. The most prominent companies in this industry are Yakult, Kyowa Hakko, Fujizawa, Tanabe, etc. As Japanese scholars involved in the early stage of guideline preparation for the FAO/WHO conference, Japanese regulatory policy on probiotics was friendly toward industrial manufacturers. In general, all producers of functional food in Japan may apply for a certification from the Ministry of Health, Labor and Welfare (MHLW) according to a "food for specialized health use" (FOSHU) system. The system was first initiated in 1984 by the Japanese government through a series of studies and recommendation from a group of scientists from the academics, industry, and government research organizations. It was suggested that with advanced age of Japanese population, the cost of health care could be controlled through proper intake of functional food, instead of offering therapeutic solutions after the person is ill. FOSHU system was therefore formally recognized and entered into Health Improvement Act, Article 26, and Food Sanitation Act, Article 11, in 1991. In fact, functional food was first coined in Japan. Probiotics were then included as part of functional food. Japan has further distinguished among functional foods so as to designate products that wish to make specific health claims against certain illness (special health protection food), general nutritional supplements, and special application food (for pregnant women, patients under recovery from surgery, or otherwise could not consume regular foods, etc.).

   There was no official requirement for the minimum viable cells in the final product. The government seemed to leave the issue to the discretion of industrial organizations such as Japanese Fermented Milks and Lactic Acid Bacteria Beverages Association. This organization stipulated that a product containing $\geq 1 \times 10^7$ viable bifidobacteria/g or ml is to be considered a probiotic food (545). Dosages also need to be considered alongside with frequency of intake, the strain involved, and the general health condition of the consumer. Food regulations in general are not concerned with such issue, as would a drug regulation. Industrial manufacturers are therefore leery of establishing a recommended dosage for fear of being driven into drug category. It was generally agreed, however, the viability of the cells needs to be indicated for the duration of product's shelf life.

There were over 69 products formerly registered through FOSHU as probiotic products in Japan by December 2005, which were allowed to claim augmentation of gastrointestinal disorder out of a total of 569 registered functional food products (546). At the time of this writing there were a total of 680 functional food products registered through FOSHU system. It should be interesting to note that more than 57% of these foods were addressed to gastrointestinal disorder, and the remedies included not only probiotics, but also fiber, prebiotics such as oligosaccharides, etc.

2. *People's Republic of China.* The country with the largest populace in the world stands to gain advantage in health care cost saving by promoting probiotics. However, Chinese culture does not have a traditional role for LAB as most western countries would have in the form of fermented dairy products. Most of the Chinese physicians received their training with a microbiology curriculum that emphasized mainly on pathogenic bacteria. The benefit of probiotics has not been widely introduced in China until early 1980s. Nevertheless, with rapid economic growth, scientific data regarding probiotics from the developed countries soon received attention in China. While regulations were drafted, Chinese Ministry of Health took into consideration that the traditional Chinese Herbal Medicine actually included a number of fungal cultures that were believed to be beneficial to health. Thus, "Fungal Health Food Review Guidelines" and "Probiotic Health Food Review Guidelines"were both issued in March 2001 in accordance to Food Sanitation Law and Regulation on Health Food, with later revisions until 2004. The basic principle is more or less similar to those practiced in Japan and other western countries. The guidelines included specific instructions on the technical capability of the parties applying for the health food registration for either fungi or bacteria. The identification of the culture should be performed by officially sanctioned research institutes that are usually government-run to ensure accuracy. There are a number of clearly defined cultures (547), which are allowed to be included in health food category without the need for extensive safety studies due to historical reason. These culture lists were attached to each guideline, effectively discouraging companies from using cultures that has not been named in the list. In case a new culture that is not listed, companies would need to go through the expense and effort to register new cultures based on these guidelines, proving safety and efficacy.

The PRC government however had been aware of the current regulation on health food were rather cumbersome, and a Provisional Health Food Registration and Management Directive went into effect on July 1, 2005. The new directive differs from the current system in the following manner:

(a) Since companies tend to concentrate on using well-known ingredients and applying for health food registration only in the 27 sanctioned functions published by the State, for fear of uncertainty in the review process if a new function is being applied, the current 27 functions will be abolished, so that new ingredients and functions will be encouraged. But benefits declared by health food for malignant tumor prevention will no longer be allowed.

(b) Parties who are allowed to apply for health food registration will now include not only corporation, but also individual persons. This is helpful for private individuals who are not attached to any official organizations to come forward with their invention.

(c) Once the documents that are required for health food registration are considered complete and accepted by the Ministry of Health, a decision must be made within 5 months by the reviewing committee, as compared to the average of 8 months at the present.

(d) Health Food License will be valid for only 5 years, and is renewable upon re-registration, which should be initiated no later than 3 months prior to the license expires. This way, the more than 8000 health food licenses already granted will be weeded out not only through marketing force but also by the economic consideration due to the cost of registration.

(e) Stricter limit was placed on the wording used for promotion of health foods. Several branded health foods are deemed as "negative examples" are named in the new regulation and will not be tolerated in the future.

(f) All advertisements relating to health foods will need to be approved first prior to publicity.

The regulation on probiotic health foods was also revised in the meantime, which clearly specified that any new culture applied for health food status would be approved by State Food and Drug Administration (SFDA) (548). The approved list of probiotic microorganisms at the time of this writing is shown in Table 1.11

There have been some arguments for the requirement of viability of probiotic products, citing examples that dead probiotics could also work in

TABLE 1.11 List of Probiotic Cultures Recommended by SFDA

Bacteria	Fungi
Bifidobacterium bifidum	*Saccharomyces cerevisiae*
Bifidobacterium infantis	*Candida utilis*
Bifidobacterium longum	*Kluyveromyces lactis*
Bifidobacterium breve	*Saccharomyces carlsbergensis*
Bifidobacterium adolescentis	*Paecilomyces hepiali* Chen et Dai, sp. Nov
Lactobacillus delbrueckii subsp. bulgaricus	*Hirsutella hepiali* Chen et Shen
Lactobacillus acidophilus	*Ganoderma lucidum*
Lactobacillus casei subsp. Casei	*Ganoderma sinensis*
Streptococcus thermophilus	*Ganoderma tsugae*
Lactobacillus reuteri	*Monacus anka*
Lactobacillus rhamnosus	*Monacus purpureus*

certain beneficial functions. Therefore, the official position is that dead organisms and/or their metabolites could also be considered as probiotics, as long as the functional ingredients and assay methods could be identified. Otherwise, a minimum of 10^6 cfu/g (mL) is required before the product expiration date.

3. *Korea.* Health/Functional Food Act of Korea was first published in August 2004. The regulation is unique in which it requires the products that come under this law must be sold in dosable form, that is, pills, tablets, capsules, etc. (549). There are also 37 categories that are recognized as generic health/functional foods that do not require special safety and efficacy studies, as shown in Table 1.12. Probiotics in the form of LAB, *Chlorella*, *Spirulina*, and *Monascus* species are also included in the table. However, specifications for a product to

TABLE 1.12 Categories of Functional Food in Korea

	Health Food Product Category by KFDA		
1	Nutritional supplement products	20	Grape seed oil products
2	Ginseng products	21	Fermented vegetable extract Products
3	Red ginseng products	22	Muco-polysaccharide protein Products
4	Eel oil products	23	Chlorophyll-containing products
5	EPA/DHA fish oil products	24	Mushroom-processed products
6	Royal jelly products	25	Aloe products
7	Yeast products	26	Plum products
8	Pollen products	27	Turtle products
9	Squalene products	28	Beta-carotene products
10	Product of digestive enzyme	29	Chitosan products
11	Edible lactic acid forming bacteria products	30	Chito-oligo-saccharide products
12	Chlorella products	31	Glucosamine products
13	Spirulina products	32	Propolis extract products
14	Edible oil containing gama-linolenic acid products	33	Green tea extracts and its their products
15	Wheat germ/rice bran oil products	34	Soyprotein and its products
16	Products with wheat germ and/or others	35	Phytosterol, Phytosterolesters and their products
17	Egg and/or soybean lecithin products	36	Fructo-oligosaccharide and its products
18	Octacosanol products	37	Red rice and its products
19	Alkoxy-glycerol products	38	Product specific

be accepted into each category are rather strict. Ministry of Health and Welfare is the agency in charge of health food legislative changes on the Health/Functional Food Act, while Korean FDA is in charge of all the other matter relating to food products. The 37 categories are listed in Table 1.12

Korea also has rather strict regulation regarding the safety and identity of probiotic products, requiring the industry to apply for a new evaluation by the Korea Food and Drug Administration each time a new culture combination is designed, even if individual cultures had been tested and approved previously. Probiotic products rank fourth in the functional food market, following nutritional supplements (such as vitamin mixes, etc.), ginseng and aloe. Furthermore, KFDA issues only health food license to companies with local registration. Overseas company can only register it health food products through an authorized domestic agent or its own subsidiary.

4. *Malaysia*. Malaysia government has been aware of the role of probiotics as a food item, and issued relative provisions in the Food Regulation (550) in 1985. Over the years Malaysia has adopted an attitude that seems closer to those being practiced in Commonwealth of United Kingdom. Lactic acid bacteria are recognized as a legitimate food ingredient, and *Bifidobacteria lactis* and *B. longum* are both mentioned specifically in the regulation with minimum of 10^6 cfu/g requirement if such cultures are to be labeled on any food item (551). The regulation also stipulates that the label must follow the following rules, which further emphasizes the requirement for viable cells.

However, Malaysian regulation does not allow any functional claim that could be construed as having therapeutic benefits, as was stipulated in the same regulation (552):

(a) In these regulations, "nutrient function claim" means a nutrition claim that describes the physiological role of the nutrient in the growth, development, and normal functions of the body.

(b) A nutrient function claim shall not imply or include any statement to the effect that the nutrient would afford a cure or treatment for or protection from a disease.

(c) Where a claim is made as to the presence of bifidobacteria in food, there shall be written in the label of a package containing such food, a statement setting out the viable bifido bacteria count present in a stated quantity of the food.

(d) There shall be written in the label on the package of food containing bifido bacteria the words "Contains viable bacteria, require special storage condition" or "Contains viable bacteria, follow instruction for storage."

The regulation specified 12 functions that could be allowed to print on the label relating to ingredients such as vitamins and minerals. But no probiotic culture was allowed to claim any benefits.

5. *Taiwan*. Taiwan has promulgated a Health Food Control Act in February 1999, focusing on the process of registration of health foods, claims, and penalties.

The Act has since been revised four times until 2002. The government also issued official guidelines regarding the efficacy and safety evaluation for certain human health conditions. At the time of writing, a total of 13 efficacy-testing protocols for health-improvement functions have been recommended. These include sero-cholesterol augmentation, osteoporosis improvement, immunity augmentation, dental health improvement, blood-sugar augmentation, liver function improvement, gastrointestinal function augmentation, antifatigue, antiaging, etc. The product may be labeled with specific claims that conform to recommended wordings by the Act, after the clinical trials have been successfully completed. The label is as follows:

The health care effects of health food shall be described in any of the following ways:

1. claiming the effect of preventing or alleviating the illness relating to nutrients when deficient in the human body if intake of the health food can make up said nutrients;
2. claiming the impact on human physiological structure and functions by the specified nutrients or specific ingredients contained in a health food or the food itself after the health food has been taken;
3. furnishing the scientific evidence to support the claim that the health food can maintain or affect human physiological structure and functions; and/or
4. describing the general advantages of taking the health food (553).

There exists specific regulation governing how a new health food registration should be applied, including requirements on the manufacturing facility, building, equipment, process control, labeling, ingredient listing, and even award for people who report to the government on violation of the specific Act. Mentioning of viable cells in the probiotic products is voluntary, with no requirement on the wording. Therefore most of the products on the market are content with claims on the viable cells at the time of the manufacturing, rather than at the end of stated shelf life, although such a requirement was recommended (not required) by the Health Food Control Act. By the end of 2005 there were 18 probiotic health foods officially registered by the Taiwan government.

6. *The Other Asian Countries.* A brief overview of other Asian countries such as India, Pakistan, Bangladesh, Nepal, Thailand, Vietnam, Philippines, Indonesia, etc. regarding functional food has been published by the UN (554). It seems that most of these countries, with the exception of India, are still in the development stage of policy and regulation regarding functional food, let alone probiotic products per se. Each country is aware of the need for special legislation for functional foods since most of them have traditional remedies that have been used for health care, but are not in the realm of western medicine. Probiotic is already recognized as having a special role in countries such as Thailand (555) and India.

7. *Impact of regulatory policy toward the development of probiotic industry.* Asia has the largest population among all of the world's continents. Economic development within Asian countries varies from one extreme to the other. It is apparent from this study that countries with endogenous microbiology industry would have more confidence in establishing more flexible guidelines and regulation on probiotic products, allowing such products to be sold as food rather than drugs. Japan is probably unique in that the industrial organizations would play important role in the initial review and certification, under the supervision and authorization of the government. Probiotic Industry has been careful in establishing the credibility of their products over the last 50 years, so that the sales of probiotic products has reached over US$5 billion in recent years, probably one of the highest per capita consumption in the world.

The world market of probiotic products has been growing at an annual rate of 15% until 2003; not only in the form of fermented dairy products but also in probiotic foods that mainly consists of viable LAB or other traditionally employed microorganisms such as *Bacillus natoensis* and *Monascus* sp.

It was remarkable that most of the countries studied in this article started to recognize the importance of regulating probiotics since 2004 probably as a response both to the market development, as well as the position taken by FAO/WHO of the UN. One of the important decision makers for purchasing probiotic foods who have been resistant to the idea of probiotics is the conventional medical service provider such as western physicians and pharmacists, especially in countries of underdevelopment economics. The main reason is probably due to the dubious reputation caused by inferior probiotic products in the past, when there was no regulation at all to control the quality of these products. It was only through the effort of major industrial players who collaborated with reputable academic institutes in various nations to perform scientific research and clinical trials, and demonstrated reproducible results through publication, were such negative attitudes among the physicians become soften. Advanced technology such as 16SRNA sequence analysis and other biochemical characterization also provided assurance of bacterial strain identification, and safety control during manufacturing. Researchers are now patenting specific cultures that have proactive physiological benefits such as allergy reduction LGG (556), cholesterol reduction *Bacillus coagulans* (557), cancer inhibition *L. acidophilus* 1–1492 (558), and cultures that have been developed in Asian countries such as LP33 (559). New patents on probiotic health functions are being filed at a rate of over 30 per year since 2002, indicating an important trend in the near future. The fact that most of the probiotic products are also qualified as food, as far as safety is concerned, provided an attractive alternative to sustain health in an aging population, with relatively low cost. Hence Asian governments have been aggressive in putting together a friendly environment to encourage the growth of the probiotic industry. Countries who have the most reasonable policy based on sound scientific knowledge would stand to gain the most advantage.

1.8.1.2 Europe

ARTHUR C. OUWEHAND AND SAMPO LAHTINEN

Health & Nutrition, Danisco, Finland

Until now, health claims on functional foods have been regulated on a national level in the EU. Regulatory bodies like the Agence Française de Sécurité Sanitaire des Aliments in France, Joint Health Claims Initiative (now closed) in the United Kingdom, Voedings Centrum in the Netherlands, and the Swedish Nutrition Foundation in Sweden have evaluated health claims for functional foods. This situation is clearly not satisfactory as the nationally approved claims may interfere with the free movement of goods in the EU internal market; a product maybe sold with a health claim in one country but this claim might not be allowed in another country. A harmonization has therefore long been desired and a new regulation on nutrition and health claims was adopted by the European parliament on January 19th, 2007 (560). The new regulation deals with nutrition claims, health claims, and reduction in disease risk claims (561, 562).

1. *Definitions*. Within the new legislation, a number of terms are defined, some of them are mentioned below.

    A "claim" is defined as any message or representation in any form, which states, suggests or implies that a food has particular characteristics. "Nutrients" are protein, carbohydrate, fat, fiber, sodium, vitamins, and minerals. "Other substances" are defined as substances other than nutrients that have a nutritional or physiological effect. Nutrition claims refer to foods with particular beneficial nutritional properties such as

    (a) the energy the food provides or does not provide; the reduced or increased levels of it.

    (b) the nutrients or other substances it contains or does not contain or contains in different amounts or proportions.

    Health claims that are based on generally accepted scientific evidence fall under Article 13.1. Health claims based on newly developed scientific evidence and/or those claims that request protection of proprietary data fall under Article 13.5. Claims regarding a reduction of disease risk and claims referring to children's development and health fall under Article 14.

2. *Health claims* other than those referring to the reduction of disease risk, Article 13.1. Generally accepted claims that were submitted under Article 13.1 did not require a full scientific dossier. Instead, member states had until January 31st, 2008 to compile a list of claims and submitted them to the European Commission (EC). Although the member states were required to submit their list of claims by that date, many had indicated that they would have deadlines for submission well prior to the 31st of January 2008. This would allow the authorities to compile and finalize the list well in advance before submitting it to the EC. After this date, the EC will consolidate the lists and the EFSA will evaluate the claims. The final list with permitted and generally accepted claims and their conditions of use are to be

adopted by January 31st, 2010, at the latest. In the mean time, health statements can be made under the assumption that they would be approved by EFSA and hence have to have been submitted under Articles 13.1, 13.5, or 14.

The following area's of health claims have been indicated:

(a) Growth, development, and the functions of the body.
(b) Psychological and behavioral functions.
(c) Slimming or weight control or a reduction in the sense of hunger or an increase in sense of satiety or reduction of available energy from the diet.

The European trade associations representing the food and food supplement industry, Confederation of Food and Drink Industries in Europe (CIAA), European Federation of Associations of Health Product Manufacturers (EHPM), European Responsible Nutrition Alliance (ERNA), and the European Botanical Forum (EBF), developed in anticipation of the need to compile a list of claims a template for registration (Table 1.13). The layout of the list has been adopted by a substantial number of EU member states as a basis for the inventory of health claims.

3. *European Trade Associations Inventory for Article 13.* The CIAA, EHPM, ERNA, and EBF have jointly developed an inventory list for submission to the Commission. The trade associations' members have included their ingredients on the inventory following the layout suggested. The list will be used as an example here to describe the requirements for the application under Article 13.

The final list contains seven major categories: vitamins, minerals, proteins, carbohydrates, fats, fiber, and probiotic ingredients. Under "fiber," seven prebiotics have been included, while under "probiotic ingredients" more than 50 strains and combinations of probiotic strains have been included.

(a) *Food Category or Food Component.* The first column in the table is used to describe the food or food component. Here, an appropriate and correct description of the active component should be given. In particular for probiotics it is important that the correct taxonomic name is used, the strain designation is given and (when available) the code under which the strain

TABLE 1.13 For Registration of Health Claims, the European Food and Food Supplement Trade Associations (CIAA, EHPM, ERNA, and EBF) Have Developed a Template Table to Summaries the Claims

	No	Food or Food Component	Health Relationship	Conditions of Use	Nature of Evidence	Ref	Example wording
Vitamins							
Minerals							
Protein							
Carbohydrates							
Fats							
Fiber							
Probiotic ingredients							

has been deposited in a public culture collection for safe deposit. Examples for probiotics are *Lactobacillus acidophilus* NCFM (ATCC SD5221), *Lactobacillus johnsonii* NCC 533 (La1) (CNCM I-1225), or *Bifidobacterium animalis* ssp. *lactis* Bb-12®. Examples for prebiotics are inulin, fructo-oligosaccharides, galacto-oligosaccharides, and polydextrose. Although the production process of prebiotics may influence the structure, and thereby the functionality, origin and production methods are not mentioned in great detail; for example, fructo-oligosaccharides from chicory or sucrose.

(b) *Health Relationship.* In the past, much attention has been paid to the relation between foods and food components and health (563). A truthful documentation of the health claim and linking it to solid scientific evidence is the key. As mentioned earlier, three areas of health claims fall out side the Article 13.1: claims related to children's health and development, claims for which new scientific data is developed or which have requested proprietary protection, and claims related to reduction of disease risk. Claims in these areas will be discussed later.

The health relationship under Article 13.1 must relate to the maintenance or improvement of healthy body functions and should refer to the healthy state of those body functions, such as metabolism, immune function, intestinal health, digestion, etc. Specific physical or chemical properties of a food or a food ingredient (e.g. pre- or probiotic) may influence these physiological functions. If the effect is not shown in the final product, it would be essential to show that the bioactivity of the ingredients are not impaired by it's inclusion into a particular food matrix and thereby the scientific substantiation of the claim.

Examples of a health relationships for pre- and probiotics could be, immune function, intestinal health, bowel function, among others.

(c) *Conditions of Use.* The health claim must refer to the food as it is ready for consumption. The amount of food or food component that has to be consumed to obtain the claimed health benefit should correlate with the dose that a consumer reasonably could be expected to consume based on portion size and frequency with which it will be consumed. So, for prebiotics for example, the amount in grams per day and for probiotics the cfu per day and possibly the format (e.g. yogurt or fermented milk) should be taken into account. Scientific studies may use high doses of pre- or probiotics in order to ascertain the likelihood of observing a health benefit. Such studies may serve as proof of principle, but would have to be replicated with lower doses feasible in a final product.

(d) *Nature of Evidence.* The idea with the current legislation is to substantiate the health claim with sound scientific evidence. The evidence to support the health claim must be based on human studies. *In vitro* and animal studies may be useful as supporting evidence, but on their own are not sufficient to substantiate a claim. The human studies should be of sufficient quality and double-blind, randomized, placebo-controlled interventions provide the strongest evidence. But, it is recognized that due to the nature of food it may not always be possible to perform such interventions, and other study designs may be sufficient to substantiate the claim. It is also recognized that not all studies give the same results. Therefore, the totality of available evidence needs to be taken into account. In this column on the nature of evidence, can also be indicated the type of documentation: individual studies, textbooks, meta-analyses, monographs, critical reviews, and opinions of authoritative bodies.

    (e) *References.* The references mentioned in the claims table should be complete and demonstrate scientific justification of the proposed claim. There are no guidelines given concerning the number of references. But, the totality of the evidence should allow an objective evaluation.

    (f) *Example Wording.* The example wording in the list is not an exhaustive list of all possible health claims. The examples should be consistent with the health relationship and the supporting evidence. They should not suggest health benefits that cannot be substantiated, are false, ambiguous, or misleading. Nor shall they doubt the safety and/or nutritional adequacy of other foods. Finally, they should be understandable by the average consumer.

4. *Article 13.5.* Any health claim based on newly developed scientific data and/or which include a request for the protection of proprietary data will have to be submitted separately to a member state. The application has to include the name of the applicant and the nutrition, substance, food or food category in question. Furthermore, references to the scientific studies with regard to the health claim and any other relevant studies, proposal of the wording of the health claim and a summary of the application have to be included. The member state will make the dossier available to EFSA. EFSA will evaluate the claim and make the summary publicly available.

5. *Article 14.* Reduction of disease risk claims and development of children. Claims on the reduction of disease risk may only be made when they have been specifically authorized. The application procedure is similar to the procedure described above for Article 13.5. When an authorization has been received for a reduction of disease risk claim, the presentation or advertising shall also include a statement indicating that the disease to which the claim is referring has multiple risk factors and that altering one of these may or may not have a beneficial effect.

6. *Labeling.* As mentioned with example wording, the health claim has to be supported by scientific evidence. Thus, labeling, presentation, and advertising should not mislead or deceive the consumer, nor should it suggest that a balanced and varied diet cannot provide appropriate quantities of the nutrient (564).

7. *Further information.* The new legislation has not been applied and, as has been mentioned, several uncertainties remain in the exact procedure that will be followed to evaluate the health claims. At the time of publication of this section, changes may have been introduced in the legislation. It is therefore advised to consult home pages of legislative and authorities in order to obtain the latest information (Table 1.14).

TABLE 1.14 Institutes for Further Information on Health Claim Legislation in the EU

Institutes	Country	Acronym	Homepage
European Food Safety Authority	EU	EFSA	http://www.efsa.europa.eu/en.html
Ministry of Health	Belgium		www.health.fgov.be/
Elintarvikevirasto	Finland	EVIRA	http://www.evira.fi/
Agence Française de Sécurité Sanitaire des Aliments	France	AFSSA	http://www.afssa.fr/
Direction Générale de la Concurrence, de la Consommation et de la Repression des Fraude	France	DGCCRF	http://www.finances.gouv.fr/DGCCRF/
Food Safety Authority of Ireland	Ireland	FSAI	http://www.fsai.ie/
Ministry of Health	Netherlands		www.minvws.nl
Agencia Española de Seguridad Alimentaria y Nutricion	Spain	AESA	www.aesa.msc.es
Swedish Nutrition Foundation	Sweden	SNF	http://www.snf.ideon.se/
Food Standards Agency	UK		http://www.food.gov.uk/foodlabelling/ull/claims/

1.8.1.3   The United States of America

JOSE M. SAAVEDRA[1] AND FRED H. DEGNAN[2]

[1] John Hopkins University School of Medicine, Baltimore, MD, USA
[2] King & Spalding LLP, washington, DC, USA

The regulation of microorganisms for human consumption in North America is guided by and is dependent on a number of multiple, complex, and interdependent factors. In the United States, the Federal Food Drug and Cosmetic Act (the FDC Act or the Act) (565) provides the Food and Drug Administration (FDA) with broad authority to classify and regulate an array of products for human consumption. Different regulatory standards govern the marketing of a product depending on how the product is classified under the Act, that is, as a "food," a "food additive," a "drug," a "new drug," a "dietary supplement" (to mention only a few possible classifications). As a result, the pathway under the Act to lawful marketing of a product—even a "probiotic" product—will be relatively easy for some products while, for others, arduous—depending on how the specific product is classified by FDA.

It is important however, to establish from the outset, that there is neither a legally recognized or regulatory definition for the term "probiotics" in North America, nor is there a standard of identity for "probiotics" for human consumption. Therefore, we will discuss the regulation of microorganisms for human consumption, some of which may be regarded by the scientific community as "probiotic" microorganisms, that is, they have been shown to provide a measurable benefit to the host when consumed.

The extent to which FDA is empowered to regulate the safety and claims of efficacy for a substance (including microorganisms) depends not only on the nature of the specific substance but also on how the specific substance is classified under the Act, that is, as a "food" or "food additive," as a "dietary supplement," as a "drug" or "new drug," etc. To determine what product classification applies, one must focus on the "intended use" of the product (most regulated products are defined in the Act according to their "intended use"). Thus, as a general rule, how a manufacturer or a purveyor of a product "intends" a product to be used will govern (565) how the product is classified under the Act and, even more importantly, (566) the data collection and substantiation requirements that must be met to achieve lawful marketing. To this end, the "intended use" of a product is generally determined by what a sponsor, manufacturer, or purveyor says about its product. As a result, "intended use" can be determined from a variety of sources, including product advertising and promotion, product labeling, representations on a company's website, and speeches or remarks by corporate officials. To help navigate through the complexity, a catalog of key statutory definitions and accompanying criteria is necessary.

1. Product classification: statutory definitions.
   Food. "Food" is defined in the Act (566) in a self-evident and circular manner as "articles used for food or drink for man or animals." The definition goes on to include chewing gum and articles "used" for components of food. Note that this least useful of the definitions focuses on actual "use," while for all the other relevant classifications, the key criterion is "intended use."

Drug. The term "drug" is defined in the Act (567) as an article "other than food" that is intended for use in the diagnosis, cure, mitigation, treatment, or prevention of disease or in other animals." This also provides that an article is a drug if it is "intended to affect the structure or function of the body of man or other animals." And, an article is also a drug if it is "intended for use as a component" of a "drug." A "new drug" is a statutory term indicating an article is not yet "generally recognized" as safe and effective for its "intended use" and must undergo extensive premarketing clearance requirements that are laid out in detail in Section 505 of the Act and FDA's implementing regulations. Product approval can, and does, take years of data collection and subsequent agency review.

Food Additive. The Act (568) broadly defines "food additive" as including any substance "the intended use of which" results or can "reasonably be expected to result, directly or indirectly, in its becoming a component or otherwise affecting the characteristics of food. ..." These substances intentionally added to food are generally subject to formal premarket review or approval by FDA. However, exempted from this requirement are "GRAS" (generally recognized as safe) substances. These are substances "generally recognized" by "qualified experts" as having been adequately shown through "scientific procedures" to be safe under "the conditions of intended use." Thus, importantly, a "GRAS" food substance or ingredient is *not* subject to the pre-market clearance requirements that accompany a "food additive" and the sponsor of a GRAS substance may proceed directly to market with it.

Dietary Supplement. The term "dietary supplement" is defined in the Act (569) as a product "other than tobacco" that contains a "dietary ingredient" (e.g., a vitamin, mineral, herb, amino acid, etc.) "intended" to "supplement the diet." Also, as defined, a dietary supplement must be intended for "ingestion," and must be in tablet, capsule, powder, softgel, gelcap, or liquid form *or*, if not intended for ingestion, in a form that is not represented as a conventional food and is not represented for use as a sole item, meal, or the diet.

Biological Product. Although not defined in the FDC Act, the term "biological product" is defined in the Public Health Service Act (PHS Act) (570) as, among other things, a "virus" "applicable" to the prevention, treatment, or cure of a disease or condition or "injuries to man." FDA is reported to be of the view that the term "virus" may, logically, include "microorganisms" and, thus, probiotics (571). Although various microorganisms, including probiotics are being studied for such type therapeutic applications, none have yet been commercialized or evaluated under this category.

Medical Food. A "medical food" is defined in an amendment to the FDC Act (572) as "a food which is formulated to be consumed or administered enterally under the supervision of a physician and which is intended for the specific dietary management of a disease or condition for which distinctive nutritional requirements, based on recognized scientific principles, are established by medical evaluation. Currently, no foods containing microorganisms have been commercialized or recognized as medical foods.

2. FDA Authority to regulate food uses of probiotics.
   The agency not only has the authority to *differentiate among types* of products bearing microorganisms, including probiotic organisms, but also to consider and evaluate *different uses according to different safety standards*. And, in every case, the agency is empowered to regulate the labeling of such products and to demand, to *varying degrees of rigor, substantiation of the claims* made on behalf of such products
3. Safety considerations with respect to the food uses of probiotics.
   Food Additive. The FDC Act (573) establishes a system in which substances that meet the definition of "food additive" must be the subject of a premarket submission containing data and information documenting the safety of the intended use of the additive. A food additive approval cannot be issued unless a petition, submitted by a sponsor, contains convincing evidence establishing that the desired use of the additive is "safe," based on proof to a "reasonable certainty" that no harm will result from the proposed use of the additive (574). The regulation contains the important acknowledgement that it is impossible to establish the absolute harmlessness of any substance.

   GRAS Substances. As discussed earlier, a key exception to the definition of "food additive" is GRAS status: if a substance is GRAS for its intended use, the substance is, by definition, not a "food additive." On the basis of this GRAS exemption, many substances, including several microorganisms, are currently lawfully marketed without a food additive regulation and have achieved GRAS status without any formal FDA review given their very long history of use in the food supply (e.g., *Lactobacillus bulgaricus* in yogurt).

   The GRAS standard focuses on two key considerations: (a) whether the data and information concerning the desired use of the substance provides a scientific basis to conclude that there is consensus among qualified experts about the safety of the substance for the intended use and (b) whether the data and information relied upon to establish safety are generally available to the scientific community. For the purposes of a GRAS opinion, information in peer reviewed scientific journals can be supplemented by (a) publication of data and information in the secondary scientific literature, (b) documentation of an opinion of an experts convened in a panel charged to consider safety, and (c) the opinion or recommendations of an "authoritative body" (for example, another federal agency or a respected scientific entity like the National Academy of Sciences).

   It is the manufacturer's responsibility to assure a product or substance is GRAS, and it should neither be assumed that any microorganism currently in the food supply qualifies it as GRAS, nor that because a bacterium is appropriate for a supplement, it is GRAS for a food, since the determination is "for an intended use." Manufacturers can conduct a "self-determination" of GRAS status for a substance. The documentation of this determination, based on the criteria mentioned above maybe kept internally, and provided to FDA if the agency requests it. Although manufacturers are not required to provide this self-determination to FDA prior to marketing a product, not working with FDA on GRAS issues before marketing a product risks adverse publicity and judicial action by FDA should the agency disagree with the self-GRAS determination and is concerned that consumers may be at risk. This is even more the case with respect to infant formulas, where FDA requires premarket notification and clearance.

The documentation of this determination, based on the criteria mentioned earlier maybe kept internally, and provided to FDA if the agency requests it. But manufacturers are not required to provide this self-determination to FDA prior to marketing a product. An exception to this is the use of a microorganism (as well as other substances) when intended for use in an infant formula, in which FDA explicitly requires premarket notification.

In some cases, the manufacturer may choose to notify FDA of its GRAS self-affirmation. To facilitate GRAS determinations, FDA for the last decade has followed a "notification" procedure under which the agency reviews a submission by a sponsor for the use (or the uses) of a given substance. The submission must contain a sufficient basis for a GRAS determination based on the criteria mentioned earlier. In reviewing a notification, FDA takes at face value the conclusions of the sponsor and the independent panel of experts. If the agency has questions with respect to the apparent adequacy of the cited data and information or the panel's conclusions, FDA will so notify the sponsor. Similarly, FDA will advise the sponsor if it has no questions. Although, a "no questions" letter is not the same as an agency conclusion that the use of a given substance is, in fact, GRAS, a "no questions" letter represents an agency acquiescence to the sponsor proceeding to market the product.

Probiotics have been the subject of at least two successful GRAS notifications to FDA. The agency has issued "no questions" letters with respect to the use of *Lactobacillus acidophilus* and *Lactobacillus lactis* for use in fresh meat for the control of pathogenic bacteria and to the use of *Streptococcus thermophilus*, and strain of *Bifidobacterium lactis* in infant formula (575). In the United States *B. lactis* is the only bacterium, which has shown probiotic effects and is currently GRAS for use in infant formula.

Dietary Supplements. Dietary supplements are not subject to the food additive provisions of the Act. As a consequence, dietary supplements not only are not subject to the demanding premarket approval requirements for food additives but also do not undergo "GRAS" scrutiny and evaluation.

From the safety point of view, a dietary supplement is "adulterated" (and thus unlawful) within the meaning of the Act if it "presents a significant or unreasonable risk of illness or injury" when used under its ordinary conditions of use (576). FDA can invoke numerous sections of the Act to declare food "adulterated" once it is on the market.

An additional safety standard applies if the dietary supplement contains a "new dietary ingredient" (NDI). An "NDI" is an ingredient "not marketed in the United

States before October 15, 1994, when the Dietary Supplement Health and Education Act (DSHEA) was passed" (577). For ingredients in supplements (including any microorganisms, with or without probiotic effects) commercialized before this date, no premarket approval is required by FDA. Thus, many microorganisms in the food supply used for a long time can be, and are neither considered NDIs nor have been subject of an NDI review by FDA. The FDA has no complete or authoritative list of dietary ingredients marketed before October 15, 1994, so it is up to the manufacturer to determine if a new product or ingredient is an NDI. If an NDI has been present in the general food supply as an article used for food "in a form in which the food has not been chemically altered," the NDI may be used in the food supply without first notifying FDA. But, if that is not the case, the NDI may only be used if (a) there is a history of use of the ingredient, or other evidence establishing that the proposed use of the ingredient "will reasonably be expected to be safe" *and* (b) the manufacturer files a premarket notification containing the information in support of safety at least 75 days before marketing the product. Once the 75-day period has expired, the sponsor is free to proceed to market. If at any time the agency disagrees with the sponsor's conclusions concerning safety, the agency may so notify the sponsor and, if need be, litigate to prevent the marketing of the product.

Good Manufacturing Practices. Regardless of whether a microorganism is in a food or a dietary supplement, and in addition to the establishment of the safety of the microorganism for its intended use, FDA will expect that both the microorganism and the food to which it is added are produced under good manufacturing practices designed to ensure safety, microbiological quality, and integrity of the probiotics.

Safety and Intended Use. It is a common misconception that "safety" evaluation is more stringent for a drug than for a supplement or a food. As mentioned earlier, from the regulatory point of view, safety depends on the intended use. While a drug may have specific documented adverse effects, its use may be approved as long as the safety of the drug has been rigorously investigated and the benefits (e.g. treating a disease) are established by "substantial evidence" and are found to outweigh the safety-related risks. On the contrary, for an infant formula, no adverse effects of any significance are permitted. Supplements, like drugs, may be labeled for use in particular subpopulations, such as adults, or children over a particular age. This is not the case for foods because, among other reasons, foods may be consumed in large amounts, by the population at large, and for very long periods of time, and may demonstrate cumulative effects that are not typical of supplement consumption patterns.

4. Labeling and claim substantiation for food uses of microorganisms.
As mentioned at the outset, there is no regulatory or legally recognized definition for the term "probiotic" in the United States or worldwide. Thus, products containing the term "probiotic" on the label currently do not meet any specific requirement relative to the term. On the contrary, there is an increasing level of information and understanding by consumers and health care providers

that the term itself conveys "some benefit" and could be itself considered a claim. For now, the regulatory framework has dealt only with express or implied claims other than those conveyed by the term "probiotic."

Content Claims. Foods or dietary supplements containing microorganisms may mention the fact they contain certain microorganisms, and sometimes indicate certain amounts. Unfortunately many commercial products, particularly supplements do not contain numbers or types of viable probiotic microbes stated on label (578–580). This is likely a consequence of multiple factors. One is the evolving and continuous change in the taxonomy of these organisms, which has evolved in parallel with the technology available to identify and differentiate different genera, species, and strains. Unless current nomenclature is used, actual contents may not reflect the label. As an example, what used to be called *L. acidophilus* until recently could in fact be one or more of six different species (*L. acidophilus, L. gasseri, L. johnsonii, L. crispatus, L. gallinarum, and L. amylovoris*). A second factor is the lack of documentation of viability and quality control throughout shelf life, which is dependent on manufacturing conditions, humidity, storage temperature, etc.

When it comes to foods, similar issues are of concern. Many manufacturers do not mention content (number) of microorganisms at all, others do. Although more recently manufacturers have been more specific regarding genus, species and strain, and amount of a particular microbe in the product (and have used more aggressively claims associated to this organism), consumers are ultimately left to relying on manufacturer's statement regarding the presence, amount of microorganisms in these products. This becomes even more important given the fact that the potential probiotic effect of these products may relate to amount of microorganisms ingested.

Lastly, since similar to the term probiotic, the term "contains active live bacteria," "active live cultures," and other similar, simply imply "some" viability, or "some" microorganisms is present. At the consumer level, this also may imply these bacteria are of some benefit (a probiotic), which of course is not the case.

Infant formula is a particular product, which in the United States is governed by the same regulatory criteria mentioned above, as well as multiple other regulatory layers, and respond to a significantly greater level of care by manufacturers as well as oversight by FDA. Therefore both specific organism and amounts are better guaranteed.

Benefit or Efficacy Claims: FDA is empowered to comprehensively regulate the use of labeling and promotional material communicating health-related information or claims to the consumer. Under the Act (Sections 403(a) and 201 (n)) (581, 582), a food is misbranded if its label or labeling is false or misleading "in any particular." Under this standard, any claim regarding the value or benefit of a food or a dietary supplement must be demonstrated by reliable information. The responsibility for documenting the validity of label claims rests with the manufacturer and the oversight by FDA. The Federal Trade Commission (FTC) has oversight of claims made in advertising.

Health Claims. FDA has special authority with respect to claims made in the labeling of food and dietary supplements, which expressly or impliedly characterize the relationship of a substance to a disease or health-related condition (583). This type of claim is, in common parlance, referred to as a "health claim." Even an implied reference to an impact on disease can constitute a "health claim."

Before a health claim may be lawfully used on food, FDA must authorize the claim upon review of a petition for the claim: FDA can only approve a claim upon finding that the claim is supported by "significant scientific agreement among qualified experts." This is a rigorous statutory standard and only a handful of claims—*none involving probiotics*—has been found by FDA to meet the standard.

Qualified Health Claims. Judicial rulings over the last decade have led to an additional category of health claims—"qualified health claims"—for foods and dietary supplements. FDA has implemented a policy of reviewing and permitting health claims to appear on foods even if the data and information in the supporting health claim submission do not meet FDA's "significant scientific agreement" standard, but is satisfied that "qualifiers" can render the claim not misleading. For example, an appropriately qualified claim might provide guidance to the effect that although there is emerging scientific evidence supporting the claim, the evidence is "not conclusive." To secure FDA's acquiescence to the use of a "qualified health claim," a sponsor must submit a premarket petition for FDA and wait for the Agency's review and conclusions. No qualified health claims have been granted to any probiotic microorganism.

Structure or Function Claims. An additional category of claims that may lawfully appear on both foods and dietary supplements is the "structure or function" claim. This type of claim presents perhaps the most significant opportunity for the exercise of marketing creativity and, commensurately, for attracting regulatory concern. As noted above, the statutory definition of "drug" includes an article "(other than food)" intended to affect the structure or function of the body of man or other animals." The parenthetical expression reflects Congress' implicit recognition that a "food" may bear a claim with respect to the effect a food or food component may have on the body, for example, "Calcium builds strong bones." This reflects the fact that certain foods quite naturally affect the structure or function of the human body and that a claim with respect to such an effect should not be regulated under the rigorous standards that govern "drugs." Thus, coffee, without being considered a "drug," can lawfully bear a claim with respect to the mild alertness effect naturally occurring caffeine can impart.

In the 1980s, some tablet and capsule dietary supplement products could not meet the definition of food and, thus, were regulated by FDA as "drugs" if their label or labeling or bore or implied structure or function-related claims. So the U.S. Congress amended the FDC Act in 1994 to expressly authorize the use of structure or function claims in the labeling of dietary supplements. These amendments, however, require that any such statement of a dietary supplement must bear the following disclaimer: "This statement has not been evaluated by the Food and Drug Administration. This product is not intended to diagnose, treat, cure, or prevent any disease." (584). Moreover, the sponsor of such a claim must notify FDA of the claim within 30 days after marketing a dietary supplement bearing the claim.

In spite of the amendment permitting supplements to bear a "structure or function" claim and not be considered a drug, FDA's primary enforcement tool against dietary supplements remains its authority (under the alternative prong of the "drug" definition found in the Act (Section 201(g)(1)(B)) to classify a supplement as a drug if product labeling and promotion imply that the product "is intended, for use in the...cure, mitigation, treatment, or prevention of disease...." Accordingly, FDA has issued detailed regulations that differentiate between (a) structure or function claims appropriate for dietary supplements and, by inference, appropriate for foods and (b) "disease" claims that may not be made on behalf of a dietary supplement or on behalf of a food, without the prior FDA authorization previously described for a "drug" or "health claim" (585).

At the heart of FDA's regulations is the concern that a structure or function claim carries the potential to imply a disease related benefit. When exactly a structure or function claim on a dietary supplement or on a food becomes a "drug" claim is not always clear. Thus, FDA's regulation contains guidance for how, in FDA's view, to assess when a structure or function claim will be treated by the agency as a "drug" claim.

With respect to probiotics, examples of appropriate structure/function claims include "helps maintain intestinal flora" and "helps replenish healthy microflora" or "helps support a healthy immune system." Notice how each claim avoids mentioning a disease or disease endpoint and each is clearly directed at healthy people who wish to remain healthy. On the contrary, attempted structure/function claims like "prevents adherence of *Candida albicans* to the intestinal mucosa" and "deters bacteria from adhering to the wall of the bladder and urinary tract" have been determined by FDA to be implied drug claims, presumably on the basis that the former claim may be viewed as suggesting the prevention of cancer and that the latter claim may be viewed as suggesting the prevention of bacterial infection of the bladder and urinary tract.

Publications as Claims. With respect to dietary supplements *but not* with respect to "foods," the FDC Act permits published scientific articles to be distributed in connection with the sale of the dietary supplement (586). The title of the publication may mention disease conditions and the relationship of the substance to such conditions. Such mentions will be exempt from any prohibition against unauthorized "disease claims" under the "health claim" or "drug" provisions of the Act, only as long as the publication (a) is not false and misleading, (b) does not promote a particular brand or manufacturer, (c) presents a balanced view of available scientific information, and (d) is reprinted in its entirety (summaries of published articles do not quality for the exemption).

Functional Foods. Although the term "functional food" is frequently employed to describe foods capable of imparting particular health benefits and is often used to describe probiotic-containing foods, the term is neither defined by FDA regulations nor to be found in the FDC Act. That said, "foods" or "dietary supplements" bearing "structure or function," "health," or "qualified health" claims—or even the entire classes of "medical foods," and "dietary supplements"—all can be regarded as "functional foods." Thus, this whole category of so-called "functional foods" is not governed by a single set of regulatory criteria, but is rather an array of possible product classifications, each with its own regulatory status and governing criteria.

5. Probiotic use as drugs.

If a probiotic falls within the definition of "drug," that is, it is commercialized for the intended use of mitigating, treating, or preventing a disease, the probiotic would be subject to an array of data collection and submission requirements beginning with those concerning the conduct of "clinical" trials under FDA's "investigational new drug" (IND) (587) procedures and regulations and ending with the submission of a new drug application containing, with respect to the desired use, (a) data and information establishing safety and (b) substantial evidence, based on adequate and well-controlled investigations, conducted by qualified researchers, establishing effectiveness.

IND regulations help guarantee the welfare of those participating in a clinical study, and include submission of data from laboratory and animal research, experience, or historical use of the drug in people, and detailed protocols of the planed trials as well as provisions for selecting qualified investigators, maintaining detailed study records, and keeping FDA informed of the progress of a given investigation and of any significant safety-related developments.

Once a probiotic successfully runs the gauntlet of clinical testing under an IND, the sponsor must submit a "new drug application" (NDA) for the desired use of the product. Only if the application is approved by FDA may the product be lawfully marketed. It should be noted that, unlike the case with "food additive" or GRAS substances, FDA accords confidential status to data submitted in a new drug application, even the fact that an IND is in effect, or that an NDA has been submitted is privileged, confidential information.

Currently, no specific probiotic product has been approved or commercialized as a drug for humans in the United States. The increase in "intended use" of specific bacteria with probiotic effects for prevention or treatment of disease will likely result ultimately in specific products marketed for this purpose, although the hurdles remain significant.

That reality, in and of itself, is not troubling—there is a general agreement, at least in the scientific community, that claims of benefit should be put to the test and proven in an ethical manner consistent with sound principles of scientific methodology. The concern that can arise, however, is that although the distinction made in the FDC Act between "drug" and nondrug uses can, in some cases, arguably be viewed as artificial or casuistic, the resulting demands on market entry for products whose "intended use" triggers regulation as a drug (or as a biological product) are stark from a product development perspective. Nevertheless, the distinction, however artificial it may in some cases be, has been and remains a touchstone of public health regulation in the United States and is ingrained in federal public health precedent and policy (588).

When a product is classified as a drug, there is an immediate and direct effect on the development of such a product, from complex and onerous investigational requirements for a "new drug" described earlier, all the way to its commercialization. As a consequence, there is little incentive for sponsors of probiotics to undertake the expense to develop probiotic products as true "biotherapeutics" subject to premarket approval. This lack of incentive has been reinforced by the fact that unlike investigational new drug studies, "food" or "dietary supplement" studies may be commenced, conducted, and terminated, without comparable FDA involvement (FDA's informed consent and IRB requirements attend any clinical trial regardless of the nature of the substance under test). The result is the understandable tendency of sponsors to clinically test "structure or function"-related indications or claims rather than indications likely to invoke new drug or biological product status.

6. Conclusions

The oral human consumption of specific microorganisms that provide health-related benefits, described elsewhere in this handbook, is considered safe and desirable by many—including researchers, food technologists, and, of course, food producers and purveyors.

The regulatory aspects of producing and commercializing such products, as summarily treated in this chapter, is complexly layered with increasingly demanding burdens of data collection and claim substantiation.

But with any decision to market probiotic products comes the understandable desire to tout the potential benefits of such products. It is in that context, that the focus of the FDC Act on "intended use" and the product classifications that derive from such a focus, force interested sponsors of such products to pause and confront the realities of public health regulation today in the United States.

Although benefit claims can trigger more demanding regulatory hurdles, common safety considerations apply to all probiotic products regardless of whether foods (including medical foods and foods for special dietary use, or dietary supplements), drugs, or biological products. Simply put, care must be taken to carefully and comprehensively document and establish the safety of any use. In the food area this is customarily accomplished in the context of (a) data results from adequately designed studies and (b) informed expert opinion. Not by coincidence, these elements square well with the key elements of any GRAS assessment.

With respect to showing the safety of the probiotic component of a dietary supplement, although food additive GRAS considerations do not apply,

TABLE 1.15 Differences Among Regulatory Categories Available for Marketing Probiotic Products in the United States

Approvals, Characteristics, and Claims	Regulatory Categories			
	Supplements	Food	Infant Formula	Drug
Premarket approval by the FDA	Not required for microorganisms used as "dietary ingredients" before October 1994	Not required for GRAS microorganisms	Premarket notification and clearance required	Required
Disease claim Describes the effect of a drug on the diagnosis, treatment, mitigation, cure, or prevention of disease	Not allowed	Not allowed	Not allowed	Allowed if approved by the FDA
Health claim Describes the effect of a dietary substance on the reduction of risk of disease by the currently healthy population	Allowed if approved by the FDA (may be unqualified or qualified)	Allowed if approved by the FDA (may be unqualified or qualified)	Allowed if approved by the FDA (may be unqualified or qualified)	Not used, although can use stronger prevention claims
Structure function claim Describes the effect of a dietary substance on the structure or function of the body	Allowed if truthful and not misleading	Allowed if truthful and not misleading	Allowed if truthful and not misleading	Required for "new drugs."
	Commonly used	FDA takes the view that effect must derive from the "nutritive value" of the food (sensible, but not clearly established).	FDA takes the view that effect must derive from the "nutritive value" of the food.	

(*continued*)

TABLE 1.15 (Continued)

Approvals, Characteristics, and Claims	Regulatory Categories			
	Supplements	Food	Infant Formula	Drug
	Label must say "this statement has not been reviewed by the FDA of intent to use this claim within 30 days of marketing the product	No requirement for label disclaimer or FDA notification	No requirement for label disclaimer or FDA notification	
Safety standards	No significant or unreasonable risk of illness or injury	Reasonable certainty of no harm under the intended conditions of use or GRAS	Food additive approval or GRAS notification done prior to, or as part of, pre-market approval	The FDA assesses safety and effectiveness and determines if benefits are established by substantial evidence and if they outweigh risk
	Target consumer group can be stipulated on the label	GRAS status can be self-determined or submitted through GRAS notification process Must be safe for general population and all subgroups		
Product examples	Capsules	Yogurt	Infant formula	No probiotic products currently are regulated as drugs for human use in the United States
	Powder sachet	Dairy drink		

Adapted from Ref. 593.

approaching the substantiation of safety in a similar way is sensible and should guarantee that the safety standards, including those regarding new dietary ingredient notifications, of the Act are met.

In conclusion, FDA's regulatory authority over the use of probiotics is comprehensive. Different regulatory standards will govern the market entry of a probiotic product depending on how the product is classified under the Act, that is, as a "food," a "drug," a "dietary supplement," etc. (Table 1.15). Numerous alternative pathways to market are, as a result, available to probiotic manufacturers. Informed care and attention need to be taken if navigating these pathways is to be successful.

1.8.2 Animal Probiotics

ALLAN LIM AND HAI-MENG TAN

Kemin Industries (Asia) Pte Ltd, Singapore

Probiotics can be defined as live microorganisms that have a beneficial effect on the animal health when used in animal nutrition. They are considered as feed additives in most countries and are therefore regulated separately from food. In some countries, they may also be considered as a veterinary chemical product and thus regulated accordingly. Probiotics used in animal nutrition comprise mainly of Gram-positive bacteria belonging to the genera of *Lactobacillus, Bifidobacterium, Pediococcus, Bacillus, Streptococcus,* and even *Enterococcus*. Others have included yeast or *Saccharomyces cerevisiae*. The approval of probiotics for use in animals follows essentially the same approach as that for humans, which is largely dependant on the efficacy and toxicity of the strains.

1.8.2.1 United States Regarded as the authority and reference on feed additive policy, the Association of American Feed Control Authority (AAFCO) published a list of microorganisms approved as direct-fed microbial products (590). Of the 45 microorganisms approved, about half belongs to the genus of *Lactobacillus* (14) and *Bifidobacterium* (16) (Table 1.16).

1.8.2.2 European Union Live microorganisms, together with enzymes and feed additives of biological origin were added to the list of feed additives regulated by the European Union in the 1980s due to the emerging market trends. The term "probiotics" have been rejected on the grounds of being too generic. In 2002, under the framework of establishing the European Food Safety Authority, a new draft regulation would group microorganisms as "zootechnical additives," defined as agents producing beneficial effect on gut microflora. This proposal was adopted in 2003, when the European Commission passed a new regulation (EC) No 1831/2003 of the European Parliament and of the Council on additives for use in animal nutrition. To be used in Europe, probiotics as additives must satisfy several criteria with regards to their identity, characteristics, and conditions for use of the additive; their safety of use in animals, humans, and environment

TABLE 1.16 Approved Direct-Fed Microbials in the United States

Genus	Species	Genus	Species
Aspergillus	niger	Lactobacillus	acidophilus
	oryzae		brevis
			bucheri[a]
Bacillus	coagulans		bulgaricus
	lentus		casei
	licheniformis		cellobiosus
	pumilus		curvatus
	subtilis		fareiminis[b]
			fermentum
Bacteroides	amylophilus		delbruckii
	capillosus		helveticus
	ruminocola		lactis
	suis		pantarum
			reuteri
Bifidobacterium	adolescentis		
	animalis	Leuconostoc	mesenteroides
	bifidum		
	infantis	Pediococcus	acidilacticii
	longum		cerevisiae (damnosus)
	thermophilum		pentosaceus
		Propionibacterium	acidipropionici[a]

[a]Cattle only.
[b]Swine only.

such as the lack of pathogenicity and production of antibiotics and antibiotic resistance; and their efficacy on animals or categories of the target animal species such as improved zootechnical performance, reduction of morbidity and mortality. To demonstrate efficacy effects, three trials conducted at a minimum of two separate sites presenting significant results ($p < 0.05$ or 0.1) are required. The main safety concerns regarding such probiotic strains, *Bacillus* and *Enterococcus* in particular, revolve around the potential production of toxins and virulence factors, detectable using modern techniques such as ELISA, cell line assays, and PCR amplification of toxin genes. Any harmful effects of overdosing are studied through administration of at least ten times the recommended maximum dose. Studies of genotoxicity and mutagenicity would also need to be carried out. Such an evaluation procedure and requirements have led to the reevaluation of some previously submitted product dossiers. Microorganisms are now regulated as zootechnical additives, which is one of the five categories of feed additives, viz:

- Technological additives (e.g. preservatives, antioxidants, emulsifiers, stabilizing agents, acidity regulators, silage additives).
- Sensory additives (e.g. flavors, colorants).
- Nutritional additives (e.g. vitamins, minerals, amino acids, trace elements).

- Zootechnical additives (e.g. digestibility enhancers, gut flora stabilizers).
- Coccidiostats and histomonostats.

Authorization of feed additives is granted by The European Food Safety Authority (EFSA), which evaluates the data submitted on efficacy, safety, and toxicology of the feed additive. Once the Commission is satisfied with the data, it prepares a draft regulation to grant authorization, following the procedure involving Member States within the Standing Committee on the Food Chain and Animal Health—Animal Nutrition. Authorizations are granted for specific animal species, specific conditions of use and for 10-year periods. Although the registration and approval can be interpreted as fairly complex, it can be argued that this is critical to ensure safety of probiotics used as feed additives that ultimately contributes to their efficacy. Approved feed additives are published in the Community Register of feed additive (591). A current list of microorganisms, together with the approved animal species, age, and dosage limits is summarized in Table 1.17.

1.8.2.3 China Feed additives are regulated by the Ministry of Agriculture of the People's Republic of China (592). A total of 16 microorganisms including bacteria and fungus are approved as probiotics in China, and they are to be used according to the application guidelines from the vendors (Table 1.18).

1.8.2.4 Japan The Food and Agricultural Materials Inspection Center (FAMIC) (593) approves a total of 11 bacteria strains for use as probiotics in animals (Table 1.18).

1.8.2.5 Korea A total of 16 microorganisms are approved for use in Korea as probiotics in animals, of which two are fungus and one is yeast (Table 1.18). In addition, there is a separate list of carriers, most of which are starch, derivatives of starch, or cereals and grains, to be used with each of the probiotics.

1.8.2.6 Thailand Feed additives including probiotics are regulated by Department of Livestock Development in Thailand (594). There are currently 42 strains of bacteria and 6 strains of fungus approved as probiotics for animals, which can be used singly or in combination, in producing finished feed at not more than 1×10^5 cfu/kg of feed (Table 1.18).

1.8.2.7 Australia Probiotics are considered a biological product, thus an import permit application must be filed with AQIS. In addition, as probiotic is considered a veterinary chemical product, registration with the Australian Pesticides and Veterinary Medicines Authority (APVMA) is required. The Australia APVMA (595) regulates probiotics as microbial agents, together with three other biological products:

- Group 1—biological chemicals (e.g., pheromones, hormones, growth regulators, enzymes and vitamins);

TABLE 1.17 Microorganisms Approved as Feed Additives in the EU

Genus	Species	Strains	Minimum Age	Species or Category of Animals	Composition in Complete Feedstuff (cfu/kg) Min.	Composition in Complete Feedstuff (cfu/kg) Max.
Aspergillus	niger					
	oryzae	AK 7001 DSM 1862				
Bacillus	badius					
	cereus	var. toyoi NCIMB 40112/CNCM I-1012		Cattles for fattening	0.2×10^9	0.2×10^9
	cereus	var. toyoi NCIMB 40112/CNCM I-1012		Rabbits for fattening	0.1×10^9	5×10^9
	cereus	var. toyoi NCIMB 40112/CNCM I-1012		Chickens for fattening	0.2×10^9	1×10^9
	cereus	var. toyoi NCIMB 40112/CNCM I-1012	From 2 to 4 months	Piglets	0.5×10^9	1×10^9
	cereus	var. toyoi NCIMB 40112/CNCM I-1012	From 4 months until slaughter	Pigs for fattening	0.2×10^9	1×10^9
	cereus	var. toyoi NCIMB 40112/CNCM I-1012	2 months	Piglets	1×10^9	1×10^9
	cereus	var. toyoi NCIMB 40112/CNCM I-1012	From service to weaning	Sows	0.5×10^9	2×10^9
	coagulans	CECT 7001				
	lentus	302				
	licheniformis	DSM 5749		Turkeys for fattening	1.28×10^9	1.28×10^9
	licheniformis	DSM 5749	3 months	Calves	1.28×10^9	1.28×10^9
	licheniformis	DSM 5749		Sows	1.28×10^9	1.28×10^9
	licheniformis	DSM 5749		Pigs for fattening	1.28×10^9	1.28×10^9
	licheniformis	DSM 5749		Piglets	1.28×10^9	1.28×10^9
	pumilus	BP288 ATCC 53682/CNCM I-3240 (NRRL B4064)/MBS-BP-01/Micron Bio-Systems Culture Collection				

	subtilis	DSM 5750	Turkeys for fattening	1.28×10^9
	subtilis	DSM 5750	Calves	1.28×10^9
	toyoi			1.28×10^9
Bacteroides	*amylophilus*			1.28×10^9
	capillosus			
	ruminocola			
	suis			
Bifidobacterium	*adolescentis*			
	animalis	subsp. *lactis* CHCC5445 (DSM15954)		
	bifidum			
	infantis			
	longum	CNCM I-3241 (ATCC 15707)		
	pseudolongum			
	thermophilum			
Candida	*glabrata*	35120		
	pinolepessi			
	utilis			
Clostridium	*butyricum*			
	sporogenes phage	P		
	tyrobutyricum phage			
Enterococcus	*cremoris*			
	diacetylactis			
	faecalis			
	faecium	NCIMB 10415	Chickens for fattening	1×10^8
	faecium	NCIMB 10415	Calves 6 months	1×10^9
	faecium	NCIMB 10415	Chickens for fattening	0.3×10^9
				1×10^8
				1×10^{10}
				2.8×10^9

(continued)

TABLE 1.17 (*Continued*)

Genus	Species	Strains	Minimum Age	Species or Category of Animals	Composition in Complete Feedstuff (cfu/kg)	
					Min.	Max.
	faecium	NCIMB 10415		Pigs for fattening	0.35×10^9	1×10^9
	faecium	NCIMB 10415		Piglets	0.35×10^9	1×10^9
	faecium	NCIMB 10415		Sows	0.7×10^9	1.25×10^9
	faecium	NCIMB 10415		Piglets	1×10^9	1×10^{10}
	faecium	NCIMB 10415	6 months	Calves	1×10^9	6.6×10^9
	faecium	DSM 7134	4 months	Calves	1×10^9	5×10^9
	faecium	DSM 7134		Piglets (weaned)	2.5×10^9	5×10^9
	faecium	DSM 10663/NCIMB 10415	6 months	Calves	1×10^9	1×10^{10}
	faecium	DSM 10663/NCIMB 10415		Piglets	1×10^9	1×10^{10}
	faecium	DSM 10663/NCIMB 10415		Chickens for fattening	1×10^9	1×10^9
	faecium	DSM 10663/NCIMB 10415		Dogs	1×10^9	1×10^{10}
	faecium	DSM 10663/NCIMB 10415		Turkeys for fattening	1×10^7	1×10^9
	faecium	DSM 10663/NCIMB 10415	6 months	Calves	5×10^8	2×10^{10}
	faecium	NCIMB 11181		Piglets	5×10^8	2×10^{10}
	faecium	NCIMB 11181		Chickens for fattening	2.5×10^8	15×10^9
	faecium	NCIMB 11181				
	faecium	1:1 mixture of ATCC 53519 and ATCC 55593		Chickens for fattening	1×10^8	1×10^8
	faecium	CECT 4515		Piglets (weaned)	1×10^9	1×10^9
	faecium	CECT 4515		Chickens for fattening	1×10^9	1×10^9
	faecium	NCIMB 10415		Dogs	4.5×10^6	2×10^9
	faecium	NCIMB 10415		Cats	5×10^6	8×10^9

	faecium	DSM 7134	Chickens for fattening	0.2×10^9	2×10^9
	faecium	DSM 7134	Piglets	0.5×10^9	4×10^9
	faecium mundtii	DSM 7134 82760	Pigs for fattening	0.2×10^9	1×10^9
Kluyveromyces	marxianus	var. lactisK1 BCCM/MUCL 39434	Dairy cows	0.25×10^6	1×10^6
	marxianus-fragilis	B0399 MUCL 41579	Piglets (weaned)	6×10^6	6×10^6
Lactobacillus	acidophilus	36587/CHCC3777 (DSM13241)/ CNCM DALA 1-1246/NCIMB 30067/NCAIM			
	acidophilus	DSM 13241	Dogs	6×10^9	2×10^{10}
	acidophilus	DSM 13241	Cats	3×10^9	2×10^{10}
	amyloliticus	CBS 116420			
	amylovorans	DSM 16251			
	brevis	DSM 12835/DSM 16570/IFA 92/KKP. 839/NCIMB 8038			
	bucheri	40177/71044/71065/BIO 73/CCM 1819/DSM 12856/DSM 13573/DSM 16774/KKP. 907/LN4637 ATCC PTA-2494/NCIMB 30137/NCIMB 30138/NCIMB 30139/NCIMB 40788/NCIMB 8007			
	bulgaricus	MA 547/3M			
	casei	32909/ATCC 7469/CCM 3775/ CHCC2115/CNCM DA LC 1-1247/ MA 67/4U/NCIMB 11970/NCIMB 30007/rhamnosus LC 705 DSM 7061			
	cellobiosus	Q1			
	collinoides	DSMZ 16680			
	curvatus				

(continued)

TABLE 1.17 (Continued)

Genus	Species	Strains	Minimum Age	Species or Category of Animals	Composition in Complete Feedstuff (cfu/kg)	
					Min.	Max.
	farciminis	CNCM MA 67/4R		Chickens for fattening	5×10^8	1×10^9
	farciminis	CNCM MA 67/4R		Turkeys for fattening	5×10^8	1×10^9
	farciminis	CNCM MA 67/4R		Laying hens	5×10^8	1×10^9
	farciminis	CNCM MA 67/4R		Piglets (weaned)	1×10^9	1×10^{10}
	fareiminis	MA27/6B				
	fermentum	DSM 16250				
	delbruckii					
	helveticus					
	lactis					
	mucosae	DSM 16246				
	paracasei	30151/DSM 16245/DSM 16572/DSM 16773/NCIMB 30151/ssp. paracasei DSM 11394/ssp.paracasei DSM 11395/ssp. paracasei CNCM I-3292 (P4126)				
	pantarum					
	pentosus	DSM 14025				
	plantarum	16627/24001/24011/252/50050/88/ Aber F1 NCIMB 41028/AK 5106 DSM 20174/AMY LMG-P22548/ ATCC 8014/C KKP/783/p/C KKP/ 788/p/CCM 3769/CNCM DALP. I-1250/CNCM I-3235/				

	(ATCC 8014)/CNCM I-820/CNCM MA 18/5U/CNCM MA 27/5M/DSM 11520/DSM 11672/DSM 12187/DSM 12836/DSM 12837/DSM 13367/DSM 13543/DSM 13544/DSM 13545/DSM 13546/DSM 13547/DSM 13548/DSM 16247/DSM 16565/DSM 16568/DSM 16571/DSM 16682/DSM 3676/DSM 3677/DSM 4748/DSM 4909/DSM 8427/DSM 8428/DSM 8862/DSM 8866/DSMZ 15683/DSMZ 16627/EU/EEC 1/24476/IFA 96/K KKP/593/p/KKP/788/p/L-256 NCIMB 30084/L43 NCIMB 30146/L44 NCIMB 30147/L54 NCIMB 30148/L58/LP286 DSM 4784 ATCC 53187/LP287 DSM 5257 ATCC 55058/LP318 DSM 4785/LP319 DSM 4786/LP329 DSM 5258 ATCC 55942/LP346 DSM 4787 ATCC 55943/LP347 DSM 5284 ATCC 55944/LSI NCIMB 30083/MA 541/2E/MBS-LP-1/Micron Bio-Systems culture collection/MiLAB 393 LMG-21295/NCIMB 12422/NCIMB 30004/NCIMB 30094/NCIMB 301			
reuteri	CNCM MA28/6E-g/CNCM MA28/6U-g/DSM 16248/DSM 16249			
rhamnosus	MA27/6R/NCIMB 30121			
rhamnosus	DSM 7133	4 months	Calves	1×10^9 5×10^9

(*continued*)

TABLE 1.17 (Continued)

Genus	Species	Strains	Minimum Age	Species or Category of Animals	Composition in Complete Feedstuff (cfu/kg)	
					Min.	Max.
	rhamnosus	DSM 7133		Piglets (weaned)	2.5×10^9	5×10^9
	sakei	DSM 16564/subsp. Sakei AK 5115 DSM 20017				
	salivarius	CNCM I-3238 (ATCC 11741)				
Lactococcus	lactis	CCM 4754, NCIMB 30117/CNCM I-3291 (ATCC 7962)/*lactis* 30044/*lactis* NCIMB 30044/CCM 4754, NCIMB 30117/NCIMB 30149/ NCIMB 30160/SR 3.54 NCIMB 30117/subsp. *Lactis biovar diacetylactis* CHCC2237/subsp. *Lactis* CHCC2871				
Leuconostoc	mesenteroides	DSM 8865				
	oeno LO1	LO1				
	pseudo-mesenteroides	CHCC2114				
Pediococcus	acidilacticii	30005/33-06 NCIMB 30086/33-11 NCIMB 30085/AK 5201 DSM 20284/CNCM I-3237 (ATCC 8042)/ CNCM MA 151/5R/CNCM MA 18/5M/DSM 10313/DSM 11673/DSM 13946/DSM 16243/ET 6/NCIMB 30005				
	acidilacticii	CNCM MA18/5M		Chickens for fattening	1×10^9	1×10^{10}

Genus	Species	Strain	Animal category	Value	Value
	acidilactici				
	cerevisiae (damnosus)				
	pentosaceus	69221/AP35/CCM 3770/CNCM MA 25/4J/DSM 12834/DSM 14021/DSM 16244/DSM 16566/DSM 16569/HTS LMG P-22549/MBSPP-1/Micron Bio-Systems culture collection/ NCIMB 12455/NCIMB 30068/ NCIMB 30089/NCIMB 30168/ NCIMB 30171	Pigs for fattening	1×10^9	1×10^9
Pediococcus	sp				
Propinibacterium	acidipropionici freudenreichii	CNCM MA 26/4U shermanii JS DSM 7067/subsp. shermanii AK 5502 DSM 4902			
	globosum shermanii	CNCM DAPB 1-1249 ATCC 9614/MBSPS-1 DSM 9576/DSM 9577			
	sp.	ATTC 17001			
Rhodopseudomonas palustris	cerevisiae	IFO 0203/37584/80566			
Scaccharomyces	cerevisiae	NCYC Sc47	Rabbits for fattening	2.5×10^9	2.5×10^9
	cerevisiae	NCYC Sc47	Rabbits for fattening	2.5×10^9	5×10^9
	cerevisiae	NCYC Sc47	Sows	5×10^9	2.5×10^{10}
	cerevisiae	NCYC Sc47	Sows	5×10^9	1×10^{10}
	cerevisiae	NCYC Sc47	Piglets (weaned)	5×10^9	1×10^{10}
	cerevisiae	NCYC Sc47	Lambs for fattening	1.4×10^9	1.4×10^{10}
	cerevisiae	NCYC Sc47	Dairy cows	4×10^8	2×10^9
	cerevisiae	CBS493.94	Dairy cows	5×10^7	3.5×10^8

(*continued*)

TABLE 1.17 (*Continued*)

Genus	Species	Strains	Minimum Age	Species or Category of Animals	Composition in Complete Feedstuff (cfu/kg) Min.	Composition in Complete Feedstuff (cfu/kg) Max.
	cerevisiae	CBS493.94		Horses	4×10^9	2.5×10^{10}
	cerevisiae	CBS493.94	6 months	Calves	2×10^8	2×10^9
	cerevisiae	CBS493.94		Cattle for fattening	1.7×10^8	1.7×10^8
	cerevisiae	CNCM I-1079		Sows	1×10^9	6×10^{10}
	cerevisiae	MUCL 39 885		Piglets (weaned)	3×10^9	3×10^9
	cerevisiae	MUCL 39 885		Cattle for fattening	9×10^9	9×10^9
	cerevisiae	MUCL 39 885		Dairy cows	1.23×10^9	2.33×10^9
	cerevisiae	CNCM I-1077		Dairy cows	4×10^8	2×10^9
	cerevisiae	CNCM I-1077		Cattle for fattening	5×10^8	1.6×10^9
Streptococcus	cremoris	CNCM DASC I-1244				
	diacetylactis					
	faecium	36 KKP.880				
	intermedius					
	lactis					
	thermophilus	CHCC3021/CNCM DAST I-1245				

- Group 2—extracts (e.g., plant extracts, oils);
- Group 3—microbial agents (e.g., bacteria, fungi, viruses, protozoa);
- Group 4—other living organisms (e.g., microscopic insects, plants and animals plus some organisms that have been genetically modified).

APVMA defines microbial agents as naturally occurring or genetically modified microorganisms, including bacteria, fungi, viruses, protozoa, microscopic nematodes, or other microbial organisms. The APVMA approves a list of microorganisms that have been reviewed by the US Food and Drug Administration, Center for Veterinary Medicine, as direct-fed microbials (Table 1.18). However, applicants must still provide up-to-date evidence that the direct-fed microbial has generally Recognized as safe (GRAS) status from the US Food and Drug Administration or the equivalent from the European Union.

1.8.2.8 New Zealand To determine whether a probiotic product is registrable, submission of class determination is necessary. Importing a probiotic product that contains microorganism would also require approval from MAF Biosecurity and ERMA. Usage of probiotic for animals in New Zealand is governed by the Agricultural Compounds and Veterinary Medicines Regulations (ACVM) 2001 (596). As an oral nutritional compound (ONC), probiotics can be exempted from registration if the Trade Name Product (TNP) are the following:

- It is not medicated;
- It contains no feed additive that is not listed in Schedule 7 Part A of the ACVM Regulations 2001;
- It makes no therapeutic of pharmacological claims attributable to a nutritional benefit;
- It contains no substance of uncertain status as either a nutrient or feed additive;
- It does not use any slow release mechanism containing high/concentrated levels of substance.

For a probiotic that requires registration, the information required by NZFSA includes information to support safety of the functional ingredient and where applicable the non-GRAS additives at the proposed feeding rate, and if appropriate, information to support the product's status as being fit for purpose and the Product Data Sheet (PDS). Details of the registration package can be found in ACVM Specified Requirements Products Standard and Guideline: Oral Nutritional Compounds Containing Nutrients with Known Therapeutic Uses (Functional Nutrients) non-GRAS ingredients (596). The GRAS status of the probiotic can be determined by the GRAS Register for ONC maintained by NZFSA (597). According to the current list, there are a total of 21 strains of microorganisms from nine genera (Table 1.18).

TABLE 1.18 Overview of Microorganisms Approved as Feed Additives in the Asia-Pacific Countries

Genus	Species	China	Japan	Korea	Thailand	Australia	New Zealand
Aspergillus	*niger*				✓	✓	✓
	oryzae				✓	✓	✓
Bacillus	*badius*		✓				
	cereus		✓		✓		
	coagulans		✓	✓	✓	✓	
	lentus				✓		
	licheniformis	✓			✓	✓	✓
	pumilus				✓	✓	
	subtilis	✓	✓	✓	✓ and strain BN, non-antibiotic producing strains only	✓	
Bacteroides	*toyoi*				✓		
	amylophilus				✓	✓	
	capillosus				✓	✓	
	ruminocola				✓	✓	
	suis				✓	✓	
Bifidobacterium	*adolescentis*				✓	✓	
	animalis				✓	✓	
	bifidum	✓			✓	✓	
	infantis				✓	✓	
	longum				✓	✓	
	pseudolongum		✓	✓	✓	✓	✓
	thermophilum		✓	✓			✓
Candida	*sp*						
	pinolepessi				✓		

Genus	Species				
Clostridium	utilis		✓		
Enterocuccus	butyricum	✓	✓		✓
	faecalis	✓	✓		✓
	faecium	✓	✓		✓
	mundtii				
Lactobacillus	acidophilus	✓			✓
	bifidus				✓
	brevis			✓	✓
	bucheri				
	bulgaricus	✓ only for pigs and poultry			✓
	casei	✓		✓	✓
	cellobiosus			✓	
	curvatus			✓	
	delbruckii			✓	✓
	fermentum			✓	✓
	helveticus			✓	✓ subspecies *lactis*
	lactis			✓	
	pantarum			✓	
	pentosus		✓	✓	
	plantarum	✓		✓	✓
	reuteri		✓		✓
	rhamnosus			✓	✓
	sakei				✓
	salivarius		✓	✓	
Lactococcus	lactis			✓	✓ subspecies *cremoris*
Leuconostoc	mesenteroides				✓
	pseudomesenteroides				✓
Pediococcus	acidilacticii	✓			✓

(continued)

TABLE 1.18 (*Continued*)

Genus	Species	China	Japan	Korea	Thailand	Australia	New Zealand
Pediocuceus	*cerevisiae (damnosus)*				✓	✓	
	pentosaceus	✓			✓	✓	✓
	sp				✓		
Propinibacterium	*freudenreichii*				✓	✓	
	globosum						
	shermanii				✓	✓	
Rhodopseudomonas	*palustris*	✓					
Scaccharomyces	*cerevisiae*	✓			✓	✓	✓
Streptococcus	*cremoris*				✓	✓	✓
	diacetylactis				✓	✓	
	faecium				✓ and *cernelle* 68	✓	
	intermedius				✓	✓	
	lactis				✓	✓	
	salivarius				✓	✓	
	thermophilus				✓		✓ subspecies *thermophilus*
Yeast							

1.8.2.9 Indonesia, Malaysia, Philippines, and Vietnam There is currently no positive list of microorganisms as feed additives for these Asian countries. Sale of probiotic products in these countries is subjected to the same registration requirements as other feed additives.

REFERENCES

1. Diplock AT, Aggett P, Ashwell M, Bornet F, Fern E, and Roberfroid M. Scientific concepts of functional foods in Europe: Consensus document. *Br. J. Nutr.* 1999; 81(Suppl. 1): S1–S27.
2. FAO/WHO Joint Expert Consultation on Evaluation of Health and Nutritional Properties of Probiotics in Food Including Powder Milk with Live Lactic Acid Bacteria, October 2001.
3. Fuller R. Probiotics in man and animals. *J. App. Bacteriol.* 1989; 66: 365–378.
4. Havenaar R and Huis Int Veld JHJ. Probiotics: A general view. In: Wood BJB, editor. *The Lactic Acid Bacteria, vol. 1: The Lactic Acid Bacteria in Health and Disease.*Elsevier Applied Science, London, 1992, pp. 151–170.
5. Lilly DM and Stillwell RH. Probiotics: Growth promoting factors produced by microorganisms. *Science* 1965; 147: 747–749.
6. Naidu AS, Biblack WR, and Clemens RA. Probiotic spectra of lactic acid bacteria (LAB). *Crit. Revs. Food Sci. Nutr.* 1999; 39: 13–126.
7. Parker RB. Probiotics, the other half of the antibiotic story. *Anim. Nutr. Health* 1974; 29: 4–8.
8. Reid G, Sander ME, Gaskins HR, Gibson GR, Mercenier A, Rastall R, Roberfroid M, Rowland I, Cherbut C, and Klaenhammer TR. New scientific paradigms for probiotics and prebiotics. *J. Clin. Gastroentrol.* 2003; 37: 105–118.
9. Salminen S, Bouley MC, Boutron-Rualt MC, Cummings J, Franck A, Gibson G, Isolauri E, Moreau MC, Roberfroid M, and Rowland I. Functional food science and gastrointestinal physiology and function. *Br. J. Nutr.* 1998; 80 (Suppl. 1): 147–171.
10. Schrezenmeir J and de Vrese M. Probiotics, prebiotics and synbiotics- approaching a definition. *Am. J. Clin. Nutr.* 2001; 73(2 Suppl.): 361S–364S.
11. Tannock GW, Munro K, Harmsen HJ, Welling GW, Smart J, and Gopal PK. Analysis of the fecal microflora of human subjects consuming a probiotic product containing *Lactobacillus rhamnosus* DR20. *Appl. Environ. Microbiol.* 2000; 66: 2578–2588.
12. Adel-Patient K, Ah-Leung S, Creminon C, Nouaille S, Chartel JM, Langella P, and Wal JM. Oral administration of recombinant *Lactococcus lactis* expressing bovine beta-lactoglobulin partially prevents mice from sensitization. *Clin. Exp. Allergy.* 2005; 35: 539–546.
13. Ahrné S, Nobaek S, Jeppsson B, Adlerberth I, Wold AE, and Molin G. The normal *Lactobacillus* flora of healthy human rectal and oral mucosa. *J. Appl. Microbiol.* 1998; 85: 88–94.
14. Alvarez-Martín P, Florez AB, and Mayo B. Screening for plasmids among human bifidobacteria species: Sequencing and analysis of pBC1 from *Bifidobacterium catenulatum* L48. *Plasmid* 2007; 57: 165–174.

15. Amann RI, Ludwingg W, and Schleifer K-F. Phylogenetic identification and *in situ* detection of individual microbial cells without cultivation. *Microbiol. Rev.* 1995; 59: 143–169.
16. Ammor MS, Delgado S, Alvarez-Martín P, Margolles A, and Mayo B. Reagentless identification of human bifidobacteria by intrinsic fluorescence. *J. Microbiol. Methods* 2007; 69: 100–106.
17. Annuk H, Shchepetova J, Kullisaar T, Songisepp E, Zilmer M, and Mikelsaar M. Characterization of intestinal lactobacilii as putative probiotic candidates. *J. Appl. Microbiol.* 2003; 94: 403–412.
18. Axelsson L, Lactic acid bacteria: classification and physiology. In: Salminen S von Wright A editors. *Lactic Acid Bacteria: Microbiology and Functional Aspects*. Marcel Dekker New York 1998, pp. 1–72.
19. Bäckhead F, Ley RE, Sonnenburg JL, Peterson DA, and Gordon JI. Host-bacterial mutualism in the human intestine. *Science* 2005; 307: 1915–1919.
20. Ballongue J, Bifidobacteria and probiotic action. In: Salminen S, von Wright A, and Ouwehand AC, editors. *Lactic Acid Bacteria: Microbiological and Functional Aspects*, 3rd edition. Marcel Dekker, New York, 2004, pp. 67–124.
21. Begley M, Gahan CGM, and Hill C. The interaction between bacteria and bile. *FEMS Microbiol. Rev.* 2005; 29: 625–651.
22. Ben-Amor K, Vaughan EE, and de Vos WM. Advanced molecular tools for the identification of lactic acid bacteria. *J. Nutr.* 2007; 137: 741S–747S.
23. Berger B, Pridmore RD, Barretto C, Delmas-Julien F, Schreiber K, Arigoni F, and Brüssow H. Similarity and differences in the *Lactobacillus acidophilus* group identified by polyphasic analysis and comparative genomics. *J. Bacteriol.* 2007; 189: 1311–1321.
24. Berthier F and Ehrlich SD. Rapid species identification within two groups of closely related lactobacilli using PCR primers that target the 16S/23S rRNA spacer region. *FEMS Microbiol. Lett.* 1998; 161: 97–106.
25. Bevilacqua L, Ovidi M, Di Mattia E, Trovatelli LD, and Canganella F. Screening of *Bifidobacterium* strains isolated from human faeces for antagonistic activities against potentially bacterial pathogens. *Microbiol. Res.* 2003; 158: 179–185.
26. Bonjoch X, Balleste E, and Blanch AR. Multiplex PCR with 16S rRNA gene-targeted primers of *Bifidobacterium* spp. to identify sources of fecal pollution. *Appl. Environ. Microbiol.* 2004; 70: 3171–3175.
27. Bourget N, Simonet JM, and Decaris B. Analysis of the genome of the five *Bifidobacterium breve* strains: plasmid content, pulsed-field gel electrophoresis genome size estimation and *rrn* loci number. *FEMS Microbiol. Lett.* 1993; 110: 11–20.
28. Boyd MA, Antonio MAD, and Hillier SL. Comparison of API 50 CH strips to whole-chromosomal DNA probes for the identification of *Lactobacillus* species. *J. Clin. Microbiol.* 2005; 43: 5309–5311.
29. Briczinski EP and Roberts RF. Technical note: a rapid pulsed-field gel electrophoresis method for analysis of bifidobacteria. *J. Dairy Sci.* 2006; 89: 2424–2427.
30. Caldini G, Trotta F, Villarini M, Moretti M, Pasquini R, Scassellati-Sforzolini G, and Cenci G. Screening of potential lactobacilli antigenotoxicity by microbial and mammalian cell-based tests. *Int. J. Food Microbiol.* 2005; 102: 37–47.

31. Camilleri M, Colemont LJ, Phillips SF, Brown ML, Thomforde GM, Chapman N, and Zinsmeister AR. Human gastric emptying and colonic filling of solids characterized by a new method. *Am. J. Physiol. Gastrointest. Liver Physiol.* 1989; 257: 284–290.
32. Canchaya C, Claesson MJ, Fitzgerald GF, van Sinderen D, and O'Toole PW. Diversity of the genus *Lactobacillus* revealed by comparative genomics of five species. *Microbiology* 2006; 152: 3185–3196.
33. Chavagnat F, Hauter M, Jimeno J, and Gasey MG. Comparison of partial *tuf* gene sequences for the identification of lactobacilli. *FEMS Microbiol. Lett.* 2002; 217: 177–183.
34. Chou LS and Weimer B. Isolation and characterization of acid- and bile-tolerant isolates from strains of *Lactobacillus acidophilus*. *J. Dairy Sci.* 1999; 82: 23–31.
35. Claesson MJ, van Sinderen D, O'Toole PW. The genus *Lactobacillus*. A genomic basis for understanding its diversity. *FEMS Microbiol. Lett.* 2007; 269: 22–28.
36. Coconnier M-H, Liévin V, Hemery E, and Servin AL. Antagonistic activity against *Helicobacter* infection *in vitro* and *in vivo* by the human *Lactobacillus acidophilus* strain LB. *Appl. Environ. Microbiol.* 1998; 64: 4573–4580.
37. Coeuret V, Gueguem M, and Vernoux JP. *In vitro* screening of potential probiotic activities of selected lactobacilli isolated from unpasteurized milk products for incorporation into soft cheese. *J. Dairy Res.* 2004; 71: 451–460.
38. Collado MC, Gueimonde M, Hernández M, Sanz Y, and Salminen S. Adhesion of selected *Bifidobacterium* strains to human intestinal mucus and the role of adhesion in enteropathogen exclusion. *J. Food Prot.* 2005; 68: 2672–2678.
39. Collado MC and Sanz Y. Method for direct selection of potentially probiotic *Bifidobacterium* strains from human feces based on their acid-adaptation ability. *J. Microbiol. Methods* 2006; 66: 560–563.
40. Collins MD, Rodrigues U, Ash C, Aguirre M, Farrow JAE, Martínez-Murcia A, Phillips BA, Williams AM, and Wallbanks S. Phylogenetic analysis of the genus *Lactobacillus* and related lactic acid bacteria as determined by reverse transcriptase sequencing of 16S rRNA. *FEMS Microbiol. Lett.* 1991; 77: 5–12.
41. Corthesy B, Boris S, Isler P, Grangette C, and Mercenier A. Oral immunization of mice with lactic acid bacteria producing *Helicobacter pylori* urease B subunit partially protect against challenge with *Helicobacter felis*. *J. Infect. Dis.* 2005; 192: 1441–1449.
42. Crittenden RG, Morris LF, Harvey ML, Tran LT, Mitchell HL, and Playne MJ. Selection of a *Bifidobacterium* strain to complement resistant starch in a synbiotic yoghurt. *J. Appl. Microbiol.* 2001; 90: 268–278.
43. Cusick SM, O'Sullivan DJ. Use of a single, triplicate arbitrarily primed-PCR procedure for molecular fingerprinting of lactic acid bacteria. *Appl. Environ. Microbiol.* 2000; 66: 2227–2231.
44. Dal Bello F and Hertel C. Oral cavity as natural reservoir for intestinal lactobacilli. *Syst. Appl. Microbiol.* 2006; 29: 69–76.
45. Daniel C, Poiret S, Goudercourt D, Dennin V, Leyer G, and Pot B. Selecting lactic acid bacteria for their safety and functional by use of a mouse colitis model. *Appl. Environ. Microbiol.* 2006; 72: 5799–5805.
46. Daud Khaled AK, Neilan BA, Henriksson A, and Conway PL. Identification and phylogenetic analysis of *Lactobacillus* using multiplex RAPD-PCR. *FEMS Microbiol. Lett.* 1997; 153: 191–197.

47. De las Rivas B, Marcobal A, and Muñoz R. Development of a multilocus sequence typing meted for analysis of *Lactobacillus plantarum* strains. *Microbiology* 2006; 52: 85–93.
48. Delcenserie V, Bechoux N, Leonard T, China B, and Daube G. Discrimination between *Bifidobacterium* species from human and animal origin by PCR-restriction fragment length polymorphism. *J. Food Prot.* 2004; 67: 1284–1288.
49. Delgado S, Ruas-Madiedo P, Suárez A, and Mayo B. Interindividual differences in microbial counts and biochemical-associated variables in the faeces of healthy Spanish adults. *Dig. Dis. Sci.* 2006; 51: 737–743.
50. Delgado S, Suarez A, and Mayo B. Bifidobacterial diversity determined by culturing and by 16S rDNA sequence analysis in feces and mucosa from ten healthy Spanish adults. *Dig. Dis. Sci.* 2006; 51: 1878–1885.
51. Dellaglio F and Felis GE, Taxonomy of lactobacilli and bifidobacteria. In: Tannock GW, editor, Probiotics and Prebiotis: Scientific Aspects. Caister Academic Press, Wymondham, UK, 2005, pp. 25–49.
52. De los Reyes-Gavilán CG, Ruas-Madiedo P, Noriega P, Cuevas I, Sánchez B, and Margolles A. Effect of acquired resistance to bile salts on enzymatic activities involved in the utilisation of carbohydrates by bifidobacteria. An overview. *Lait* 2005; 85: 113–123.
53. Deplancke B and Gaskins HR. Microbial modulation of innate defence: goblet cells and the intestinal mucus layer. *Am. J. Clin. Nutr.* 2001; 73: 1131S–1141S.
54. Diancourt L, Passet V, Chervaux C, Garault P, Smokvina T, and Brisse S. Multilocus sequence typing of *Lactobacillus casei* reveals a clonal population structure with low levels of homologous recombination. *Appl. Environ. Microbiol.* doi:10.1128/AEM.01095-07 2007; 73: 6601–6611
55. Dong X, Cheng G, and Jian W. Simultaneous identification of five *Bifidobacterium* species isolated from human beings using multiple PCR primers. *Syst. Appl. Microbiol.* 2000; 23: 386–390.
56. Dunne C, Murphy L, Flynn S, O'Mahohy L, O'Halloran S, Feeney M, Morrissey D, Thornton G, Fitzgerld G, Daly Ch Kiely B, Quigley EMM, O'Sullivan GC, Shanahan F, and Collins JK. Probiotics: from myth to reality. Demonstration of functionality in animal models of disease and in human clinical trials. *Antonie van Leeuwenhoek* 1999; 76: 279–292.
57. Dunne C, O'Mahony L, Murphy L, Thornton G, Morrissey D, O'Halloran S, Feeny M, Flynn S, Fitzgeral G, Daly C, Kiely B, O'Sullivan GC, Shanahan F, and Collins JK. *In vitro* selection criteria for probiotic bacteria of human origin: correlation with *in vivo* findings. *Am. J. Clin. Nutr.* 2001; 73: 386S–392S.
58. Du Plessis EM and Dicks LM. Evaluation of random amplified polymorphic DNA (RAPD)-PCR as a method to differentiate *Lactobacillus acidophilus, Lactobacillus crispatus, Lactobacillus amylovorus, Lactobacillus gallinarum, Lactobacillus gasseri*, and *Lactobacillus johnsonii*. *Curr. Microbiol.* 1995; 31: 114–118.
59. EFSA. Opinion of the Scientific Committee on a request from EFSA related to a generic approach to the safety assessment by EFSA of microorganisms used in food/feed and the production of food/feed additives. *EFSA J.* 2005; 226: 1–12.
60. FAO/WHO. Probiotic in foods. Health and nutritional properties and guidelines for evaluation. In *FAO Food and Nutrition Paper 85*, 2006, ISBN 92-5-105513-0. Also available at ftp://ftp.fao.org/docrep/fao/009/a0512e/a0512e00.pdf.

61. Fasoli S, Marzotto M, Rizzotti L, Rossi F, Dellaglio F, and Torriani S. Bacterial composition of commercial probiotics products as evaluated by PCR-DGGE analysis. *Int. J. Food Microbiol.* 2003; 82: 59–70.
62. Favier CF, Vaughan EE, de Vos WM, and Akkermans AD. Molecular monitoring of succession of bacterial communities in human neonates. *Appl. Environ. Microbiol.* 2002; 68: 219–226.
63. FDA. Substances Generally Recognized As Safe. *Fed. Reg. FDA* 1997; 62: 18937–18964.
64. Felis GE, Dellaglio L, Mizzi L, and Torriani S. Comparative sequence analysis of *recA* gene fragment brings new evidence for a change in the taxonomy of the *Lactobacillus casei* group. *Int. J. Syst. Microbiol.* 2001; 51: 2113–2117.
65. Felis GE and Dellaglio F. Taxonomy of lactobacilli and bifidobacteria. *Curr. Issues Intest. Microbiol.* 2007; 8: 44–61.
66. Garrigues C, Stuer-Lauridsen B, and Johansen E. Characterisation of *Bifidobacterium animalis* subsp. *lactis* BB-12 and other probiotic bacteria using genomics, transcriptomics and proteomics. *Aust. J. Dairy Technol.* 2005; 60: 84–92.
67. Germond JE, Mamin O, and Mollet B. Species specific identification of nine human *Bifidobacterium* spp. in feces. *Syst. Appl. Microbiol.* 2002; 25: 536–543.
68. Gilarova R, Voldrich M, Demnerova K, Cerovsky M, and Dobias J. Cellular fatty acids analysis in the identification of lactic acid bacteria. *Int. J. Food Microbiol.* 1994; 24: 315–319.
69. Giraffa G, Lazzi C, Gatti M, Rossetti L, Mora D, and Neviani E. Molecular typing of *Lactobacillus delbrueckii* of dairy origin by PCR-RFLP of protein-coding genes. *Int. J. Food Microbiol.* 2003; 82: 163–172.
70. Gueimonde M, Noriega L, Margolles A, de los Reyes-Gavilán CG, and Salminen S. Ability of *Bifidobacterium* strains with acquired resistance to bile to adhere to human intestinal mucus. *Int. J. Food Microbiol.* 2005; 101: 341–346.
71. Gueimonde M, Jaloe L, He F, Hiramatsu M, and Salminen S. Adhesion and competitive inhibition and displacement of human enteropathogens by selected lactobacilli. *Food Res. Int.* 2006; 39: 467–471.
72. Gürtler V and Stanisish VA. New approaches to typing and identification of bacteria using the 16S-23S rDNA spacer region. *Microbiology* 1996; 142: 3–16.
73. Ha CG, Cho JK, Lee CH, Chai YG, Ha YA, and Shin SH. Cholesterol lowering effect of *Lactobacillus plantarum* isolated from human feces. *J. Microbiol. Biotechnol.* 2006; 16: 1201–1209.
74. Hamaji Y, Fujimori M, Sasaki T, Matsuhashi H, Matsui-Seki K, Shimatani-Shibata Y, Kano Y, Amano J, and Taniguchi S. Strong enhancement of recombinant cytosine deaminase activity in *Bifidobacterium longum* for tumor-targeting enzyme/prodrug therapy. *Biosci. Biotech. Biochem.* 2007; 71: 874–883.
75. Han KS, Kim Y, Choi S, Oh S, Park S, Kim SH, and Whang KY. Rapid identification of *Lactobacillus acidophilus* by restriction analysis of the 16S-23S rRNA intergenic spacer region and flanking 23S rRNA gene. *Biotechnol. Lett.* 2005; 27: 1183–1188.
76. Han W, Mercenier A, Ait-Belgnaoui A, Pavan S, Lamine F, van Swam IL, Kleerebezem M, Salvador-Cartier C, Hisbergues M, Bueno L, Theodorou V, and Fioramonti J. Improvement of an experimental colitis in rats by lactic acid bacteria producing superoxide dismutase. *Inflamm. Bowel Dis.* 2006; 12: 1044–1052.

77. Hazebrouck S, Oozeer R, Adel-Patient K, Langella P, Rabot S, Wal JM, and Corthier G. Constitutive delivery of bovine beta-lactoglobulin to the digestive tracts of gnobiotic mice by engineered *Lactobacillus casei*. *Appl. Environ. Microbiol.* 2006; 72: 7460–7467.
78. HazebrouckS Pothelune L, Azevedo V, Corthier G, Wal JM, and Langella P. Efficient production and secretion of bovine beta-lactoglobulin by *Lactobacillus casei*. *Microbial Cell Factories* 2007; 6: 12–20.
79. Heikkilä MP and Saris PEJ. Inhibition of *Staphylococcus aureus* by commensal bacteria of human milk. *J. Appl. Microbiol.* 2003; 95: 471–478.
80. Heilig GHJ, Zoetendal EG, Vaughan EE, Marteau P, Akkermans ADL, and de Vos WM. Molecular diversity of *Lactobacillus* spp. and other lactic acid bacteria in the human intestine as determined by specific amplification of 16S ribosomal DNA. *Appl. Environ. Microbiol.* 2002; 68: 114–123.
81. Hudault S, Liévin V, Bernet-Camard M-F, and Servin AL. Antagonistic activity exerted *in vitro* and *in vivo* by *Lactobacillus paracasei* (strain GG) against *Salmonella typhimurium* C5 infection. *Appl. Environ. Microbiol.* 1997; 63: 513–518.
82. Hutt P, Shchepetova J, Loivukene K, Kullisaar T, and Mikelsaar M. Probiotic lactobacilli enhance eradication on *Salmonella* Typhimurium in animal model. *J. Appl. Microbiol.* 2006; 100: 1324–1332.
83. Huys G, Vancanneyt M, D'Haene K, Vankerckhoven V, Goossens H, and Swings J. Accuracy of species identity of commercial bacterial cultures intended for probiotic or nutritional use. *Res. Microbiol.* 2006; 157: 803–810.
84. Jacobsen CN, Nielsen VR, Hayford AE, Moller PL, Michaelsen KF, Paerregaard A, Sandstrom B, Tvede M, and Jakobsen M. Screening of probiotic activities of forty-seven strains of *Lactobacillus* spp. by *in vitro* techniques and evaluation of the colonization ability of five selected strains in humans. *Appl. Environ. Microbiol.* 1999; 65: 4949–4956.
85. Johansson M-L, Molin G, Jeppsson B, Nobaek S, Ahrné S, and Bengmark S. Administration of different *Lactobacillus* strains in fermented oatmeal soup: *In vivo* colonization of human intestinal mucosa and effect on the indigenous flora. *Appl. Environ. Microbiol.* 1993; 59: 15–20.
86. Kandlder O and Weiss N, Genus Lactobacillus Beijerinck 1901, 212[AL]. In: Sneath PHA, Mair NS, Sharpe ME, and Holt JG, editors, Bergey's Manual of Systematic Bacteriology Vol. 2. Williams and Wilkins, Baltimore, MD, 1986, pp. 1208–1234.
87. Kaushal G and Shao J. Oral delivery of beta-lactamase by *Lactococcus lactis* subsp. *lactis* transformed with plasmid ss80. *Int. J. Pharm.* 2006; 312: 90–95.
88. Kimura K, McCartney AL, McConnell MA, and Tannock GW. Analysis of fecal populations of bifidobacteria and lactobacilli and investigation of the immunological responses of their human hosts to the predominant strains. *Appl. Environ. Microbiol.* 1997; 63: 3394–3398.
89. Krizova J, Spanova A, and Rittich B. Evaluation of amplified ribosomal DNA restriction analysis (ARDRA) and species-specific PCR for identification of *Bifidobacterium* species. *Syst. Appl. Microbiol.* 2006; 29: 36–44.
90. Kwon HS, Yang EH, Yeon SW, Kang BH, and Kim TY. Rapid identification of probiotic *Lactobacillus* species by multiplex PCR using species-specific primers based on the region extending from 16S rRNA through 23S rRNA. *FEMS Microbiol. Lett.* 2004; 239: 267–275.
91. Kwon HS, Yang EH, Lee SH, Yeon SW, Kang BH, and Kim TY. Rapid identification of potentially probiotic *Bifidobacterium* species by multiplex PCR using species-specific

primers based on the region extending from 16S rRNA through 23S rRNA. *FEMS Microbiol. Lett.* 2005; 250: 55–62.

92. Lay C, Rigottier-Gois L, Holmstrøm K, Rajilic M, Vaughan EE, de Vos WM, Collins MD, Thiel R, Namsolleck P, Blaut M, and Doré J. Colonic microbiota signatures across five northern European countries. *Appl. Environ. Microbiol.* 2005; 71: 4153–4155.

93. Lee KY, So JS, and Heo TR. Thin layer chromatographic determination of organic acids for rapid identification of bifidobacteria at genus level. *J. Microbiol. Methods* 2001; 45: 1–6.

94. Leblond-Bourget N, Philippe H, Mangin I, and Decaris B. 16S rRNA and 16S to 23S internal transcribed spacer sequence analyses reveal inter- and intraspecific *Bifidobacterium* phylogeny. *Int. J. Syst. Bacteriol.* 1999; 46: 102–111.

95. Maiden MC, Bygraves JA, Feil E, Morelli G, Russell JE, Urwin R, Zhang Q, Zhou J, Zurth K, Caugant DA, Feavers IM, Achtman M, and Spratt BG. Multilocus sequence typing: A portable approach to the identification of clones within populations of pathogenic microorganisms. *Proc. Natl. Acad. Sci. USA* 1998; 95: 3140–3145.

96. Makarova K, Slesarev A, Wolf Y, Sorokin A, Mirkin B, Koonin E, Pavlov A, Pavlova N, Karamychev V, Polouchine N, Shakhova V, Grigoriev I, Lou Y, Rohksar D, Lucas S, Huang K, Goodstein DM, Hawkins T, Plengvidhya V, Welker D, Hughes J, Goh Y, Benson A, Baldwin K, Lee JH, Diaz-Muñiz I, Dosti B, Smeianov V, Wechter W, Barabote R, Lorca G, Altermann E, Barrangou R, Ganesan B, Xie Y, Rawsthorne H, Tamir D, Parker C, Breidt F, Broadbent J, Hutkins R, O'Sullivan D, Steele J, Unlu G, Saier M, Klaenhammer T, Richardson P, Kozyavkin S, Weimer B, and Mills D. Comparative genomics of the lactic acid bacteria. *Proc. Natl. Acad. Sci. USA* 2006; 103: 15611–15616.

97. Mangin I, Bourget N, Bouhnik Y, Bisetti N, Simonet JM, and Decaris B. Identification of *Bifidobacterium* strains by rRNA gene restriction patterns. *Appl. Environ. Microbiol.* 1994; 60: 1451–1458.

98. Mangin I, Bourget N, and Decaris B. Ribosomal DNA polymorphism in the genus *Bifidobacterium*. *Res. Microbiol.* 1996; 147: 183–192.

99. Mangin I, Suau A, Magne F, Garrido D, Gotteland M, Neut C, and Pochart P. Characterization of human intestinal bifidobacteria using competitive PCR and PCR-TTGE. *FEMS Microbiol. Ecol.* 2006; 55: 28–37.

100. Margolles A, García L, Sánchez B, Gueimonde M, and de los Reyes-Gavilán C.G. Characterisation of *Bifidobacterium* strains with acquired resistance to cholate–a preliminary study. *Int. J. Food Microbiol.* 2003; 82: 191–198.

101. Martin R, Langa S, Reviriego C, Jiménez E, Marín ML, Xaus J, Fernández L, and Rodríguez JM. Human milk is a source of lactic acid bacteria for the infant gut. *J. Pediatr.* 2003; 143: 754–758.

102. Masco L, Huys G, Gevers D, Verbrugghen L, and Swings J. Identification of *Bifidobacterium* species using rep-PCR fingerprinting. *Syst. Appl. Microbiol.* 2003; 26: 557–563.

103. Masco L, Ventura M, Zink R, Huys G, and Swings J. Polyphasic taxonomic analysis of *Bifidobacterium animalis* and *Bifidobacterium lactis* reveals relatedness at the subspecies level: reclassification of *Bifidobacterium animalis* as *Bifidobacterium animalis* subsp. *animalis* subsp. nov. and *Bifidobacterium lactis* as *Bifidobacterium animalis* subsp. *lactis* subsp. nov. *Int. J. Syst. Evol. Microbiol.* 2004; 54: 1137–1143.

104. Masco L, Huys G, De Brandt E, Temmerman R, and Swings J. Culture-dependent and culture-independent qualitative analysis of probiotic products claimed to contain bifidobacteria. *Int. J. Food Microbiol.* 2005; 102: 221–230.

105. Matijasic BB, Narat M, Zoric Peternel MZ, and Rogelj I. Ability of *Lactobacillus gasseri* K7 to inhibit *Escherichia coli* adhesion *in vitro* on Caco-2 cells and *ex vivo* on pigs' jejunal tissue. *Int. J. Food Microbiol.* 2006; 107: 92–96.
106. Matsuki T, Watanabe K, Tanaka R, Fukuda M, and Oyaizu H. Distribution of bifidobacterial species in human intestinal microflora examined with 16S rRNA-gene-targeted species-specific primers. *Appl. Environ. Microbiol.* 1999; 65: 4506–4512.
107. Matsuki T, Watanabe K, Fujimoto J, Miyamoto Y, Takada T, Matsumoto K, Oyaizu H, and Tanaka R. Development of 16S rRNA-gene-targeted group-specific primers for the detection and identification of predominant bacteria in human feces. *Appl. Environ. Microbiol.* 2002; 68: 5445–5451.
108. Matsuki T, Watanabe K, and Tanaka R. Genus- and species-specific PCR primers for the detection and identification of bifidobacteria. *Curr. Issues Intest. Microbiol.* 2003; 4: 61–99.
109. Matsumura A, Saito T, Arakuni M, Kitazawa H, Kawai Y, and Itoh T. New binding assay and preparative trial of cell-surface lectin from *Lactobacillus acidophilus* group lactic acid bacteria. *J. Dairy Sci.* 1999; 82: 2525–2529.
110. Matto J, Malinen E, Suihko ML, Alander M, Palva A, and Saarela M. Genetic heterogeneity and functional properties of intestinal bifidobacteria. *J. Appl. Microbiol.* 2004; 97: 459–470.
111. Maus JE and Ingham SC. Employment of stressful conditions during culture production to enhance subsequent cold- and acid tolerance of bifidobacteria. *J. Appl. Microbiol.* 2003; 95: 146–154.
112. Molenaar D, Bringel F, Schuren FH, de Vos WM, Siezen RJ, and Kleerebezem M. Exploring *Lactobacillus plantarum* genome diversity by using microarrays. *J. Bacteriol.* 2005; 187: 6119–6127.
113. Molin G, Jeppsson B, Johansson ML, Ahrné S, Nobaek S, Stahl M, and Bengmark S. Numerical taxonomy of *Lactobacillus* spp. associated with healthy and diseased mucosa of the human intestines. *J. Appl. Bacteriol.* 1993; 74: 314–323.
114. Moon GS, Pyun YR, Park MS, Ji GE, and Kim WJ. Secretion of recombinant pediocin PA-1 by *Bifidobacterium longum*, using the signal sequence for bifidobacterial alpha-amylase. *Appl. Environ. Microbiol.* 2005; 71: 5630–5632.
115. Mori K, Yamazi K, and Ishiyama T. Comparative sequence analyses of the genes coding for 16S rRNA of *Lactobacillus casei*-related taxa. *Int. J. Syst. Bacteriol.* 1997; 47: 54–57.
116. Mullie C, Odou MF, Singer E, Romond MB, and Izard D. Multiplex PCR using 16S rRNA gene-targeted primers for the identification of bifidobacteria from human origin. *FEMS Microbiol. Lett.* 2003; 222: 129–136.
117. Nigatu A, Ahrne S, and Molin G. Randomly amplified polymorphic DNA (RAPD) profiles for the distinction of *Lactobacillus* species. *Antonie van Leeuwenhoek.* 2001; 79: 1–6.
118. Noriega L, Gueimonde M, Sánchez B, Margolles A, and de los Reyes-Gavilán CG. Effect of the adaptation of high bile salts concentrations on glycosidic activity, survival at low pH and cross-resistance to bile in *Bifidobacterium*. *Int. J. Food Microbiol.* 2004; 94: 79–86.
119. Nour M. 16S-23S and 23S-5S intergenic spacer regions of lactobacilli: nucleotide sequence, secondary structure and comparative analysis. *Microbiol. Rev.* 1998; 59: 143–169.
120. Ogier J-C, Lafarge V, Girard V, Rault A, Maladen V, Gruss A, Leveau J-Y, and Delacroix-Buchet A. Molecular fingerprint of dairy microbial ecosystems by use of temporal

temperature denaturing gradient gel electrophoresis. *Appl. Environ. Microbiol.* 2004; 70: 5628–5643.
121. O'Sullivan DJ. Screening of intestinal microflora for effective probiotic bacteria. *J. Agric. Food Chem.* 2001; 49: 1751–1760.
122. Ouwehand AC, Isolauri E, Kirjavainen PV, and Salminen SJ. Adhesion of four *Bifidobacterium* strains to human intestinal mucus from subjects in different age groups. *FEMS Microbiol. Lett.* 1999; 172: 61–64.
123. Ouwehand AC, Kirjavainen PV, Grönlund MM, Isolauri E, and Salminen SJ. Adhesion of probiotic micro-organisms to intestinal mucus. *Int. Dairy J.* 1999; 9: 623–630.
124. Ouwehand AC, Hashimoto H, Isolauri E, Benno Y, and Salminen S. Adhesion of *Bifodobacterium* spp. to human intestinal mucus. *Microbiol. Immunol.* 2001; 45: 259–262.
125. Ouwehand AC, Tuomola EM, Tölkkö S, and Salminen S. Assessment of adhesion properties of novel probiotic strains to human intestinal mucus. *Int. J. Food Microbiol.* 2001; 64: 119–126.
126. Ouwehand AC, Salminen S, Tölkkö S, Roberts P, Ovaska J, and Salminen E. Resected human colonic tissue: new model for characterizing adhesion of lactic acid bacteria. *Clin. Diagn. Lab. Immunol.* 2002; 9: 184–186.
127. Palys T, Nakamura LK, and Cohan FM. Discovery and classification of ecological diversity in the bacterial world: the role of DNA sequence data. *Int. J. Syst. Bacteriol.* 1997; 47: 1145–1156.
128. Park SH and Itoh K. Species-specific oligonucleotide probes for the detection and identification of *Lactobacillus* isolated from mouse faeces. *J. Appl. Microbiol.* 2005; 99: 51–57.
129. Pavan S, Desreumaux PD, and Mercenier A. Use of mouse models to evaluate the persistence, safety, and immune modulation capacities of lactic acid bacteria. *Clin. Diagn. Lab. Immunol.* 2003; 10: 696–701.
130. Pennacchia C, Ercolini D, Blaiotta G, Pepe O, Mauriello G, and Villani F. Selection of *Lactobacillus* strains for fermented sausages for their potential use as probiotics. *Meat Sci.* 2004; 67: 309–317.
131. Peña JA, Li SY, Wilson PH, Thibodeau AA, Szary AJ, and Versalovic J. Genotypic and phenotypic studies on murine intestinal lactobacilli: species differences in mice with and without colitis. *Appl. Environ. Microbiol.* 2004; 70: 558–5568.
132. Pereira DIA and Gibson GR. Cholesterol assimilation by lactic acid bacteria and bifidobacteria isolated from the human gut. *Appl. Environ. Microbiol.* 2002; 68: 4689–4693.
133. Pot B, Hertel C, Ludwig W, Descheemaeker P, Kersters K, and Schleifer KH. Identification and classification of *Lactobacillus acidophilus*, *L. gasseri* and *L. johnsonii* strains by SDS-PAGE and rRNA-targeted oligonucleotide probe hybridisation. *J. Gen. Microbiol.* 1993; 139: 513–517.
134. Prasad J, Gill H, Smart J, and Gopa PK. Selection and characterization of *Lactobacillus* and *Bifidobacterium* strains for use as probiotics. *Int. Dairy J.* 1998; 8: 993–1002.
135. Randazzo CL, Torriani S, Akkermans AL, de Vos WM, and Vaughan EE. Diversity, dynamics, and activity of bacterial communities during production of an artisanal Sicilian cheese as evaluated by 16S rRNA analysis. *Appl. Environ. Microbiol.* 2002; 68: 1882–1892.

136. Reid G. The scientific basis for probiotic strains of lactobacilli. *Appl. Environ. Microbiol.* 1999; 65: 3763–3766.
137. Requena T, Burton J, Matsuki T, Munro K, Simon MA, Tanaka R, Watanabe K, and Tannock GW. Identification, detection, and enumeration of human *Bifidobacterium* species by PCR targeting the transaldolase gene. *Appl. Environ. Microbiol.* 2002; 68: 2420–2427.
138. Rettger LF, Levy MN, Weinstein L, and Weiss JE, Lactobacillus acidophilus and its therapeutic application. Yale University Press, New Haven, CT, 1935.
139. Riedel CU, Foata F, Goldstein DR, Blue S, and Eikmanns BJ. Interaction of bifidobacteria with Caco-2 cells–adhesion and impact on expression profiles. *Int. J. Food Microbiol.* 2006; 110: 62–68.
140. Rinkinen M, Westemarck E, Salminen S, and Ouwehand AC. Absence of host specificity for *in vitro* adhesion of probiotic lactic acid bacteria to intestinal mucus. *Vet. Microbiol.* 2003; 97: 55–61.
141. Rodtong S and Tannock GW. Differentiation of *Lactobacillus* strains by ribotyping. *Appl. Environ. Microbiol.* 1993; 59: 3480–3484.
142. Roy D, Ward P, and Champagne G. Differentiation of bifidobacteria by use of pulsed-field gel electrophoresis and polymerase chain reaction. *Int. J. Food Microbiol.* 1996; 29: 11–29.
143. Roy D and Sirois S. Molecular differentiation of *Bifidobacterium* species with amplified ribosomal DNA restriction analysis and alignment of short regions of the *ldh* gene. *FEMS Microbiol. Lett.* 2000; 191: 17–24.
144. Roy D, Ward P, Vincent D and Mondou F Molecular identification of potentially probiotic lactobacilli. *Curr. Microbiol.* 2000; 40: 40–46.
145. Roy D and Ward P. Comparison of fructose-1,6-biphosphatase gene (*fbp*) sequences for the identification of *Lactobacillus rhamnosus*. *Curr. Microbiol.* 2004; 49: 313–320.
146. Saarela M, Rantala M, Hallamaa K, Nohynek L, Virkajarvi I, and Matto J. Stationary-phase acid and heat treatments for improvement of the viability of probiotic lactobacilli and bifidobacteria. *J. Appl. Microbiol.* 2004; 96: 1205–1214.
147. Sakamoto M, Hayashi H, and Benno Y. Terminal restriction fragment length polymorphism analysis for human fecal microbiota and its application for analysis of complex bifidobacterial communities. *Microbiol. Immunol.* 2003; 47: 133–142.
148. Sakata S, Ryu CS, Kitahara M, Sakamoto M, Hayashi H, Fukuyama M, and Benno Y. Characterization of the genus *Bifidobacterium* by automated ribotyping and 16S rRNA gene sequences. *Microbiol. Immunol.* 2006; 50: 1–10.
149. Salminen S, Nurmi J, and Gueimonde M. The genomics of probiotic intestinal microorganisms. *Genome Biol.* 2005; 6: 255. Also available at http://genomebiology.com/2005/6/7/225.
150. Satokari RM, Vaughan EE, Smidt H, Saarela M, Mättö J, and de Vos WM. Molecular approaches for the detection and identification of bifidobacteria and lactobacilli in the human gastrointestinal tract. *Syst. Appl. Microbiol.* 2003; 26: 572–584.
151. Scardovi V, Genus *Bifidobacterium*. Orla-Jensen 1924, 472^{AL}. In: Sneath PHA, Mair NS, Sharpe ME, and Holt JG, editors. Bergey's Manual of Systematic Bacteriology, Vol. 2. Williams and Wilkins, Baltimore, MD, 1986, 1418–1434.
152. Schell MA, Karmirantzou M, Snel B, Vilanova D, Berger B, Pessi G, Zwahlen MC, Desiere F, Bork P, Delley M, Pridmore RD, and Arigoni F. The genome sequence of *Bifidobacterium longum* reflects its adaptation to the human gastrointestinal tract. *Proc. Natl. Acad. Sci. USA.* 2002; 99: 14422–14427.

153. Schillinger U, Yousif NM, Sesar L, and Franz CM. Use of group-specific and RAPD-PCR analyses for rapid differentiation of *Lactobacillus* strains from probiotic yogurts. *Curr. Microbiol.* 2003; 47: 453–456.

154. Settanni L, van Sinderen D, Rossi J, and Corsetti A. Rapid differentiation and *in situ* detection of 16 sourdough *Lactobacillus* species by multiplex PCR. *Appl. Environ. Microbiol.* 2005; 71: 3049–3059.

155. Sghir A, Antonopoulos D, and Mackie RI. Design and evaluation of a *Lactobacillus* group-specific ribosomal RNA-targeted hybridization probe and its application to the study of intestinal microecology in pigs. *Syst. Appl. Microbiol.* 1998; 21: 291–296.

156. Sghir A, Chow JM, and Mackiw RI. Continuous culture selection of bifidobacteria and lactobacilli from human faecal samples using fructooligosaccharides as selective substrate. *J. Appl. Microbiol.* 1998; 85: 769–777.

157. Shanahan F. Immunology. Therapeutic manipulation of gut flora. *Science* 2000; 289: 1311–1312.

158. Shuhaimi MA, Ali M, Saleh NM, and Yazid AM. Utilisation of enterobacterial repetitive intergenic consensus (ERIC) sequence-based PCR to fingerprint the genomes of *Bifidobacterium* isolates and other bacteria. *Biol. Technol. Lett.* 2001; 23: 731–736.

159. Simpson PJ, Stanton C, Fitzgerald GF, and Ross RP. Genomic diversity and relatedness of bifidobacteria isolated from a porcine cecum. *J. Bacteriol.* 2003; 185: 2571–2581.

160. Song Y-L, Kato N, Liu C-X, Matsumiya Y, Kato H, and Watanabe K. Rapid identification of 11 human intestinal *Lactobacillus* species by multiplex PCR assays using group- and species-specific primers derived from the 16S-23S rRNA intergenic spacer region and its flanking 23S rRNA. *FEMS Microbiol. Lett.* 2000; 187: 167–173.

161. Stackebrandt E and Goebel BM. Taxonomic note: A place for DNA-DNA reassociation and 16S rRNA sequence analysis in the present species definition in bacteriology. *Int. J. Syst. Bacteriol.* 1994; 44: 846–849.

162. Stahl M, Molin G, Persson A, Ahrné S, and Stahl S. Restriction endonuclease patterns and multivariate analysis as a classification tool for *Lactobacillus* spp. *Int. J. Syst. Bacteriol.* 1990; 40: 189–193.

163. Staley JT. The bacterial species dilemma and the genomic-phylogenetic species concept. *Phil. Trans. R. Soc. Lond. B. Biol. Sci.* 2006; 361: 1899–1909.

164. Steidler L, Hans W, Schotte L, Neirynck S, Obermeir F, Falk W, Fiers W, and Remaut E. Treatment of murine colitis by *Lactococcus lactis* secreting interleukin-10. *Science* 2000; 289: 1352–1355.

165. Steidler L. Genetically engineered probiotics. *Best Pract. Res. Clin. Gastroenterol.* 2003; 17: 861–876.

166. Stiles ME. Biopreservation by lactic acid bacteria. *Antonie van Leeuwenhoek* 1996; 70: 331–345.

167. Strahinic I, Busarcevic M, Pavlica D, Milasin J, Golic N, and Toposirovic L. Molecular and biochemical characterization of human oral lactobacilli as putative probiotic candidates. *Oral Microbiol. Immunol.* 2007; 22: 111–117.

168. Takata T, Shirakawa T, Kawasaki Y, Kinoshita S, Gotoh A, Kano Y, and Kawabata M. Genetically engineered *Bifidobacterium animalis* expressing the *Salmonella* flagellin gene for the mucosa immunization in a mouse model. *J. Gene Med.* 2006; 8: 1341–1346.

169. Tallon R, Arias S, Bressollier P, and Urdaci MC. Strain- and matrix-dependent adhesion of *Lactobacillus plantarum* mediated by proteinaceous bacterial compounds. *J. Appl. Microbiol.* 2007; 102: 442–451.
170. Tannock GW. Identification of lactobacilli and bifidobacteria. *Curr. Issues Mol. Biol.* 1999; 1: 53–64.
171. Tannock GW, Tilsala-Timisjarvi A, Rodtong S, Munro Ng J, Munro K, and Alatossava T. Identification of *Lactobacillus* isolates from the gastrointestinal tract, silage, and yogurt by 16S-23S rRNA gene intergenic spacer region. *Appl. Environ. Microbiol.* 1999; 65: 4264–4276.
172. Theunissen J, Britz TJ, Torriani S, and Witthuhn RC. Identification of probiotic microorganisms in South African products using PCR-based DGGE analysis. *Int. J. Food Microbiol.* 2005; 98: 11–21.
173. Tissier MH, Réchérches sur la flore intestinale normale et pathologique du nourisson (thesis). University of Paris, France, 1900, pp. 1–253.
174. Tobes R and Ramos JL. REP code: defining bacterial identity in extragenic space. *Environ. Microbiol.* 2005; 7: 225–228.
175. Torriani S, Felix GE, and Dellaglio F. Differentiation of *Lactobacillus plantarum, L. pentosus*, and *L. paraplantarum* by *recA* gene sequence analysis and multiplex PCR assay with *recA*-derived primers. *Appl. Environ. Microbiol.* 2001; 67: 3450–3454.
176. Torriani S, Clementi F, Vancanneyt M, Hoste B, Dellaglio F, and Kersters K. Differentiation of *Lactobacillus plantarum, L. pentosus* and *L. paraplantarum* species by RAPD-PCR and AFLP. *Syst. Appl. Microbiol.* 2001; 24: 554–560.
177. Tuomola EM and Salminen S. Adhesion of some probiotic and dairy *Lactobacillus* strains to Caco-2 cell cultures. *Int. J. Food Microbiol.* 1998; 41: 45–51.
178. Tuomola E, Crittenden R, Playne M, Isolauri E, and Salminen S. Quality assurance criteria for probiotic bacteria. *Am. J. Clin. Nutr.* 2001; 73: 393S–398S.
179. Tynkkynen S, Satokari R, Saarela M, Mattila-Sandholm T, and Saxelin M. Comparison of ribotyping, randomly amplified polymorphic DNA analysis, and pulsed-field gel electrophoresis in typing of *Lactobacillus rhamnosus* and *L. casei* strains. *Appl. Environ. Microbiol.* 1999; 65: 3908–3914.
180. Ugarte MB, Guglielmotti D, Giraffa G, Reinheimer J, and Hynes E. Nonstarter lactobacilli isolated from soft and semihard Argentinean cheeses: genetic characterization and resistance to biological barriers. *J. Food Prot.* 2006; 69: 2983–2991.
181. Vandamme P, Pot B, Gillis M, de Vos P, Kersters K, and Swings J. Polyphasic taxonomy, a consensus approach to bacterial systematics. *Microbiol. Rev.* 1996; 60: 407–438.
182. Vaughan EE, Heilig HG, Ben-Amor K, and de Vos WM. Diversity, vitality and activities of intestinal lactic acid bacteria and bifidobacteria assessed by molecular approaches. *FEMS Microbiol. Rev.* 2005; 29: 477–490.
183. Vaugien L, Prevots F, and Roques C. Bifidobacteria identification based on 16S rRNA and pyruvate kinase partial gene sequence analysis. *Anaerobe* 2002; 8: 341–344.
184. Vásquez A, Ahrné S, Pettersson B, and Molin G. Temporal temperature gradient gel electrophoresis (TTGE) as a tool for identification of *Lactobacillus casei, Lactobacillus paracasei, Lactobacillus zeae* and *Lactobacillus rhamnosus*. *Lett. Appl. Microbiol.* 2001; 32: 215–219.

185. Venema K and Maathuis AJ. A PCR-based method for identification of bifidobacteria from the human alimentary tract at the species level. *FEMS Microbiol. Lett.* 2003; 224: 143–149.
186. Ventura M, Casas IA, and Morelli L. Rapid amplified ribosomal DNA restriction analysis (ARDRA) identification of *Lactobacillus* spp. isolated from fecal and vaginal samples. *Syst. Appl. Microbiol.* 2000; 23: 504–509.
187. Ventura M, Reniero R, and Zink R. Specific identification and targeted characterization of *Bifidobacterium lactis* from different environmental isolates by a combined multiplex-PCR approach. *Appl. Environ. Microbiol.* 2001; 67: 2760–2765.
188. Ventura M, Elli M, Reniero R, and Zink R. Molecular microbial analysis of *Bifidobacterium* isolates from different environments by the species-specific amplified ribosomal DNA restriction analysis (ARDRA). *FEMS Microbiol. Ecol.* 2001; 36: 113–121.
189. Ventura M and Zink R. Rapid identification, differentiation, and proposed new taxonomic classification of *Bifidobacterium lactis*. *Appl. Environ. Microbiol.* 2002; 68: 6429–6434.
190. Ventura M and Zink R. Specific identification and molecular typing analysis of *Lactobacillus johnsonii* by using PRC-based methods and pulsed-field gel electrophoresis. *FEMS Microbiol. Lett.* 2002; 217: 141–154.
191. Ventura M and Zink R. Comparative sequence analysis of the *tuf* and *recA* genes and restriction fragment length polymorphism of the internal transcribed spacer region sequences supply additional tools for discriminating *Bifidobacterium lactis* from *Bifidobacterium animalis*. *Appl. Environ. Microbiol.* 2003; 69: 7517–7522.
192. Ventura M, Canchaya C, Meylan V, Klaenhammer TR, and Zink R. Analysis, characterization, and loci of the *tuf* genes in *Lactobacillus* and *Bifidobacterium* species and their direct application for species identification. *Appl. Environ. Microbiol.* 2003; 69: 6908–6922.
193. Ventura M, Meylan V, and Zink R. Identification and tracing of *Bifidobacterium* species by use of enterobacterial repetitive intergenic consensus sequences. *Appl. Environ. Microbiol.* 2003; 69: 4296–4301.
194. Ventura M, Canchaya C, Zink R, Fitzgerald GF, and van Sinderen D. Characterization of the *groEL* and *groES* loci in *Bifidobacterium breve* UCC 2003: genetic, transcriptional, and phylogenetic analyses. *Appl. Environ. Microbiol.* 2004; 70: 6197–6209.
195. Ventura M, Canchaya C, van Sinderen D, Fitzgerald GF, and Zink R. *Bifidobacterium lactis* DSM 10140: identification of the *atp* (atpBEFHAGDC) operon and analysis of its genetic structure, characteristics, and phylogeny. *Appl. Environ. Microbiol.* 2004; 70: 3110–3121.
196. Ventura M, van Sinderen D, Fitzgerald GF, and Zink R. Insights into the taxonomy, genetics and physiology of bifidobacteria. *Antonie Van Leeuwenhoek* 2004; 86: 205–223.
197. Ventura M, Canchaya C, Bernini V, Del Casale A, Dellaglio F, Neviani E, Fitzgerald GF, and van Sinderen D. Genetic characterization of the *Bifidobacterium breve* UCC 2003 *hrcA* locus. *Appl. Environ. Microbiol.* 2005; 71: 8998–9007.
198. Ventura M, Zhang Z, Cronin M, Canchaya C, Kenny JG, Fitzgerald GF, and van Sinderen D. The ClgR protein regulates transcription of the *clpP* operon in *Bifidobacterium breve* UCC 2003. *J. Bacteriol.* 2005; 187: 8411–8426.
199. Ventura M, Zink R, Fitzgerald GF, and van Sinderen D. Gene structure and transcriptional organization of the *dnaK* operon of *Bifidobacterium breve* UCC 2003 and application of the operon in bifidobacterial tracing. *Appl. Environ. Microbiol.* 2005; 71: 487–500.

200. Ventura M, Canchaya C, Del Casale A, Dellaglio F, Neviani E, Fitzgerald GF, and van Sinderen D. Analysis of bifidobacterial evolution using a multilocus approach. *Int. J. Syst. Evol. Microbiol.* 2006; 56: 2783–2792.
201. Vincent D, Roy D, Mondou F, and Dery C. Characterization of bifidobacteria by random DNA amplification. *Int. J. Food Microbiol.* 1998; 43: 185–193.
202. Walker DC, Girgis HS, and Klaenhammer TD. The *groESL* chaperone operon of *Lactobacillus johnsonii. Appl. Environ. Microbiol.* 1999; 65: 3033–3041.
203. Wang RF, Beggs ML, Erickson BD, and Cerniglia CE. DNA microarray analysis of predominant human intestinal bacteria in fecal samples. *Mol. Cell Probes* 2004; 18: 223–234.
204. Xanthopoulos V, Ztaliou I, Gaier W, Tzanetakis N, and Litopoulou-Tzanetaki E. Differentiation of *Lactobacillus* isolates from infant faeces by SDS-PAGE and rRNA-targeted oligonucleotide probes. *J. Appl. Microbiol.* 1999; 87: 743–749.
205. Yaeshima T, Takahashi S, Ishibashi N, and Shimamura S. Identification of bifidobacteria from dairy products and evaluation of a microplate hybridization method. *Int. J. Food Microbiol.* 1996; 30: 303–313.
206. Yeung PSM, Sanders ME, Kitts CL, Cano R, and Tong PS. Species-specific identification of commercial probiotic strains. *J. Dairy Sci.* 2002; 85: 1039–1051.
207. Yeung PS, Kitts CL, Cano R, Tong PS, and Sanders ME. Application of genotypic and phenotypic analyses to commercial probiotic strain identity and relatedness. *J. Appl. Microbiol.* 2004; 97: 1095–1104.
208. Yin X, Chambers JR, Barlow K, Park AS, and Wheatcroft R. The gene encoding xylulose-5-phosphate/fructose-6-phosphate phosphoketolase (xfp) is conserved among *Bifidobacterium* species within a more variable region of the genome and both are useful for strain identification. *FEMS Microbiol. Lett.* 2005; 246: 251–257.
209. Zavaglia AG, Kociubinski G, Pérez P, and de Antoni G. Isolation and characterization of *Bifidobacterium strains* from probiotic formulations. *J. Food Prot.* 1998; 61: 865–873.
210. Zhong W, Millsap K, Bialkowska-Hobrzanska H, and Reid G. Differentiation of *Lactobacillus* species by molecular typing. *Appl. Environ. Microbiol.* 1998; 64: 2418–2423.
211. Zhu L, Li W, and Dong X. Species identification of genus *Bifidobacterium* based on partial HSP60 gene sequences and proposal of *Bifidobacterium thermacidophilum* subsp. *porcinum* subsp. nov. *Int. J. Syst. Evol. Microbiol.* 2003; 53: 1619–1623.
212. Ahrne S, Lonnermark E, Wold AE, Aberg N, Hesselmar B, et al. Lactobacilli in the intestinal microbiota of Swedish infants. *Microbes Infect.* 2005; 7: 1256–1262.
213. Apajalahti JHA, Kettunen A, Nurminen PH, Jatila H, and Holben WE. Selective plating understimates abundance and shows differential recovery of bifidobacterial species from human feces. *Appl. Environ. Microbiol.* 2003; 69: 5731–5735.
214. Arroyo L, Cotton LN, and Martin JH. AMC Agar—a composite medium for selective enumeration of *Bifidobacterium longum. Cult. Dairy Prod. J.* 1995; 30: 12–15.
215. Barcina I, Arana I, Santorum P, Iriberri J, and Egea L. Direct viable count of gram-positive and gram-negative bacteria using ciprofloxacin as inhibitor of cellular division. *J. Microbiol. Methods* 1995; 22: 139–150.
216. Bartosch S, Fite A, Macfarlane GT, and McMurdo ME. Characterization of bacterial communities in feces from healthy elderly volunteers and hospitalized elderly patients by

using real-time PCR and effects of antibiotic treatment on the fecal microbiota. *Appl. Environ. Microbiol.* 2004; 70: 3575–3581.

217. Belanger SD, Boissinot M, Clairoux N, Picard FJ, and Bergeron MG. Rapid detection of *Clostridium difficile* in feces by real-time PCR. *J. Clin. Microbiol.* 2003; 41: 730–734.

218. Ben-Amor K, Breeuwer P, Verbaarschot P, Rombouts FM, and Akkermans AD, et al. Multiparametric flow cytometry and cell sorting for the assessment of viable, injured, and dead *Bifidobacterium* cells during bile salt stress. *Appl. Environ. Microbiol.* 2002; 68: 5209–5216.

219. Ben-Amor K, Heilig H, Smidt H, Vaughan EE, Abee T, and de Vos WM. Genetic diversity of viable, injured, and dead fecal bacteria assessed by fluorescence-activated cell sorting and 16S rRNA gene analysis. *Appl. Environ. Microbiol.* 2005; 71: 4679–4689.

220. Bibiloni R, Gómez Zavaglia A, and de Anthony G. Enzyme-based most probable number method for the enumeration of *Bifidobacterium* in dairy products. *J. Food Prot.* 2001; 64: 2001–2006.

221. Bonaparte C, Klein G, Kneifel W, and Reuter G. Development of a selective culture medium for the enumeration of bifidobacteria in fermented milks [French]. *Lait* 2001; 81: 227–235.

222. Brandt K and Alatossava T. Specific identification of certain probiotic *Lactobacillus rhamnosus* strains with PCR primers based on phage related sequences. *Int. J. Food Microbiol.* 2003; 84: 189–196.

223. Camaschella P, Mignot O, Pirovano F, and Sozzi T. Method for differentiated enumeration of mixed cultures of thermophilic lactic acid bacteria and bifidobacteria by using only one culture medium. *Lait* 1998; 78: 461–467.

224. Collado MC, Moreno Y, Cobo JM, Mateos JA, and Hernandez M. Molecular detection of *Bifidobacterium animalis* DN-173010 in human feces during fermented milk administration. *Food Res. Int.* 2006; 39: 530–535.

225. Conte MP, Schippa S, Zamboni I, Penta M, and Chiarini F, et al. Gut-associated bacterial microbiota in paediatric patients with inflammatory bowel disease. *Gut* 2006; 55: 1760–1767.

226. Corthier G, Muller MC, L'Haridon R. Selective enumeration of *Bacteroides vulgatus* and *Bacteriodes distasonis* organisms in the predominant human fecal flora by using monoclonal antibodies. *Appl. Environ. Microbiol.* 1996; 62: 735–738.

227. Dave RI and Shah NP. Viability of yoghurt and probiotic bacteria in yoghurts made from commercial starter cultures. *Int. Dairy J.* 1997; 7: 31–41.

228. Delgado S, Ruas-Madiedo P, Suárez A, and Mayo B. Interindividual differences in microbial counts and biochemical-associated variables in the feces of healthy Spanish adults. *Digest Dis. Sci.* 2006; 51: 737–743.

229. Delgado S, Suárez A, and Mayo B. Bifidobacterial diversity determined by culturing and by 16S rDNA sequence analysis in feces and mucosa from ten healthy Spanish adults. *Digest Dis, Sci.* 2006; 51: 1878–1885.

230. Delgado S, Suárez A, and Mayo B. Identification of dominant bacteria in feces and colonia mucosa from healthy Spanish adults by culturing and by 16S rDNA sequence analysis. *Digest Dis, Sci.* 2006; 51: 744–751.

231. Delgado S, Suárez A, Otero L, and Mayo B. Variation of microbiological and biochemical parameters in the feces of two healthy people over a 15 day period. *Eur. J. Nutr.* 2004; 43: 375–380.

232. Dore J, Sghir A, Hannequart-Gramet G, Corthier G, and Pochart P. Design and evaluation of a 16S rRNA-targeted oligonucleotide probe for specific detection and quantitation of human faecal *Bacteroides* populations. *Syst. Appl. Microbiol.* 1998; 21: 65–71.
233. Eckburg PB, Bik EM, Bernstein CN, Purdom E, and Dethlefsen L, et al. Diversity of the human intestinal microbial flora. *Science* 2005; 308: 1635–1638.
234. Eerola E and Lehtonen OP. Optimal data processing procedure for automatic bacterial identification by gas-liquid chromatography of cellular fatty acids. *J. Clin. Microbiol.* 1988; 26: 1745–1753.
235. Egert M, de Graaf AA, Maathuis A, de Waard P, and Plugge CM, et al. Identification of glucose fermenting bacteria present in an in vitro model of the human intestine by RNA-stable isotope probing. *FEMS Microbiol. Ecol.* 2007; 60: 126–135.
236. Favier CF, Vaughan EE, de Vos WM, and Akkermans ADL. Molecular monitoring of succession of bacterial communities in human neonates. *Appl, Environ, Microbiol.* 2002; 68: 219–226.
237. Favier CF, de Vos WM, and Akkermans ADL. Development of bacterial and bifidobacterial communities in feces of newborn babies. *Anaerobe* 2003; 9: 219–229.
238. Fernandez L, Marin ML, Langa S, Martin R, and Reviriego C, et al. A novel genetic label for detection of specific probiotic lactic acid bacteria. *Food Sci. Tech. Int.* 2004; 10: 101–108.
239. Fite A, Macfarlane GT, Cummings JH, Hopkins MJ, and Kong SC, et al. Identification and quantification of mucosal and faecal desulfovibrios using real time polymerase chain reaction. *Gut* 2004; 53: 523–529.
240. Fitzsimons NA, Akkermans ADL, de Vos WM, and Vaughan EE. Bacterial gene expresion detected in human faeces by reverse transcription PCR. *J. Microbiol. Methods* 2003; 55: 133–140.
241. Franks AH, Harmsen HJM, Raangs GC, Jansen GJ, Schut F, and Welling GW. Variations of bacterial populations in human feces measured by fluorescent *in situ* hybridisation with group-specific 16S rRNA-targeted oligonucleotido probes. *Appl. Environ. Microbiol.* 1998; 64: 3336–3345.
242. Fujiwara S, Seto Y, Kinura A, and Hashiba H. Establishment of orally-administered *Lactobacillus gasseri* SBT2055SR in the gastrointestinal tract of humans and its influence on intestinal microbiota and metabolism. *J. Appl. Microbiol.* 2001; 90: 343–352.
243. Fukushima Y, Yamano T, Kusano A, Takada M, Amano M, and Iino H. Effect of fermented milk containing *Lactobacillus johnsonii* La1 (LC1) on defecation in healthy Japanese adults—a double blind placebo controlled study. *Biosci. Microflora* 2004; 23: 139–147.
244. Furrie E. A molecular revolution in the study of intestinal microflora. *Gut* 2006; 55: 141–143.
245. Gill SR, Pop M, DeBoy RT, Eckburg PB, Turnbaugh PJ, et al. Metagenomic analysis of the human distal gut microbiome. *Science* 2006; 312: 1355–1359.
246. Goldenberg O, Herrmann S, Marjoram G, Noyer-Weidner M, and Hong G, et al. Molecular monitoring of the intestinal flora by denaturing high performance liquid chromatography. *J. Microbiol. Methods* 2007; 68: 94–105.
247. Goossens DAM, Jonkers DMAE, Russel MGVM, Stobberingh EE, and Stockbrügger RW. The effect of a probiotic drink with *Lactobacillus plantarum* 299V on the bacterial composition of faeces and mucosal biopsies of rectum and ascending colon. *Aliment Pharmacol. Ther.* 2006; 23: 255–263.

248. Gueimonde M, Tolkko S, Korpimaki T, and Salminen S. New real-time quantitative PCR procedure for quantification of bifidobacteria in human faecal samples. *Appl. Environ. Microbiol.* 2004; 70: 4165–4169.

249. Gueimonde M, Delgado S, Mayo B, Ruas-Madiedo P, Margolles A, and de los Reyes-Gavilán CG. Viability and diversity of probiotic *Lactobacillus* and *Bifidobacterium* populations included in commercial fermented milks. *Food Res. Int.* 2004; 37: 839–850.

250. Gueimonde M, Kalliomäki M, Isolauri E, and Salminen S. Probiotic intervention in neonates—Will permanent colonisation ensue? *J. Pediatr. Gastroenterol. Nutr.* 2006; 42: 604–606.

251. Gueimonde M, Debor L, Tölkkö S, Jokisalo E, and Salminen S. Quantitative assessment of faecal bifidobacterial populations by using novel real-time PCR assays. *J. Appl. Microbiol.* 2007; 102: 1116–1122.

252. Gueimonde M, Ouwehand A, Huhtinen H, Salminen E, and Salminen S. Qualitative and quantitative analyses of the bifidobacterial microbiota in the colonic mucosa of patients with colorectal cancer, inflammatory bowel disease and diverticulitis. *World J. Gastroenterol.* 2007b; 13: 3985–3989.

253. Haarman M and Knol J. Quantitative real-time PCR assays to identify and quantify fecal *Bifidobacterium* species in infants receiving a prebiotic infant formula. *Appl. Environ. Microbiol.* 2005; 71: 2318–2324.

254. Haarman M and Knol J. Quantitative real-time PCR analysis of fecal *Lactobacillus* species in infants receiving a prebiotic infant formula. *Appl. Environ. Microbiol.* 2006; 72: 2359–2365.

255. Hadadji M, Benama R, Saidi N, Henni DE, and Kihal M. Identification of cultivable *Bifidobacterium* species isolated from breast-fed infants feces in West-Algeria. *African J. Biotechnol.* 2005; 4: 422–430.

256. Harmsen HJ, Wildeboer-Veloo AC, Raangs GC, Wagendorp AA, Klijn N, et al. Analysis of intestinal flora development in breast-fed and formula-fed infants by using molecular identification and detection methods. *J. Pediatr. Gastroenterol. Nutr.* 2000; 30: 61–67.

257. Harmsen HJM, Gibson GR, Elfferich P, Raangs GC, Wildeboer-Veloo ACM, et al. Comparison of viable cell counts and fluorescence in situ hybridisation using specific rRNA-based probes for the quantification of human fecal bacteria. *FEMS Microbiol. Lett.* 1999; 183: 125–129.

258. Harmsen HJM, Raangs GC, He T, Degener JE, and Welling GW. Extensive set of 16S rRNA-based probes for detection of bacteria in human feces. *Appl. Environ. Microbiol.* 2002; 68: 2982–2990.

259. Hayashi H, Sakamoto M, and Benno Y. Phylogenetic analysis of the human gut microbiota using 16S rDNA clone libraries and strictly anaerobic culture based methods. *Microbiol. Immunol.* 2002; 46: 535–548.

260. Hayashi H, Sakamoto M, and Benno Y. Fecal microbial diversity in a strict vegetarian as determined by molecular analysis and cultivation. *Microbiol. Immunol.* 2002; 46: 819–831.

261. Hayashi H, Sakamoto M, Kitahara M, and Benno Y. Molecular analysis of fecal microbiota in elderly individuals using 16S rDNA library and T-RFLP. *Microbiol. Immunol.* 2003; 47: 557–570.

262. Heilig HGHJ, Zoetendal EG, Vaughan EE, Marteau P, Akkermans ADL, and de Vos WM. Molecular diversity of *Lactobacillus* spp. and other lactic acid bacteria in the human intestine as determined by specific amplification of 16S ribosomal DNA. *Appl. Environ. Microbiol.* 2002; 68: 114–123.

263. Hopkins MJ and MacFarlane GT. Changes in predominant bacterial populations in human faeces with age and with *Clostridium difficile* infection. *J. Med. Microbiol.* 2002; 51: 448–454.

264. Hopkins MJ, Sharp R, and Macfarlane GT. Age and disease related changes in intestinal bacterial populations assessed by cell culture, 16S rRNA abundance, and community cellular fatty acid profiles. *Gut* 2001; 48: 198–205.

265. Huijsdens XW, Linskens RK, Mak M, Meuwissen SGM, Vandenbroucke-Grauls CMJE, and Savelkoul PHM. Quantification of bacteria adherent to gastrointestinal mucosa by real-time PCR. *J. Clin. Microbiol.* 2002; 40: 4423–4427.

266. IDF. *Detection and enumeration of* Lactobacillus acidophilus. Bulletin No. 306. International Dairy Federation, Brussels, Belgium, 1995.

267. Jansen GJ, Wildeboer-Veloo ACM, Tonk RHJ, Franks AH, and Welling GW. Development and validation of an automated, microscopy-based method for enumeration of groups of intestinal bacteria. *J. Microbiol. Methods* 1999; 37: 215–221.

268. Jones CM and Thies JE. Soil microbial community analysis using two-dimensional polyacrilamide get electrophoresis of the bacterial ribosomal internal transcribed spacer regions. *J. Microbiol. Methods* 2007; 69: 256–267.

269. Kalliomäki M, Kirjavainen P, Eerola E, Kero P, Salminen S, and Isolauri E. Distinct patterns of neonatal gut microflora in infants in whom atopy was and was not developing. *J. Allergy Clin. Immunol.* 2001; 107: 129–134.

270. Kaprelyants AS and Kell DB. Dormancy in stationary-phase cultures of *Micrococcus luteus*: flow cytometric analysis of starvation and resuscitation. *Appl. Environ. Microbiol.* 1993; 59: 3187–3196.

271. Kaufmann P, Pfefferkorn A, Teuber M, and Meile L. Identification and quantification of bifidobacterium species isolated from food with genus-specific 16S rRNA-targeted probes by colony hybridisation and PCR. *Appl. Environ. Microbiol.* 1997; 63: 1268–1273.

272. Khan AA, Nawaz MS, Robertson L, Khan SA, and Cerniglia CE. Identification of predominant human and animal anaerobic intestinal bacterial species by terminal restriction fragment patterns (TRFP's): a rapid, PCR-based method. *Mol. Cell Probe* 2001; 15: 349–355.

273. Kirjavainen PV, Arvola T, Salminen SJ, and Isolauri E. Aberrant composition of gut microbiota of allergic infants: a target of bifidobacterial therapy at weaning? *Gut* 2002; 51: 51–55.

274. Klaassens ES, de Vos WM, and Vaughan EE. Metaproteomics approach to study the functionality of the microbiota in the human gastrointestinal tract. *Appl. Environ. Microbiol.* 2007; 73: 1388–1392.

275. Klingberg TD and Budde BB. The survival and persistence in the human gastrointestinal tract of five potential probiotic lactobacilli consumed as freeze-dried cultures or as probiotic sausage. *Int. J. Food Microbiol.* 2006; 109: 157–159.

276. Lahtinen SJ. New insights into the viability of probiotic bacteria. PhD Thesis. Functional Foods Forum, Department of Biochemistry and Food Chemistry, University of Turku, Finland, 2007.

277. Lahtinen SJ, Gueimonde M, Ouwehand AC, Reinikainen JP, and Salminen SJ. Probiotic bacteria may become dormant during storage. *Appl.Environ. Microbiol.* 2005; 71: 1662–1663.
278. Lahtinen SJ, Gueimonde M, Ouwehand AC, Reinikainen JP, and Salminen SJ. Comparison of four methods to enumerate probiotic bifidobacteria in a fermented food product. *Food Microbiol.* 2006; 23: 571–577.
279. Lahtinen SJ, Ouwehand AC, Reinikainen JP, Korpela JM, Sandholm J, and Salminen SJ. Intrinsic properties of so-called dormant probiotic bacteria, determined by flow cytometric viability assays. *Appl. Environ. Microbiol.* 2006; 72: 5132–5134.
280. Langendijk PS, Schut F, Jansen GJ, Raangs GC, and Kamphuis GR, et al. Quantitative fluorescence in situ hybridisation of *Bifidobacterium* ssp. with genus-specific 16S rRNA-targeted probes and its application in fecal samples. *Appl. Environ. Microbiol.* 1995; 61: 3069–3075.
281. Lankaputhra WEV and Shah NP. A simple method for selective enumeration of *Lactobacillus acidophilus* in yogurt supplemented with *L. acidophilus* and *Bifidobacterium* spp. *Milchwissenschaft.* 1996; 5: 446–451.
282. Lapierre L, Undeland P, and Cox LJ. Lithium chloride-sodium propionate agar for the enumeration of bifidobacteria in fermented dairy products. *J. Dairy Sci.* 1992; 75: 1192–1196.
283. Larsen CN, Nielsen S, Kaestel P, Brockmann E, and Bennedsen M, et al. Dose-response study of probiotic bacteria *Bifidobacterium animalis* subsp *lactis* BB-12 and *Lactobacillus paracasei* subsp *paracasei* CRL-341 in healthy young adults. *Eur. J. Clin. Nutr.* 2006; 60: 1284–1293.
284. Lyons SR, Griffen AL, and Leys EJ. Quantitative real-time PCR for *Porphyromonas gingivalis* and total bacteria. *J. Clin. Microbiol.* 2000; 38: 2362–2365.
285. Malinen E, Matto J, Salmitie M, Alander M, Saarela M, and Palva A. PCR-ELISA II: Analysis of *Bifidobacterium* populations in human faecal samples from a consumption trial with *Bifidobacterium lactis* Bb-12 and a galacto-oligosaccharide preparation. *Syst. Appl. Microbiol.* 2002; 25: 249–2458.
286. Manichanh C, Rigottier-Gois L, Bonnaud E, Gloux K, and Pelletier E, et al. Reduced diversity of faecal microbiota in Cronh's disease revealed by a metagenomic approach. *Gut* 2006; 55: 205–211.
287. Marteau P, Pochart P, Doré J, Béra-Maillet C, Bernalier A, and Corthier G. Comparative study of bacterial groups within the human cecal and fecal microbiota. *Appl. Environ. Microbiol.* 2001; 67: 4939–4942.
288. Matijasic BB, Stojkovic S, Salobir J, Malovrh S, and Rogelj I. Evaluation of the *Lactobacillus gasseri* K7 and LF221 strains in weaned piglets for their possible probiotic use and their detection in the faeces. *Anim. Res.* 2004; 53: 35–44.
289. Matsuki T, Watanabe K, Fujimoto J, Miyamoto Y, and Takada T, et al. Development of 16S rRNA-gene-targeted group-specific primers for the detection and identification of predominant bacteria in human feces. *Appl. Environ. Microbiol.* 2002; 68: 5445–5451.
290. Matsuki T, Watanabe K, Fujimoto J, Takada T, and Tanaka R. Use of 16S rRNA-gene-targeted group-specific primers for real-time PCR analysis of predominant bacteria in human feces. *Appl. Environ. Microbiol.* 2004; 70: 7220–7228.
291. Matsuki T, Watanabe K, and Tanaka R. Genus- and species-specific PCR primers for the detection and identification of bifidobacteria. *Curr. Issues Intest. Microbiol.* 2003; 4: 61–69.

292. Matsuki T, Watanabe K, Tanaka R, Fukuda M, and Oyaizu H. Distribution of bifidobacterial species in human intestinal microflora examined with 16S rRNA-gene targeted species-specific primers. *Appl. Environ. Microbiol.* 1999; 65: 4506–4512.

293. Mätto J, Fondén R, Tolvanen T, von Wright A, Vilpponen-Salmela T, et al. Intestinal survival and persistence of probiotic *Lactobacillus* and *Bifidobacterium* strains administered in triple-strain yoghurt. *Int. Dairy J.* 2006; 16: 1174–1180.

294. Maukonen J, Mätto J, Satokari R, Söderlund H, Mattila-Sandholm T, and Saarela M. PCR DGGE and RT-PCR DGGE show diversity and short-term temporal stability in the *Clostridium coccoides-Eubacterium rectale* group in the human intestinal microbiota. *FEMS Microbiol. Ecol.* 2006; 58: 517–528.

295. Midtvedt AC, Carlstedt-Duke B, Norin H, Saxerholt H, and Midtvedt T. Development of five metabolic activities associated with the intestinal flora in healthy infants. *J. Pediatr. Gastroenterol. Nutr.* 1998; 7: 559–567.

296. Mohan R, Koebnick C, Schildt J, Schmidt S, Mueller M, et al. Effects of *Bifidobacterium lactis* Bb12 supplementation on intestinal microbiota of preterm infants: a double-blind, placebo-controlled, randomized study. *J. Clin. Microbiol.* 2006; 44: 4025–4031.

297. Morelli L, Garbagna N, Rizzello F, Zonenschain D, and Grossi E. *In vivo* association to human colon of *Lactobacillus paracasei* B21060: Map from biopsies. *Digest Liver Dis.* 2006; 38: 894–898.

298. Moreno Y, Collado MC, Ferrús MA, Cobo JM, Hernández E, and Hernández M. Viability assessment of lactic acid bacteria in commercial dairy products stored at 4°C using LIVE/DEAD® BacLight™ staining and conventional plate counts. *Int. J. Food Sci. Technol.* 2006; 41: 275–280.

299. Nadkarni MA, Martin FE, Jacques NA, and Hunter N. Determination of bacterial load by real-time PCR using a broad range (universal) probe and primers set. *Microbiology* 2002; 148: 257–266.

300. Nebra Y and Blanch AR. A new selective medium for *Bifidobacterium* spp. *Appl. Environ. Microbiol.* 1999; 65: 5173–5176.

301. Nystrom T. Aging in bacteria. *Curr. Opin. Microbiol.* 2002; 5: 596–601.

302. Oozeer R, Leplingard A, Mater DDG, Mogenet A, and Michelin R, et al. Survival of *Lactobacillus casei* in the human digestive tract after consumption of fermented milk. *Appl. Environ. Microbiol.* 2006; 72: 5615–5617.

303. Palmer C, Bik EM, Eisen MB, Eckburg PB, and Sana TR, et al. Rapid quantitative profiling of complex microbial populations. *Nucl. Acids Res.* 2006; 10(34): e5.

304. Ravula RR and Shah NP. Effect of acid casein hydrolysate and cysteine on the viability of yogurt and probiotic bacteria in fermented frozen dairy desserts. *Aust. J. Dairy Technol.* 1998; 53: 175–179.

305. Requena T, Burton J, Matsuki T, Munro K, Simon MA, et al. Identification, detection and enumeration of human *Bifidobacterium* species by PCR targeting the transaldolase gene. *Appl. Environ. Microbiol.* 2002; 68: 2420–2427.

306. Rollins DM and Colwell RR. Viable but non-culturable stage of *Campylobacter jejuni* and its role in survival in the natural aquatic environment. *Appl. Environ. Microbiol.* 1986; 52: 531–538.

307. Roy D. Media for the isolation and enumeration of bifidobacteria in dairy products. *Int. J. Food Microbiol.* 2001; 69: 167–182.

308. Sakamoto M, Hayashi H, and Benno Y. Terminal restriction fragment length polymorphism analysis for human fecal microbiota and its application for analysis of complex bifidobacterial communities. *Microbiol. Immunol.* 2003; 47: 133–142.

309. Salminen S, Gibson GR, McCartney AL, and Isolauri E. Influence of mode of delivery on gut microbiota composition in seven year old children. *Gut* 2004; 53: 1388–1389.

310. Salminen S, Nurmi J, and Gueimonde M. The genomics of probiotic intestinal microorganisms. *Genome Biol.* 2005; 6: 225.

311. Satokari RM, Vaughan EE, Akkermans ADL, Saarela M, and de Vos WM. Bifidobacterial diversity in human feces detected by genus specific PCR and denaturing gradient gel electrophoresis. *Appl. Environ. Microbiol.* 2001; 67: 504–513.

312. Schmalenberger A, Schwieger F, and Tebbe CC. Effect of primers hybridising to different evolutionary conserved regions of the small-subunit rRNA gene in PCR-based microbial community analyses and genetic profiling. *Appl. Environ. Microbiol.* 2001; 67: 3557–3563.

313. Schwiertz A, Le Blay G, and Blaut M. Quantitation of different *Eubacterium* spp. in human fecal samples with species-specific 16S rRNA-targeted oligonucloeotide probes. *Appl. Environ. Microbiol.* 2000; 66: 375–382.

314. Sghir A, Gramet G, Suau A, Rochet V, Pochart P, and Dore J. Quantification of bacterial groups within human fecal flora by oligonucleotide probe hybridisation. *Appl. Environ Microbiol.* 2000; 66: 2263–2266.

315. Shah NP. Probiotic bacteria: Selective enumeration and survival in dairy foods. *J. Dairy Sci.* 2000; 83: 894–907.

316. Shah NP, Lankaputhra WEV, Britz ML, and Kyle WSA. Survival of *Lactobacillus acidophilus* and *Bifidobacterium bifidum* in commercial yoghurt during refrigerated storage. *Int. Dairy J.* 1995; 5: 515–521.

317. Sneath P (editor). *Bergey's Manual of Systematic Bacteriology.* Williams & Wilkins Co, Baltimore, MD, 1986.

318. Su P, Henriksson A, and Mitchell H. Survival and retention of the probiotic *Lactobacillus casei* LAFTI L26 in the gastrointestinal tract of the mouse. *Lett. Appl. Microbiol.* 2007; 44: 120–125.

319. Suau A, Bonnet R, Sutren M, Godon J-J, and Gibson G, et al. Direct analysis of genes encoding 16S RNA from complex communities reveals many novel molecular species within the human gut. *Appl. Environ. Microbiol.* 1999; 65: 4799–4807.

320. Takada T, Matsumoto K, and Nomoto K. Develpment of multi-color FISH method for análisis of seven *Bifidobacterium* species in human feces. *J. Microbiol. Methods* 2004; 58: 413–421.

321. Talwalkar A and Kailasapathy K. Comparison of selective and differential media for the accurate enumeration of strains of *Lactobacillus acidophilus, Bifidobacterium* spp. and *Lactobacillus casei* complex from commercial yoghurts. *Int. Dairy J.* 2004; 14: 143–149.

322. Tannock GW, Munro K, Bibiloni R, Simon MA, and Hargreaves P, et al. Impact of consumption of oligosaccharide-containing biscuits on the fecal microbiota of humans. *Appl. Environ. Microbiol.* 2004; 70: 2129–2136.

323. Tannock GW, Munro K, Harmsen HJM, Welling GW, Smart J, and Gopal PK. Analysis of the fecal microflora of human subjects consuming a probiotic product containing *Lactobacillus rhamnosus* DR20. *Appl. Environ. Microbiol.* 2000; 66: 2578–2588.

324. Tapiainen T, Ylitalo S, Eerola E, and Uhari M. Dynamics of gut colonization and source of intestinal flora in healthy newborn infants. *APMIS* 2006; 114: 812–817.
325. Tuohy KM, Finlay RK, Wynne AG, and Gibson GR. A human volunteer study on the prebiotic effects of HP-Inulin—Fecal bacteria enumerated using fluorescent *in situ* hybridisation (FISH). *Anaerobe* 2001; 7: 113–118.
326. Van de Casteele S, Vanheuverzwijn T, Ruyssen T, Van Assche P, Swings J, and Huys G. Evaluation of culture media for selective enumeration of probiotic strains of lactobacilli and bifidobacteria in combination with yoghurt or cheese starters. *Int. Dairy J.* 2006; 16: 1470–1476.
327. Vanhoutte T, Huys G, De Brandt E, and Swings J. Temporal stability analysis of the microbiota in human feces by denaturing gradient gel electrophoresis using universal and group-specific 16S rRNA gene primers. *FEMS Microb. Ecol.* 2004; 48: 437–446.
328. Vinderola CG and Reinheimer JA. Culture media for the enumeration of *Bifidobacterium bifidum* and *Lactobacillus acidophilus* in the presence of yoghurt bacteria. *Int. Dairy J.* 1999; 9: 497–505.
329. Vinderola CG, Bailo N, and Reinheimer JA. Survival of probiotic microflora in Argentinian yoghurts during refrigerated storage. *Food Res. Int.* 2000; 33: 97–102.
330. Vlková E, Nevoral J, Jencikova B, Kopecny J, and Godefrooij J, et al. Detection of infant faecal bifidobacteria by enzymatic methods. *J. Microbiol. Methods* 2005; 60: 365–373.
331. Vulevic J, McCartney AL, Gee JM, Johnson IT, and Gibson GR. Microbial species involved in production of 1,2-*sn*-diacylglycerol and effects of phosphatidylcholine on human fecal microbiota. *Appl. Environ. Microbiol.* 2004; 70: 5659–5666.
332. Walter J, Hertel C, Tannock GW, Lis CM, Munro K, and Hammes WP. Detection of *Lactobacillus*, *Pediococcus*, *Leuconostoc*, and *Weissella* species in human feces by using group-specific PCR primers and denaturing gradient gel electrophoresis. *Appl. Environ. Microbiol.* 2001; 67: 2578–2585.
333. Wang RF, Beggs ML, Erikson BD, and Cerniglia CE. DNA microarray analysis of predominant human intestinal bacteria in fecal samples. *Mol. Cell Probe* 2004; 18: 223–234.
334. Wang RF, Beggs ML, Robertson LH, and Cerniglia CE. Design and evaluation of oligonucleotide-microarray method for the detection of human intestinal bacteria in fecal samples. *FEMS Microbiol. Lett.* 2002; 213: 175–182.
335. Wang RF, Cao WW, and Cerniglia CE. PCR detection and quantitation of predominant anaerobic bacteria in human and animal fecal samples. *Appl. Environ. Microbiol.* 1996; 62: 1242–1247.
336. Wang RF, Kim SJ, Robertson LH, and Cerniglia CE. Development of a membrane-array method for the detection of human intestinal bacteria in fecal samples. *Mol. Cell Probe* 2002; 16: 341–350.
337. Wei G, Pan L, Du H, Chen J, and Zhao L. ERIC-PCR fingerprinting-based community DNA hybridization to pinpoint genome-specific fragments as molecular markers to identify and track populations common to healthy human guts. *J. Microbiol. Methods* 2004; 59: 91–108.
338. Wildeboer-Veloo ACM, Harmsen HJM, Degener JE, and Welling GW. Development of a 16S rRNA-based probe for *Clostridium ramosum*. *C. spiroforme* and *C. cocleatum* and its application for the quantification in human faeces from volunteers of different age groups *Microbiol. Ecol. Health Dis.* 2003; 15: 131–136.

339. Wilson KH and Blitchington RB. Human colonic biota studied by ribosomal DNA sequence analysis. *Appl. Environ. Microbiol.* 1996; 62: 2273–2278.
340. Wintzingerode FV, Göbel UB, and Stackebrandt E. Determination of microbial diversity in environmental samples: pitfalls of PCR-based rRNA analysis. *FEMS Microbiol. Rev.* 1997; 21: 213–229.
341. Woodmansey E, McMurdo MET, Macfarlane GT, and Macfarlane S. Comparison of compositions and metabolic activities of fecal microbiotas in young adults and in antibiotic-treated and non-antibiotic-treated elderly subjects. *Appl. Environ. Microbiol.* 2004; 70: 6113–6122.
342. Xiao W and Oefner PJ. Denaturing high-performace liquid chromatography: a review. *Hum. Mutat.* 2001; 17: 439–474.
343. Yamamoto T, Morotomi M, and Tanaka R. Species-specific oligonucleotide probes for five *Bifidobacterium* species detected in human intestinal microflora. *Appl. Environ. Microbiol.* 1992; 58: 4076–4079.
344. Yamano T, Iino H, Takada M, Blue S, Rochat F, and Fukushima Y. Improvement of the human intestinal flora by ingestion of the probiotic strain *Lactobacillus johnsonii* La1. *Br. J. Nutr.* 2006; 95: 303–312.
345. Yaqub S, Anderson JG, MacGregor SJ, and Rowan NJ. Use of a fluorescent viability stain to assess lethal and sublethal injury in food-borne bacteria exposed to high-intensity pulsed electric fields. *Lett. Appl. Microbiol.* 2004; 39: 246–251.
346. Yuki N, Watanabe K, Mike A, Tagami Y, and Tanaka R, et al. Survival of probiotic, *Lactobacillus casei* strain Shirota, in the gastrointestinal tract: selective isolation from faeces and identification using monoclonal antibodies. *Int. J. Food Microbiol.* . 48 : 1999; 51–57.
347. Zinkevich V and Beech IB. Screening of sulphate-reducing bacteria in colonoscopy samples from healthy and colitic human gut mucosa. *FEMS Microbiol. Ecol.* 2000; 34: 147–155.
348. Zoetendal EG, Akkermans ADL, and de Vos WM. Temperature gradient gel electrophoresis analysis of 16S rRNA from human fecal samples reveals stable and host-specific communities of active bacteria. *Appl. Environ. Microbiol.* 1998; 64: 3854–3859.
349. Zoetendal EG, Ben-Amor K, Harmsen HJ, Schut F, Akkermans AD, and de Vos WM. Quantification of uncultured Runinococcus obeum-like bacteria in human fecal samples by fluorescent in situ hybridisation and flow cytometry using 16S rRNA-targeted probes. *Appl. Environ. Microbiol.* 2002; 68: 4225–4232.
350. Zoetendal EG, von Wright A, Vilpponen-Salmela T, Ben-Amor K, Akkermans ADL, and de Vos WM. Mucosa-associated bacteria in the human gastrointestinal tract are uniformly distributed along the colon and differ from the community recovered from feces. *Appl. Environ. Microbiol.* 2002; 68: 3401–3407.
351. Abdo Z, Schuette U, Bent S, Williams C, Forney L, and Joyce P. Statistical methods for characterizing diversity of microbial communities by analysis of terminal restriction fragment length polymorphisms of 16S rRNA genes. *Environ. Microbiol.* 2006; 8: 929–938.
352. Bernhard AE and Field KG. A PCR assay to discriminate human and ruminant feces on the basis of host differences in Bacteroides-Prevotella genes encoding 16S rRNA. *Appl. Environ. Microbiol.* 2000; 66: 4571–4574.
353. Braid MD, Daniels L, and Kitts CL. Removal of PCR inhibitors from soil DNA by chemical flocculation. *J. Microbiol. Methods* 2003; 52: 389–393.

354. Danovaro R, Luna GM, Dell'Anno A, and Pietrangeli B. Comparison of two fingerprinting techniques, terminal restriction fragment length polymorphism and automated ribosomal intergenic spacer analysis, for determination of bacterial diversity in aquatic environments. *Appl. Environ. Microbol.* 2006; 72: 5982–5989.

355. Egert M and Friedrich M. Formation of pseudo-terminal restriction fragments, a PCR-related bias affecting terminal restriction fragment length polymorphism analysis of microbial community structure. *Appl. Environ. Microbiol.* 2003; 69: 2555–2562.

356. Gianninò V, Santagati M, Guardo G, Cascone C, Rappazzo G, and Stefani S. Conservation of the mosaic structure of the four internal transcribed spacers and localization of the *rrn* operons on the *Streptococcus pneumoniae* genome. *FEMS Microbiol. Lett.* 2003; 223: 245–252.

357. Hayashi K. PCR-SSCP: A simple and sensitive method for detection of mutations in the genomic DNA. *Genome Res.* 1991; 1: 34–38.

358. Horz HP, Vianna M, Gomes B, and Conrads G. Evaluation of universal probes and primer sets for assessing total bacterial load in clinical samples: General implications and practical use in endodontic antimicrobial therapy. *J. Clin. Microbiol.* 2005; 43: 5332–5337.

359. Jernberg C, Sullivan A, Edlund C, and Jansson JK. Monitoring of antibiotic-induced alterations in the human intestinal microflora and detection of probiotic stains by use of terminal restriction fragment length polymorphism. *Appl. Environ. Microbiol.* 2005; 71: 501–506.

360. Kibe R, Sakamoto M, Hayashi H, Yokota H, and Benno Y. Maturation of the murine cecal microbiota as revealed by terminal restriction fragment length polymorphism and 16S rRNA gene clone libraries. *FEMS Microbiol. Lett.* 2004; 235: 139–146.

361. Kitts CL. Terminal restriction fragment patterns: a tool for comparing microbial communities and assessing community dynamics. *Curr. Issues Intest. Microbiol.* 2001; 2: 17–25.

362. Lee D-H, Zo Y-G, and Kim S-J. Nonradioactive method to study genetic profiles of natural bacterial communities by PCR-single strand-conformation polymorphism. *Appl. Environ. Microbiol.* 1996; 62: 3112–3120.

363. Li F, Hullalr MAJ, and Lampe JW. Optimization of terminal restriction fragment polmorphorphism (TRFLP) analysis of human gut microbiota. *J. Microbiol. Methods* 2007; 68: 303–311.

364. Liu W, Marsh T, Cheng H, and Forney LJ. Characterization of microbial diversity by determining terminal restriction fragment length polymorphisms of genes encoding 16S rRNA. *Appl. Environ. Microbiol.* 1997; 63: 4516–4522.

365. Maidak BL, Cole JR, Lilburn TG, Parker CT, Saxman PR, Farris RJ, Garrity GM, Olsen GJ, Schmidt TM, and Tiedje JM. The RDP-II (Ribosomal Database Project). *Nucl. Acids Res.* 2001; 29: 173–174.

366. Marsh TL. Terminal restriction fragment length polymorphism (T-RFLP): A merging method for characterizing diversity among homologous populations of amplification products. *Curr. Opin. Microbiol.* 1999; 2: 323–327.

367. Marsh TL, Saxman P, Cole J, and Tiedje T. Terminal restriction fragment length polymorphism analysis program, a web-based research tool for microbial community analysis. *Appl. Environ. Microbiol.* 2000; 66: 3616–3620.

368. Mills DK, Entry J, Gillevet P, and Mathee K. Assessing microbial community diversity using amplicon length heterogeneity polymerase chain reaction. *Soil Sci. Soc. Am. J.* 2007; 71: 572–578.

369. Muyzer G. DGGE/TGGE: a method for identifying genes from natural ecosystems. *Curr. Opin. Microbiol.* 1999; 2: 317–322.

370. Osborn AM, Moore ERB, and Timmis KN. An evaluation of terminal-restriction fragment length polymorphism (T-RFLP) analysis for the study of microbial community structure and dynamics. *Environ. Microbiol.* 2000; 2(1): 39–50.

371. Ranjard L, Poly F, Lata J, Thioulouse J, and Nazaret S. Characterization of bacterial and fungal soil communities by automated ribosomal intergenic spacer analysis fingerprints: biological and methodological variability. *Appl. Environ. Microbiol.* 2001; 67: 4479–4487.

372. Sauer P, Gallo J, Kesselova M, Kolar M, and Koukalova D. Universal primers for detection of common bacterial pathogens causing prosthetic joint infection. *Biomed. Pap. Med. Fac. Univ. Palacky Olomouc Czech. Repub.* 2005; 14(2): 285–288.

373. Zhu H, Qu F, and Zhu L. Isolation of genomic DNAs from plants and bacteria using benzyl chloride. *Nucl. Acids Res.* 1993; 21: 5279–5280.

374. Zoetendal EG, Ben-Amor K, Akkermann A, Abee T, and deVos WM. DNA isolation protocols affect the detection limit of PCR approaches of bacteria in samples from the human gastrointestinal tract. *Syst. Appl. Microbiol.* 2001; 24: 405–410.

375. Khan AA, Nawaz MS, Khan SA, and Cerniglia CE. Identification of predominant human and animal anaerobic intestinal bacterial species by terminal restriction fragment patterns (TRFPs): a rapid, PCR-based method. *Mol. Cell Probes* 2001; 15(6): 349–355.

376. Nadkarni MA, Martin F, Jacques N, and Hunter N. Determination of bacterial load by real-time PCR using a broad-range probe and primers set. *Microbiology* 2002; 148: 257–266.

377. Nagashima K, Hisada T, Sato M, and Mochizuki J. Application of new primer-enzyme combinations to terminal restriction fragment length polymorphism profiling of bacterial populations in human feces. *Appl. Environ. Microbiol.* 2003; 69(2): 1251–1262.

378. Sakamoto M, Hayashi H, and Benno Y. Terminal restriction fragment length polymorphism analysis for human fecal microbiota and its application for analysis of complex bifidobacterial communities. *Microbiol. Immunol.* 2003; 47: 133–142.

379. Takai K and Horikoshi K. Rapid detection and quantification of members of the archael community by quantitative PCR using fluorogenic probes. *Appl. Environ. Microbiol.* 2000; 66: 5066–5072.

380. Lee YK, Mathematical modeling of intestinal bacteria-host interaction. In: Salminen S, von Wright A, and Ouwehand A, editors. *Lactic Acid Bacteria, Microbiological & Functional Aspects*, 3rd ed. Marcel Dekker, New York, 2004, 351–363.

381. Alamprese C, Foschino R, Rossi M, Pompei C, and Corti S. Effects of *Lactobacillus rhamnosus* GG addition in ice cream. *Int. J. Dairy Technol.* 2005; 58: 200–206.

382. Alamprese C, Foschino R, Rossi M, Pompei C, and Savani L. Survival of *Lactobacillus johnsonii* La1 and influence of its addition in retail-manufactured ice cream produced with different sugar and fat concentrations. *Int. Dairy J.* 2002; 12: 2–3.

383. Aragon-Alegro LC, Alegro JHA, Cardarelli HR, Chiu MC, and Saad SMI. Potentially probiotic and synbiotic chocolate mousse. *Food Sci. Technol.* 2007; 40: 669–675.

384. Ammor MS and Mayo B. Selection criteria for lactic acid bacteria to be used as functional starter cultures in dry sausage production: an update. *Meat Sci.* 2007; 76: 138–146.

385. Anal AK and Singh H. Recent advances in microencapsulation of probiotics for industrial applications and targeted delivery. *Trends Food Sci. Technol.* 2007; 18: 240–251.

386. Angelov A, Gotcheva V, Kuncheva R, and Hristozova T. Development of a new oat-based probiotic drink. *Int. J. Food Microbiol.* 2006; 112: 75–80.
387. Arihara K, Ota H, Itoh M, Kondo Y, Sameshima T, Yamanaka H, Akimoto M, Kanai S, and Miki T. *Lactobacillus acidophilus* group lactic acid bacteria applied to meat fermentation. *J. Food Sci.* 1998; 63: 544–547.
388. Bâati L, Fabre-Gea C, Auriol D, and Blanc PJ. Study of the cryotolerance of *Lactobacillus acidophilus*: effect of culture and freezing conditions on the viability and cellular protein levels. *Int. J. Food Microbiol.* 2000; 59: 241–247.
389. Başyiğit G, Kuleaşan H, and Karahan AG. Viability of human-derived probiotic lactobacilli in ice cream produced with sucrose and aspartame. *J. Ind. Microbiol. Biotechnol.* 2006; 33: 796–800.
390. Bautista-Garfias CR, Ixta-Rodríguez O, Martínez-Gómez F, López MG, and Aguilar-Figueroa BR. Effect of viable or dead *Lactobacillus casei* organisms administered orally to mice on resistance against *Trichinella spiralis* infection. *Parasite* 2001; 8(Suppl.): S226–S228.
391. Bergamini CV, Hynes ER, Quiberoni A, Suarez VB, and Zalazar CA. Probiotic bacteria as adjunct starters: influence of the addition methodology on their survival in a semi-hard Argentinean cheese. *Food Res. Int.* 2005; 38: 597–604.
392. Bergamini CV, Hynes ER, and Zalazar CA. Influence of probiotic bacteria on the proteolysis profile of a semi-hard cheese. *Int. Dairy J.* 2006; 16: 856–866.
393. Boirivant M and Strober W. The mechanism of action of probiotics. *Curr. Opin. Gastroenterol.* 2007; 23: 679–692.
394. Bolduc MP, Raymond Y, Fustier P, Champagne CP, and Vuillemard JC. Sensitivity of bifidobacteria to oxygen and redox potential in non-fermented pasteurized milk. *Int. Dairy J.* 2006; 16: 1038–1048.
395. Boylston TD, Vinderola CG, Ghoddusi HB, and Reinheimer JA. Incorporation of bifidobacteria into cheeses: challenges and rewards. *Int. Dairy J.* 2004; 14: 375–387.
396. Corbo MR, Albenzio M, De Angelis M, Sevi A, and Gobbetti M. Microbiological and biochemical properties of canestrato pugliese hard cheese supplemented with bifidobacteria. *J. Dairy Sci.* 2001; 84: 551–561.
397. Corcoran BM, Stanton C, Fitzgerald GF, and Ross RP. Survival of probiotic lactobacilli in acidic environments is enhanced in the presence of metabolizable sugars. *Appl. Environ. Microbiol.* 2005; 71: 3060–3067.
398. Crittenden R. An update on probiotic Bifidobacteria. Salminen S, von Wright A, and Ouwerhand A, editors. Lactic Acid Bacteria: Microbiological and Functional Aspects. Marcel Dekker, New York, 2004, 125–157.
399. Crittenden R, Sanguansri L, and Augustin MA. *Probiotic Storage and Delivery.* Patent WO2005030229-A1, 2005.
400. Crittenden R, Weerakkody R, Sanguansri L, and Augustin MA. Synbiotic microcapsules that enhance microbial viability during non-refrigerated storage and gastrointestinal transit. *Appl. Environ. Microbiol.* 2006; 72: 2280–2282.
401. Dave RI and Shah NP. Effectiveness of ascorbic acid as an oxygen scavenger in improving viability of probiotic bacteria in yoghurts made with commercial starter cultures. *Int. Dairy J.* 1997; 7: 435–443.

402. Dehlink E, Domig KJ, Loibichler C, Kampl E, Eiwegger T, Georgopoulos A, Kneifel W, Urbanek R, and Szépfalusi Z. Heat- and formalin-inactivated probiotic bacteria induce comparable cytokine patterns in intestinal epithelial cell-leucocyte cocultures. *J. Food Protection* 2007; 70: 2417–2421.
403. Doleyres Y and Lacroix C. Technologies with free and immobilised cells for probiotic bifidobacteria production and protection. *Int. Dairy J.* 2005; 15: 973–988.
404. Donkor ON and Shah NP. Production of β-Glucosidase and hydrolysis of isoflavone phytoestrogens by *Lactobacillus acidophilus*, *Bifidobacterium lactis*, and *Lactobacillus casei* in soymilk. *J. Food Sci.* 2008; 73: M15–M20.
405. Erkkilä S, Petäjä E, Eerola S, Lilleberg L, Mattila-Sandholm T, and Suihko ML. Flavour profiles of dry sausages fermented by selected novel meat starter cultures. *Meat Sci.* 2001; 58: 111–116.
406. Erkkilä S, Suihko ML, Eerola S, Petäjä E, and Mattila-Sandholm T. Dry sausage fermented by *Lactobacillus rhamnosus* strains. *Int. J. Food Microbiol.* 2001; 64: 205–210.
407. FAO/WHO. Guidelines for the evaluation of probiotics in Food. Report of a Joint FAO/WHO Working Group on Drafting Guidelines for the Evaluation of probiotics in Food. London Ontario, Canada. April 30 and May 1. 2002. [Online.] http://www.who.int/foodsafety/fs_management/en/probiotic_guidelines.pdf
408. Farnworth ER, Mainville I, Desjardins MP, Gardner N, Fliss I, and Champagne C. Growth of probiotic bacteria and bifidobacteria in a soy yogurt formulation. *Int. J. Food Microbiol.* 2007; 116: 174–181.
409. Favaro-Trindade CS, Bernardi S, Bodini RB, Balieiro JC, and Almeida E. Sensory acceptability and stability of probiotic microorganisms and vitamin C in fermented acerola (*Malpighia emarginata* DC.) ice cream. *J. Food Sci.* 2006; 71: S492–S495.
410. Fernandez Murga ML, de Ruiz Holgado AP, and de Valdez GF. Survival rates and enzyme activities of *Lactobacillus acidophilius* following frozen storage. *Cryobiology 1998*, 1998; 36: 315–319.
411. Gardiner GE, Bouchier P, O'Sullivan E, Kelly J, Collins JK, Fitzgerald G, Ross RP, and Stanton C. A spray-dried culture for probiotic Cheddar cheese manufacture. *Int. Dairy J.* 2002; 12: 749–756.
412. Gibson GR and Roberfroid MB. Dietary modulation of the human colonic microbiota—introducing the concept of prebiotics. *J. Nutr.* 1995; 125: 1401–1412.
413. Gokavi S, Zhang LW, Huang MK, Zhao X, and Guo MR. Oat-based symbiotic beverage fermented by *Lactobacillus plantarum*, *Lactobacilus paracasei* ssp. *casei*, and *Lactobacillus acidophilus*. *J. Food Sci.* 2005; 70: M216–M223.
414. Gopal P, Dekker J, Prasad J, Pillidge C, Delabre ML, and Collett M. Development and commercialisation of Fonterra's probiotic strains. *Aust. J. Dairy Technol.* 2005; 60: 173–182.
415. Heenan CN, Adams MC, Hosken RW, and Fleet GH. Survival and sensory acceptability of probiotic microorganisms in a nonfermented frozen vegetarian dessert. *Food Sci. Technol.* 2004; 37: 461–466.
416. Hekmat S and McMahon DJ. Survival of *Lactobacillus acidophilus* and *Bifidobacterium bifidum* in ice cream for use as a probiotic food. *J. Dairy Sci.* 1992; 75: 1415–1422.
417. Heller KJ. Probiotic bacteria in fermented foods: product characteristics and starter organisms. *Am. J. Clin. Nutr.* 2001; 73(2 Suppl.): 374S–379S.

418. Kailasapathy K. Microencapsulation of probiotic bacteria: technology and potential applications. *Curr. Issues Intest. Microbiol.* 2002; 3: 39–48.
419. Kawasaki S, Mimura T, Satoh T, Takeda K, and Niimura Y. Response of the microaerophilic *Bifidobacterium* species, *B. boum* and *B. thermophilum*, to oxygen. *Appl. Environ. Microbiol.* 2006; 72: 6854–6858.
420. Kaya M and Aksu MI. Effect of modified atmosphere and vacuum packaging on some quality characteristics of sliced sucuk produced using probiotics culture. *J. Sci. Food Agric.* 2005; 85: 2281–2288.
421. Kirjavainen PV, Salminen S, and Isolauri E. Probiotic bacteria in the management of atopic disease: underscoring the importance of viability. *J. Pediatr. Gastroenterol. Nutr.* 2003; 36: 223–237.
422. Klingberg TD and Budde BB. The survival and persistence in the human gastrointestinal tract of five potential probiotic lactobacilli consumed as freeze-dried cultures or as probiotic sausage. *Int. J. Food Microbiol.* 2006; 109: 157–159.
423. Kourkoutas Y, Bosnea L, Taboukos S, Baras C, Lambrou D, and Kanellaki M. Probiotic cheese production using *Lactobacillus casei* cells immobilized on fruit pieces. *J. Dairy Sci.* 2006; 89: 1439–1451.
424. Krasaekoopt W, Bhandari B, and Deeth H. The influence of coating materials on some properties of alginate beads and survivability of microencapsulated probiotic bacteria. *Int. Dairy J.* 2004; 14: 737–743.
425. Landuyt A. Probiotic chocolate: restoring the balance. *Food Eng. Ingredients* 2007; 32: 20–23.
426. Leroy F, Verluyten J, and de Vuyst L. Functional meat starter cultures for improved sausage fermentation. *Int. J. Food Microbiol.* 2006; 106:270–285.
427. Lopez-Rubio A, Gavara R, and Lagaron JM. Bioactive packaging: turning foods into healthier foods through biomaterials. *Trends Food Sci. Technol.* 2006; 17: 567–575.
428. Lucklow T, Sheehan V, Delahunty C, and Fitzgerald G. Determining the odor and flavour characteristics of probiotic, health promoting ingredients and the effects of repeated exposure on consumer acceptance. *J. Food Sci.* 2005; 70: S53–S59.
429. Magarinos H, Selaive S, Costa M, Flores M, and Pizarro O. Viability of probiotic microorganisms (*Lactobacillus acidophilus* La-5 and *Bifidobacterium animalis* subsp. *lactis* Bb-12) in ice cream. *Int. J. Dairy Technol.* 2007; 60: 128–134.
430. Mattila-Sandholm T, Myllärinen P, Crittenden R, Morgensen G, Fondén R, and Saarela R. Technological challenges for future probiotic foods. *Int. Dairy J.* 2002; 12: 173–182.
431. Mättö J, Alakomi HL, Vaari A, Virkajärvi I, and Saarela M. Influence of processing conditions on *Bifidobacterium animalis* subsp. *lactis* functionality with a special focus on acid tolerance and factors affecting it. *Int. Dairy J.* 2006; 16: 1029–1037.
432. Meile L, Ludwig W, Rueger U, Gut C, Kaufmann P, Dasen G, Wenger S, and Teuber M. *Bifidobacterium lactis* sp. nov., a moderately oxygen tolerant species isolated from fermented milk. *Syst. Appl. Microbiol.* 1997; 20: 57–64.
433. Mortazavian AM, Ehsani MR, Mousavi SM, Reinheimer JA, Emamdjomeh Z, Sohrabvandi S, and Rezaei K. Preliminary investigation of the combined effect of heat treatment and incubation temperature on the viability of the probiotic micro-organisms in freshly made yogurt. *Int. J. Dairy Technol.* 2006; 59: 8–11.

434. Mugnier J and Jung G. Survival of bacteria and fungi in relation to water activity and the solvent properties of water in biopolymer gels. *Appl. Environ. Microbiol.* 1985; 50: 108–114.
435. Muthukumarasamy P and Holley RA. Survival of *Escherichia coli* O157:H7 in dry fermented sausages containing micro-encapsulated probiotic lactic acid bacteria. *Food Microbiol.* 2007; 24: 82–88.
436. Nebesny E, Żyżelewicz D, Motyl I, and Libudzisz Z. Dark chocolates supplemented with *Lactobacillus* strains. *Eur. Food Res. Technol.* 2007; 225: 33–42.
437. Nebesny E, Zyzelewicz D, Motyl I, and Libudzisz Z. Properties of sucrose-free chocolates enriched with viable lactic acid bacteria. *Eur. Food Res. Technol.* 2005; 220: 3–4.
438. NutraIngredients. *BioGaia to Extend TopLife Product.* 29 April 2003. Available at www.nutraingredients.com.
439. NutraIngredients. *BioGaia Develops "Healthy" Bottle Top.* 01 April 2004. Available at www.nutraingredients.com.
440. O'Brien J, Crittenden R, Ouwerhand AC, and Salminen S. Safety evaluation of probiotics. *Trends Food Sci. Technol.* 1999; 10: 418–424.
441. Ong L, Henriksson A, and Shah NP. Chemical analysis and sensory evaluation of Cheddar cheese produced with *Lactobacillus acidophilus, Lb. casei, Lb. paracasei or Bifidobacterium* sp. *Int. Dairy J.* 2007; 17: 937–945.
442. Ong L, Henriksson A, and Shah NP. Proteolytic pattern and organic acid profiles of probiotic Cheddar cheese as influenced by probiotic strains of *Lactobacillus acidophilus, Lb. paracasei, Lb. casei or Bifidobacterium* sp. *Int. Dairy J.* 2007; 17: 67–78.
443. Pennacchia C, Vaughan EE, and Villani F. Potential probiotic *Lactobacillus* strains from fermented sausages: further investigations on their probiotic properties. *Meat Sci.* 2006; 73: 90–101.
444. Pérez-Conesa D, López G, and Ros G. Fermentation capabilities of bifidobacteria using nondigestible oligosaccharides, and their viability as probiotics in commercial powder infant formula. *J. Food Sci.* 2005; 70: M279–M285.
445. Phillips M, Kailasapathy K, and Tran L. Viability of commercial probiotic cultures (*L. acidophilus, Bifidobacterium* sp., *L. casei, L. paracasei* and *L. rhamnosus*) in cheddar cheese. *Int. J. Food Microbiol.* 2006; 108: 276–280.
446. Rebucci R, Sangalli L, Fava M, Bersani C, Cantoni C, and Baldi A. Evaluation of functional aspects in *Lactobacillus* strains isolated from dry fermented sausages. *J. Food Qual.* 2007; 30: 187–201.
447. Rodgers S and Odongo R. Survival of *Lactobacillus acidophilus, L casei* and *Bifidobacterium lactis* in coleslaw during refrigerated storage. *Food Aust.* 2002; 54: 185–188.
448. Rodgers S. Incorporation of probiotic cultures in foodservice products: an exploratory study. *J. Foodserv.* 2007; 18: 108i–118.
449. Ross RP, Desmond C, Fitzgerald GF, and Stanton C. Overcoming the technological hurdles in the development of probiotic foods. *J. Appl. Microbiol.* 2005; 98: 1410–1417.
450. Saarela M, Virkajärvi I, Alakomi H-L, Sigvart-Mattila P, and Mättö J. Stability and functionality of freeze-dried probiotic *Bifidobacterium* cells during storage in juice and milk. *Int. Dairy J.* 2006; 16: 1477–1482.

451. Saarela M, Virkajärvi I, Nohynek L, Vaari A, and Mättö J. Fibres as carriers for *Lactobacillus rhamnosus* during freeze-drying and storage in apple juice and chocolate-coated breakfast cereals. *Int. J. Food Microbiol.* 2006; 112: 171–178.
452. Salminen S, von Wright A, Morelli L, Marteau P, Brassart D, de Vos WM, Fondén R, Saxelin M, Collins K, Mogensen G, Birkeland SE, and Mattila-Sandholm T. Demonstration of safety of probiotics—a review. *Int. J. Food Microbiol.* 1998; 44: 93–106.
453. Sameshima T, Magome C, Takeshita K, Arihara K, Itoh M, and Kondo Y. Effect of intestinal *Lactobacillus* starter cultures on the behaviour of *Staphylococcus aureus* in fermented sausage. *Int. J. Food Microbiol.* 1998; 41: 1–7.
454. Savoie S, Champagne CP, Chiasson S, and Audet P. Media and process parameters affecting the growth, strain ratios and specific acidifying activities of a mixed lactic starter containing aroma-producing and probiotic strains. *J. Appl. Microbiol.* 2007; 103: 163–174.
455. Saxelin M, Grenov B, Svensson U, Fondén R, Reniero R, and Mattila-Sandholm T. The technology of probiotics. *Trends Food Sci. Technol.* 1999; 10: 387–392.
456. Schmidt G and Zink R. Basic features of the stress response in three species of bifidobacteria: *B. longum*, *B. adolescentis*, and *B. breve*. *Int. J. Food Microbiol.* 2000; 55: 41–45.
457. Shimamura S, Abe F, Ishibashi N, Miyakawa H, Yaeshima T, Araya T, and Tomita M. Relationship between oxygen sensitivity and oxygen metabolism of *Bifidobacterium* species. *J. Dairy Sci.* 1992; 75: 3296–3306.
458. Sloan AE. The top 10 functional food trends 2004. *Food Technol.* 2004; 58: 28–51.
459. Talwalkar A, Miller CW, Kailasapathy K, and Nguyen MH. Effect of packaging materials and dissolved oxygen on the survival of probiotic bacteria in yoghurt. *Int. J. Food Sci. Technol.* 2004; 39: 605–611.
460. Tang AL, Shah NP, Wilcox G, Walker KZ, and Stojanovska L. Fermentation of calcium-fortified soymilk with *Lactobacillus*: effects on calcium solubility, isoflavone conversion, and production of organic acids. *J. Food Sci.* 2007; 72: M431–M436.
461. Tharmaraj N and Shah NP. Survival of *Lactobacillus acidophilus*, *Lactobacillus paracasei* subsp. *paracasei*, *Lactobacillus rhamnosus*, *Bifidobacterium animalis* and *Propionibacterium* in cheese-based dips and the suitability of dips as effective carriers of probiotic bacteria. *Int. Dairy J.* 2004; 14: 1055–1066.
462. Tuomola E, Crittenden R, Playne M, Isolauri E, and Salminen S. Quality assurance criteria for probiotic bacteria. *Am. J. Clin. Nutr.* 2001; 73(Suppl.): 393S–398S.
463. von Wright A. Regulating the safety of probiotics–The European approach. *Curr. Pharm. Design* 2005; 11: 17–23.
464. Yilmaztekin M, Oezer BH, and Atasoy F. Survival of *Lactobacillus acidophilus* LA-5 and *Bifidobacterium bifidum* BB-02 in white-brined cheese. *Int. J. Food Sci. Nutr.* 2004; 55: 53–60.
465. Senok AC, Ismaeel AY, and Botta GA. Probiotics: facts and myths. *Clin. Microbiol. Infect.* 2005; 11(12): 958–966.
466. Huang Y, Kotula L, and Adams MC. The in vivo assessment of safety and gastrointestinal survival of an orally administered novel probiotic, *Propionibacterium jensenii* 702, in a male Wistar rat model. *Food Chem. Toxicol.* 2003; 41(12): 1781–1787.
467. Fernandez MF, Boris S, and Barbes C. Safety evaluation of *Lactobacillus delbrueckii* subsp. *lactis* UO 004, a probiotic bacterium. *Res. Microbiol.* 2005; 156(2): 154–160.

468. Hong HA, Le HD, and Cutting SM. The use of bacterial spore formers as probiotics. *FEMS Microbiol. Rev.* 2005; 29(4): 813–835.
469. de Vries MC, Vaughan EE, Kleerebezem M, and de Vos WM. *Lactobacillus plantarum*— survival, functional and potential probiotic properties in the human intestinal tract. *Int. Dairy J.* 2006; 16(9): 1018–1028 (4th NIZO Dairy Conference—Prospects for Health, Well-Being and Safety).
470. Guglielmotti DM, Marco MB, Golowczyc M, Reinheimer JA, and Quiberoni AdL. Probiotic potential of *Lactobacillus delbrueckii* strains and their phage resistant mutants. *Int. Dairy J.* 2007; 17(8): 916–925.
471. Reid G. Safe and efficacious probiotics: what are they? *Trends Microbiol.* 2006; 14(8): 348–352.
472. Benchimol EI and Mack DR. Safety issues of probiotic ingestion. *Pract. Gastroenterol.* 2005; 29(11): 23–34.
473. Ezendam J and Van Loveren H. Probiotics: Immunomodulation and evaluation of safety and efficacy. *Nutr. Rev.* 2006; 64(1): 1–14.
474. von Wright A. Regulating the safety of probiotics—the European approach. *Curr. Pharm. Design* 2005; 11(1): 17–23.
475. FAO/WHO. Guidelines for the evaluation of probiotics in food. Report of a Joint FAO/ WHO Working Group on Drafting Guidelines for the Evaluation of Probiotics in Food. London, Ontario, Canada: FAO/WHO, April 30 and May 1, 2002.
476. Cummins J and Ho M-W. Genetically modified probiotics should be banned. *Microb. Ecol. Health Dis.* 2005; 17(2): 66–68.
477. Henriksson A, Borody T, and Clancy R. Probiotics under the regulatory microscope. *Expert Opin. Drug Safety* 2005; 4(6): 1135–1143.
478. Reid G. The importance of guidelines in the development and application of probiotics. *Curr. Pharm. Design* 2005; 11(1): 11–16.
479. Whittaker PJ. Re: All nutritional supplements should be classified as drugs. *BMJ*. August 9, 2007; DOI:10.1136/bmj.39272.581736.55.
480. European Food Safety Authority (EFSA). EFSA Scientific Colloquium 2 Summary Report QPS. Qualified Presumption of Safety of Micro-organisms in Food and Feed. Brussels, Belgium, 2005.
481. Bernardeau M, Guguen M, and Vernoux JP. Beneficial lactobacilli in food and feed: long-term use, biodiversity and proposals for specific and realistic safety assessments. *FEMS Microbiol. Rev.* 2006; 30(4): 487–513.
482. Sanders ME, Tompkins T, Heimbach JT, and Kolida S. Weight of evidence needed to substantiate a health effect for probiotics and prebiotics Regulatory considerations in Canada, E.U. and U.S. *Eur. J. Nutr.* 2005; 44(5): 303–310.
483. Gueimonde M, Frias R, and Ouwehand A. Assuring the continued safety of lactic acid bacteria used as probiotics. *Biologia* 2006; 61(6): 755–760.
484. Yeung PSM, Sanders ME, Kitts CL, Cano R, and Tong PS. Species-specific identification of commercial probiotic strains. *J. Dairy Sci.* 2002; 85(5): 1039–1051.
485. Lourens-Hattingh A and Viljoen BC. Survival of probiotic bacteria in South African commercial bio-yogurt. *South African J. Sci.* 2002; 98(5–6): 298–300.
486. Weese JS. Microbiologic evaluation of commercial probiotics. *J. Am. Vet. Med. Assoc.* 2002; 220(6): 794–797.

487. Temmerman R, Pot B, Huys G, and Swings J. Identification and antibiotic susceptibility of bacterial isolates from probiotic products. *Int. J. Food Microbiol.* 2003; 81(1): 1–10.

488. Huys G, Vancanneyt M, D'Haene K, Vankerckhoven V, Goossens H, and Swings J. Accuracy of species identity of commercial bacterial cultures intended for probiotic or nutritional use. *Res. Microbiol.* 2006; 157(9): 803–810.

489. Hamilton-Miller JMT and Shah S. Deficiencies in microbiological quality and labelling of probiotic supplements. *Int. J. Food Microbiol.* 2002; 72(1–2): 175–176.

490. Playne M. Classification and identification of probiotic bacterial strains. *Probiotica* 1999: 1–2, 4.

491. Vesterlund S, Vankerckhoven V, Saxelin M, Goossens H, Salminen S, and Ouwehand AC. Safety assessment of *Lactobacillus* strains: Presence of putative risk factors in faecal, blood and probiotic isolates. *Int. J. Food Microbiol.* 2007; 116(3): 325–331.

492. Salminen S, Isolauri E, and vonWright A, Safety of probiotic bacteria. In: Preedy V Watson R, editors. Reviews in Food and Nutrition Toxicity. London, UK: Taylor and Francis; 2003, pp. 271–283.

493. Rautio M, Jousimies-Somer H, Kauma H, Pietarinen I, Saxelin M, Tynkkynen S, et al. Liver abscess due to a *Lactobacillus rhamnosus* strain indistinguishable from *L. rhamnosus* strain GG. *Clinical Infectious Diseases* 1999; 28: 1159–1160.

494. Mackay A, Taylor M, Kibbler C, and Hamilton-Miller J. Lactobacillus endocarditis caused by a probiotic organism. *Clin. Infect. Dis.* 1999; 5: 290–292.

495. Wolf B, Wheeler K, Ataya D, and Garleb K. Safety and tolerance of *Lactobacillus reuteri* supplementation to a population infected with the human immunodeficiency virus. *Food Chem. Toxicol.* 1998; 36: 1085–1094.

496. Borriello SP, Hammes WP, Holzapfel W, Marteau P, Schrezenmeir J, and Vaara M, et al. Safety of probiotics that contain *Lactobacilli* or *Bifidobacteria*. *Clin. Infect. Dis.* 2003; 36: 775–780.

497. LeDoux D, Labombardi VJ, and Karter D. *Lactobacillus acidophilus* bacteraemia after use of a probiotic in a patient with AIDS and Hodgkin's disease. *Int. J. STD AIDS* 2006; 17(4): 280–282.

498. Land MH, Rouster-Stevens K, Woods CR, Cannon ML, Cnota J, and Shetty AK. Lactobacillus sepsis associated with probiotic therapy. *Pediatrics* 2005; 115(1): 178–181.

499. Srinivasan R, Meyer R, Padmanabhan R, and Britto J. Clinical safety of *Lactobacillus casei* shirota as a probiotic in critically ill children. *J. Pediatr. Gastroenterol. Nutr.* 2006; 42(2): 171–173.

500. Hammerman C, Bin-Nun A, and Kaplan M. Safety of probiotics: comparison of two popular strains. *BMJ* 2006; 333(7576): 1006–1008.

501. Marteau P. Living drugs for gastrointestinal diseases: the case for probiotics. *Digest. Dis.* 2006; 24(1–2): 137–147.

502. Donohue D, Salminen S, and Marteau P. Safety of probiotic bacteria. In: Salminen S, Wright Av, editors. Lactic Acid Bacteria, 2nd ed. New York: Marcel Dekker, 1998, pp. 369–383.

503. Hew CM, Korakli M, and Vogel RF. Expression of virulence-related genes by *Enterococcus faecalis* in response to different environments. *Syst. Appl. Microbiol.* 2007; 30(4): 257–267.

504. Salminen MK, Tynkkynen S, Rautelin H, Saxelin M, Vaara M, and Ruutu P, et al. Lactobacillus bacteremia during a rapid increase in probiotic use of *Lactobacillus rhamnosus* GG in Finland. *Clin. Infect. Dis.* 2002; 35(10): 1155–1160.
505. Sullivan A and Nord CE. Probiotic lactobacilli and bacteraemia in Stockholm. *Scand. J. Infect. Dis.* 2006; 38(5): 327–331.
506. Cannon JP, Danziger LH, Lee TA, and Bolanos JT. Pathogenic relevance of *Lactobacillus*: a retrospective review of over 200 cases. *Eur. J. Clin. Microbiol. Infect. Dis.* 2005; 24(Jan): 31–40.
507. Danielsen M, Wind A, Leisner J, and Arpi M. Antimicrobial susceptibility of human blood culture isolates of *Lactobacillus* spp. *European J. Clin. Microbiol. Infect. Dis.* 2007; 26(4): 287–289.
508. Mathur S and Singh R. Antibiotic resistance in food lactic acid bacteria—a review. *Int. J. Food Microbiol.* 2005; 105(3): 281–295.
509. Franz CMAP, Stiles ME, Schleifer KH, and Holzapfel WH. Enterococci in foods—a conundrum for food safety. *Int. J. Food Microbiol.* 2003; 88(2–3): 105–122.
510. Kastner S, Perreten V, Bleuler H, Hugenschmidt G, Lacroix C, and Meile L. Antibiotic susceptibility patterns and resistance genes of starter cultures and probiotic bacteria used in food. *Syst. Appl. Microbiol.* 2006; 29(2): 145–155.
511. Hummel AS, Hertel C, Holzapfel WH, and Franz CMAP. Antibiotic resistances of starter and probiotic strains of lactic acid bacteria. *Appl. Environ. Microbiol.* 2007; 73(3): 730–739.
512. Saarela M, Maukonen J, von Wright A, Vilpponen-Salmela T, Patterson AJ, Scott KP, et al. Tetracycline susceptibility of the ingested *Lactobacillus acidophilus* LaCH-5 and Bifidobacterium animalis subsp. lactis Bb-12 strains during antibiotic/probiotic intervention. *Int. J. Antimicrob. Agents* 2007; 29(3): 271–280.
513. Courvalin P, Antibiotic resistance: the pros and cons of probiotics. *Digest. Liver Dis.* Papers from the 3rd International Congress on Probiotics, Prebiotics and New Foods 2006; 38(Suppl. 2): S261–S265.
514. Pohjavuori E, Viljanen M, Korpela R, Kuitunen M, Tiittanen M, Vaarala O, et al. *Lactobacillus* GG effect in increasing IFN-γ production in infants with cow's milk allergy. *J. Allergy Clin. Immunol.* 2004; 114(1): 131–136.
515. Reid G, Bruce AW, Fraser N, Heinemann C, Owen J, and Henning B. Oral probiotics can resolve urogenital infections. *FEMS Immunol. Med. Microbiol.* 2001; 30: 49–52.
516. Kalliomaki M, Salminen S, Arvilommi H, Kero P, Koskinen P, and Isolauri E. Probiotics in primary prevention of atopic disease; a randomised placebo-controlled trial. *Lancet* 2001; 357: 1076–1079.
517. Heyman M and Heuvelin E. Probiotic micro-organisms and immune regulation: the paradox. *Nutr. Clin. Metabol.* 2006; 20(2): 85–94.
518. Isolauri E. Probiotics in human disease. *Am. J. Clin. Nutr.* 2001; 73(6): 1142S–1146.
519. Kirjavainen PV, El-Nezami HS, Salminen SJ, Ahokas JT, and Wright PFA. The effect of orally administered viable probiotic and dairy lactobacilli on mouse lymphocyte proliferation. *FEMS Immunol. Med. Microbiol.* 1999; 26(2): 131–135.
520. Wagner R, Warner T, Roberts L, Farmer J, and Balish E. Colonization of congenitally immunodeficient mice with probiotic bacteria. *Infect. Immun.* 1997; 65(8): 3345–3351.

521. Wagner R, Pierson C, Warner T, Dohnalek M, Farmer J, Roberts L, et al. Biotherapeutic effects of probiotic bacteria on candidiasis in immunodeficient mice. *Infect. Immun.* 1997; 65(10): 4165–4172.

522. Timmerman HM, Niers LEM, Ridwan BU, Koning CJM, Mulder L, Akkermans LMA, et al. Design of a multispecies probiotic mixture to prevent infectious complications in critically ill patients. *Clin. Nutr.* 2007; 26(4): 450–459.

523. Baken KA, Ezendam J, Gremmer ER, de Klerk A, Pennings JLA, Matthee B, et al. Evaluation of immunomodulation by *Lactobacillus casei* Shirota: Immune function, autoimmunity and gene expression. *Int. J. Food Microbiol.* 2006; 112(1): 8–18.

524. Steidler L. Genetically engineered probiotics. Bailliere's best practice and research. *Clin. Gastroenterol.* 2003; 17(5): 861–876.

525. Frossard CP, Steidler L, and Eigenmann PA. Oral administration of an IL-10-secreting *Lactococcus lactis* strain prevents food-induced IgE sensitization. *J. Allergy Clin. Immunol.* 2007; 119(4): 952–959.

526. Salminen MK, Tynkkynen S, Rautelin H, Poussa T, Saxelin M, Ristola M, et al. The efficacy and safety of probiotic *Lactobacillus rhamnosus* GG on prolonged, noninfectious diarrhea in HIV patients on antiretroviral therapy: A randomized, placebo-controlled, crossover study. *HIV Clin. Trials* 2004; 5(4): 183–191.

527. Weizman Z, Asli G, and Alsheikh A. Effect of a probiotic infant formula on infections in child care centers: comparison of two probiotic agents. *Pediatrics* 2005; 115(1): 5–9.

528. Hickson M, D'Souza AL, Muthu N, Rogers TR, Want S, Rajkumar C, et al. Use of probiotic Lactobacillus preparation to prevent diarrhoea associated with antibiotics: randomised double blind placebo controlled trial. *BMJ* 2007; 335(7610): 80.

529. Saeed BAK. Re: Probiotics are they the answer for *Clostridium difficile* associated diarrhoea? *BMJ* . Available at http://www.bmj.com/cgi/eletters/335/7610/80#172144, Accessed on 13 Aug 2007.

530. Connolly E, Abrahamsson T, and Bjorksten B. Safety of D(−)-lactic acid producing bacteria in the human infant. *J. Pediatr. Gastroenterol. Nutr.* 2005; 41(4): 489–492.

531. Besselink MG, van Santvoort HC, Buskens E, Boermeester MA, van Goor H, Timmerman HM, et al. Probiotic prophylaxis in predicted severe acute pancreatitis: a randomised, double-blind, placebo-controlled trial. *Lancet* 2008; 371(9613): 651–659.

532. Saxelin M, Chuang N-H, Chassy B, Rautelin H, Mäkelä P, Salminen S, et al. Lactobacilli and bacteremia in southern Finland 1989–1992. *Clin. Infect. Dis.* 1996; 22: 564–566.

533. Saxelin M, Rautelin H, Salminen S, and Mäkelä P. The safety of commercial products with viable Lactobacillus strains. *Infect. Dis. Clin. Pract.* 1996; 5: 331–335.

534. Steidler L, Hans W, Schotte L, Neirynck S, Obermeier F, and Falk W et al. Treatment of murine colitis by *Lactococcus lactis* secreting interleukin-10. *Science* 2000; 289: 1352–1355.

535. Kaur IP, Chopra K, and Saini A. Probiotics: potential pharmaceutical applications. *Eur. J. Pharm. Sci.* 2002; 15(1): 1–9.

536. Braat H, Rottiers P, Hommes DW, Huyghebaert N, Remaut E, and Remon J-P, et al. A Phase I Trial With Transgenic Bacteria Expressing Interleukin-10 in Crohn's Disease. *Clin. Gastroenterol. Hepatol.* 2006; 4(6): 754–759.

537. Reid G, Gibson GR, Gill HS, Klaenhammer TR, Rastall RA, and Rowland IR, et al. Use of genetically modified microbes for human health. *Microbial Ecol. Health Dis.* 2006; 18(2): 75–76.

538. Datamonitor. Insights into Tomorrow's Nutraceutical Consumers, 2005. Available at http://www.marketresearch.com/product/display.asp?productid=1173944. Accessed on August 8, 2007.
539. BCC Research. *Probiotics: ingredients, supplements, foods* BCC Inc, Norwalk, CT, 2005.
540. FAO/WHO. *Joint Expert Consultation on Evaluation of Health and Nutritional Properties of Probiotics in Food Including Powder Milk with Live Lactic Acid Bacteria*, October 2001.
541. Ta Kung Pao (Chinese Newspaper, HK), November 28, 2004.
542. Rubin R, A bug for what's bugging you? 7 August 2003, USA TODAY.
543. Frost & Sullivan. End-User Analysis of the European Probiotics Market. NYC 2004.
544. FAO/WHO. *Joint FAO/WHO Working Group Guidelines for the Evaluation of Probiotics in Food.* Food and Agriculture Organization, United Nations, 2002.
545. Ishibashi N and Shimamur S. Bifidobacteria: research and development in Japan. *Food Technol.* 1993; 47: 126–135.
546. Japanese Health Food and Nutrition Food Association. Available at http://www.jhnfa.org/.
547. Notification by the Ministry of Health regarding Guidelines on the review of Fungal and Probiotic Health foods, Wei-Fa-Jian-Fa, 2001, No. 84, Ministry of Health, People's Republic of China.
548. Regulation on Registration and Review of Probiotic Health Foods, 2005; Ministry of Health, PRC. Available at http://www.sfda.gov.cn/cmsweb/webportal/W945325/A64003018_1.html.
549. Kim JY, Kim DB, and Lee HJ. Regulation on Health/Functional Foods in Korea. *Toxicology* 2006; 221: 112–118.
550. Food Safety Information System, Malaysia. Available at http://fsis.moh.gov.my/fqc/.
551. Food Regulation, Malaysia Ministry of Health, part V. *Food Additive and Nutritional Supplement*, 26A Bifidobacteria, Table 12A, 1985.
552. Food Regulation, Malaysia Ministry of Health, part V. *Food Additive and Nutritional Supplement*, 18E Nutrient Function Claim, 1985.
553. Health Food Control Act, Department of Health, Executive Yuan, Taiwan, ROC, Article 4. Available at http://food.doh.gov.tw/chinese/ruler/ruler_2_1.asp?chieng=2&lawsidx=417.
554. Report of the regional expert consultation of the Asia-Pacific network for food and nutrition on functional foods and their implications in the daily diet, 2004, Available at http://www.fao.org/docrep/007/ae532e/ae532e02.htm#bm02.2.
555. Thai Medical and Health Product Bulletin, Vol. 5, No. 4, pp. 8–12. Available at http://www.fda.moph.go.th/fda-net/html/product/apr/about/bulletin%205-4.pdf.
556. Isolauri Erika, Metsaniitty Leena, Korhonen Hannu, Salminen Seppo, and Syvaoja Eeva-Liisa. *Methods of Preventing or Treating Allergies*, US patent no 6,506,380. Assignee: Valio Oy, 2003.
557. Farmer S and Lefkowitz Andrew R, *Methods for Reducing Cholesterol Using Bacillus coagulans Spores, Systems and Compositions*, US Patent no 6,811,786. Assignee: Ganeden Biotech, Inc., 2004.
558. Luquet F-M, Bladwin C, and Lacroix M, Lactic acid bacteria and their use for treating and preventing cancer. Eu patent No.145082, Assignee: B:o k plus International Inc. 2006.

559. Wang MF, Lin HC, Wang YY, and Hsu CH. Treatment of perennial allergic rhinitis with lactic acid bacteria. *Pediatr. Allergy Immunol.* 2004; 15(2): 152–158.

560. European Parliament and the Council of the European Union. Regulation (EC) No 1924/2006 of the European Parliament and the Council of 20 December 2006 on nutrition and health claims made on foods. *Off. J. Eur. Union* 2006; L404: 9–25.

561. Richardson DP, Binns NM, and Viner P. Guidelines for an evidence-based review system for the scientific justification of diet and health relation ships under Article 13 of the new European legislation on nutrition and health claims. *Food. Sci. Technol. Bull.* 2007; 3(8): 83–97.

562. Ruffell M. EU nutrition and health claims regulation. *Nutrfoods* 2007; 6(1): 39–41.

563. Aggett PJ, Antoine JM, Asp NG, Bellisle F, Contor L, Cummings JH, et al. PASSCLAIM: consensus on criteria. *Eur. J. Nutr.* 2005; 44(Suppl. 1): 5–30.

564. European Parliament and the Council of the European Union. Regulation (EC) No 1925/2006 of the European Parliament and the Council of 20 December 2006 on the addition of vitamins and minerals and of certain other substances to foods. *Off. J. Eur. Union* 2006; L404: 26–38.

565. US Federal Food, Drug, and Cosmetic Act, Sections 301, et. seq. (21 USC 301, et. seq.). Available at http://frwebgate.access.gpo.gov/cgi-bin/getdoc.cgi?dbname=browse_usc&docid=Cite:+21USC301.

566. US Federal Food, Drug, and Cosmetic Act, Section 321(f) (21 USC 321(f)). Available at http://frwebgate.access.gpo.gov/cgi-bin/getdoc.cgi?dbname=browse_usc&docid=Cite:+21USC321.

567. US Federal Food, Drug, and Cosmetic Act, Section 321(g) (21 USC 321(g)). Available at http://frwebgate.access.gpo.gov/cgi-bin/getdoc.cgi?dbname=browse_usc&docid=Cite:+21USC321.

568. US Federal Food, Drug, and Cosmetic Act, Section 321(s) (21 USC 321(s)). Available at http://frwebgate.access.gpo.gov/cgi-bin/getdoc.cgi?dbname=browse_usc&docid=Cite:+21USC321.

569. US Federal Food, Drug, and Cosmetic Act, Section 321(ff) (21 USC 321(ff)). Available at http://frwebgate.access.gpo.gov/cgi-bin/getdoc.cgi?dbname=browse_usc&docid=Cite:+21USC321.

570. US Code, The Public Health and Welfare, Public Health Service Sections 201, et seq. (42 USC 201, et seq.). Available at http://frwebgate.access.gpo.gov/cgi-bin/getdoc.cgi?dbname=browse_usc&docid=Cite:+42USC201.

571. The Changing Agenda for Unraveling the Host-Microbe Relationship. Workshop Summary. Forum on Microbial Threats. Board on Global Health 2006. Institute of Medicine, The National Academies Press, Washington, DC. Available at www.nap.edu.

572. US Federal Food, Drug, and Cosmetic Act, Section 360ee(b)(3) (21 USC 360ee(b)(3)). Available at http://frwebgate.access.gpo.gov/cgi-bin/getdoc.cgi?dbname=browse_usc&docid=Cite:+21USC360ee.

573. US Federal Food, Drug, and Cosmetic Act, Section 348 (21 USC 348). Available at http://frwebgate.access.gpo.gov/cgi-bin/getdoc.cgi?dbname=browse_usc&docid=Cite:+21USC348.

574. US Code of Federal Regulations 21 CFR 180.1(a) Available at http://a257.g.akamaitech.net/7/257/2422/26mar20071500/edocket.access.gpo.gov/cfr_2007/aprqtr/21cfr180.1.htm.

REFERENCES 175

575. FDA GRAS Notifications 0049 and 0171. Available at http://www.cfsan.fda.gov/~dms/eafus.html.
576. US Federal Food, Drug, and Cosmetic Act, Section 342(f) (21 USC 342(f)). Available at http://frwebgate.access.gpo.gov/cgi-bin/getdoc.cgi?dbname=browse_usc&docid= Cite:+21USC342.
577. US Federal Food, Drug, and Cosmetic Act, Section 350b (21 USC 350b). Available at http://frwebgate.access.gpo.gov/cgi-bin/getdoc.cgi?dbname=browse_usc&docid= Cite:+21USC350b.
578. Yeung PS, Sanders ME, Kitts CL, Cano R, and Tong PS. Species-specific identification of commercial probiotic strains. *J. Dairy Sci.* 2002; 85(5): 1039–1051.
579. Temmerman R, Scheirlinck I, Huys G, and Swings J. Culture-independent analysis of probiotic products by denaturing gradient gel electrophoresis. *Appl. Environ. Microbiol.* 2003; 69(1): 220–226.
580. Drisko J, Bischoff B, Giles C, Adelson M, Rao RVS, and McCallum, R. Evaluation of five probiotic products for label claims by DNA extraction and polymerase chain reaction analysis. *Dig. Dis. Sci.* 2005; 50(6): 1113–1117.
581. US Federal Food, Drug, and Cosmetic Act, Section 343(a) (21 USC 343(a)). Available at http://frwebgate.access.gpo.gov/cgi-bin/getdoc.cgi?dbname=browse_usc&docid= Cite:+21USC343.
582. US Federal Food, Drug, and Cosmetic Act, Section 321(n) (21 USC 321(n)). Available at http://frwebgate.access.gpo.gov/cgi-bin/getdoc.cgi?dbname=browse_usc&docid= Cite:+21USC321.
583. US Federal Food, Drug, and Cosmetic Act, Section 343(r) (21 USC 343(r)). Available at http://frwebgate.access.gpo.gov/cgi-bin/getdoc.cgi?dbname=browse_usc&docid= Cite:+21USC343.
584. US Federal Food, Drug, and Cosmetic Act, Section 343(r)(6) (21 USC 343(r)(6)). Available at http://frwebgate.access.gpo.gov/cgi-bin/getdoc.cgi?dbname=browse_usc&docid=Cite:+21USC343.
585. US Code of Federal Regulations, Title 21, Part 101.93(f) (21 CFR 101.93(f)). Available at http://a257.g.akamaitech.net/7/257/2422/26mar20071500/edocket.access.gpo.gov/cfr_2007/aprqtr/21cfr101.93.htm.
586. US Federal Food, Drug, and Cosmetic Act, Section 343-2 (21 USC 343-2). http://frwebgate.access.gpo.gov/cgi-bin/getdoc.cgi?dbname=browse_usc&docid=Cite: +21USC343-2.
587. US Code of Federal Regulations, Title 21, Part 312 (21 CFR Part 312). Available at http://www.access.gpo.gov/nara/cfr/waisidx_07/21cfr312_07.html.
588. Draft guidance Complimentary and Alternative Medicine Products and their Regulation by the Food and Drug Administration December 2006. Available at http://www.fda.gov/OHRMS/DOCKETS/98fr/06d-0480-gld0001.pdf.
589. Sanders ME et al. Probiotics: their potential to impact human health. Council for Agricultural Science and Technology (CAST) Issue Paper, 2007. Available at www.cast-science.org.
590. 36.14, Direct-fed microorganisms. Feed Ingredients Definitions. Official Publication, 2007. Association of American Feed Control Officials Incorporated. Available at http://www.aafco.org.
591. Food and Feed Safety, European Commission. Available at http://ec.europa.eu/food/food/animalnutrition/feedadditives/legisl_en.htm.

592. Ministry of Agriculture, the People's Republic of China. Available at http://www.agri.gov.cn.
593. Food and Agricultural Materials Inspection Center (FAMIC), Japan. Available at http://www.famic.go.jp.
594. Department of Livestock Development, Ministry of Agriculture and Cooperative, Thailand. Available at http://www.dld.go.th/webenglish/index.html#home.
595. Australian Pesticides and Veterinary Medicines Authority, Australia. Available at http://www.apvma.gov.au/index.asp.
596. Agricultural Compounds and Veterinary Medicines Regulations 2001 SR 2001/101, New Zealand. Available at http://www.nzfsa.govt.nz/site/exit/legislation/index.htm.
597. GRAS register for Oral Nutritional Compounds, New Zealand. Available at http://www.nzfsa.govt.nz/acvm/registers-lists/index.htm.

2

SELECTION AND MAINTENANCE OF PROBIOTIC MICROORGANISMS

YUAN KUN LEE
Department of Microbiology, National University of Singapore, Singapore

2.1 ISOLATION OF PROBIOTIC MICROORGANISMS

Various potential probiotic microorganisms could be isolated from the mouth, gastrointestinal (GI) content, and feces of animal and human by subculturing the respective microorganisms on the appropriate enrichment or selective agar media as suggested below.

Isolation of Probiotic Microorganisms[a,b]

Organisms	Nonselective Medium	Selective Medium	Supplements	Culture Conditions
Lactobacilli	MRS medium	SL medium	Cycloheximide[c] (100 mg/L), cysteine[d] (0.05% w/v), growth factors[e], other carbohydrates[f]	10% O_2 + 90% N_2 or H_2
Bifidobacteria	MRS medium	TPY medium[g]	Antibiotics[h]	37–40°C in anaerobic jar, 10% CO_2 + 90% H_2

(continued)

Handbook of Probiotics and Prebiotics, Second Edition Edited by Yuan Kun Lee and Seppo Salminen
Copyright © 2009 John Wiley & Sons, Inc.

(Continued)

Organisms	Nonselective Medium	Selective Medium	Supplements	Culture Conditions
Streptococci	Strep Base medium[i]	TYC medium[j]	—	5% CO_2 in air
Enterococci	Brain–heart infusion medium[k]	Kanamycin aesculin medium[l]	—	>3% CO_2 in air
Yeast	YPD medium[m]	Rose Bengal medium[n]	Chloramphenical (100 mg/L)	Air

[a] After Butler JP, editor. *Bergey's Manual of Systematic Bacteriology*, Vol. 2. Williams & Wilkins, Baltimore, 1986.
[b] After Balows A, Trüper H, Devorkin M, Harder W, and Schleifer K-H, editors. *The Prokaryotes. A Handbook on the Biology of Bacteria: Ecophysiology, Isolation, Identification, Applications*, Vol. II, Springer Verlag, New York, 1992.
[c] To eliminate yeasts.
[d] To isolate anaerobic lactobacilli from intestinal sources.
[e] Examples are meat extract, tomato juice, fresh yeast extract, malt extract, ethanol, mevalonic acid (sake), beer, and juices, to improve the isolation of lactobacilli, which are adapted to a particular ecological niche.
[f] Replacement of glucose by maltose, fructose, sucrose, or arabinose is recommended for the isolation of heterofermentative lactobacilli.
[g] TPY medium: trypticase (BBL), 10 g; phytone (BBL), 5 g; glucose, 5 g; yeast extract (Difco), 2.5 g; Tween 80, 1 mL; cysteine hydrochloride, 0.5 g; K_2HPO_4, 2 g; $MgCl_2 \cdot 6H_2O$, 0.5 g; $ZnSO_4 \cdot 7H_2O$, 0.25 g; $CaCl_2$, 0.15 g; $FeCl_3$, a trace; agar, 15 g; distilled water, 1000 mL. Final pH value is about 6.5 after autoclaving at 121°C for 25 min; dilutions can be made with the same liquid medium.
[h] Use of kanamycin, neomycin, paramomycin, sodium propionate, lithium chloride, sorbic acid, or sodium azide could improve selectivity in some cases.
[i] Strep Base: proteose peptone, 20 g/L; yeast extract, 5 g/L; NaCl, 5 g/L; Na_2HPO_4, 1 g/L; glucose, 5 g/L. Dissolve ingredients in distilled water, adjust pH to 7.6. Autoclave at 121°C for 15 min.
[j] TYC medium: trypticase, 15.0 g/L; yeast extract, 5.0 g/L; L-cysteine, 0.2 g/L; Na_2SO_3, 0.1 g/L; NaCl, 1.0 g/L; $NaHCO_3$, 2.0 g/L; $Na_2HPO_4 \cdot 12H_2O$, 2.0 g/L; sodium acetate ($3H_2O$), 20.0 g/L; sucrose, 50.0 g/L. Adjust pH to 7.3. Autoclave at 121°C for 15 min.
[k] Brain–heart infusion medium: calf brain infusion solids, 12.5 g/L; beef heart infusion solids, 5.0 g/L; proteose peptone, 10.0 g/L; dextrose, 2.0 g/L; NaCl, 5.0 g/L; Na_2HPO_4, 2.5 g/L; pH 7.4. Sterilize by autoclaving at 121°C for 15 min.
[l] Kanamycin aesculin medium: tryptone, 20 g/L; yeast extract, 5 g/L; NaCl, 5 g/L; sodium citrate, 1 g/L; aesculin, 1 g/L; ferric ammonium citrate, 0.5 g/L; sodium azide, 0.15 g/L; kanamycin sulfate, 0.02 g/L. Bring to boil to dissolve completely. Sterilize by autoclaving at 121°C for 15 min.
[m] YPD medium: yeast extract, 3 g/L; peptone, 5 g/L; dextrose, 10 g/L; malt extract, 3 g/L; final pH 6.2. Bring to boil to dissolve. Sterilize by autoclaving at 121°C for 15 min.
[n] Rose Bengal medium: mycological peptone, 5 g/L; glucose, 10 g/L; dipotassium phosphate, 1 g/L; magnesium sulfate, 0.5 g/L; Rose Bengal, 0.05 g/L; pH 7.2. Bring to boil to dissolve completely. Add chloramphenical and mix gently. Sterilize by autoclaving at 121°C for 15 min.

2.2 SELECTION OF PROBIOTIC MICROORGANISMS

Common criteria for the selection and development of a new probiotic strain are mentioned below.

2.2.1 Manufacturing Criteria (General Criteria)

A pleasant aroma and flavor profiles are of importance in the formulation of probiotic functional foods. More importantly, probiotic strains should be hardy and stable enough to withstand conventional industrial production processes. These criteria may include the following:

(a) ease in maintaining in storage without loss of viability;
(b) ability to revitalize and grow quickly to the maximum concentration in a simple and cheap fermentation medium;
(c) ability to grow and survive in microaerophilic or aerobic conditions;
(d) ability to withstand physical handling without significant loss of viability;
(e) ability to survive in the food matrices and during processing.

In general, lactobacilli are insensitive to cold storage at $-12°C$ or lower (5), but sensitive to freeze–drying process. A 5% glycerol in suspension medium (1) and chopped meat broth with 12% sucrose (10) increase survival during lyophilization.

Most lactic acid bacteria can be cultured in conventional stirred tank fermentor and are rather resistant to centrifugation.

During animal feed pellet production, heat treatment at 90°C, over a very short period of 14 s, kills most of the lactic acid bacteria and yeasts, except for bacterial spores.

2.2.2 Shelf Life and Gut Transit (General Criteria)

If the desirable probiotic properties are dependent on the metabolic activities and the close vicinity of probiotics to host cells for a prolonged period of time, sufficient numbers of viable cells must be present at the time of consumption and the probiotics must be able to colonize on the intestinal mucosal surface.

2.2.2.1 Shelf Life of Viable Probiotics Under Different Storage Conditions
The shelf life of viable probiotic microorganisms is determined largely by their storage temperature.

Cell Survival During Cold Storage of Cultured Milk at 4°C

Organisms	Initial Cell Concentration (cfu/mL)	Viable Cell Concentration (% Initial)			
		Day 0	Day 14	Day 28	Day 49
L. casei strain Shirota	1.6×10^8	100	93.8	93.8	87.5
L. casei NCIMB 8822	1.3×10^8	100	43.1	6.8	2.5
L. casei NCIMB 11970	1.5×10^8	100	43.3	8.0	4.6

(continued)

(*Continued*)

Organisms	Initial Cell Concentration (cfu/mL)	Viable Cell Concentration (% Initial)			
		Day 0	Day 14	Day 28	Day 49
L. paracasei NCIMB 8001	3.6×10^8	100	100	94.4	83.3
L. paracasei NCIMB 9709	1.4×10^8	100	100	100	100
L. paracasei NCIMB 9713	3.3×10^8	100	100	75.8	42.4
L. rhamnosus NCIMB 6375	4.8×10^8	100	64.5	10.4	4.0
L. rhamnosus GG	2.0×10^8	100	90	50	49
L. acidophilus NCIMB 8690	1.7×10^8	100	25.9	1.9	0.008
L. acidophilus CH 5	7.0×10^7	100	0.03	0.0014	0.001
L. bulgaricus NCIMB 11778	5.2×10^7	100	67.3	1.2	0.06
S. thermophilus NCIMB 10387	3.2×10^6	100	10.3	0.25	0.072

Data provided by Lee YK, National University of Singapore.

Cell Survival During Storage of Cultured Milk at 25°C (Room Temperature)

Organisms	Initial Cell Concentration (cfu/mL)	Viable Cell Concentration (% Initial)			
		Day 0	Day 14	Day 28	Day 49
L. casei strain Shirota	1.6×10^8	100	87.5	62.5	57.5
L. casei NCIMB 8822	1.3×10^8	100	60.8	57.7	55.4
L. casei NCIMB 11970	1.5×10^8	100	66.7	42	10
L. paracasei NCIMB 8001	3.6×10^8	100	72.2	20.8	3.9
L. paracasei NCIMB 9709	1.4×10^8	100	71.4	12.9	4.1
L. paracasei NCIMB 9713	3.3×10^8	100	93.9	17.9	4.8
L. rhamnosus NCIMB 6375	4.8×10^8	100	25	7.1	1.9
L. rhamnosus GG	2.0×10^8	100	65	46.5	16
L. acidophilus NCIMB 8690	1.7×10^8	100	2.3	0.03	$<10^{-6}$
L. acidophilus CH 5	7.0×10^7	100	0.005	$<10^{-5}$	$<10^{-5}$
L. bulgaricus NCIMB 11778	5.2×10^7	100	21	$<10^{-5}$	$<10^{-5}$
S. thermophilus NCIMB 10387	3.2×10^6	100	0.013	$<10^{-4}$	$<10^{-4}$

Data provided by Lee YK, National University of Singapore.

2.2.2.2 Tolerance to Digestive Juices This is to determine the possibility of a microorganism to survive in various parts of the GI tract. Lysozyme (4), gastric juice (2), and jejunal fluid (7) were being used for testing the tolerance of passage through the GI tract. The following data show the stability of lactic acid bacteria in a synthetic gastric juice (0.2% bile salt w/v in MRS) at 37°C. It appears that the tolerance of lactic acid bacteria to bile and acidity is strain dependent and needs to be evaluated individually. Lactobacilli are more acid tolerant in a food carrier and the food matrix determines their pH tolerance (2, 7).

Organisms	Viable Cell Concentration (cfu/mL)	
	Initial	After 3 h
L. casei strain Shirota	2.0×10^8	1.0×10^8
L. casei NCIMB 8822	2.9×10^8	4.0×10^2
L. casei NCIMB 11970	5.8×10^8	5.7×10^2
L. paracasei NCIMB 8001	8.7×10^8	1.2×10^7
L. paracasei NCIMB 9709	3.2×10^8	<100
L. paracasei NCIMB 9713	5.2×10^8	1.1×10^7
L. rhamnosus NCIMB 6375	4.5×10^8	6.9×10^6
L. rhamnosus GG	2.3×10^8	1.0×10^8
L. acidophilus NCIMB 8690	1.5×10^7	5.4×10^4
L. acidophilus CH 5	2.0×10^7	2.1×10^5
L. delbrueckii subsp. bulgaricus NCIMB 11778	9.6×10^6	<100
S. thermophilus NCIMB 10387	2.6×10^6	6.1×10^3

Data provided by Lee YK, National University of Singapore.

2.2.2.3 Adhesion and Colonization onto Specific Site of Body Surface (See Also Section 5.1) If a long residence time and multiplication on the host mucosal surface are required to achieve maximum probiotic effects, the probiotic microorganism needs to colonize the surface and achieve a relatively short generation time to prevent from being removed together with the mucus and lumen content. Adhesion can be considered as the first step of colonization. A reason for most commercial probiotics not able to colonize the human GI tract permanently has been attributed to their doubling rate in GI tract being slower than the washout rate (6).

2.2.3 Health Properties (Specific Criteria)

(a) Ability to utilize prebiotics for growth (see Chapter 7).
(b) Ability to synthesize vitamins (see Section 4.3.1).
(c) Ability to inhibit or exclude pathogens (see Section 5.1).
(d) Antibiotic resistance across a wide range of antibiotics.
   Antibiotic resistance would allow the probiotics to be used with antibiotic administration to prevent antibiotic-associated diarrhea (see Section 4.3.3). However, there is a concern on transfer of antibiotic resistance genes.
(e) Ability to synthesize β-galactosidase (see Section 4.3.1).
(f) Ability to deconjugate bile acid.
(g) Ability to produce antimicrobial substances (see Section 5.2).
(h) Ability to modulate immune reactions (see Section 5.3).
   Immune activation by probiotics has been suggested to be responsible for the prevention and treatment of acute gastroenteritis in humans. Immune

effects have also been related to probiotic effects seen in colon and bladder cancer studies. Some target cells and biomarkers of immune responses of specific diseases are listed below.

Disorders	Target Cells, Cytokines, and Antibody
Allergies	Th1/Th2, Treg, Tr-1, IL-12, IFN-γ, IL-10, TGF-β, IL-4, IL-5, IgE
Autoimmune diseases	Treg, Tr-1, Th3, IL-10, TGF-β, IL-1β, IL-6, TNF-α
Cancer	NK cell, cytotoxic T cell, NKT cell, IL-1β, IL-12, TNF-α
Infections	Macrophage, NK cell, neutrophil, IL-12, IL-18, IFN-γ, TNF-α
Inflammatory bowel disease	Treg, Tr-1, IL-10, IL-6, TNF-α

(i) Specific induction of death of tumor cells.

2.2.4 Safety (See Also Section 1.7)

The safety of lactic acid bacteria used in clinical and functional food is of great importance. In general, lactic acid bacteria have a good record in safety, and no major problems have occurred although cases of infection have been reported with several strains in immunocompromised patients. Safety has been documented with dairy strains (2, 7, 9). It is most important for novel probiotics to have their safety assured and to conform to all local regulations (see Section 1.7.1).

2.2.5 Identification (See Section 1.2)

An example of a decision tree for the selection of a probiotic strain for a defined probiotic property.

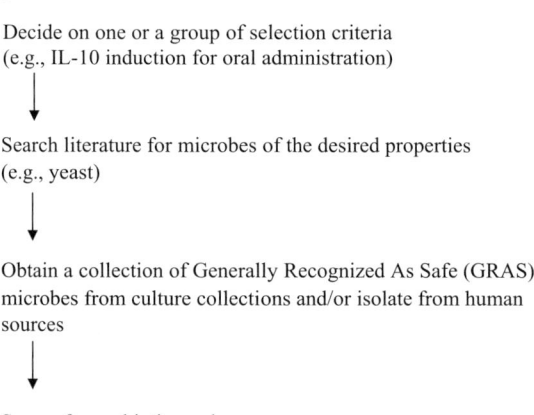

(e.g., induction of IL10 production in HTP1 cells)

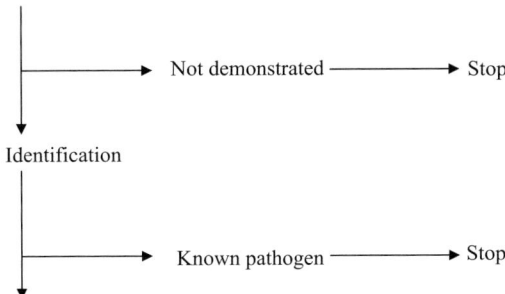

Identification

Biosafety
(e.g., GRAS status of newly isolated microbes, short- and long-term toxicology tests)

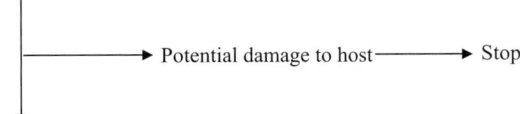

Growth properties
(e.g., ease of cultivation, specific growth rate, growth yield, tolerance to processing stress)

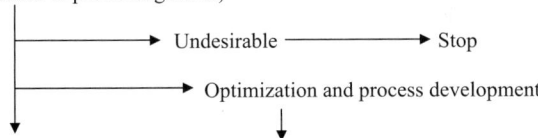

Stability and shelf life in product matrix

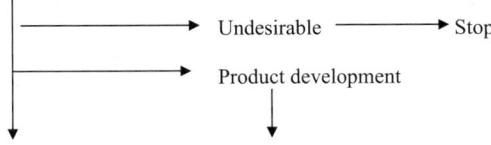

184 SELECTION AND MAINTENANCE OF PROBIOTIC MICROORGANISMS

Tolerance to GI passage (if viable probiotic is essential)
(e.g., low pH, gastric juice, bile, pancreas juice, intestinal juice)

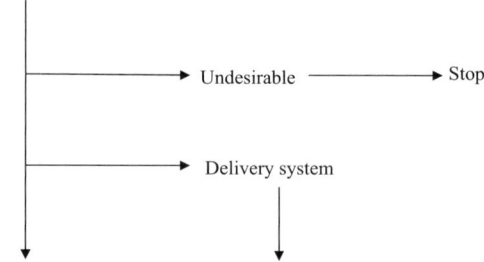

Colonization of GI tract (if persistence in GI tract is desirable)

(e.g., adhesion to intestinal tissue cells, resident time in GI tract of intended host)

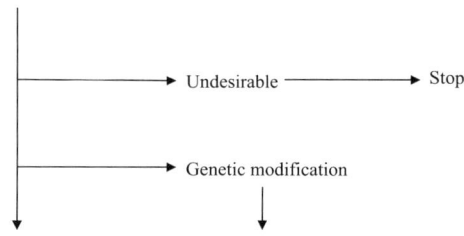

Animal models for the above criteria

Human/clinical study

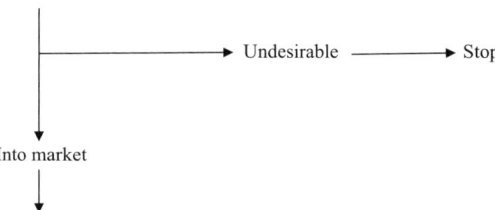

Into market

Follow up and refinement of product

2.3 MAINTENANCE OF PROBIOTIC MICROORGANISMS

Organism	Period of Storage	Culture Medium/Method	Growth Phase	Storage Temperature (°C)	Cryoprotectant	Frequency of Transfer
Lactobacilli	Short term	MRS agar stab	Early exponential	4–7	MRS agar	Weekly
	Long term	MRS agar spread	Early exponential	−20	MRS	Several months
		MRS	Late exponential	Lyophilized, 5–8	Skim milk or horse serum + 7.5% w/v glucose	10–20 years
Bifidobacteria	Short term	MRS 0.5% agar stab	Early exponential	3–4	MRS 0.5% agar stab	2 weeks
B. thermophilum	Short term	MRS agar slant in 10% CO_2 in air	Early exponential	3–4	MRS agar slant	2 weeks
Streptococci	Short term	Appropriate medium slope/stab	Early exponential	4	Culture medium	Weekly
		Litmus milk + 10% chalk + 0.3% yeast extract + 10% glucose	Early exponential	10	Litmus milk	Weekly
	Long term	Appropriate medium	Late exponential	−80, deep frozen or in liquid N_2	1% tryptone, 0.5% yeast extract, 0.1% glucose, 0.1% cysteine HCl, 2% bovine serum	Several months

(*continued*)

185

(Continued)

Organism	Period of Storage	Culture Medium/Method	Growth Phase	Storage Temperature (°C)	Cryoprotectant	Frequency of Transfer
Enterococci	Short term	Litmus milk + 1% chalk	Early exponential	Lyophilized, 5–8	Skim milk or serum	10–20 years
				4	Litmus milk + 1% chalk	3 months
	Long term	Appropriate medium	Late exponential	Lyophilized, −20	Nutrient broth + inactivated horse serum + glucose	2 years
Yeast	Short term	YPD agar	Late exponential	4	YPD	1 year
	Long term	YPD medium	Late exponential	−20	1 part culture + 1 part glycerol	10 years

After Butler JP, editor. *Bergey's Manual of Systematic Bacteriology*, Vol. 2, Williams and Wilkins, Baltimore, 1986. After Balows A, Trüper H, Devorkin M, Harder W, Schleifer K-H, editors. *The Prokaryotes. A Handbook on the Biology of Bacteria: Ecophysiology, Isolation, Identification, Applications*, Vol. II, Springer Verlag, New York, 1992.

REFERENCES

1. Bozoglu TF and Gurakan GC. Freeze−drying injury of *Lactobacillus acidophilus*. *J. Food Protect.* 1989; 52: 259−260.
2. Conway PL, Gorbach SL, and Goldin BR. Survival of lactic acid bacteria in the human stomach and adhesion to intestinal cells. *J. Diary Sci.* 1987; 70: 1−12.
3. Donohue DC and Salminen S. Safety of probiotic bacteria. *Asia Pacific J. Clin. Nutr.* 1996; 5: 25−28.
4. Gilliland SE. Beneficial interrelationships between certain microorganisms and humans: candidate microorganisms for use as dietary adjuncts. *J. Food Protect.* 1979; 42: 164.
5. Klaenhammer TR and Kleeman EG. Growth characteristics, bile sensitivity and freeze damage in colonial variants of *Lactobacillus acidophilus*. *Appl. Environ. Microbiol.* 1981; 41: 1461−1467.
6. Lee YK, Ho PS, Low CS, Arvilommi H, and Salminen S. Permanent colonization by *Lactobacillus casei* is hindered by the low rate of cells division in mouse gut. *Appl. Environ. Microbiol.* 2004; 70: 670−674.
7. Robins-Browne RM, Path FF, and Levine MM. The fate of ingested lactobacilli in the proximal small intestine. *Am. J. Clin. Nutr.* 1981; 34: 514−519.
8. Salminen S, Isolauri E, and Salminen E. Clinical uses of probiotics for stabilizing the gut mucosal barrier: successful stains and future challenges. *Anton. Leeuwenhoek.* 1996; 70: 347−358.
9. Saxelin M, Rautelin H, Salminen S, and Mäkelä H. The safety of commercial products with viable *Lactobacillus* strains. *Infect. Dis. Clin. Pract.* 1996; 5: 331−335.
10. Staab JA and Ely JK. Viability of lyophilized anaerobes in two media. *Microbiology* 1987; 24: 174−178.

3

GENETIC MODIFICATION OF PROBIOTIC MICROORGANISMS

Koji Nomoto, Mayumi Kiwaki, and Hirokazu Tsuji
Yakult Central Institute for Microbiology Research, Tokyo, Japan

3.1 MUTANTS OBTAINED FROM PROBIOTIC MICROORGANISMS BY RANDOM MUTAGENESIS

Handbook of Probiotics and Prebiotics, Second Edition Edited by Yuan Kun Lee and Seppo Salminen
Copyright © 2009 John Wiley & Sons, Inc.

Spontaneous Mutagenesis

Selection of Mutants	Character	Sites of Mutation	Species	References
Growth in medium containing sodium cholate	Cholate resistance, changes in fermentation pattern and membrane protein profile	—	BFan, BFbi	(33)
Growth in medium containing bile salt	Bile resistance, tolerance to low pH, changes in glycosidase activities, F6PPK activity, fermentation pattern, membrane protein profile, fatty acid composition, higher production of F_1F_0-ATPase in the absence of bile at low pH	—	BFan, BFbi, BFin, BFflo	(44, 51, 52, 54, 55)
Mutants from Ref 33, 44	Changes in the antibiotic resistance pattern, higher adhesion to mucus, changes in hydrophobicity	—	BFan, BFbi, BFlo	
Growth in medium containing rifaximin	Rrifaximin resistance, changes in protein profile	$rpoB$	BFin	(67)
Loss of halo in agar containing glycodeoxycholic acid	BSH negative	bsh	BFlo	(61)
Survival of low pH condition	Resistance to acid and bile, changes in protein profile	—	BFlo	(53)
Growth in medium containing bacteriocin, mundticin KS	Mun^r, changes in fatty acid composition	—	ENfm	(56)
Growth in medium containing rifampicin	Rif^r, at various MIC	$rpoB$	ENfm	(9)
Growth in medium containing fluoroquinolone	SPX^r, NOR^r at various MIC	$gyrA$, $parC$	ENfm	(46)

Condition	Phenotype	Gene	Species	Ref
Growth in medium containing quinolones	Quinolone resistance at various MIC	gyrA, parC	ENfm	(71)
Survival after repeated freezing and thawing	Improved cryotolerance, higher acidifying activity after thawing	—	LBdb	(36)
Growth in medium containing neomycin	Acid sensitive, decrease in membrane-bound H^+-ATPase activity	—	LBdb	(45)
Growth under low pH condition	Higher lactic acid production on hydrolyzed cane sugar	—	LBde	(27)
Survival after infection with lytic phage	No plaque-forming ability, absence of lysogeny	—	LBde	(19)
Survival after infection with lytic phage	Stable phage resistance, similar or improved immunomodulating capacity	—	LBde	(66)
Growth in medium containing bile salt of mutants form Ref. 66	Phage resistance, bile resistance	—	LBde	(18)
Growth under low pH condition	Improved acid tolerance	—	LBsp	(48)
Growth on agar containing lacticin 3147	Decreased adsorption of lacticin 3147, cross-resistance to other antimicrobes, NaCl, lysozyme	—	LCla	(20)
Color production on agar containing TTC	Production of excess CO_2	—	LCld	(8)
Growth in medium containing galactose	Gal$^+$, higher GalK activity	galK promoter	STth	(65)
Loss of the ability to solidify milk medium	Nonropy, no EPS production	espF*	STth	(14)

(continued)

(Continued)

Selection of Mutants	Character	Sites of Mutation	Species	References
Growth in medium containing thymidine and TMP	Growth dependence on thymidine	thyA	STth	(58)
Loss of halos in agar containing urea and pH indicator	Urease negative, higher acidifying activity in milk	urease operon	STth	(39)
Growth in medium containing α-ketobutyrate and leucine	α-Acetolactate decarboxylase deficiency, higher production of diacetyl	aldC	STth	(38)

Chemical Mutagenesis

Conditions	Selections	Character	Species	References
EMS				
120 μL/mL, 25°C, 2 h	Color production on agar containing X-gal	Overproduction of α-galactosidase	BFbr, BFlo, LBdb, STth	(24)
20 μL/mL, 37°C, 45 min	Poor growth on glucose–milk agar, loss of the ability to hydrolyze L-leucine-β-naphthylamide	Proteinase deficiency, aminopeptidase deficiency	LBca	(10)

NTG				
50 mg/mL, 25°C, 2 h	Color production on agar containing X-gal	Overproduction of α-galactosidase	BFbr, BFlo, LBdb, STth	(24)
1 mg/mL, 1.5 h	Loss of halo in agar containing taurodeoxycholate	Partial reduction of CBSH activity, decrease in growth rate in the presence of bile salt	LBam	(15)
37°C, 60 min, a two to three log reduction in cell number	Less biomass and more acid production as lower A_{600}/A_{650} ratio in medium containing pH indicator	Overproduction of EPS, ropy, or mucoid phenotype	LBdb	(69)
170 μg/mL, 37°C, 15 min	Loss of the ability to grow on agar containing low concentration of Mn^{2+}	Mn^{2+} dependency, loss of the high-affinity Mn^{2+}, and Cd^{2+} uptake system	LBpl	(23)
200 μg/mL, 37°C, 60 min	Production of α-acetolactate on agar containing pyruvate	α-Acetolactate decarboxylase deficiency, no increase in production of diacetyl	LBrh	(37)
	Production of larger halos on agar containing $CaCO_3$ at acidic pH	Growth on lower pH, increase in lactic acid production	LBrh	(68)
	Growth on acidic region of pH gradient plate	Improved acid tolerance	LBsp	(48)
Deposition of a crystal of NTG on agar plate, 30°C, 5 days	Expression of reporter *cat* gene under p15 promoter-like sequence from LBac	Deletion and mutation in P15 sequence, change of promoter activity in various species	LCla	(2)
20–500 μg/mL, 30°C, 60 min	Production of α-acetolactate on agar containing citrate	α-Acetolactate decarboxylase deficiency	LCld	(40)
50 μg/mL, 30°C, 60 min	Production of acetoin and diacetyl on agar excluding citrate	Attenuation of LDH activity, lower production of lactate	LCld	(5)
20 or 200 μg/mL, 30°C, 60 min	Color production resulting from low acid formation on agar containing TTC	Production of excess CO_2	LCld	(8)

(continued)

(*Continued*)

Conditions	Selections	Character	Species	References
Two cycles of NTG treatment with concentration adjusted to 90% lethality	Production of α-acetolactate, low acid formation on agar containing TTC	α-Acetolactate decarboxylase deficiency, low LDH activity, overproduction of α-acetolactate, diacetyl, acetoin	LCld	(35)
5–50 μg/mL, 37°C, 60 min	Loss of halos in agar containing urea and pH indicator	Urease negative, less acidifying activity than parent strains	STth	(39)
Hydroxylamine				
5 μg of plasmid containing P170 promoter region, 0.4 M hydroxylamine in 200 μL, 70°C, 2 h	Transformation to *L. lactis*, color production on agar containing X-Gal	Mutations in promoter P170, increased promoter activity retaining pH-responsiveness	LCla	(31)

UV Irradiation

Conditions	Selections	Character	Species	References
250 nm, at a distance of 25 cm, 1, 1.5, 2, and 2.5 min	Loss of halo in agar containing glycodeoxycholic acid	BSH negative, mutation in *bsh*	BFlo	(61)
254 nm, at a distance of 6 cm, 30 s	Higher acid production on agar containing hydrolyzed cane sugar	Efficient growth on and higher production of fructose and lactic acid from hydrolyzed cane sugar, efficient production of lactose from cellobiose and cellotriose	LBde	(1, 27, 47)

Tools and Methods	Selections	Character	Insertion Sites	Species	References
254 nm, at a distance of 20 cm, 1–40 min	Production of acid on agar containing a high concentration of lactate calcium	Overproduction of lactic acid, resistance to feedback repression		LBla	(4)
	Production of larger halos on agar containing $CaCO_3$ at acidic pH	Growth on lower pH, increase in lactic acid production		LBrh	(68)
254 nm at 44.7 J/m²	Color production resulting from low acid formation on agar containing TTC	Production of excess CO_2		LCld	(8)

Transposon Mutagenesis

Tools and Methods	Selections	Character	Insertion Sites	Species	References
Tn916					
Transformation with the plasmid carrying Tn916	Loss of the intrageneric coaggregation ability	Cog negative, absence of a 100-kDa surface protein	Upstream region of 33.6 kDa unknown protein	STgo	(70)
Transformation with the plasmid carrying Tn916	Loss of the intrageneric coaggregation ability	Cog negative	*dltB*	STgo	(7)
Plate mating with *E. faecalis* carrying Tn916 on plasmid	Color production resulting from loss of ropy character on agar containing ruthenium red	Lost the ability to produce EPS	*epsK*	STth	(60)
Plate mating with *E. faecalis* carrying Tn916 on plasmid	Color production resulting from loss of ropy character on agar containing ruthenium red	Lost the ability to produce EPS, impaired growth, abnormal cell morphology	*pbp2b*	STth	(59)

(continued)

(Continued)

Tools and Methods	Selections	Character	Insertion Sites	Species	References
Tn917					
Transposition from ts plasmid by temperature shift	Difference in killing activities against nematodes in aerobic or anaerobic condition	Lower H_2O_2 production with lower killing activity, higher H_2O_2 production with higher killing activity	nox, upstream region of npr	ENfm	(42)
Transposition of Tn917 carrying promoter-less lacZ gene from ts plasmid without temperature shift	Replication twice on agar containing Em for selection of Tn917, color production on agar containing X-gal	Random insertion into chromosome, β-Gal$^+$ promoter fusant	—	LCla	(25)
Transposition of Tn917 carrying promoter-less lacZ gene from ts plasmid without temperature shift	Color production on agar containing X-gal	β-Gal$^+$ with promoter regulated by pH and/or temperature	—	LCla	(26)
Transposition of Tn917 carrying promoter-less lacZ gene and nuclease gene lacking signal sequence without temperature shift	Color production on agar containing X-gal, development of orange halo in agar containing DNA and TB	Nuc$^+$ and Lac$^+$, fusion of nuclease with secretion protein	CluA, DexB, HAS, membrane transporter	LCla	(50)
ISS1					
Transposition of pG$^+$h9: ISS1 into chromosome by temperature shift	Expression of reporter CAT without p-coumaric acid induction	padA promoter losing regulation by phenolic acids	rhe3, lp_0385	LBpl	(22)

Method	Phenotype	Gene(s)	Strain	Ref.
Transposition of pG⁺h9: IS*S1* by temperature shift	Growth on agar under the condition of low pH and high temperature	Acid tolerance, multi-stress resistance	RBS of *pstS*, promoter of *deoB*, intergenic region of *glnP-glnQ hpt, relA, guaA, recN*	LCla	(49)
Transposition of pGh: IS*S1* by temperature shift	Growth on agar containing Ab under high temperature for integration, temperature shift without Ab for plasmid-excision	Random insertion into chromosome, leaving a IS*S1* copy after plasmid-excision	—	LCla, STth	(32)
Transposition of pG⁺h9: IS*S1* by temperature shift	Poor growth on milk-based agar	Slower growth in milk, BCAA requirement for growth	*ilvB, ilvC*	STth	(13)
Transposition of pG⁺h9: IS*S1* by temperature shift	Survival after challenge with lytic phage	No or delayed synthesis of intracellular phage DNA	orf 90, orf 269	STth	(30)
Transposition of pG⁺h9: IS*S1* by temperature shift	Modifications of temperature, growth phase, Ab concentration	Improved frequency of single copy insertion	—	STth	(3)
Transposition of pG⁺h9: IS*S1* by temperature shift	Poor growth on milk-based agar, from library in ref. 13	Significant lower growth rate in milk	*amiA1*	STth	(12)
Transposition of pG⁺h9: IS*S1* by temperature shift	Growth on agar containing Em	Undetectable level of recombination events between endogenous IS*S1*	—	STth	(63)
Transposition of pG⁺h9: IS*S1* by temperature shift	Lower tolerance to menadione, growth, and survival at lower concentration of menadione	Sensitive phenotype to oxidative agents, change in cell morphology	*mreD, rodA, pbp2b, cpsX, tgt, ossF, ossG, fatD, sufD, iscU*	STth	(62, 64)

(continued)

(*Continued*)

Tools and Methods	Selections	Character	Insertion Sites	Species	References
Transposition of pG+h9: IS*S1* by temperature shift	Higher tolerance to menadione, growth, and survival at higher concentration of menadione	Resistance to oxidative agents	*deoB, gst, rggC, osrD, osrE, osrF, osrG, orfP*	STth	(11)
Tn10					
Transposition of mini-Tn*10* from ts plasmid to chromosome by temperature shift	Resistance to Spec, difference of colony morphotype on agar	Riboflavin auxotrophy, growth response to exogenous vitamin B_2	*ribD*	BAce	(57)

Insertional Mutagenesis

Tools and Methods	Selections	Character	Insertion Sites	Species	References
Integration through homologous recombination of the RepA⁻ plasmid pORI19-based genomic library using ts helper plasmid	Loss of a halo on agar containing autoclaved bacterial cells, loss of color formation on agar containing maltose, and pH indicator	Autolysin negative, maltose negative	*acmA, malK*	LCla	(29)
Integration through homologous recombination of the RepA⁻ plasmid pORI19-based genomic library using ts helper plasmid	Growth on phage-containing soft agar, colonies exhibiting a reduced growth on sucrose containing agar	Delay of lysis caused by phage infection, reduced acidification capacity in milk	Protein implicating folate metabolism, subunits of oligopeptide transport system	STth	(28)

| | Integration through homologous recombination of the RepA⁻ plasmid pORI19-based genomic library using ts helper plasmid | Growth on agar containing lacticin 3147 | Resistance to lacticin 3147 and nisin | *mleS, yjjC, ymcF, pi322, tra9811* | LCla | (21) |

PCR-Mediated Mutagenesis

Conditions	Template	Selections	Character	Species	References
PCR under random mutagenesis condition with low GTP content	*mesYI* Gene cloned on plasmid	Cloning, sequencing, transformation into *Leuconostoc* strain, bacteriocin assay using well-diffusion method	Amino acid substitution in bacteriocin peptide, loss of the bacteriocin production, or weaker bactericidal activity	LEme	(41)
Successive PCR with manganese or dITP as mutagen	*sppK* Gene cloned on plasmid	Cloning, transformation into *Lactobacillus* strain, constitutive expression of reporter gene from peptide pheromone promoter	Amino acid substitution in histidine protein kinase, no or weakend response to peptide pheromone	LBpl, LBsk	(34)

Mutator Strain

Methods	Selections	Character	Species	Reference
Passage of the plasmid carrying *Lactobacillus* gusA gene through ESco mutator strain	Recloning of gusA gene from ESco producing blue color on agar plates containing X-glu at pH 8	Amino acid substitution in GusA, higher β-glucuronidase activity over the pH range of 4 to 8 in LBga	LBga	(6)

Genome Shuffling

Methods	Selections	Character	Species	References
Serial protoplast fusion of acid-tolerant populations generated by adaptation and NTG treatment	Growth on acidic region of pH gradient plate	Successive improvement of population, growth at lower pH, higher production of lactate	LBsp	(48)
Serial fusion of lethal protoplasts obtained by heat and UV treatment using acid-tolerant mutants generated by UV and NTG treatment as initial strains	Growth on agar at low pH, larger halos on agar containing $CaCO_3$	Successive improvement of population, growth at lower pH, higher production of lactate	LBrh	(68)

Abbreviations—Ab: antibiotics; *acmA*: peptidoglycan hydrolase gene; *aldC*: α-acetolactate decarboxylase gene; *amiA1*: oligopeptide-binding protein gene; BAce: *Bacillus cereus*; BCAA: branched-chain amino acid; BFan: *Bifidobacterium animalis*; BFbi: *Bifidobacterium bifidum*; BFbr: *Bifidobacterium breve*`; BFin: *Bifidobacterium infantis*; BFlo: *Bifidobacterium longum*; β-Gal$^+$: β-galactosidase expression; BSH: bile salt hydrolase; *bsh*: bile salt hydrolase gene; CAT: chloramphenicol acetyl transferase; CBSH: conjugated bile salt hydrolase; CluA: cell-wall-associated protein involved in cell aggregation; Cog: coaggregation; *cpsX*: gene encoding membrane protein putatively involved in exopolysaccharide translocation; *deoB*: phosphopentomutase gene for purine/pyrimidine salvage; DexB: exoglycosylate; dITP: deoxyinosine triphosphate; *dltB*: gene of putative membrane protein involved in D-alanine incorporation; Em: erythromycin; EMS: ethyl methane sulfonate; ENfm: *Enterococcus faecium*;

EPS: exopolysaccharide; *epsK*: gene putatively involved in polymerization and export of exopolysaccharide; ESco: Escherichia coli; *espF**: gene encoding unknown protein locating in the eps gene cluster; F6PPK: fructose-6-phosphate phosphoketolase; *fatD*: iron ABC transporter gene; Gal$^+$: galactose fermenting; *galK*: galactokinase gene; GalK: galactokinase; *glnP*: gene encoding glutamate/glutamine ABC transporter; *glnQ*: gene encoding glutamate/glutamine ABC transporter; *gst*: proton/sodium-dicarboxylate symporter gene; *guaA*: GMP synthetase gene; *gusA*: β-glucuronidase gene; GusA: β-glucuronidase; *gyrA*: DNA gyrase subunit gene; HAS: hyaluronate synthase; *hpt*: hypoxanthine guanine phosphoribosyltransferase gene; *ilvB*: gene encoding large subunit of acetolactate synthase; LBac: *Lactobacillus acidophilus*; LBam: *Lactobacillus amylovorus*; *iscU*: gene encoding protein putatively involved in assembly of Fe/S cluster; Lac$^+$: LacZ expression; LBac: *Lactobacillus acidophilus*; LBam: *Lactobacillus amylovorus*; LBca: *Lactobacillus casei*; LBdb: *Lactobacillus delbrueckii* subsp. *bulgaricus*; LBde: *Lactobacillus delbrueckii*; LBga: *Lactobacillus gasseri*; LBla: *Lactobacillus lactis*; LBpl: *Lactobacillus plantarum*; LBrh: *Lactobacillus rhamnosus*; LBsk: *Lactobacillus sakei*; LBsp: *Lactobacillus* Sp; LCla: *Lactococcus lactis*; LCld: *Lactococcus lactis* subsp. *lactis* biovar *diacetylactis*; LDH: lactate dehydrogenase; lp_0385: putative histidine-binding protein; *malK*: ATP-binding protein gene; *mesY*: gene encoding pre-Mesentericin Y105 and immunity protein; *mleS*: malate-inducible malolactic enzyme gene; *mreD*: gene encoding putative rod-shaped determining protein; Mun^r: mundticin KS resistance; NOR^r: norfloxacin resistance; *nox*: NADH oxidase gene; *npr*: NADH peroxidase gene; NTG: *N*-methyl-*N'*-nitro-*N'*-nitrosoguanidine; Nuc$^+$: nuclease expression; rf 269: putative oxido-reductase; orf 90: putative chorismate mutase chain A; *orfP*: gene encoding protein with no known homologue; *osrD*: gene encoding protein with no known homologue; *osrF*: RNA methyltransferase gene; *osrG*: gene encoding protein with unknown function; *padA*: phenolic acid decarboxylase gene; *parC*: DNA topoisomerase IV subunit gene; *pbp2b*: gene encoding the penicillin-binding protein 2b; *pi322*: gene encoding phage-like protein; *pstS*: gene encoding phosphate ABC transporter; RBS: ribosome-binding site; *recN*: gene encoding DNA repair protein; *relA*: pppGpp synthetase gene; *rggC*: glycosyltransferase regulatory determinant gene; *rhe3*: putative ATP-dependant RNA halicase gene; *ribD*: riboflavin deaminase/reductase gene; Rif^r: rifampicin resistance; *rodA*: gene encoding putative rod-shaped determining protein; *rpoB*: β-subunit of RNA polymerase gene; Spec: spectinomycin; *sppK*: histidine kinase gene of two-component signal transduction system; SPX^r: sparfloxacin resistance; STgo: *Streptococcus gordonii*; STth: *Streptococcus thermophilus*; *sufD*: gene encoding protein putatively involved in assembly of Fe/S cluster; TB: Toluidine Blue O; *tgt*: tRNA guanine transglycosylase gene; *thyA*: thymidylate synthase gene; TMP: trimethoprim; *tra981l*: transposase gene of IS981; ts: temperature sensitive; TTC: 2,3,5-triphenyl tetrazolium; X-gal: 5-bromo-4-chloro-3-indolyl-D-D-galactopyranoside; X-glu: 5-bromo-4-chloro-3-indolyl-D-D-glucuronide; *yjjC*: gene encoding putative ATP binding domain of an ABC transporter; *ymcF*: gene encoding putative peptidoglycan-bound protein.

3.2 PLASMIDS

Strains	Plasmid	Size	Identified Gene	Phenotypes	References
Bifidobacterium asteroids					
DSM20089	pCIBA089	2.1 kb	*rep*	Cryptic	(93)
Bifidobacterium bifidum					
NCFB 1454	—	ca. 8 kb	—	Bac$^+$	(207)
Bifidobacterium breve					
NCFB 2258	pCIBb1	5,750 bp	*rep*	Cryptic	(146)
Bifidobacterium catenulatum					
L48	pBC1	2,540 bp	*repB, copG-like*	Cryptic	(76, 77)
Bifidobacterium longum					
BK51	pTB6	3,624 bp	*repB, membB, mobA*	—	(185)
DJO0A	pDOJH10L	10,073 bp	*rep, mob*	—	(128)
DJO0A	pDOJH10S	3,661 bp	*rep, mob*	—	(128)
KJ	pKJ50	4,960 bp	*mob, rep*	—	(153)
NAL8	pNAL8H	10 kb	—	—	(116)
NAL8	pNAL8M	4.9 kb	—	—	(116)
NCC293	pNCC293	1.8 kb	—	—	(124)
RW041	pNAC2	3,684 bp	*repB*	—	(91)
RW041	pNAC3	10,224 bp	*repB*	—	(91)
RW048	pNAC1	3,538 bp	*repB*	—	(91, 116)
A24	pBIFA24	4,892 bp	*rep, mob*	Cryptic	(155)
Bifidobacterium pseudomonas subsp. Globosum					
DPC479	pASV479	4.8 kb	*rep, ftsK-like*	—	(167)
Enterococcus faecium					
C264	pC264V	40 kb	*vanA*	Vm^r	(208)

Strain	Plasmid	Size	Genes	Phenotype	Ref.
FH1	pHTa	65.9 kb	*traA, vanA*	Vm^r, Cnj$^+$	(190, 191)
FH4	pHTb	63.7 kb	*traA, vanA*	Vm^r, Cnj$^+$	(190, 191)
FH7	pHTg	66.5 kb	*traA, vanA*	Vm^r, Cnj$^+$	(190, 191)
GF112	pMG1	65.1 kb	*traA, aacA-aphD*	Cnj$^+$, Gm^r, Km^r	(186, 191)
I125	pI125V	370 kb	*vanA*	Vm^r	(208)
L50	pCIZ1	50 kb	*entL50A, entL50B*	Bac$^+$	(92)
L50	pCIZ2	7,383 bp	*entqA, entqB, entqC*	Bac$^+$	(92)
NIAI 157	pEN100	49.0 kb	—	Bac$^+$	(150)
NIAI 157	pEN200	43.7 kb	—	—	(150)
T8	plasmid T8	7 kb	*mob, bac, imm*	Bac$^+$	(97)
ZB18	pZB18	67 kb	*vanA*	Vm^r, Cnj$^+$	(209)
ZB22	pZB22	200 kb	*vanA*	Vm^r, Cnj$^+$	(209)
Lactobacillus acidophilus					
I16	—	4.4 kb, 6.1 kb, 11.5 kb	—	Em^r	(87)
Lactobacillus brevis					
ABBC45	pRH45II	12,605 bp	*orf5*	Hop^r	(182)
ABBC45	pRH45	—	*horA*	Hop^r	(164, 184)
Lactobacillus casei					
A23	pSMA23	3,497 bp	*rep mob*	—	(181)
CRL 705	—	ca. 35 kb	*lac705b, lac705a*	Bac$^+$	(94)
IMPC LC34	—	8.8 kb	—	ABS	(169)
L-49	pLC494	8,846 bp	*repA, repB*	—	(78)
Lactobacillus delbrueckii subsp. *bulgaricus*					
B36	pDOJ1	6220 bp	*mob, tnp*	—	(127)
CFR 2028	—	9.4 kb, 6.5 kb	—	—	(81)
JCL414	pJBL2	8,716 bp	*hsdS*	—	(85)

(*continued*)

(Continued)

Strains	Plasmid	Size	Identified Gene	Phenotypes	References
NCC88	pN42	8,140 bp	*hsdS*	—	(85)
B1	pLBB1	6,127 bp	—	Cryptic	(80)
Lactobacillus fermentum					
KC5b	pKC5b	4,392 bp	—	Cryptic	(156)
ROT1	pLME300	19,398 bp	*erm(LF)*, *vat(E)*	Em^r, Df^r	(114)
VKM1311	pLF1311	2,389 bp	*repA*, *repB*	Cryptic	(74)
Lactobacillus helveticus					
ATCC15009	pLH1	19,360 bp	rep	—	(189)
ATCC15009	pLH2, pLH3	5.7 kb, 3.4 kb	—	Cryptic	(189)
CP53	pCP53	11.5 kb	—	Cryptic	(171)
GRA2	pLHg2	2.5 kb	—	Cryptic	(162)
GRA4	pLHg4	2.5 kb	—	Cryptic	(162)
NBIMCC 2727	pLBL4	2,556 bp	—	Cryptic	(111)
PRO1	pLHp1	6.4 kb	—	Cryptic	(162)
ROB1	pLHr1	2.3 kb	—	Cryptic	(162)
Lactobacillus hilgardii					
IOEB 0006	pHDC	80 kb	*hdcP, hdcA, hdcB, hisRS*	HDC$^+$	(136)
Lactobacillus johnsonii					
FI9785	p9785S	3,471 bp	*mob, rep*	Cryptic	(118)
FI9785	—	25.6 kb	—	Cryptic	(118)
Lactobacillus paracasei					
NFBC338	—	80 kb	—	—	(101)
Lactobacillus paracasei subsp. *paracasei*					
BGBUK2-16	—	ca. 80 kb	—	Bac$^+$	(135)

Lactobacillus paraplantarum				
C7	pC7	2,134 bp	*repA*	Cryptic	(129, 154)
Lactobacillus plantarum				
423	pPLA4	8,135 bp	*rep, mob, plaA, plaB, plaC, plaD*	Bac$^+$	(201, 202)
5057	pMD5057	10,877 bp	*tet(M)*	Tc^r	(96)
AS1.2986	pLP2000		*rep*	—	(95)
AS1.2986	pLP9000		*corA*	—	(95)
BIFI-38	pPB1	2,899 bp	*repA, repB, repC, mob*	Cryptic	(98)
CCUG 43738	—	ca. 14 kb	*tet(S)*	Tc^r	(119)
DG 507	ptet(M)$_{DG\ 507}$	ca. 10 kb	*tet(M)*	Tc^r	(120)
DG 507	perm(B)$_{DG\ 507}$	ca. 8.5 kb	*erm(B)*	Em^r	(120)
DG 522	ptet(M)$_{DG\ 522}$	ca. 40 kb	*tet(M)*	Tc^r	(120)
FB335	pLp3	ca. 36 kb	*ISLpl1*	—	(141)
L137	pLTK2	2,295 bp	*repA*	—	(122)
LL441	—	30 kb	*lacL, lacM*	Lac$^+$	(109)
LL441	—	40 kb, 80 kb		Bac$^+$	(99)
LL441	—	70 kb	*melA,*		(99)
NC7	p256	7,222 bp	*orf2, orf3*	TA$^+$	(179)
NGRI0101	pLKL	6.8 kb	—	Phage^r	(106)
NGRI0101	pLKS	2,025 bp	—		(106)
WCFS1	pWCFS101	1,917 bp	*rep*	Cryptic	(198)
WCFS1	pWCFS102	2,365 bp	*rep*	Cryptic	(198)
WCFS1	pWCFS103	36,069 bp	*arsR, arsD1, arsA, arsB, arsD2*	Cnj$^+$, AS^r	(198)
WHE92	HE92	ca. 11 kb	*papABCD*	Bac$^+$	(139)
					(*continued*)

(Continued)

Strains	Plasmid	Size	Identified Gene	Phenotypes	References
Lactobacillus reuteri					
100-23	pGT232	5.1 kb	*repA*	—	(117)
L1	pTE80	7.0 kb	*ermB* (*ermAM*)	Em^r	(131)
N16	pTR15	15 kb	*ermB* (*ermAM*)	Em^r	(131)
Lactobacillus sakei					
L45	pCIM1	50 kb	*lasAMNTUVPJW, lasXY*	Bac$^+$	(161)
RV332	pRV500	12,959 bp	*repA, hsdM, hsdR, hsdS*	—	(75)
Lactobacillus salivarius subsp. *salivarius*					
UCC118	pMP118	242,436 bp	*abp118, tktA, mipB, bsh*	Bac$^+$, Pen$^+$	(89, 130)
UCC118	pSF118-20	20.4 kb	—	—	(89)
UCC118	pSF118-44	44 kb	—	—	(89)
Lactobacillus sp.					
121B	p121BS	4,232 bp	*ermT*	Em^r, Ty^r	(206)
Lactococcus lactis					
C5	pAJ01	4 kb	—	Em^r	(160)
DPC5552	pCBG104	ca. 68 kb	—	Bac$^+$, Phage^r	(140)
NCDO 1867	—	ca. 70 kb	*gdh, cadA, cadC,*	GDH$^+$	(187)
NIZO B40	pNZ4000	42,180 bp	*epsRXABCDEFGHIJKL, repB1, repB2, repB3, repB4, mobA, corA*	Eps$^+$, Cnj$^+$	(197, 199, 200)
W-37	pSRQ900	10,836 bp	*abiQ*	Phage^r	(84)
Strain	pGdh442	68,319 bp	*gdh, cadA, cadC*	GDH$^+$, Cd/Zn^r	(188)

Lactococcus lactis subsp. *cremoris*

Strain	Plasmid	Size	Genes	Phenotype	Ref
DCH-4	pSRQ700	7,784 bp	*llaDCHIA, llaDCHIB, llaDCHIC*	Phage^r	(84)
HO2	pCI605	4.5 kb	—	—	(110)
HO2	pCI623	22.5 kb	—	—	(110)
HO2	pCI642	42 kb	*abiB*	—	(110)
HO2	pCI646	46 kb	—	Lac$^+$, Phage^r	(110)
HO2	pCI658	58 kb	—	Phage^r, Ads$^-$	(110)
HP	pHP003	13,433 bp	*prtP, prtM*	Prt$^+$	(88)
IL420	pIL2614	14 kb	*hsdR, hsdM, hsdS, abiP*	Phage^r	(102, 168)
P8-2-47	pBM02	3.85 kb	*copG, repB*	—	(166)
SK11	pSK11P	75,814 bp	*lacRABCDFEGX*	Lac$^+$	(172)
SK11	pSK11L	47,165 bp	*prtP, prtM*	Prt$^+$	(172)
SK11	pSK11A, pSK11B	10,372 bp, 13,332 bp	—	—	(172)
UC509.9	pCIS3	6,159 bp	*hsdS*	Phage^r	(170)
UC509.9	pCIS1	4 kb	*hsdS*	—	(170)
UC509.9	pCIS5			Lac$^+$	(170)
UC653	pCI750	65 kb	*abiGi, abiGii*	Phage^r	(143)
W10	pEW104	12.1 kb	*llagi*	Phage^r	(137)
W12	pAW122	8 kb	*hsdS*	—	(138)
W56	pJW566	24.4 kb	*llaBIII*	Phage^r	(126)

Lactococcus lactis subsp. *lactis*

Strain	Plasmid	Size	Genes	Phenotype	Ref
DPC3147	pMRC01	60,232 bp	*ltnA, ltnB, ltnM1, ltnT, ltnM2, tra*	Bac$^+$, Cnj$^+$, Phage^r	(90, 103, 163)
DPC4268	pMT60	60 kb	—	Prt$^+$	(192)
DPC4275	pMRC02	80 kb	*itnE, itnA, itnM, itnT, itnD*	Bac$^+$, Prt$^+$, Phage^r	(192)

(continued)

(Continued)

Strains	Plasmid	Size	Identified Gene	Phenotypes	References
IPLA 972	pBL1	10.9 kb	—	Bac$^+$	(165)
K214	pK214	29,871 bp	mdt(A), str, cat, tet(S)	MLS^r, Sm^r, Cm^r, Tc^r	(157)
KR2	pKR223	16,174 bp	llaKR2IR, llaKR2IM, abiR	Phage^r	(196)
LL57-1	pND324	3.6 kb	rep	—	(105)
LL58-1	pND306	54kb	lcoA, lcoB, lcoC	Cu^r	(132)
M14	pAR141	1,594 bp	repB, copG	—	(159)
M71	pND302	8.8 kb	repB, cadA, cadC	Cd^r	(133)
M71	pND302	8.8 kb	—	Cd^r	(134)
M189	pND300	60 kb	nisR, nisK, nicF, nisE, nisG	Bac$^+$	(104)
MJC15	pCD4	6094 bp	repB, hsdS	—	(108)
NCDO 275	pCI2000	60 kb	repA, parA	—	(123)
W1	pSRQ800	7,858 bp	abiK	Phage^r	(84)
BCRC10791	pL2	5,299 bp	repB	Cryptic	(86)
Lactococcus lactis subsp. *lactis* biovar *diacetylactis*					
CRL264	pCIT264		citQRP	Cit$^+$	(112)
DRC3	pNP40	64,980 bp	abiE, abiF, cadA, llaJIM1, llaJIM2, llaJIR1, llaJIR2	Phage^r, Cd^r	
(144, 145, 193)					
DPC220	pAH33	6,159 bp	hsdS	—	(147, 148)
DPC220	pAH82	20.3 kb	hsdR, hsdM, hsdS, cadA, cadC	Cd^r, Phage^r	(147, 148)
DPC721	pAH90	26,490 bp	hsdR, hsdM, hsdS, cadA, cadC, corA	Phage^r, Cd^r, Co^r, Ads$^-$	(147, 148)

77NIAI N-7	pCM1	8,280 bp	*citP*	Cit$^+$	(79)
S50	pS6, pS7a, pS7b, pS80	6.3 kb, 7.31 kb, 7.27 kb, 80 kb	—	—	(125)
S50	pS140	140 kb	—	Prt$^+$, Bac$^+$, Cnj$^+$	(125)
UK12922	pND859	16 kb	*abi-859*	Phage^r	(100)
Leuconostoc citreum					
IH3	pIH01	1.8 kb	—	Cryptic	(152)
Leuconostoc mesenteroides					
SY2	pFIMBL1	4,661 bp	—	Cryptic	(121)
Leuconostoc mesenteroides subsp. *mesenteroides*					
FR52	pFR18	1,828 bp	*rep18*	Cryptic	(82)
Y110	pTXL1	2,665 bp	—	Cryptic	(83)
Pediococcus acidilactici					
NCIMB 6990	pEOC01	11,661 bp	*ermB, aadE,* ω-ε-ζ operon	Cl^r, Em^r, Sm^r	(142)
Pediococcus damnosus					
ABBC478	pRH478	14,567 bp	*horA*	Hop^r	(183)
IOEB8801	pF8801	5.5 kb	*rep, mob, dps*	Rop$^+$	(203)
Pediococcus parvulus					
2.6	pPP2	35 kb	*gtf*	Glu$^+$	(204)
ATO77	pATO77	ca. 9 kb	*papABCD*	Bac$^+$	(139)
Pediococcus pentosaceus					
Pep1	pHD1	ca. 13.6 kb	—	Bac$^+$	(151)
RS4	pRS4	3,550 bp	*mob, rep*	Cryptic	(73)
S34	pS34	ca. 8.9 kb	*papABCD*	Bac$^+$	(139)
Streptococcus salivarius					
K12		ca. 190 kb	—	Bac$^+$	(205)

(*continued*)

(Continued)

Strains	Plasmid	Size	Identified Gene	Phenotypes	References
Streptococcus thermophilus					
NDI-6	pCI65st	6,499 bp	*repA, hsp1, hsp2, eno, hsdS*	HS^r, Phage^r	(149)
S4	pSt04		*shsp*	Hsp$^+$	(107)
ST113	pER13	4,139 bp	*copG, repB, mob*	—	(178)
ST116	pER16	4.5 kb	*rep, shsp, hsdS*	—	(174, 175)
ST134	pER341	2,798 bp	*rep, hsp16.4*	—	(176, 177)
ST135	pER35	10 kb	*rep, shsp, hsdR, hsdM, hsdS*	Phage^r	(174, 175)
ST136	pER36	3.7 kb	*rep, shsp*	—	(174, 175)
ST137	pER371		*rep371*	—	(173)
ST2-1	pND103	3.5 kb	*rep, shsp*	—	(180)
ST2783	pt38	2,911 bp	*rep, shsp*	—	(158)
SMQ312	pSMQ-312b	6,710 bp	—	—	(115)
SMQ316	pSMQ-316	6,710 bp	—	—	(115)
SMQ308	pSMQ-308	8,144 bp	—	—	(115, 194)
ST4	pSt04	3.1 kb	*repA, shsp*	—	(113)
ERI	pER1-1	3.4 kb	*repA, shsp*	—	(113)
J34	pJ34	3.4 kb	*repA*	—	(113)
ERI	pER1-2	4.4 kb	—	—	(113)
ST22	pSt22-2	4.2 kb	—	—	(113)
ST106	pSt106	5.3 kb	—	—	(113)

ST8	pSt08	7.5 kb	repA		Phage^r	(113)
ST0	pST0	8.1 kb	—		Phage^r	(113)
SMQ172	pSMQ172	4,230 bp	—		—	(195)
SMQ173	pSMQ173b	4,449 bp	—		—	(194)

Tetragenococcus halophila

D10	pD1	25 kb	aspD, aspT		AspD$^+$	(72)

Abbreviations——abi: abortive infection; ABS: antibiotic substance; Ads: phage adsorption; AspD$^+$: aspartate-decarboxylating; AS^r: arsenate—arsenite resistance; *bsh*: bile salt hydrolase; *citP*: citrate permease; *copG*: transcriptional repression protein; *corA*: magnesium transporter with affinity for cobalt; Df^r: dalfopristin resistance; *dps*: glucosyltransferase; *eno*: enolase; Eps$^+$: exopolysaccharide production; *erm*: erythromycin ribosome methylase; *ftsK*: DNA translocate; *gdh*: glutamate dehydrogenase; Glu$^+$: β-D-glucan production; Gm^r: gentamycin resistance; *gtf*: glycosyltransferase; HDC$^+$: histamine producing; *hdcA*: histidine decarboxylase; *hdcP*: histidine/histamine exchanger; *hisRS*: histidyl-tRNA synthetase; Hop^r: hop resistance; *hsd*: subunits for type IC R/M system; *hsdM*: methylase; *hsdR*: restriction; *hsdS*: specificity; Hsp$^+$: heat shock protein production; HS^r: heat-shock resistance; *imm*: immunity protein; Km^r: kanamycin resistance; Lac$^+$: lactose fermentation; *llaDCHI*: typeII R/M system; *mdt*(A): macrolide efflux protein; *mipB*: transaldolase; MLS^r: macrolide—lincosamide—streptogramin resistance; *mob*: mobilization protein; *orA*: ATP-dependent multidrug transporter; Pen$^+$: pentose fermentation; Phage^r: bacteriophage resistance; Prt$^+$: proteolytically active; R/M: restriction-modification system; *rep*: replication protein; Rop$^+$: ropy phenotype; *shsp*: small heat shock protein; Sm^r: streptomycin resistance; *str*: streptomycin adenylase; TA: toxin—antitoxin system for plasmid maintenance; Tc^r: tetracyclin resistance; *tet(S)*: tetracycline resistance; *tktA*: transketolase; *tnp*: transposase; Ty^r: tylosin resistance; Vm^r: vancomycin resistance; ω–ε–ζ: post-sagregational killing system.

3.3 VECTORS FOR LACTOBACILLI AND BIFIDOBACTERIA

Vector	Replicon	Source	Size (kb)	Marker	Characteristics	Host	References
pAM401	pGB354	STag	10.4	Cm^r, Tc^r	Shuttle	ESco, ENfa, ENfm, ENdu, LCla, LBca	(235)
pAM5	pBC1	BFca	—	Tc^r	Shuttle	BFpsca, BFbr, ESco	(212)
pAV001	pTB6	BFlo	3.6	Sp^r	Shuttle	BFbr, BFlo, ESco	(246, 249)
pBC1.2	pBC1	BFca	—	Cm^r	Shuttle	BFpsca, BFbr, ESco	(212)
pBES2	pMG1	BFlo	7.6	Cm^r, Ap^r	Shuttle	BFlo, ESco	(299, 316)
pBKJ36F	pKJ36	BFlo	6.775	Cm^r	Shuttle	ESco, BFan, BFin	
pBKJ50F	pKJ50	BFan	8.115	Cm^r	Shuttle	ESco, BFan	
pBLES100	pTB6	BFlo	9.1	Sp^r	Shuttle	ESco, BFan, BFlo	(246, 341)
pBRASTA101	pTB6	BFlo	5.0	Sp^r	Shuttle	ESco, BFlo	(342)
pCLJ15	pMB1	BFlo	5.9	Em^r	Shuttle	ESco, BFan, BFbi, BFin, BFlo, BFma	(317)
pCW4	pC7	LBpapl	5.3	Em^r	Shuttle	LEme, LBbu, LBpapl, ESco	(301)
pDOJ4	pDOJ1	LBde	13.317	Cm^r	Shuttle, lacZ	LBde, ESco, STth, LCla	(267)
pDOJHR	pDOJH10S	BFlo	8.596	Cm^r	Shuttle, lacZ	BFlo, ESco	(268)
pDSO44Sp	pTB6	BFlo	4.3	Sp^r	Shuttle	ESco, BFlo	(342)
pEM1110	p8014-2	LBpl	ca. 7.4^a	Em^r	Shuttle	ESco, LBca	(284)
pFBYC050E	pTXL1	LEme	7.72	Ap^r, Em^r	Shuttle	ESco, LEme, LEcr, PDac, LBsk	(221)
pFBYC18E	pFR18	LEme	3.5	Em^r	—	LEcr, LEme, LBsk	(220)
pFI2431	p9785S	LBjo	4.871	Cm^r	—	LBjo, LBga	(252)
pFXL03	pWV01	LCla	3.860	Thy^+	Shuttle	ESco, LBac	(238)
pGKV210	pWV01	LCla	4.4	Em^r	Shuttle	ESco, LBpl	(309)
PGOSBif33	pNAL8L	BFlo	6.691	Em^r	Shuttle	LBlo, BFan, ESco	(244)
pHO304	pHI01	LEci	4.9	Em^r	Shuttle	LEmeme, LEci, LBpl, LCla	(297)

pIAb1	ENfa	8.375	Em^r	Shuttle	LBca, LCla, BAsu, ESco	(307)
pIAb5	ENfa	7.250	Em^r	Shuttle	LBca, LCla, BAsu, ESco	(306, 307)
pIAV1	LCla	6.110	Em^r, Cm^r	Shuttle	LBca, LCla, BAsu, ESco	(307)
pIK650	LBdela	8.96	Em^r	—	LBdebul	(327)
pILE	LBca	7.390	Em^r	Shuttle	LBca, ESco	(215)
pILE4942	LBca	5.318	Cm^r	Shuttle	LBca, LBcaala, LBga, LBac, ESco	(214)
pJRS233	pWV01	6.026	Em^r	ts-shuttle	LBfe, ESco	(346–348)
pKRV3	p256	6.0	Em^r	Shuttle	ESco, LBsk, LBpl	(218)
pL120	pSMA23	7.5	Em^r	Shuttle	LBca, LBga, ESco	(338)
pL142	pSMA23	6.4	Em^r	Shuttle	LBca, LBga, ESco	(338)
pL157	pSMA23	6.3	Cm^r	Shuttle	LBca, LBga, ESco	(338)
pLEB590	LCla	3.1	Nis^r	Food-grade	LCla, LBpl	(339)
pLEM415	LBre	6.3	Em^r	Shuttle	ESco, LBre, LBdebu	(327, 356)
pLF5	BFlo	5.7	Cm^r	Shuttle	ESco, BFan, BFbi, BFin, BFlo, BFma	(317)
pLP3537	LBpe	6.305	Em^r, Ap^r	Shuttle	LBsf, ESco	(251)
pLPV111	LBpl	4.2	Em^r	Shuttle, lacZ	ESco, LBsk, LBpl	(335)
pNZ123	LCla	2.8	Cm^r	Shuttle	ESco, LBac, LBhe, LBbr, LBga, LBjo	(228, 261)
pPKCm1	BFas	6.200	Ap^r, Cm^r	Shuttle	ESco, BFbr, BFanla, BFlo, BFpslo, BFde	(308)
pRN14	LBpl	6.89	Em^r	Shuttle	LBpl, ESco	(226, 258)
pRS4C1	PDpe	7.8	Ap^r, Cm^r	Shuttle	ESco, LBpl, LBca, PDpe, PDac	(210)
pRV500	LBsk	7.289	Em^r, Ap^r	Shuttle	LBcu, LBca, LBpl, LBsk, ESco	(211)
pSJ33E	LEme	7.7	Em^r	Shuttle	LBbr, ESco, LEme	(256)

(continued)

(*Continued*)

Vector	Replicon	Source	Size (kb)	Marker	Characteristics	Host	References
pSKEm	pCIBAO89	BFas	4.400	Em^r	Shuttle	ESco, BFbr	(232)
pSP1	pKC5b	LBfe	9.4	Em^r, Cm^r	Shuttle	LBfe, LBje, LBga, LBcr, LBjo, LBsa, STmu, STgo, STsn	(303)
pSPEC1	pMB1	BFlo	5.9	Sp^r	Shuttle	ESco, BFan, BFbi, BFin, BFlo, BFma	(317)
pSTE32	rep32	LBre	5.5	Em^r	Shuttle	LBre, ESco	(353)
pTC82-RO	pTC82	LBre	7.0	Em^r	Shuttle	ESco, LBre	(274)
pTE15-RO	pTE15	LBre	6.7	Em^r	Shuttle	ESco, LBre, LBfe	(273)
pTE80-RO	pTE80	LBre	6.9	Em^r	Shuttle	ESco, LBre	(273)
pTRE3	pMB1	BFlo	2.8	Em^r	—	BFan, BFbi, BFin, BFlo, BFma	(317)
pTRK563	pWV01	LCla	3.56	Em^r	Shuttle	ESco, LBga, LBjo, LBac, LBre	(224, 227)
pTRK669	pWV01	LCla	—	Cm^r	ts shuttle	ESco, LBac, LBga	(319)
pTRKH2	pAMb1	ENfa	6.9	Em^r	Shuttle	LBbr, ESco, LBca, LBpl	(222, 277, 278)
pVI1055	pIP501	STag	6.7	Em^r	ts shuttle	ESco, LBbu	(326)
pX3	pBUL1	LBdebu	9.063	Em^r	—	LBdela, LBdebu	(313, 327)

Replication Origin Screening Vector

Vector	Replicon	Source	Size (kb)	Marker	Host	References
pBif	pBluescript II KS	ESco	6.5	Cm^r	BFbr, BFpsca	(212)
pUE80$^{+/-}$	pUC18/19	ESco	3.8	Em^r	LBre	(273)

Promoter Screening Vector

Vector	Replicon	Source	Size (kb)	Marker	Reporter Gene (Supplementation)	Host	References
pEH100	pRV500	LBsk	6.83	Cm^r	*ermGT, gusA* (IVET vector)	LBsk, ESco	(253)
pGKV110	pWV01	LCla	2.9	Em^r	Cat	ESco, LBpl	(343)
pIL4242	pWV01	LCla	4.8	Em^r	LuxAB (integrative vector)	ESco, LBca	(295)
pJW100	pGT232	LBre	7.4	Cm^r	*bglM, ermGT*	ESco, LBre	(349)
pMDY23	pNCC293	BFlo	6.092	Sp^r	*gusA*	ESco, BFlo	(262)
pNZ272	pSH71	LCla	4.7	Cm^r	*gusA*	LBrh, ESco, LBfe, LBpl	(288)
pNZ7120	pAMb1	LCla	6.215	Em^r	*alr*	ESco, LCla, LBpl	(223)
pSH85350	p353-2	LBpe	8.916	Em^r, Ap^r	*melA*	LBsf, ESco	(251)

Secretion Signal Screening Vector

Vector	Replicon	Source	Size (kb)	Marker	Reporter gene	Host	References
pFUN	pAMβ1	LCla	8.072	Em^r, Ap^r	*nuc*	ESco, BFbr, LCla	(281)
pNICE-gfpSP	rep32	LBre	8.9	Em^r	*gfp*	LBre, ESco	(351)

Conjugative Vector

Vector	Replicon	Source	Size (kb)	Marker	Source of Tra	Host	References
pIL-BS::2.2	pAMb1	ENfa	10.1	Em^r, Ap^r	Helper plasmid, pIL205	ESco, LCla, LBhe	(345)
pSA3b6::pVA797	pIP501	STag	44	Em^r, Cm^r	Cointegrated plasmid, pVA797	LCla, LBhe	(344)

(continued)

Integration Vector

Vector	Replicon	Source	Size (kb)	Markers	Region involved in Integration (Supplementation)	Host	References
pBluescript	ColE1	ESco	3.204	Ap^r	Chromosome DNA	LBpl	(276)
pEM39	pWV01	LCla	5.4	Em^r	attP from phage A2 (ts)	ESco, LBca, LBpl, LBpaca, LCla	(213)
pEM40	Puc	ESco	5.407	Ap^r, Em^r	attP from phage A2	LBca, LBrh	(259, 266)
pEM68	p8014-2	LBpl	ca. 9	Ap^r, Cm^r	(Exision-vector coding β-recombinase for six system)	LBca	(283)
pEM76	pUC	ESco	6.3	Ap^r, Em^r	attP from phage A2 (six site for exision)	LBca	(235, 283, 284)
pEM94	pWV01	LCla	8.450	Ap^r, Cm^r	(ts, exision vector coding β-recombinase for six system)	LBca	(284)
pGh9:ISS1	pWV01	LCla	4.6	Em^r	IS$S1$ (ts)	LBpl, LCla, STth, ENfa, LBrh	(245, 347)
pGID023	pE194	SAau	—	Em^r	Chromosome DNA (unstable vector)	LBpl	(219)
pIlac	pUC	ESco	4.895	Em^r, Ap^r	$lacG'$-$lacF$	LBca	(241, 355)
pJB34-S	pWV01	LCla	4.1	Em^r	$lacII$	LBga	(225)
pJC4	p15A	ESco	7.6	Em^r	IS1223 (Cel$^+$)	LBjo, LBga, LBbu, LBpl	(255)
pJDC9	pMB9	ESco	6.95	Em^r	Chromosome DNA	LBpl	(243)
pLBS-GFP-EmR	pUC	ESco	6.671	Em^r, Km^r, Ap^r	lbs locus (expression vector from P$_{lbs}$)	LBcr, LBjo, LBre, LBde	(290)

pMEC10	pUC	ESco	ca. 8.1	Ap^r, Em^r	attP from phage mv4	LBpl, LBca, LBrh	(233, 248, 302)
pMSK742	p15A	ESco	4.033	Em^r	attP from phage φFSW	ESco, LBca	(331)
pMSK761	p15A	ESco	5.491	Em^r	attP from phage φFSW (integration-excision vector using Δcat gene)	ESco, LBca	(330)
pNZ5319	p15A	ESco	—	Cm^r, Em^r	Chromosome DNA (lox66, lox71 for exision)	ESco, LBpl	(265)
pNZ5348	pE194	SAau	—	Em^r	(Unstable exision vector coding Cre for lox system)	ESco, LBpl	(265)
pORI28	pWV01	LCla	1.65	Em^r	Chromosome DNA (repA deleted)	LBac, LBga, LBre	(287, 319, 325, 337)
pORI280	pWV01	LCla	5.3	Em^r, Xgal$^+$	Chromosome DNA (repA deleted)	LCla, LBbr	(217)
pRV300	pUC	ESco	3.539	Em^r	Chromosome DNA	LBsk	(270)
pRV80	pUC	ESco	ca. 4.6	Em^r	Δ$lacLM$ gene	LBsk	(240, 336)
pSKE-IN	ColE1	ESco	5.5	Ap^r, Em^r	attP from phage φAT3	LBrh, LBca	(279)
pTN1	pLC2	LBcu	3.299	Em^r	chromosome DNA (ts)	LCla, LBga	(292)
pVI49	pIP501	STag	ca. 8	Em^r	IS1223 (ts)	LBbu	(326)
pVI52	pIP501	STag	ca. 8	Em^r	IS1201 (ts)	LBbu	(326)

(*continued*)

Expression Vector

Vector	Replicon	Source	Size (kb)	Marker	Promoter/Secdretion/Anchor signal (Supplementation)	Host	References
p123(1–5N)	pSH71	LCla	3.5	Cm^r	P_{Slayer}, S_{Slayer}	ESco, LBac	(229)
pBES16PR	pMG1	BFlo	11.1	Cm^r	P_{16P}	BFlo, ESco	(298)
pBESAF2	pMG1	BFlo	ca. 8.5	Cm^r, Ap^r	P_{amy}, S_{amy}	ESco, BFlo	(300)
pBIFRIBO	pASV479	BFpslo	9	Cm^r	P_{rRNA}	BFbr, ESco	(321)
pBLES100-S-eCD	pTB6	BFlo	12.7	Sp^r	P_{HU}	ESco, BFlo	(291, 322)
pBV220	pUC	ESco	3.666	Ap^r	P_{PR}, P_{PL} (inducible)	ESco, BFad, BFlo	(237, 271)
pBV222	pUC	ESco	3.684	Ap^r	P_{PR}, P_{PL} (unstable vector, inducible, His6-tag)	ESco, BFlo	(354)
pBV22210-ENDO	pMB1	BFlo	6.493	Cm^r	P_{PR}, P_{PL} (inducible, His6-tag-ENDO$^+$)	ESco, BFlo	(354)
pCW5	pC7	LBpapl	—	Em^r	P_{32} (GFP$^+$)	LBbu, LBpp, LBpl, ESco, LEme	(269)
pCWAIp:Usp-E7mm	pWV01	LCla	—	Cm^r	P_{nisA}, S_{Usp45}, A_{CWAIp}	ESco, LCla, LBpl	(231)
pFBYC069	pTXL1	LEme	8.39	Ap^r, Em^r	P_{59}	ESco, LEme, LEcr, PDac, LBsk	(221)
pGM4	p256	LBpl	—	Em^r	P_{sppA} (regulated expression, PepN$^+$)	LBpl, ESco,	(286)
pIAb5lacamy	pAMb1	ENfa	9.250	Em^r	P_{lac}	ESco, LBca	(242)
pIAlac	pAMb1	ENfa	ca. 7.53	Em^r	P_{lac}	ESco, LBca	(230, 305, 306)
pIL1020	pWV01	LCla	5.198	Em^r, Cm^r	P_{59}	LBpl, ESco	(293)
pKTH2119	pWV01	LCla	5.2	Em^r, Cm^r	$P1_{slpA}$, $P2_{slpA}$	LCla, LBpl, LBga	(257)
pKTH2121	pWV01	LCla	6.0	Cm^r, Em^r	P_{slpA}, S_{slpA}	LCla, LBbr, LBjo, LBmu, LBpl, ESco	(294)

Plasmid	Replicon	Size (kb)	Markers	Features	Host	Reference
pLacTd-Ter	pWV01	3.021	Em^r	P$_{lacTd}$ (integration vector)	LCla	(247)
pLEB600	pSH71	3.6	Lac$^+$	P$_{pepR}$ (food-grade vector)	LBca,	(340)
pLEM3	pLEM3	—	Em^r	P$_{ldhL}$ (GFP$^+$)	LBre	(356)
pLEM415::gfp	p353-2	8.2–9.8^b	Em^r, Ap^r	P$_{amy}$, S$_{amy}$, A$_{prt}$ (inducible, Gus$^+$)	LBdela, LBfr, ESco ESco, LBca, LBpaca, LBpl	(254, 263, 280, 310, 311)
pLP401						
pLP402	p353-2	9.5	Em^r, Ap^r	P$_{amy}$, S$_{amy}$ (inducible, Gus$^+$)	ESco, LBca	(280, 311)
pLP403	p353-2	9.3	Em^r, Ap^r	P$_{amy}$ (inducible, Gus$^+$)	ESco, LBca	(311)
pLP501	p353-2	8.5–9.7	Em^r, Ap^r	P$_{ldh}$, S$_{prt}$, A$_{prt}$ (Gus$^+$)	ESco, LBca, LBpaca, LBpl	(263,280,282, 296,310,311)
pLP502	p353-2	9.3	Em^r, Ap^r	P$_{ldh}$, S$_{prt}$ (Gus$^+$)	ESco, LBca	(280, 311)
pLP503	p353-2	9.1–9.5	Em^r, Ap^r	P$_{ldh}$ (Gus$^+$)	ESco, LBca	(311, 357)
pLP601	pSH71	6.5	Cm^r	P$_{amy}$, S$_{amy}$, A$_{prt}$	ESco, LBac, LBcr	(311)
pLP602	pSH71	6.15	Cm^r	P$_{amy}$, S$_{amy}$	ESco, LBac, LBcr	(311)
pLP701	pSH71	5.7	Cm^r	P$_{xyl}$, S$_{amy}$, A$_{prt}$	ESco, LBac, LBcr	(311)
pLP702	pSH71	5.4	Cm^r	P$_{xyl}$, S$_{amy}$	ESco, LBac, LBcr	(311)
pLP803	pSH71	5.8	Cm^r	P$_{ldh}$	ESco, LBac, LBcr	(311)
pLPM11	p353-2	ca. 7.6	Em^r	P$_{amy}$ (inducible)	LBca, ESco, LBpl	(259, 285)
pLPMSSA3	p353-2	—	Em^r	P$_{amy}$ (inducible)	LBca, ESco	(216)
pLR	pMB1	8.0	Cm^r	P$_{HU}$, S$_{BIF3}$	BFlo, ESco	(315)
pMD112	pSH71	—	Cm^r	P$_{prt}$	LBjo, LCla,	(324)
pMEC127	pLAB1000	7.498	Em^r, Ap^r	P$_{ldh}$ (TTFC$^+$)	ESco, LBpl	(314)
pMEC17	pSH71	7.2	Km^r, Cm^r	P$_{25}$ (GFP$^+$)	ESco, LBpl, LCla	(239)
pMEC30	pLAB1000	6.6	Ap^r, Em^r	P$_{ldh}$ (GFP$^+$)	ESco, LBpl, LCla, LBrh	(233, 239)
pMEC45	pSH71	4.1	Cm^r	P$_{nisA}$ (inducible, GFP$^+$)	ESco, LBpl, LCla, LBrh	(233, 239)
pMSP3535	pAMb1	8.353	Em^r	P$_{nisA}$ (inducible)	LCla, LBpaca	(234)
pNICE	rep32	8.05	Em^r	P$_{nisA}$ (inducible)	LBre, ESco	(353)

(continued)

(Continued)

Vector	Replicon	Source	Size (kb)	Marker	Promoter/Secdretion/Anchor signal (Supplementation)	Host	References
pNIES	rep32	LBre	8.2	Em^r	P$_{nisA}$, S$_{amy}$ (inducible)	LBre, ESco	(353)
pNIES-GFP	rep32	LBre	9	Em^r	P$_{nisA}$, S$_{amy}$ (inducible, GFP$^+$)	LBre, ESco	(352)
pNZ3004	pWV01	LCla	4.9	Em^r	P$_{lacA}$	ESco, LBga	(228)
pNZ7017	pSH71	LCla	—	Cm^r	P$_{pepN}$, P$_{nisA}$	LCla, LBga	(350)
pNZ8032	pSH71	LCla	4.8	Cm^r	P$_{nisA}$ (Gus$^+$)	ESco, LCla, LBbr	(217)
pNZ8037	pSH71	LCla	3.1—4.4	Cm^r	P$_{nisA}$ (inducible)	LBpl, ESco	(236, 275, 285, 304)
pNZ8048	pSH71	LCla	3	Cm^r	P$_{nisA}$ (inducible)	ESco, BFbr, LBsa, LBplLCla	(234, 328, 329, 337)
pPSAB1	pMG1	BFlo	8.6	Cm^r, Ap^r	P$_{amy}$, S$_{Amy}$	ESco, BFlo	(289)
pRc/CMV2-VP1-Rep. 8014	p8014-2	LBpl	8.3	Ap^r, Nm^r	P$_{cmv}$ (eukaryotic expression vector)	LBac, ESco, PK15	(272)
pRV85	pWV01	LCla	ca. 6.25	Em^r	P$_{ldh}$ (ts, GFP$^+$)	ESco, LBsk	(240)
pSAK332	pAMb1	ENfa	6.995	Em^r	P$_{erm}$, S$_{SAK}$, Ap$_{763}$	ESco, LBca	(264)
pSEC:LEISS-Nuc-BLG	pWV01	LCla	4.264	Cm^r	P$_{nisA}$, S$_{Usp45\text{-}LEISS}$	ESco, LBca	(248)
pSECE1	2.76kb plasmid	LCla	4.66	Em^r	P$_{amy}$, S$_{amy}$	ESco, STth, LCla, LBdebu	(323)
pSIP300	p256	LBpl	5.4	Em^r	P$_{sapA}$ (inducible)	ESco, LBsk, LBpl	(333)
pSIP304	pSH71	LCla	—	Em^r	P$_{sapA}$ (inducible, Gus$^+$)	ESco, LBsk, LBpl	(334)
pSIP400	p256	LBpl	5.5	Em^r	P$_{sppA}$ (inducible)	ESco, LBsk, LBpl	(333)
pSIP401	p256	LBpl	5.6	Em^r	P$_{sppA}$, P$_{spplP}$ (inducible)	ESco, LBsk, LBpl	(333)
pSIP409	pSH71	LCla	—	Em^r	P$_{sppA}$, P$_{spplP}$ (inducible, Gus$^+$)	ESco, LBsk, LBpl	(318, 334)
pSIP500	p256	LBpl	5.4	Em^r	P$_{nisA}$ (inducible)	ESco, LBsk, LBpl	(333)

pTG2247	pSH71	LCla	6.5	Km^r, Cm^r	P$_{25}$	ESco, LBpl, LBpaca	(250, 332)
pTKR664	pWV01	LCla	—	Em^r	P$_6$ (Gus$^+$)	ESco, LBga, LBac, LBjo, LCla	(320)
pTSV2	pAMb1	LCla	—	Em^r	P$_{LPS2}$, S$_{Usp45}$	ESco, LBpl, LBga	(312)
pWK7	pLTK2	LBpl	9.6	Em^r, Km^r	P$_{acc}$ (CHO$^+$)	ESco, LBpl, LBca	(260)

^{a} Estimated from the description in text.
^{b} Vary with the report.

Abbreviations.—BAsu: *Bacillus subtilis*; BFad: *Bifidobacterium adolescentis*; BFan: *Bifidobacterium animalis*; BFanla: *Bifidobacterium animalis* subsp. *lactis*; BFas: *Bifidobacterium asteroids*; BFbr: *Bifidobacterium breve*; BFca: *Bifidobacterium catenulatum*; BFde: *Bifidobacterium dentium*; BFin: *Bifidobacterium infantis*; BFlo: *Bifidobacterium longum*; BFma: *Bifidobacterium magnum*; BFpsca: *Bifidobacterium pseudocatenulatum*; BFpslo: *Bifidobacterium pseudolongum*; ENdu: *Enterococcus durans*; ENfa: *Enterococcus faecalis*; LBac: *Lactobacillus acidophilus*; LBbr: *Lactobacillus brevis*; LBbu: *Lactobacillus bulgaricus*; LBcaal: *Lactobacillus casei* subsp. *alactosus*; LBcr: *Lactobacillus crispatus*; LBcu: *Lactobacillus curvatus*; LBdebu: *Lactobacillus delbrueckii* subsp. *bulgaricus*; LBdela: *Lactobacillus delbrueckii* subsp. *lactis*; LBfe: *Lactobacillus fermentum*; LBfr: *Lactobacillus fructosus*; LBga: *Lactobacillus gasseri*; LBhe: *Lactobacillus helveticus*; LBhi: *Lactobacillus hilgardii*; LBje: *Lactobacillus jensenii*; LBjo: *Lactobacillus johnsonii*; LBpapl: *Lactobacillus paraplantarum*; LBpe: *Lactobacillus pentosus*; LBrh: *Lactobacillus rhamnosus*; LBsa: *Lactobacillus salivarius*; LBsf: *Lactobacillus sanfranciscensis*; LBsk: *Lactobacillus sakei*; LBspp: *Lactobacillus spp.*; LCla: *Lactococcus lactis*; LEci: *Leuconostoc citreum*; LEcr: *Leuconostoc cremoris*; LEme: *Leuconostoc mesenteroides*; LEmeme: *Leuconostoc mesenteroides* subsp *mesenteroides*; PDac: *Pediococcus acidilactici*; PDpe: *Pediococcus pentosaceus*; SAau: *Staphylococcus aureus*; SCspp: *Saccharomyces spp.*; STag: *Streptococcus agalactiae*; STgo: *Streptococcus gordonii*; STmu: *Streptococcus mutans*; STsn: *Streptococcus sanguis*; *alr*: alanine racemase gene; Ap^r: ampicillin resistance; attP: attachment site of the phage; bglM: promoterless β-glucanase gene; β-recombinase: resolvase for deletion of DNA between six site; Cel$^+$: endoglucanase production; CHO$^+$: cholesterol oxidase production; Cm^r: chloramphenicol resistance; Cre: site-specific recombinase; CWAlp: cell-wall anchor of LBpl protein, lp 2940; ΔlacLM: deleted lacLM operon for β-galactosidase; ENDO$^+$: endostatin production; ermGT: promoterless erythromycin resistance determinant; *gfp*: green fluorescent protein gene; GFP$^+$: green fluorescent protein production; Gus$^+$: β-glucuronidase production; IVET: *in vivo* expression technology; Km^r: kanamycin resistance; *lacG'-lacF*: lactose operon genes of LBca; *lacII*: phospho-β-galactosidase gene; *lbs*: laminin-binding S-layer protein gene; LEISS: synthetic peptide LEISSTCDA for improved secretion of proteins; lox66: mutated loxP site for Cre recognition; lox71: mutated loxP site for Cre recognition; *luxAB*: luciferase gene; melA: α-galactosidase gene; Nis^r: nisin resistance; *nuc*: nuclease gene lacks transcriptional and translational signals; P$_{16P}$: rRNA promoter; P$_{25}$: promoter25 from STth; p$_{32}$: promoter32 from pMG36e; P$_{59}$: lactococcal promoter; P$_6$: P6 promoter from LBac; P$_{acc}$: acetyl coenzyme A carboxylase promoter; P$_{amy}$: α-amylase promoter; P$_{cmv}$: human cytomegalovirus promoter; P$_{bs}$: lbs promoter; P$_{ldh}$: lactate dehydrogenase promoter; P$_{LPS2}$: histone-like protein promoter; P$_{rRNA}$: rRNA promoter; P$_{xyl}$: xylose promoter; rep32: replication origin from LBre plasmid; S$_{Amy}$: α-amylase signal sequence; S$_{BIF3}$: β-galactosidase signal peptide; six: target site of β-recombinase; Sp^r: spectinomycin resistance; S$_{Pri}$: proteinase signal sequence; S$_{Slayer}$: surface-layer protein signal sequence; S$_{Usp45}$: lactococcal USP45 protein signal sequence; Thy$^+$: thymidylate synthase production; ts: thermosensitive replication.

3.4 GENETIC RECOMBINATION

Organism	Characteristics Acquired	Introduced Sequence Involved	Targeted Sequence in the Organism	Gene Transfer	References
Insertion of Plasmid DNA by Homologous Recombination					
Ec. faecalis	Disrupted gls24 gene	gls24 gene fragments on nonreplicative plasmid	Chromosomal gls24 gene	Ep	(373)
Lb. acidophilus	Inactivated putative adhesion factors	fbpA, mub, slpA, ORF1633 and 1634 gene fragments on a temperature sensitive plasmid	chromosomal fbpA, mub, slpA, ORF1633, and 1634 gene	Ep	(367)
Lb. acidophilus	Inactivated β-galactosidase activitiy	lacL gene fragments on a temperature sensitive plasmid	Chromosomal lacL gene	Ep	(392)
Lb. acidophilus	Inactivated glutamate/GABA antiporter, transcriptional regulator, amino acid permiase, and ornithine decarboxylase	ORF La57, La867, La995, and La996 gene fragments on a temperature sensitive plasmid	Chromosomal ORF La57, La867, La995, and La996 gene	Ep	(361)
Lb. casei	Inactivated lactose operon antiterminator	lacT gene fragment on a nonreplicative plasmid	Chromosomal lacT gene	Ep	(375)
Lb. casei	β-glucronidase activity	gusA gene on a nonreplicative plasmid	Chromosomal lacF-G gene	Ep	(374)
Lb. casei	Acetohydroxy acid synthase activity	ilvBN gene on a nonreplicative plasmid	Chromosomal lacF-G gene	Ep	(374)
Lb. gasseri	Inactivated β-glucronidase activity	gusA gene fragments on a temperature sensitive plasmid	Chromosomal gusA gene	Ep	(392)

Organism	Function	Construct	Type	Ref.
Lb. gasseri	Nisin controlled expression genes and erythromycin resistance	nisP, nisR, and nisK genes on a temperature sensitive plasmid	Chromosomal pepN gene downstream	Ep	(388)
Lb. helveticus	Inactivated D(+)-lactate dhydrogenase activity	ldhD gene fragment on a temperature sensitive plasmid	Chromosomal ldhD gene	Ep	(363)
Lb. plantarum	α-Amylase and endoglucanase	Genomic DNA fragment on a nonreplicative plasmid with amyA, celA	Chromosomal DNA	Ep	(394)
Lb. sake	Inactivated PTS carbohydrates fermenting ability	ptsI gene fragment on a nonreplicative plasmid	Chromosomal ptsI gene	Ep	(383)
Lb. sake	Inactivated lactose fermenting ability	lacL gene fragment on a nonreplicative plasmid	Chromosomal lacL gene	Ep	(383)
Lc. lactis	Inactivated oligopeptide transport system and endpeptidase pepO	oppA and pepO gene on nonreplicative plasmid	Chromosomal oppA and pepO gene	Ep	(369)
Lc. lactis ss. cremoris	Lactose fermenting ability	Lac gene on a temperature sensitive plasmid	Chromosomal DNA	Ep	(372)
Lc. lactis ss. lactis	Inactivated enzyme II^lac and phospho-β-galactosidase activity	lacE, lacG gene fragments on a nonreplicative plasmid with Cm^r gene	Chromosomal lacE and lacG gene	Ep	(398)
Lc. lactis ss. lactis	Erythromycin resistance	Random chromlsamal fragments and Em^r gene on a temperature sensitive plasmid	Chromosomal DNA	Ep	(380)

(continued)

(Continued)

Organism	Characteristics Acquired	Introduced Sequence Involved	Targeted Sequence in the Organism	Gene Transfer	References
Lc. lactis ss. lactis	Erythromycin and chloramphenicol resistance	Em^r and Cm^r gene on a nonreplicative plasmid	Chromosomal DNA	Ep	(381)
Replacement of DNA Fragment by Homologous Recombination					
Lb. acidophilus	β-galactosidase activity	Genomic DNA fragment having β-*gal* gene on a nonreplicative plasmid	Chromosomal DNA	Ep	(384)
Lb. brevis	Expressed c-Myc epitope	*slpA* gene carrying the c-*myc* gene on a temperature sensitive plasmid	Chromosomal *slpA* gene	Ep	(360)
Lb. fermentum	Inactivated D(+), L(−)-lactate dehydrogenase activity	*ldhD* and *ldhL* gene fragment on a temperature sensitive plasmid	Chromosomal *ldhD, L* gene	Ep	(358)
Lb. fermentum	Inactivated D(+)-lactate dehydrogenase activity	*ldhD* gene fragment on a temperature sensitive plasmid	Chromosomal *ldhD* gene	Ep	(358)
Lb. gasseri	Superoxide dismutase activity	*sodA* gene on a temperature sensitive plasmid	Chromosomal *lacII* gene	Ep	(365)
Lb. helveticus	Inactivated the X-propyl dipeptidyl aminopeptidase activity	Deleted *pepXP* gene on a temperature sensitive plasmid	Chromosomal *pepXP* gene	Ep	(362)

Lb. helveticus	Inactivated D(+)-lactate dehydrogenase activity	Deleted *ldhD* gene on a temperature sensitive plasmid	Chromosomal *ldhD* gene	Ep	(378)
Lb. helveticus	L(−)-lactose dehydrogenase activity	*ldhL* gene on a temperature sensitive plasmid	Chromosomal *ldhD* gene	Ep	(378)
Lb. helveticus	Inactivated the aminopeptidases and X-propyl dipeptidyl aminopeptidase	*pepC*, *pepN*, and *pepX* gene fragments on a temperature sensitive plasmid	Chromosomal *pepC*, *pepN*, and *pepX* gene	Ep	(368)
Lb. johnsonii	Inactivated D(+)-lactate dehydrogenase activity	Deleted *ldhD* gene on a conjugative plasmid	Chromosomal *ldhD* gene	Cnj	(379)
Lb. plantarum	Inactivated conjugated bile acid hydrolyses the activity	*cml*-containing *cbh* gene on a nonreplicative plasmid	Chromosomal *cbh* gene	Ep	(382)
Lb. sakei	Inactivated the β-galactosidase activity	Deleted *lacLM* gene on a nonreplicative plasmid	Chromosomal *lacLM* gene	Ep	(400)
Lc. lactis	Nuclease activity	*his* gene fragment with *nucA* and Pnis gene on a nonreplicative plasmid	Chromosomal *his* gene	Ep	(397)
Lc. lactis	Inactivated the X-propyl dipeptidyl aminopeptidase activity	Deleted *pepXP* gene on a conjugative plasmid	Chromosomal *pepXP* gene	Cnj	(386)
Lc. lactis ss. lactis	Inactivated phospho-β-galactosidase activity	*lacG* gene fragments on a nonreplicative plasmid with Em^r gene	Chromosomal *lacG* gene	Ep	(398)
Str. thermophilus	Amylase activity	Deleted *lacZ* gene on a temperature sensitive plasmid with *thyA* gene	Chromosomal *lacZ* gene	Ep	(393)

(*continued*)

(Continued)

Organism	Characteristics Acquired	Introduced Sequence Involved	Targeted Sequence in the Organism	Gene Transfer	References
Str. thermophilus	Inactivated the lactose fermenting ability	Deleted *lacZ* gene on a nonreplicative plasmid	Chromosomal *lacZ* gene	Ep	(387)
Str. thermophilus	Chloramphenicol resistance	Promoterless *cat* gene inserted between *lacS* and *lacZ* genes on a nonreplicative plasmid	Chromosomal *lacS-lacZ* genes	Ep	(387)
Insertion of Transporsable Element					
Lb. casei Lb. helveticus Lc. lactis	Conversion of a temperate phage into a lytic phage	Endogenous ISL1	Temperate phage genome on chromosome		(395)
	Inactivated lactose fermenting ability	Endogenous ISL2	*lacL-lacM* locus on chromosome		(403)
Lb. plantarum	Erythromycin resistance	Tn917 on a temperature sensitive plasmid	Resident plasmid	Ep	(371)
Lb. sake	Abolished bacteriocin production	Endogenous IS1163	*las* operon on resident plasmid		(399)
Ec. faecalis	Erythromycin resistance	Conjugative chromosomal element Ω6001 bearing *erm* gene	Chromosomal DNA	Cnj	(390)
Insertion of plasmid DNA via transposition of transporsable element					
Ec. faecalis Str. thermophilus	Erythromycin resistance	ISS1 on a temperature sensitive plasmid	Chromosomal DNA	Ep	(385)
Ec. faecalis	Erythromycin resistance	*att* site of Tn1545 on a nonreplicative plasmid with a provision of integrase in trans	Chromosomal DNA	Ep, Cnj	(401)

Lb. gasseri	Erythromycin resistance	IS1223 on a nonreplicative plasmid having *erm* gene	Chromosomal DNA	Ep	(402)
Insertion of plasmid DNA by phage function					
Ec. faecalis Lb. casei Lc. lactis	Erythromycin resistance	*attP* site of phage mv4 on a nonreplicative plasmid with *int* gene	Chromosomal DNA	Ep	(359)
Lb. gasseri	Erythromycin resistance	*attP* site of phage φadh on a nonreplicative plasmid with *int* gene	Chromosomal DNA	Ep	(391)
Lb. casei	Expressed bovine β-lactogloblin	BLG gene and *attP* site of phage φmv4 on a nonreplicative plasmid with *int* gene	Chromosomal tRNA^{ser} locus	Ep	(376)
Lb.plantarum	Erythromycin resistance	*attP* site of phage mv4 on a nonreplicative plasmid with *int* gene	Chromosomal DNA	Ep	(389)
Lc. lactis ss. cremoris	β-Glucronidase activity	*gusA* gene and *attP* site of phage φTP901-1 on a nonreplicative plasmid	Chromosomal *attB* site	Ep	(364)
Lc. lactis ss. cremoris	β-Galactosidase activity	*lacLM* gene and *attP* site of phage φTP901-1 on a nonreplicative plasmid	Chromosomal *attB* site	Ep	(364)
Lc. lactis ss. cremoris	Erythromycin resistance	*attP* site of phage φTP901-1 on a nonreplicative plasmid	Chromosomal *attB* site	Ep	(369)
Str. thermophilus	Chloramphenicol resistance	*attP* site of phage φSfi21 on a nonreplicative plasmid with *int* gene	Chromosomal DNA	Ep	(366)

(*continued*)

(Continued)

Organism	Characteristics Acquired	Introduced Sequence Involved	Targeted Sequence in the Organism	Gene Transfer	References
Unidentified Mechanism					
Lb. casei	Double resistance to both rifampicin and streptomycin	Chromosomal DNA of Rif^r strain	Chromosomal DNA of St^r strain	Pf	(377)
Lb. fermentum	Trehalose-fermenting ability	Chromosomal DNA of trehalose-positive *Lc. lactis* harboring plasmid pAMβ1	Chromosomal DNA	Pf	(370)
Ec. faecalis *Lb. plantarum*	Double resistance to both chloramphenicol and erythromycin	Cointegrant with conjugative plasmid pAMβ1 formed in *Ec. faecalis*	Successful or unsuccessful resolution of cointegrant in *Lb. plantarum*	Mob	(396)

Abbreviations—Ec: Enterococcus; Lb: Lactobacillus; Lc: Lactococcus; Str: Streptococcus; ss: subspecies; amyA: α-amylase gene; BLG: bovine β-lactoglobulin; cbh: conjugated bile acid hydrolase; celA: endoglucanase gene; cml: chloramphenicol-resistance gene; fbpA: fibronectin-binding protein gene; gftD: glucosyltransferase gene; gls24: putative stress protein gene; gusA: β-glucronidase gene; his: histidine biosynthesis gene; ilvBN: acetohydroxy acid synthase gene; Int: integrase gene; lacII: phospho-β-galactosidase II gene; lacE: enzyme II^{lac}; lacF: enzyme II A^{lac}; lacG: phospho-β-galactosidase gene; lacM-lacL or lacL: β-galactosidase gene; lacZ: β-galactosidase gene; lacS: lactose permease gene; lacT: lactose operon antiterminator gene; las operon: lactocin S operon; mub: mucinbinding protein; ldhD: D(+)-lactate dehydrogenase gene; ldhL: L(−)-lactate dehydrogenase gene; c-Myc: human c-Mic oncogene; nisP: Nicin promotor gene; nisR: Nicin transcriptional regulator gene; nisK: histidine protein kinase gene; nucA: nuclease gene; oppA: oligopeptide transport system gene; pepC: broad specificity aminopeptidase gene; pepN: broad specificity aminopeptidase gene; pepO: endopeptidase gene; pepX: X-propyl dipeptidyl aminopeptidase gene; pepXP: X-propyl dipeptidyl aminopeptidase gene; ptsH: HPr gene; ptsI: enzyme I gene; recA: recA protein gene; scrA: sucrose-specific EII permease gene; slpA: S-layer protein; sodA: superoxide dismutase gene; Rif^r: rifampicin resistance; St^r: streptomycin resistance; thyA: thymidylate synthase gene; Tn: transposon; Cnj: conjugation; Ep: electroporation; Mob: mobilization; Pf: protoplast fusion.

REFERENCES

1. Adsul M, Khire J, Bastawde K, and Gokhale D. Production of lactic acid from cellobiose and cellotriose by *Lactobacillus delbrueckii* mutant Uc-3. *Appl. Environ. Microbiol.* 2007; 73: 5055–5057.
2. Arsenijevic S and Topisirovic L. Molecular analysis of mutated *Lactobacillus acidophilus* promoter-like sequence P15. *Can. J. Microbiol.* 2000; 46: 938–945.
3. Baccigalupi L, Naclerio G, De Felice M, and Ricca E. Efficient insertional mutagenesis in *Streptococcus thermophilus*. *Gene* 2000; 258: 9–14.
4. Bai D-M, Zhao X-M, Li X-G, and Xu S-M. Strain improvement and metabolic flux analysis in the wild-type and a mutant *Lactobacillus lactis* strain for L(+)-lactic acid production. *Biotechnol. Bioeng.* 2004; 88: 681–689.
5. Boumerdassi H, Monnet C, Desmazeaud M, and Corrieu G. Isolation and properties of *Lactococcus lactis* subsp. *lactis* biovar *diacetylactis* CNRZ 483 mutants producing diacetyl and acetoin from glucose. *Appl. Environ. Microbiol.* 1997; 63: 2293–2299.
6. Callanan MJ, Russell WM, and Klaenhammer TR. Modification of *Lactobacillus* β-glucuronidase activity by random mutagenesis. *Gene* 2007; 389: 122–127.
7. Clemans DL, Kolenbrander PE, Debabov DV, Zhang Q, Lunsford RD, Sakone H, Whittaker CJ, Heaton MP, and Neuhaus FC. Insertional inactivation of genes responsible for the D-alanylation of lipoteichoic acid in *Streptococcus gordonii* DL1 (Challis) affects intrageneric coaggregations. *Infect. Immun.* 1999; 67: 2464–2474.
8. El Attar A, Monnet C, Aymes F, and Corrieu G. Method for the selection of *Lactococcus lactis* mutants producing excess carbon dioxide. *J. Dairy Res.* 2000; 67: 641–646.
9. Enne VI, Delsol AA, Roe JM, and Bennett PM. Rifampicin resistance and its fitness cost in *Enterococcus faecium*. *J. Antimicrob. Chemother.* 2004; 53: 203–207.
10. Fernández de Palencia P, Martin-Hernández MC, Joosten HM, and Peláez C. Isolation and characterization of proteinase- and aminopeptidase-deficient mutants of *Lactobacillus casei* subsp. *casei* IFPL 731. *Lett. Appl. Microbiol.* 1997; 25: 215–219.
11. Fernandez A, Thibessard A, Borges F, Gintz B, Decaris B, and Leblond-Bourget N. Characterization of oxidative stress-resistant mutants of *Streptococcus thermophilus* CNRZ368. *Arch. Microbiol.* 2004; 182: 364–372.
12. Garault P, Le Bars D, Besset C, and Monnet V. Three oligopeptide-binding proteins are involved in the oligopeptide transport of *Streptococcus thermophilus*. *J. Biol. Chem.* 2002; 277: 32–39.
13. Garault P, Letort C, Juillard V, and Monnet V. Branched-chain amino acid biosynthesis is essential for optimal growth of *Streptococcus thermophilus* in milk. *Appl. Environ. Microbiol.* 2000; 66: 5128–5133.
14. Germond J-E, Delley M, D'Amico N, and Vincent SJ. Heterologous expression and characterization of the exopolysaccharide from *Streptococcus thermophilus* Sfi39. *Eur. J. Biochem.* 2001; 268: 5149–5156.
15. Grill JP, Cayuela C, Antoine JM, and Schneider F. Isolation and characterization of a *Lactobacillus amylovorus* mutant depleted in conjugated bile salt hydrolase activity: relation between activity and bile salt resistance. *J. Appl. Microbiol.* 2000; 89: 553–563.
16. Gueimonde M, Margolles A, de los Reyes-Gavilán CG, and Salminen S. Competitive exclusion of enteropathogens from human intestinal mucus by *Bifidobacterium* strains

with acquired resistance to bile–a preliminary study. *Int. J. Food Microbiol.* 2007; 113: 228–232.

17. Gueimonde M, Noriega L, Margolles A, de los Reyes-Gavilan CG, and Salminen S. Ability of *Bifidobacterium* strains with acquired resistance to bile to adhere to human intestinal mucus. *Int. J. Food Microbiol.* 2005; 101: 341–346.

18. Guglielmotti D, Marcó MB, Vinderola C, de Los Reyes Gavilán CG, Reinheimer J, and Quiberoni A. Spontaneous *Lactobacillus delbrueckii* phage-resistant mutants with acquired bile tolerance. *Int. J. Food Microbiol.* 2007; 119: 236–242.

19. Guglielmotti DM, Reinheimer JA, Binetti AG, Giraffa G, Carminati D, and Quiberoni A. Characterization of spontaneous phage-resistant derivatives of *Lactobacillus delbrueckii* commercial strains. *Int. J. Food Microbiol.* 2006; 111: 126–133.

20. Guinane CM, Cotter PD, Hill C, and Ross RP. Spontaneous resistance in *Lactococcus lactis* IL1403 to the lantibiotic lacticin 3147. *FEMS Microbiol. Lett.* 2006; 260: 77–83.

21. Guinane CM, Cotter PD, Lawton EM, Hil C, and Ross RP. Insertional mutagenesis to generate lantibiotic resistance in *Lactococcus lactis*. *Appl. Environ. Microbiol.* 2007; 73: 4677–4680.

22. Gury J, Barthelmebs L, and Cavin J-F. Random transposon mutagenesis of *Lactobacillus plantarum* by using the pGh9:IS*S1* vector to clone genes involved in the regulation of phenolic acid metabolism. *Arch. Microbiol.* 2004; 182: 337–345.

23. Hao Z, Reiske HR, and Wilson DB. Characterization of cadmium uptake in *Lactobacillus plantarum* and isolation of cadmium and manganese uptake mutants. *Appl. Environ. Microbiol.* 1999; 65: 4741–4745.

24. Ibrahim SA, O'Sullivan DJ. Use of chemical mutagenesis for the isolation of food grade β-galactosidase overproducing mutants of bifidobacteria, lactobacilli and *Streptococcus thermophilus*. *J. Dairy Sci.* 2000; 83: 923–930.

25. Israelsen H and Hansen EB. Insertion of Transposon Tn*917* Derivatives into the *Lactococcus lactis* subsp. *lactis* Chromosome. *Appl. Environ. Microbiol.* 1993; 59: 21–26.

26. Israelsen H, Madsen SM, Vrang A, Hansen EB, and Johansen E. Cloning and partial characterization of regulated promoters from *Lactococcus lactis* Tn*917-lacZ* integrants with the new promoter probe vector, pAK80. *Appl. Environ. Microbiol.* 1995; 61: 2540–2547.

27. Kadam SR, Patil SS, Bastawde KB, Khire JM, and Gokhale DV. Strain improvement of *Lactobacillus delbrueckii* NCIM 2365 for lactic acid production. *Process Biochem.* 2006; 41: 120–126.

28. Labarre C, Schirawski J, van der Zwet A, Fitzgerald GF, and van Sinderen D. Insertional mutagenesis of an industrial strain of *Streptococcus thermophilus*. *FEMS Microbiol. Lett.* 2001; 200: 85–90.

29. Law J, Buist G, Haandrikman A, Kok J, Venema G, and Leenhouts K. A system to generate chromosomal mutations in *Lactococcus lactis* which allows fast analysis of targeted genes. *J. Bacteriol.* 1995; 177: 7011–7018.

30. Lucchini S, Sidoti J, and Brüssow H. Broad-range bacteriophage resistance in *Streptococcus thermophilus* by insertional mutagenesis. *Virology* 2000; 275: 267–277.

31. Madsen SM, Arnau J, Vrang A, Givskov M, and Israelsen H. Molecular characterization of the pH-inducible and growth phase-dependent promoter P170 of *Lactococcus lactis*. *Mol. Microbiol.* 1999; 32: 75–87.

32. Maguin E, Prévost H, Ehrlich SD, and Gruss A. Efficient insertional mutagenesis in lactococci and other gram-positive bacteria. *J. Bacteriol.* 1996; 178: 931–935.
33. Margolles A, Garcia L, Sánchez B, Gueimonde M, and de los Reyes-Gavilán CG. Characterisation of a *Bifidobacterium* strain with acquired resistance to cholate–a preliminary study. *Int. J. Food Microbiol.* 2003; 82: 191–198.
34. Mathiesen G, Axelsen GW, Axelsson L, and Eijsink VG. Isolation of constitutive variants of a subfamily 10 histidine protein kinase (SppK) from *Lactobacillus* using random mutagenesis. *Arch. Microbiol.* 2006; 184: 327–334.
35. Monnet C, Aymes F, and Corrieu G. Diacetyl and α-acetolactate overproduction by *Lactococcus lactis* subsp. *lactis* biovar *diacetylactis* mutants that are deficient in α-acetolactate decarboxylase and have a low lactate dehydrogenase activity. *Appl. Environ. Microbiol.* 2000; 66: 5518–5520.
36. Monnet C, Béal C, and Corrieu G. Improvement of the resistance of *Lactobacillus delbrueckii* ssp. *bulgaricus* to freezing by natural selection. *J. Dairy Sci.* 2003; 86: 3048–3053.
37. Monnet C and Corrieu G. Selection and Properties of *Lactobacillus* Mutants Producing α-Acetolactate. *J. Dairy Sci.* 1998; 81: 2096–2102.
38. Monnet C and Corrieu G. Selection and properties of α-acetolactate decarboxylase-deficient spontaneous mutants of *Streptococcus thermophilus*. *Food Microbiol.* 2007; 24: 601–606.
39. Monnet C, Pernoud S, Sepulchre A, Fremaux C, and Corrieu G. Selection and properties of *Streptococcus thermophilus* mutants deficient in urease. *J. Dairy Sci.* 2004; 87: 1634–1640.
40. Monnet C, Schmitt P, and Diviès C. Development and Use of a Screening Procedure for Production of αAcetolactate by *Lactococcus lactis* subsp. *lactis* biovar diacetylactis Strains. *Appl. Environ. Microbiol.* 1997; 63: 793–795.
41. Morisset D, Berjeaud J-M, Marion D, Lacombe C, and Frère J. Mutational analysis of mesentericin y105, an anti-*Listeria* bacteriocin, for determination of impact on bactericidal activity, in vitro secondary structure, and membrane interaction. *Appl. Environ. Microbiol.* 2004; 70: 4672–4680.
42. Moy TI, Mylonakis E, Calderwood SB, and Ausubel FM. Cytotoxicity of hydrogen peroxide produced by *Enterococcus faecium*. *Infect. Immun.* 2004; 72: 4512–4520.
43. Noriega L, de los Reyes-Gavilán CG, and Margolles A. Acquisition of bile salt resistance promotes antibiotic susceptibility changes in *bifidobacterium*. *J. Food Prot.* 2005; 68: 1916–1919.
44. Noriega L, Gueimonde M, Sánchez B, Margolles A, and de los Reyes-Gavilán CG. Effect of the adaptation to high bile salts concentrations on glycosidic activity, survival at low PH and cross-resistance to bile salts in *Bifidobacterium*. *Int. J. Food Microbiol.* 2004; 94: 79–86.
45. Ongol MP, Sawatari Y, Ebina Y, Sone T, Tanaka M, Tomita F, Yokota A, and Asano K. Yoghurt fermented by *Lactobacillus delbrueckii* subsp. *bulgaricus* H^{+}-ATPase-defective mutants exhibits enhanced viability of *Bifidobacterium breve* during storage. *Int. J. Food Microbiol.* 2007; 116: 358–366.
46. Oyamada Y, Ito H, Fujimoto K, Asada R, Niga T, Okamoto R, Inoue M, and Yamagishi J. Combination of known and unknown mechanisms confers high-level resistance to fluoroquinolones in *Enterococcus faecium*. *J. Med. Microbiol.* 2006; 55: 729–736.

47. Patil SS, Kadam SR, Bastawde KB, Khire JM, and Gokhale DV. Production of lactic acid and fructose from media with cane sugar using mutant of *Lactobacillus delbrueckii* NCIM 2365. *Lett. Appl. Microbiol.* 2006; 43: 53–57.
48. Patnaik R, Louie S, Gavrilovic V, Perry K, Stemmer WP, Ryan CM, and del Cardayre S. Genome shuffling of *Lactobacillus* for improved acid tolerance. *Nat. Biotechnol.* 2002; 20: 707–712.
49. Rallu F, Gruss A, Ehrlich SD, and Maguin E. Acid- and multistress-resistant mutants of *Lactococcus lactis*: Identification of intracellular stress signals. *Mol. Microbiol.* 2000; 35: 517–528.
50. Ravn P, Arnau J, Madsen SM, Vrang A, and Israelsen H. The development of Tn*Nuc* and its use for the isolation of novel secretion signals in *Lactococcus lactis*. *Gene* 2000; 242: 347–356.
51. Ruas-Madiedo P, Hernández-Barranco A, Margolles A, and de los Reyes-Gavilán CG. A bile salt-resistant derivative of *Bifidobacterium animalis* has an altered fermentation pattern when grown on glucose and maltose. *Appl. Environ. Microbiol.* 2005; 71: 6564–6570.
52. Ruiz L, Sánchez B, Ruas-Madiedo P, de Los Reyes-Gavilán CG, and Margolles A. Cell envelope changes in *Bifidobacterium animalis* ssp. *lactis* as a response to bile. *FEMS Microbiol. Lett.* 2007; 274: 316–322.
53. Sánchez B, Champomier-Vergès M-C, Del Carmen Collado M, Anglade P, Baraige F, Sanz Y, de Los Reyes-Gavilán CG, Margolles A, and Zagorec M. Low pH adaptation and the acid tolerance response of *Bifidobacterium longum* biotype *longum*. *Appl. Environ. Microbiol.* 2007; 73: 6450–6459.
54. Sánchez B, de los Reyes-Gavilán CG, and Margolles A. The F_1F_0-ATPase of *Bifidobacterium animalis* is involved in bile tolerance. *Environ. Microbiol.* 2006; 8: 1825–1833.
55. Sánchez B, Noriega L, Ruas-Madiedo P, de los Reyes-Gavilán CG, and Margolles A. Acquired resistance to bile increases fructose-6-phosphate phosphoketolase activity in *Bifidobacterium*. *FEMS Microbiol. Lett.* 2004; 235: 35–41.
56. Sakayori Y, Muramatsu M, Hanada S, Kamagata Y, Kawamoto S, and Shima J. Characterization of *Enterococcus faecium* mutants resistant to mundticin KS, a class IIa bacteriocin. *Microbiology* 2003; 149: 2901–2908.
57. Salvetti S, Celandroni F, Ghelardi E, Baggiani A, and Senesi S. Rapid determination of vitamin B_2 secretion by bacteria growing on solid media. *J. Appl. Microbiol.* 2003; 95: 1255–1260.
58. Sasaki Y, Ito Y, and Sasaki T. *ThyA* as a selection marker in construction of food-grade host-vector and integration systems for *Streptococcus thermophilus*. *Appl. Environ. Microbiol.* 2004; 70: 1858–1864.
59. Stingel F and Mollet B. Disruption of the gene encoding penicillin-binding protein 2b (*pbp2b*) causes altered cell morphology and cease in exopolysaccharide production in *Streptococcus thermophilus* Sfi6. *Mol. Microbiol.* 1996; 22: 357–366.
60. Stingele F, Neeser J-R, and Mollet B. Identification and characterization of the *eps* (exopolysaccharide) gene cluster from *Streptococcus thermophilus* Sfi6. *J. Bacteriol.* 1996; 178: 1680–1690.
61. Tanaka H, Hashiba H, Kok, and Mierau I. Bile salt hydrolase of *Bifidobacterium longum*-biochemical and genetic characterization. *Appl. Environ. Microbiol.* 2000; 66: 2502–2512.

62. Thibessard A, Borges F, Fernandez A, Gintz B, Decaris B, and Leblond-Bourget N. Identification of *Streptococcus thermophilus* CNRZ368 genes involved in defense against superoxide stress. *Appl. Environ. Microbiol.* 2004; 70: 2220–2229.
63. Thibessard A, Fernandez A, Gintz B, Decaris B, and Leblond-Bourget N. Transposition of pGh9:IS*S1* is random and efficient in *Streptococcus thermophilus* CNRZ368. *Can. J. Microbiol.* 2002; 48: 473–478.
64. Thibessard A, Fernandez A, Gintz B, Leblond-Bourget N, and Decaris B. Effects of *rodA* and *pbp2b* disruption on cell morphology and oxidative stress response of *Streptococcus thermophilus* CNRZ368. *J. Bacteriol.* 2002; 184: 2821–2826.
65. Vaughan EE, van den Bogaard PT, Catzeddu P, Kuipers OP, and de Vos WM. Activation of silent *gal* genes in the *lac-gal* regulon of *Streptococcus thermophilus*. *J. Bacteriol.* 2001; 183: 1184–1194.
66. Vinderola G, Marcó MB, Guglielmotti DM, Perdigón G, Giraffa G, Reinheimer J, and Quiberoni A. Phage-resistant mutants of *Lactobacillus delbrueckii* may have functional properties that differ from those of parent strains. *Int. J. Food Microbiol.* 2007; 116: 96–102.
67. Vitali B, Turron S, Dal Piaz F, Candela M, Wasinger V, and Brigidi P. Genetic and proteomic characterization of rifaximin resistance in *Bifidobacterium infantis* BI07. *Res. Microbiol.* 2007; 158: 355–362.
68. Wang Y, Li Y, Pei X, Yu L, and Feng Y. Genome-shuffling improved acid tolerance and L-lactic acid volumetric productivity in *Lactobacillus rhamnosus*. *J. Biotechnol.* 2007; 129: 510–515.
69. Welman AD, Maddox IS, and Archer RH. Screening and selection of exopolysaccharide-producing strains of *Lactobacillus delbrueckii* subsp. *bulgaricus*. *J. Appl. Microbiol.* 2003; 95: 1200–1206.
70. Whittaker CJ, Clemans DL, and Kolenbrander PE. Insertional inactivation of an intrageneric coaggregation-relevant adhesin locus from *Streptococcus gordonii* DL1 (Challis). *Infect. Immun.* 1996; 64: 4137–4142.
71. Wickman PA, Black JA, Smith Moland E, Thomson KS, and Hanson ND. *In vitro* development of resistance to DX-619 and other quinolones in enterococci. *J. Antimicrob. Chemother.* 2006; 58: 1268–1273.
72. Abe K, Ohnishi F, Yagi K, Nakajima T, Higuchi T, Sano M, Machida M, Sarker RI, and Maloney PC. Plasmid-encoded *asp* operon confers a proton motive metabolic cycle catalyzed by an aspartate-alanine exchange reaction. *J. Bacteriol.* 2002; 184: 2906–2913.
73. Alegre MT, Rodriguez MC, and Mesas JM. Nucleotide sequence, structural organization and host range of pRS4, a small cryptic *Pediococcus pentosaceus* plasmid that contains two cassettes commonly found in other lactic acid bacteria. *FEMS Microbiol. Lett.* 2005; 250: 151–156.
74. Aleshin VV, Semenova EV, Doroshenko VG, Jomantas YV, Tarakanov BV, and Livshits VA. The broad host range plasmid pLF1311 from *Lactobacillus fermentum* VKM1311. *FEMS Microbiol. Lett.* 1999; 178: 47–53.
75. Alpert CA, Crutz-Le Coq AM, Malleret C, and Zagorec M. Characterization of a theta-type plasmid from *Lactobacillus sakei*: a potential basis for low-copy-number vectors in lactobacilli. *Appl. Environ. Microbiol.* 2003; 69: 5574–5584.
76. Álvarez-Martín P, Flórez AB, and Mayo B. Screening for plasmids among human bifidobacteria species: sequencing and analysis of pBC1 from *Bifidobacterium catenulatum* L48. *Plasmid* 2007; 57: 165–174.

77. Álvarez-Martín P, O'Connell-Motherway M, van Sinderen D, and Mayo B. Functional analysis of the pBC1 replicon from *Bifidobacterium catenulatum* L48. *Appl. Microbiol. Biotechnol.* 2007; 76: 1395–1402.
78. An HY and Miyamoto T. Cloning sequencing of plasmid pLC494 isolated from human intestinal *Lactobacillus casei*: construction of an *Escherichia coli-Lactobacillus* shuttle vector. *Plasmid* 2006; 55: 128–134.
79. An HY, Tsuda H, and Miyamoto T. Expression of citrate permease gene of plasmid pCM1 isolated from *Lactococcus lactis* subsp. *lactis* biovar diacetylactis NIAI N-7 in *Lactobacillus casei* L-49-4. *Appl. Microbiol. Biotechnol.* 2007; 74: 609–616.
80. Azcárate-Peril MA and Raya RR. Sequence analysis of pLBB1, a cryptic plasmid from *Lactobacillus delbrueckii* subsp. *bulgaricus*. *Can. J. Microbiol.* 2002; 48: 105–112.
81. Balasubramanyam BV and Varadaraj MC. Cultural conditions for the production of bacteriocin by a native isolate of *Lactobacillus delbrueckii* ssp. *bulgaricus* CFR 2028 in milk medium. *J. Appl. Microbiol.* 1998; 84: 97–102.
82. Biet F, Cenatiempo Y, and Fremaux C. Characterization of pFR18, a small cryptic plasmid from *Leuconostoc mesenteroides* ssp. *mesenteroides* FR52, and its use as a food grade vector. *FEMS Microbiol. Lett.* 1999; 179: 375–383.
83. Biet F, Cenatiempo Y, and Fremaux C. Identification of a replicon from pTXL1, a small cryptic plasmid from *Leuconostoc mesenteroides* subsp. *mesenteroides* Y110, and development of a food-grade vector. *Appl. Environ. Microbiol.* 2002; 68: 6451–6456.
84. Boucher I, Émond É, Parrot M, and Moineau S. DNA sequence analysis of three *Lactococcus lactis* plasmids encoding phage resistance mechanisms. *J. Dairy Sci.* 2001; 84: 1610–1620.
85. Bourniquel AA, Casey MG, Mollet B, and Pridmore RD. DNA sequence and functional analysis of *Lactobacillus delbrueckii* subsp. *lactis* plasmids pN42 and pJBL2. *Plasmid* 2002; 47: 153–157.
86. Chang SM and Yan TR. DNA sequence analysis of a cryptic plasmid pL2 from *Lactococcus lactis* subsp. *lactis*. *Biotechnol. Lett.* 2007; 29: 1519–1527.
87. Chin SC, Abdullah N, Siang TW, and Wan HY. Plasmid profiling and curing of *Lactobacillus* strains isolated from the gastrointestinal tract of chicken. *J. Microbiol.* 2005; 43: 251–256.
88. Christensson C, Pillidge CJ, Ward LJ, O'Toole PW. Nucleotide sequence and characterization of the cell envelope proteinase plasmid in *Lactococcus lactis* subsp. *cremoris* HP. *J. Appl. Microbiol.* 2001; 91: 334–343.
89. Claesson MJ, Li Y, Leahy S, Canchaya C, van Pijkeren JP, Cerdeño-Tárraga AM, Parkhill J, Flynn S, O'Sullivan GC, Collins JK, Higgins D, Shanahan F, Fitzgerald GF, van Sinderen D, and O'Toole PW. Multireplicon genome architecture of *Lactobacillus salivarius*. *Proc. Natl. Acad. Sci. USA.* 2006; 103: 6718–6723.
90. Coakley M, Fitzgerald G, and Ros RP. Application and evaluation of the phage resistance- and bacteriocin-encoding plasmid pMRC01 for the improvement of dairy starter cultures. *Appl. Environ. Microbiol.* 1997; 63: 1434–1440.
91. Corneau N, Émond É, and LaPointe G. Molecular characterization of three plasmids from *Bifidobacterium longum*. *Plasmid* 2004; 51: 87–100.
92. Criado R, Diep DB, Aakra Å, Gutiérrez J, Nes IF, Hernández PE, and Cintas LM. Complete sequence of the enterocin Q-encoding plasmid pCIZ2 from the multiple

bacteriocin producer *Enterococcus faecium* L50 and genetic characterization of enterocin Q production and immunity. *Appl. Environ. Microbiol.* 2006; 72: 6653–6666.
93. Cronin M, Knobel M, O'Connell-Motherway M, Fitzgerald GF, and van Sinderen D. Molecular dissection of a bifidobacterial replicon. *Appl. Environ. Microbiol.* 2007; 73: 7858–7866.
94. Cuozzo SA, Sesma F, Palacios JM, de Ruiz Holgado AP, and Raya RR. Identification and nucleotide sequence of genes involved in the synthesis of lactocin 705, a two-peptide bacteriocin from *Lactobacillus casei* CRL 705. *FEMS Microbiol. Lett.* 2000; 185: 157–161.
95. Daming R, Yinyu W, Zilai W, Jun C, Hekui L, and Jingye Z. Complete DNA sequence and analysis of two cryptic plasmids isolated from *Lactobacillus plantarum*. *Plasmid* 2003; 50: 70–73.
96. and Danielsen M. Characterization of the tetracycline resistance plasmid pMD5057 from *Lactobacillus plantarum* 5057 reveals a composite structure. *Plasmid* 2002; 48: 98–103.
97. De Kwaadsteniet M, Fraser T, Van Reenen CA, DicksLM. Bacteriocin T8, a novel class IIa sec-dependent bacteriocin produced by *Enterococcus faecium* T8, isolated from vaginal secretions of children infected with human immunodeficiency virus. *Appl. Environ. Microbiol.* 2006; 72: 4761–4766.
98. de las Rivas B, Marcobal A, and Muñoz R. Complete nucleotide sequence and structural organization of pPB1, a small *Lactobacillus plantarum* cryptic plasmid that originated by modular exchange. *Plasmid* 2004; 52: 203–211.
99. Delgado S and Mayo B. Development of *Lactobacillus plantarum* LL441 and its plasmid-cured derivatives in cheese. *J. Ind. Microbiol. Biotechnol.* 2003; 30: 216–219.
100. Deng YM, Harvey ML, Liu CQ, and Dunn NW. A novel plasmid-encoded phage abortive infection system from *Lactococcus lactis* biovar. *diacetylactis*. *FEMS Microbiol. Lett.* 1997; 146: 149–154.
101. Desmond C, Ross RP, Fitzgerald G, and Stanton C. Sequence analysis of the plasmid genome of the probiotic strain *Lactobacillus paracasei* NFBC338 which includes the plasmids pCD01 and pCD02. *Plasmid* 2005; 54: 160–175.
102. Domingues S, Chopin A, Ehrlich SD, and Chopin MC. The Lactococcal abortive phage infection system AbiP prevents both phage DNA replication and temporal transcription switch. *J. Bacteriol.* 2004; 186: 713–721.
103. Dougherty BA, Hill C, Weidman JF, Richardson DR, Venter JC, and Ross RP. Sequence and analysis of the 60 kb conjugative, bacteriocin-producing plasmid pMRC01 from *Lactococcus lactis* DPC3147. *Mol. Microbiol.* 1998; 29: 1029–1038.
104. Duan K, Harvey ML, Liu CQ, and Dunn NW. Identification and characterization of a mobilizing plasmid, pND300, in *Lactococcus lactis* M189 and its encoded nisin resistance determinant. *J. Appl. Bacteriol.* 1996; 81: 493–500.
105. Duan K, Liu CQ, Liu YJ, Ren J, and Dunn NW. Nucleotide sequence and thermostability of pND324, a 3.6-kb plasmid from *Lactococcus lactis*. *Appl. Microbiol. Biotechnol.* 1999; 53: 36–42.
106. Eguchi T, Doi K, Nishiyama K, Ohmomo S, and Ogata S. Characterization of a phage resistance plasmid, pLKS, of silage-making *Lactobacillus plantarum* NGRI0101. *Biosci. Biotechnol. Biochem.* 2000; 64: 751–756.
107. El Demerdash HA, Oxmann J, Heller KJ, and Geis A. Yoghurt fermentation at elevated temperatures by strains of *Streptococcus thermophilus* expressing a small heat-shock

protein: application of a two-plasmid system for constructing food-grade strains of *Streptococcus thermophilus*. *Biotechnol. J.* 2006; 1: 398–404.
108. Émond É, Lavallée R, Drolet G, Moineau S, and LaPointe G. Molecular characterization of a theta replication plasmid and its use for development of a two-component food-grade cloning system for *Lactococcus lactis*. *Appl. Environ. Microbiol.* 2001; 67: 1700–1709.
109. Fernández M, Margolles A, Suárez JE, and Mayo B. Duplication of the β-galactosidase gene in some *Lactobacillus plantarum* strains. *Int. J. Food Microbiol.* 1999; 48: 113–123.
110. Forde A, Daly C, and Fitzgerald GF. Identification of four phage resistance plasmids from *Lactococcus lactis* subsp. *cremoris* HO2. *Appl. Environ. Microbiol.* 1999; 65: 1540–1547.
111. Gancheva AG, Miteva VI, and Stefanova TT. A *Lactobacillus helveticus* plasmid detects restriction fragment length polymorphism in different bacterial species. *FEMS Microbiol. Lett.* 2000; 190: 335–339.
112. García-Quintáns N, Magni C, de Mendoza D, and López P. The citrate transport system of *Lactococcus lactis* subsp. *lactis* biovar diacetylactis is induced by acid stress. *Appl. Environ. Microbiol.* 1998; 64: 850–857.
113. Geis A, El Demerdash HA, and Heller KJ. Sequence analysis and characterization of plasmids from *Streptococcus thermophilus*. *Plasmid* 2003; 50: 53–69.
114. Gfeller KY, Roth M, Meile L, and Teuber M. Sequence and genetic organization of the 19.3-kb erythromycin- and dalfopristin-resistance plasmid pLME300 from *Lactobacillus fermentum* ROT1. *Plasmid* 2003; 50: 190–201.
115. Girard SL and Moineau S. Analysis of two theta-replicating plasmids of *Streptococcus thermophilus*. *Plasmid* 2007; 58: 174–181.
116. Guglielmetti S, Karp M, Mora D, Tamagnini I, and Parini C. Molecular characterization of *Bifidobacterium longum* biovar *longum* NAL8 plasmids and construction of a novel replicon screening system. *Appl. Microbiol. Biotechnol.* 2007; 74: 1053–1061.
117. Heng NC, Bateup JM, Loach DM, Wu X, Jenkinson HF, Morrison M, and Tannock GW. Influence of different functional elements of plasmid pGT232 on maintenance of recombinant plasmids in *Lactobacillus reuteri* populations in vitro and in vivo. *Appl. Environ. Microbiol.* 1999; 65: 5378–5385.
118. Horn N, Wegmann U, Narbad A, and Gasson MJ. Characterisation of a novel plasmid p9785S from *Lactobacillus johnsonii* FI9785. *Plasmid* 2005; 54: 176–183.
119. Huys G, D'Haene K, and Swings J. Genetic basis of tetracycline and minocycline resistance in potentially probiotic *Lactobacillus plantarum* strain CCUG 43738. *Antimicrob. Agents Chemother.* 2006; 50: 1550–1551.
120. Jacobsen L, Wilcks A, Hammer K, Huys G, Gevers D, and Andersen SR. Horizontal transfer of *tet*(M) and *erm*(B) resistance plasmids from food strains of *Lactobacillus plantarum* to *Enterococcus faecalis* JH2-2 in the gastrointestinal tract of gnotobiotic rats. *FEMS Microbiol. Ecol.* 2007; 59: 158–166.
121. Jeong SJ, Park JY, Lee HJ, and Kim JH. Characterization of pFMBL1, a small cryptic plasmid isolated from *Leuconostoc mesenteroides* SY2. *Plasmid* 2007; 57: 314–323.
122. Kaneko Y, Kobayashi H, Kiatpapan P, Nishimoto T, Napitupulu R, Ono H, and Murooka Y. Development of a host-vector system for *Lactobacillus plantarum* L137 isolated from a traditional fermented food produced in the Philippines. *J. Biosci. Bioeng.* 2000; 89: 62–67.
123. Kearney K, Fitzgerald GF, and Seegers JF. Identification and characterization of an active plasmid partition mechanism for the novel *Lactococcus lactis* plasmid pCI2000. *J. Bacteriol.* 2000; 182: 30–37.

124. Klijn A, Moine D, Delley M, Mercenier A, Arigoni F, and Pridmore RD. Construction of a reporter vector for the analysis of *Bifidobacterium longum* promoters. *Appl. Environ. Microbiol.* 2006; 72: 7401–7405.

125. Kojic M, Strahinic I and Topisirovic L. Proteinase PI, and lactococcin A genes are located on the largest plasmid in *Lactococcus lactis* subsp. *lactis* bv. diacetylactis S50. *Can. J. Microbiol.* 2005; 51: 305–314.

126. Kong J and Josephsen J. The ability of the plasmid-encoded restriction and modification system *Lla*BIII to protect *Lactococcus lactis* against bacteriophages. *Lett. Appl. Microbiol.* 2002; 34: 249–253.

127. Lee JH, Halgerson JS, Kim JH, O'Sullivan DJ. Comparative sequence analysis of plasmids from *Lactobacillus delbrueckii* and construction of a shuttle cloning vector. *Appl. Environ. Microbiol.* 2007; 73: 4417–4424.

128. Lee JH, O'Sullivan DJ. Sequence analysis of two cryptic plasmids from *Bifidobacterium longum* DJO10A and construction of a shuttle cloning vector. *Appl. Environ. Microbiol.* 2006; 72: 527–535.

129. Lee KH, Park WJ, Kim JY, Kim HG, Lee JM, Kim JH, Park JW, Lee JH, Chung SK, and Chung DK. Development of a monitoring vector for *Leuconostoc mesenteroides* using the green fluorescent protein gene. *J. Microbiol. Biotechnol.* 2007; 17: 1213–1216.

130. Li Y, Canchaya C, Fang F, Raftis E, Ryan KA, van Pijkeren JP, van Sinderen D, O'Toole PW. Distribution of megaplasmids in *Lactobacillus salivarius* and other lactobacilli. *J. Bacteriol.* 2007; 189: 6128–6139.

131. Lin CF and Chung TC. Cloning of erythromycin-resistance determinants and replication origins from indigenous plasmids of *Lactobacillus reuteri* for potential use in construction of cloning vectors. *Plasmid* 1999; 42: 31–41.

132. Liu CQ, Charoechai P, Khunajakr N, Deng YM, Widodo, and Dunn NW. Genetic and transcriptional analysis of a novel plasmid-encoded copper resistance operon from *Lactococcus lactis*. *Gene* 2002; 297: 241–247.

133. Liu CQ, Khunajakr N, Chia LG, Deng YM, Charoenchai P, and Dunn NW. Genetic analysis of regions involved in replication and cadmium resistance of the plasmid pND302 from *Lactococcus lactis*. *Plasmid* 1997; 38: 79–90.

134. Liu CQ, Leelawatcharamas V, Harvey ML, and Dunn NW. Cloning vectors for lactococci based on a plasmid encoding resistance to cadmium. *Curr. Microbiol.* 1996; 33: 35–39.

135. Lozo J, Vukasinovic M, Strahinic I, and Topisirovic L. Characterization and antimicrobial activity of bacteriocin 217 produced by natural isolate *Lactobacillus paracasei* subsp. *paracasei* BGBUK2-16. *J. Food Prot.* 2004; 67: 2727–2734.

136. Lucas PM, Wolken WA, Claisse O, Lolkema JS, and Lonvaud-Funel A. Histamine-producing pathway encoded on an unstable plasmid in *Lactobacillus hilgardii* 0006. *Appl. Environ. Microbiol.* 2005; 71: 1417–1424.

137. Madsen A and Josephsen J. The *Lla*GI restriction and modification system of *Lactococcus lactis* W10 consists of only one single polypeptide. *FEMS Microbiol. Lett.* 2001; 200: 91–96.

138. Madsen A, Westphal C, and Josephsen J. Characterization of a novel plasmid-encoded HsdS subunit, S.*Lla*W12I, from *Lactococcus lactis* W12. *Plasmid* 2000; 44: 196–200.

139. Miller KW, Ray P, Steinmetz T, Hanekamp T, and Ray B. Gene organization and sequences of pediocin AcH/PA-1 production operons in *Pediococcus* and *Lactobacillus* plasmids. *Lett. Appl. Microbiol.* 2005; 40: 56–62.

140. Mills S, Coffey A, O'Sullivan L, Stokes D, Hill C, Fitzgerald GF, and Ross RP. Use of lacticin 481 to facilitate delivery of the bacteriophage resistance plasmid, pCBG104 to cheese starters. *J. Appl. Microbiol.* 2002; 92: 238–246.

141. Nicoloff H and Bringel F. IS*Lpl1* is a functional IS*30*-related insertion element in *Lactobacillus plantarum* that is also found in other lactic acid bacteria. *Appl. Environ. Microbiol.* 2003; 69: 6032–6040.

142. O'Connor EB, O'Sullivan O, Stanton C, Danielsen M, Simpson PJ, Callanan MJ, Ross RP, and Hill C. pEOC01: a plasmid from *Pediococcus acidilactici* which encodes an identical streptomycin resistance (*aadE*) gene to that found in *Campylobacter jejuni*. *Plasmid* 2007; 58: 115–126.

143. O'Connor L, Coffey A, Daly C, and Fitzgerald GF. AbiG, a genotypically novel abortive infection mechanism encoded by plasmid pCI750 of *Lactococcus lactis* subsp. *cremoris* UC653. *Appl. Environ. Microbiol.* 1996; 62: 3075–3082.

144. O'Driscoll J, Glynn F, Cahalane O, O'Connell-Motherway M, Fitzgerald GF, and Van Sinderen D. Lactococcal plasmid pNP40 encodes a novel, temperature-sensitive restriction-modification system. *Appl. Environ. Microbiol.* 2004; 70: 5546–5556.

145. O'Driscoll J, Glynn F, Fitzgerald GF, and van Sinderen D. Sequence analysis of the lactococcal plasmid pNP40: a mobile replicon for coping with environmental hazards. *J. Bacteriol.* 2006; 188: 6629–6639.

146. O'Riordan K and Fitzgerald GF. Molecular characterisation of a 5.75-kb cryptic plasmid from *Bifidobacterium breve* NCFB 2258 and determination of mode of replication. *FEMS Microbiol. Lett.* 1999; 174: 285–294.

147. O'Sullivan D, Ross RP, Twomey DP, Fitzgerald GF, Hill C, and Coffey A. Naturally occurring lactococcal plasmid pAH90 links bacteriophage resistance and mobility functions to a food-grade selectable marker. *Appl. Environ. Microbiol.* 2001; 67: 929–937.

148. O'Sullivan D, Twomey DP, Coffey A, Hill C, Fitzgerald GF, and Ross RP. Novel type I restriction specificities through domain shuffling of HsdS subunits in *Lactococcus lactis*. *Mol. Microbiol.* 2000; 36: 866–875.

149. O'Sullivan T, van Sinderen D, and Fitzgerald G. Structural and functional analysis of pCI65st, a 6.5 kb plasmid from *Streptococcus thermophilus* NDI-6. *Microbiology* 1999; 145 (Pt. 1): 127–134.

150. Ohmomo S, Murata S, Katayama N, Nitisinprasart S, Kobayashi M, Nakajima T, Yajima M, and Nakanishi K. Purification and some characteristics of enterocin ON-157, a bacteriocin produced by *Enterococcus faecium* NIAI 157. *J. Appl. Microbiol.* 2000; 88: 81–89.

151. Osmanağaoğlu O, Beyatli Y, Gündüz U, and Saçilik SC. Analysis of the genetic determinant for production of the pediocin P of *Pediococcus pentosaceus* Pep1. *J. Basic Microbiol.* 2000; 40: 233–241.

152. Park J, Lee M, Jung J, and Kim J. pIH01, a small cryptic plasmid from *Leuconostoc citreum* IH3. *Plasmid* 2005; 54: 184–189.

153. Park MS, Shin DW, Lee KH, and Ji GE. Sequence analysis of plasmid pKJ50 from *Bifidobacterium longum*. *Microbiology* 1999; 145(Pt. 3): 585–592.

154. Park WJ, Lee KH, Lee JM, Lee HJ, Kim JH, Lee JH, Chang HC, and Chung DK. Characterization of pC7 from *Lactobacillus paraplantarum* C7 derived from Kimchi and development of lactic acid bacteria–*Escherichia coli* shuttle vector. *Plasmid* 2004; 52: 84–88.

155. Park YS, Kim KH, Park JH, Oh IK, and Yoon SS. Isolation and molecular characterization of a cryptic plasmid from *Bifidobacterium longum*. *Biotechnol. Lett.* 2008; 30: 145–151.

156. Pavlova SI, Kilic AO, Topisirovic L, Miladinov N, Hatzos C, and Tao L. Characterization of a cryptic plasmid from *Lactobacillus fermentum* KC5b and its use for constructing a stable *Lactobacillus* cloning vector. *Plasmid* 2002; 47: 182–192.

157. Perreten V, Schwarz FV, Teuber M, and Levy SB. Mdt(A), a new efflux protein conferring multiple antibiotic resistance in *Lactococcus lactis* and *Escherichia coli*. *Antimicrob. Agents Chemother.* 2001; 45: 1109–1114.

158. Petrova P, Miteva V, Ruiz-Masó JA, and del Solar G. Structural and functional analysis of pt38, a 2.9 kb plasmid of *Streptococcus thermophilus* yogurt strain. *Plasmid* 2003; 50: 176–189.

159. Raha AR, Hooi WY, Mariana NS, Radu S, Varma NR, and Yusoff K. DNA sequence analysis of a small cryptic plasmid from *Lactococcus lactis* subsp. *lactis* M14. *Plasmid* 2006; 56: 53–61.

160. Raha AR, Ross E, Yusoff K, Manap MY, and Ideris A. Characterisation and molecular cloning of an erythromycin resistance plasmid of *Lactococcus lactis* isolated from chicken cecum. *J. Biochem. Mol. Biol. Biophy.* 2002; 6: 7–11.

161. Rawlinson EL, Nes IF, and Skaugen M. LasX, a transcriptional regulator of the lactocin S biosynthetic genes in *Lactobacillus sakei* L45, acts both as an activator and a repressor. *Biochimie* 2002; 84: 559–567.

162. Ricci G, Borgo F, and Fortina MG. Plasmids from *Lactobacillus helveticus*: distribution and diversity among natural isolates. *Lett. Appl. Microbiol.* 2006; 42: 254–258.

163. Ryan MP, Rea MC, Hill C, and Ross RP. An application in cheddar cheese manufacture for a strain of *Lactococcus lactis* producing a novel broad-spectrum bacteriocin, lacticin 3147. *Appl. Environ. Microbiol.* 1996; 62: 612–619.

164. Sami M, Suzuki K, Sakamoto K, Kadokura H, Kitamoto K, and Yoda K. A plasmid pRH45 of *Lactobacillus brevis* confers hop resistance. *J. Gen. Appl. Microbiol.* 1998; 44: 361–363.

165. Sánchez C, Hernández de Rojas A, Martínez B, Argüelles ME, Suárez JE, Rodríguez A, and Mayo B. Nucleotide sequence and analysis of pBL1, a bacteriocin-producing plasmid from *Lactococcus lactis* IPLA 972. *Plasmid* 2000; 44: 239–249.

166. Sánchez C and Mayo B. Sequence and analysis of pBM02, a novel RCR cryptic plasmid from *Lactococcus lactis* subsp *cremoris* P8-2-47. *Plasmid* 2003; 49: 118–129.

167. Sangrador-Vegas A, Stanton C, van Sinderen D, Fitzgerald GF, and Ross RP. Characterization of plasmid pASV479 from *Bifidobacterium pseudolongum* subsp. *globosum* and its use for expression vector construction. *Plasmid* 2007; 58: 140–147.

168. Schouler C, Clier F, Lerayer AL, Ehrlich SD, and Chopin MC. A type IC restriction-modification system in *Lactococcus lactis*. *J. Bacteriol.* 1998; 180: 407–411.

169. Scolari G, Torriani S, and Vescovo M. Partial characterization and plasmid linkage of a non-proteinaceous antimicrobial compound in a *Lactobacillus casei* strain of vegetable origin. *J. Appl. Microbiol.* 1999; 86: 682–688.

170. Seegers JF, van Sinderen D, and Fitzgerald GF. Molecular characterization of the lactococcal plasmid pCIS3: Natural stacking of specificity subunits of a type I restriction/modification system in a single lactococcal strain. *Microbiology* 2000; 146(Pt. 2): 435–443.

171. Shinoda T, Kusuda D, Ishida Y, Ikeda N, Kaneko K, Masuda O, and Yamamoto N. Survival of *Lactobacillus helveticus* strain CP53 in the human gastrointestinal tract. *Lett. Appl. Microbiol.* 2001; 32: 108–113.
172. Siezen RJ, Renckens B, van Swam I, Peters S, van Kranenburg R, Kleerebezem M, and de Vos WM. Complete sequences of four plasmids of *Lactococcus lactis* subsp. *cremoris* SK11 reveal extensive adaptation to the dairy environment. *Appl. Environ. Microbiol.* 2005; 71: 8371–8382.
173. Solaiman DK and Somkuti GA. Characterization of a novel *Streptococcus thermophilus* rolling-circle plasmid used for vector construction. *Appl. Microbiol. Biotechnol.* 1998; 50: 174–180.
174. Solow BT and Somkuti GA. Comparison of low-molecular-weight heat stress proteins encoded on plasmids in different strains of *Streptococcus thermophilus*. *Curr. Microbiol.* 2000; 41: 177–181.
175. Solow BT and Somkuti GA. Molecular properties of *Streptococcus thermophilus* plasmid pER35 encoding a restriction modification system. *Curr. Microbiol.* 2001; 42: 122–128.
176. Somkuti GA, Solaiman DK, and Steinberg DH. Structural and functional properties of the *hsp16.4*-bearing plasmid pER341 in *Streptococcus thermophilus*. *Plasmid* 1998; 40: 61–72.
177. Somkuti GA and Steinberg DH. Promoter activity of the pER341-borne ST_{Phsp} in heterologous gene expression in *Escherichia coli* and *Streptococcus thermophilus*. *FEMS Microbiol. Lett.* 1999; 179: 431–436.
178. Somkuti GA and Steinberg DH. Molecular organization of plasmid pER13 in *Streptococcus thermophilus*. *Biotechnol. Lett.* 2007; 29: 1991–1999.
179. Sørvig E, Skaugen M, Naterstad K, Eijsink VG, and Axelsson L. Plasmid p256 from *Lactobacillus plantarum* represents a new type of replicon in lactic acid bacteria, and contains a toxin-antitoxin-like plasmid maintenance system. *Microbiology* 2005; 151: 421–431.
180. Su P, Jury K, Allison GE, Wong WY, Kim WS, Liu CQ, Vancov T, and Dunn NW. Cloning vectors for *Streptococcus thermophilus* derived from a native plasmid. *FEMS Microbiol. Lett.* 2002; 216: 43–47.
181. Sudhamani M, Ismaiel E, Geis A, Batish V, and Heller KJ. Characterisation of pSMA23, a 3.5kbp plasmid of *Lactobacillus casei*, and application for heterologous expression in *Lactobacillus*. *Plasmid* 2008; 59: 11–19.
182. Suzuki K, Koyanagi M, and Yamashita H. Genetic characterization of non-spoilage variant isolated from beer-spoilage *Lactobacillus brevis* ABBC45. *J. Appl. Microbiol.* 2004; 96: 946–953.
183. Suzuki K, Sami M, Iijima K, Ozaki K, and Yamashita H. Characterization of *horA* and its flanking regions of *Pediococcus damnosus* ABBC478 and development of more specific and sensitive *horA* PCR method. *Lett. Appl. Microbiol.* 2006; 42: 392–399.
184. Suzuki K, Sami M, Kadokura H, Nakajima H, and Kitamoto K. Biochemical characterization of *horA*-independent hop resistance mechanism in *Lactobacillus brevis*. *Int. J. Food Microbiol.* 2002; 76: 223–230.
185. Tanaka K, Samura K, and Kano Y. Structural and functional analysis of pTB6 from *Bifidobacterium longum*. *Biosci. Biotechnol. Biochem.* 2005; 69: 422–425.

REFERENCES 241

186. Tanimoto K and Ike Y. Analysis of the conjugal transfer system of the pheromone-independent highly transferable *Enterococcus* plasmid pMG1: identification of a *tra* gene (*traA*) up-regulated during conjugation. *J. Bacteriol.* 2002; 184: 5800–5804.
187. Tanous C, Chambellon E, Sepulchre AM, and Yvon M. The gene encoding the glutamate dehydrogenase in *Lactococcus lactis* is part of a remnant Tn*3* transposon carried by a large plasmid. *J. Bacteriol.* 2005; 187: 5019–5022.
188. Tanous C, Chambellon E, and Yvon M. Sequence analysis of the mobilizable lactococcal plasmid pGdh442 encoding glutamate dehydrogenase activity. *Microbiology* 2007; 153: 1664–1675.
189. Thompson JK, Foley S, McConville KJ, Nicholson C, Collins MA, and Pridmore RD. Complete sequence of plasmid pLH1 from *Lactobacillus helveticus* ATCC15009: analysis reveals the presence of regions homologous to other native plasmids from the host strain. *Plasmid* 1999; 42: 221–235.
190. Tomita H and Ike Y. Genetic analysis of transfer-related regions of the vancomycin resistance *Enterococcus* conjugative plasmid pHTa: identification of *oriT* and a putative relaxase gene. *J. Bacteriol.* 2005; 187: 7727–7737.
191. Tomita H, Tanimoto K, Hayakawa S, Morinaga K, Ezaki K, Oshima H, and Ike Y. Highly conjugative pMG1-like plasmids carrying Tn*1546*-like transposons that encode vancomycin resistance in *Enterococcus faecium*. *J. Bacteriol.* 2003; 185: 7024–7028.
192. Trotter M, McAuliffe OE, Fitzgerald GF, Hill C, Ross RP, and Coffey A. Variable bacteriocin production in the commercial starter *Lactococcus lactis* DPC4275 is linked to the formation of the cointegrate plasmid pMRC02. *Appl. Environ. Microbiol.* 2004; 70: 34–42.
193. Trotter M, Mills S, Ross RP, Fitzgerald GF, and Coffey A. The use of cadmium resistance on the phage-resistance plasmid pNP40 facilitates selection for its horizontal transfer to industrial dairy starter lactococci. *Lett. Appl. Microbiol.* 2001; 33: 409–414.
194. Turgeon N, Frenette M, and Moineau S. Characterization of a theta-replicating plasmid from *Streptococcus thermophilus*. *Plasmid* 2004; 51: 24–36.
195. Turgeon N and Moineau S. Isolation and characterization of a *Streptococcus thermophilus* plasmid closely related to the pMV158 family. *Plasmid* 2001; 45: 171–183.
196. Twomey DP, De Urraza PJ, McKay LL, O'Sullivan DJ. Characterization of AbiR, a novel multicomponent abortive infection mechanism encoded by plasmid pKR223 of *Lactococcus lactis* subsp. *lactis* KR2. *Appl. Environ. Microbiol.* 2000; 66: 2647–2651.
197. van Kranenburg R and de Vos WM. Characterization of multiple regions involved in replication and mobilization of plasmid pNZ4000 coding for exopolysaccharide production in *Lactococcus lactis*. *J. Bacteriol.* 1998; 180: 5285–5290.
198. van Kranenburg R, Golic N, Bongers R, Leer RJ, de Vos WM, Siezen RJ, and Kleerebezem M. Functional analysis of three plasmids from *Lactobacillus plantarum*. *Appl. Environ. Microbiol.* 2005; 71: 1223–1230.
199. van Kranenburg R, Kleerebezem M, and de Vos WM. Nucleotide sequence analysis of the lactococcal EPS plasmid pNZ4000. *Plasmid* 2000; 43: 130–136.
200. van Kranenburg R, Marugg JD, van S II, Willem NJ, and de Vos WM. Molecular characterization of the plasmid-encoded *eps* gene cluster essential for exopolysaccharide biosynthesis in *Lactococcus lactis*. *Mol. Microbiol.* 1997; 24: 387–397.

201. Van Reenen CA, Chikindas ML, Van Zyl WH, and Dicks LM. Characterization and heterologous expression of a class IIa bacteriocin, plantaricin 423 from *Lactobacillus plantarum* 423, in *Saccharomyces cerevisiae*. *Int. J. Food Microbiol.* 2003; 81: 29–40.

202. Van Reenen CA, Van Zyl WH, and Dicks LM. Expression of the immunity protein of plantaricin 423, produced by *Lactobacillus plantarum* 423, and analysis of the plasmid encoding the bacteriocin. *Appl. Environ. Microbiol.* 2006; 72: 7644–7651.

203. Walling E, Gindreau E, and Lonvaud-Funel A. A putative glucan synthase gene *dps* detected in exopolysaccharide-producing *Pediococcus damnosus* and *Oenococcus oeni* strains isolated from wine and cider. *Int. J. Food Microbiol.* 2005; 98: 53–62.

204. Werning ML, Ibarburu I, Dueñas MT, Irastorza A, Navas J, and López P. *Pediococcus parvulus gtf* gene encoding the GTF glycosyltransferase and its application for specific PCR detection of β-D-glucan-producing bacteria in foods and beverages. *J. Food Prot.* 2006; 69: 161–169.

205. Wescombe PA, Burton JP, Cadieux PA, Klesse NA, Hyink O, Heng NC, Chilcott CN, Reid G, and Tagg JR. Megaplasmids encode differing combinations of lantibiotics in *Streptococcus salivarius*. *Antonie Van Leeuwenhoek* 2006; 90: 269–280.

206. Whitehead TR and Cotta MA. Sequence analyses of a broad host-range plasmid containing *ermT* from a tylosin-resistant *Lactobacillus* sp. Isolated from swine feces. *Curr. Microbiol.* 2001; 43: 17–20.

207. Yildirim Z, Winters DK, and Johnson MG. Purification, amino acid sequence and mode of action of bifidocin B produced by *Bifidobacterium bifidum* NCFB 1454. *J. Appl. Microbiol.* 1999; 86: 45–54.

208. Zheng B, Tomita H, Xiao YH, and Ike Y. The first molecular analysis of clinical isolates of VanA-type vancomycin-resistant *Enterococcus faecium* strains in Mainland China. *Lett. Appl. Microbiol.* 2007; 45: 307–312.

209. Zheng B, Tomita H, Xiao YH, Wang S, Li Y, and Ike Y. Molecular characterization of vancomycin-resistant *Enterococcus faecium* isolates from mainland China. *J. Clin. Microbiol.* 2007; 45: 2813–2818.

210. Alegre MT, Rodríguez MC, and Mesas JM. Nucleotide sequence, structural organization and host range of pRS4, a small cryptic *Pediococcus pentosaceus* plasmid that contains two cassettes commonly found in other lactic acid bacteria. *FEMS Microbiol. Lett.* 2005; 250: 151–156.

211. Alpert CA, Crutz-Le Coq AM, Malleret C, and Zagorec M. Characterization of a theta-type plasmid from *Lactobacillus sakei*: a potential basis for low-copy-number vectors in lactobacilli. *Appl. Environ. Microbiol.* 2003; 69: 5574–5584.

212. Álvarez-Martín P, O'Connell-Motherway M, van Sinderen D, and Mayo B. Functional analysis of the pBC1 replicon from *Bifidobacterium catenulatum* L48. *Appl. Microbiol. Biotechnol.* 2007; 76: 1395–1402.

213. Alvarez MA, Herrero M, and Suárez JE. The site-specific recombination system of the *Lactobacillus* species bacteriophage A2 integrates in gram-positive and gram-negative bacteria. *Virology* 1998; 250: 185–193.

214. An HY and Miyamoto T. Cloning and sequencing of plasmid pLC494 isolated from human intestinal *Lactobacillus casei*: construction of an *Escherichia coli-Lactobacillus* shuttle vector. *Plasmid* 2006; 55: 128–134.

215. An HY, Tsuda H, and Miyamoto T. Expression of citrate permease gene of plasmid pCM1 isolated from *Lactococcus lactis* subsp. *lactis* biovar *diacetylactis* NIAI N-7 in *Lactobacillus casei* L-49-4. *Appl. Microbiol. Biotechnol.* 2007; 74: 609–616.

216. Antikainen J, Anton L, Sillanpaa J, and Korhonen TK. Domains in the S-layer protein CbsA of *Lactobacillus crispatus* involved in adherence to collagens, laminin and lipoteichoic acids and in self-assembly. *Mol. Microbiol.* 2002; 46: 381–394.

217. Åvall-Jääskeläinen S, Kylä-Nikkilä K, Kahala M, Miikkulainen-Lahti T, and Palva A. Surface display of foreign epitopes on the *Lactobacillus brevis* S-layer. *Appl. Environ. Microbiol.* 2002; 68: 5943–5951.

218. Axelsson L, Lindstad G, and Naterstad K. Development of an inducible gene expression system for *Lactobacillus sakei*. *Lett. Appl. Microbiol.* 2003; 37: 115–120.

219. Barthelmebs L, Divies C, and Cavin JF. Knockout of the *p*-coumarate decarboxylase gene from *Lactobacillus plantarum* reveals the existence of two other inducible enzymatic activities involved in phenolic acid metabolism. *Appl. Environ. Microbiol.* 2000; 66: 3368–3375.

220. Biet F, Cenatiempo Y, and Fremaux C. Characterization of pFR18, a small cryptic plasmid from *Leuconostoc mesenteroides* ssp. *mesenteroides* FR52, and its use as a food grade vector. *FEMS Microbiol. Lett.* 1999; 179: 375–383.

221. Biet F, Cenatiempo Y, and Fremaux C. Identification of a replicon from pTXL1, a small cryptic plasmid from *Leuconostoc mesenteroides* subsp. *mesenteroides* Y110, and development of a food-grade vector. *Appl. Environ. Microbiol.* 2002; 68: 6451–6456.

222. Broadbent JR, Gummalla S, Hughes JE, Johnson ME, Rankin SA, and Drake MA. Overexpression of *Lactobacillus casei* D-hydroxyisocaproic acid dehydrogenase in cheddar cheese. *Appl. Environ. Microbiol.* 2004; 70: 4814–4820.

223. Bron PA, Hoffer SM, Van S II, De Vos WM, and Kleerebezem M. Selection and characterization of conditionally active promoters in *Lactobacillus plantarum*, using alanine racemase as a promoter probe. *Appl. Environ. Microbiol.* 2004; 70: 310–317.

224. Bruno-Bárcena JM, Andrus JM, Libby SL, Klaenhammer TR, and Hassan HM. Expression of a heterologous manganese superoxide dismutase gene in intestinal lactobacilli provides protection against hydrogen peroxide toxicity. *Appl. Environ. Microbiol.* 2004; 70: 4702–4710.

225. Bruno-Bárcena JM, Azcarate-Peril MA, Klaenhammer TR, and Hassan HM. Marker-free chromosomal integration of the manganese superoxide dismutase gene (*sodA*) from *Streptococcus thermophilus* into *Lactobacillus gasseri*. *FEMS Microbiol. Lett.* 2005; 246: 91–101.

226. Cahyanto MN, Kawasaki H, Nagashio M, Fujiyama K, and Seki T. Construction of *Lactobacillus plantarum* strain with enhanced L-lysine yield. *J. Appl. Microbiol.* 2007; 102: 674–679.

227. Callanan MJ, Russell WM, and Klaenhammer TR. Modification of *Lactobacillus* β-glucuronidase activity by random mutagenesis. *Gene* 2007; 389: 122–127.

228. Cho JS, Choi YJ, and Chung DK. Expression of *Clostridium thermocellum* endoglucanase gene in *Lactobacillus gasseri* and *Lactobacillus johnsonii* and characterization of the genetically modified probiotic *lactobacilli*. *Curr. Microbiol.* 2000; 40: 257–263.

229. Chu H, Kang S, Ha S, Cho K, Park SM, Han KH, Kang SK, Lee H, Han SH, Yun CH, and Choi Y. *Lactobacillus acidophilus* expressing recombinant K99 adhesive fimbriae has

an inhibitory effect on adhesion of enterotoxigenic *Escherichia coli. Microbiol. Immunol.* 2005; 49: 941–948.

230. Colombi D, Oliveira ML, Campos IB, Monedero V, Pérez-Martinez G, and Ho PL. Haemagglutination induced by *Bordetella pertussis* filamentous haemagglutinin adhesin (FHA) is inhibited by antibodies produced against $FHA_{430-873}$ fragment expressed in *Lactobacillus casei. Curr. Microbiol.* 2006; 53: 462–466.

231. Cortes-Perez NG, Azevedo V, Alcocer-González JM, Rodriguez-Padilla C, Tamez-Guerra RS, Corthier G, Gruss A, Langella P, and Bermudez-Humarán LG. Cell-surface display of E7 antigen from human papillomavirus type-16 in *Lactococcus lactis* and in *Lactobacillus plantarum* using a new cell-wall anchor from lactobacilli. *J. Drug Target* 2005; 13: 89–98.

232. Cronin M, Knobel M, O'Connell-Motherway M, Fitzgerald GF, and van Sinderen D. Molecular dissection of a bifidobacterial replicon. *Appl. Environ. Microbiol.* 2007; 73: 7858–7866.

233. De Keersmaecker SC, Braeken K, Verhoeven TL, Perea Vélez M, Lebeer S, Vanderleyden J, and Hols P. Flow cytometric testing of green fluorescent protein-tagged *Lactobacillus rhamnosus* GG for response to defensins. *Appl. Environ. Microbiol.* 2006; 72: 4923–4930.

234. Desmond C, Fitzgerald GF, Stanton C, and Ross RP. Improved stress tolerance of GroESL-overproducing *Lactococcus lactis* and probiotic *Lactobacillus paracasei* NFBC 338. *Appl. Environ. Microbiol.* 2004; 70: 5929–5936.

235. Fernández M, Martínez-Bueno M, Martín MC, Valdivia E, and Maqueda M. Heterologous expression of enterocin AS-48 in several strains of lactic acid bacteria. *J. Appl. Microbiol.* 2007; 102: 1350–1361.

236. Fernández M, van Doesburg W, Rutten GA, Marugg JD, Alting AC, van Kranenburg R, and Kuipers OP. Molecular and functional analyses of the *metC* gene of *Lactococcus lactis*, encoding cystathionine β-lyase. *Appl. Environ. Microbiol.* 2000; 66: 42–48.

237. Fu GF, Li X, Hou YY, Fan YR, Liu WH, and Xu GX. *Bifidobacterium longum* as an oral delivery system of endostatin for gene therapy on solid liver cancer. *Cancer Gene Ther.* 2005; 12: 133–140.

238. Fu X and Xu JG. Development of a chromosome-plasmid balanced lethal system for *Lactobacillus acidophilus* with *thyA* gene as selective marker. *Microbiol. Immunol.* 2000; 44: 551–556.

239. Geoffroy MC, Guyard C, Quatannens B, Pavan S, Lange M, and Mercenier A. Use of green fluorescent protein to tag lactic acid bacterium strains under development as live vaccine vectors. *Appl. Environ. Microbiol.* 2000; 66: 383–391.

240. Gory L, Montel MC, and Zagorec M. Use of green fluorescent protein to monitor *Lactobacillus sakei* in fermented meat products. *FEMS Microbiol. Lett.* 2001; 194: 127–133.

241. Gosalbes MJ, Esteban CD, Galán JL, and Pérez-Martínez G. Integrative food-grade expression system based on the lactose regulon of *Lactobacillus casei. Appl. Environ. Microbiol.* 2000; 66: 4822–4828.

242. Gosalbes MJ, Pérez-Arellano I, Esteban CD, Galán JL, Pérez-Martínez G. Use of *lac* regulatory elements for gene expression in *Lactobacillus casei. Lait* 2001; 81: 29–35.

243. Grangette C, Nutten S, Palumbo E, Morath S, Hermann C, Dewulf J, Pot B, Hartung T, Hols P, and Mercenier A. Enhanced antiinflammatory capacity of a *Lactobacillus*

plantarum mutant synthesizing modified teichoic acids. *Proc. Natl. Acad. Sci. USA.* 2005; 102: 10321–10326.

244. Guglielmetti S, Karp M, Mora D, Tamagnini I, and Parini C. Molecular characterization of *Bifidobacterium longum* biovar *longum* NAL8 plasmids and construction of a novel replicon screening system. *Appl. Microbiol. Biotechnol.* 2007; 74: 1053–1061.

245. Gury J, Barthelmebs L, and Cavin JF. Random transposon mutagenesis of *Lactobacillus plantarum* by using the pGh9:IS*S1* vector to clone genes involved in the regulation of phenolic acid metabolism. *Arch. Microbiol.* 2004; 182: 337–345.

246. Hamaji Y, Fujimori M, Sasaki T, Matsuhashi H, Matsui-Seki K, Shimatani-Shibata Y, Kano Y, Amano J, and Taniguchi S. Strong enhancement of recombinant cytosine deaminase activity in *Bifidobacterium longum* for tumor-targeting enzyme/prodrug therapy. *Biosci. Biotechnol. Biochem.* 2007; 71: 874–883.

247. Hazebrouck S, Oozeer R, Adel-Patient K, Langella P, Rabot S, Wal JM, and Corthier G. Constitutive delivery of bovine β-lactoglobulin to the digestive tracts of gnotobiotic mice by engineered *Lactobacillus casei*. *Appl. Environ. Microbiol.* 2006; 72: 7460–7467.

248. Hazebrouck S, Pothelune L, Azevedo V, Corthier G, Wal JM, and Langella P. Efficient production and secretion of bovine β-lactoglobulin by *Lactobacillus casei*. *Microb. Cell Fact.* 2007; 6: 12.

249. Hidaka A, Hamaji Y, Sasaki T, Taniguchi S, and Fujimori M. Exogenous cytosine deaminase gene expression in *Bifidobacterium breve* I-53-8w for tumor-targeting enzyme/prodrug therapy. *Biosci. Biotechnol. Biochem.* 2007; 71: 2921–2926.

250. Hols P, Slos P, Dutot P, Reymund J, Chabot P, Delplace B, Delcour J, and Mercenier A. Efficient secretion of the model antigen M6-gp41E in *Lactobacillus plantarum* NCIMB 8826. *Microbiology* 1997; 143(Pt. 8): 2733–2741.

251. Hörmann S, Vogel RF, and Ehrmann M. Construction of a new reporter system to study the NaCl-dependent *dnaK* promoter activity of *Lactobacillus sanfranciscensis*. *Appl. Microbiol. Biotechnol.* 2006; 70: 690–697.

252. Horn N, Wegmann U, Narbad A, and Gasson MJ. Characterisation of a novel plasmid p9785S from *Lactobacillus johnsonii* FI9785. *Plasmid* 2005; 54: 176–183.

253. Hüfner E, Markieton T, Chaillou S, Crutz-Le Coq AM, Zagorec M, and Hertel C. Identification of *Lactobacillus sakei* genes induced during meat fermentation and their role in survival and growth. *Appl. Environ. Microbiol.* 2007; 73: 2522–2531.

254. Hultberg A, Tremblay DM, de Haard H, Verrips T, Moineau S, Hammarström L, and Marcotte H. Lactobacillli expressing llama VHH fragments neutralise *Lactococcus* phages. *BMC Biotechnol* 2007; 7: 58.

255. Jang SJ, Ham MS, Lee JM, Chung SK, Lee HJ, Kim JH, Chang HC, Lee JH, and Chung DK. New integration vector using a cellulase gene as a screening marker for *Lactobacillus*. *FEMS Microbiol. Lett.* 2003; 224: 191–195.

256. Jeong SJ, Park JY, Lee HJ, and Kim JH. Characterization of pFMBL1, a small cryptic plasmid isolated from *Leuconostoc mesenteroides* SY2. *Plasmid* 2007; 57: 314–323.

257. Kahala M and Palva A. The expression signals of the *Lactobacillus brevis slpA* gene direct efficient heterologous protein production in lactic acid bacteria. *Appl. Microbiol. Biotechnol.* 1999; 51: 71–78.

258. Kaneko Y, Kobayashi H, Kiatpapan P, Nishimoto T, Napitupulu R, Ono H, and Murooka Y. Development of a host-vector system for *Lactobacillus plantarum* L137 isolated from a

traditional fermented food produced in the Philippines. *J. Biosci. Bioeng.* 2000; 89: 62–67.

259. Kerovuo J and Tynkkynen S. Expression of *Bacillus subtilis* phytase in *Lactobacillus plantarum* 755. *Lett. Appl. Microbiol.* 2000; 30: 325–329.

260. Kiatpapan P, Yamashita M, Kawaraichi N, Yasuda T, and Murooka Y. Heterologous expression of a gene encoding cholesterol oxidase in probiotic strains of *Lactobacillus plantarum* and *Propionibacterium freudenreichii* under the control of native promoters. *J. Biosci. Bioeng.* 2001; 92: 459–465.

261. Kim YH, Han KS, Oh S, You S, and Kim SH. Optimization of technical conditions for the transformation of *Lactobacillus acidophilus* strains by electroporation. *J. Appl. Microbiol.* 2005; 99: 167–174.

262. Klijn A, Moine D, Delley M, Mercenier A, Arigon F, and Pridmore RD. Construction of a reporter vector for the analysis of *Bifidobacterium longum* promoters. *Appl. Environ. Microbiol.* 2006; 72: 7401–7405.

263. Kruger C, Hultberg A, van Dollenweerd C, Marcotte H, and Hammarström L. Passive immunization by lactobacilli expressing single-chain antibodies against *Streptococcus mutans*. *Mol. Biotechnol.* 2005; 31: 221–231.

264. Kushiro A, Takahashi T, Asahara T, Tsuji H, Nomoto K, and Morotomi M. *Lactobacillus casei* acquires the binding activity to fibronectin by the expression of the fibronectin binding domain of *Streptococcus pyogenes* on the cell surface. *J. Mol. Microbiol. Biotechnol.* 2001; 3: 563–571.

265. Lambert JM, Bongers RS, and Kleerebezem M. Cre-*lox*-based system for multiple gene deletions and selectable-marker removal in *Lactobacillus plantarum*. *Appl. Environ. Microbiol.* 2007; 73: 1126–1135.

266. Lebeer S, De Keersmaecker SC, Verhoeven TL, Fadda AA, Marchal K, and Vanderleyden J. Functional analysis of *luxS* in the probiotic strain *Lactobacillus rhamnosus* GG reveals a central metabolic role important for growth and biofilm formation. *J. Bacteriol.* 2007; 189: 860–871.

267. Lee JH, Halgerson JS, Kim JH, O'Sullivan DJ. Comparative sequence analysis of plasmids from *Lactobacillus delbrueckii* and construction of a shuttle cloning vector. *Appl. Environ. Microbiol.* 2007; 73: 4417–4424.

268. Lee JH, O'Sullivan DJ. Sequence analysis of two cryptic plasmids from *Bifidobacterium longum* DJO10A and construction of a shuttle cloning vector. *Appl. Environ. Microbiol.* 2006; 72: 527–535.

269. Lee KH, Park WJ, Kim JY, Kim HG, Lee JM, Kim JH, Park JW, Lee JH, Chung SK, and Chung DK. Development of a monitoring vector for *Leuconostoc mesenteroides* using the green fluorescent protein gene. *J. Microbiol. Biotechnol.* 2007; 17: 1213–1216.

270. Leloup L, Ehrlich SD, Zagorec M, and Morel-Deville F. Single-crossover integration in the *Lactobacillus sake* chromosome and insertional inactivation of the *ptsI* and *lacL* genes. *Appl. Environ. Microbiol.* 1997; 63: 2117–2123.

271. Li X, Fu GF, Fan YR, Liu WH, Liu XJ, Wang JJ, and Xu GX. *Bifidobacterium adolescentis* as a delivery system of endostatin for cancer gene therapy: selective inhibitor of angiogenesis and hypoxic tumor growth. *Cancer Gene Ther.* 2003; 10: 105–111.

272. Li YG, Tian FL, Gao FS, Tang XS, and Xia C. Immune responses generated by *Lactobacillus* as a carrier in DNA immunization against foot-and-mouth disease virus. *Vaccine* 2007; 25: 902–911.

273. Lin CF and Chung TC. Cloning of erythromycin-resistance determinants and replication origins from indigenous plasmids of *Lactobacillus reuteri* for potential use in construction of cloning vectors. *Plasmid* 1999; 42: 31–41.
274. Lin CF, Ho JL, and Chung TC. Characterization of the replication region of the *Lactobacillus reuteri* plasmid pTC82 potentially used in the construction of cloning vector. *Biosci. Biotechnol. Biochem.* 2001; 65: 1495–1503.
275. Lindholm A, Ellmén U, Tolonen-Martikainen M, and Palva A. Heterologous protein secretion in *Lactococcus lactis* is enhanced by the *Bacillus subtilis* chaperone-like protein PrsA. *Appl. Microbiol. Biotechnol.* 2006; 73: 904–914.
276. Liu S. A simple method to generate chromosomal mutations in *Lactobacillus plantarum* strain TF103 to eliminate undesired fermentation products. *Appl. Biochem. Biotechnol.* 2006; 129–132: 854–863.
277. Liu S, Dien BS, Nichols NN, Bischoff KM, Hughes SR, and Cotta MA. Coexpression of pyruvate decarboxylase and alcohol dehydrogenase genes in *Lactobacillus brevis*. *FEMS Microbiol. Lett.* 2007; 274: 291–297.
278. Liu S, Nichols NN, Dien BS, and Cotta MA. Metabolic engineering of a *Lactobacillus plantarum* double *ldh* knockout strain for enhanced ethanol production. *J. Ind. Microbiol. Biotechnol.* 2006; 33: 1–7.
279. Lo TC, Shih TC, Lin CF, Chen HW, and Lin TH. Complete genomic sequence of the temperate bacteriophage PhiAT3 isolated from *Lactobacillus casei* ATCC 393. *Virology* 2005; 339: 42–55.
280. Maassen CB, Laman JD, den Bak-Glashouwer MJ, Tielen FJ, van Holten-Neelen JC, Hoogteijling L, Antonissen C, Leer RJ, Pouwels PH, Boersma WJ, and Shaw DM. Instruments for oral disease-intervention strategies: recombinant *Lactobacillus casei* expressing tetanus toxin fragment C for vaccination or myelin proteins for oral tolerance induction in multiple sclerosis. *Vaccine* 1999; 17: 2117–2128.
281. MacConaill LE, Fitzgerald GF, and Van Sinderen D. Investigation of protein export in *Bifidobacterium breve* UCC2003. *Appl. Environ. Microbiol.* 2003; 69: 6994–7001.
282. Marcotte H, Kõll-Klais P, Hultberg A, Zhao Y, Gmür R, Mändar R, Mikelsaar M, and Hammarström L. Expression of single-chain antibody against RgpA protease of *Porphyromonas gingivalis* in *Lactobacillus*. *J. Appl. Microbiol.* 2006; 100: 256–263.
283. Martín MC, Alonso JC, Suárez JE, and Alvarez MA. Generation of food-grade recombinant lactic acid bacterium strains by site-specific recombination. *Appl. Environ. Microbiol.* 2000; 66: 2599–2604.
284. Martín MC, Fernández M, Martín-Alonso JM, Parra F, Boga JA, and Alvarez MA. Nisin-controlled expression of Norwalk virus VP60 protein in *Lactobacillus casei*. *FEMS Microbiol. Lett.* 2004; 237: 385–391.
285. Martínez B, Sillanpää J, Smit E, Korhonen TK, and Pouwels PH. Expression of *cbsA* encoding the collagen-binding S-protein of *Lactobacillus crispatus* JCM5810 in *Lactobacillus casei* ATCC 393(T). *J. Bacteriol.* 2000; 182: 6857–6861.
286. Mathiesen G, Sørvig E, Blatny J, Naterstad K, Axelsson L, and Eijsink VG. High-level gene expression in *Lactobacillus plantarum* using a pheromone-regulated bacteriocin promoter. *Lett. Appl. Microbiol* 2004; 39: 137–143.
287. McAuliffe O, Cano RJ, and Klaenhammer TR. Genetic analysis of two bile salt hydrolase activities in *Lactobacillus acidophilus* NCFM. *Appl. Environ. Microbiol.* 2005; 71: 4925–4929.

288. McCracken A, Turner MS, Giffard P, Hafner LM, and Timms P. Analysis of promoter sequences from *Lactobacillus* and *Lactococcus* and their activity in several *Lactobacillus* species. *Arch. Microbiol.* 2000; 173: 383–389.
289. Moon GS, Pyun YR, Park MS, Ji GE, and Kim WJ. Secretion of recombinant pediocin PA-1 by *Bifidobacterium longum*, using the signal sequence for bifidobacterial alpha-amylase. *Appl. Environ. Microbiol.* 2005; 71: 5630–5632.
290. Mota RM, Moreira JL, Souza MR, Horta MF, Teixeira SM, Neumann E, Nicol JR, and Nunes AC. Genetic transformation of novel isolates of chicken *Lactobacillus* bearing probiotic features for expression of heterologous proteins: a tool to develop live oral vaccines. *BMC Biotechnol.* 2006; 6: 2.
291. Nakamura T, Sasaki T, Fujimori M, Yazawa K, Kano Y, Amano J, and Taniguchi S. Cloned cytosine deaminase gene expression of *Bifidobacterium longum* and application to enzyme/pro-drug therapy of hypoxic solid tumors. *Biosci. Biotechnol. Biochem.* 2002; 66: 2362–2366.
292. Neu T and Henrich B. New thermosensitive delivery vector and its use to enable nisin-controlled gene expression in *Lactobacillus gasseri*. *Appl. Environ. Microbiol.* 2003; 69: 1377–1382.
293. Noonpakdee W, Sitthimonchai S, Panyim S, and Lertsiri S. Expression of the catalase gene *katA* in starter culture *Lactobacillus plantarum* TISTR850 tolerates oxidative stress and reduces lipid oxidation in fermented meat product. *Int. J. Food Microbiol.* 2004; 95: 127–135.
294. Oh Y, Varmanen P, Han XY, Bennett G, Xu Z, Lu T, and Palva A. *Lactobacillus plantarum* for oral peptide delivery. *Oral Microbiol. Immunol.* 2007; 22: 140–144.
295. Oozeer R, Furet JP, Goupil-Feuillerat N, Anba J, Mengaud J, and Corthier G. Differential activities of four *Lactobacillus casei* promoters during bacterial transit through the gastrointestinal tracts of human-microbiota-associated mice. *Appl. Environ. Microbiol.* 2005; 71: 1356–1363.
296. Pant N, Hultberg A, Zhao Y, Svensson L, Pan-Hammarström Q, Johansen K, Pouwels PH, Ruggeri FM, Hermans P, Frenken L, Borén T, Marcotte H, and Hammarström L. Lactobacilli expressing variable domain of llama heavy-chain antibody fragments (lactobodies) confer protection against rotavirus-induced diarrhea. *J. Infect. Dis.* 2006; 194: 1580–1588.
297. Park J, Lee M, Jung J, and Kim J. pIH01, a small cryptic plasmid from *Leuconostoc citreum* IH3. *Plasmid* 2005; 54: 184–189.
298. Park MS, Kwon B, Shim J, Huh CS, and Ji GE. Heterologous expression of cholesterol oxidase in *Bifidobacterium longum* under the control of 16S rRNA gene promoter of bifidobacteria. *Biotechnol. Lett.* 2008; 30: 165–172.
299. Park MS, Moon HW, and Ji GE. Molecular Characterization of Plasmid from *Bifidobacterium longum*. *J Microbiol. Biotechnol.* 2003; 13: 457–462.
300. Park MS, Seo JM, Kim JY, and Ji GE. Heterologous gene expression and secretion in *Bifidobacterium longum*. *Lait* 2005; 85: 1–8.
301. Park WJ, Lee KH, Lee JM, Lee HJ, Kim JH, Lee JH, Chang HC, and Chung DK. Characterization of pC7 from *Lactobacillus paraplantarum* C7 derived from Kimchi and development of lactic acid bacteria–*Escherichia coli* shuttle vector. *Plasmid* 2004; 52: 84–88.

302. Pavan S, Hols P, Delcour J, Geoffroy MC, Grangette C, Kleerebezem M, and Mercenier A. Adaptation of the nisin-controlled expression system in *Lactobacillus plantarum*: a tool to study in vivo biological effects. *Appl. Environ. Microbiol.* 2000; 66: 4427–4432.

303. Pavlova SI, Kiliç AO, Topisirovi L, Miladinov N, Hatzos C, and Tao L. Characterization of a cryptic plasmid from *Lactobacillus fermentum* KC5b and its use for constructing a stable *Lactobacillus* cloning vector. *Plasmid* 2002; 47: 182–192.

304. Peltoniemi K, Vesanto E, and Palva A. Genetic characterization of an oligopeptide transport system from *Lactobacillus delbrueckii* subsp. *bulgaricus*. *Arch. Microbiol.* 2002; 177: 457–467.

305. Pérez-Arellano I and Pérez-Martínez G. Structural features of the *lac* promoter affecting *gus*A expression in *Lactobacillus casei*. *Curr. Microbiol.* 2002; 45: 191–196.

306. Pérez-Arellano I and Pérez-Martínez G. Optimization of the green fluorescent protein (GFP) expression from a lactose-inducible promoter in *Lactobacillus casei*. *FEMS Microbiol. Lett.* 2003; 222: 123–127.

307. Pérez-Arellano I, Zúñiga M, and Pérez-Martínez G. Construction of compatible wide-host-range shuttle vectors for lactic acid bacteria and *Escherichia coli*. *Plasmid* 2001; 46: 106–116.

308. Perez-Casal J, Price JA, Maguin E, and Scott JR. An M protein with a single C repeat prevents phagocytosis of *Streptococcus pyogenes*: use of a temperature-sensitive shuttle vector to deliver homologous sequences to the chromosome of *S. pyogenes*. *Mol. Microbiol.* 1993; 8: 809–819.

309. Phumkhachorn P, Rattanachaikunsopon P, and Khunsook S. Use of the gfp gene in monitoring bacteriocin-producing *Lactobacillus plantarum* N014, a potential starter culture in nham fermentation. *J. Food Prot.* 2007; 70: 419–424.

310. Pouwels PH, Leer RJ, and Boersma WJ. The potential of *Lactobacillus* as a carrier for oral immunization: development and preliminary characterization of vector systems for targeted delivery of antigens. *J. Biotechnol.* 1996; 44: 183–192.

311. Pouwels PH, Vriesema A, Martinez B, Tielen FJ, Seegers JF, Leer RJ, Jore J, and Smit E. Lactobacilli as vehicles for targeting antigens to mucosal tissues by surface exposition of foreign antigens. *Methods Enzymol.* 2001; 336: 369–389.

312. Pusch O, Kalyanaraman R, Tucker LD, Wells JM, Ramratnam B, and Boden D. An anti-HIV microbicide engineered in commensal bacteria: secretion of HIV-1 fusion inhibitors by lactobacilli. *AIDS* 2006; 20: 1917–1922.

313. Ravin V, Sasaki T, Räisänen L, Riipinen KA, and Alatossava T. Effective plasmid pX3 transduction in *Lactobacillus delbrueckii* by bacteriophage LL-H. *Plasmid* 2006; 55: 184–193.

314. Reveneau N, Geoffroy MC, Locht C, Chagnaud P, and Mercenier A. Comparison of the immune responses induced by local immunizations with recombinant *Lactobacillus plantarum* producing tetanus toxin fragment C in different cellular locations. *Vaccine* 2002; 20: 1769–1777.

315. Reyes Escogido ML, De León Rodríguez A, and Barba de la Rosa AP. A novel binary expression vector for production of human IL-10 in *Escherichia coli* and *Bifidobacterium longum*. *Biotechnol. Lett.* 2007; 29: 1249–1253.

316. Rhim SL, Park MS, and Ji GE. Expression and secretion of *Bifidobacterium adolescentis* amylase by *Bifidobacterium longum*. *Biotechnol.* 2006; 28: 163–168.

317. Rossi M, Brigid P, and Matteuzzi D. Improved cloning vectors for *Bifidobacterium* spp. *Lett. Appl. Microbiol.* 1998; 26: 101–104.
318. Rud I, Jensen PR, Naterstad K, and Axelsson L. A synthetic promoter library for constitutive gene expression in *Lactobacillus plantarum*. *Microbiology* 2006; 152: 1011–1019.
319. Russell WM and Klaenhammer TR. Efficient system for directed integration into the *Lactobacillus acidophilus* and *Lactobacillus gasseri* chromosomes via homologous recombination. *Appl. Environ. Microbiol.* 2001; 67: 4361–4364.
320. Russell WM and Klaenhammer TR. Identification and cloning of gusA, encoding a new beta-glucuronidase from *Lactobacillus gasseri* ADH. *Appl. Environ. Microbiol.* 2001; 67: 1253–1261.
321. Sangrador-Vegas A, Stanton C, van Sinderen D, Fitzgerald GF, and Ross RP. Characterization of plasmid pASV479 from *Bifidobacterium pseudolongum* subsp. *globosum* and its use for expression vector construction. *Plasmid* 2007; 58: 140–147.
322. Sasaki T, Fujimori M, Hamaji Y, Hama Y, Ito K, Amano J, and Taniguchi S. Genetically engineered *Bifidobacterium longum* for tumor-targeting enzyme-prodrug therapy of autochthonous mammary tumors in rats. *Cancer Sci.* 2006; 97: 649–657.
323. Satoh E, Ito Y, Sasaki Y, and Sasaki T. Application of the extracellular alpha-amylase gene from *Streptococcus bovis* 148 to construction of a secretion vector for yogurt starter strains. *Appl. Environ. Microbiol.* 1997; 63: 4593–4596.
324. Scheppler L, Vogel M, Zuercher AW, Zuercher M, Germond JE, Miescher SM, and Stadler BM. Recombinant *Lactobacillus johnsonii* as a mucosal vaccine delivery vehicle. *Vaccine* 2002; 20: 2913–2920.
325. Schwab C, Walter J, Tannock GW, Vogel RF, and Gänzle MG. Sucrose utilization and impact of sucrose on glycosyltransferase expression in *Lactobacillus reuteri*. *Syst. Appl. Microbiol.* 2007; 30: 433–443.
326. Serror P, Ilami G, Chouayekh H, Ehrlich SD, and Maguin E. Transposition in *Lactobacillus delbrueckii* subsp. *bulgaricus*: identification of two thermosensitive replicons and two functional insertion sequences. *Microbiology* 2003; 149: 1503–1511.
327. Serror P, Sasak T, Ehrlich SD, and Maguin E. Electrotransformation of *Lactobacillus delbrueckii* subsp. *bulgaricus* and *L. delbrueckii* subsp. *lactis* with various plasmids. *Appl. Environ. Microbiol.* 2002; 68: 46–52.
328. Sheehan VM, Sleator RD, Fitzgerald GF, and Hill C. Heterologous expression of BetL, a betaine uptake system, enhances the stress tolerance of *Lactobacillus salivarius* UCC118. *Appl. Environ. Microbiol.* 2006; 72: 2170–2177.
329. Sheehan VM, Sleator RD, Hill C, and Fitzgerald GF. Improving gastric transit, gastrointestinal persistence and therapeutic efficacy of the probiotic strain *Bifidobacterium breve* UCC2003. *Microbiology* 2007; 153: 3563–3571.
330. and Shimizu-Kadota M. A method to maintain introduced DNA sequences stably and safely on the bacterial chromosome: application of prophage integration and subsequent designed excision. *J. Biotechnol.* 2001; 89: 73–79.
331. Shimizu-Kadota M, Kiwaki M, Sawaki S, Shirasawa Y, Shibahara-Sone H, and Sako T. Insertion of bacteriophage phiFSW into the chromosome of *Lactobacillus casei* strain Shirota (S-1): characterization of the attachment sites and the integrase gene. *Gene* 2000; 249: 127–134.

332. Slos P, Dutot P, Reymund J, Kleinpeter P, Prozzi D, Kieny MP, Delcour J, Mercenier A, and Hols P. Production of cholera toxin B subunit in *Lactobacillus*. *FEMS Microbiol. Lett.* 1998; 169: 29–36.

333. Sørvig E, Grönqvist S, Naterstad K, Mathiesen G, Eijsink VG, and Axelsson L. Construction of vectors for inducible gene expression in *Lactobacillus sakei* and *L plantarum*. *FEMS Microbiol. Lett.* 2003; 229: 119–126.

334. Sørvig E, Mathiesen G, Naterstad K, Eijsink VG, and Axelsson L. High-level, inducible gene expression in *Lactobacillus sakei* and *Lactobacillus plantarum* using versatile expression vectors. *Microbiology* 2005; 151: 2439–2449.

335. Sørvig E, Skaugen M, Naterstad K, Eijsink VG, and Axelsson L. Plasmid p256 from *Lactobacillus plantarum* represents a new type of replicon in lactic acid bacteria, and contains a toxin-antitoxin-like plasmid maintenance system. *Microbiology* 2005; 151: 421–431.

336. Stentz R, Loize C, Malleret C, and Zagorec M. Development of genetic tools for *Lactobacillus sakei*: disruption of the beta-galactosidase gene and use of *lacZ* as a reporter gene To study regulation of the putative copper ATPase, AtkB. *Appl. Environ. Microbiol.* 2000; 66: 4272–4278.

337. Sturme MH, Nakayama J, Molenaar D, Murakami Y, Kunugi R, Fujii T, Vaughan EE, Kleerebezem M, and de Vos WM. An *agr*-like two-component regulatory system in *Lactobacillus plantarum* is involved in production of a novel cyclic peptide and regulation of adherence. *J. Bacteriol.* 2005; 187: 5224–5235.

338. Sudhamani M, Ismaiel E, Geis A, Batish V, and Heller KJ. Characterisation of pSMA23, a 3.5kbp plasmid of *Lactobacillus casei*, and application for heterologous expression in *Lactobacillus*. *Plasmid* 2008; 59: 11–19.

339. Takala TM and Saris PE. A food-grade cloning vector for lactic acid bacteria based on the nisin immunity gene *nisI*. *Appl. Microbiol. Biotechnol.* 2002; 59: 467–471.

340. Takala TM, Saris PE, and Tynkkynen SS. Food-grade host/vector expression system for *Lactobacillus casei* based on complementation of plasmid-associated phospho-beta-galactosidase gene *lacG*. *Appl. Microbiol. Biotechnol.* 2003; 60: 564–570.

341. Takata T, Shirakawa T, Kawasaki Y, Kinoshita S, Gotoh A, Kano Y, and Kawabata M. Genetically engineered *Bifidobacterium animalis* expressing the *Salmonella flagellin* gene for the mucosal immunization in a mouse model. *J. Gene Med.* 2006; 8: 1341–1346.

342. Tanaka K, Samura K, and Kano Y. Structural and functional analysis of pTB6 from *Bifidobacterium longum*. *Biosci. Biotechnol. Biochem.* 2005; 69: 422–425.

343. Thompson JK, McConville KJ, McReynolds C, and Collins MA. Electrotransformation of *Lactobacillus plantarum* using linearized plasmid DNA. *Lett. Appl. Microbiol.* 1997; 25: 419–425.

344. Thompson JK, McConville KJ, McReynolds C, and Collins MA. Potential of conjugal transfer as a strategy for the introduction of recombinant genetic material into strains of *Lactobacillus helveticus*. *Appl. Environ. Microbiol.* 1999; 65: 1910–1914.

345. Thompson JK, McConville KJ, Nicholson C, and Collins MA. DNA cloning in *Lactobacillus helveticus* by the exconjugation of recombinant *mob*-containing plasmid constructs from strains of transformable lactic acid bacteria. *Plasmid* 2001; 46: 188–201.

346. Turner MS and Giffard PM. Expression of *Chlamydia psittaci*- and human immunodeficiency virus-derived antigens on the cell surface of *Lactobacillus fermentum* BR11 as fusions to bspA. *Infect. Immun.* 1999; 67: 5486–5489.

347. Turner MS, Hafner LM, Walsh T, and Giffard PM. Identification and characterization of the novel LysM domain-containing surface protein Sep from *Lactobacillus fermentum* BR11 and its use as a peptide fusion partner in *Lactobacillus* and *Lactococcus*. *Appl. Environ. Microbiol.* 2004; 70: 3673–3680.

348. Turner MS, Woodberry T, Hafner LM, and Giffard PM. The *bspA* locus of *Lactobacillus fermentum* BR11 encodes an L-cystine uptake system. *J. Bacteriol.* 1999; 181: 2192–2198.

349. Walter J, Heng NC, Hamme WP, Loach DM, Tannock GW, and Hertel C. Identification of *Lactobacillus reuteri* genes specifically induced in the mouse gastrointestinal tract. *Appl. Environ. Microbiol.* 2003; 69: 2044–2051.

350. Wegkamp A, Starrenburg M, de Vos WM, Hugenholtz J, and Sybesma W. Transformation of folate-consuming *Lactobacillus gasseri* into a folate producer. *Appl. Environ. Microbiol.* 2004; 70: 3146–3148.

351. Wu CM and Chung TC. Green fluorescent protein is a reliable reporter for screening signal peptides functional in *Lactobacillus reuteri*. *Microbiol. Methods* 2006; 67: 181–186.

352. Wu CM and Chung TC. Mice protected by oral immunization with *Lactobacillus reuteri* secreting fusion protein of *Escherichia coli* enterotoxin subunit protein. *FEMS Immunol. Med. Microbiol.* 2007; 50: 354–365.

353. Wu CM, Lin CF, Chang YC, and Chung TC. Construction and characterization of nisin-controlled expression vectors for use in *Lactobacillus reuteri*. *Biosci. Biotechnol. Biochem.* 2006; 70: 757–767.

354. Xu YF, Zhu LP, Hu B, Fu GF, Zhang HY, Wang JJ, and Xu GX. A new expression plasmid in *Bifidobacterium longum* as a delivery system of endostatin for cancer gene therapy. *Cancer Gene Ther.* 2007; 14: 151–157.

355. Yao XY, Wang HM, Li DJ, Yuan MM, Wang XL, Yu M, Wang MY, Zhu Y, and Meng Y. Inoculation of *Lactobacillus* expressing hCG beta in the vagina induces an anti-hCG beta antibody response in murine vaginal mucosa. *J. Reprod. Immunol.* 2004; 63: 111–122.

356. Yu QH, Dong SM, Zhu WY, and Yang Q. Use of green fluorescent protein to monitor *Lactobacillus* in the gastro-intestinal tract of chicken. *FEMS Microbiol. Lett.* 2007; 275: 207–213.

357. Zegers ND, Kluter E, van Der Stap H, van Dura E, van Dalen P, Shaw M, and Baillie L. Expression of the protective antigen of *Bacillus anthracis* by *Lactobacillus casei*: towards the development of an oral vaccine against anthrax. *J. Appl. Microbiol.* 1999; 87: 309–314.

358. Aarnikunnas JN, Von Weymarn K, Ronnholm M, Leisola, and Palva A. Metabolic engineering of *Lactobacillus fermentum* for production of mannitol and pure L-lactic acid or pyruvate. *Biotechnol. Bioeng.* 2003; 82: 653–663.

359. Alvarez MA, Herrero M, and Suarez JE. The site-specific recombination system of the Lactobacillus species bacteriophage A2 integrates in gram-positive and gram-negative bacteria. *Virology* 1998; 250: 185–193.

360. Avall-Jaaskelainen S, Kyla-Nikkila K, Kahala M, Miikkulainen-Lahti T, and Palva A. Surface display of foreign epitopes on the *Lactobacillus brevis* S-layer. *Appl. Environ. Microbiol.* 2002; 68: 5943–5951.

361. Azcarate-Peril MA, Altermann E, Hoover-Fitzula RL, Cano RJ, and Klaenhammer TR. Identification and inactivation of genetic loci involved with *Lactobacillus acidophilus* acid tolerance. *Appl. Environ. Microbiol.* 2004; 70: 5315–5322.

362. Bhowmik T, Fernandez L, and Steele JL. Gene replacement in *Lactobacillus helveticus*. *J. Bacteriol.* 1993; 175: 6341–6344.
363. Bhowmik T and Steele JL. Cloning, characterization and insertional inactivation of the *Lactobacillus helveticus* D(-) lactate dehydrogenase gene. *Appl. Microbiol. Biotechnol.* 1994; 41: 432–439.
364. Brondsted L and Hammer K. Use of the integration elements encoded by the temperate lactococcal bacteriophage TP901-1 to obtain chromosomal single-copy transcriptional fusions in *Lactococcus lactis*. *Appl. Environ. Microbiol.* 1999; 65: 752–758.
365. Bruno-Barcena JM, Azcarate-Peril MA, Klaenhammer TR, and Hassan HM. Marker-free chromosomal integration of the manganese superoxide dismutase gene (sodA) from *Streptococcus thermophilus* into *Lactobacillus gasseri*. *FEMS Microbiol. Lett.* 2005; 246: 91–101.
366. Bruttin A, Foley S, and Brussow H. The site-specific integration system of the temperate *Streptococcus thermophilus* bacteriophage phiSfi21. *Virology* 1997; 237: 148–158.
367. Buck BL, Altermann E, Svingerud T, and Klaenhammer TR. Functional analysis of putative adhesion factors in *Lactobacillus acidophilus* NCFM. *Appl. Environ. Microbiol.* 2005; 71: 8344–8351.
368. Christensen JE and Steele JL. Impaired growth rates in milk of *Lactobacillus helveticus* peptidase mutants can be overcome by use of amino acid supplements. *J. Bacteriol.* 2003; 185: 3297–3306.
369. Christiansen B, Johnsen MG, Stenby E, Vogensen FK, and Hammer K. Characterization of the lactococcal temperate phage TP901-1 and its site-specific integration. *J. Bacteriol.* 1994; 176: 1069–1076.
370. Cocconcelli PS, Morelli L, Vescovo M, and Bottazzi V. Intergeneric protoplast fusion in lactic acid bacteria. *FEMS Microbiol. Lett.* 1986; 35: 211–214.
371. Cosby WM, Axelsson LT, and Dobrogosz WJ. Tn917 transposition in *Lactobacillus plantarum* using the highly temperature-sensitive plasmid pTV1Ts as a vector. *Plasmid* 1989; 22: 236–243.
372. Feirtag JM, Petzel JP, Pasalodos E, Baldwin KA, and McKay LL. Thermosensitive plasmid replication, temperature-sensitive host growth, and chromosomal plasmid integration conferred by *Lactococcus lactis* subsp. *cremoris* lactose plasmids in *Lactococcus lactis* subsp. *lactis*. *Appl. Environ. Microbiol.* 1991; 57: 539–548.
373. Giard JC, Rince A, Capiaux H, Auffray Y, and Hartke A. Inactivation of the stress- and starvation-inducible gls24 operon has a pleiotrophic effect on cell morphology, stress sensitivity, and gene expression in *Enterococcus faecalis*. *J. Bacteriol.* 2000; 182: 4512–4520.
374. Gosalbes MJ, Esteban CD, Galan JL, and Perez-Martinez G. Integrative food-grade expression system based on the lactose regulon of *Lactobacillus casei*. *Appl. Environ. Microbiol.* 2000; 66: 4822–4828.
375. Gosalbes MJ, Esteban CD, and Perez-Martinez G. In vivo effect of mutations in the antiterminator LacT in *Lactobacillus casei*. *Microbiology* 2002; 148: 695–702.
376. Hazebrouck S, Pothelune L, Azevedo V, Corthier G, Wal JM, and Langella P. Efficient production and secretion of bovine beta-lactoglobulin by *Lactobacillus casei*. *Microb. Cell Fact.* 2007; 6: 12.
377. Kang Y, Kim JH, and Ryu DDY. Protoplast fusion of *Lactobacillus casei*. *Agric. Biol. Chem.* 1987; 51: 2221–2227.

378. Kyla-Nikkila K, Hujanen M, Leisola M, and Palva A. Metabolic engineering of *Lactobacillus helveticus* CNRZ32 for production of pure L-(+)-lactic acid. *Appl. Environ. Microbiol.* 2000; 66: 3835–3841.

379. Lapierre L, Germond JE, Ott A, Delley M, and Mollet B. D-Lactate dehydrogenase gene (ldhD) inactivation and resulting metabolic effects in the *Lactobacillus johnsonii* strains La1 and N312. *Appl. Environ. Microbiol.* 1999; 65: 4002–4007.

380. Law J, Buist G, Haandrikman A, Kok J, Venema G, and Leenhouts K. A system to generate chromosomal mutations in *Lactococcus lactis* which allows fast analysis of targeted genes. *J. Bacteriol.* 1995; 177: 7011–7018.

381. Leenhouts KJ, Kok J, and Venema G. Campbell-like integration of heterologous plasmid DNA into the chromosome of *Lactococcus lactis* subsp. *lactis*. *Appl. Environ. Microbiol.* 1989; 55: 394–400.

382. Leer RJ, Christiaens H, Verstraete W, Peters L, Posno M, and Pouwels PH. Gene disruption in *Lactobacillus plantarum* strain 80 by site-specific recombination: isolation of a mutant strain deficient in conjugated bile salt hydrolase activity. *Mol. Gen. Genet.* 1993; 239: 269–272.

383. Leloup L, Ehrlich SD, Zagorec M, and Morel-Deville F. Single-crossover integration in the *Lactobacillus sake* chromosome and insertional inactivation of the ptsI and lacL genes. *Appl. Environ. Microbiol.* 1997; 63: 2117–2123.

384. Lin MY, Harlander S, and Savaiano D. Construction of an integrative food-grade cloning vector for *Lactobacillus acidophilus*. *Appl. Microbiol. Biotechnol.* 1996; 45: 484–489.

385. Maguin E, Prevost H, Ehrlich SD, and Gruss A. Efficient insertional mutagenesis in lactococci and other gram-positive bacteria. *J. Bacteriol.* 1996; 178: 931–935.

386. Mayo B, Kok J, Bockelmann W, Haandrikman A, Leenhouts KJ, and Venema G. Effect of X-Prolyl Dipeptidyl Aminopeptidase Deficiency on *Lactococcus lactis*. *Appl. Environ. Microbiol.* 1993; 59: 2049–2055.

387. Mollet B, Knol J, Poolman B, Marciset O, and Delley M. Directed genomic integration, gene replacement, and integrative gene expression in *Streptococcus thermophilus*. *J. Bacteriol.* 1993; 175: 4315–4324.

388. Neu T and Henrich B. New thermosensitive delivery vector and its use to enable nisin-controlled gene expression in *Lactobacillus gasseri*. *Appl. Environ. Microbiol.* 2003; 69: 1377–1382.

389. Pavan S, Hols P, Delcour J, Geoffroy MC, Grangette C, Kleerebezem M, and Mercenier A. Adaptation of the nisin-controlled expression system in *Lactobacillus plantarum*: A tool to study in vivo biological effects. *Appl. Environ. Microbiol.* 2000; 66: 4427–4432.

390. Pozzi G, Musmanno RA, Renzoni EA, Oggioni MR, and Cusi MG. Host-vector system for integration of recombinant DNA into chromosomes of transformable and nontransformable streptococci. *J. Bacteriol.* 1988; 170: 1969–1972.

391. Raya RR, Fremaux C, De Antoni GL, and Klaenhammer TR. Site-specific integration of the temperate bacteriophage phi adh into the Lactobacillus gasseri chromosome and molecular characterization of the phage (attP) and bacterial (attB) attachment sites. *J. Bacteriol.* 1992; 174: 5584–5592.

392. Russell WM and Klaenhammer TR. Efficient system for directed integration into the *Lactobacillus acidophilus* and *Lactobacillus gasseri* chromosomes via homologous recombination. *Appl. Environ. Microbiol.* 2001; 67: 4361–4364.

393. Sasaki Y, Ito Y, and Sasaki T. ThyA as a selection marker in construction of food-grade host-vector and integration systems for *Streptococcus thermophilus*. *Appl. Environ. Microbiol.* 2004; 70: 1858–1864.

394. Scheirlinck T, Mahillon J, Joos H, Dhaese P, and Michiels F. Integration and expression of alpha-amylase and endoglucanase genes in the *Lactobacillus plantarum* chromosome. *Appl. Environ. Microbiol.* 1989; 55: 2130–2137.

395. Shimizu-Kadota M, Kiwaki M, Hirokawa H, and Tsuchida N. ISL1: a new transposable element in *Lactobacillus casei*. *Mol. Gen. Genet.* 1985; 200: 193–198.

396. Shrago AW and Dobrogosz WJ. Conjugal transfer of group B streptococcal plasmids and comobilization of *Escherichia coli*-Streptococcus shuttle plasmids to *Lactobacillus plantarum*. *Appl. Environ. Microbiol.* 1988; 54: 824–826.

397. Simoes-Barbosa A, Abreu H, Silva Neto A, Gruss A, and Langella P. A food-grade delivery system for *Lactococcus lactis* and evaluation of inducible gene expression. *Appl. Microbiol. Biotechnol.* 2004; 65: 61–67.

398. Simons G, Nijhuis M, and de Vos WM. Integration and gene replacement in the *Lactococcus lactis* lac operon: induction of a cryptic phospho-beta-glucosidase in LacG-deficient strains. *J. Bacteriol.* 1993; 175: 5168–5175.

399. Skaugen M and Nes IF. Transposition in *Lactobacillus sake* and its abolition of lactocin S production by insertion of IS1163, a new member of the IS3 family. *Appl. Environ. Microbiol.* 1994; 60: 2818–2825.

400. Stentz R, Loizel C, Malleret C, and Zagorec M. Development of genetic tools for *Lactobacillus sakei*: disruption of the beta-galactosidase gene and use of lacZ as a reporter gene to study regulation of the putative copper ATPase, AtkB. *Appl. Environ. Microbiol.* 2000; 66: 4272–4278.

401. Trieu-Cuot P, Carlier C, Poyart-Salmeron C, and Courvalin P. An integrative vector exploiting the transposition properties of Tn1545 for insertional mutagenesis and cloning of genes from gram-positive bacteria. *Gene* 1991; 106: 21–27.

402. Walker DC and Klaenhammer TR. Isolation of a novel IS3 group insertion element and construction of an integration vector for *Lactobacillus* spp. *J. Bacteriol.* 1994; 176: 5330–5340.

403. Zwahlen MC and Mollet B. ISL2, a new mobile genetic element in *Lactobacillus helveticus*. *Mol. Gen. Genet.* 1994; 245: 334–348.

4

ROLE OF PROBIOTICS IN HEALTH AND DISEASES

M. CARMEN COLLADO
Functional Foods Forum, University of Turku, Turku, Finland

A probiotic has been defined as a "live microorganism which when administered in adequate amounts confers health benefits to the host" (5). Probiotics were originally used to improve the health of both animals and humans through the modulation of the intestinal microbiota. At present, the specific live microbial food ingredients and their effects on human health are studied both within food matrices and as single or mixed culture preparations (20, 4). Several well-characterized strains of lactobacilli and Bifidobacteria are available for human use to reduce the risk of gastrointestinal infections or treat such infections (16, 17).

Some of the beneficial effects of probiotic consumption include: the improvement of intestinal tract health by means of regulation of microbiota and stimulation and development of the immune system, synthesizing and enhancing the bioavailability of nutrients, reducing symptoms of lactose intolerance and reducing risk of certain diseases. The mechanisms by which probiotics exert their effects are largely unknown, but may involve modifying gut pH, antagonizing pathogens through production of antimicrobial compounds, competing for pathogen binding, receptor sites, nutrients and growth factors, stimulating immunomodulatory cells, and producing lactase.

The primary clinical interest in the application of probiotics has been in the prevention and treatment of gastrointestinal infections and diseases (14). Several proposed health effects of probiotics are summarized in Fig. 4.1. The general mechanisms by which probiotics may have an effect can be divided into different categories: normalization of microbiota, modulation of immune response, and metabolic functions.

Handbook of Probiotics and Prebiotics, Second Edition   Edited by Yuan Kun Lee and Seppo Salminen
Copyright © 2009 John Wiley & Sons, Inc.

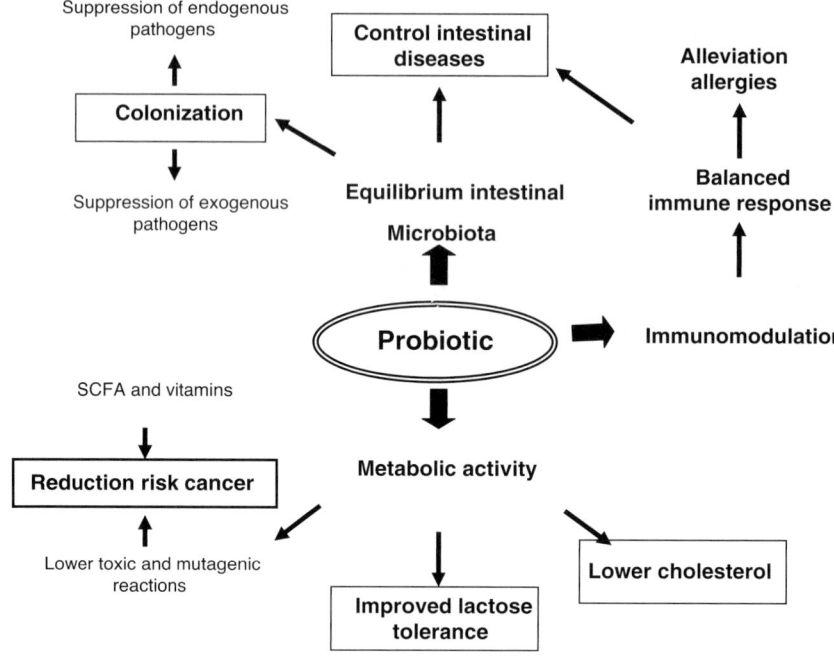

FIGURE 4.1 Beneficial effects of probiotics on human health. Adapted from Parvez et al. (14).

Microbiota deviations have been associated with enhanced risk of specific diseases as well as with the presentation of these diseases (11, 9, 21). Therefore, modulation of an unbalanced indigenous microbiota forms the rationale of probiotic therapy. During the last few decades, a number of studies have been carried out on the effects of probiotic microorganisms using different formulas and with numerous purposes of preventing or treating diseases. The most extensive researches and clinical applications of probiotics have been related to the management of gastrointestinal infections caused by pathogenic microorganisms.

The development of adjuvant or alternative therapies based on bacterial replacement is becoming important due to the rapid emergence of antibiotic-resistant pathogenic strains and the adverse consequences of antibiotic therapies on the protective flora, which enhances the risk of infection (6). Probiotic strains have been shown to exert a protective effect against acute diarrhea, rotavirus diarrhea, and antibiotic-associated diarrhea (3, 7, 13, 18, 19), as well as *Helicobacter pylori* (*H. pylori*) infection (8), and they alleviate symptoms of gastrointestinal diseases, such as irritable bowel syndrome (IBS) (10, 12, 2). In addition, probiotics have also shown other health benefits related to pathogen infection and immune system stimulation (15, 18). The use of probiotics should be further investigated for its possible benefits and its side effects if any (1, 14).

Knowledge about intestinal microbiota, nutrition, immunity, and genetics in health and disease has increased in the past years. This information helps to develop new probiotic strains with disease-specific functions. Recent advances facilitate the understanding of when to use probiotics and how they affect specific pathological states. The effects of probiotics in the gut are well documented and they include upregulation of immunoglobulin, downregulation of inflammatory cytokines, and enhancement of gut barrier function among others. In addition, new evidences support the use of probiotics in the prevention and treatment of a number of diseases including atopic diseases, immune disorders, obesity, diabetes, and some cancers.

In vitro studies are important to assess the safety of probiotic strains, and they are useful in expanding the knowledge of specific properties of tested strains. However, it is realized that the currently available tests, including identification of strains and characterization of some activities, are not accurate to predict the potential use and functionality of probiotic strains *in vivo*. Rational selection and validation of probiotic strains should be based on the evidence obtained in *in vitro* and *in vivo* models with a reliable predicted value or function, and followed by studies in humans. Thus, *in vitro* data are not enough to assure probiotic characteristics to specific strains. Hence, it is crucial that the probiotic strains for human use should undergo *in vivo* studies, and animal studies should be followed by the clinical trials on human studies. The principal outcome of efficacy studies on probiotics should have proven benefits in human studies, such as statistically and biologically significant improvement in condition, alleviate symptoms, improved well-being or quality of life; reduced risk of disease or prolonged period for a relapse or faster recovery from illness. A definite evidence for efficacy should be proven by *in vivo* tests in humans. Despite considerable progress in "probiotic" research over the last 5 years, not all probiotic bacteria available in the market have a solid scientific record and further *in vitro* and *in vivo* studies are still needed.

4.1 CELL LINE MODELS IN RESEARCH

J.M. LAPARRA

Department of Food Science, Cornell University, Ithaca, New York, USA

In recent decades, many efforts have been directed to understand several of the processes accounting in the gut after oral food intake. A major handicap in this field has been the lack of a homogenous cell model that could form a polarized barrier mimicking the physiological and functional morphology of the intestinal epithelia. Presently, several *in vitro* models provide an effective approximation to *in vivo* situation and offer the advantage of good reproducibility in improving the understanding of underlying mechanism(s) in absorption processes and the relationship between food components in gut environment. The *in vitro* models have been proved to be successful while evaluating the key stages in the absorption of nutrients and cross talking between intestinal bacteria and intestinal cells.

The small intestinal epithelium is the first physiological barrier to exogenous molecules from the lumen (e.g., nutrients, toxins) to reach the blood compartment.

This epithelium is composed of several cell types: enterocyte, globet cell, paneth cell, endocrine, and stem cell (28, 53). This tissue is affected by a host of factors brought into play by the ingestion of food and the chemical and physical presence of food in the lumen of the gut. Furthermore, the intestinal media is a complex environment where enteric bacteria are also a critical component in the development of gut function. Otherwise, harmful bacteria can counteract host defense mechanism(s) or can allow tissue and organ invasion, impairing the intestinal barrier by causing the disruption of membrane integrity.

Human studies are the definitive tool to evaluate the physiological processes; however, *in vitro* models are increasingly developed, which are being successfully applied to study drug and nutrient transport across intestinal epithelium. *In vitro* cultures offer several advantages that include (a) rapid assessment of the potential permeability and metabolism; (b) elucidation of the underlying molecular mechanism(s) in absorption processes; (c) rapid evaluation of strategies for pharmaceutical development (i.e., drug targeting, enhancing transport, and minimizing first-pass metabolism); (d) the opportunity to use human rather than animal tissues; and (e) the opportunity to minimize time-consuming, expensive, and sometimes controversial animal studies (24). Besides, when compared to the *in vivo* situation, the lack of (neuro)endocrine regulators, blood flow, and other physiological factors, the *in vitro* systems constitute a relatively simple approach.

A major handicap in the study of processes at intestinal epithelia has been the lack of an *in vitro* culture model that could form a polarized barrier mimicking the mature human enterocytes. Presently, the use of cell lines (i.e., Caco-2, HT29, SW116, LS174T, SW-480) obtained from human colorectal cancer allowed the development of model systems representing features of the intestinal epithelium (51, 29, 46). In order to mimic the biological barrier at intestinal level with an *in vitro* cell culture model, the selection of the cell line becomes particularly important that will condition the transport and metabolic characteristics of the system. The aspects that must be considered are the ones related to cell seeding density, confluency, and the stage of cellular differentiation (24, 58). The intestinal cell lineage, Caco-2 is the cell model, which is most commonly used to reproduce the features of the human small intestine (56, 35). The primary clone that is available through ATCC is identified as HTB-37; however, numerous other clones are also known to exist in various laboratories worldwide, some of which are more useful or unique to the study of certain nutrients or pharmaceuticals (57, 27). Under appropriate conditions, Caco-2 cells differentiate spontaneously to polarized enterocyte-like monolayers exhibiting uptake and barrier characteristics similar to the small intestinal epithelial layer (51, 40). They develop microvilli and tight junctions that in many ways act similar to small intestinal epithelial cells (51, 40). Caco-2 cells are known to express carrier-mediated transport systems for several nutrients, amino acids and vitamins (33, 41), minerals (23, 42), and have also been suggested for toxic metals (22, 45). The Caco-2 cells are very often cultured on permeable supports, which closely resemble the *in vivo* situation, where the cells can take up and secrete molecules on both sides of the monolayer (58). Proper selection of the diameter, pore size, and nature of the permeable support may influence the barrier characteristics of the culture (24). Nevertheless, the Caco-2 model differs from the small intestinal epithelia in several aspects and owes to the lack of an adequate mucus layer.

The HT29-MTX cell line resulted from the isolation of HT29 cells adapted to metotrexate (MTX) (47). The HT29 cell line cannot differentiate spontaneously under standard conditions representing undifferentiated colonic epithelial cells (61). The differentiation of globet cells is characterized by the secretion of several mucins (62). The MUC5AC mucin gene, usually expressed in the stomach, accounts for the major expressed mucin gene in HT29-MTX cells. The presence of nutrients such as glucose or galactose, in the media also conditions their ability to undergo differentiation (61). In the intestinal epithelium, among the varieties of different cell types, enterocytes and globet cells represent the two major cell phenotypes. Therefore, cocultures of Caco-2 cells and mucin-secreting HT29-MTX cells would provide a system that exhibits the two main important cellular types encountered in the human intestinal epithelium. Recent efforts have been focused in developing cocultures of the two human intestinal cell lines to raise a more physiological model mimicking the intestinal epithelia (28, 42, 43). Novellaux et al. (50) grew (up to 21 days) mixed cultures of these cell lines in different concentrations (25%, 50%, and 75%), showing an increased permeability to molecules such as Lucifer yellow (453 Da) or 20 kDa-dextran (50). This observation is in accordance with lower transepithelial electrical resistance (TEER) values caused by increasing proportions of HT29 in the system (58, 52, 50). The presence of a mucous gel adhering loosely to the monolayers could be observed (58, 52); however, it should be pointed out as previously reported that HT29-MTX cells may only produce a fraction of the mucin molecule species found in the normal human intestine (60). It is important to mention that, although cocultures generally grew in well-defined monolayers, some anomalies, such as multilayered areas have been reported in this model (58). It is possible that the discrepancy between growth rates of the two cell lines produced these overgrown regions in the cultures.

Another important aspect that should be taken into account when using *in vitro* cell cultures is that these cell lines, which originated from tumor tissues have altered the cellular metabolism compared to physiological conditions. In addition, once the cell lines are established, they may cause an adaptation of their metabolism to the growth conditions in the culture media with a rather simple composition than the complex mixture in nutrients provided by foods in the gut. In this sense, a recent proteomic approach provided better insights in the expression patterns of human intestinal epithelial cells and the cell lines Caco-2 and HT29 (46). Lenaerts et al. (46) demonstrated that Caco-2 cells as well as HT29 cells express proteins that are characteristic to human intestinal epithelium *in vivo*. Furthermore, several of these biologically significant proteins are expressed at comparable levels in Caco-2 cells and small intestinal scrapings. Interestingly, these similarities in expression are not observed between Caco-2 cells and cells from the large intestine (46). One of the main conclusions on this report states that, although they originated from colon adenocarcinoma, after differentiation Caco-2 cells exhibit the phenotype of enterocytes supporting their usability in *in vitro* studies to mimic the small intestine.

The development of monolayer tissue cultures has proven to be a significant advancement in the efforts to better understand the factors affecting absorption (i.e., bioavailability) of nutrients. The incorporation of Caco-2 cells grown on solid or microporous supports, coupled to *in vitro* gastrointestinal digestion, allowing mineral uptake and/or transport to be estimated, improves the systems used for bioavailability studies (29, 30). The earliest publications on coupling *in vitro* digestion with culture of Caco-2 cell monolayers appeared in the mid 1990s (38, 37). These combined systems were generated from the need to evaluate the extent of absorption and those factors that can affect the bioavailabilty of essential micronutrients in foods. Though bioavailability depends largely on the ability to cross the intestinal barrier of the ultimate soluble physicochemical forms in which nutrients, after gastrointestinal digestion, reach the small intestine. In addition to the nutritional relevance of this concept, it has also been recently applied to evaluate the potential toxicological risk associated to the consumption of foods contaminated with toxic metals (44). Antinutrients and promoter substances within foods can either inhibit or enhance the absorption and/or utilization of food-derived components solubilized in the gut.

Presently, interest has grown in relation to the functional food-derived components. The latter presenting so many different chemical structures and properties. *In vitro* models have offered a useful tool to evaluate or perform a primary screening of their potential effects on intestinal epithelia. The combined *in vitro* digestion and Caco-2 culture model has been used to evaluate the effects of several functional food components. Most of these studies monitored the antioxidant effects of casein phosphopeptides (30), carotenoids (26), flavonoids (61), and phenolic extracts (54) upon the accumulation of reactive oxygen species. In addition to evaluate these beneficial effects, research related to antioxidants has also been focused on the ability of individual polyphenolics to reduce cell growth and proliferation (43) using the fact that Caco-2 cells are originated from tumor tissues.

Within this context of functional foods, we cannot rule out probiotics, which are nonpathogenic microorganisms that exert a positive effect on their host's health. They should also exhibit properties such as (a) resistance to gastric acidity and bile toxicity, (b) to be able to colonize the gastrointestinal tract, (c) inhibit intestinal pathogens, (d) stimulate the immune system, and (e) show good technological properties (49). The ability of probiotics to adhere and colonize mucosal surfaces in the gut is a prerequisite for bacterial maintenance. In this sense, monolayers of Caco-2 or HT29 cells have been successfully used to evaluate the adhesion ability of the main groups of bacteria representative of the human normal microbiota; the *Bifidobacterium* (25, 32, 39); and *Lactobacillus* (34). The mechanism(s) whereby the adhering strains interacted with intestinal epithelia were investigated. These studies showed that Bifidobacteria (*B. breve, B. longum, B. bifidum, B. infantis*) exhibit different adhesive capacities by interacting with the brush border membrane of the differentiated Caco-2 cells, which resulted in high calcium-independent for *B. breve* and was mediated by an adhesion-promoting factor (25). Furthermore, the inhibition by Bifidobacteria strains of Caco-2 cell colonization by harmful bacteria such as *Escherichia coli* (H10407 and JPN15), *Yersinia pseudotuberculosis* (YPIII pYV$^-$), and *Salmonella serovar typhimurium* (SL1344) has been evaluated (25). When comparing the *in vitro* data with *in vivo* adhesion ability of Bifidobacteria (*B. adolescentis, B. angulatum, B. animalis, B. breve, B. infantis, B. longum, B. pseudolongum*), the values obtained by using Caco-2 cells were in good agreement with the results obtained in humans fed with fermented milks (32). Delgado et al. (34) reported that *Lactobacillus* (*L. gasseri*, F71 and L1; and *L. paracasei*, BA3 and F76) exhibit similar adhesion values to Caco-2 or HT29 cultures. It is important to mention that there are no studies related to the bacterial adhesion ability by using a combined *in vitro* digestion and Caco-2/HT29 coculture system. Otherwise, the ability of *Lactobacillus* and *Bifidobacterium* strains to grow in culture media (MRSC) in the presence of different concentrations of bovine bile (0.25–2%) or low pH values (3.5–5.0) has been evaluated (34, 31).

Cell monolayers may also be used to investigate the harmful effect of pathogens such as *E. coli* through the production of its enterohemolysin (EHly) (36). The latter authors evidenced that essential biological processes for cell survival (i.e., mitochondrial function, disorganization of cytoskeleton) or death (i.e., apoptosis) were affected when Caco-2 cultures were challenged to EHly. On the contrary, the cross talking between bacteria-intestinal cells to monitor the probiotic-induced stabilization of gut barrier function has also been successfully evaluated by using the Caco-2 cell model (59, 55). The beneficial effects of *Lactobacillus* and the probiotic mixture VSL#3 through the induction of antimicrobial peptides such as human β-defensin (hBD)-2 by Caco-2 cultures has been reported (55). In the same way, Wehkamp et al. (59) demonstrated that the probiotic bacterium *E. coli Nissle* 1917 and other probiotic strains might exert their beneficial effects by inducing the synthesis of hBD-2.

In summary, taking into account that cultures of human cell lines of tumor origin are not always comparable to normal tissue cells, the *in vitro* models provide an effective approximation to the *in vivo* situation and offer several advantages of good reproducibility. The strong correlations shown in many aspects of the human studies indicate that cell line models are useful and valuable tools in understanding the underlying mechanism(s) of many of the processes occurring at the intestinal level. Furthermore, cell line models constitute a promising tool for studying the potential relationship between food-derived components, gut microbiota, and intestinal epithelia.

4.2 LABORATORY ANIMAL MODELS IN RESEARCH

Rafael Frias

Central Animal Laboratory, University of Turku, Finland

In addition to *in vitro* and *ex vivo* assays, preclinical testing with probiotics can be performed in *in vivo* models by using laboratory animals. There are many hypotheses and theories, which need to be examined on certain probiotics, in which complex physiological interactions provided by the *in vivo* model are still required. Experiments using animals are strictly regulated by legislation, and scientists should aim at using nonanimal methods whenever possible. If laboratory animals are to be used, scientists must be competent, that is, they must hold an appropriate academic qualification, possess sufficient skills, and be obliged to conduct animal experiments in accord with the highest scientific, humane, and ethical principles. The implementation of the 3 Rs for *replacement* (use of alternative methods), *reduction* (use of minimum number of animals), and *refinement* (use of improved experimental procedures, high standards of animal welfare, etc.) are indispensable prerequisites not only for public acceptance but also for good science (63–66).

Laboratory animals are used in a wide range of biological and medical experiments including research with probiotics. *In vivo* models have allowed the examination and evaluation of various properties of different probiotic strains, and have considerably contributed to our knowledge on their function, and on their efficacy and safety for both human and veterinary use. There are many published examples where health data generated from evaluating probiotics in animal models has given rise to similar outcomes in the model and in the target (67–76).

Animal models are nonhuman living organisms, vertebrate, or invertebrate that are scientifically used in the investigation of biological phenomena and physiological function between the model and the target species. The target species are not only humans but also other species. Mice and rats are the most popular laboratory animal models used in research with probiotics, but several other species have been used as well (Fig. 4.2). The choice of an appropriate and a high quality animal model is of crucial importance. Significant and valid data that is to be extrapolated in the target species strongly depends on a suitable model only. A sound knowledge of comparative anatomy, physiology, and genetics is an advantage when selecting an animal model. Nonetheless, phylogenetic closeness is not always a guarantee for valid extrapolation (e.g., chimpanzees are not good models to study human AIDS virus as they do not acquire the infection), and the ultimate selection of the animal model should be primarily based on how well the model explains the specific aims rather than how well it represents the target. New animal models are continuously developed for use in the investigation of mechanisms of action, measurement of pharmacokinetics, diagnostic and therapeutic procedures, nutrition and metabolic diseases, and the safety and efficacy of test substances, such as novel probiotics for

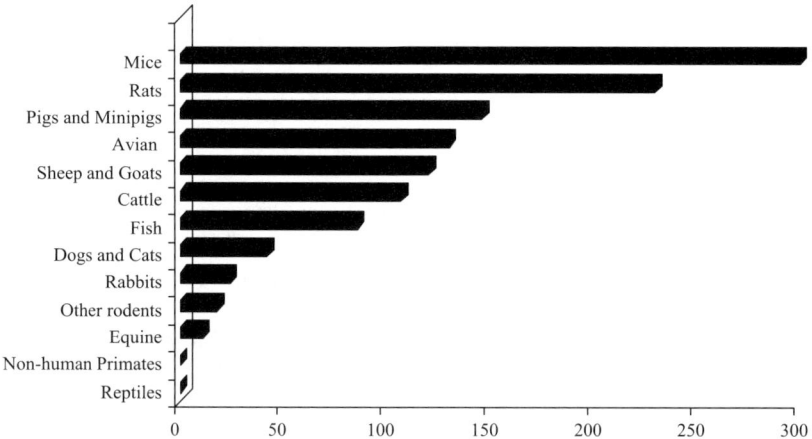

FIGURE 4.2 Number of citations in pubMed containing different animal models and probiotics, March 2008.

human and veterinary use. In particular, animal models can be divided into five different disease categories (77):

1. *Induced models* are healthy animals in which the disease condition to be investigated is experimentally induced by a physical, chemical, or biological method. Usually, this class allows a free choice of species, although the etiology and the characterization of the disease in the model are different from the corresponding disease in the target species.
2. *Spontaneous models* are animals with a naturally occurring disease whose manifestations are similar to the corresponding disease in the target species. The disease as well as the responses is usually comparable between the model and the target. Spontaneous models include inbred, congenic, hybrid, and mutant murine strains with hereditary disease or genetic anomalies (e.g., nude mouse).
3. *Genetically modified (GM) models* are animals whose genome has been manipulated to produce a disease. Genetically engineered animals are further subdivided in transgenic animals (when DNA is inserted into its genome) and knock-outs (when specific gene has been deleted from animal's genome to produce a specific genotype). A wide number of disease models have been developed with GM animals and the number seems to be constantly increasing. Currently, GM animals are possibly the most important class of disease models and they provide exceptional opportunities for biomedical research. The mouse is the most common genetically altered animal used in biomedical research, but other animals have also been manipulated. The production of GM animals is not an accurate science and sometimes unpredictable outcomes in terms of scientific results and welfare problems have been obtained.
4. *Negative models* are laboratory animals in which a certain disease does not develop. These animals naturally demonstrate a lack of response to a specific stimulus (e.g., infection), and are commonly used in studies that are aimed at understanding the physiology and the mechanisms of resistance to a disease.
5. *Orphan models* are animals that suffer from a species-specific disease, which is not yet described in the target species. These models are of value in the investigation of a similar disease that is being recognized in the target species, for example, bovine spongiform encephalopathy in cows and Creutzfeldt-Jakob in humans.

There are numerous events and their interactions that may influence the suitability of an animal model, such as the species, breed, strain, genotype, and other various exogenous factors. Microbial factors are one of the most important concerns to scientists working with probiotics because of their potential to confound and invalidate results and conclusions from the animal model. Several groups of microorganisms (viruses, mycoplasmas, bacteria, fungi, and parasites) are responsible for infections and diseases in laboratory animals. It is of crucial importance that the laboratory animal model is free from unwanted microbial agents (e.g., specific pathogen free) in order to achieve reliable, valid, and reproducible experimental outcomes. Likewise, it is important to remark that an infection is not synonymous with disease, and for instance most rodent infections are subclinical (latent) and do not lead to overt clinical symptoms. Hence, all microbial infections, clinical or subclinical, are likely to cause various degrees of abnormalities in experimental results and increase biological variability, which in turn leads to increased animal use having a considerable impact on the project costs as well as on the animal welfare (77–79).

The term extrapolation describes how data obtained from an *in vivo* model can be reliably used in the target species. At times, data generated from the model are not applicable in the target and *vice versa*. This is particularly true in microbiology models, where the high variability in species-specific responses does not allow the correct extrapolation of the results. Each experiment must be individually interpreted and cautious rather than rigorous extrapolation should be exercised. A test substance (e.g., probiotic strain) may be effective, ineffective, or even generates side effects in the animal model, and sometimes the inverse response may be observed in the target species. Body size and metabolic rate should be also taken into account when selecting an animal model, as there are usually large differences between the model and the target species (e.g., mouse vs. human). In mammals, the organ size is generally proportional to the body size and it is expressed as percentage of body weight (BW), for example, the weight of the heart is about 5–6 g/kg (0.5–0.6%) BW, and the total blood volume is approximately 70–80 ml/kg (7–8%) of BW. In contrast, the metabolic rate is much higher in smallest than in large mammals. For instance, the normal heart rate in a mouse may go over 700 beats/min, and the respiration goes over 200 breaths/min. In small mammals, oxygen supply to body tissues is limited by the low stroke volume (i.e., volume of blood ejected by single heart contraction), and rodents must considerably increase both the heart and respiration rates to compensate. As a consequence, other physiological variables like food and water intake are similarly affected by their high metabolic rate. The metabolism and excretion of a test substance is directly correlated to animal's metabolic rate rather than BW. The metabolic rate (M) is expressed by oxygen consumption (mL) per BW (g) per hour, and is related to BW as follows: $M = 3.8 \times BW^{-0.25}$. Therefore, scaling is of crucial importance in studies where an accurate dose of a test substance must be calculated and administered to the laboratory animal. If the aim is to achieve similar test substance concentrations in the model and the target, the dose should be calculated in proportion to animal's BW. If the dose of test substance is known in the animal model or in the target, then the dosage for a subject with a different BW can be calculated as follows:

$$\text{Dose}_1 = \frac{\text{Dose}_2 \times \text{BW}_1^{-0.25}}{\text{BW}_2^{-0.25}}$$

This formula and similar ones have been successfully used for calculating dosages and for establishing safety data from the model to determine the levels of exposure to the target in several pharmacological and toxicological studies, but care should be taken with too broad generalization. $BW^{-0.50}$ should be used in animals weighing <100 g.

Marked variations may occur depending on test substances, species, genotype, microbiological quality, and other environmental conditions (77, 80–83).

The reader is referred to published literature on probiotics and animal examples, as well as monographs, reviews, and other relevant resources on laboratory animal science (symposia, websites) that may be very useful in collecting additional information on specific animal models.

4.3 EFFECTS ON HUMAN HEALTH AND DISEASES

YUAN KUN LEE[1] AND SEPPO SALMINEN[2]

[1]*Department of Microbiology, National University of Singapore, Singapore*
[2]*Functional Foods Forum, University of Turku, Finland*

Probiotics are live microbial food supplements which when taken in adequate amounts exert a health benefit to the human or animal host (88, 89), also see Section 1.1). However, probiotics are not a uniform group of microorganisms with health benefits; hence, the efficacy and safety of each specific species and strain should be assessed individually rather than as a group of probiotics. It is also important to recognize that not all lactic acid bacteria or Bifidobacteria are probiotics as their health benefits need to be scientifically demonstrated. As in the case of *Lactobacillus rhamnosus*, one of the strains had been demonstrated to be effective in treatment of rotavirus diarrhea in children, while the other strain had no such clinical effect (86). Similarly, a nonviable strain of *Lactobacillus rhamnosus* was not effective in atopic eczema symptom alleviation while a viable strain was demonstrated to be effective (85, 84).

As probiotics by definition impart health benefits to the host, it is important to review the scientific documentation and basis for such benefits. Based on research reports and mainly clinical intervention studies it has been demonstrated that all probiotic health effects are strain, dose, disease, and possibly host dependent. Human placebo-controlled clinical intervention trials to elucidate the health effects of specific probiotics or probiotic combinations are also cited here.

4.3.1 Nutritional Effects

Examples of nutritional effects of probiotics are summarized in the following table:

Effects	Specific Probiotics	Mechanisms	References
Produce water-soluble vitamins: thiamine, nicotin, folic acid, pyridoxin, Vit. B_{12}	*B. bifidum* *B. infantis* *B. breve* *B. adolescentis* *B. longum*	— — — — —	(90) — — — —

(continued)

(*Continued*)

Effects	Specific Probiotics	Mechanisms	References
Produce biotin	B. adolescentis M101-4 B. bifidum A234-4 B. breve I-53-8 B. infantis I-10-5 B. longum M101-2		(91)
Increase bioavailability of iron	L. acidophilus SBT2062		(92)
Deconjugation of bile salts (taurocholic acid and taurodeoxycholic acid)	L. reuteri 100-23 L. delbrueckii 100-18 L. fermentum 100-20 L. delbrueckii 100-21 E. faecium E. faecalis	Reduced bile salt hydrolase activity in ilea	(93)
Partial digestion of lactose in milk	Yogurt	Increased lactase activity in milk and in duodenum. Reduce lactose maldigestion	(89)

4.3.1.1 Lactose Maldigestion Lactose maldigestion is defined by an increase in blood glucose concentration of <1.12 mmol/L or breath hydrogen of >20 ppm after ingestion of 1 g/kg body wt. or 50 g lactose (94).

Hydrogen is the end product of anaerobic microbial metabolism of lactose in human gut. Studies on the influence on hydrogen exhalation of natural and heat-denatured yogurt compared with milk in lactose-intolerant persons (94–104) are summarized in the following figure.

Natural and denatured (pasteurized, heated) yogurts improved lactose digestion in lactose maldigesters. The effect of natural yogurt was higher than that of heat-treated yogurt. Natural yogurt almost abolished the symptom of lactose maldigestion in lactose maldigesting person, for lactose intake of up to 20 g. These observations suggest that in yogurt, beside microbial β-galactosidase (active only in natural yogurt but inactivated by heat treatment), other factors, such as the delay of lactose in gastrointestinal transit (prolongs the action of residual β-galactosidase and decreases osmotic load of the lactose) (97, 104) positive effects on intestinal functions, and colonic microbiota may contribute to lactose digestion in the intestine.

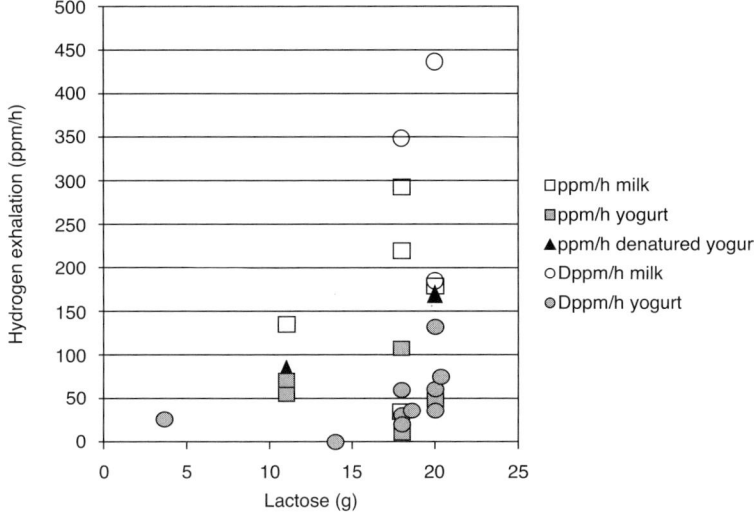

ppm/h = Part per million total breath hydrogen per hour;
Dppm/h = Part per million breath hydrogen above that of the baseline per hour

4.3.1.2 β-Galactosidase in Fermented Milk Products The effect of the enzyme β-galactosidase in yogurt on the digestion of lactose in lactose maldigesting people is summarized in the following figure (data extracted from References (105–109) cited below).

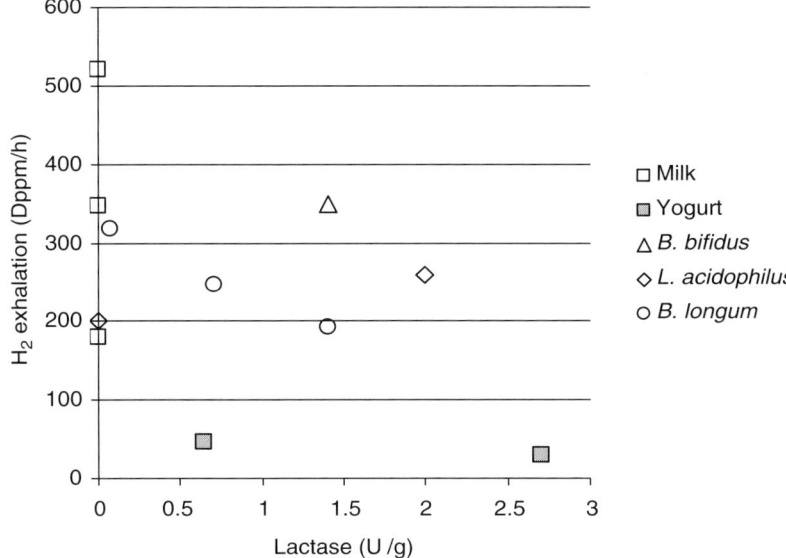

The poor correlation between lactose maldigestion (hydrogen exhalation) and β-galactosidase (lactase) content in fermented milk observed in these studies implies that β-galactosidase may not be the sole and major factor involves in the digestion of lactose in milk products. Other factors, such as intracellular β-galactosidase released in the small intestine and the yogurt matrix may play an important role in improving lactose digestion. It is also noteworthy that there are 100-fold differences in the β-galactosidase activities between the different types of common starter cultures used in yogurt production (111). Such activity differences may result in significantly different lactose levels in the final products with consequent impact on symptoms in subjects highly sensitive to small amounts of lactose in their diet.

4.3.2 Prevention and Treatment of Oral Infection and Dental Caries

Acidogenic oral bacteria such as Streptococci and Lactobacilli are associated with the presence and onset of dental caries (110, 111). The caries bacteria colonize acidic environments, such as fissures and interdental spaces, and produce acids from fermentable sugars. On the contrary there was a suggestion that temporary colonization of a nonlactose and slow sugar fermenting bacteria, which are able to inhibit the caries pathogens, may prevent caries in children (112, 113).

At the other end of the age spectrum, overgrowth of oral yeast is a common problem among the elderly. Temporary colonization of probiotic bacteria in the oral cavity may competitively exclude the oral yeast (114).

Probiotics	Dose	Disease	Subjects	Clinical Trials	Observations	References
L. rhamnosus GG	$5-10 \times 10^5$ cfu/mL milk	Dental caries	594, 1–6 years from 18 municipal day-care centers, Helsinki, Finland	5 days/week for 7 months; age 1–2 years: control: average 218 mL/day, probiotic: 248 mL/day; age 3–4 years: control: 240 mL/day, probiotic: 232 mL/day; age 5–6 year: control: 269 mL/day, probiotic: 257 mL/day	3–4 year-old: occlusal surface carries baseline-adjusted odd ratios (OR) = 0.22 ($p = 0.059$); cumulative caries OR = 0.34 ($p = 0.057$); active caries OR = 0.34 ($p = 0.059$); risk of caries OR = 0.56 ($p = 0.01$); protective effect for 1–2 years and 5–6 years not significant	(114)
L. rhamnosus GG	50 g cheese	Oral candida	276 elderly from Helsinki, Finland	16 weeks	High salivary yeast count ($>10^4$ cfu/mL), reduction OR = 0.25 ($p = 0.004$); hyposalivation reduction OR = 0.44 ($p = 0.05$)	(112)

General Conclusion: The 7-month intervention was too short for caries to progress; nevertheless, a protective effect of probiotic bacteria in 3–4 year-old children was observed and warrants further study. The protective effect in the 3–4 year-old children in the probiotic group could indeed be due to an effect on *S. mutants*, as initial caries is known to correlate with mutant streptococcus counts (110).

One study suggests that probiotic bacteria can be effective in controlling oral Candida and hyposalivation in the elders and warrants further clinical confirmation.

4.3.3 Prevention and Treatment of Diarrhea

Each year about 4 billion diarrhea episodes occur worldwide, accounting for almost 4% of all deaths and 5% of days lost in disability. Prevention of diarrhea is also a public-health issue. Environmental hygiene, food storage and handling, and clean portable water go a long way in the prevention of diarrhea. However several of these factors could not be easily improved due to local conditions, especially in the developing countries. The effects of probiotics in the prevention and treatment of diarrhea are well studied. Several recent meta-analyses of data from clinical trials provide a detail analysis on the evidences on the efficacy of probiotics in preventing diarrhea of various causes (115–118).

4.3.3.1 Acute (Rotavirus) Diarrhea in Children   Acute diarrhea is a common pediatric disease, which contributes to substantial morbidity and mortality worldwide, especially in developing countries where up to seven episodes could be experienced per year (122, 125). Even in the developed countries, such as the United States, 16.5 million children develop acute diarrhea annually. The current treatment of diarrhea is mainly supportive and primarily involves symptomatic care. Recently, probiotics have been proposed as an adjunctive therapy in the treatment of acute diarrhea in children.

Probiotics	Pathogen (% Rotavirus)	Subject No.	Age, Months	Dose/Day	Days of Therapy	Duration of Diarrhea (Days ± SD)	P value	References
L. acidophilus	Overall: 49	P: 38, C: 33	1–54	1 packet QD	Variable	P: 1.1 ± 0.9, C: 1.1 ± 0.7	>0.05	(120)
	P: 51.4, C: 44.4	P: 37, C: 36	3–24	1×10^{10} BID	2	P: 1.8 ± 1.1, C: 2.4 ± 1.5	0.034	(140)
L. acidophilus + L. bulgaricus + S. thermophilus		P: 53, C: 41	<36	1×10^8 per capsule, four to eight capsules daily	Variable	P: 2.7 ± 2.5, C: 2.1 ± 1.6	>0.05	(133)
L. acidophilus + B. infantis	100	P: 50, C: 50	6–60	L. acidophilus: 1×10^9, B. infantis: 1×10^9, TID + IV hydration	4	P: 3.1 ± 0.7, C: 3.6 ± 0.8	<0.01	(130)
1. L. reuteri	P: 63, C: 86	P: 19, C: 21	6–36	$1 \times 10^{10-11}$ QD	5	P: 1.7 ± 1.6, C: 2.9 ± 2.3	0.07	(137)
	100	P: 21, C: 20	6–36	$1 \times 10^{10-11}$ QD, 1×10^7 QD	5	P: 1.5 ± 1.1, 1.9 ± 0.9, C: 2.5 ± 1.5	0.01, >0.05	(138)
L. rhamnosus GG	P: 92, 74; C: 79	P: 24, 23; C: 24	4–45	$1 \times 10^{10-11}$ QD fermented milk, $1 \times 10^{10-11}$ QD freeze dried	5	P: 1.4 ± 0.8, 1.4 ± 0.8; C: 2.4 ± 1.1	<0.001 ANOVA	(126)

(continued)

(Continued)

Probiotics	Pathogen (% Rotavirus)	Subject No.	Age, Months	Dose/Day	Days of Therapy	Duration of Diarrhea (Days ± SD)	P value	References
	100	P: 22, C: 17	7–37	$1 \times 10^{10-11}$ BID, pasteurized	5	P: 1.1 ± 0.6, C: 2.5 ± 1.4	0.001	(128)
	100	P: 21, C: 21	5–28	1×10^{10} BID, Freeze-dried	5	P: 1.5 ± 0.7, C: 2.3 ± 0.8	0.002	(127)
	P: 10, C: 26	P: 20, C: 19	1–24	$1 \times 10^{10-11}$ BID	2	P: 1.9 ± 0.6, C: 3.3 ± 2.3	<0.05	(132)
	P: 60, C: 62.5	P: 52, C: 48	3–36	3×10^9 BID	5	P: 3.3 ± 2.9, C: 5.9 ± 2.8	<0.01	(124)
	28	P: 59, C: 64	1–36	5×10^9 BID in ORS	5	P: 2.7 ± 2.2, C: 3.8 ± 2.8	0.02	(139)
	P: 38, C: 32	P: 147, C: 140	1–36	$1 \times 10^{10}/$ 250 mL ORS	Variable	P: 2.4 ± 1.2, C: 3.0 ± 1.5	<0.03	(123)
	50	P: 61, C: 63	1–24	5×10^9	Not specified	P: 1.59 ± 0.16, C: 1.63 ± 0.19	>0.05	(121)
	P: 24.4, C: 39.3	P: 52, C: 61	3–36	1×10^9 mL^{-1} milk, 150 mL/kg/day	Ad libitum	P: 2.44 ± 1.26, C: 2.10 ± 1.17	>0.05	(136)
L. rhamnosus GG	100	LGG: 16	6–35	LGG 6.25×10^9 BID, freeze-dried	5	LGG: 1.8 ± 0.8	0.04 ANOVA	(131)
L. rhamnosus		*L. rhamnosus*: 14		*L. rhamnosus* 2.75×10^8 BID		*L. rhamnosus*: 2.8 ± 1.2		(131)
L. rhamnosus + *L. delbruckii* + *L. bulgaricus*		Mixture: 19		Mixture 3.5×10^9 BID		Mixture: 2.6 ± 1.4		(131)

L. rhamnosus + L. reuteri	P: 54, C: 74	P: 24, C: 19	9–44	2×10^{10} BID	5	P: 3.2 ± 1.7, C: 4.8 ± 3.5	0.05	(135)
	P: 60, C: 71	P: 30, C: 39	5–36	2×10^{10} BID	5	P: 3.4 ± 1.6, C: 4.2 ± 2.0	0.07	(134)
Saccharomyces boulardii	P: 39, C: 44	P: 100, C: 100	3–84	250 mg in ORS	5	P: 4.7 ± 2.5, C: 5.5 ± 3.2	0.03	(129)
	P: 16, C: 20	P: 50, C: 50	2–144	150 mg BID in ORS	5	P: 3.5, C: 4.8	0.001	(119)
		P: 44, C: 44	3–24	<1 year 250 mg, >1 year 500 mg, in ORS	6	P: 4.70 (2–10), C: 6.16 (2–13)	<0.05	(141)

P: Probiotics; C: control; BID: twice a day; QD: once a day; TID: three times a day; ORS: oral rehydration salt/solution; IV: intravenous; ANOVA: Analysis of Variance.

General Conclusion: The efficacy of various probiotics as adjunctive therapy in the treatment of acute diarrhea in more than 1800 children was evaluated. The evidences in general (1340 out of 1800 children were evaluated) appear to support the efficacy of probiotic therapy in the setting of acute diarrhea in children. The *L. rhamnosus* and *S. boulardii* are more effective than *L. acidophilus* and *L. reuteri*.

4.3.3.2 Antibiotic-Associated Diarrhea Antibiotic-associated diarrhea is a common complication of antibiotic treatment. The frequency of antibiotic-associated diarrhea ranges from 26% to 60% during hospital outbreak to 0.1% in outpatient settings (151, 153, 164). The risk factors include broad-spectrum antibiotics, host physiological factors, and hospital environment (153, 142). The disease usually occurs 2–8 weeks after antibiotic treatment, due to the disturbance of intestinal microbiota, which serves as a mucosal barrier for opportunistic pathogens (154).

Current strategy in the treatment of antibiotic diarrhea includes discontinuation of the antibiotics and use of second line and specific (if etiology is known) antibiotics. Clinical trials have been conducted to evaluate the effects of probiotics in the prevention and treatment of antibiotic-associate diarrhea, and the results are summarized in the following table.

Probiotics	Antibiotic	Subjects	Dose/Day	Test Duration	Follow-Up	Risk Ratio (95% CI)	References
Bacillus clausii	3 antibiotics	100 adults	6×10^9	2 weeks	0	0.88 (0.50, 1.57)	(157)
Bacillus longum	Clindamycin	20 adults	5×10^{10}	21 days	0	0.57 (0.24, 1.35)	(158)
Clostridium butyricum MIYAIRI	Varied	110, 1 month to 15 years	1–4×10^7	6 days	0	0.12 (0.05, 0.28)	(159)
Enterococcus faecium SF68	Varied	45 adults	1.5×10^7	7 days	0	0.32 (0.07, 1.41)	(167)
L. acidophilus	Amoxicillin	27 adults	1.2×10^8	Varied	0	1.25 (0.81, 1.94)	(166)
L. rhamnosus GG	3 antibiotics	42 adults	6×10^9	7 days	0	0.17 (0.02, 1.27)	(147)
	3 antibiotics	120 adults	1×10^{10}	2 weeks	0	0.13 (0.03, 0.52)	(144)
	Varied	81, 1 to 36 months	1×10^{10}	Varied	0	0.20 (0.06, 0.66)	(161)
	Varied	188, 6 months to 10 years	1–2×10^{10}	10 days	0	0.29 (0.13, 0.63)	(164)
	β-Lactam/varied	267 adults	2×10^{10}	2 weeks	1 week	0.98 (0.68, 1.42)	(163)
	Varied	119, 2 weeks to 13 years	4×10^{10}	2 weeks	12 weeks	0.32 (0.09, 1.11)	(145)
L. acidophilus + *L. bulgaricus*	Ampicillin	79 adults	2×10^9	5 days	0	0.40 (0.12, 1.36)	(148)
	Amoxicillin	38, 5 months to 6 years	2×10^9	10 days	0	0.96 (0.61, 1.50)	(162)
L. acidophilus + *B. infantis*	Varied	18, 1 to 36 months	6×10^9	7 days	0	0.47 (0.18, 1.21)	(149)
L. acidophilus + *B. lactis*	Clindamycin	20 adults	1×10^{11}	21 days	0	0.29 (0.08, 1.05)	(158)
L. acidophilus + *B. longum*	3 antibiotics	42 adults	5×10^9	7 days	0	0.17 (0.02, 1.27)	(147)

(*continued*)

(Continued)

Probiotics	Antibiotic	Subjects	Dose/Day	Test Duration	Follow-Up	Risk Ratio (95% CI)	References
B. lactis + S. thermophilus	Varied	77 adults	1×10^7	15 days	15 days	0.51 (0.21, 1.23)	(146)
L. sporogenes + fructooligosaccharide	Varied	98 children	$5.5 \times 10^8 + 250$ mg	10 days	0	0.48 (0.29, 0.77)	(151)
Saccharomyces boulardii	β-Lactam/ tetracycline	388 adults	4×10^9	1 week	0	0.26 (0.13, 0.53)	(143)
	Varied	69 adults	4×10^9	2 weeks	0	1.53 (0.54, 4.35)	(153)
	3 antibiotics	43 adults	5×10^9	7 days	0	0.16 (0.02, 1.21)	(147)
	Varied	246 adults	1×10^{10}	1–2 weeks	2 weeks	0.19 (0.07, 0.55)	(150)
	Varied	180 adults	2×10^{10}	With antibiotics	2 weeks	0.43 (0.21, 0.90)	(160)
	β-Lactam	193 adults	2×10^{10}	4 weeks	7 weeks	0.49 (0.21, 1.17)	(156)

General Conclusion: The clinical trials conducted suggest that most of the tested probiotics (except *L. acidophilus* alone in one study and *S. boulardii* in another study) can significantly reduce the incidence of antibiotic-associated diarrhea with varied efficacy.

4.3.3.3 Clostridium difficile Associated Diarrhea Patients under treatment with antibiotics may develop *Clostridium difficile* colitis (168). It was reported that 80% of the patients responded well to the initial treatment of vancomycin or metronidazole. The remaining 20% may develop recurrent episodes of diarrhea, which may persist over several years, despite repeated antibiotic treatments (169, 172). There is a growing interest in the use of probiotics for the treatment of *Clostridium*-associated diarrhea due to the potential effects of probiotics in re-establishing the mucosal integrity, intestinal commensal microbiota, and modulating of immune responses.

Probiotics	Subject No.	Dose/Day	Test Duration	Follow-Up	Risk Ratio (95% CI)	References
L. acidophilus + *B. bifidum*	138	2×10^{10}	20 days	0	0.59 (0.35, 0.98)	(173)
L. plantarum 299v + metronidazole	20	5×10^{10}	38 days	0	0.33 (0.10, 1.06)	(175)
L. rhamnosus GG + vancomycin or metronidazole	15	6×10^{11}	3 weeks	4 weeks	0.55 (0.22, 1.35)	(170)
Saccharomyces boulardii +	124	2×10^{10}	4 weeks	4 weeks	2.63 (0.35, 19.85)	(172)
vancomycin or metronidazole	32	2×10^{10}	4 weeks	4 weeks	0.33 (0.07, 1.59)	(174)

General Conclusion: The few studies conducted have demonstrated the potential of probiotics in the treatment of *Clostrdium difficile*-associated diarrhea. However, one study showed the adverse effects when antibiotics and probiotics were combined. Further studies are also necessary.

4.3.3.4 Radiation-Induced Diarrhea Radiation therapy disturbs the indigenous intestinal microbiota, which may lead to acute enteritis and colitis (177). Attempts to treat this complication with antibiotics, sucralfate, and anti-inflammatory drugs have met with inconclusive clinical results. Probiotics could be an alternative treatment.

Probiotics	Dose	Subject No.	Clinical Trials	Reference
L. casei + L. plantarum + L. acidophilus + L. delbruekii spp. Bulgaricus + B. longum + B. breve + B. Infantis + S. salivarius spp. thermophilus; 4.5×10^{11} cfu/g lyophilized	1 g (sachet) twice daily	Probiotic: 243, placebo: 239	51.8% placebo versus 31.6% probiotic patients had diarrhea ($p < 0.01$); 55.4% placebo versus 1.4% probiotic suffered grade 3 or 4 diarrhea ($p = 0.001$); 14.7 ± 6 placebo versus 5.1 ± 3 daily bowel movements ($p < 0.05$); mean time to use loperamide 86 ± 6 h placebo versus 122 ± 8 h probiotics ($p = 0.001$)	(176)

General Conclusion: The above clinical observations clearly indicate that probiotics are potentially useful to protect cancer patients against the risk of radiation-induced diarrhea.

4.3.3.5 Traveler's Diarrhea Traveler's diarrhea is a common health complaint among travelers, every year 12 million cases of traveler's diarrhea are reported (180, 186, 189). Rates of traveler's diarrhea can range from 5% to 50%, depending on the destination and origin of the travelers. Most cases (80–85%) of traveler's diarrhea are due to bacterial pathogens, such as *Aeromonas hydrophila, Campylobacter jejuni,* Enterotoxigenic *E. coli*, Enteroaggregative *E. coli*, *Plesiomonas shigelloides, Shigella* species, *Salmonella s*pecies, *Vibrio cholera, Vibrio parahemolyticus,* and *Yersinia enterocolitica* (178, 181). Other less frequent causes of traveler's diarrhea are viruses (Norwalk or Rotavirus) and parasites (*Cyclospora, Cryptosporidium, Entamoeba histolytica, Giardia lamblia*). In some cases the cause of diarrhea cannot be determined.

The summary compares the efficacy of probiotics for the prevention of traveler's diarrhea based on published clinical trials. Diarrhea is defined as 4–6 loose, watery, or bloody bowel movements per day. Traveler's diarrhea usually lasts for 2–6 days if untreated. In up to 15% of cases, diarrhea may be prolonged (1 week to 1 month, rarely up to 1 year) and associated with repeated bouts of abdominal cramping, malaise, nausea, fever, or muscle pain.

Probiotics	Subject No.	Dose/ Day	Test Duration	Follow-Up	Risk Ratio (95% CI)	References
L. acidophilus	319	$2 \times 10^{8-9}$	3 weeks	0	1.13 (0.91, 1.40)	(184)
	202	2×10^{11}	3 weeks	1 week	1.08 (0.67, 1.75)	(185)
L. rhamnosus GG	245	2×10^9	1–3 weeks	0	0.52 (0.18, 1.52)	(182)
	756	2×10^9	2 weeks	0	0.88 (0.75, 1.04)	(187)
L. fermentum KLD	181	2×10^{11}	3 weeks	1 week	1.00 (0.59, 1.69)	(183)
L. acidophilus + L. bulgaricus	50	$4-7 \times 10^9$	8 days	3 weeks	1.19 (0.52, 2.69)	(188)

(*Continued*)

Probiotics	Subject No.	Dose/ Day	Test Duration	Follow-Up	Risk Ratio (95% CI)	References
L. acidophilus + *L. bulgaricus* + *B. bifidum* + *S.thermophilus*	94	3×10^9	2 weeks	0	0.61 (0.41, 0.89)	(179)
Heat killed *Salmonella* + *Shigella* + *E. coli*	310	Nonviable	3 weeks	0	1.01 (0.81, 1.27)	(184)
Saccharomyces boulardii	832	5×10^9	3 weeks	0	0.79 (0.66, 0.94)	(184)
	713	5×10^9	3 weeks	0	0.97 (0.72, 1.06)	(185)
	805	1×10^{10}	3 weeks	0	0.75 (0.62, 0.90)	(184)
	664	2×10^{10}	3 weeks	0	0.74 (0.60, 0.92)	(185)

General Conclusion: The efficacy of probiotics in the prevention of traveler's diarrhea appears to be species dependent. *L. rhamnosus*, *Saccharomyces boulardii*, and a mixture of *Lactobacillus*, *Bifidobacterium*, and *Streptococcus* showed positive effect, where as *L. acidophilus*, *L. fermentum*, and *L. bulgaricus* showed no effect even at very high dose.

4.3.3.6 Diarrhea in Tube-Fed Patients Enteral tube feeding (ETF) is an established method of nutritional support for both hospital and community patients (192, 193). Despite the benefits and widespread use of ETF, some patients experience complications, which include diarrhea (2–63% of patients) (194). Manipulation of the colonic microbiota by probiotics may reduce the incidence of ETF diarrhea via suppression of enteropathogens.

Probiotics	Dose	Subject No.	Clinical Trials	References
L. acidophilus + *L. bulgaricus*	1 g TD	41	Diarrhea defined as >200 g stool/ 24 h or >3 liquid stools/day. 17% developed diarrhea during first 5 days of ETF. No statistically significant difference in the incidence of diarrhea between patient groups	(191)
S. boulardii	500 mg QD	128 (64 each group)	Diarrhea defined as score >12 on Hart and Dobb diarrhea score. Frequency of diarrhea days was 18.9% placebo versus 14.2% probiotic (OR 0.67, 95% CI 0.50–0.90; $p = 0.0069$).	(190)

TD: thrice daily; QD: once a day.

General Conclusion: The two prospective trials suggest that probiotics may not prevent ETF diarrhea; however, probiotics may reduce the number of days that patients suffer from diarrhea. The trials are too few to be conclusive.

4.3.4 Treatment of Irritable Bowel Syndrome

Approximately 5–20% of the world population is estimated to suffer from IBS (196, 198). Its main clinical symptoms include abdominal discomfort or pain, diarrhea, constipation, bloating, and flatulence. The current therapies for IBS are considered to be only moderately effective, and new approaches in treatment are being constantly sought. The pathogenesis of IBS remains unclear, but available evidence suggests that altered gut motility, visceral hypersensitivity, and dysregulation of the brain–gut axis are important mechanisms (196). There is accumulating clinical evidence to suggest that an imbalance intestinal microbial profile (202, 208) and enteric bacteria mediated mucosal inflammation (195, 203) may be associated with IBS. Some of the key studies on the effect of probiotic bacteria in relieving IBS symptoms are summarized in the following table.

Probiotics	Dose/Day	Subjects	Treatment	Clinical Trials	References
L. casei Shirota	1×10^9 cfu/mL in 65 mL beverage/day	70 chronic idiopathic constipation patients of either sex, aged 18–70	4 weeks	Degree of constipation, flatulence and bloating on a four-point scale, stool consistency assessed by Bristol Stool form scale; 89% of probiotic group versus 56% of placebo showed positive effect on constipation and stool consistency ($p = 0.003$); no change in degree flatulence and bloating sensation	(202)
L. rhamnosus GG	1×10^{10} cfu as enterocoated tablet	19 female patients (Rome criteria), mean age 40 (24–60), mean duration symptoms 4.9 years (0.5–18)	2 weeks run-in, 8 weeks treatment, then 2 weeks washout, 8 weeks crossover	No significant difference probiotic versus placebo in mean symptom scores for pain, urgency, or bloating	(206)
L. plantarum 299v	5×10^7 cfu/mL in 125 mL rose-hip drink/day	12 patients (Rome criteria) either sex, aged 18–65	4 weeks, then 4 weeks switched over	No significant difference in the median symptom score between placebo and probiotic in pain severity, stool urgency, abdominal distension, wind, stool frequency, and pain frequency	(207)

(continued)

(Continued)

Probiotics	Dose/Day	Subjects	Treatment	Clinical Trials	References
	5×10^7 cfu/mL in 400 mL rose-hip drink/day	52 patients (Rome criteria): placebo: 27; probiotic: 25	4 weeks	Visual analog scale: 0 = no flatulence, 10 = worst possible; flatulence significantly reduced in test group (6.5–3.1 per day, $p = 0.004$) versus placebo (7.4–5.6 per day, $p = 0.03$); no significant difference between test and placebo in abdominal pain and defecation function	(205)
	In liquid suspension	40 patients: probiotic: 20; placebo: 20	4 weeks	95% probiotic versus 15% placebo showed symptom score improvement ($p < 0.0001$); all probiotic showed resolution of abdominal pain versus 11 placebo ($p = 0.0012$); no significant difference in normalization of stool frequency in constipated patients ($p = 0.17$)	(204)
B. animalis DN-173 010	1.25×10^{10} cfu/ pot (125 g) yogurt	274 primary care adults with constipation predominant IBS (Rome II)	6 weeks	At week 3, probiotic group Functional Digestive Disorders Quality of Life discomfort score (0–100) was higher (65.2% versus 47.7%, $p < 0.005$), bloating score lower (0.56 versus 0.31, $p = 0.03$)	(197)

L. rhamnosus GG + L. rhamnosus LC705 + B. breve Bb99 + Proprionibacterium freudenreichii ssp. Shermanii JS	8–9 × 10^9 cfu each/day in capsule	103 either sex fulfilling Rome I or II criteria (100%) (68%): probiotic: 52 (mean age 46); control (mean age 45)	6 months	Total symptom score = abdominal pain + distension + flatulence + borborygmi, 0 = absence of symptom, 4 = severe symptom; probiotic total symptom score 7.7 points lower than placebo ($p = 0.015$), or medium reduction of 42% in probiotic versus 6% in placebo; borborygmi probiotic baseline-adjusted symptom score 2.8 versus placebo score 5.0 ($p = 0.008$); no significant trend in other symptoms	(199)

(*continued*)

(Continued)

Probiotics	Dose/Day	Subjects	Treatment	Clinical Trials	References
B. longum + B. infantis + B. breve Bb99 + L. acidophilus + L. casei + L. delbrueckii spp. bulgaricus + L. plantarum + S. salivarius spp. thermophilus	Total 4.5×10^9 viable lyophilized bacteria/day, in 6 oz yogurt twice daily	48 patients (Rome II criteria) of either sex, aged 18–75: placebo: 24, probiotic: 24; colonic transit measurements using scintigraphy with ^{111}In charcoal. Symptoms assessed by a composite score of abdominal bloating, sensation of flatulence, pain, fecal urgency, stool frequency, stool consistency, ease of passage, and sense of complete evacuation	Two weeks baseline run-in, 8-week treatment	Colonic transit measured by scintigraphy with ^{111}In charcoal; symptoms assessed by composite score of abdominal bloating, sensation of flatulence, pain, fecal urgency, stool frequency, stool consistency, ease of passage, and sense of complete evacuation; probiotic flatulence 29.7 ± 2.6 versus placebo 39.5 ± 2.6 ($p = 0.011$); probiotic colonic transit geometric center 2.83 ± 0.19 versus placebo 2.27 ± 0.2 ($p = 0.05$); no difference in relief of bloating, stool-related symptoms, abdominal pain, and bloating score	(200)

General Conclusion: Some probiotic strains applied either singly or in combination appeared effective in relieving some of the IBS symptoms, such as constipation, flatulence, and borborygmi. The effects and efficacy varied widely between studies and between strains of probiotics.

4.3.5 Prevention and Treatment of Inflammatory Bowel Diseases

The chronic idiopathic inflammatory bowel diseases (IBD) such as Crohn's disease, ulcerative colitis, and pouchitis may be caused by a hyper-responsive cell-mediated immune response to intestinal commensal bacteria or microbiota aberrancies in genetically susceptible individuals (217). Therefore, it is rational to consider therapeutic approaches that correct the aberrancies in microbiota and eliminate the inflammation inducing bacteria and adjuvants for the treatment of IBD in conjunction with anti-inflammatory and immunosuppressant agents (217).

Disease	Probiotics	Dose/Day	Subjects	Treatment	Clinical Trials	References
Crohn's disease	*Saccharomyces boulardii*	1 g	32 patients (age 23–49), Crohn's disease activity index (CDAI) <150: probiotic: 16, placebo: 16	6 months with either mesalamine 1 g three time daily or mesalamine 1 g two times a day plus 1 g probiotics	37.5% placebo and 6.25% probiotic suffered clinical relapses as assessed by CDAI values ($p = 0.04$)	(213)
	L. rhamnosus GG	6×10^9 cfu/ 2.46 g bag, twice daily	45 (age 22–71) operated on for Crohn's disease: probiotic: 23, placebo: 22	1 year	Probiotics endoscopic recurrence 60% versus placebo 35.3% ($p = 0.297$); no significant differences in severity of lesions between groups	(215)
Ulcerative colitis		2×10^9 cfu	11 patients with moderate to active Crohn's disease (CDAI 150–300): probiotic: 5, placebo: 6	Received a tapering steroid regime and antibiotics for 1 week, followed by 6 months treatment	Remission defined as freedom from relapse at the 6 months follow-up. Relapse defined as increase CDAI >100; two patients in each group sustained remission. Median time to relapse 16 ± 4 weeks in probiotic versus 12 ± 4.3 weeks in placebo ($p = 0.5$)	(218)

Ulcerative colitis	*Escherichia coli* serotype O6:K5:H1	2.5×10^{10} cfu/capsule, two capsules twice daily	120 consecutive patients (age 30–60): probiotic: 57, placebo: 59	Received either 800 mg three times daily mesalazine or *E. coli*. All patients received standard medical therapy and 1 week course oral gentamycin. After remission, patients maintained on either mesalazine or *E. coli* and follow up for 12 months	75% patients on mesalazine attained remission versus 68% in *E. coli* group; mean time to remission 44 days (median 42) in mesalazine group versus 42 days (median 37) for *E. coli* group ($p = 0.0092$); 73% relapsed versus 67% in *E. coli* group ($p = 0.0059$); mean duration of remission 206 days in mesalazine group (median 175) versus 221 days (median 185) in *E. coli* group ($p = 0.0174$)	(216)
		2.5×10^{10} cfu/capsule, two capsules twice daily	120 consecutive patients: probiotic: 57, placebo: 59	Received either 500 mg three times daily mesalazine or *E. coli* for 12 weeks	No significant difference in CAI ($p = 0.12$) between the two groups; relapse rate 11.3% under mesalazine versus 16.0% under *E. coli*; comparable relapse-free time 103 ± 4 days for mesalazine versus 106 ± 5 days for *E. coli*	(214)

(*continued*)

(Continued)

Disease	Probiotics	Dose/Day	Subjects	Treatment	Clinical Trials	References
	L. casei + L. plantarum + L. acidophilus + L. delbrueckii spp. bulgaricus + B. longum + B. breve + B. infantis + S. salivarius spp. thermophilus	9×10^{11} cfu, viable lyophilized per sachet, four sachets daily in two divided oral doses	34 (age 18–65) ambulatory patients with mild to moderate active UC, not responded to mesalamine therapy	6 weeks, with continued treatment of oral or rectal mesalamine	Intent to treat analysis, total remission/response rate 77%; remission (UCDAI <2) in 53%; response (decrease in UCDAI, but final score >3) in 24%; no response in 9%; two components of probiotics (S. salivarius spp. thermophilus, B. infantis) detected by PCR/DGGE in biopsies collected from three patients in remission, but not in remaining eight patients	(209)
Pouchitis	L. casei + L. plantarum + L. acidophilus + L. delbrueckii spp. bulgaricus + B. longum + B. breve + B. infantis + S. salivarius spp. thermophilus	5×10^{11} cfu, viable lyophilized per gram, 6 g daily	40 patients in clinical and endoscopic remission of pouchitis: probiotic: 20, placebo: 20	9 months	Probiotic 15% relapsed versus 100% in placebo ($p < 0.001$); fecal concentration of Lactobacilli, Bifidobacteria, S. thermophilus increased significantly from baseline in probiotic group ($p < 0.01$)	(212)

L. casei + L. plantarum + L. acidophilus + L. delbrueckii spp. bulgaricus + B. longum + B. breve + B. infantis + S. salivarius spp. thermophilus	9×10^{11} cfu, viable lyophilized per packet	40 consecutive patients (age 18–65) underwent ileal pouch-anal anastomosis for ulcerative colitis: probiotic: 20, placebo: 20	Received treatment immediately after ileostomy closure for 1 year	19% probiotics versus 40% placebo had one episode acute pouchitis (log-rank test, $z = 2.273$; $p = 0.05$)	(210, 211)

General Conclusion: The clinical trials suggest that high number ($>5 \times 10^{10}$ cfu daily) of specific probiotic strains or specific probiotic combinations can prevent recurrent intestinal inflammation and may possibly treat active IBD, with best results in pouchitis and to a lesser extent, ulcerative colitis.

4.3.6 Treatment of *H. pylori* Infection

Helicobacter pylori infection of gastrointestinal tract was observed in 70–90% of the population in developing countries and 25–50% in developed countries (227), and it was linked to the development of gastric and duodenal ulcers, gastric cancer, atrophic gastritis, mucosa-associated lymphoma tissue lymphoma, and other gastric complications (222, 223, 233) and a Class 1 carcinogen in the list of the International Cancer Research Institute. The recommended treatment for the eradication of *H. pylori* is a combination of proton-pump inhibitor and antibiotics. The regimens achieve high eradication rate; however, there is approximately 10–35% of patients failed in the treatment. Antibiotic-associated gastrointestinal side effects such as diarrhea, nausea, vomiting, bloating, and abdominal pain are the drawbacks of anti-*H. pylori* therapies, often resulting in discontinuation of therapy (221). These manifestations are related to the changes in the intestinal microbiota profile. Several studies reported that certain probiotic bacteria exhibit inhibitory effect against *H. pylori in vitro* and *in vivo* (219, 232). The following table summarizes clinical reports on the supplementation of probiotics in *H. pylori* eradication treatment (proton pump inhibitor plus two or three antibiotics). Only randomized controlled trials are included.

Probiotics	Dose/Day	Subject No.	Treatment	Test Method	Follow-Up Period (Weeks)	Eradication Odd Ratios (95% CI)	Overall Side Effects (95% CI)	References
Bacillus clausii	2×10^9 spores TDS	P: 54, C: 52	14 days	UBT	6	1.05 (0.45–2.45)		(231)
Clostridium butyrium	40 mg TDS	P: 47, C: 50	7 days	UBT	4	2.00 (0.47–8.51)	0.34 (0.12–0.97)	(229)
Lactobacillus acidophilus LB	5×10^9 heat-killed TDS	P: 60, C: 60	10 days	UBT, H	6	2.79 (1.10–7.04)	1.00 (0.30–3.30)	(224)
L. casei DN-114 001	1×10^{10} cfu in 100 mL fermented milk	P: 39, C: 47	14 days	UBT, SAT	4	4.07 (1.43–11.58)	0.92 (0.31–2.76)	(235)
L. rhamnosus GG	6×10^9 cfu	P: 60, C: 60	14 days	UBT	6	1.22 (0.51–2.91)	0.48 (0.23–0.99)	(220)
L. rhamnosus GG, L. rhamnosus LC705, Propionibacterium freudenreichii ssp. Shermanii JS, Bifidobacterium breve Bb99	1×10^9 cfu/mL milk-base, fruit drink twice daily	P: 23, C: 24	4 weeks	UBT	16	2.76 (0.48–15.95)	0.61 (0.09–4.01)	(230)
L. rhamnosus GG, L. acidophilus, Saccharomyces boulardii	6×10^9, 5×10^9, 6×10^9 cfu	P: 64, C: 21	14 days	UBT	5–6	1.23 (0.38–3.97)	0.16 (0.05–0.46)	(226)

(continued)

(Continued)

Probiotics	Dose/Day	Subject No.	Treatment	Test Method	Follow-Up Period (Weeks)	Eradication Odd Ratios (95% CI)	Overall Side Effects (95% CI)	References
Lactobacillus + *Bifidobacterium*	5×10^9 cfu in 200 mL yogurt, BD	P: 80, C: 80	5 weeks	UBT	4–8	2.81 (1.10–7.22)		(234)
B. animalis + *L. casei*	1×10^7 cfu	P: 33, C: 32	3 months	UBT	4–12	1.39 (0.52–3.74)		(228)
Bifidobacterium longum, Faecal streptococci, L. acidophilus	One capsule	P: 64, C: 64	7 days	UBT, ELISA	4	2.63 (0.49–14.07)		(225)
Total		P: 524, C: 490				1.84 (1.34–2.54)	0.44 (0.30–0.66)	

P: probiotic; C: control; UBT: ^{13}C-urea breath test; ELISA: enzyme-linked immunosorbent assay for specific antibody; H: histology; SAT: stool antigen test; BD: twice daily; TDS: three time a day. Overall side effects: diarrhea + epigastric pain + nausea + taste disturbance.

General Conclusion: The odds ratio of the pooled *H. pylori* eradication (1.84, 95% CI 1.34–2.54) and the occurrence of side effects (0.44, 95% CI 0.30–0.66) are in favor of patients receiving probiotic supplement. Both the eradication rate and prevention of side effects of treatment appeared to be strain dependent rather than dose dependent. The *Bacillus clausii* supplement seems to have no positive effect on the anti-*H. pylori* treatment (OR 1.05) despite of a comparable daily dosage (6×10^9 cfu) being given for 14 days, while *L. rhamnosus* GG (OR 1.22), *L. rhamnosus* GG + *L. acidophilus S. boulardii* (OR 1.23), and *B. animalis* + *L. cacei* (OR 1.39) have little effect in preventing the side effects. *L. casei* DN-114 001 showed a high (OR 4.07) success rate of eradication at a high daily dosage of 1×10^{10} cfu.

4.3.7 Prevention of Postoperative Infections

Septic complications remain a major cause of morbidity and mortality in patients undergoing elective abdominal surgery despite the routine use of antibiotic prophylaxis (239, 241). In many cases the patients were immune compromised and surgical stress exacerbated immune dysfunction, causing the patients susceptible to infections. There are reports suggesting that infections in immunocompromised surgical patients often arise from their intestinal microbiota (242). If this is the case, it may be possible to reduce infection by correcting intestinal microbial imbalance induced by surgical stress. The few blinded randomized controlled studies that investigated the clinical efficacy of symbiotics (combination of probiotics and prebiotics) are summarized in the following tables.

Positive Effects

Probiotics	Dose/Day	Subjects	Treatment	Clinical Trials	References
B. breve strain Yakult, L. casei strain Shirota	1×10^8 cfu + 1×10^8 cfu/g	54 patients (average age 63) with biliary cancer undergone combined liver and extrahepatic bile duct resection with hepaticojejunostomy: probiotic: 21, placebo: 23	Enteral or oral feeding of 1 g probiotic or placebo with 12 g galactooligosaccharides for 14 days	19% infectious in symbiotic group versus 52% in controls ($p < 0.05$); postoperative changes in lactulose–mannitol ratios (assay for intestinal permeability) and serum diamine oxidase (marker of intestinal integrity) activities were identical between the two groups. Thus infections were not due to alteration of intestinal permeability and mucosal integrity.	(237)

L. plantarum 299	172 patients following major abdominal surgery or liver transplantation	Given either: (A) conventional parenteral or enteral nutrition, (B) enteral nutrition with fiber and live probiotics, or (C) nutrition with fiber and heat inactivated Lactobacilli (placebo)	After liver, gastric and pancreas resection, bacterial infection 31% in conventional group A versus 4% in probotic group B, 13% in placebo group C; In the 95 liver transplant recipients, 13% group B patients developed infection versus 48% group a, 34% group C patients; Difference between groups A and B was statistically significant in both cases; Cholangitis and pneumonia: most frequent infection; Enterococci: most frequently isolated bacteria	(240)

No Effects

Probiotics	Dose/Day	Subjects	Treatment	Clinical Trials	References
L. plantarum 299v	5×10^7 cfu/mL oatmeal based drink, 500 mL daily	129 patients (age 58–77) undergoing elective major abdominal surgery: probiotic: 64, placebo: 65	Given treatment at least 1 week preoperative and in postoperative period	No significant difference between the two groups in bacterial translocation (12% versus 12%, $p = 0.82$), gastric colonization with enteric organisms (11% versus 17%, $p = 0.42$), or septic morbidity (135% versus 15%, $p = 0.74$)	(238)
L. acidophilus La5 + *L. bulgaricus* + *B. Lactis* Bb-12 + *S. thermophilus*	4×10^9 cfu/capsule	137 patients (age 47–80) undergoing elective abdominal surgery: probiotic: 72, placebo: 65	Given either placebo or probiotics with oligofructose for 2 weeks preoperative and continue in postoperative until discharge	No significant difference between the two groups in bacterial translocation (12.1% versus 10.7%, $p = 0.808$), gastric colonization (41% versus 44%, $p = 0.719$), systemic inflammation, or septic complications (32% versus 31%, $p = 0.882$)	(236)

General Conclusion: The few clinical reports on the effects of probiotics in the prevention of postoperative infection are conflicting and a further evaluation is required.

4.3.8 Prevention and Treatment of Respiratory Tract Infections

Common virus respiratory infections such as common cold and influenza continue to cause considerable health and economic burden in children and adults alike. Several strains of probiotic microorganisms have been reported to enhance mucosal immunity against pathogens. It is reasonable to assume that probiotics may have positive effects in the prevention and treatment of respiratory tract infections.

Probiotics	Disease	Subjects	Dose	Treatment	Clinical Trials	References
Enterococcus faecalis and autolysate (Symbioflor 1)	Recurrence hypertrophic sinusitis	157 chronic patients: probiotic: 78, placebo: 79	30 drops Symbioflor three times a day	6 months treatment, 8 months follow-up	During treatment, occurrence of relapses: 17 incidents in probiotic group versus 33 in placebo ($p = 0.019$); during follow-up: probiotic 33 incidents versus placebo 57 ($p = 0.013$); time interval to first relapse in probiotic group 513 days versus 311 days in placebo; severity and clinical parameters of the acute relapses was comparable in both group	(246)
	Relapse of chronic recurrent bronchitis	136 chronic patients: probiotic: 70, placebo: 66	30 drops Symbioflor three times a day	6 months treatment, 8 months follow-up	During treatment, occurrence of relapses: 12 incidents in probiotic group versus 27 in placebo ($p = 0.013$); during follow-up: 27 in probiotic versus 39 in placebo ($p = 0.127$); time interval to first relapse: 699 days in probiotic versus 334 days in placebo; at end of observation, 91% probiotic experienced 1 relapse versus 62% in placebo ($p = 0.01$); four patients in probiotic group required antibiotic versus 13 in placebo	(245)

Probiotic	Condition	Subjects	Dose	Duration	Results	Ref.
L. casei DN-114001, yogurt	Winter respiratory infection	180 elderly		3 weeks	No difference in incidence of infections; duration infections: 7.0 ± 3.2 days probiotic group versus 8.7 ± 3.7 days in placebo group ($p = 0.024$); maximum temperature of probiotic group $38.3 \pm 0.5°C$ versus control group $38.5 \pm 0.6°C$ ($p = 0.01$)	(248)
L. rhamnosus GG	Respiratory infection	571 healthy children, (1–6 years): probiotic: 282, placebo: 289	$5–10 \times 10^5$ cfu/mL milk, 260 mL daily, three times feeding a day, 5 days a week. Mean compliance 60%	7 months	A relative 17% (17.2–0.1%) reduction in number respiratory infection ($p = 0.05$); age adjusted odds ratio 0.75 ($p = 0.13$); a relative 19% reduction in antibiotic treatments for respiratory infection ($p = 0.03$); adjusted odds ratio 0.72 ($p = 0.08$)	(247)
L. gasseri PA16/8, B. longum SP07/03, B. bifidum MF20/5	Common cold	454 healthy adults (18–67 years) not vaccinated	5×0^5 cfu spray dried/ tablet/day	2 Winter/ Spring (3 and 5 months each)	No effect on incidence of common cold (placebo 153 episodes versus probiotic 158); shorten duration of episode by almost 2 days (placebo 8.9 ± 1.0 versus probiotic 7.0 ± 0.5 days, $p = 0.045$); days with fever placebo 1.0 ± 0.3 versus probiotic 0.24 ± 0.1 ($p = 0.017$); increased in cytotoxic T cells plus T suppressor cells in 14 days (placebo $19 \pm 19 \times 10^6 L^{-1}$ blood versus probiotic $64 \pm 15 \times 10^6 L^{-1}$ blood, $p = 0.035$)	(243, 244)

General Conclusion: The intake of probiotics in general, reduced the incidents and duration of respiratory infections among children and adults.

4.3.9 Prevention and Treatment of Allergic Diseases

Allergy, in the form of atopic diseases, such as atopic eczema, allergic rhinitis, and asthma, is a chronic disorder of increasing prevalence in the more developed countries (251). A survey conducted in 1998 in Bangkok area, Thailand as compared to a similar survey in 1990 showed that the period prevalence of asthma increased fourfold, allergic rhinitis increased nearly threefolds, whereas eczema remained stable (271). The 1998 survey of 3628, 6–7 years old and 3713, 13–14 years children reported that the cumulative 12-month period prevalence of wheezing was between 18.3% and 12.7%; rhinitis 44.2% and 39.7%; and eczema 15.4% and 14.0%, respectively. Surveys conducted in 1994–1995 in various regions of Finland reported that among the children of ages 13–14, 10–20% of the children had symptoms of asthma, 15–23% allergic rhinitis and 15–19% atopic eczema (262, 265). There appears to be an inverse association between atopy and infection in early life, exposure to farm life, animals, and elder siblings, which implies that the recent rapid rise in atopy might be a result of improved hygiene (260, 267–269). Other reports, however, did not support the "Hygiene Hypothesis" (261) or indicated variation in different time points of near history (253). There was a proposal that specific microbes in the commensal gut microbiota are more important than sporadic infections in atopic disease prevention and that atopic diseases can be predicted by specific intestinal microbiota aberrancies in infants prior to the occurrence of atopic symptoms or atopic disease (254). Based on the microbiota aberrancies and the hygiene hypothesis several treatment and prevention studies have been conducted. The following tables summarize human clinical trials in the investigation of probiotic consumption and prevention of atopic diseases. Only double-blind randomized placebo-controlled trials are included.

Positive Effects

Probiotics	Dose/Day	Subjects	Treatment	Clinical Trials	References
B. lactis Bb-12 or *L. rhamnosus* GG (ATCC53103)	*B. lactis* 1 × 10⁹ cfu/g whey formula, LGG 3 × 10⁸ cfu/g whey formula	27 infants (mean age 4.6 months) with atopic eczema during exclusive breast feeding	Weaned to probiotic supplemented hydrolyzed whey formulas or without probiotics	Median (interquartile range) SCORAD score 16 (7–25) during breast feeding; after 2 months, median SCORAD in *Bifidobacterium* group 0 (0–3.8), in *Lactobacillus* group 1 (0.1–8.7), in unsupplemented group 13.4 (4.5–18.2) ($p = 0.01$); In parallel a reduction in soluble CD4 in serum ($p = 0.005$) and eosinophilic protein X in urine (*Bifidobacterium* $p = 0.01$, *Lactobacillus* $p = 0.04$), suggesting alleviation of allergic inflammation; at 2 months, serum TGF-β1 decreased in *Bifidobacterium* group ($p = 0.04$) but tended to increase with *Lactobacillus* ($p = 0.07$); serum IL-1ra, TNF-α, GM-CSF, sICAM-1, RANTES, MCP-1α not modified by probiotics; after 6 months the median SCORAD score 0 (0–6.6) in all groups alike	(252)

(*continued*)

(Continued)

Probiotics	Dose/Day	Subjects	Treatment	Clinical Trials	References
L. rhamnosus GG	1×10^{10} cfu/capsule	159 mothers with family history of atopic disease	Received two capsules daily for 2–4 weeks before expected delivery; after delivery breastfeeding mothers or bottle feeding children took probiotic for 6 months	At 2 years, frequency of atopic eczema probiotics group 23% versus placebo 46% ($p = 0.008$), relative risk (95% CI) 0.51 (0.32–0.84); Preventive effects not depend on mode of administration, in probiotic group atopic eczema diagnosed in 25% infants who consumed probiotic themselves and 21% whose breastfeeding mothers took capsules ($p = 0.74$); Concentration of total IgE, frequencies of increased antigen-specific IgE concentration and positive reaction in skin-prick tests were similar between two groups	(255)
		Probiotic: 53, placebo: 54	Follow-up of Kalliomaki et al., 2001	At 4 years, 26.4% *Lactobacillus* group versus 46.3% placebo developed atopic eczema; relative risk 0.57, 95% CI (0.33–0.97); Skin prick test reactivity same in both group	(255, 256)

	Probiotic: 53, placebo: 62	Follow-up of Kalliomaki et al., 2001	At 7 years, 42.6% *Lactobacillus* group versus 66.1% placebo developed atopic eczema; relative risk 0.64, 95% CI (0.45–0.92)	(255, 257)
3×10^8 cfu/g whey formula	35 infants (mean age 5.5 months) with atopic eczema and cow's milk allergy (CMA)	Received hydrolyzed whey formula supplemented with viable or heat-inactivated probiotic	SCORAD scores (interquartile range) decreased from 13 (range 4–29) to 8 (0–29) units in placebo group; from 19 (4–47) to 5 (0–18) units in viable probiotic group; from 15 (0–29) to 7 (0–26) units in heat-inactivated probiotic group; Heat-inactivated probiotic group showed adverse gastrointestinal symptoms and diarrhea	(258)
L. rhamnosus 19070-2 or *L. reuteri* DSM12246	Lyophilized 10^{10} cfu each/g	41 children (median age 4 years) with moderate (SCORAD score 16–40) and severe (SCORAD score >40) atopic dermatitis(AD)	Orally fed 1 g each strain or placebo in 2.5–5 mL water twice daily for 6 weeks, 6-week washout between each intervention period	Frequency of gastrointestinal symptoms (diarrhea, vomiting, abdominal pain) placebo 39% versus 10% probiotic treatment ($p = 0.02$); positive association between lactulose:mannitol ratio (small intestinal permeability) and eczema severity ($r = 0.61$, $p = 0.02$ after placebo; $r = 0.53$, $p = 0.05$ after probiotic treatment)	(266)

(*continued*)

(Continued)

Probiotics	Dose/Day	Subjects	Treatment	Clinical Trials	References
L. rhamnosus GG (ATCC53103) or LGG 5×10^9 cfu + *L. rhamnosus* LC705 5×10^9 cfu + *Bifidobacterium breve* Bb99 2×10^8 cfu + *Propionibacterium freudenreichii* ssp. Shermanii JS 2×10^9 cfu/capsule	LGG 5×10^9 cfu/capsule	230 infant (mean age 6.4 months) with suspected CMA: LGG: 80, mixture: 76, placebo: 74	One capsule content mixed with food twice daily for 4 weeks	Four weeks after treatment, CMA was diagnosed in 120 infants. No differences in SCORAD index and CMA between groups; in IgE-sensitized infants, LGG group SCORAD 26.1 versus placebo 19.8 from baseline ($p = 0.036$)	(272)
L. fermentum VRI-033 PCC	1×10^9 cfu	53 children (aged 6–18 months) with moderate or severe AD	Twice daily for 8 weeks; a final assessment at 16 weeks	Reduction in SCORAD index was significant in probiotic group ($p = 0.03$) but not placebo group; significantly more children receiving probiotics (92%) had a SCORAD index better than baseline versus placebo group (63%); at completion, more children in probiotic group had mild AD (54%) versus placebo group (30%)	(273)
B. longum BB536	5×10^{10} cfu/2 g dextrin powder	44 adults with clinical history (>2 years, and serum JCP-specific IgE) of Japanese cedar pollinosis (JCPsis)	2 g twice daily with 100 mL milk for 13 weeks during pollen season	Subjective symptom scores indicated significant decreased in rhinorrhea ($p = 0.0167$), nasal blockage ($p = 0.0118$), and composite scores	(274)

Probiotic	Dose	Subjects	Protocol	Results	Ref.
				($p = 0.0339$) in probiotic group versus placebo; no significant difference in scores for sneezing, nasal itching, eye, and throat symptoms; probiotic significantly suppressed increases in plasma thymus- and activation-regulated chemokine (a Th2 marker) and tended to suppress elevation of JCP-specific IgE ($p = 0.067$)	
L. rhamnosus GG (ATCC 53103) 5×10^9 cfu + *L. rhamnosus* LC705 (DSM 7061) 5×10^9 cfu + *B. breve* Bb99 (DSM 13692) 2×10^8 cfu + *Propionibacterium freudenreichii* spp. Shermanii JS (DSM 7076) 2×10^9 cfu/capsule	One capsule	1223 pregnant mothers carrying children in risk for allergy; Infants: probiotic 461, placebo 464	Mothers took one capsule twice daily during 2–4 weeks before delivery. Their newborn infants received one opened probiotic capsule mixed with 20 drops sugar syrup + 0.8 g galactooligosaccharides once daily for 6 months after birth; The placebo group received microcrystalline cellulose without galactooligosaccharides	At 2 years, no effect on the cumulative incidence of allergic diseases (food allergy, eczema + asthma + allergic rhinitis) between the groups, but probiotic group tended to reduce IgE-associated (atopic) diseases (odds ratio, 0.71; 95% CI 0.50–1.00; $p = 0.052$); probiotic treatment reduced eczema (OR 0.74; 95% CI 0.55–0.98; $p = 0.035$) and atopic eczema (OR 0.66; 95% CI 0.46–0.95; $p = 0.025$); Lactobacilli and Bifidobacteria more frequently ($p = 0.001$) colonized guts of supplemented infants	(259)

No Effects

Probiotics	Dose/Day	Subjects	Treatment	Clinical Trials	References
L. rhamnosus GG (ATCC 53103)	55×10^9 cfu/capsule	36 young adults and teenagers allergic to birch pollen: probiotic 18, placebo 18	Two capsules twice daily for 5.5 months (2.5 months before pollen season, 1 month during season, 2 months after)	Probiotic treatment did not alleviate symptoms (nasal, eye, lung) and skin prick tests or reduce their use of medication during the birch-pollen season or subsequent 2 months; Treatment did not significantly affect symptoms caused by apple in oral challenge tests (60 g slice chewed 1 min and swallowed)	(250)
L. rhamnosus or L. rhamnosus GG	3×10^8 cfu/g hydrolyzed whey formula (ca. 5×10^9 cfu/100 mL formula)	50 infants below 5 months fulfilling Hanifin criteria for AD and diagnosed with CMA: L. rhamnosus 17, LGG 16, placebo 17	4–6 weeks whey formula, followed by treatment for 3 months	No statistically significant effect of probiotic on SCORAD, sensitization (total IgE and food-specific IgEs, skin prick testing for cow's milk), inflammatory parameters (blood eosinophils, eosinophil protein X in urine, fecal α-1-antitrypsin) and cytokine production (IL-4, IL-5, IFN-δ production by peripheral blood mononuclear cells after polyclonal stimulation)	(249)

L. acidophilus LAVRI-A1 (also identified as LAFTI-L10)	3×10^9 cfu	177 newborns of women with allergy: probiotic 89, placebo 88	6 months	Probiotic group showed significantly higher rates of Lactobacilli colonization ($p = 0.039$); at 6 months, AD rates similar in probiotic (25.8%) versus placebo (22.7%), $p = 0.629$; no difference in AD rates at 12 months, but proportion of children with allergen skin prick test + AD significantly higher in probiotic group ($p = 0.045$); at 12 months, rate of sensitization significantly higher in probiotic group ($p = 0.03$); The presence of culturable Lactobacilli in stools at 6 months associated with increased risk of subsequent cow's milk sensitization ($p = 0.012$)	(270)

General Conclusion: The prevention studies suggest that there is a clear effect on specific strains to prevent atopic eczema in children with intestinal and food involvement. Also positive demonstrations have been reported for specific probiotics in treatment of atopic eczema in infants. The studies, which showed positive effect of probiotics suggest that impairment of intestinal mucosal barrier is involved in pathogenesis of atopic diseases, and probiotics stabilize intestinal barrier function and decrease gastrointestinal symptom in children with atopic diseases. In addition, immunomodulation may be involved in the probiotic effects as noted, for example, in the TGF-β levels in breast milk of probiotic-treated mothers (264). Supplementation of infant formula with viable but not heat-inactivated probiotics is beneficial to young children with suffering from atopic diseases with intestinal involvement. Probiotics were equally effective among children receiving probiotics in whey/milk formula, in water suspension, and through mothers' milk where mothers took probiotic-containing capsules. Probiotics have shown some efficacy in relieving or delaying IgE-mediated type 1 allergy to pollen.

Other studies using the same strains of probiotics; however, could not confirm the impact of these probiotics on the symptoms of infant allergic diseases and pollen allergy in adults. The differences could not be explained by dosage of probiotics, temporary intestinal colonization of probiotics, age of children, disease history, and length of study.

Why are there differences in the study results with different probiotics?

A study suggested that early probiotic supplementation was associated with increased allergen sensitization. However, this study was conducted using a previously uncharacterized probiotics *L. acidophilus* LAVRI. This has later been characterized as *L. acidophilus* LAFTI-L10. Additionally, the preparation was given postnatally and the previous prevention studies have all used prenatal and postnatal administration of the probiotics. Another recent studies (264a, 264b) clearly demonstrated protection against sensitization when a combination of well-characterized probiotics (*Lactobacillus rhamnosus* GG and *Bifidobacterium lactis* Bb12) were used. Other sources of differences in clinical efficacy may include probiotic factors (strains, dose, viability, properties, prenatal/postnatal administration), host/mother factors (genetics, type of microbiota, colonization, health status, compliance), and environment factors (mode of delivery, microbiotic composition, transfer of microbiota, breast feeding, feeding regimens, antibiotics, immunomodulatory drugs). The allergy studies are ongoing but not yet completed. The reasons for differences in clinical efficacy have been reviewed by Prescott and Björksten (263).

4.3.10 Antitumor Effects

There are numerous claims on the association of intestinal microecology and cancers (278, 280, 282). There are also several mechanistic studies on the effects of probiotics binding to carcinogens and reducing the mutagenicity of urine and intestinal contents, and their inhibitory properties on tumor in *in vitro* and in animal models (277, 279, 281). There is, however, few randomized controlled clinical trials and epidemiological studies demonstrating the antitumor effects of probiotics in human, these are summarized in the following table.

Cancer	Probiotics	Dose/Day	Subjects	Clinical Trials	References
Superficial bladder cancer	L. casei Shirota	1×10^{10} cfu/g, three times daily	58 patients with superficial bladder cancer (age 50–69); 16 primary cases with multiple lesions (6 in probiotic group/10 in placebo), 9 recurrent cases with single lesion (5/4), 23 recurrent cases with multiple lesions (12/11); after transurethral resection of bladder tumor (TUR-Bt), subjects given oral doses for a year	50% recurrence-free interval after TUR-Bt determined by Kaplan–Meier method, was longer in probiotic treatment (350 days) versus placebo (195 days) (long rank test $p = 0.03$); no adverse side effect	(275)
		1×10^{10} cfu/g, three times daily	138 patients (under 80 years) had transitional cell carcinoma of bladder; after transurethral resection of tumor, patients were divided into: group A multiple lesions without history of recurrent; group B single lesion with a history of recurrence; group C multiple lesions with a history of recurrence; patients received treatment for a year	50% recurrence-free interval 688 days in probiotic group versus 543 days placebo group; risk of tumor recurrence 2.58-fold higher in placebo than probiotic group in Cox's multiple life table analysis; Comparison of recurrence-free curve showed a shorter recurrence-free period in group C compare to groups A and B; mild transient adverse reactions, including diarrhea, constipation, hepatic transaminases, occurred in three patients in placebo group and three in probiotic group	(276)

(*continued*)

(Continued)

Cancer	Probiotics	Dose/Day	Subjects	Clinical Trials	References
		1×10^{10} cfu/bottle	180 cases (mean age 67 ± 10) and 445 population-based controls matched by gender and age	Compare to population-based controls, odds ratio (smoking adjusted) of probiotic group drank for previous 10–15 years were 0.46 (0.25–0.84, $p = 0.01$) for 1–2 times/week and 0.65 (0.38–1.13, $p = 0.13$) for 3–4 or more times/week	(277)
Breast cancer	Yogurt (probiotic strains not indicated)		Population-based case-control (for age, education, age at menarche, age at first pregnancy), white female patients 172 and controls 190, age 20–54 years	Yogurt reduced odds ratio in both premenopausal (OR = 0.4, CI = 0.1–1.1 for highest versus lowest quartile) and postmenopausal women (OR = 0.2, CI = 0.0–0.8)	(284)
	Yogurt and buttermilk (probiotic strains not indicated)		Case-control Caucasian women (age 25–44 or 55–64 years); 133 incident breast cancer cases versus 289 population control	Significantly lower consumption of fermented milk products (yogurt and buttermilk) among breast cancer cases versus population control (116 ± 100 versus 157 ± 144 g/day, $p < 0.01$); age adjusted odds ratio of daily consumption of 1.5 glasses (>225 g) fermented milk versus none was 0.5 (95% confidence interval 0.23–1.08); odds ratio expressed per 225 g fermented milk was 0.63 (multivariate-adjusted 95% confidence interval 0.41–0.96); daily intake of milk showed no significant differences between cases and controls	(285)

General Conclusion: The few clinical studies do provide evidences that dietary probiotics interact with the host and possibly with the intestinal microbiota and dietary content to exert protective effects in the etiology of some cancers such as superficial bladder cancer and breast cancer.

4.3.11 Reduction of Serum Cholesterol

Hypercholesterol is a risk factor for cardiovascular diseases, which is a leading cause of death in many countries (294). A 1% reduction in serum cholesterol is estimated to result in 2–3% reduction in the risk of coronary artery disease (297). It is suggested that intestinal Lactobacilli may reduce serum cholesterol level through bacterial assimilation in the intestine (289) and deconjugation of bile salts (288, 293). Short-chain fatty acids produced by Lactobacilli may also inhibit hepatic cholesterol synthesis and distribution of cholesterol in the plasma and liver (298). Human clinical studies using various probiotics have given mixed results and do not allow any general conclusion.

Positive Effects

Probiotics	Dose	Subjects	Treatment	Clinical Trials	References
L. acidophilus		30 healthy men (age 33–64 years), mean serum total cholesterol 5.23 ± 1.03 mmol/L	Three times daily 125 mL milk products + 2.5% fructooligosaccharides; 2-way crossover with 3 weeks treatment period and 1 week wash-out in between	Probiotic milk significantly lower serum total cholesterol 4.4% ($p < 0.001$), LDL cholesterol 5.4% ($p < 0.005$), LDL/HDL ratio 5.3% ($p < 0.05$); Serum HDL cholesterol, triglycerides, blood glucose unchanged	(300)
L. acidophilus 145 + *B. longum* 913	*L. acidophilus* 1×10^6–10^8 cfu; *B. longum* 1×10^5 cfu/g yogurt	29 healthy women (age 19–56 years), 15 normocholesterolaemic, 14 hypercholesterolaemic	300 g yogurt + 1% oligofructose daily, crossover study 7 weeks each	No difference in total and LDL cholesterol; HDL increased significantly by 0.3 mmol/L ($p = 0.002$); LDL/HDL decreased from 3.24 to 2.48 ($p = 0.001$)	(293)
Enterococcus faecium M-74	2×10^9 cfu/capsule	43 volunteers: probiotic 20, mean age 75.4 ± 1.5 years; placebo 18, mean age 78.1 ± 1.7 years	1 capsule + 50 mg selenium for 60 weeks daily	Probiotic reduced total cholesterol by 12%, achieved by fall in LDL cholesterol (3.85 ± 0.27 week 0 versus 3.09 ± 0.21 mmol/L week 56); no significant change in HDL and triglycerides	(291)

E. faecium + S. termophilus	1×10^8–10^{11} cfu/L	58 healthy, nonobese, normocholesterolaemic males (age 44 years)	200 mL daily fermented milk for 6 weeks	Total cholesterol reduced significantly in probiotic group (−0.37 mmol/L, CI −0.51 to −0.23), no change in placebo group (−0.02 mmol/L) ($p = 0.01$); LDL cholesterol reduced by 0.42 mmol/L (10%); HDL and triglyceride unchanged in both group	(286)

No Effects

Probiotics	Dose	Subjects	Treatment	Clinical Trials	References
L. acidophilus L-1	5×10^9 cfu/L	78 volunteer with normal to borderline high cholesterol levels (5.4 ± 0.7 mmol/L)	500 mL daily control yogurt for 2 weeks, followed by 500 mL daily treatment for 6 weeks	No significant difference in serum total cholesterol, LDL, HDL and triglycerol level	(287)
L. acidophilus LA-1 freeze-dried, reduce cholesterol *in vitro*	3×10^{10} cfu/capsule	80 volunteers (mean age 46.5 years) with elevated cholesterol	Two capsules three times daily for 6-week, followed by 6 weeks washout, crossover for 6 weeks on capsules	No change in serum total cholesterol, HDL, LDL and triglycerides from baseline	(295)
L. acidophilus + L. bulgaricus	2×10^6 cfu/tablet	354 nonfasting volunteers	One tablet four times daily for 6 weeks, 3 weeks washout, crossover for 6 weeks	No significant difference in total lipoprotein, LDL and HDL concentration in both groups	(296)
L. acidophilus DDS Plus + B. longum	3×10^8 cfu/capsule	37 postmenopausal women with total cholesterol ≥ 5 mmol/L	3 capsules daily + 15–30 mg fructooligosaccharide in milk or soy protein isolate for 6 weeks	No significant difference in the reduction of total cholesterol, HDL, LDL, and triglycerides between the groups	(290)

L. fermentum PCC	2×10^9 cfu/capsule	46 volunteers with total cholesterol ≥ 4 mmol/L: probiotic 23, age 50 ± 12 years; placebo 21, age 53 ± 11 years	Two capsules twice daily for 10 weeks	No significant difference in the reduction of LDL cholesterol (probiotic 7.0% versus placebo 5.2%, $p > 0.05$), total cholesterol, HDL % triglycerides, liver enzymes	(301)
E. faecium + S. thermophilus	1×10^8–10^{11} cfu/L	87 nonobese, normocholesterolemic volunteers (age 50–70 years)	200 mL daily for 6 months	Reduction of LDL by 0.21 mmol/L in probiotic group ($p < 0.02$) after 1 month and 0.32 mmol/L after 3 months ($p < 0.001$); at 6 months, no difference in LDL, HDL, triglyceride in both group	(299)

4.3.12 Enhancement of Vaccine Responses

The evidence on the immunomodulatory impact of probiotics on the host is accumulating rapidly. Probiotics are increasingly administered in infancy, raising the question of their effect on the immunological responses to vaccines.

Probiotics	Dose/Day	Vaccine	Subjects	Treatment	Clinical Trials	References
L. rhamnosus GG (ATCC 53103)	5×10^{10} cfu/0.1 g dry powder in 5 mL water	Rhesus-human reassortant rotavirus vaccine strain DXRRV, corresponding to human rotavirus VP7 serotype 1	2–5 months old healthy infants: probiotic 28, placebo 27	5 mL given orally before vaccination; administration continued at home for 5 days, two doses daily	Rotavirus IgM seroconversion detected in 96% probiotic group versus 85% placebo ($p = 0.15$); rotavirus IgA seroconversion detected in 93% probiotic group versus 74% placebo ($p = 0.05$)	(304)
L. rhamnosus GG or L. acidophilus CRL431	10^{10} cfu in 100 mL fermented milk	Live attenuated poliomyelitis viruses type 1 strain LSc_1, type 2 strain P2712 and type 3 strain Leon $12a_1b$	64 healthy male volunteers (aged 20–30 years): LGG 21, L. acidophilus 21, placebo 22	5 weeks, in second week, subjects were vaccinated	Probiotics increased poliovirus neutralizing antibody titers by 3.9 ($p = 0.047$), 2.2 ($p = 0.083$) and 3.9 ($p = 0.036$) folds, respectively, for IgM, IgG and IgA; no consistent difference noted between probiotic strains	(303)

(continued)

(Continued)

Probiotics	Dose/Day	Vaccine	Subjects	Treatment	Clinical Trials	References
L. rhamnosus GG (ATCC 53103) 5×10^9 cfu + L. rhamnosus LC705 5×10^9 cfu + B. breve Bbi99 2×10^8 cfu + Propionibacterium freudenreichii spp. Shermanii JS 2×10^9 cfu/capsule	One capsule	DTwP vaccine: diphtheria toxoid, tetanus toxoid, inactivated Bordetella pertussis; Hib conjugate vaccine: Haemophilus influenzae type b, tetanus toxoid, Corynebacterium diphtheria toxin CRM_{197}	1223 allergy high risk infants: probiotic 32, placebo 29	4 weeks before expected delivery, mothers took one capsule twice daily; Their infants received one opened capsule in 20 drops sugar syrup containing 0.8 g galactooligosaccharides once daily for 6 months; The infants received a DTwP vaccine at 3, 4, and 5 months, and a Hib conjugate at 4 months	Protective Hib antibody concentrations (>1 μg/mL) occurred more frequently in probiotic group (50%) than placebo (21%), $p = 0.02$; Geometric mean (interquartile range) Hib IgG tended to be higher in probiotic group, 0.75 (0.15–2.71) μg/mL than placebo, 0.40 (0.15–0.92) μg/mL ($p = 0.064$); diphtheria IgG in probiotic 0.38 (0.14–0.78) versus placebo 0.47 (0.19–1.40) IU/mL ($p = 0.449$); tetanus IgG in probiotic 1.01 (0.47–1.49) versus placebo 0.81 (0.56–1.39) IU/mL ($p = 0.310$)	(305)

General Conclusion: There is no evidence that suggests impairment of vaccine responses. The probiotics may improve response to some of the vaccines, such as *Haemophilus influenzae*, rotavirus, and polio vaccines.

4.4 EFFECTS ON FARM ANIMALS

ALLAN LIM AND HAI-MENG TAN
Kemin Industries (Asia) Pte Ltd Singapore, Singapore

Probiotics are useful in the treatment of disturbed intestinal microbiota and increased gut permeability, which are characteristics of many intestinal disorders. Such bacteria are able to survive gastric conditions to colonize the intestine, at least temporarily, by adhering to the epithelium. They have been reported to improve the growth rate and feed utilization in pigs, chicken, and calves, and to improve their feed conversion ratio. There is a significant decrease in the occurrence of diarrhea as observed in pigs and calves fed with probiotics (352). Probiotics are also believed to neutralize the effect of enterotoxin from *E. coli*, which is pathogenic for pigs (326). The beneficial effects of probiotics ranged from displacement of harmful bacteria, such as *Clostridium perfringens* to reduction of bacterial urease activity, to synthesis of vitamins, stimulatory effects on the immune system, maintenance of a healthy normal balanced microbiota (eubiosis), and contribution to digestion (339, 358, 391–393).

The positive effects of probiotics have been harnessed by humans since biblical times when fermented milk and dairy products were widely consumed by different cultures. In contrast, probiotics were only introduced to farm animals in the last two decades. As might be expected, most of the effects of probiotics in animals are associated with improved performance and to a lesser extent improved immunity against infections. The interest in using probiotics was also accelerated by the movement in developed countries especially the European Union to ban antibiotic growth promoters (AGP) from animal diets. The use of probiotics in place of growth promoting antibiotics has also been viewed by many leading animal meat producers as a major differentiation factor in highly sophisticated markets such as Japan.

The effects of probiotics on animal performance can be influenced by the variability in farm practices, species, age, method of application, strains of microorganisms, and diet. It is now widely accepted that probiotics can improve animal performance through competitive exclusion with pathogens in the digestive systems, and that animals generally benefit from probiotic microorganisms isolated from their own digestive tracts (352, 401). There is also an evidence to suggest that probiotics compete with pathogens for specific receptors at the intestinal surfaces. In many cases, the propensity for pathogens to adhere to the intestinal surface is subjected to the dynamic microbial–host interactions in the gut, which can be further modulated by species, genotype, and age. The effects of probiotics on the gut immune system has been the subject of intense research in recent years. It is believed that probiotic supplementation to young animals will accelerate the maturation of the gut immune system resulting in lower morbidity and mortality. As the use of synbiotics, which is a combination of pre- and probiotics, is gaining scientific credibility as functional food ingredients at nutritional and therapeutic levels (355, 377), there is also an increasing evidence to show its effectiveness in helping young animals to achieve a better growth performance (371, 332).

4.4.1 Poultry

Due to the higher sensitivity of poultry production to profitability, the impact of the ban on growth promoting antibiotics in Europe with effect from 1 January 2006 has a profound impact on poultry farming. Many alternatives to AGP have been evaluated in poultry production with mixed results. One of the more successful probiotic bacteria used in poultry production is *Bacillus subtilis*, from which proprietary strains have been offered (e.g., under the CloSTAT™ trade name (Kemin Industries, Inc., USA)). Apart from improving the growth benefits, *B. subtilis* is also known to inhibit the growth of pathogens in the digestive tract of chickens, which can lead to a considerable economic loss. The unique ability of a patented *B. subtilis* strain PB6 to kill *Clostridium perfringens*, *Campylobacter jejuni*, and *Streptococcus pneumoniae* has been demonstrated (354, 391) (Figs. (4.3–4.5). The spore-forming property of *B. subtilis* is also a huge advantage in their survival during pellet formation, which is now widely used for broiler feed production. Other commercial probiotics contain either a single strain of *Enterococcus faecium* (Protexin® (Protexin, UK)) or a combination of strains from *Lactobacillus*, *Streptococcus*, and *Enterococcus* (Lactina, Lactina Ltd., Bulgaria). As the protective effect of probiotics on poultry is believed to be by competitive exclusion of pathogens, the adhesion of probiotics to the gastrointestinal tracts of poultry is a key performance parameter. Extensive studies have indicated that lactic acid bacteria (344, 322, 352), *B. subtilis* (391), and *Enterocuccus faecium* (363) have strong ability to colonize the intestines of animals. Many beneficial effects of probiotics are suggested such as improved immune system, modification of gut microbiota, reduced inflammatory reactions, decreased ammonia and urea

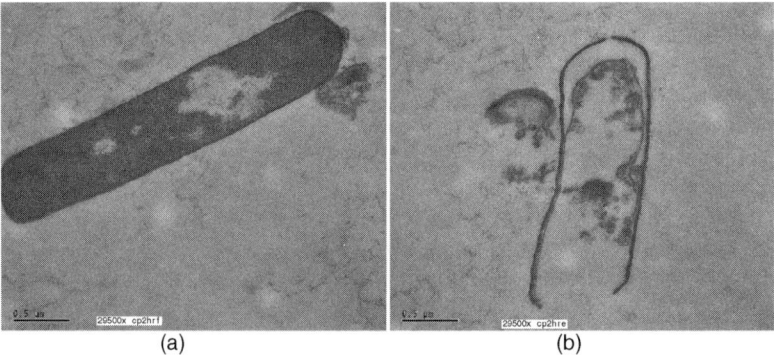

FIGURE 4.3 The effect of *B. subtilis* PB6 (CloSTAT) on *Clostridium perfringens*: (a) intact cell of *C. perfringens* and (b) rupture and death of the *C. perfringens* cell after 4 h of exposure to *B. subtilis* PB6 (CloSTAT) (transmission electron micrograph magnification: 29,000×).

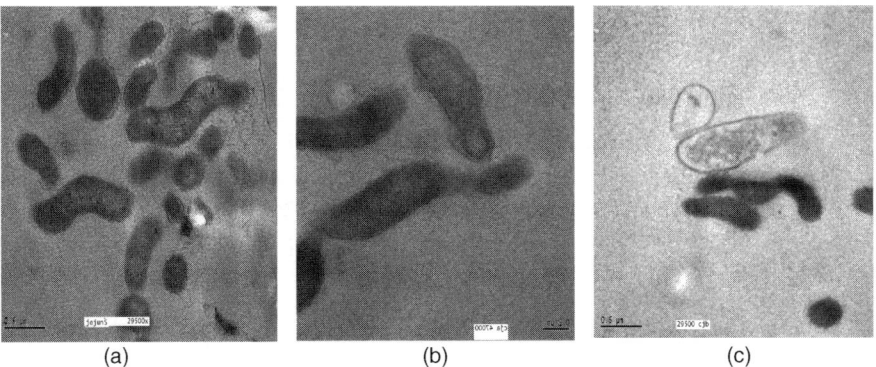

FIGURE 4.4 The effect of *B. subtilis* PB6 (CloSTAT) on *Campylobacter jejuni* at 37°C: (a) normal cells (magnification 29,500×), (b) pertubation of cell wall after 1 h (magnification 47,000×), and (c) rupture and cell death after 4 h (magnification 47,000×).

excretion, lower serum cholesterol, improved mineral adsorption (393, 331, 372) that may have an indirect positive impact on the profitability of poultry production. There are other studies that document the tendency of probiotics to improve the growth performance of layers and egg quality. Table 4.1 summarizes the effects of probiotics on growth and other aspects of poultry production.

4.4.2 Swine

Pigs in the postweaning period are most vulnerable to enteric bacterial diseases, and are therefore subject to extensive antibiotic treatments to reduce mortality and morbidity. With the ban on growth promoting antibiotics in EU and more stringent use of therapeutic antibiotics, the use of probiotics to improve growth performance and resistance against bacterial diseases in pigs has been widely studied. The most notable project was the HEALTHYPIGUT sponsored by the EU between 2001 and 2004, which allowed scientists from nine research institutes to evaluate alternatives to

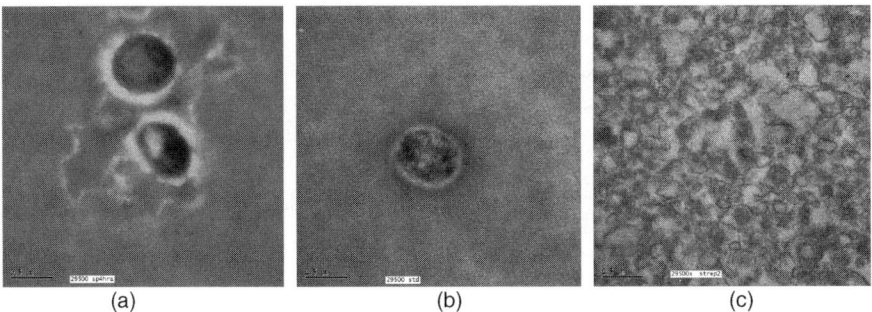

FIGURE 4.5 The effect of *B. subtilis* PB6 (CloSTAT) on *Streptococcus pneumoniae* at 37°C: (a) normal cells, (b) and (c) rupture and cell death after 4 h (magnification 29,500×).

TABLE 4.1 Effects of Probiotics on Poultry

Probiotics	Dosage	Animals	Growth Effects	Other Effects	References
B. subtilis	0.25×10^6, 0.5×10^6, and 1×10^6 cfu/g feed	Turkeys	Significantly increased ($p < 0.01$) BW gain after 12 weeks and improved ($p < 0.05$) feed efficiency after 20 weeks	Increased crop and cecal B. sultilis counts but no effects on intestinal E. coli and Lactobacillus counts	(344)
Lactobacillus (Lacto)	8.8×10^8 cfu/g feed	Layers	Significantly improved ($p < 0.05$) BW gain for corn–soybean meal and barley–corn–soybean meal diets by 24% and 21%, respectively; no effects on feed consumption and conversion	Significantly improved ($p < 0.05$) egg weight, egg mass and egg size. Significantly increased ($p < 0.05$) passage rates of digesta. Significantly increased ($p < 0.05$) retentions of fat, calcium, phosphorus, copper, and manganese; no significant ($p > 0.05$) effects on internal egg quality and egg specific gravity	(361)
B. subtilis (FPBS)	20 mg/g diet	Female broilers	Significant reductions ($p < 0.05$) in feed efficiency and nitrogen utilization	Significant reductions ($p < 0.05$) in the ratio of abdominal fat or liver to BW, acetyl coenzyme A carboxylase activity, liver, and serum cholesterol contents	(383)

Lactobacillus	4.84×10^7 cfu/g feed	Layers	Significantly increased ($p < 0.05$) daily feed consumption by 2.5% during the laying phase	Significantly increased ($p < 0.05$) egg size by 2.1%; significantly improved ($p < 0.05$) nitrogen and calcium retention	(362)
L. casei	2.4×10^5 cfu/g feed	Broiler chicks	Significantly increased ($p < 0.05$) average daily weight gain by 7.0% after 3 weeks	Significantly decreased ($p < 0.05$) urease activity in the small intestine during the first 3 weeks	(404)
B. coagulans	1.6×10^7 (days 1–7), 4.0×10^6 (days 8–40) cfu/g diet	Broilers	Numerical improvement in feed conversion (33–39 days), significantly higher ($p < 0.054$) mean BW and daily weight gain after 33 and 49 days	—	(322)
Culture of *L. acidophilus*, *L. fermentum*, *L. crispattus*, and *L. brevis*	0.5×10^6, 1.0×10^6, 1.5×10^6 cfu/g; 0.1% and 0.15% culture	Broilers	Supplementation at 0.5×10^6 and 1.0×10^6 cfu/g significantly improved ($p < 0.05$) feed conversion by 6% and 13%, respectively, and increased ($p < 0.05$) BW from 1 to 42 days	Significantly reduced ($p < 0.05$) cecal coliform and Lactobacilli counts after 10 days Significantly reduced ($p < 0.05$) serum cholesterol after 30 days	(343)

(*continued*)

TABLE 4.1 (*Continued*)

Probiotics	Dosage	Animals	Growth Effects	Other Effects	References
Bacillus, Lactobacillus, Streptococcus, Clostridium. Saccharomyces, and Candida	—	Male broilers	Improved productivity	Significantly reduced ($p < 0.05$) water and increased ($p < 0.05$) lipid content in liver. Significantly decreased ($p < 0.05$) cholesterol content in thigh meat and liver, and significantly increased ($p < 0.05$) the linolenic acid and the unsaturated fatty acid/saturated fatty acid ratio in pectoral and thigh meat. Significantly reduced ($p < 0.05$) the number of E. coli and Salmonella in ceca, and significantly reduced ($p < 0.05$) the pH value of cecal contents. Significantly reduced ($p < 0.05$) the concentration of ammonia and increased ($p < 0.05$) the concentration of acetic acid in ceca	(330)

B. subtilis C-3102	3×10^5 viable spores (cfu)/g feed	Broilers	Significant increase ($p < 0.05$) in BW after 21 days. Significant improvement ($p < 0.05$) in FCR by 1.9–3.0% between 21 and 42 days	Significant reductions in aerobic plate count ($p < 0.001$) coliforms ($p < 0.001$), and Campylobacter ($p < 0.05$) in carcass	(333)
Lactobacillus culture	1×10^6 cfu/g feed	Female broilers	Significantly increased ($p < 0.05$) BW at day 21 by 22%; significantly increased ($p < 0.05$) weight gain and feed intake from day 1 to 42. Feed efficiency significantly improved by 42% from day 1 to 21	Significantly increased antibody titer against Newcastle disease vaccine by 36% after 42 days	(409)
E. faecium	0.5×10^6, 1.0×10^6, 1.5×10^6 cfu/g feed	Layers	Significantly lower ($p < 0.05$) feed consumption and feed conversion for 500 ppm supplementation only	Significantly lower damaged eggs for dosage at 1.0×10^6 cfu/g only	(312)
B. subtilis (FPBS)	0.5%, 1.0%, 2.0%	Female broilers	FPBS at 0.5% and 1.0% significantly reduced ($p < 0.05$) feed intake and improved feed efficiency by 9.8% and 8.6%, respectively	FPBS at 1% and 2% significantly reduced ($p < 0.05$) ammonia gas excretion. Significantly reduced ($p < 0.05$) hepatic triglyceride and cholesterol contents	(384)

(*continued*)

TABLE 4.1 (Continued)

Probiotics	Dosage	Animals	Growth Effects	Other Effects	References
B. subtilis var. Natto	0.5×10^6, 1×10^6, 3×10^6/g feed	Chickens	Numerical increase in feed intake and BW gain after 28 days. Numerical improvement in feed efficiency after 28 days	Significantly higher ($p < 0.05$) villus in duodenum and ileum. Significantly broader ($p < 0.05$) cells in the duodenum and jejunum. Significant increase ($p < 0.05$) in cell mitosis in jejunum. Significantly reduced ($p < 0.05$) serum ammonia after 28 days	(382)
L. acidophilus + L. casei, B. bifidum + A. oryzae + S. faecium + Torulopsis spp. (Probiolac)	2.7×10^5, 5.4×10^5 cfu/g feed	Layers	No effects on feed conversion and BW gain	Significantly increased ($p < 0.05$) hen-housed egg production, shell weight and thickness. Significantly higher antibody titer against sheep red blood cells, and significantly higher cutaneous basophilic hypersensitivity CBH response phytohaemagglutinin	(370)

E. faecium (Cernelle) 68 SF68	0.035 mg/g diet	Broilers	Numerical improvements in feed consumption and feed efficiency	Significantly reduced ($p < 0.05$) serum aspartate aminotransferase (AST) and alanine aminotransfera (ALT) levels, and serum cholesterol. No effects on small intestine weight and ileum pH. No effect on blood parameters	(369)
Protexin® Boost (L. plantarum, L. bulgaricus, L. acidophilus, L. rhamnosus, B. bifidum, S. thermophilus, E. faecium, A. oryzae, C. pintolopessi)	—	Broilers	Significantly increased ($p < 0.01$) live weight gains at 2nd, 4th, 5th and 6th weeks	Significantly increased ($p < 0.01$) carcass yield, breast and leg weights. Significantly higher ($p < 0.01$) antibody titer to sheep red blood cells antigen, and significantly increased ($p < 0.05$) weights of bursa and spleen	(346)
Lactina® (S. thermophilus, E. faecium, Lactobacillus)	0.3 mg/g feed	Mule ducklings	Significantly increased ($p < 0.05$) average BW by 3.4%, improved feed conversion by 3.74%, and reduced mortality after 93 days	Significantly increased ($p < 0.05$) weight of internal organs after 54 and 93 days. Significantly reduced ($p < 0.05$) number of E. coli, Lactobacilli, and Salmonella in cecum. No effects on blood parameters and serum cholesterol	(328)

(continued)

TABLE 4.1 (*Continued*)

Probiotics	Dosage	Animals	Growth Effects	Other Effects	References
B. subtilis PB6	10^3 cfu/g feed	Male broilers	Significantly improved ($p < 0.05$) BW of unchallenged and challenged broilers by 4.6% and 7.4%, respectively, after 42 days. Significantly reduced ($p < 0.05$) feed conversion of the challenged broilers by 7.6% or 15% points after 42 days	Numerically increase in *Lactobacillus* cell counts in chime in jejunum of broilers, unchallenged and challenged with a pathogenic strain of *E. coli*	(392)
B. subtilis PB6	10^2–10^3 cfu/g feed	Male broilers	Significantly decreased ($p < 0.05$) feed intake by 5.9% and improved ($p < 0.05$) feed conversion 2.6% from 21 to 42 days. Significantly increased ($p < 0.05$) in weight gain and feed intake from 1 to 42 days	—	(393)

antibiotics in improving gut health in pigs, especially during the postweaning period. In one particular study, a combination of live yeast and *Pediococccus acidilactici* in piglet diets was found to improve the piglet's intestinal mucosa, which is critical to nutrient absorption and immunity against pathogens. As can be seen in Table 4.2, the effects of probiotics is highly variable but it is with piglets that most of the significant success was observed.

4.4.3 Ruminants

The rumen is the site of active metabolism of carbohydrates and protein, accounting for almost all the energy and two-thirds of amino acids available to the ruminants. The condition in the rumen, which is characterized by constant temperature and pH (5.6 and 6.8), and lack of oxygen, is well suited for many strains of anaerobic bacteria, protozoa, and fungus.

One of the earliest applications of probiotics was to treat ruminal acidosis by yeast. Ruminal acidosis is a condition characterized by the overproduction of volatile organic acids such as propionic acid and acetates when nonfiber carbohydrates are fed to cows and cattle. If left untreated, ruminal acidosis can lead to decreased appetite, lameness, diarrhea, and decreased milk fat content. Feeding yeasts to these animals often alleviated the effects of ruminal acidosis, probably due to lowering of redox potential in the rumen, increase in cellulolytic bacteria (325), and general improvement in ruminal digestion (399).

Apart from treating ruminal acidosis, fungus, yeasts, and bacteria have been used in ruminants with varying success since the 1970s to increase milk production and weight gain, improve health status, and resistance to diseases. The high variability in trial results is in part attributed to the rumen microbiota, age, and diet of the animals. As the bacteria, yeast, fungus, and protozoa in the rumen of adult ruminants are formidable barriers to exogenous microorganisms, most of the success in probiotic application has been with calves. Since the microbiota of the small and large intestines of these immature ruminants are similar to the rumen except the absence of protozoa, many digestive disorders in calves can be treated by probiotics (380). The rumen of calves is populated first by cellulotytic and methanogenic bacteria, followed by cellulotytic and lactate fermenting bacteria by the 3rd week (308). The population of lactate fermenting bacteria will decrease gradually and by the 13th week, the ruminal flora of the calf will be similar to that of the adult ruminant (319). Protozoa are usually found after week 13 (319), although their establishment could be accelerated by switching milk to dry feeds (365).

One of the main objectives of feeding calves with probiotics is to reduce morbidity and mortality caused by intestinal and respiratory diseases, and this was confirmed by a recent report on veal calves treated with multi-species probiotic (MSPB) (395). In beef cattle, the effects of probiotics is heavily dependent on the type of diet, age, and type of probiotics. Wallace and Newbold (400) pointed out that improvement in average daily gain is higher for beef cattle fed high forage diets than those on 50% concentrate. Among the different probiotics tested so far, several strains of lactic acid

TABLE 4.2 Effects of Probiotics on the Performance of Pigs from Selected Studies Carried Out Since 1980

Probiotics	Dosage	Animals	Growth Effects	Other Effects	References
L. fermentum, *S. salivarius*	—	Pigs	—	Decreased *E. coli* in stomach	(309)
L. acidophilus (Probios), *S. faecium* type Cernelle 68 (Feed-Mate)	0.750 mg/g feed	Pigs (4 weeks old)	*L. acidophilus* tended to improve average daily gain and feed conversion with and without antibiotic supplementation. *S. faecium* type Cernelle 68 tended to decrease average daily gain	—	(372)
L. acidophilus (Probios), *S. faecium* type Cernelle 68 (Feed-Mate)	0.500 mg/g feed, 0.500 mg/g feed	Pigs (growing-finishing)	Numerical reduction in average daily gain and feed conversion	—	(372)
L. acidophilus	2×10^{12} cfu/milk feed	Gnotobiotic piglets	—	Significantly increased ($p < 0.001$) population of *L. acidophilus* in tissues and feces. Increased ($p = 0.06$) white blood cell counts but had no effect on hemotocrit	(373)

L. acidophilus	2×10^{12} cfu/milk feed	Conventional piglets	Slight reduction ($p < 0.1$) in average daily gains	Significant increased ($p < 0.001$) population of Lactobacilli in the intestines. Significantly reduced ($p < 0.01$) coliform counts	(373)
L. acidophilus (DDS 1), L. acidophilus (Probios)	2×10^{12} cfu/day, 4×10^{3} cfu/g feed	Piglets	Numerical improvement in average daily gain and feed conversion by 11% and 1.5%, respectively, by L. acidophilus (DDS 1). Numerical improvement in average daily gain and feed conversion by L. acidophilus (Prebios) + lactose. Numerical increase in average daily gain by L. acidophilus (Probios)	L. acidophilus significantly increased ($p < 0.05$) fecal Lactobacilli counts. No effect on fecal coliform counts	(374)
L. acidophilus	4×10^{3} cfu/g feed	Piglets	No effects on daily gain, feed intake, and feed efficiency in three trials	—	(336)
L. acidophilus	4×10^{3} cfu/g feed (pigs $\leq$ 20kg), 2×10^{3} cfu/g feed	Growing-finishing	Significantly reduced ($p < 0.05$) daily gain by 5.8% in two trials. No effects on feed intake and feed efficiency	—	(336)

(continued)

TABLE 4.2 (*Continued*)

Probiotics	Dosage	Animals	Growth Effects	Other Effects	References
L. acidophilus	2×10^3 cfu/g feed	Pigs (average weight of 31.2 kg)	Numerically improved ($p > 0.05$) daily gain and feed intake by 1.4% and 2.3%, respectively, in three trials	—	(336)
S. faecium	—	Gnotobiotic pigs (*E. coli* challenged)	No significant effect on growth	Reduced diarrhea and recovered faster from *E. coli* challenge. Reduced mortality from 18% to 8.5%	(396)
S. faecalis	10^6 cfu/g feed	Small piglets	—	Significantly increased ($p < 0.05$) fecal coliforms and aerobic spore-forming bacilli counts. Inhibition of *Salmonella* sp. in the intestines of piglets	(368)
Lactobacillus fermentation product	0.5 mL/day	Sows	Numerical increase in average daily gain	Numerical reduction in *E. coli* in stomach. Jejunal coliform cell counts increased when challenged with *E. coli*. No effect on histopathology	(375)
L. acidophilus P47, *L. acidophilus* RP32	5×10^{10} cfu/day, 5×10^{10} cfu/day	Pigs	—	*L. acidophilus* RP32 significantly reduced ($p < 0.05$) serum cholesterol	(334)

L. reuteri	Supplied in fermented milk	Piglets (weaned at day 2)	No effect on growth		(378)
L. bulgaricus, S. thermophilus	Supplied in fermented milk	Piglets (weaned at day 2)	Significantly increased ($p < 0.05$) feed gain		(378)
Lactobacillus sp.	0.1%	Piglets	Increased feed intake and weight gain	—	(353)
L. plantarum, L. acidophilus, L. casei, S. faecium (Probios®)	1×10^7 cfu/pig on days 1 and 7, 2×10^7 cfu/pig at weaning	Pigs	—	Significantly increased ($p < 0.05$) lactase and dipeptidase activities during preweaning stage. Significantly increased sucrase ($p < 0.001$), lactase ($p < 0.01$), tripeptidase ($p < 0.05$) at 17 days. Significant difference in the activities of sucrase ($p < 0.01$), dipeptidase ($p < 0.05$) and tripeptidase ($p < 0.05$) between the proximal and distal sites of the small intestines	(323)

(continued)

TABLE 4.2 (*Continued*)

Probiotics	Dosage	Animals	Growth Effects	Other Effects	References
B. pseudolongum, *L. acidophilus*	3×10^8–3×10^9 cfu/piglet (suckling–weaning), 3×10^8–3×10^9 cfu/piglet (suckling–weaning)	Piglets	Significantly increased ($p < 0.005$) BW gain. *B. pseudolongum* significantly increased feed conversion by 8.3%	—	(305)
B. thermophilum + *B. animalis*	1×10^{10} cfu + 1×10^{10} cfu/piglet in milk replacer without antibiotics	Piglets	Significantly increased ($p < 0.05$) mean BW gain	Significantly increased ($p < 0.05$) survival rate from 75% to 95%	(305)
B. subtilis, *B. licheniformis*, *B. pumilus*	3×10^6 cfu/g feed	Pigs	—	Reduced fecal coliform counts, increased fecal lactic acid bacteria count. No effects on dry matter, NDF, ADF, ash and N digestibility	(348)
B. cereus (CenBiot)	0.5×10^6 to 1×10^6 spores/g feed	Sows, piglets	Significantly increased ($p < 0.05$) average daily weight gain by 24%. Significantly improved ($p < 0.05$) feed conversion during nursery phase	Significantly reduced ($p < 0.05$) diarrhea from 35% to 18% in piglets. *E. coli* K88 was not detected in piglets fed *B. cereus* during suckling phase as compared to 18% in control	(406)

B. licheniformis, B. toyoi	10^7 viable spores/g feed, 10^6 viable spores/g feed	Piglets	—	Reduced incidence of diarrhea, no infection of enterogenic E. coli	(350)
E. faecium	—	Piglets	—	Reduced infection of Chlamydiaceae from 85% to 60% ($p<0.05$)	(376)
E. faecium SF68	1.6×10^6–1.2×10^6 cfu/g feed (sows), 1.7×10^5–1.2×10^5 cfu/g feed (piglets)	Sows, piglets	—	Significantly reduced total IgG in piglets at 8 weeks ($p<0.01$) and T cell (CD8+) in the jejunal epithelium of piglets ($p<0.001$). Reduction of up to 50% in frequency of β-hemolytic and O141 serovars of E. coli in piglets. No effects on fecal anaerobic and coliform bacteria counts of piglets and sows, and no effect on immune system of sows	(386)
A. oryzae (Amaferm®)	1.3 g/day	Sows	No significant effects on feed intake, body condition, piglet weaning weight or return to estrus	No significant effects on piglet and nursery pig growth	(341)

(continued)

TABLE 4.2 (*Continued*)

Probiotics	Dosage	Animals	Growth Effects	Other Effects	References
	1.5 mg/g feed	Piglets	No significant effects on average daily gain, feed intake, and feed-to-grain ratio	—	(341)
E. faecium NCIMB 10415 (SF68), *B. cereus* var. *toyoi* NCIMB 40112	1.6×10^6, 1.2×10^6, 1.7×10^5, 2×10^5 cfu/g dry weight feed; 2.6×10^5, 4×10^5, 1.3×10^6, 1.4×10^6 cfu/g dry weight feed	Gestating sows, lactating sows, nursed piglets, weaned piglets	—	Supplementation with *B. cereus* significantly increased fecal IgA of sows ($p = 0.045$) and piglets before weaning ($p = 0.004$). Serum IgG was significantly increased ($p < 0.001$) by *E. faecium* and *B. cereus* in weaned piglets	(387)

bacteria and propionibacteria appear to have reproducible positive effects on daily gain and feed efficiency (349). These effects are probably the cumulative result of alterations to the chemical and biological environment of the rumen and the intestines (349). Since feedlot cattle have been recognized as a host for *E. coli* O157:H7, much attention has been focused on the reduction of this pathogen by competitive exclusion of probiotics. There are already some trials showing growth inhibition of *E. coli* O157:H7 by certain strains of *E. coli* (407) and reduced shedding of *E. coli* O157:H7 during slaughter (349).

The effects of probiotics on the performance of dairy cattle has also attracted considerable attention due to the potential economic returns. Cultures of *S. cerevisiae* and *A. oryzae* have been used with mixed results on lactating cows. Some studies have reported significant improvements in dry matter intake, milk yield, and milk composition in cows fed probiotics. Most of the variability in the results can be attributed to farm environments, physiological status of animals, diets, and types of probiotics. In general, probiotics of fungal origin tended to be more effective in cows fed with medium to high amounts of concentrates or during early lactation (340). Interestingly, some studies have shown tendencies for improved growth and lactating performance in ruminants fed a combination of probiotics and prebiotics, such as lactic acid bacteria and mannanoligosaccharide on cows (335), and *E. faecium* and lactulose on calves (332). It is, however, not conclusive whether the positive effects of synbiotics (pro and prebiotic) are additive or synergistic.

The effects of probiotics on health, growth, and performance of ruminants are summarized in Table 4.3.

4.4.4 Rabbits

Rabbits have been regularly used as test models for human probiotics (360), mainly due to the similarity of the digestive systems. As such, most of the studies concentrated on the ability of potential human probiotics to adhere and colonize the gastrointestinal tract of rabbits. There are a limited number of studies focused on improving the growth and well being of these animals. Yu and Tsen (405) reported that due to the limited resistance to the gastric juice and lack of adhesive capacity in the intestinal tract, the use of lactobacillus as probiotics in rabbits maybe limited. Table. 4.4 summarize the effects of probiotics on rabbits.

4.4.5 Pets

In 2006, the world's dog and cat food imports by major importing countries were estimated at US$700, an increase of 11% over the previous year (410). One of the fastest growing markets for pet food is China, largely due to the growing affluence of the urban population and the perceived prestige associated with pet ownership (410). As pets are now regarded as human companions, their diets have also evolved from table scraps that fulfill the basic metabolic needs to highly formulated foods. In addition, providing nutritionally balanced diets to dogs and cats is now regarded as part of the pet owners' responsibility to maintain the health and

TABLE 4.3 Effects of Probiotics on Calves from Studies Since 1980

Probiotics	Dosage	Animals	Growth Effects	Other Effects	References
L. bulgaricus	1.5×10^8–24×10^8 cfu/feed	Calves (0–9 weeks)	Significantly increased ($p < 0.05$) average daily gain by 43% in one of three trials for calves fed with Lactobacillus at 6×10^8 cfu/feed	No significant effects on fecal lactobacillus and coliform counts, and no significant effects on nutrient digestibilities and nitrogen intake	(388)
Yeast culture (Diamond V Mills, Inc., USA)	1.85% in feed	Steers, lambs	No significant effect on dry matter intake	Significantly increased ($p < 0.05$) liquid flow rate in steers. No significant effects on liquid turnover time and ruminal fermentation of steers. No significant effects on dry matter intake, digestibility, water intake and urine excretion in lambs	(306)
S. faecalis	$0.3 \times 10^{8.7}$ cfu/day	Cows	—	Significantly increased ($p < 0.05$) Bifidobacteria in the feces	(368)
S. faecium SF-68	—	Female calves	Improved fattening performance	No significant effect on carcass quality	(317)
S. faecium M 74	1×10^7 cfu/g in milk replacer	Female calves	Numerically improved average daily gain	No significant effect on carcass quality	(318)
L. acidophilus	10^9–10^{10} cfu/d/calve	Calves	No significant difference in growth performance and health	Significantly increased ($p < 0.05$) fecal Lactobacilli count	(345)

Supplement	Dose	Animal	Effect	Ref.
Yeast culture (Diamond V Mills, Inc., USA), Yeast + A. oryzae fermentation extract (Vitaferm®), A. oryzae fermentation extract (Amaferm®)	90 g/day, 90 g/day, 2.63 g/day	Cows	Crude protein and hemicellulose digestibility, and ruminal cellulolytic bacteria count were significantly increased ($p < 0.01$) by all fungal supplements. Dry matter digestibility was significantly increased ($p < 0.05$) by supplements containing A. oryzae. Numerical increase in acetate:propionate ratio by all fungal supplements	(402)
A. oryzae	—	Calves	Increased feed intake and weight gains	(307)
A. oryzae (Amaferm®)	0.92–5.5 g/kg starter before weaning, 0.28–1.65 g/kg starter after weaning	Calves	Significantly reduced the number of weeks to weaning in heifer calves ($p < 0.05$) and heifer subset ($p < 0.01$). Amaferm at higher dosage significantly increased ($p < 0.05$) feed intake from week 5 to 10	(314)
L. acidophilus, L. lactis, B. subtilis (Biomate FG), B. subtilis (Biomate 2B)	2.2×10^9, 2.2×10^6, 1.1×10^9 cfu/day; 1.24×10^{10} cfu/day	Calves	No significant effect on growth by both probiotics. Calves fed B. subtilis concentrate tended to have highest feed efficiency. Numerically higher fecal Lactobacilli counts in calves fed mixed concentrate	(342)

(continued)

TABLE 4.3 (Continued)

Probiotics	Dosage	Animals	Growth Effects	Other Effects	References
S. cerevisiae, A. oryzae extract (Diamond V Mills, Inc., USA)	1.68×10^9 cfu of S. cerevisiae/day in TMR	Cows	—	Significantly reduced ($p < 0.05$) lactose and SNF (solids-nonfat) contents. No effects on milk yield, milk fat, and protein contents	(337)
B. pseudolongum, L. acidophilus, B. thermophilum + E. faecium + L acidophilus	10^9 cfu/day, 10^9 cfu/day, 1×10^{10} cfu + 1×10^{10} cfu + 1×10^9 cfu/day in milk replacer	Calves	BW gain was significantly increased by B. pseudolongum and L. acidophilus by 25% and 22%, respectively. Feed conversion was significantly improved by B. pseudolongum and L. acidophilus by 11% and 13%, respectively	No significant effect on fecal score	(305)
B. thermophilum + E. faecium + L acidophilus	1×10^{10} cfu + 1×10^{10} cfu + 1×10^9 cfu/day in milk replacer	Calves	Numerical increased ($p > 0.05$) BW gain and improved ($p > 0.05$) feed conversion	Reduced diarrhea from 78% to 10%	(305)
L. acidophilus, L. lactis, B. subtilis (Biomate FG)	5.5×10^5, 5.5×10^2, 2.75×10^5, cfu/g milk replacer	Bull calves	No significant effect, on BW gain and starter consumption	Significantly increased mean capsular volume of blood after 10 days	(359)
L. acidophilus	5×10^7 cfu/feed/ calve	Calves	Significantly improved ($p < 0.01$) BW gain during first 2 weeks. No significant difference in total BW gain and feed efficiency	No significant effect on occurrence of diarrhea	(324)

S. cerevisiae (Biomate Yeast Plus®)	5×10^{11} cfu/day	Cows	No significant effect on dry matter intake ($p > 0.10$)	Numerically increased milk urea nitrogen ($p > 0.10$). No effect, on milk yield and milk composition ($p < 0.10$)	(389)
E. coli (17 strains), *Proteus mirabilis*	10^{10} cfu/calve in 2% skimmed milk	Calves	—	*E. coli* O157:H7 was not recovered from the rumen of the calves challenged with the pathogen	(407)
P. freudenreichii (PF24) + *L. acidophilus* (LA45), *P. freudenreichii* (PF24) + *L. acidophilus* (LA45+LA51)	$10^9 + 10^6$ cfu/g feed, $10^9 + 10^6$ cfu/g feed, $10^9 + 10^8$ cfu/g feed	Beef steers	Significantly increased ($p < 0.05$) average daily gain and feed efficiency by 6.9% and 4.8%, respectively	Tended to increase carcass quality ($p > 0.05$)	(381)
P. acidipropionici	$1 \times 10^7 – 1 \times 10^9$ cfu/day	Beef steers	—	Acetate:propionate ratio was significantly reduced ($p < 0.05$) except at 10^8 cfu/day dosage. Significantly decreased ($p < 0.01$) butyrate with increasing dosage	(347)
L. acidophilus (NP45), *L. acidophilus* (NP51), *P. freudenreichii*	$1 \times 10^6, 1 \times 10^9, 1 \times 10^9$ cfu/animal/day	Beef steers	No significant difference to growth performance when in treatments with single or combination of probiotics	Significant reduction in lamina propria thickness ($p = 0.02$) and decreased fecal *E. coli* shedding ($p = 0.06$) by *L. acidophilus*	(329)
A. oryzae fermentation extract (Amaferm®)	5 g/day in TMR	Cows	—	No effects on lactation performance and rumen parameters	(338)

(*continued*)

TABLE 4.3 (*Continued*)

Probiotics	Dosage	Animals	Growth Effects	Other Effects	References
E. faecium, *S. cerevisiae* (Probios® TC)	5×10^9, 5×10^9 cfu/ cow/day	Cows	No significant effects on BW change	Significantly reduced in situ undegraded forage dry matter at 12 h ($p < 0.05$), 24, 28, and 72 h ($p < 0.01$). Increased dry matter intake at prepartum ($p < 0.1$) and postpartum periods. Significantly increased milk lactose content ($p < 0.05$). Numerically increased milk protein yield ($p > 0.1$). Significantly increased ($p < 0.05$) blood glucose 7 days postpartum and reduced β-hydroxybutyrate at −1 day ($p < 0.01$), +1 ($p < 0.09$) and +7 ($p > 0.05$) days postpartum	(366)
P. freudenreichii (NP24), *L. acidophilus* (NP51)	1×10^9 cfu/steer/ day, 1×10^7, 1×10^8, 1×10^9 cfu/steer/ day	Beef steers	No significant effects ($p > 0.10$) on final BW and dry matter intake. Mixture of *P. freudenreichii* and *L. acidophilus* significantly improved ($p < 0.05$) gain efficiency by approximately 2%	Numerical increase ($p = 0.10$) in marbling score and percentage of USDA choice cattle	(398)

TABLE 4.4 The Effects of Probiotics on Rabbits

Probiotics	Dosage	Animals	Growth Effects	Other Effects	References
L. delbrueckii subsp. *Bulgaricus*, *S. thermophilus*, *L. acidophilus*	10^8 cfu/kg BW	Rabbits	No effect on growth in rabbits on casein diets	Rabbits fed with *L. delbrueckii* subsp. *Bulgaricus* and *S. thermophilus* showed increased in intestinal *Enterococcus* population. Reduced cecal *C. septicum* and *C. clostridiforme* from 80.7% and 12.7%, respectively, to 0. Significantly reduced ($p < 0.05$) plasma cholesterol	(320)
L. casei Shirota	—	Infant rabbits	—	Decreased severity of diarrhea and lowered the level of STEC colonization in the gastrointestinal tract	(367)
S. thermophilus (Sc. t.), *L. delbrueckii* ssp. *bulgaricus* (Lb. b.)	5.8×10^{10} cfu/day, 3.6×10^{10} cfu/day	Rabbits	—	Probiotic strains were recovered in wet feces for the stomach, the duodenum and large intestine at 10^5, 10^7 cells/mL, and 10^7 cells/g, respectively, after supplementation	(327)

(continued)

TABLE 4.4 (*Continued*)

Probiotics	Dosage	Animals	Growth Effects	Other Effects	References
B. cereus var. toyoi	0.2×10^6 spores/g feed	Doe rabbits	No effect on daily feed intake during lactation, feed efficiency increased significantly ($p = 0.01$) by 15% to 0.314. No effect on doe BW at parturition or at weaning	Significantly shortened ($p = 0.05$) interval from parturition to effective mating and between parturitions, and tended to improve litter size at weaning ($p = 0.09$). Litter weight at weaning tended to be higher ($p = 0.10$)	(364)
B. cereus var. toyoi	$2 \times 10^5 – 1 \times 10^6$ spores/g feed	Rabbits	Significant increase ($p = 0.002$) in final live weight by 2.5%, significant increase ($p = 0.01$) in daily weight gain by 4.2%, and significantly improved ($p = 0.01$) feed conversion by 3.7%	Significantly reduced ($p = 0.03$) morbidity by 43.4%	(394)

TABLE 4.5 The Effects of Probiotics on Cats and Dogs from Recent Studies

Probiotics	Dosage	Animals	Growth Effects	Other Effects	References
Bacillus CIP 5832	7.5×10^6 cfu/day	Dogs	Numerical improvement in dry matter, protein, and energy digestibility	—	(315)
L. acidophilus, *E. faecium*, *S. cerevisiae*	10^9 cells/mL	Gnotobiotic mice	—	Mice fed with *E. faecium* survived challenge with S. typhimurium ($p < 0.05$). By comparison, survival time was longer ($p < 0.05$) for mice fed with the three probiotic than those fed with *L. acidophilus* or *S. cerevisiae*	(356)
E. faecium SF68	5×10^8 cfu/day	8 weeks old puppy	No significant effects on feed intake and weight gain	Significantly increased mature B cells (CD21+/MHCII+) by 7.4% and 4.5% after 31 and 44 weeks, respectively	(311)
E. faecium NCIB 10415	9.2×10^9 cfu/day	Dogs	—	Significant reduction ($p < 0.05$) in fecal counts of *Clostridium* spp. in 10 out of 12 dogs; numerical increase in *Salmonella* spp. and *Campylobacter* spp. in all dogs	(397)

(continued)

TABLE 4.5 (*Continued*)

Probiotics	Dosage	Animals	Growth Effects	Other Effects	References
L. rhamnosus GG, B. lactis Bb12, L. pentosus UK1A, Sk2A, E. faecium M74, SF273	10^7–10^8 cfu/mL	Canine jejunal mucosa	—	Significantly reduced ($p < 0.05$) the adhesion of *C. perfringens*. Both strains of *Enterococci* significantly enhanced ($p < 0.05$) the adhesion of canine pathogen *Campylobacter jejuni*	(379)
L. acidophilus	>10^9 cfu/day	Dogs	—	Significant increases ($p < 0.05$) in red blood cells, hematocrit, hemoglobin concentration, neutrophils, monocytes, and serum immunoglobin G concentration, and significant reductions ($p < 0.05$) in RBC fragility and serum nitric oxide concentration	(310)

L. acidophilus (DSM 13241)	2×10^8 cfu/day	Cats	—	Significant reduction in fecal population of Clostridium spp ($p = 0.011$) and E. faecalis ($p < 0.001$). Significant increased ($p < 0.001$) fecal population of Lactobacilli and L. acidophilus. Reduced plasma endotoxin concentration, improved white blood cell counts and phagocytic capacity	(357)
L. fermentum AD1	—	Dogs	—	Significant increase in fecal Lactobacilli and Enterococci counts. Significant increase in serum total protein and total lipid, and significant reduction of serum glucose	(390)
L. acidophilus NCC2628, NCC2766, L. johnsonii NCC2767	10^9 cfu/(g day)	Dogs	No significant effect on BW	Significant reduction ($p < 0.0001$) in Canine Inflammatory Bowel Disease Activity Index in all dogs suffering from food responsive diarrhea	(385)
L. animalis LA4	10^9 cfu/day	Adult dogs	—	Increased fecal Lactobacilli and reduced fecal Enterococci ($p = 0.08$)	(313)

well being of these animals (316). The correlation between the gastrointestinal tract and overall health of the animals has led to a deluge of probiotics and prebiotics into pet food in recent years (316). Many such products contain mixtures of probiotic bacteria strains from *Pediococcus*, *Lactobacillus*, *Bifidobacterium*, *Bacillus*, *Streptococcus*, *Enterococcus*, or *Saccharomyces* yeast with claims of improved digestion, increased appetite, reduced diarrhea, increased firmness of stool, reduction in vomiting, reduction of body odor, reduction of flatulence, or improved swallowing. A more recent patent claimed improved health when pets were fed on a mixture of lactic acid bacteria preparation from *Lactobacillus reuteri* NCC2581 (CNCM I-2448), *Lactobacillus reuteri* NCC2592 (CNCM I-2450), *Lactobacillus rhamnosus* NCC2583 (CNCM I-2449), *Lactobacillus reuteri* NCC2603 (CNCM I-2451), *Lactobacillus reuteri* NCC2613 (CNCM I-2452), *Lactobacillus acidophilus* NCC2628 (CNCM I-2453), and *Enterococcus faecium* SF 68 (NCIMB 10415) (408). The scientific basis for many such claims remains to be verified by independent studies. Weese and Arroyo (403) reviewed 19 commercial pet foods containing probiotics and concluded that the average bacterial count was 1.8×10^5 cfu/g, and no relevant bacteria growth was present in five of the products. This implies that a significant number of products may not have enough viable bacteria to colonize the digestive tracts of the pets, and could also explain the mixed response seen in a small number of studies cited by Myers (360). A summary of more recent outcomes on supplementation of the diets of cats and dogs with probiotics is presented in Table 4.5.

REFERENCES

1. Bengmark S, Bacteria for optimal health. *Nutrition* 2000; 16: 611–615.
2. Camilleri M. Probiotics and irritable bowel syndrome: rationale, putative mechanisms, and evidence of clinical efficacy. *J. Clin. Gastroenterol.* 2006; 40(3): 264–269.
3. Chouraqui JP, Van Egroo LD, and Fichot MC. Acidified milk formula supplemented with bifidobacterium lactis: impact on infant diarrhea in residential care settings. *J. Pediat. Gastroenterol. Nutr.* 2004; 38(3): 288–292.
4. Collado MC, Meriluoto J, and Salminen S. In vitro Analysis of probiotic strains combinations to inhibit pathogen adhesion to human intestinal mucus. *Food Res. Int.* 2007; 40(5): 629–636.
5. FAO/WHO. Guidelines for the evaluation of probiotics in food. Food and Agricultural Organization of the United Nations and World Health Organization. *Working Group Report*, 2002.
6. Forestier C, De Champs C, Vatoux C, and Joly B. Probiotic activities of *Lactobacillus casei rhamnosus*: in vitro adherence to intestinal cells and antimicrobial properties. *Res. Microbiol.* 2001; 152: 167–173.
7. Gaudier E, Michel C, Segain JP, Cherbut C, and Hoebler C. The VSL#3 probiotic mixture modifies microflora but does not heal chronic dextran-sodium sulfate-induced colitis or reinforce the mucus barrier in mice. *J. Nutr.* 2005; 135(12): 2753–2761.

8. Gotteland M, Brunser O, and Cruchet S. Systematic review: are probiotics useful in controlling gastric colonization by *Helicobacter pylori*? *Aliment. Pharmacol. Ther.* 2006; 23(8): 1077–1086.
9. Juntunen M, Kirjavainen PV, Ouwehand AC, Salminen SJ, and Isolauri E. Adherence of probiotic bacteria to human intestinal mucus in healthy infants and during rotavirus infection. *Clin. Diagn. Lab. Immunol.* 2001; 8(2): 293–296.
10. Kajander K, Hatakka K, Poussa T, Farkkila M, and Korpela R. A probiotic mixture alleviates symptoms in irritable bowel syndrome patients: a controlled 6-month intervention. *Aliment. Pharmacol. Ther.* 2005; 22(5): 387–394.
11. Kalliomäki M, Salminen S, Arvilommi H, Kero P, Koskinen P, and Isolauri E. Probiotics in primary prevention of atopic disease: a randomised placebo-controlled trial. *Lancet* 2001; 357: 1076–1079.
12. Kim HJ, Vazquez Roque MI, Camilleri M, Stephens D, Burton DD, Baxter K, Thomforde G, and Zinsmeister AR. A randomized controlled trial of a probiotic combination VSL#3 and placebo in irritable bowel syndrome with bloating. *Neurogastroenterol. Motil.* 2005; 17(5): 687–696.
13. McFarland LV. Meta-analysis of probiotics for the prevention of antibiotic associated diarrhea and the treatment of *Clostridium difficile* disease. *Am. J. Gastroenterol.* 2006; 101: 812–822.
14. Parvez S, Malik KA, Ah Kang S, and Kim HY. Probiotics and their fermented food products are beneficial for health. *J. Appl. Microbiol.* 2006; 100(6): 1171–1185.
15. Rautava S, Arvilommi H, and Isolauri E. Specific probiotics in enhancing maturation of IgA responses in formula-fed infants. *Pediatr. Res.* 2006; 60(2): 221–224.
16. Salminen S, Bouley C, Boutron-Ruault M-C, Cummings JH, Franck A, Gibson GR, Isolauri E, Moreau M-C, Roberfroid M, and Rowland I. Functional food science and gastrointestinal physiology and function. *Br. J. Nutr.* 1998; 80: S147–S171.
17. Salminen S, Ouwehand AC, Benno Y, and Lee YK. Probiotics: how should they be defined? *Trends Food Sci. Technol.* 1999; 10: 107–110.
18. Santosa S, Farnworth E, and Jones PJ. Probiotics and their potential health claims. *Nutr. Rev.* 2006; 64(6): 265–274.
19. Sazawal S, Hiremath G, Dhingra U, Malik P, Deb S, and Black RE. Efficacy of probiotics in prevention of acute diarrhoea: a meta-analysis of masked, randomised, placebo-controlled trials. *Lancet Infect. Dis.* 2006; 6(6): 374–382.
20. Timmerman HM, Koning CJ, Mulder L, Rombouts FM, and Beynen AC. Monostrain, multistrain, and multispecies probiotics—A comparison of functionality and efficacy. *Int. J. Food Microbiol.* 2004; 96: 219–233.
21. Turnbaugh PJ, Ley RE, Mahowald MA, Magrini V, Mardis ER, and Gordon JI. An obesity-associated gut microbiome with increased capacity for energy harvest. *Nature* 2006; 444(7122): 1027–1031.
22. Aduayom I, Campbell PGC, Denizeau F, and Jumarie C. Different transport mechanisms for cadmium and mercury in Caco-2 cells: inhibition of Cd uptake by Hg without evidence for reciprocal effects. *Tox. Appl. Pharmacol.* 2003; 189: 56.
23. Arredondo M, Orellana A, Gárate MA, and Nuúñez MT. Intracellular iron regulates iron absorption and IRP activity in intestinal epithelial (Caco-2) cells. *Am. J. Physiol.– Gastroenterol. L.* 1997; 273: G275–G280.

24. Audus KL, Bartel RL, Hidalgo IJ, and Borchardt RT. The use of cultured epithelial and endothelial cells for drug transport and metabolism studies. *Pharm. Res.* 1990; 7: 435–451.
25. Bernet M-F, Brassart D, Neeser J-R, Servin AL. Adhesion of human Bifidobacterial strains to cultured human intestinal epithelial cells and inhibition of enteropathogen-cell interactions. *Appl. Environ. Microbiol.* 1993; 59: 4121–4128.
26. Bestwick CS and Milne L. Effects of β-carotene on antioxidant enzyme activity, intracellular reactive oxygen, and membrane integrity within post confluent Caco-2 intestinal cells. *Biochim. Biophys. Acta.* 2000; 1474: 47–55.
27. Caro I, Boulenc X, Rousset M, Meunier V, Bourrie M, Julian B, Joyeux H, Roques C, Berger Y, Zweibaum A, and Fabre G. Characterization of a newly isolated Caco-2 clone (TC-7), as a model of transport processes and biotransforming of drugs. *Int. J. Pharma.* 1995; 116: 147–158.
28. Cheng H and Leblond CP. Origin, differentiation, and renewal of the four main epithelial cell types in the mouse small intestine. V. Unitarian Theory of the origin of the four epithelial cell types. *Am. J. Anat.* 1974; 141: 537–561.
29. Chung YS, Song IS, Erickson RH, Sleisenger MH, and Kim YS. Effect of growth and sodium butyrate on brush border membrane-associated hydrolases in human colorectal cancer cell lines. *Cancer Res.* 1985; 45: 2976–2982.
30. Cilla A, Laparra JM, Alegria A, Barbera R, and Farre R. Antioxidant effect derived from bioaccessible fractions of fruit beverages against H_2O_2-induced oxidative stress in Caco-2 cells. *Food Chem.* 2008; 106: 1180–1187.
31. Collado MC and Sanz Y. Method for direct selection of potentially probiotic *Bifidobacterium* strains from human feces based on their acid-adaptation ability. *J. Microbiol. Meth.* 2006; 66(3): 560–563.
32. Crociani J, Grill J-P, Huppert M, and Ballongue J. Adhesion of different Bifidobacteria strains to human enterocyte-like Caco-2 cells and comparison with *in vivo* study. *Lett. Appl. Microbiol.* 1995; 21: 146–148.
33. Dantzig AH and Bergin L. Uptake of the cephalosporin, cephalexin, by a dipeptide transport carrier in the human intestinal cell line, Caco-2. *Biochim. Biophys. Acta.* 1990; 1027: 211–217.
34. Delgado S, O'Sullivan E, Fitzgerald G, and Mayo B. Substractive screening for probiotic properties of *Lactobacillus* species from the human gastrointestinal tract. *J. Food Sci.* 2007; 72: 310–313.
35. Ekmekcioglu C. A physiological approach for preparing and conducting intestinal bioavailability studies using experimental systems. *Food Chem.* 2002; 76: 225–230.
36. Figueiredo PMS, Furumura MT, Aidar-Ugrinovich L, Pestana de Castro AF, Pereira IL, and Yano T. Induction of apoptosis in Caco-2 and HT29 human intestinal epithelial cells by enterohemolysin produced by classic enteropathogenic *Escherichia coli*. *Lett. Appl. Microbiol.* 2007; 45: 358–363.
37. Gangloff MB, Glahn RP, Miller DD, Norwell WA, and Van Campen DR. Assessment of iron availability using combined in vitro digestion and Caco-2 cell culture. *Nutr. Res.* 1996; 16: 479–487.
38. Glahn RP, Gangloff MB, Miller DD, Wien EM, Kapsokefalou M, and Van Campen DR. Use of In vitro digestion and Caco-2 cell culture to study iron uptake/bioavailability. *FASEB J.* 1994; 8: A712.

39. Gueimonde M, Jalonen L, He F, Hiramatsu M, and Salminen S. Adhesion and competitive inhibition and displacement of human enteropathogens by selected lactobacilli. *Food Res. Int.* 2006; 39(4): 467–471.
40. Hidalgo IJ, Raub TJ, and Borchardt RT. Characterization of the human-colon carcinoma cell-line (Caco-2) as a model system for intestinal epithelial permeability. *Gastroenterol.* 1989; 96: 736–749.
41. Hu M and Borchardt RT. Transport of a large neutral amino acid in a human intestinal epithelial cell line (Caco-2): uptake and efflux of phenylalanine. *Biochim. Biophys. Acta.* 1992; 1135: 233–244.
42. Johnson DM, Yamaji S, Tennant J, Srai SK, and Sharp PA. Regulation of divalent metal transporter expression in human intestinal epithelial cells following exposure to non-haem iron. *FEBS Lett.* 2005; 579: 1923–1929.
43. Kaur M, Singh RP, Gu M, Agarwal R, and Agarwal C. Grape seed extract inhibits In vitro and In vivo growth of human colorectal carcinome cells. *Clin. Cancer Res.* 2006; 12: 6194–6202.
44. Laparra JM, Vélez D, Montoro R, Barberá R, and Farré R. Bioavailability of inorganic arsenic in cooked rice: practical aspects for human health risk assessment. *J. Agric. Food Chem.* 2005; 53: 8829–8833.
45. Laparra JM, Vélez D, Montoro R, Barberá R, and Farré R. Evaluation of the human intestinal permeability of As(III) and its effects upon intracellular GSH levels using Caco-2 cells. *Toxicol. In vitro* 2006; 20: 658–663.
46. Lenaerts K, Bouwman FG, Lamers WH, Renes J, and Mariman EC. Comparative proteomic analysis of cell lines and scrapings of the human intestinal epithelium. *BMC Genomics* 2007; 8: 1–14.
47. Lesuffleur T, Barbat A, Luccioni C, Beaumatin J, Clair M, Kornowski A, Dussaulx E, Dutrillaux B, and Zweibaum A. Dihydrofolate reductase gene amplification-associated shift of differentiation in methotrexate-adapted HT-29 cells. *J. Cell Biol.* 1991; 115: 1409–1418.
48. Lesuffleur T, Porchet N, Aubert JP, Swallow D, Gum JR, Kim YS, Real FX, and Zweibaum A. Differential expression of the human mucin genes MUC1 to MUC5 in relation to growth and differentiation of different mucus-secreting HT-29 cell subpopulations. *J. Cell Sci.* 1993; 106: 771–783.
49. Mattila-Sandholm T, Myllärinen P, Crittenden R, Mogensen G, Fondén R, and Saarela M. Technological challenges for future probiotic foods. *Int. Dairy J.* 2002; 12: 173–182.
50. Novellaux G, Devillé C, El Moualij B, Zorzi W, Deloyer P, Schenider YJ, Peulen O, and Dandrifosse G. Development of a serum-free co-culture of human intestinal epithelium cell-lines (Caco-2/HT29-5M21). *BMC Cell Biol.* 2006; 7: 1–11.
51. Pinto M, Robine-Leon S, Appay MD, Kedinger M, Triadou N, Dussaulx E, Lacroix B, Simon-Assmann P, Haffen K, Fogh J, and Zweibaum A. Enterocyte-like differentiation and polarization of the human colon carcinoma cell line caco-2 in culture. *Biology of the Cell* 1983; 47: 323–330.
52. Pontier C, Pachot J, Botham R, Lenfant B, and Arnaud P. HT29-MTX and Caco-2/TC7 monolayers as predictive models for human intestinal absorptioin: role of the mucus layer. *J. Pharma. Sci.* 2001; 90: 1608–1619.
53. Potten CS. Stem cells in gastrointestinal epithelium: numbers, characteristics and death. *Phil. Trans. Royal Soc. Lond. B-Biol. Sci.* 1998; 353: 821–830.

54. Schaefer S, Baum M, Eisenbrand G, and Janzowski C. Modulation of oxidative cell damage by reconstituted mixtures of phenolic apple juice extracts in human colon cell lines. *Mol. Nutr. Food Res.* 2006; 50: 413–417.
55. Schlee M, Harder J, Koten B, Stange EF, Wehkamp J, and Fellerman K. Probiotic lactobacilli and VSL#3 induce enterocyte beta-defensin 2. *Clin. Exp. Immunol.* 2008; 151: 528–535.
56. Van Campen DR and Glahn RP. Micronutrient bioavailability techniques: accuracy, problems and limitations. *Field Crops Res.* 1999; 60: 93–113.
57. Walter E and Kissel T. Transepithelial transport and metabolism of thyrotropin-releasing hormone (TRH) in monolayers of a human intestinal cell line (Caco-2): evidence for an active transport component? *Pharm. Res.* 1994; 11: 1575–1580.
58. Walter E, Janich S, Roessler BJ, Hilfinger JM, and Amidon GL. HT29-MTX/Caco-2 cocultures as an *in vitro* model for the intestinal epithelium: *In vitro–In vivo* correlation with permeability data from rats and humans. *J. Pharm. Sci.* 1996; 85: 1070–1076.
59. Wehkamp J, Harder J, Wehkamp K, Meissner W, Schlee M, Enders C, Sonneborn U, Nuding S, Bengmark S, Fellerman K, Schröder JM, and Stange EF. NF-κB- and AP-1-mediated induction of human beta defensin-2 in intestinal epithelia cells by *Eschericia coli* Nissle 1917: a novel effect of probiotic bacterium. *Infect. Immun.* 2004; 72: 5750–5758.
60. Wikman A, Karlsson J, Carlstedt I, and Artursson P. A drug absorption model based on the mucus layer producing human intestinal globet cell line HT29-H. *Pharm. Res.* 1993; 10: 843–852.
61. Yokomizo A and Moriwaki M. Effects of flavonoids on oxidative stress induced by hydrogen peroxide in human intestinal Caco-2 cells. *Biosci. Biotech. Biochem.* 2006; 70: 1317–1324.
62. Zweibaum A, Pinto M, Chevalier G, Dussalux E, Triadou N, Lacroix B, Haffen K, Brun JL, and Rousset M. Enterocytic differentiation of a subpopulation of the human colon tumour cell line HT-29 selected for growth in sugar-free medium and its inhibition by glucose. *J. Cell Biol.* 1985; 122: 21–29.
63. European, Convention for the Protection of Vertebrate Animals Used for Experimental and Other Scientific Purposes (European Treaty Series 123), Council of Europe, Strasbourg, 1986.
64. European Commission, Provisions of the European Union Member States Regarding the Protection of Animals Used for Experimental and Other Scientific Purposes, Directive 86/609/EEC. *Off. J. Eur. Comm.* 1986; 1: L358.
65. National Research Council, Institute of Laboratory Animal Resources, *Guide for the Care and Use of Laboratory Animals*, National Academy Press, Washington D.C., 1996.
66. Russell WMS and Burch RL. *The Principles of Humane Experimental Technique*, Methuen & Co. Ltd., London, 1959.
67. Gueimonde M, Frias R, and Ouwehand AC. Assuring the continued safety of lactic acid bacteria used as probiotics. *Biol. Bratislava.* 2006; 61(6): 755–760.
68. Pelletier C, Bouley C, Bourliov P, Carbon C. Evaluation of safety properties of *Lactobacillus* strains by using an experimental model of endocarditis in rabbit. SOMED Meeting, Paris 1996.
69. Nonaka Y, Izumo T, Izumi F, Maekawa T, Shibata H, Nakano A, Kishi A, Akatani K, and Kiso Y. Antiallergic effects of *Lactobacillus pentosus* strain S-PT84 mediated by

modulation of Th1/Th2 immunobalance and induction of IL-10 production. *Int. Arch. Allergy Immunol.* 2008; 145(3): 249–257.

70. Amital H, Gilburd B, and Shoenfeld Y. Probiotic supplementation with *Lactobacillus casei* (Actimel) induces a Th1 response in an animal model of antiphospholipid syndrome. *Ann. N.Y. Acad. Sci.* 2007; 1110: 661–669.

71. Blümer N, Sel S, Virna S, Patrascan CC, Zimmermann S, Herz U, Renz H, and Garn H. Perinatal maternal application of *Lactobacillus rhamnosus* GG suppresses allergic airway inflammation in mouse offspring. *Clin. Exp. Allergy* 2007; 37(3): 348–357.

72. Bauer E, Williams BA, Smidt H, Verstegen MW, and Mosenthin R. Influence of the gastrointestinal microbiota on development of the immune system in young animals. *Curr. Issues Intest. Microbiol.* 2006; 7(2): 35–51.

73. McVay MR, Boneti C, Habib CM, Keller JE, Kokoska ER, Jackson RJ, and Smith SD. Formula fortified with live probiotic culture reduces pulmonary and gastrointestinal bacterial colonization and translocation in a newborn animal model. *J. Pediatr. Surg.* 2008; 43(1): 25–29.

74. Mogilner JG, Srugo I, Lurie M, Shaoul R, Coran AG, Shiloni E, and Sukhotnik I. Effect of probiotics on intestinal regrowth and bacterial translocation after massive small bowel resection in a rat. *J. Pediatr. Surg.* 2007; 42(8): 1365–1371.

75. Ruan X, Shi H, Xia G, Xiao Y, Dong J, Ming F, and Wang S. Encapsulated Bifidobacteria reduced bacterial translocation in rats following hemorrhagic shock and resuscitation. *Nutrition* 2007; Oct 23(10): 754–761.

76. Daniel C, Poiret S, Goudercourt D, Dennin V, Leyer G, and Pot B. Selecting lactic acid bacteria for their safety and functionality by use of a mouse colitis model. *Appl. Environ. Microbiol.* 2006; 72(9): 5799–5805.

77. Hau J and van Hoosier GL. In: *Animal Models*, Vol. II, 2nd ed. Laboratory Animal Science, 2004, pp. 1–9.

78. Nicklas W, Baneux P, Boot R, Decelle T, Deeny AA, Fumanelli M, and Illgen-Wilcke B. FELASA (Federation of European Laboratory Animal Science Associations Working Group on Health Monitoring of Rodent and Rabbit Colonies). Recommendations for the health monitoring of rodent and rabbit colonies in breeding and experimental units. *Lab. Anim.* 2002; 36(1): 20–42.

79. Nicklas W, Homberger F, Illgen-Wilcke B, Jacobi K Kraft V, Kunstyr I, Mähler M, Meyer H, and Phlmeyer-Esch G. GV-SOLAS (Working Group on Hygiene). Implications of infectious agents on results of animal experiments. *Lab Anim*. 1999; Jan 33 (Suppl. 1) S1: 39–S2: 87.

80. Schmidt-Nielsen K, *How Animals Work*. Cambridge University Press, London, 1972.

81. Schmidt-Nielsen K. *Animal Physiology, Adaptation and Environment*. Cambridge University Press, London, 1975.

82. Kleiber M. Body size and metabolism. *Hilgardia*. 1932; 6: 135.

83. Hau J and Poulsen OM. Doses for laboratory animals based on metabolic rate. *Scand. J. Lab. Anim. Sci.* 1986; 15: 81.

84. Isolauri E, Arvola T, Sutas Y, Moilanen E, and Salminen S. Probiotics in the management of atopic eczema. *Clin. Exp. Allergy* 2000; 30(11): 1604–1610.

85. Kirjavainen PV, Salminen SJ, and Isolauri E. Probiotic bacteria in the management of atopic disease underscoring the importance of viability. *J. Pediatr. Gastroenterol. Nutr.* 2003; 36: 223–227.

86. Majamaa H, Isolauri E, Saxelin M, and Vesikari T. Lactic acid bacteria in the treatment of acute rotavirus gastroenteritis. *J. Pediatr. Gastroenterol. Nutr.* 1995; 20: 333–338.
87. Salminen S, Bouley C, Boutron-Ruault MC, et al. Gastrointestinal physiology and function-targets for functional food development. *Br. J. Nutr.* 1998; 80: S147–S171.
88. WHO, 2002. Available at http://www.who.int/foodsafety/fs_management/en/probiotic_guidelines.pdf
89. De Vres M, Stegelmann A, Richter B, Fenselau S, Laue C, and Schrezenmeir J. Probiotics compensation for lactase insufficiency. *Am. J. Clin. Nutr.* 2001; 73 (Suppl.): 421S–429S.
90. Deguchi Y, Morishita T, and Mutai M. Comparative studies on synthesis of water-soluble vitamins among human species of *Bifidobacteria*. *Agric. Biol. Chem.* 1985; 49: 13–19.
91. Noda H, Akasaka N, and Ohsug M. Biotin production by *Bifidobacteria*. *J. Nutr. Sci. Vitaminol.* 1994; 40: 181–188.
92. Oda T, Kado-oka Y, and Hashiba H. Effect of *Lactobacillus acidophilus* on iron bioavailability in rats. *J. Nutr. Sci. Vitaminol.* 1994; 40: 613–616.
93. Tannock GW, Dashkevicz MP, and Feighner SD. Lactobacilli and bile salt hydrolase in the murine intestinal tract. *Appl. Environ. Microbiol.* 1989; 55: 1848–1851.
94. Bayless TM. Lactose malabsorption, milk intolerance, and symptom awareness in adults. In: Paige DM, Bayless TM, editors. *Lactose Digestion: Clinical and Nutritional Implications*. Johns Hopkins University Press, Baltimore, 1981, pp. 117–123.
95. Kolars JC, Levitt MD, Aouji M, and Savaiano DA. Yogurt-an autodigesting source of lactose. *N. Engl. J. Med.* 1984; 310: 1–3.
96. Lerebours E, N'Djitoyap Ndam C, Lavoine A, Hellot MF, Antoine JM, and Colin R. Yogurt and fermented-then-pasteurized milk: effects of short-term and long-term ingestion on lactose absorption and mucosal lactase activity in lactase-deficient subjects. *Am. J. Clin. Nutr.* 1989; 49: 823–827.
97. Marteau P, Flourie B, Pochart P, Chastang C, Desjeux J-F, and Rambaud J-C. Effect of the microbial lactase (EC 3.2.1.23) activity in yogurt on the intestinal absorption of lactose: an in vivo study on lactase-deficient humans. *Br. J. Nutr.* 1990; 64: 71–79.
98. Martini MC, Kukielka D, and Savaiano DA. Lactose digestion from yogurt: influence of a meal and additional lactose. *Am. J Clin Nutr.* 1991; 53: 1253–1258.
99. Martini MC, Lerebours EC, Lin WJ, et al. Strains and species of lactic acid bacteria in fermented milks (yogurts): effect on in vivo lactose digestion. *Am. J. Clin. Nutr.* 1991; 54: 1041–1046.
100. Martini MC, Smith DE, and Savaiano DA. Lactose digestion from flavored and frozen yogurts, ice milk, and ice cream by lactase-deficient persons. *Am. J. Clin. Nutr.* 1987; 46: 636–640.
101. Rosado JL, Solomons NW, and Allen LH. Lactose digestion from unmodified, low-fat and lactose-hydrolyzed yogurt in adult lactose-maldigesters. *Eur. J. Clin. Nutr.* 1992; 46: 61–68.
102. Savaiano DA, Abou ElAnouar A, Smith DE, and Levitt MD. Lactose malabsorption from yogurt, pasteurized yogurt, sweet acidophilus milk, and cultured milk in lactase-deficient individuals. *Am. J. Clin. Nutr.* 1984; 40: 1219–1223.
103. Varela-Moreiras G, Antoine JM, Ruiz-Roso B, Varela G, Effects of yogurt and fermented-then-pasteurized milk on lactose absorption in an institutionalized elderly group. *J. Am. Coll. Nutr.* 1992; 11: 168–171.

104. Vesa TH, Marteau PR, Briet FB, Boutron-Ruault MC, and Rambaud JC. Raising milk energy content retards gastric emptying of lactose in lactose-intolerant humans with little effect on lactose digestion. *J. Nutr.* 1997; 127: 2316–2320.

105. Jiang T, Mustapha A, and Savaiano DA. Improvement of lactose digestion in humans by ingestion of unfermented milk containing *Bifidobacterium longum*. *J. Dairy Sci.* 1996; 79: 750–757.

106. Martini MC, Lerebours EC, Lin WJ, et al. Strains and species of lactic acid bacteria in fermented milks (yogurts): effect on in vivo lactose digestion. *Am. J. Clin. Nutr.* 1991; 54: 1041–1046.

107. Montes RG, Bayless TM, Saavedra JM, and Perman JA. Effect of milks inoculated with *Lactobacillus acidophilus* or a yogurt starter culture in lactose-maldigesting children. *J. Dairy Sci.* 1995; 78: 1657–1664.

108. Rosado JL, Solomons NW, and Allen LH. Lactose digestion from unmodified, low-fat and lactose-hydrolyzed yogurt in adult lactose-maldigesters. *Eur. J. Clin. Nutr.* 1992; 46: 61–68.

109. Sanders M, Walker D, Walker K, Aoyama K, and Klaenhammer T. Performance of commercial cultures in fluid milk applications. *J. Dairy Sci.* 1996; 79: 943–955.

110. Babaahmady KG, Challacombe SJ, Marsh PD, and Newman HN. Ecological study of *Streptococcus mutans*. *Streptococcus sobrinus* and *Lactobacillus* spp. at subsites from approximal dental plaque from children. *Caries Res.* 1998; 32: 51–58.

111. Granath L, Cleaton-Jones P, Fatti LP, and Grossman ES. Salivary Lactobacilli explain dental caries better than salivary mutans Streptococci in 4–5-year-old children. *Scand. J. Dent. Res.* 1994; 102: 319–323.

112. Nase L, Hatakka K, Savilaht E, Saxelin M, Ponka A, Poussa T, Korpela R, and Meurman JH. Effect of long-term consumption of a probiotic bacterium, *Lactobacillus rhamnosus* GG, in milk on dental caries and caries risk in children. *Caries Res.* 2001; 35: 412–420.

113. Meurman JH, Antila H, and Salminen S. Recovery of *Lactobacillus* strain GG (ATCC 53103) from saliva of healthy volunteers after consumption of yogurt prepared with the bacterium. *Microbiol. Ecol. Health Dis.* 1994; 7: 295–298.

114. Hatakka K, Ahola AJ, Yli-Knuuttila H, Richardson M, Poussa T, Meurman JH, and Korpela R. Probiotics reduce the prevalence of oral candida in the elderlya randomized controlled trial. *J. dent. Res.* 2007; 86: 125–130.

115. Johnston BC, Supina AL, and Vohra S. Probiotics for pediatric antibiotic-associated diarrhea: a meta-analysis of randomized placebo-controlled trials. *CMAJ* 2006; 175: 377–383.

116. McFarland LV, Meta-analysis of probiotics for the prevention of antibiotic associated diarrhea and the treatment of *Clostridium difficile* disease. *Am. J. Gastroenterol.* 2006; 101: 812–822.

117. Sazawal S, Hiremath G, Dhingra U, Malik P, Deb S, and Black RE. Efficacy of probiotics in prevention of acute diarrhoea: a meta-analysis of masked, randomised, placebo-controlled trials. *Lancet Inf. Dis.* 2006; 6: 374–382.

118. Szajewska H, Ruszczynski M, and Radzikowski A. probiotics in the prevention of antibiotic-associated diarrhea in children: a meta-analysis of randomized controlled trials. *J. Pediatr.* 2006; 149: 367–372.

119. Billoo AG, Memon MA, Khaskheli SA, Murtaza G, Iqbal K, Shekhani MS, and Siddiqi AQ. Role of a probiotic (*Saccharomyces boulardii*) in management and prevention of diarrhea. *World J. Gastroenterol.* 2006; 12: 4557–4560.

120. Boulloche J, Mouterde O, and Mallet E. Traitment des diarrhe'es aigues chez le nourrisson et le jeune enfant. *Ann. Pediatr. (Paris)* 1994; 41: 457–463.
121. Costa-Ribeiro H, Ribeiro TC, Mattos AP, et al. Limitations of probiotic therapy in acute, severe dehydrating diarrhea. *J. Pediatr. Gastroenterol. Nutr.* 2003; 36: 112–115.
122. Glass RI, Lew JF, Gangarosa RE, et al. Estimates of morbidity and mortality rates for diarrheal diseases in American children. *J. Pediatr.* 1991; 118: S27–S33.
123. Guandalini S, Pensabene L, Zikri MA, et al. *Lactobacillus* GG administered in oral rehydration solution to children with acute diarrhea: a multicenter European trial. *J. Pediatr. Gastroenterol. Nutr.* 2000; 30: 54–60.
124. Guarino A, Canani RB, Spagnuolo MI, Albano F, and Di Benedetto L. Oral bacterial therapy reduces the duration of symptoms and of viral excretion in children with mild diarrhea. *J. Pediatr. Gastroenterol. Nutr.* 1997; 25(5): 516–519.
125. Guerrant RL, Hughes JM, Lima NL, and Crane J. Diarrhea in developed and developing countries: magnitude, special settings, and etiologies. *Rev. Infect. Dis.* 1990; 12 (Suppl. 1): S41–S50.
126. Isolauri E, Juntunen M, Rautanen T, Sillanaukee P, and Koivula T. A human *Lactobacillus* strain (*Lactobacillus casei* sp strain GG) promotes recovery from acute diarrhea in children. *Pediatrics* 1991; 88: 90–97.
127. Isolauri E, Kaila M, Mykkanen H, Ling W, and Salminen S. Oral bacteriotherapy for viral gastroenteritis. *Dig. Dis. Sci.* 1994; 39: 2595–2600.
128. Kaila M, Isolauri E, Soppi E, Virtanen E, Laine S, and Arvilommi H. Enhancement of the circulating antibody secreting cell response in human diarrhea by a human *Lactobacillus* strain. *Pediatr. Res.* 1992; 32: 141–144.
129. Kurugol Z and Koturoglu G. Effects of *Saccharomyces boulardii* in children with acute diarrhea. *Acta Paediatr.* 2005; 94: 44–47.
130. Lee MC, Lin LH, Hung KL, and Wu HY. Oral bacterial therapy promotes recovery from acute diarrhea in children. *Acta. Paediatr. Tw.* 2001; 42: 301–305.
131. Majamaa H, Isolauri E, Saxelin M, and Vesikari T. Lactic acid bacteria in the therapy of acute gastroenteritis. *J. Pediatr. Gastroenterol. Nutr.* 1995; 20: 333–338.
132. Pant AR, Graham M, Allen SJ, Harikul S, Sabchareon A, Cuevas L, and Hart CA. *Lactobacillus GG* and acute diarrhea in young children in the tropics. *J. Trop. Pediatr.* 1996; 42: 162–165.
133. Pearce JL and Hamilton JR. Controlled trial of orally administered Lactobacilli in acute infantile diarrhea. *J. Pediatr.* 1974; 84: 261–26.
134. Rosenfeldt V, Michaelsen KF, Jakobsen M, et al. Effect of probiotic *Lactobacillus* strains in young children hospitalized with acute diarrhea. *Pediatr. Infect. Dis. J.* 2002a; 21: 411–416.
135. Rosenfeldt V, Michaelsen KF, Jakobsen M, et al. Effect of probiotic *Lactobacillus* strains on acute diarrhea in a cohort of nonhospitalized children attending day-care centers. *Pediatr. Infect. Dis. J.* 2002b; 21: 417–419.
136. Salazar-Lindo E, Miranda-Langschwager P, Campos-Sanchez M, et al. *Lactobacillus casei* strain GG in the treatment of infants with acute watery diarrhea: a randomized, double-blind, placebo controlled clinical trial. *BMC Pediatr.* 2004; 4: 18.
137. Shornikova AV, Casas IA, Isolauri E, Mykkanen H, and Vesikari T. *Lactobacillus reuteri* as a therapeutic agent in acute diarrhea in young children. *J. Pediatr. Gastroenterol. Nutr.* 1997a; 24: 399–404.

138. Shornikova A, Casas I, Mykkanen H, Salo E, and Vesikari T. Bacteriotherapy with *Lactobacillus reuteri* in rotavirus gastroenteritis. *Pediatr Infect Dis J.* 1997b; 16: 1103–1107.

139. Shornikova AV, Isolauri E, Burkanova L, Lukovnikova S, and Vesikari T. A trial in the Karelian Republic of oral rehydration and *Lactobacillus* GG for treatment of acute diarrhoea. *Acta Paediatr.* 1997c; 86: 460–465.

140. Simakachorn N, Pichaipat V, Rithipornpaisarn P, Kongkaew C, Tongpradit P, and Varavithya W. Clinical evaluation of the addition of lyophilized, heat-killed *Lactobacillus acidophilus* LB to oral rehydration therapy in the treatment of acute diarrhea in children. *J. Pediatr. Gastroenterol. Nutr.* 2000; 30: 68–72.

141. Villarruel G, Rubio DM, Lopez F, Cintioni J, Gurevech R, Romero G, and Vandenplas Y. *Saccharomyces boulardii* in acute childhood diarrhea: a randomized placebo-controlled study. *Acta Paediatr.* 2007; 96: 538–541.

142. Ackermann G, Thomalla S, Achermann F, et al. Prevalence and characteristics of bacteria and host factors in an outbreak situation of antibiotic-associated diarrhea. *J. Med. Microbiol.* 2005; 54 (Pt. 2): 149–153.

143. Adam J, Barret C, Barret-Bellet A, et al. Controlled double-blind clinical trials of Ultra-Levure: multicentre study by 25 physicians in 388 cases. *Gaz. Med. Fr.* 1977; 84: 2072–2078.

144. Armuzzi A, Cremonini F, Bartolozzi F, et al. The effect of oral administration of *Lactobacillus* GG on antibiotic-associated gastrointestinal side-effects during *Helicobacter pylori* eradication therapy. *Aliment. Pharmacol. Ther.* 2001; 15: 163–169.

145. Arvola T, Laiho K, Torkkeli S, et al. Prophylactic *Lactobacillus* GG reduces antibiotic-associated diarrhea in children with respiratory infections: a randomized study. *Pediatrics* 1999; 104: e64.

146. Correa NB, Filho P, Luciano A, et al. A randomized formula controlled trial of *Bifidobacterium lactis* and *Streptococcus thermophilus* for prevention of antibiotic-associated diarrhea in infants. *J. Clin. Gastroenterol.* 2005; 39: 385–389.

147. Cremonini F, Di Caro S, Covino M, et al. Effect of different probiotic preparations on anti-*Helicobacter pylori* therapy-related side effects: a parallel group, triple blind, placebo-controlled study. *Am. J. Gastroenterol.* 2002; 97: 2744–2749.

148. Gotz V, Romankiewicz JA, Moss J, et al. Prophylaxis against ampicillin-associated diarrhea with a *Lactobacillus* preparation. *Am. J. Hosp. Pharm.* 1979; 36: 754–757.

149. Jirapinyo P, Thamonsiri N, Densupsoontorn N, et al. Prevention of antibiotic-associated diarrhea in infants by probiotics. *J. Med. Assoc. Thai.* 2002; 85 (Suppl. 2): S739–S742.

150. Kotowska M, Albrecht P, and Szajewska H. *Saccharomyces boulardii* in the prevention of antibiotic-associated diarrhoea in children: a randomized double-blind placebo-controlled trial. *Aliment. Pharmacol. Ther.* 2005; 21: 583–590.

151. LaRosa M, Bottaro G, Gulino N, et al. Prevention of antibiotic-associated diarrhea with *Lactobacillus sporogens* and fructo-oligosaccharides in children. A multicentric double-blind vs placebo study. *Minerva Pediatr.* 2003; 55(5): 447–452.

152. Levy DG, Stergachis A, McFarland LV, et al. Antibiotics and *Clostridium difficile* diarrhea in the ambulatory care setting. *Clin. Ther.* 2000; 22: 91–102.

153. Lewis SJ, Potts LF, and Barry RE. The lack of therapeutic effect of *Saccharomyces boulardii* in the prevention of antibiotic-related diarrhoea in elderly patients. *J. Infect.* 1998; 36: 171–174.

154. McFarland LV, Epidemiology, risk factors and treatments for antibiotic-associated diarrhea. *Dig. Dis.* 1998; 16: 292–307.
155. McFarland LV, Normal flora: diversity and functions. *Microb. Ecol. Health Dis.* 2000; 12: 193–207.
156. McFarland LV, Surawicz CM, Greenberg RN, et al. Prevention of beta-lactam-associated diarrhea by *Saccharomyces boulardii* compared with placebo. *Am. J. Gastroenterol.* 1995; 90: 439–448.
157. Nista EC, Candelli M, Cremonini F, et al. *Bacillus clausii* therapy to reduce side-effects of anti-*Helicobacter pylori* treatment: randomized, double-blind, placebo controlled trial. *Aliment. Pharmacol. Ther.* 2004; 20: 1181–1188.
158. Orrhage K, Brismar B, and Nord CE. Effects of supplements of *Bifidobacterium longum* and *Lactobacillus acidophilus* on the intestinal microbiota during administration of clindamycin. *Microb. Ecol. Health Dis.* 1994; 7: 17–25.
159. Seki H, Shiohara M, Matsumura T, et al. Prevention of antibiotic-associated diarrhea in children by *Clostridium butyricum* MIYAIRI. *Pediatr. Int.* 2003; 45: 86–90.
160. Surawicz CM, Elmer GW, Speelman P, McFarland LV, Chinn J, and van Belle G. Prevention of antibiotic-associated diarrhea by *Saccharomyces boulardii*: a prospective study. *Gastroenterology* 1989; 96: 981–988.
161. Szajewska H, Kotowska M, Mrukowicz JZ, et al. Efficacy of *Lactobacillus GG* in prevention of nosocomial diarrhea in infants. *J. Pediatr.* 2001; 138: 361–365.
162. Tankanow RM, Ross MB, Ertel IJ, et al. A double-blind, placebo-controlled study of the efficacy of Lactinex in the prophylaxis of amoxicillin-induced diarrhea. *DICP* 1990; 24: 382–384.
163. Thomas MR, Litin SC, Osmon DR, Corr AP, Weaver AL, and Lohse CM. Lack of effect of *Lactobacillus* GG on antibiotic-associated diarrhea: a randomized, placebo-controlled trial. *Mayo Clin. Proc.* 2001; 76: 883–889.
164. Vanderhoof JA, Whitney DB, Antonson DL, Hanner TL, Lupo JV, and Young RJ. *Lactobacillus* GG in the prevention of antibiotic-associated diarrhea in children. *J. Pediatr.* 1999; 135: 564–568.
165. Winston DJ, Ho WG, Bruckner DA, et al. Beta-lactam antibiotic therapy in febrile granulocytopenic patients. A randomized trial comparing cefoperazone plus piperacillin, ceftazidime plus piperacillin, and imipenem alone. *Ann. Intern. Med.* 1991; 115: 849–859.
166. Witsell DL, Garrett G, Yarbrough WG, et al. Effect of *Lactobacillus acidophilus* on antibiotic-associated gastrointestinal morbidity: a prospective randomized trial. *J. Otolarynogol.* 1995; 24: 231–233.
167. Wunderlich PF, Braun L, Fumagalli I, et al. Double-blind report on the efficacy of lactic acid-producing *Enterococcus* SF68 in the prevention of antibiotic-associated diarrhoea and in the treatment of acute diarrhoea. *J. Int. Med. Res.* 1989; 17: 333–338.
168. Barbut F and Petit JC. Epidemiology of *Clostridium difficile*-associated infections. *Clin. Microbiol. Infect.* 2001; 7: 405–410.
169. Barbut F, Richard A, Hamadi K, Chomette V, Burghoffer B, and Petit JC. Epidemiology of recurrences or reinfections of *Clostridium difficile*-associated diarrhea. *J. Clin. Microbiol.* 2000; 38: 2386–2388.
170. Lawrence SJ, Korzenik JR, and Mundy LM. Probiotics for recurrent *Clostridium dificile* disease. *J. Med. Microbiol.* 2005; 54: 905–906.

171. McFarland LV, Surawicz CM, Greenberg RN, et al. A randomized placebo-controlled trial of *Saccharomyces boulardii* in combination with standard antibiotics for *Clostridium difficile* disease. *JAMA* 1994; 271: 1913–1918.

172. McFarland LV, Surawicz CM, Rubin M, et al. Recurrent *Clostridium difficile* disease: epidemiology and clinical characteristics. *Infect. Control Hosp. Epidemiol.* 1999; 20: 43–50.

173. Plummer S, Weaver MA, Harris JC, et al. *Clostridium difficile* pilot study: effects of probiotic supplementation on the incidence of *Clostridium difficile* diarrhea. *Int. Microbiol.* 2004; 7: 59–62.

174. Surawicz CM, McFarland LV, Greenberg RN, et al. The search for a better treatment for recurrent *Clostridium difficile* disease: use of high dose vancomycin combined with *Saccharomyces boulardii*. *Clin. Infect. Dis.* 2000; 31: 1012–1017.

175. Wullt M, Hagslatt ML, and Odenholt I. *Lactobacillus plantarum* 299v for the treatment of recurrent *Clostridium difficile*-associated diarrhoea: a double-blind, placebo-controlled trial. *Scand. J. Infect. Dis.* 2003; 35: 365–367.

176. Delia P, Sansotta G, Donato V, Frosina P, Messina G, De Renzis C, and Famularo G. Use of probiotics for prevention of radiation-induced diarrhea. *World J. Gastroenterol.* 2007; 13: 912–915.

177. Donner CS, Pathophysiology and therapy of chronic radiation-induced injury to the colon. *Dig. Dis.* 1998; 16: 253–261.

178. Adachi JA, Ericsson CD, Jiang ZD, et al. Azithromycin found to be comparable to levoflozacin for the treatment of US travelers with acute diarrhea acquired in Mexico. *Clin. Infect. Dis.* 2003; 37: 1165–1171.

179. Black F, Anderson P, Orskov J, Gaarslev K, and Laulund S. Prophylactic efficacy of *Lactobacilli* on travelers' diarrhea. *Travel Med.* 1989; 7: 333–335.

180. Cheng AC and Thielman NM. Update on traveler's diarrhea. *Curr. Infect. Dis. Rep.* 2002; 4: 70–77.

181. Clarke SC, Diarrhoeagenic *Escherichia coli* an emerging problem? *Diagn. Microbiol. Infect. Dis.* 2001; 41: 93–98.

182. Hilton E, Kolakowski P, Singer C, and Smith M. Efficacy of *Lactobacillus* GG as a diarrheal preventive in travelers. *J. Travel. Med.* 1997; 4: 41–43.

183. Katelaris PH, Salam I, and Farthing MJ. *Lactobacilli* to prevent traveler's diarrhea? *N. Engl. J. Med.* 1995; 333: 1360–1361.

184. Kollaritsch H, Kremsner P, Wiedermann G, and Scheiner O. Prevention of traveler's diarrhea: comparison of different nonantibiotic preparations. *Travel Med. Int.* 1989: 9–17.

185. Kollaritsch H, Holst H, Grobara P, and Wiedermann G. Prophylaxe der reisediarrhoe mit *Saccharomyces boulardii* [Prevention of traveler's diarrhea with *Saccharomyces boulardii*. Results of a placebo controlled double-blind study]. *Fortschr. Med.* 1993; 111: 152–156.

186. McFarland LV. Meta-analysis of probiotics for the prevention of traveler's diarrhea. *Travel Med. Infect. Dis.* 2007; 5(2): 97–105.

187. Oksanen PJ, Salminen S, Saxelin M, et al. Prevention of travelers diarrhea by lactobacillus GG. *Ann. Med.* 1990; 22: 53–56.

188. Pozo-Olano JD, Warram Jr. JH, Gomez RG, and Cavazos MG. Effect of a Lactobacilli preparation on traveler's diarrhea. A randomized, double blind clinical trial. *Gastroenterology* 1978; 74: 829–830.

189. Sanders JW and Tribble DR. Diarrhea in the returned traveler. *Curr Gastroenterol. Rep.* 2001; 3: 304–314.

190. Bleichner G, Blehaut H, Mentec H, and Moyse D. *Saccharomyces boulardii* prevents diarrhea in critically ill tube-fed patients. A multicenter, randomized, double-blind placebo-controlled trial. *Intensive. Care Med.* 1997; 23: 517–523.

191. Heimburger DC, Sockwell DG, and Geels WJ. Diarrhea with enteral feeding: prospective reappraisal of putative causes. *Nutrition* 1994; 10: 392–396.

192. Hill SA, Nielsen MS, and Lennard-Jones JE. Nutritional support in intensive care units in England and Wales: a survey. *Eur. J. Clin. Nutr.* 1995; 49: 371–378.

193. Silk DBA, Cottam TK, Nielsen MS, Elcoat C, Furness KM, Howard JP, Lennard-Jones JE, and Plester CE. Organization of Nutritional Support in Hospitals. British Association Parenteral & Enteral Nutrition Working Party Report. BAPEN, Maidenhead, 1994.

194. Smith CE, Marien L, Brogdon C, Faust-Wilson P, Lohr G, Gerald KB, and Pingleton S. Diarrhea associated with tube feeding in mechanically ventilated critically ill patients. *Nurs. Res.* 1990; 39: 148–152.

195. Chadwick VS, Chen W, Shu D, et al. Activation of the mucosal immune system in irritable bowel syndrome. *Gastroenterology* 2002; 122: 1778–1783.

196. Drossman DA, Camilleri M, Mayer EA, and Whitehead WE. AGA technical review on irritable bowel syndrome. *Gastroenterology* 2002; 123: 2108–2131.

197. Guyonnet D, Chassany O, Ducrotte P, Picard C, Mouret M, Mercier CH, and Matuchansky C. Effects of fermented milk containing *Bifidobacterium animalis* DN-173 010 on the health-related quality of life and symptoms in irritable bowel syndrome in adults in primary care: a multicentre, randomized, double-blind, controlled trial. *Aliment. Pharmacol. Ther.* 2007; 26: 475–486.

198. Hillila MT and Farkkila MA. Prevalence of irritable bowel syndrome according to different diagnostic criteria in a nonselected adult population. *Aliment. Pharmacol. Ther.* 2004; 20: 339–345.

199. Kajander K, Hatakka K, Poussa T, Farkkila M, and Korpela R. A probiotic mixture alleviates symptom in irritable bowel syndrome patients: a controlled 6-month intervention. *Aliment. Pharmacol. Ther.* 2005; 222: 387–394.

200. Kim HJ, Vazquez Roque MI, Camilleri M, et al. A randomized controlled trial of a probiotic combination VSL.3 and placebo in irritable bowel syndrome with bloating. *Neurogastroenterol. Motil.* 2005; 17: 687–696.

201. King TS, Elia M, and Hunter JO. Abnormal colonic fermentation in irritable bowel syndrome. *Lancet* 1998; 352: 1187–1189.

202. Koebnick C, Wagner I, Leitzmann P, Stern U, and Zunft HJF. Probiotic beverage containing *Lactobacillus casei* Shirota improves gastrointestinal symptoms in patients with chronic constipation. *Can. J. Gastroenterol.* 2003; 17: 655–659.

203. Linskens RK, Huijsdens XW, Savelkoul PH, Vandenbroucke-Grauls CM, Meuwissen SG, The bacterial flora in inflammatory bowel disease: current insights in pathogenesis and the influence of antibiotics and probiotics. *Scand. J. Gastroenterol. Suppl.* 2001; 36: 29–40.

204. Niedzielin K, Kordecki H, and Birkenfeld B. A controlled, double-blind, randomized study on the efficacy of *Lactobacillus plantarum* 299v in patients with irritable bowel syndrome. *Eur. J. Gastroenterol. Hepatol.* 2001; 13: 1143–1147.

205. Nobaek S, Johansson ML, Molin G, Ahrne S, and Jeppsson B. Alteration of intestinal microflora is associated with reduction in abdominal bloating and pain in patients with irritable bowel syndrome. *Am. J. Gastroenterol.* 2000; 95: 1231–1238.
206. O'Sullivan MA and O'Morain CA. Bacterial supplementation in the irritable bowel syndrome. A randomized double-blind placebo-controlled crossover study. *Dig. Liver Dis.* 2000; 32: 294–301.
207. Sen S, Mullan MM, Parker TJ, Woolner JT, Tarry SA, and Hunter JO. Effect of *Lactobacillus plantarum* 299v on colonic fermentation and symptoms of irritable bowel syndrome. *Dig. Dis. Sci.* 2002; 47: 2615–2620.
208. Treem WR, Ahsan N, Kastoff G, and Hyams JS. Fecal short-chain fatty acids in patients with diarrhea-predominant irritable bowel syndrome: *in vitro* studies of carbohydrate fermentation. *J. Pediatr. Gastroenterol. Nutr.* 1996; 23: 280–286.
209. Bibiloni R, Fedorak RN, Tannock GW, Madsen KL, Gionchetti P, Campieri M, de Simone C, and Sartor RB. VSL# 3 probiotic-mixture induces remission in patients with active ulcerative colitis. *Am. J. Gastroenterol.* 2005; 100: 1539–1546.
210. Gionchetti P, Amadini C, Rizzello F, Venturi A, Poggioli G, and Campieri M. Probiotics for the treatment of postoperative complications following intestinal surgery. *Best Pract. Res. Clin. Gastroenterol.* 2003; 17: 821–831.
211. Gionchetti P, Rizzello F, Helwig U, et al. Prophylaxis of pouchitis onset with probiotic therapy: a double-blind, placebo-controlled trial. *Gastroenterology* 2003; 124: 1202–1209.
212. Gionchetti P, Rizzello F, Venturi A, et al. Oral bacteriotherapy as maintenance treatment in patients with chronic pouchitis: a double-blind, placebo-controlled trial. *Gastroenterology* 2000; 119: 305–309.
213. Guslandi M, Mezzi G, Sorghi M, and Testoni PA. *Saccharomyces boulardii* in maintenance treatment of Crohn's disease. *Dig. Dis. Sci.* 2000; 45: 1462–1464.
214. Kruis W, Schutz E, Fric P, Fixa B, Judmaier G, and Stolte M. Double-blind comparison of an oral *Escherichia coli* preparation and mesalazine in maintaining remission of ulcerative colitis. *Aliment. Pharmacol. Ther.* 1997; 11: 853–858.
215. Prantera C, Scribano ML, Falasco G, Andreoli A, and Luzi C. Ineffectiveness of probiotics in preventing recurrence after curative resection for Crohn's disease: a randomised controlled trial with *Lactobacillus* GG. *Gut* 2002; 51: 405–409.
216. Rembacken BJ, Snelling AM, Hawkey PM, Chalmers DM, and Axon AT. Nonpathogenic *Escherichia coli* versus mesalazine for the treatment of ulcerative colitis: a randomised trial. *Lancet* 1999; 354: 635–639.
217. Sartor RB, Therapeutic manipulation of the enteric microflora in inflammatory bowel diseases: antibiotics, probiotics, and prebiotics. *Gastroenterology* 2004; 126: 1620–1633.
218. Schultz M, Timmer A, Herfarth HH, Sartor RB, Vanderhoof JA, and Rath HC. *Lactobacillus* GG in inducing and maintaining remission of Crohn's disease. *BMC Gastroenterol.* 2004; 4: 5–8.
219. Aiba Y, Suzuki N, Kabir AM, et al. Lactic acid-mediated suppression of *Helicobacter pylori* by the oral administration of *Lactobacillus salivarius* as a probiotic in a gnotobiotic murine model. *Am. J. Gstroenterol.* 1998; 93: 2097–2101.
220. Armuzzi A, Cremonini F, Bartolozzi F, et al. The effect of oral administration of *Lactobacillus* GG on antibiotic-associated gastrointestinal side-effects during *Helicobacter pylori* eradication therapy. *Aliment. Pharmacol. Ther.* 2001; 15: 163–169.

221. Bell GD, Powell K, Burridge SM, et al. Experience with 'triple' anti-*Helicobacter pylori* eradication therapy: side effects and the importance of testing the pre-treatment bacterial isolate for metronidazole resistance. *Aliment. Pharmacol. Ther.* 1992; 6: 427–435.
222. Boot H, de Jong D, van Heerde P, et al. Role of *Helicobacter pylori* eradication in high grade MALT lymphoma. *Lancet* 1995; 346: 448–449.
223. Brenes F, Ruiz B, Correa P, et al. *Helicobacter pylori* causes hyperproloferation of the gastric epithelium: pre- and posteradication indices of proliferating cell nuclear antigen. *Am. J. Gastroenterol.* 1993; 88: 1870–1975.
224. Canducci F, Armuzzi A, Cremonini F, et al. A lyophilized and inactivated culture of *Lactobacillus acidophilus* increases *Helicobacter pylori* eradication rates. *Aliment. Pharmacol. Ther.* 2000; 14: 1625–1629.
225. Cao YJ, Qu CM, Yuan Q, et al. Control of intestinal flora alteration induced by eradication therapy of *Helicobacter pylori* infection in the elders. *Chin. J. Gastroenterol. Hepatol.* 2005; 14: 195–199.
226. Cremonini F, Di Caro S, Covino M, et al. Effect of different probiotic preparations on anti-*Helicobacter pylori* therapy-related side effects: a parallel group, triple blind, placebo-controlled study. *Am. J. Gastroenterol.* 2002; 97: 2744–2749.
227. Go MF, Natural history and epidemiology of *Helicobacter pylori* infection. *Aliment. Pharmacol. Ther.* 2002; 16 (Suppl. 1): 3–15.
228. Goldman CG, Barrado DA, Balcarce N, Rua EC, Oshiro M, Calcagno ML, Janjetic M, Fuda J, Weill R, Salgueiro MJ, Valencia ME, Zubillaga MB, and Boccio JR. Effect of a probiotic food as an adjunvant to triple therapy for eradication of *Helicobacter pylori* infection in children. *Nutrition* 2006; 22: 984–988.
229. Guo JB, Yang PF, Wang MT, et al. The application of clostridium to the eradication of *Helicobacter pylori*. *Chin. J. Celiopathy.* 2004; 4: 163–165.
230. Myllyluoma E, Veijola L, Ahlroos T, et al. Probiotic supplementation improves tolerance to *Helicobacter pyroli* eradication therapya placebo-controlled, double-blind randomized pilot study. *Aliment. Pharmacol. Ther.* 2005; 21: 1263–1272.
231. Nista EC, Candelli M, Cremonini F, et al. *Bacillus clausii* therapy to reduce side effects of anti-*Helicobacter pylori* treatment: randomized, double-blind, placebo controlled trial. *Aliment. Pharmacol. Ther.* 2004; 20: 1181–1188.
232. Pinchuk IV, Bressollier P, Verneuil B, et al. In vitro anti-*Helicobacter pylori* activity of the probiotic strain *Bacillus subtilis* 3 is due to secretion of antibiotics. *Antimicrob. Agents Chemother.* 2001; 45: 3156–3161.
233. Rauws EA and Tytgat GN. Cure of duodenal ulcer associated with eradication of *Helicobacter pylori*. *Lancet* 1990; 335: 1233–1235.
234. Sheu BS, Wu JJ, Lo CY, et al. Impact of supplement with *Lactobacillus*- and *Bifidobacterium*-containing yogurt on triple therapy for *Helicobacter pylori* eradication. *Aliment. Pharmacol. Ther.* 2002; 16: 1669–1675.
235. Sykora J, Valeckova K, Amlerova J, et al. Effects of a specially designed fermented milk product containing probiotic *Lactobacillus casei* DN-114 001 and the eradication of *H. pylori* in children: a prospective randomized double-blind study. *J. Clin. Gastroenterol.* 2005; 39: 692–698.
236. Anderson ADG, McNaught CE, Jain PK, and MacFie J. Randomized clinical trial of symbiotic therapy in elective surgical patients. *Gut* 2004; 53: 241–245.

237. Kanazawa H, Nagino M, Kamiya S, Komatsu S, Mayumi T, Takagi K, Asahara T, Nomoto K, Tanaka R, and Nimura Y. Synbiotics reduce postoperative infectious complications: a randomized controlled trial in biliary cancer patients undergoing hepatectomy. *Langenbecks Arch. Surg.* 2005; 390: 104–113.

238. McNaught CE, Woodcock NP, MacFie J, and Mitchell CJ. A prospective randomized study of the probiotic *Lactobacillus platarum* 299V on indices of gut barrier function in elective surgical patients. *Gut* 2005; 51: 827–831.

239. Nagino M, Kamiya J, Uesaka K, Sano T, Yamamoto H, Hayakawa N, Kanai M, and Nimura Y. Complications of hepatectomy for hilar cholangiocarcinoma. *World J. Surg.* 2001; 27: 1277–1283.

240. Rayes N, Seehofer D, Muller AR, Hansen S, Bengmark S, and Neuhaus P. Influence of probiotics and fiber on the incidence of bacterial infections following major abdominal surgeryresults of a prospective trial. *Z. Gastroenterol.* 2002; 40: 869–876.

241. Shigeta H, Nagino M, Kamiya J, Uesaka K, Sano T, Yamamoto H, Hayakawa N, Kanai M, and Nimura Y. Bacteremia after hepatectomy: an analysis of single center. 10-year experience with 407 patients. *Langenbecks Arch. Surg.* 2002; 387: 117–124.

242. Tancrede CH and Andremont AO. Bacterial translocation and Gram-negative bacteremia in patients with hematological malignancies. *J. Infect. Dis.* 1985; 152: 99–103.

243. De Vrese M, Winkler P, Rautenberg P, Harder T, Noah C, Laue C, Ott S, Hampe J, Schreiber S, Heller K, and Jurgen S. Effect of *Lactobacillus gaseri* PA 16/8, *Bifidobacterium longum* SP 07/3, *B. bifidum* MF 20/5 on common cold episodes: a double blind, randomized, controlled trial. *Clin. Nutr.* 2005; 24: 481–491.

244. De Vrese M, Winkler P, Rautenberg P, Harder T, Noah C, Laue C, Ott S, Hampe J, Schreiber S, Heller K, and Jurgen S. Probiotic bacteria reduced duration and severity but not the incidence of common cold episodes in a double blind, randomized, controlled trial. *Vaccine* 2006; 24: 6670–6674.

245. Habermann W, Zimmermann K, Skarabis H, Kunze R, and Rusch V. Influence of a bacterial immunostimulant (human *Enterococcus faecalis* bacteria) on the recurrence of relapses in patient with chronic bronchitis. *Arzneimittelforschung* 2001; 51: 931–937.

246. Habermann W, Zimmermann K, Skarabis H, Kunze R, and Rusch V. Reduction of acute relapses in patients with chronic recurrent hypertrophic sinusitis during treatment with a bacterial immunostimulant (*Enterocucus faecalis* bacteria of human origina medical probiotic). *Arzneimittelforschung* 2002; 52: 622–627.

247. Hatakka K, Savilahti E, Ponka A, Meurman JH, Poussa T, Nase L, et al. Effect of long term consumption of probiotic milk on infections in children attending day care centers: double blind, randomized trial. *BMJ* 2001; 322: 1327–1329.

248. Turchet P, Laurenzano M, Auboiron S, and Antoine JM. Effect of fermented milk containing the probiotic *Lactobacillus casei* DN-114001 on winter infections in free-living elderly subjects: a randomized, controlled pilot study. *Nutr. Health Aging* 2003; 7: 75–77.

249. Brouwer ML, Wolt-Plompen SAA, Dubois AE, ven der Heide S, Jansen DF, Hoijer MA, and Holgate ST. The epidemic of allergy and asthma. *Nature* 1999; 402 (6760 Suppl.): B2–B4.

250. Helin T, Haahtela S, and Haahtela T. No effect of oral treatment with an intestinal bacterial strain, *Lactobacillus rhamnosus* (ATCC 53103), on birch-pollen allergy: a placebo-controlled double-blind study. *Allergy* 2002; 57: 243–246.

251. Holgate ST. The epidemic of allergy and asthma. *Nature* 1999; 402 (6760 Suppl.): B2–B4.

252. Isolauri E, Arvola T, Suta Y, Moilanen E, and Salminen S. Probiotics in the management of atopic ecema. *Clin. Expt. Allergy* 2000; 30: 1604–1610.

253. Isolauri E, Huurre A, Salminen S, and Impivaara O. The allergy epidemic extends beyond the past few decades. *Clin. Exp. Allergy* 2004 Jul; 34(7): 1007–1010.

254. Kalliomaki M, Kirjavainen P, Eerola E, Kero P, Salminen S, and Isolauri E. Distinct patterns of neonatal gut microflora in infants in whom atopy was and was not developing. *J. Allergy Clin. Immunol.* 2001 Jan; 107(1): 129–134.

255. Kalliomaki M, Salminen S, Arvilommi H, Kero P, Koskinen P, and Isolauri E. Probiotics in primary prevention of atopic disease: a randomized placebo-controlled trial. *Lancet* 2001; 357: 1076–1079.

256. Kalliomaki M, Salminen S, Poussa T, Arvilommi H, and Isolauri E. Probiotics and prevention of atopic disease: 4-year follow-up of a randomized placebo-controlled trial. *Lancet* 2003; 361: 1869–1871.

257. Kalliomaki M, Salminin S, Poussa T, and Isolauri E. Probiotics during the first 7 years of life: a cumulative risk reduction of eczema in a randomized, placebo-controlled trial. *J. Allergy Clin. Immunol.* 2007; 119: 1019–1021.

258. Kirjavainen PV, Salminen SJ, and Isolauri E. Probiotic bacteria in the management of atopic disease: underscoring the importance of viability. *J. Pediatr. Gastroenterol. Nutr.* 2003; 36: 223–227.

259. Kukkonen K, Savilahti E, Haahtela T, Juntunen-Backman K, Korpela R, Poussa T, Tuure T, and Kuitunen M. Probiotics and prebiotic galacto-oligosaccharides in the prevention of allergic diseases: a randomized, double-blind, placebo-controlled trial. *J. Allergy Clin. Immunol.* 2007; 119: 192–198.

260. Matricardi PM, Rosmini F, Riondino S, et al. Exposure to foodborne and orofecal microbes versus airborne viruses in relation to atopy and allergic asthma: epidemiological study. *BMJ* 2000; 320: 412–417.

261. Paunio M, Heinonen OP, Virtanen M, Leinikki P, Patja A, and Peltola H. Measles history and atopic diseases: a population-based cross-sectional study. *JAMA* 2000; 283: 343–346.

262. Pekkanen J, Remes ST, Husman T, Lindberg M, Kajosaari M, Koivikko A, and Soininen L. Prevalence of asthma symptoms in video and written questionnaires among children in four regions of Finland. *Eur. Respir. J.* 1997; 10: 1787–1794.

263. Prescott SL and Bjorksten B. Probiotics for the prevention or treatment of allergic diseases. *J. Allergy Clin. Immunol.* 2007 Aug; 120(2): 255–262.

264. Rautava S, Kalliomaki M, and Isolauri E. Probiotics during pregnancy and breast-feeding might confer immunomodulatory protection against atopic disease in the infant. *J. Allergy Clin. Immunol.* 2002 Jan; 109(1): 119–121.

264a. Wickens K, Black PN, Stanley TV, Mitchell E, Fitzharris P, Tannock GW, Purdie G, Crane J, Probiotic Study Group. A differential effect of 2 probiotics in the prevention of eczema and atopy: A double-blind, randomized, placebo-controlled trial. *J. Allergy Clin. Immunol.* 2008 Aug 31 [Epub ahead of print].

264b. Huarre A, Laitinen K, Rautava S, Korkeamäki M, Isolauri E. Impact of maternal atopy and probiotic supplementation during pregnancy on infant sensitization: A double-blind placebo-controlled study. *Clin. Exp. Allergy.* 2008; 38: 1342–1348.

265. Remes ST, Korppi M, Kajossari M, Koivikko A, Soininen L, and Pekkanen J. Prevalence of allergic rhinitis and atopic dermatitis among children in four regions of Finland. *Allergy* 1998; 53: 682–689.

266. Rosenfeldt V, Benfeldt E, Valerius NH, Paerregaard A, and Michaelsen KF. Effect of probiotics on gastrointestinal symptoms and small intestinal permeability in children with atopic dermatitis. *J. Pediatr.* 2004; 145: 612–616.
267. Shaheen SO, Aaby P, Hall AJ, et al. Measles and atopy I Guinea-Bissau. *Lancet* 1996; 347: 1792–1796.
268. Shirakawa T, Enomoto T, Shimazu S, and Hopkin JM. The inverse association between tuberculin responses and atopic disorder. *Science* 1997; 275: 77–79.
269. Strachan DP, Hay feve, hyfiene, and household size. *BMJ* 1989; 299: 1259–1260.
270. Taylor AL, Dunstan JA, and Prescott SL. Probiotic supplementation for the first 6 months of life fails to reduce the risk of atopic dermatitis and increases the risk of allergen sensitization in higher-risk children: a randomized controlled trial. *J. Allergy Clin. Immunol.* 2007; 119: 184–191.
271. Vichyanond P, Jirapongsananuruk O, Visitsuntorn N, and Tuchinda M. Prevalence of asthma, rhinitis and eczema in children from the Bangkok area using the ISAAC (International Study for Asthma and Allergy in Children) questionnaires. *J. Med. Assoc. Thai.* 1998; 81: 175–184.
272. Viljanen M, Savilahti E, Haahtela T, Juntunen-Backman K, Korpela R, Poussa T, Tuure T, and Kuitunen M. Probiotics in the treatment of atopic eczema/dermatitis syndrome in infants: a double-blind placebo-controlled trial. *Allergy* 2005; 60: 494–500.
273. Weston S, Halbert A, Richmond P, and Prescott SL. Effects of probiotics on atopic dermatitis: a randomized controlled trial. *Arch. Dis. Childhood* 2005; 90: 892–897.
274. Xiao JZ, Kondo S, Yanagisawa N, Takahashi N, Odamaki T, Iwabuchi N, Miyaji K, Iwatsuki K, Togashi H, Enomoto K, and Enomoto T. Probiotics in the treatment of Japanese cedar pollinosis: a double-bland placebo-controlled trial. *Clin. Exp. Allergy* 2006; 36: 1425–1435.
275. Aso Y, and Akazan H. Prophylactic effects of a *Lactobacillus casei* preparation on the recurrence of superficial bladder cancer. *Urol. Int.* 1992; 49: 125–129.
276. Aso Y, Akaza H, Tsukamoto T, Imai K, and Naito S. Preventive effect of a *Lactobacillus casei* preparation on the recurrence of superficial bladder cancer in a double-blind trial. *Eur. Urol.* 1995; 27: 104–109.
277. De Roos N and Katan M. Effects of probiotic bacteria on diarrhea, lipid metabolism, and carcinogenesis: a review of papers published between 1988 and 1998. *Am. J. Clin. Nutr.* 2000; 71: 405–411.
278. Hill MJ, Draser BS, Hawksworth G, et al. Bacteria and aetiogy of cancer of the large bowel. *Lancet* 1971; 1: 95–100.
279. Hirayama K, and Rafter J. The role of lactic acid bacteria in colon cancer prevention: mechanistic considerations. *Ant. Leeu.* 1999; 76: 391–294.
280. Kanazawa K, Konishi F, Mitsuoka T, et al. Factors influencing the development of sigmoid colon cancer. Bacteriology and biochemical studies. *Cancer* 1996; 77: 1701–1706.
281. McFarland LA, A review of the evidence of health claims for biotherapeutic agents. *Microb. Ecol. Health Dis.* 2000; 12: 65–76.
282. Moore WE and Moore LH. Intestinal flora of population that have a high risk of colon cancer. *Appl. Environ. Microbiol.* 1995; 61: 3202–3207.
283. Ohashi Y, Nakai S, Tsukamoto T, Masumori N, Akaza H, Miyanaga N, Kitamura T, Kawabe K, Kotake T, Kuroda M, Naito S, Koga H, Saito Y, Nomata K, Kitagawa M, and

Aso Y. Habitual Intake of lactic acid bacteria and risk reduction of bladder cancer. *Urol Int.* 2002; 68: 273–280.

284. Pryor M, Slattery ML, Robinson LM, and Egger M. Adolescent diet and breast cancer in Utah. *Cancer Res.* 1989; 49: 2161–2167.

285. Van't Veer P, Dekker JM, Lamers JWJ, Kok FJ, Schouten EG, Brants HAM, Sturmans F, and Hermus RJJ. Consumption of fermented milk products and breast cancer: a case-control study in the Netherlands. *Cancer Res.* 1989; 49: 4020–4023.

286. Agerbaek M, Gerdes LU, and Richelsen B. Hypocholesterolaemic effect of a new fermented milk product in healthy middle-aged men. *Eur. J. Clin. Nutr.* 1995; 49: 346–352.

287. De Roos NM, Schouten G, and Katan MB. Yogurt enriched with *Lactobacillus acidophilus* does not lower blood lipids in healthy men and women with normal to borderline high serum cholesterol levels. *Eur. J. Clin. Nutr.* 1999; 53: 277–280.

288. De Smet I, De Boever P, and Verstraete W. Cholesterol lowering in pigs through enhanced bacterial bile hydrolase activity. *Br. J. Nutr.* 1998; 79: 185–194.

289. Gilliland SE, Nelson CR, and Maxwell C. Assimilation of cholesterol by *Lactobacillus acidophilus*. *Appl. Environ. Microbiol.* 1985; 49: 377–381.

290. Greany KA, Nettleton JA, Wangen KE, Thomas W, and Kurzer MS. Probiotic consumption does not enhance the cholesterol-lowering effect of soy in postmenopausal women. *J. Nutr.* 2004; 134: 3277–3283.

291. Hlivak P, Odraska J, Ferencik M, Ebringer L, Jahnova E, and Mikes Z. One-year application of probiotic strain *Enterococcus faecium* M-74 decreases serum cholesterol levels. *Bratisl. Lek. Listy.* 2005; 106: 67–72.

292. Kießling G, Schneider J, and Jahreis G. Long-term consumption of fermented dairy products over 6 months increases HDL cholesterol. *Eur. J. Clin. Nutr.* 2002; 56: 843–849.

293. Klaver FA and Meer RVD. The assumed assimilation of cholesterol by Lactobacilli and *Bifidobacterium bifidum* is due to their bile salt-deconjugating activity. *Appl. Environ. Microbiol.* 1993; 59: 1120–1124.

294. Law MR, Wald NJ, Wu T, Hackshaw A, and Bailey A. Systematic underestimation of association between serum cholesterol concentration and ischaemic heart disease in observational studies: data from BUPA study. *Bret. Med. J.* 1994; 308: 363–366.

295. Lewis SJ and Burmeister S. A double-blind placebo-controlled study of the effects of *Lactobacillus acidophilus* on plasma lipids. *Eur. J. Clin. Nutr.* 2005; 59: 776–780.

296. Lin SY, Ayres JW, Winkler W, and Sandine WE. *Lactobacillus* effects on cholesterol: *in vitro* and *in vivo* results. *J. Dairy Sci.* 1989; 72: 2885–2899.

297. Manson JE, Tosteson H, Ridker PM, Satterield S, Hebert P, and O'Connor GT. The primary prevention of myocardial infaction. *N. Engl. J. Med.* 1992; 326: 1406–1416.

298. Pereira DI and Gibson GR. Effects of consumption of probiotics and prebiotics on serum lipid levels in humans. *Crit. Rev. Biochem. Mol. Biol.* 2002; 37(4): 259–281.

299. Richelsen B, Kristensen K, and Pedersen SB. Long-term (6 months) effect of a new fermented milk product on the level of plasma lipoproteinsa placebo-controlled and double blind study. *Eur. J. Clin. Nutr.* 1996; 50: 811–815.

300. Schaafsma G, Meuling WJ, Dokkum W, and Bouley C. Effects of a milk product, fermented by *Lactobacillus acidophilus* and with fructo-oligosaccharides added, on blood lipids in male volunteers. *Eur. J. Clin. Nutr.* 1998; 52: 436–440.

301. Simons LA, Armansec SG, and Conway P. Effect of *Lactobacillus fermentum* on serum lipids in subjects with elevated serum. *Nutr. Met. Cardio. Dis.* 2006; 16: 531–535.
302. De Vrese M, Rautenberg P, Laue C, Koopmans M, Herremans T, and Schrezenmeir J. Probiotic bacteria stimulate virus-specific neutralizing antibodies following a booster polio vaccination. *Eur. J. Nutr.* 2005; 44: 406–413.
303. Isolauri E, Joensuu J, Suomalainen H, Luomala M, and Nesikari T. Improved immunogenicity of oral DXRRV reassortant rotavirus vaccine by *Lactobacillus casei* GG. *Vaccine* 1995; 13: 310–312.
304. Kukkonen K, Nieminen T, Poussa T, Savilahti E, and Kuitunen M. Effect of probiotics on vaccine antibody responses in infancya randomized placebo-controlled double-blind trial. *Pediatr. Allergy Immunol.* 2006; 17: 416–421.
305. Abe F, Ishibashi N, and Shimamura S. Effect of administration of Bifidobacteria and Lactic Acid Bacteria to newborn calves and piglets. *J. Dairy Sci.* 1995; 78: 2838–2846.
306. Adams DC, Gaylean ML, and Kiesling HE. Influence of viable yeast culture, sodium bicarbonate and monensin on liquid dilution rate, rumen fermentation and feedlot performance of growing steers and digestibility in lambs. *J. Anim. Sci.* 1981; 53(3): 780–789.
307. Allison BC and McGraw RL. Efficacy of Vita-ferm formula for stocker calves. Anim. Sci. Newsletter A&T and NC State Uni. 1989, November, pp. 12–18.
308. Anderson KL, Nagaraja TG, Morril JL, Avery TB, Galitzer SJ, and Boyer JE. Ruminal microbial development in conventionally or early-weaned calves. *J. Anim. Sci.* 1987; 64: 1215–1226.
309. Barrow PA, Brooker BE, Fuller R, and Newport MJ. The attachment of bacteria to the epithelium of the pig and its importance in the microecology of the intestine. *J. Appl. Bacteriol.* 1980; 48: 147–154.
310. Baillon M-LA Marshall-Jones ZV, and Butterwick RF. Effects of probiotic *Lactobacillus acidophilus* strain DSM13241 in healthy adult dogs. *Am. J. Vet. Res.* 2004; 65(3): 338–343.
311. Benyacoub J, Gail L, Czarnecki-Maulden GL, Cavadini C, Sauthier T, Anderson RE, Schiffrin EJ, and von der Weid T. Supplementation of food with *Enterococcus faecium* (SF68) stimulates immune functions in young dogs. *J. Nutr.* 2003; 133: 1158–1162.
312. Belavi T, Ucan US, Coscedilun B, Kurtogbreveu V, Idot, and Cetingul S. Effect of dietary probiotic on performance and humoral immune response in layer hens. *Br. Poult. Sci.* 2001; 42(4): 456–461.
313. Biagi G, Cipollini I, Pompei A, Zaghini G, and Matteuzzi D. Effect of a *Lactobacillus animalis* strain on composition and metabolism of the intestinal microflora in adult dogs. *Vet. Microbiol.* 2007; 124(1–2): 160–165.
314. Beharka AA, Nagaraja TG, and Morrill JL. Performance and ruminal function development of young calves fed diets with *Aspergillus oryzae* fermentation extract. *J. Dairy Sci.* 1991; 74: 4326–4336.
315. Biourge V, Vallet C, Levesque A, Sergheraert R, Chevalier S, and Jean-Luc Roberton J-L. The use of probiotics in the diet of dogs. *J. Nutr.* 1998; 128: 2730S–2732S.
316. Bontempo1 V, Nutrition and health of dogs and cats: evolution of pet food. *Vet. Res. Commun.* 2005; 29(2): 45–50.
317. Burgstaller G, Ferstl R, and Alps H. Addition of lactic acid bacteria (*Streptococcus faecium* SF-68) to a milk replacer for calf feeding. *Zuechtungskunde* 1984; 56(2–3): 156–162.

318. Burgstaller G, Ferstl R, and Alps H. Lactic acid bacteria (*Streptococcus faecium* M 74) in combination with avoparcin in milk replacers for fattening calves. *Zuechtungskunde* 1985; 57(4): 278–283.

319. Bryant MP, Small N, Bouma C, and Robinson I. Studies on the composition of the ruminal flora and vauna of young calves. *J. Dairy Sci.* 1958; 41(12): 1747–1767.

320. Canzi E, Zanchi R, Camaschella P, Cresci A, Greppi GF, Orpianesi C, Serratoni M, and Ferrari A. Modulation by lactic-acid bacteria of the intestinal ecosystem and plasma cholesterol in rabbits fed a casein diet. *Nutr. Res.* 2000; 20(9): 1329–1340.

321. Carlos G, Gonzalez SN, and Oliver G. Some probiotic properties of chicken Lactobacilli. *Can. J. Microbiol.* 1999; 45: 981–987.

322. Cavazzoni V, Adami A, and Castrovilli C. Performance of broiler chickens supplemented with *Bacillus coagulans* as probiotic. *Br. Poult. Sci.* 1998; 39: 526–529.

323. Collington GK, Parker DS, and Armstron DG. The influence of inclusion of either an antibiotic or a probiotic in the diet on the development of digestive enzyme activity in the pig. *Br. J. Nutr.* 1990; 64(1): 59–70.

324. Cruywagen CW, Jordaan I, and Venter L. Effect of *Lactobacillus acidophilus* supplementation of milk replacer on preweaning performance of calves. *J. Dairy Sci.* 1996; 79: 486–486.

325. Dawson KA, Newman KE, and Boling JA. Effects of microbial supplements containing yeast and Lactobacilli on roughe-fed ruminal microbial activities. *J. Anim. Sci.* 1990; 68: 3392–3398.

326. De Mitchell I and Kenworhty R. Investigations on a metabolite from *Lactobacillus bulgaricus* which neutralizes the effect of enterotoxin from *Escherichia coli* pathogenic for pigs. *J. Appl. Bacteriol.* 1976; 41: 163–174.

327. Dilmi-Bouras A and Sadoun D. Survival of starters of yogurt in the digestive tract of rabbits. *Lait* 2002; 82: 247–253.

328. Djouvinov D, Boicheva S, Simeonova T, and Vlaikova T. Effect of feeding Lactina® probiotic on performance, some blood parameters and caecal microflora of mule ducklings. *Trakia J. Sci.* 2005; 3(2): 22–28.

329. Elam NA, Gleghorn JF, Rivera JD, Gaylean MJ, Defoor PJ, Brashears MM, and Younts-Dahl SM. Effects of live cultures of *Lactobacillus acidophilus* and *Propionibacterium freudenreichii* on performance of finishing beef steers. *J. Anim. Sci.* 2003; 81: 2686–2698.

330. Endo T and Nakano M. Influence of a probiotic on productivity, meat components, lipid metabolism, caecal flora and metabolites, and raising environment in broiler production. *Anim. Sci. J.* 1999; 70(4): 207–218.

331. Farnell MA, Donoghue AM, Solios de los Santos F, Blore PJ, Hargis BM, and Donoghue DJ. Up-regulation of oxidative burst and degranulation in chicken heterophils stimulated with probiotic bacteria. *Poult. Sci.* 2006; 85: 1900–1906.

332. Fleige S. Lactulose in combination with *Enterococcus faecium*: protective role in calves. Thesis submitted for the Doctoral Degree of Natural Science, Munich University of Technology, 2007.

333. Fritts CA, Kersn JH, Motl MA, Kroger EC, Yan E, Si J, Jiang Q, Campos MM, Waldroup AL, and Waldroup PW. *Bacillus subtilis* C-3102 (Calsporin) improves live performance and microbiological status of broiler chickens. *J. Appl. Poult. Res.* 2000; 9: 149–155.

334. Gilliland SE, Nelson CR, and Maxwell C. Assimilation of cholesterol by *Lactobacillus acidophilus*. *Appl. Environ. Microbiol.* 1985; 49(2): 377–381.

335. Gomez-Basauri J, de Ondarza MB, and Siciliano-Jones J. Intake and milk production of dairy cows fed lactic acid bacteria and mannonoligosaccharide. *J. Dairy Sci.* 2001; 84(Suppl. 1): 283.

336. Harper AF, Kornegay ET, Bryant KL, and Thomas HR. Efficacy of virginiamycin and a commercially-available Lactobacillus probiotic in swine diets. *Anim. Feed Sci. Technol.* 1983; 8: 69–76.

337. Higginbotham GE, Collar CA, Aseltine MS, and Bath DL. Effect of yeast culture and *Aspergillus oryzae* extract on milk yield in a commercial dairy herd. *J. Dairy Sci.* 1994; 77: 343–348.

338. Higginbotham GE, Santos JEP, Juchem SO, and DePeters EJ. Effect of feeding *Aspergillus oryzae* extract on milk production and rumen parameters. *Livestock Prod. Sci.* 2004; 86: 55–59.

339. Hofacre CL, Froyman R, Gautrias B, George B, Goodwin MA, and Brown J. Use of Aviguard and other intestinal bioproducts in experimental *Clostridium perfringens*-associated necrotizing enteritis in broiler chickens. *Avian Dis.* 1998; 42: 579–584.

340. Huber JT. Probiotics in cattles. In: Fuller R editor. Probiotics 2: Applications and Practical Aspects. Chapman and Hall, London, 1997, pp. 162–186.

341. Jackson JE, Lambert BD, Cadle JM, Kaiser GE, and Stafford JA. The effect of *Aspergillus oryzae* on performance of swine. *Texas J. Agric. Nat. Res.* 2006; 19: 1–7.

342. Jenny BF, Vandijik HJ, and Collins JA. Performance and fecal flora of calves fed a *Bacillus subtilis* concentrate. *J. Dairy Sci.* 1991; 74: 1968–1973.

343. Jin LZ, Ho YW, Abdullah N, and Jalaludin S. Growth performance, intestinal microbial populations, and serum cholesterol of broilers fed diets containing *Lactobacillus cultures*. *Poult. Sci.* 1998; 77(9): 1259–1265.

344. Jiraphocakul S, Sullivan TW, and Shahani KM. Influence of a dried *Bacillus subtilis* culture and antibiotics on performance and intestinal microflora in turkeys. *Poult. Sci.* 1990; 69(11): 1966–1973.

345. Jonsson E and Olsson I. The effect on performance, health and faecal microflora of feeding Lactobacillus strains to neonatal calves. *Swedish J. Agric. Res.* 1985; 15(2): 71–76.

346. Kabir SML, Rahman MM, Rahman MB, Rahman MM, and Ahmed SU. The dynamics of probiotics on growth performance and immune response in broilers. *Int. J. Poult. Sci.* 2004; 3(5): 361–364.

347. Kim SL, Standorf D, Roman-Rosario H, Yokoyama MT, and Rust SR. Potential use of propionibacterium acidipropionici, strain DH42 as a direct-fed microbial for cattle. Michigan State University Beef Cattle, Sheep and Forage System Research and Demonstration Report 2001, pp. 45–55. Downloaded from http://beef.ans.msu.edu/Extension/Publications/Beef__Sheep_and_Forage_Researc/beef__sheep_and_forage_researc.html on 20th December 2007.

348. Kornegay ET and Risley CR. Nutrient digestibilities of a corn–soybean meal diet as influenced by Bacillus products fed to finishing swine. *J. Anim. Sci.* 1996; 74: 799–805.

349. Krehbiel CR, Rust SR, Zhang G, and Gilliland SE. Bacterial direct-fed microbials in ruminant diets: performance response and mode of action. *J. Anim. Sci.* 2003; 81(Suppl. 2): E120–E132.

350. Kyriakis SC, Tsiloyiannis VK, Vlemmas J, Sarris K, Tsinas AC, Alexopoulos C, and Jansegers L. The effect of probiotic LSP 122 on the control of postweaning diarrhoea syndrome of piglets. *Res. Vet. Sci.* 1999; 67: 223–228.

351. Lee YK, Lim CY, Teng WL, Ouwehand AC, Tuomola EM, and Salminen S. Quantitative approach in the study of adhesion of lactic acid bacteria to intestinal cells and their competition with Enterobacteria. *Appl. Environ. Microbiol.* 2000; 66(9): 3692–3697.

352. Lee YK, Nomoto K, Salminen S, and Gorbach SL. *Handbook of Probiotics.* John Wiley & Sons, NY, 1999.

353. Lessard M and Brisson GJ. Effect of a Lactobacillus fermentation product on growth, immune response and fecal enzyme activity in weaned pigs. *Can. J. Anim. Sci.* 1987; 67 (2): 509–516.

354. Lin ASH, Teo AYL, and Tan HM. Antimicrobial compounds from *Bacillus subtilis* for use against animal and human pathogens. US Patent No: 7,247,299B2, 2007.

355. Losada MA and Olleros T. Towards a healthier diet for the colon: the influence of fructooligosaccharides and Lactobacilli on intestinal health. *Nutr. Res.* 2002; 22: 71–84.

356. Maia OB, Duarte R, Silva AM, Cara DC, and Nicoli JR. Evaluation of the components of a commercial probiotic in gnotobiotic mice experimentally challenged with *Salmonella enterica* subsp. *enterica* ser. *Typhimurium. Vet. Microbiol.* 2001; 79: 183–189.

357. Marshall-Jones ZV, Baillon M-LA Julie M, Croft JM, and Butterwick RF. Effects of *Lactobacillus acidophilus* DSM13241 as a probiotic in healthy adult cats. *Am. J. Vet. Res.* 2006; 67(6): 1005–1012.

358. Matarese LE, Seidner DL, and Steiger E. The role of probiotics in gastrointestinal disease. *Nutr. Clin. Practice* 2003; 18(6): 507–516.

359. Morrill JL, Morrill JM, Feyerharm AM, and Laster JF. Plasma proteins and probiotic as ingredients in milk replacer. *J. Dairy Sci.* 1995; 78: 902–907.

360. Myers D and Probiotics. *J. Exotic Pet Med.* 2007; 16(3): 195–197.

361. Nahashon SN, Nakaue HS, Snyder SP, and Mirosh LW. Performance of single comb White Leghorn layers fed corn–soybean meal and barley–corn–soybean meal diets supplemented with a direct-fed microbial. *Poult. Sci.* 1994; 73(11): 1712–1723.

362. Nahashon SN, Nakaue HS, and Mirosh LW. Performance of single Comb White Leghorn fed a diet supplemented with a live microbial during the growth and egg laying phases. *Anim. Feed Sci. Technol.* 1996; 57(1–2): 25–38.

363. Netherwood T, Gilbert HJ, Parker DS, and O'Donnell AG. Probiotics shown to change bacterial community structure in the avian gastrointestinal tract. *Appl. Environ, Microbiol.* 1999; 65(11): 5134–5138.

364. Nicodemus N, Carabano R, Garcia J, and de Blas JC. Performance response of doe rabbits to Toyocerin® (*Bacillus cereus* var. toyoi) supplementation. *World Rabbit Sci.* 2004; 12(2): 109–118.

365. Nieto N, Cabarello AG, and Martinez A. Comparative study of the ruminal protozoal fauna in calves under different methods of rearing. *Revista de Salud Anim.* 1985; 7(3): 281–286.

366. Nocek JE and Kautz WP. Direct-fed microbial supplementation on ruminal digestion, health, and performance of pre-postpartum dairy cattle. *J. Dairy Sci.* 2006; 89: 260–266.

367. Ogawa M, Shimizu K, Nomoto K, Takahashi M, Watanuki M, Tanaka R, Tanaka T, Hamabata T, Yamasaki S, and Takeda Y. Protective effect of *Lactobacillus casei* strain Shirota on Shiga toxin-producing *Escherichia coli* O157:H7 infection in infant rabbits. *Infect. Immun.* 2001; 69(2): 1101–1108.

368. Ozawa K, Yabu-uchi K, Yamanaka K, Yamashita Y, Nomura S, and Oku I. Effect of *Streptococcus faecalis* BIO-4R on intestinal flora of weanling piglets and calves. *Appl. Environ. Microbiol.* 1983; 45(5): 1513–1518.

369. Ozcan M, Arslan M, Matur E, Cotelioglu U, IAkyazi I, and Erarslan E. The effects of *Enterococcus faecium* Cernelle 68 (SF 68) on output properties and some haematological parameters in broilers. *Med. Wet.* 2003; 59(6): 496–500.

370. Panda AK, Reddy MR, Rama Rao SV, and Praharaj NK. Production performance, serum/yolk cholesterol and immune competence of white Leghorn layers as influenced by dietary supplementation with probiotic. *Trop. Anim. Health Prod.* 2003; 35(1): 85–94.

371. Patterson JA and Burkholder KM. Application of prebiotics and probiotics in poultry production. *Poult. Sci.* 2003; 82: 627–631.

372. Pollmann DS, Danielson DM, and Peo Jr ER. Effects of microbial feed additives on performance of starter and growing-finishing pigs. *J. Anim. Sci.* 1980a; 51(3): 577–581.

373. Pollmann DS, Danielson DM, Wren WB, Peo Jr ER, and Shahani KM. Influence of *Lactobacillus acidophilus* inoculum on gnotobiotic and conventional pigs. *J. Anim. Sci.* 1980b; 51(3): 629–637.

374. Pollmann DS, Danielson DM, and Peo Jr ER. Effects of *Lactobacillus acidophilus* on starter pigs fed a diet supplemented with lactose. *J. Anim. Sci.* 1980c; 51(3): 638–644.

375. Pollmann DS, Kennedy GA, Koch BA, and Allee GL. Influence of nonviable Lactobacillus fermentation product on artificially reared pigs. *Nutr. Rep. Int.* 1984; 29(4): 977–982.

376. Pollamann M, Nordhoff M, Pospischil A, Tedin K, and Wieler LH. Effects of a probiotic strain of *Enterococcus faecium* on the rate of natural chlamydia infection in Swine. *Infect. Immun.* 2005; 73(7): 4346–4353.

377. Rastall RA and Maitin V. Prebiotics and synbiotics: towards the next generation. *Curr. Opin. Biotechnol.* 2002; 13: 490–496.

378. Ratcliffe B, Cole CB, Fuller R, and Newport MJ. The effect of yogurt and milk fermented with a porcine intestinal strain of *Lactobacillus reuteri* on the performance and gastrointestinal flora of pigs weaned at 2 days of age. *Food Microbiol.* 1986; 3: 203–211.

379. Rinkinen M, Jalava K, Westermarck E, Salminen S, and Ouwehand AC. Interaction between probiotic lactic acid bacteria and canine enteric pathogens: a risk factor for intestinal *Enterococcus faecium* colonization? *Vet. Microbiol.* 2003; 92(1–2): 111–119.

380. Russell JB and Rychlik JL. Factors that alter rumen microbial ecology. *Science* 2001; 292 (5519): 1119–1122.

381. Rust SR, Metz K, and Ware DR. Effects of Bovamine™ rumen culture on the performance and carcass characteristics of feedlot steers. Michigan State University Beef Cattle, Sheep and Forage System Research and Demonstration Report 2000, pp. 22–26. Downloaded from http://beef.ans.msu.edu/Extension/Publications/Beef__Sheep_and_Forage_Researc/beef__sheep_and_forage_researc.html on 20 December 2007.

382. Samanya M and Yamaguchi K. Histological alterations of intestinal villi in chickens fed dried *Bacillus subtilis* var. natto. *Comp. Biochem. Physiol. Part A* 2002; 133: 95–104.

383. Santoso U, Tanaka K, and Ohtani S. Effect of dried *Bacillus subtilis* culture on growth, body composition and hepatic lipogenic enzyme activity in female broiler chicks. *Br. J. Nutr.* 1995; 74(4): 523–529.

384. Santoso U, Tanaka K, Ohtani S, and Sakaida M. Effect of fermented product from *Bacillus subtilis* on feed conversion efficiency, lipid accumulation and ammonia production in broiler chicks. *Asia-Australasian J. Anim. Sci.* 2001; 14(3): 333–337.

385. Sauter SN, Benyacoub J, Allenspach K, Gaschen F, Ontsouka E, Reuteler G, Cavadini C, Knorr R, and Blum JW. Effects of probiotic bacteria in dogs with food responsive diarrhoea treated with an elimination diet. *J. Anim. Physiol. Anim. Nutr.* 2006; 90(7–8): 269–277.

386. Scharek L, Guth J, Reiter K, Weyrauch KD, Taras D, Schwwerk P, Schierack P, Schmidt MFG, Wieler LH, and Tedin K. Influence of a probiotic *Enterococcus faecium* strain on development of the immune system of sows and piglets. *Vet. Immunol. Immunopathol.* 2005; 105: 151–161.

387. Scharek L, Guth J, Filter M, and Schmidt MFG. Influence of the probiotic bacteria *Enterococcus faecium* NCIMB 10415 (SF68) and *Bacillus cereus* var. *toyoi* NCIMB 40112 on the development of serum IgG and fecal IgA of sows and piglets. *Arch. Anim. Nutr.* 2007; 61(4): 223–234.

388. Schwab CG, Moore JJ, Hoyt PM, and Pretience JL. Performance and fecal flora of calves fed a nonviable *Lactobacillus bulgiaricus* fermentation product. *J. Dairy Sci.* 1980; 63: 1412–1423.

389. Soder KJ and Holden LA. Dry matter intake and milk yield and composition of cows fed yeast prepartum and postpartum. *J. Dairy Sci.* 1999; 82: 605–610.

390. Strompfová V, Marcináková M, Simonová M, Bogovic-Matijasić B, and Lauková A. Application of potential probiotic *Lactobacillus fermentum* AD1 strain in healthy dogs. *Anaerobe* 2006; 12(2): 75–79.

391. Teo A and Tan HM. Inhibition of *Clostridium perfringens* by a novel strain of *Bacillus subtilis* isolated from the gastrointestinal tracts of healthy chickens. *Appl. Environ. Microbiol.* 2005; 71: 4185–4190.

392. Teo A and Tan HM. Effect of *Bacillus subtilis* PB6 (CloSTAT) on broilers infected with a pathogenic strain of *Escherichia coli*. *J. Appl. Poult. Res.* 2006; 15: 229–235.

393. Teo A and Tan HM. Evaluation of the performance and intestinal gut microflora of broilers fed on corn-soydiets supplemented with *Bacillus subtilis* PB6 (CloSTAT). *J. Appl. Poult. Res.* 2007; 16: 296–303.

394. Thocino A, Xiccato G, Carraro L, and Jimenez G. Effect of diet supplementation with Toyocerin® (*Bacillus cereus* var. toyoi) on performance and health of growing rabbits. *World Rabbit Sci.* 2005; 13(1): 17–28.

395. Timmerman H. Probiotics reduce costly problems in calves. *Feed Mix.* 2007; 15(1): 1517.

396. Underdahl NR. The effect of feeding *Streptococcus faecium* upon *Escherichia coli* induced diarrhea in gnotobiotic pigs. *Prog. Food Nutr. Sci.* 1983; 7(3–4): 5–12.

397. Vahjen W and Männer K. The effect of a probiotic *Enterococcus faecium* product in diets of healthy dogs on bacteriological counts of *Salmonella* spp., *Campylobacter* spp., and *Clostridium* spp. in faeces. *Arch. Anim. Nutr.* 2003; 57(3): 229–233.

398. Vasconcelos JT, Elam NA, Brashears MM, and Gaylean ML. Effects of increasing dose of live cultures of *Lactobacillus acidophilus* (Strain NP51) combined with a single dose of *Propionibacterium freudenreichii* (Strain NP24) on performance and carcass characteristics of finishing beef steers. *J. Anim. Sci.* 2008; 86(3): 756–762.

399. Wallace RJ. Ruminal microbiology, biotechnology, and ruminant nutrition: progress and problems. *J. Anim. Sci.* 1994; 72: 2992–3003.

400. Wallace RJ and Newbold CJ. Probiotics for ruminants. In: Fuller R, editors. *Probiotics: The Scientific Basis*. Chapman and Hall, London, 1992, pp. 317–353.
401. Walter J. The microecology of Lactobacilli in the gastrointestinal tract. In: Tannock G, editor. *Probiotics and Prebiotics: Scientific Aspects*. Caister Academic Press, UK, 2005.
402. Wiedmeier RD, Arambel MJ, and Walters JL. Effect of yeast culture and *Aspergillus oryzae* fermentation extract on ruminal characteristics and nutrient digestibility. *J. Dairy Sci.* 1987; 70: 2063–2068.
403. Weese JS and Arroyo L. Bacteriological evaluation of dog and cat diets that claim to contain probiotics. *Can. Vet. J.* 2003; 44(3): 212–215.
404. Yeo J and Kim KI. Effect of feeding diets containing an antibiotic, a probiotic, or yucca extract on growth and intestinal urease activity in broiler chicks. *Poult. Sci.* 1997; 76(2): 381–385.
405. Yu B and Tsen HY. Lactobacillus cells in the rabbit digestive tract and the factors affecting their distribution. *J. Appl. Bacteriol.* 1993; 75(3): 269–275.
406. Zani JL, Weykamp da Cruz F, Freitas dos Santos A, and Gil-Turnes C. Effect of probiotic CenBiot on the control of diarrhea and feed efficiency in pigs. *J. Appl. Microbiol.* 1998; 84(1): 68–71.
407. Zhao T, Doyle MP, Harmon BG, Brown CA, Mueller PO, and Parks AH. Reduction of carriage of enterohemorrhagic *Escherichia coli* O157:H7 in cattle by inoculation with probiotic bacteria. *J. Clin. Microbiol.* 1998; 36(3): 641–647.
408. Zink R, Roniero R, Rochat F, Cavadini C, Der Weid T, Schiffrin E, Benyacoub J, Rousseau V, and Perez P. Probiotics for pet food applications. US Patent No: 7,189,390, 2007.
409. Zulkifli I, Abdullah N, Azrin NM, and Ho YW. Growth performance and immune response of two commercial broiler strains fed diets containing. *Br. Poult. Sci.* 2000; 41(5): 593–597.
410. USDA. Pet Food Update. Foreign Agricultural Service, United States Department of Agriculture, October 2007 (http://www.fas.usda.gov/scriptsw/attacherep/default.asp).

5

MECHANISMS OF PROBIOTICS

5.1 ADHESION TO INTESTINAL MUCUS AND EPITHELIUM BY PROBIOTICS

SAMPO LAHTINEN AND ARTHUR OUWEHAND

Health & Nutrition, Danisco, Finland

Adhesion of probiotics to intestinal mucus and epithelial cells has long been considered one of the most important selection criteria for probiotic microorganisms. Adhesion to the intestinal mucosa may prevent probiotic cells from being washed out; and therefore, enabling temporary colonization, immune modulation, and competitive exclusion of pathogens. The question whether externally administered probiotics really adhere to epithelial cells or to mucus *in vivo* has been somewhat controversial (7). Although the evidence of the adhesion of probiotics to mucus *in vivo* is still limited, recent studies indicate that such adhesion may indeed occur and may thus serve as a mechanism for probiotic action. A number of different methods have been used in the adhesion assays of probiotics (8, 74). The most common are the tests assessing the adhesion to epithelial cell lines and to intestinal mucus. In addition, assessment of the adhesion to resected colonic tissue has been demonstrated (58). This method takes into account epithelial cells and intestinal mucosa as well as the endogenous bacteria attached to mucosa.

378 MECHANISMS OF PROBIOTICS

TABLE 5.1 Adhesion of Probiotics to Gastrointestinal Epithelial Cell Lines

Probiotics	Study Design	Result	References
Lactobacillus isolates from various sources	Adhesion and detection of adherent bacteria from birds	Adherence to epithelial cells; species specific	(20)
Lactobacillus GG and other dairy strains	Adhesion to Caco-2 cell cultures	*Lactobacillus* GG more adherent than other tested dairy strains	(18)
Human *Lactobacillus* isolates	Adhesion to Caco-2 cell cultures	Adhesion strain specific; dependent on age of the cell culture	(12)
Thirteen human strains of *Bifidobacterium*	Adhesion to Caco-2 cells and mucus from HT-29-MTX cells	Adhesion calcium independent	(5)
L. acidophilus LA1 and three other strains	Adhesion to Caco-2 and HT-29-MTX cell cultures	Adhesion calcium independent	(6)
Commercial probiotic and dairy strains	Adhesion to Caco-2 cell cultures	Adhesion strain specific	(45, 72)
Thirteen *B. longum* isolates	Adhesion to Caco-2, HT-29 and KATO III cell cultures	Adhesion correlated with autoaggregation properties	(17)
Five commercial probiotic strains	Adhesion to Caco-2, HT-29 and HT-29-MTX cells	All strains showed strong adhesion	(21)
Various lactobacilli and bifidobacteria	Adhesion to Caco-2 cell cultures; induction of cytokine production	Adhesion does not affect cytokine production by epithelial cells	(49)
Eight *Bifidobacterium* strains	Adhesion to Caco-2 cell cultures and determination of global expression profiles	Caco-2 cells respond differentially to *E. coli* and *Bifidobacterium*; response independent from adhesion	(64)

5.1.1 Adhesion to Gastrointestinal Epithelial Cell Lines

A classic method for assessing adhesive properties of probiotic bacteria is the use of cultures of intestinal epithelial cells, mostly Caco-2, HT-29, or mucus-secreting HT-29-MTX cells (Table 5.1).

5.1.2 Adhesion to Intestinal Mucus

The ability to adhere to intestinal mucosa has been suggested as a potential mechanism for temporary colonization, and the adhesion to mucus is commonly used in *in vitro* adhesion assays (Table 5.2).

TABLE 5.2 Adhesion of Probiotics to Intestinal Mucus

Probiotics	Study Design	Result	Reference
Commercial probiotic strains	*In vitro* adhesion to mucus from healthy infants and adults	Probiotics more adherent to adult mucus compared to infant mucus	(41)
Commercial probiotics, dairy strains, fecal isolates	*In vitro* adhesion to mucus from healthy adults; analysis of cell surface hydrophobicity	Adhesion variable (2–43%); hydrophobicity does not correlate with adhesion	(55)
Bifidobacteria from human and animal hosts	*In vitro* adhesion to human and bovine mucus	Adhesion strain specific; host specificity may play a role	(28)
Bifidobacterium lactis Bb-12 and other strains	*In vitro* adhesion to mucus isolated from three age groups	Bb-12 most adherent; adhesion highest to adult mucus	(46)
Commercial probiotics and fecal isolates	*In vitro* adhesion to human, canine, possum, bird, and fish mucus	No host specificity observed, strain selection more important	(65)

5.1.3 Colonization of Probiotics in Human Intestine as Assessed by Biopsies

Adhesion to intestinal mucus or epithelial cells is thought to be a prerequisite for temporary colonization of probiotics in the intestine. While colonization of externally administered probiotics in humans is thus far not well documented, increasing evidence suggests that temporary colonization of probiotics is possible (Table 5.3).

5.1.4 Comparisons Between *In Vitro* and *In Vivo* Results

The validity of *in vitro* adhesion tests and the correlation to *in vivo* colonization have been assessed in several studies. Indirect evidence suggests that adhesion to intestinal cells and mucus is associated with temporary colonization. However, *in vitro* and *in vivo* results do not always correlate (Table 5.4).

5.1.5 Adhesins

Several bacterial components, including cell wall proteins, carbohydrates, and lipoteichoic acids, have been suggested to be involved in the adhesion of probiotics to intestinal contents (Table 5.5).

5.1.6 Factors Affecting the Adhesion Properties of Probiotics

Adhesion properties of probiotics are strain specific, and the factors such as cell wall properties and composition (26) and possibly also host specificity are the most

TABLE 5.3 Colonization of Probiotics in Human Intestine as Assessed by Biopsies

Probiotics	Study design	Result	Reference
Nineteen strains of *Lactobacillus*	Strains administered in combination to 13 volunteers	At 11 days postadministration, 5/19 strains were isolated from biopsies	(37)
L. rhamnosus GG	Strain administered to volunteers, followed by colonic biopsies between 7 and 14 days postadministration	LGG detected in biopsies after administration; temporary colonization	(2, 3)
L. paracasei B21060	Strain administered to seven volunteers, followed by colonic biopsies at 2 days postadministration	Most volunteers had at least one positive biopsy sample	(48)

important determinants of adhesion properties. However, several other factors have also been reported to affect the adhesion properties (Table 5.6).

5.1.7 Adhesive and Inhibitory Properties of Nonviable Probiotics

Studies on inactivated probiotics have suggested that nonviable probiotic bacteria may also be capable of adhesion. This is because adhesion properties are related to specific

TABLE 5.4 Comparisons Between *In Vitro* and *In Vivo* Results

Probiotics	Study Design	Result	Reference
Twelve strains of *Bifidobacterium*	*In vitro* adherence to Caco-2 cells compared to *in vivo* human study	Good *in vitro* adhesion was observed for strains capable of colonization	(16)
Forty-seven *Lactobacillus* strains	Strains selected based on *in vitro* assays and administered to human volunteers	Survival and temporary colonization related to *in vitro* adhesion	(35)
Wild-type and mutant *L. crispatus*	*In vitro* and *in vivo* (human) properties of aggregating and nonaggregating variants compared	Only wild-type aggregating *L. crispatus* associated with *in vitro* adhesion and *in vivo* colonization	(10)
L. johnsonii and *L. paracasei*	*In vitro* properties compared to colonization in mice	Strains showed similar *in vitro* properties but different *in vivo* colonization	(34)

TABLE 5.5 Adhesins Associated with Probiotic Adhesion

Adhesin	Probiotics	Target of Adhesion	References
Proteins	L. fermentum 104-R	Porcine mucin, mucus and epithelium	(31, 66)
	L. acidophilus BG2FO4	Intestinal cell lines and mucus	(14)
	Human Lactobacillus isolates	Intestinal epithelial cell line (Caco-2)	(24)
	L. plantarum	Mannose and HT-29 cells	(1, 62)
	L. acidophilus and L. agilis	Intestinal 407 cells	(40)
	Commercial probiotic strains	Mucus isolated from healthy volunteers	(71)
	L. reuteri 1063	Mucus and mucin from pigs and hens	(67)
	L. brevis	Intestinal cell lines and fibronectin	(33)
	L. johnsonii La1	Intestinal cell lines and mucus	(4, 22)
	L. acidophilus NCFM	Intestinal epithelial cell line (Caco-2)	(9)
	L. acidophilus M92	Intestinal epithelium from mice	(19)
	L. reuteri 100-23	Mouse intestine (in vivo)	(73)
	L. reuteri 104-R (formerly L. fermentum)	Intestinal epithelial cell line (Caco-2)	(47)
	31 L. plantarum strains	Porcine mucin, reconstituted basement membranes and Caco-2 cells	(70)
	Lactobacillus isolates from pigs	Pig and human intestinal cells and collagen, laminin, and fibronectin	(36)
Lipoteichoic acids	Bifidobacterium bifidum	Isolated colonocytes	(50)
	Various Lactobacillus strains	Hemagglutination inhibition by strains adherent to murine gastric epithelium	(69)
	L. johnsonii La1	Intestinal epithelial cell line (Caco-2)	(23)
Carbohydrates	L. fermentum 104-R	Porcine gastric squamous epithelium	(31)
	Human Lactobacillus isolates	Intestinal epithelial cell line (Caco-2)	(24)

TABLE 5.6 Factors Affecting the Adhesion Properties of Probiotics

Factor	Study Design	Effect	References
Low pH	Adhesion of probiotics to Caco-2 epithelial cells	Increased adhesion Changes in adhesion strain dependent	(8, 24) (64)
Presence of bile	Adhesion of probiotics to mucus	Reduced adhesion	(25, 52)
Growth phase	Adhesion of *L. fermentum* to keratinizing gastric epithelium in mice	Adherence higher during stationary phase compared to exponential phase	(68)
	Adhesion of probiotics to Caco-2 epithelial cells	Decreased adhesion during logarithmic phase, increased during stationary phase	(8)
Growth medium	Adhesion of potential probiotic strains to ileostomy glycoproteins	Culture medium affects adhesion	(60)
Acquired acid resistance	Adhesion of acid-resistant variants of *Bifidobacterium* to mucus	Acquired acid resistance may in some cases increase adhesion	(15)
Acquired bile resistance	Adhesion of bile-resistant variants of *Bifidobacterium* to mucus	Acquired bile resistance may increase adhesion	(25)
Endogenous microbiota	Adhesion to mucus in the presence of fecal or mucosal microbiota	Endogenous microbiota does not affect adhesion to mucus *in vitro*	(51, 56)
Presence of other probiotic strains	Adhesion of probiotic combinations to ileostomy glycoproteins	Combinations of probiotics may have synergistic effects on adhesion	(54)
	Adhesion of probiotics to porcine intestinal epithelial IPEC-J2 cells	Probiotic strains may reduce adhesion of other strains by competing for same adhesion sites	(44)
Digestive enzymes	Adhesion of probiotics to mucus	Reduced adhesion in most cases	(52)
	Adhesion of *Propionibacterium acidipropionici* to intestinal cells	No effect on adhesion	(76)
Binding of mycotoxins	Adhesion of *Lactobacillus* GG to Caco-2 epithelial cells	Decreased adhesion when cell is bound to mycotoxin	(39)
Presence of lignans	Adhesion of commercial probiotics to mucus	Increased adhesion in some strains	(43)

TABLE 5.6 (*Continued*)

Factor	Study Design	Effect	References
Presence of calcium	Adhesion of probiotics and human isolates to Caco-2 epithelial cells	Adhesion calcium independent in some cases and calcium dependent in others	(5, 6, 12, 14)
	Adhesion of probiotics to porcine intestinal epithelial IPEC-J2 cells	Adhesion of most strains increased by the presence of calcium	(44)
Presence of Mg^{++} and Zn^{++} ions	Adhesion of probiotics to porcine intestinal epithelial IPEC-J2 cells	Adhesion not affected by Mg^{++} and Zn^{++} ions	(44)

cell surface properties rather than metabolic activity or cell division. Different inactivation methods may cause strain-specific changes in the adhesion properties of the cells. For example, heating may improve the adhesion of certain probiotic strains, but may be detrimental to the adhesive properties of other strains (Table 5.7).

5.1.8 Role of Age and Diseases on Adhesion

Aberrant gastrointestinal microbiota has been linked to a number of gastrointestinal and other diseases. In addition, age-related differences in the composition of gastrointestinal microbiota have been suggested, especially in the case of bifidobacteria. In

TABLE 5.7 Adhesive and Inhibitory Properties of Nonviable Probiotics

Probiotics	Study Design	Result	References
L. acidophilus	In vitro adhesion to human intestinal cells	Viable and nonviable cells equally adherent	(32)
L. fermentum 104-R	Adhesion of sonicated cells to porcine gastric squamous epithelium	No major changes in the adhesion compared to untreated controls	(31)
L. acidophilus LB	Adhesion of heat-killed probiotic to epithelial cell cultures; pathogen inhibition	Heat-killed probiotic is highly adherent to epithelial cell cultures and inhibits pathogen adhesion	(11, 13)
Commercial probiotic strains	Effects of different inactivation methods on the adhesion to mucus in vitro	Changes in adhesion dependent on the method of inactivation	(61)
Streptococcus thermophilus and L. acidophilus	Inhibition of E. coli infection on epithelial cell line by viable or killed probiotics	Viable probiotics most effective, antibiotic-killed cells also effective, heat-killed cells ineffective	(63)

TABLE 5.8 Role of Age and Diseases on Adhesion

Probiotics	Study Design	Result	References
Bifidobacteria isolated from healthy seniors and adults	*In vitro* adhesion to human mucus	Reduced adhesion of bifidobacteria from seniors compared to adults	(30)
Commercial bifidobacteria	*In vitro* adhesion to mucus from infants, adults, and elderly	Reduced adhesion to mucus from elderly subjects compared to mucus from adults and infants	(53)
Bifidobacteria from allergic infants	*In vitro* adhesion to mucus from healthy volunteers	Reduced adhesion to mucus from allergic infants compared to mucus from healthy infants	(27, 29)
Commercial probiotics	*In vitro* adhesion to mucus and resected colonic tissue from patients with major intestinal disorders	Adhesion is disease dependent	(57, 59)
Commercial probiotics	*In vitro* adhesion to mucus from infants with rotavirus infection	Rotavirus infection does not affect adhesion	(38)
Bifidobacterium infantis	Administration of probiotic to patients with ulcerative colitis	Probiotic strain recovered from biopsies, suggests temporary colonization	(75)
L. plantarum 299v	Administration of probiotic to critically ill patients	Probiotic strain recovered from 3/9 biopsies, suggesting temporary colonization	(42)

certain cases, the adherence properties of the intestinal mucus may be altered during the disease. The ability of certain probiotics to colonize subjects with a disease has also been assessed (Table 5.8).

5.2 COMBINED PROBIOTICS AND PATHOGEN ADHESION AND AGGREGATION

M. Carmen Collado and Seppo Salminen
Functional Foods Forum, University of Turku, Finland

The human intestinal microbiota constitutes a complex ecosystem that plays an important role in health and diseases in which various pathogens alter the intestinal bacterial homeostasis. There are also food probiotic bacteria that exhibit specific beneficial properties through microbiota modulation. The protective role of probiotic bacteria against gastrointestinal pathogens and the underlying mechanisms have received special attention. Pathogen inhibition by lactic acid bacteria might provide significant human health protection against pathogens either as a natural barrier against exposure in the gastrointestinal tract or as a method to decontaminate drinking water or food. This would enhance human health and have a positive economic impact especially in developing countries where the people suffer from frequent gastrointestinal infections.

The development of adjuvant or alternative therapies based on bacterial replacement is considered important due to the rapid emergence of antibiotic-resistant pathogenic strains and the adverse consequences of antibiotic therapies on the protective microbiota (100). Probiotic strains have shown to exert a protective effect at least against traveler's diarrhea, rotavirus diarrhea, and antibiotic-associated diarrhea (89, 109, 116).

The possible mechanisms underlying these antagonistic effects include competition for adhesion sites and nutritional sources, secretion of antimicrobial substances, toxin inactivation, and immune stimulation (99). Probiotics exhibit antagonistic effects against pathogens belonging to the genera *Listeria*, *Clostridium*, *Salmonella*, *Shigella*, *Escherichia*, *Helicobacter*, *Campylobacter*, and *Candida* (91, 107, 120, 141, 143). There are an increasing number of scientific reports on the effects of combined probiotics on the health of the host (98, 106, 115, 123, 125, 137). The best known probiotic combination consists of a mixture of eight lactic acid bacterial species (VSL # 3) and is reported to be effective in several human diseases (82, 86, 104, 106, 117). Combinations of well-known probiotics strains such as *L. rhamnosus* GG and *B. lactis* Bb12 (90, 92, 93, 127).

5.2.1 Aggregation

In order to manifest beneficial effects, probiotic bacteria need to achieve an adequate mass through aggregation. Consequently, the ability of probiotics to aggregate is a desirable property. Organisms with the ability to coaggregate with other bacteria such as pathogens may have an advantage over noncoaggregating organisms that are more easily removed from the intestinal environment.

5.2.2 Adhesion

Adherence to the intestinal epithelium and mucus is associated with stimulation of the immune system (80, 114), and adhesion to the intestinal mucosa is also crucial for transient colonization (114), an important prerequisite for probiotics to control the balance of the intestinal microbiota. Intestinal mucus has a dual role; it protects the mucosa from certain microorganisms while providing an initial binding site, nutrient source, and matrix on which bacteria can proliferate. Further, adherence of bacteria to intestinal epithelium is required for both colonization and infection of the gastrointestinal tract by many pathogens (101). Adhesion is a complex process involving nonspecific (hydrophobicity) and specific ligand–receptor mechanisms. Adherence of bacterial cells is usually related to cell surface characteristics (84, 87). Many authors have reported that the coaggregation abilities of *Lactobacillus* species might enable it to form a barrier that prevents colonization by pathogenic bacteria (85, 137, 139, 140).

5.2.3 Assay for Adhesion

Cell adhesion is a complex process involving contact between the bacterial cell membrane and interacting surfaces. The ability to adhere to epithelial cells and mucosal surfaces has been suggested to be an important property of many bacterial strains used as probiotics. Several researchers have reported investigations on composition, structure, and forces of interaction related to bacterial adhesion to intestinal epithelial cells (97, 130, 132) and mucus (90, 129, 137). The bacterial adhesion to hydrocarbons test (BATH) has been extensively used for measuring cell surface hydrophobicity in lactic acid bacteria (119, 148) and bifidobacteria (87, 107, 132). Usually, bacterial adhesion to hydrocarbons test or microbial adhesion to hydrocarbons test (MATH has been analyzed (112). Affinity to hydrocarbons (hydrophobicity) was reported as adhesion before and after extraction with organic solvents, respectively. In general, hydrophobicity results demonstrated a great heterogeneity among probiotic strains in adhesion to hydrocarbons, although generally lactobacilli strains showed the highest adhesion percentages (96). Many studies on the microbial cell surface chemistry showed that the presence of (glycol-) proteinaceous material at the cell surface results in higher hydrophobicity, whereas hydrophilic surfaces are associated with the presence of polysaccharides (108, 130, 136).

5.2.4 Assay for Aggregation

Bacterial aggregation is related to cell-to-cell adherence (85, 97, 146) between bacteria of the same strain (autoaggregation) or between genetically different strains (coaggregation) and is of considerable importance in several ecological niches, especially in the human gut where probiotics have to be active (113). Bacterial aggregation was analyzed largely in oral, dental, and biofilm environments (79, 121, 122, 124, 134) but there are only a few studies using probiotic strains and pathogens (85, 94, 96, 133, 139, 140). Coaggregation with potential gut pathogens may contribute to the probiotic properties ascribed to specific probiotic strains. Since detection of bacterial aggregation between probiotics and pathogens is important for human health, such studies are a priority to characterize and assess different methods.

In general, specific probiotic strains show higher autoaggregation abilities than pathogen strains (96). A positive relationship between autoaggregation and adhesion ability has been reported for some bifidobacterial species (97, 132) and also, there is a correlation between adhesion ability and hydrophobicity in some lactobacilli (96, 97, 149). However, these correlations have not been found by other authors (148). Nevertheless, these studies indicated that specific strains have highest percentages of adhesion to hydrocarbons and also showed higher autoaggregation abilities (Fig. 5.1). For bifidobacterial strains, a positive correlation was observed between aggregation ability and hydrophobicity (Fig. 5.1).

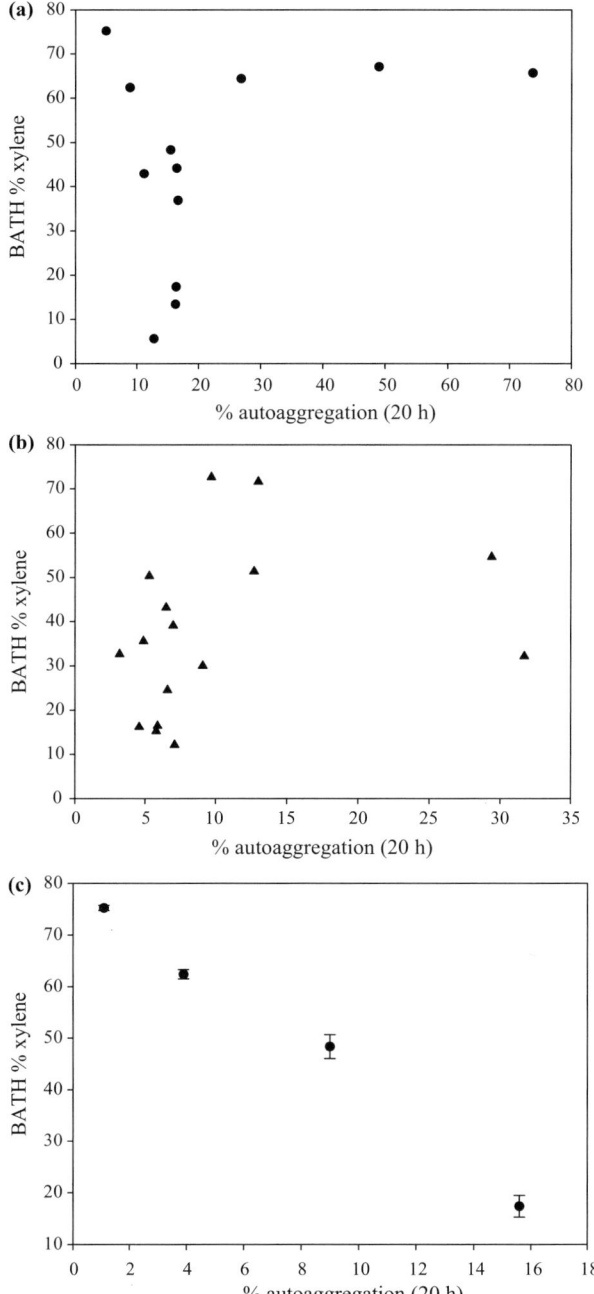

FIGURE 5.1 Autoaggregation index of the probiotic and pathogenic strains as a function of adhesion to xylene ($n \geq 3$). (A) Specific probiotic strains, (B) model pathogens, and (C) common *Bifidobacterium* strains (adapted from Refs 94 and 96).

Coaggregation phenomenon has been observed in bacteria from the human oral cavity, urogenital tract, mammalian gut, and potable water supply systems (134). It has been suggested that bacteriocin-producing lactic acid bacteria, which coaggregate with pathogens, may constitute an important host defense mechanism (133, 142). Several reports have suggested that the cellular aggregation could be important in promoting the colonization of beneficial lactobacilli in the gastrointestinal or vaginal tract (88, 113).

Recent reports (96) showed coaggregation between commercial probiotic strains and potential pathogens such as B. vulgatus, C. histolyticum, and St. aureus.

Based on the aggregation abilities, bacteria could be divided into three distinct groups. The first group showed high aggregation abilities, the second lower aggregation, and the third high coaggregation and low autoaggregation abilities (Fig. 5.2).

The findings suggest that the ability to autoaggregate, together with cell surface hydrophobicity as measured by BATH and coaggregation abilities with pathogen strains, can be used for preliminary screening in order to identify potentially probiotic bacteria suitable for human or animal use. Focused effort is needed to identify and characterize bacterial cell wall components, compositions, and properties to understand their role in adhesion to hydrocarbons, autoaggregation, and relation to coaggregation mechanisms.

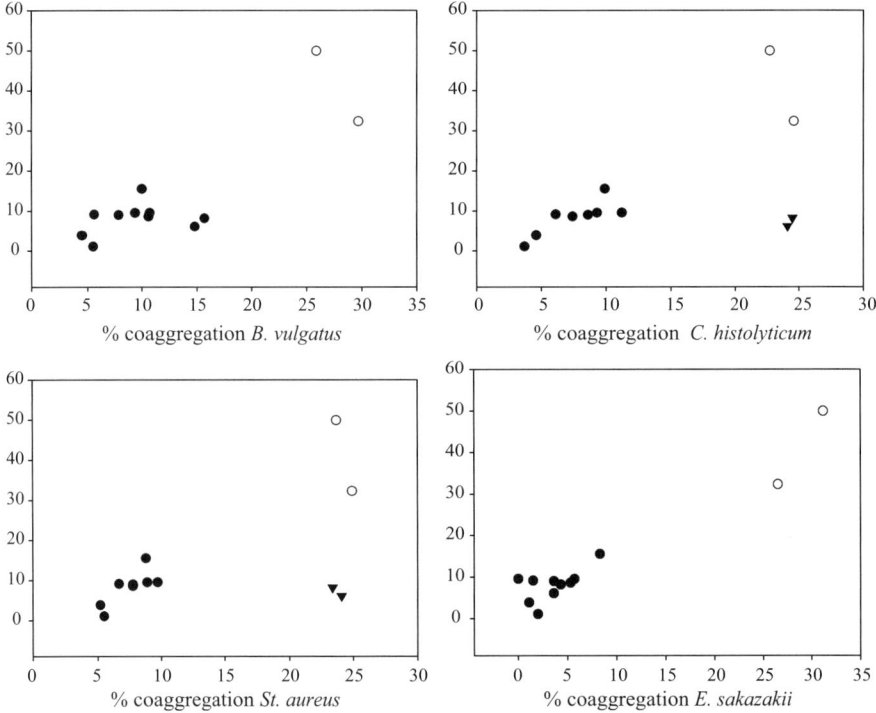

FIGURE 5.2 Coaggregation abilities (4 h incubation at 20°C) of probiotic strains ($n \geq 3$). (○) Higher aggregation and coaggregation, (●) lower aggregation and coaggregation, and (▼) lower aggregation and higher coaggregation (adapted from Ref 96).

5.2.5 Factors that Determine Adhesion

Adhesion to intestinal mucosa is regarded as a prerequisite for probiotic microorganisms. Adhesion allows the colonization, although transient, of the human intestinal tract (78) and it has been related to many of the health benefits attributed to specific probiotics. Thus, the ability to adhere to epithelial cells and mucosal surfaces has been suggested to be an important property of many probiotic bacterial strains (90, 95, 126, 128) and combination of probiotics. Several researchers have reported on the roles of composition, structure, and forces of interaction in bacterial adhesion to intestinal epithelial cells (97, 132) and mucus (90, 145). It has been shown that some lactobacilli and bifidobacteria share carbohydrate-binding specificities with some pathogens (81, 102, 110) providing the rationale for the use of these microorganisms to prevent infection at an early stage, by inhibiting the adhesion of these pathogens. The adhesion mechanisms involved passive forces, electrostatic and hydrophobic interactions, lipoteichoic acids, and specific structures such as polysaccharides and lectins. In general, it is assumed that probiotic strains are able to inhibit the attachment of pathogenic bacteria by means of steric hindrance at enterocyte pathogen receptors. It has been suggested that proteinaceous components are involved in the adhesion of probiotic strains to intestinal cells (81, 90, 102, 111).

5.2.6 *In Vitro* Models

The difficulties involved in studying bacterial adhesion *in vivo*, specifically in humans, have led to the development of *in vitro* model systems for the preliminary selection of potentially adherent strains. Many different intestinal mucosa models have been used to assess the adhesive ability of probiotics; among them the adhesion to human intestinal mucus has been widely used (90, 92, 95, 111, 118, 127, 135, 145). Good correlations have been previously reported between this mucus model and the enterocyte-like Caco-2 model (77, 128, 146). These intestinal models have also been used to demonstrate that probiotic bacteria can competitively inhibit the adhesion of pathogenic microorganisms and displace the previously adhered pathogens, such as *Salmonella*, *Escherichia coli*, *Listeria monocytogenes*, *Staphylococcus aureus*, *Bacteroides vulgatus*, *Clostridium difficile*, and *C. perfringens* (81, 83, 90, 92, 95, 110, 111).

The adhesion levels of the probiotic and pathogens strains showed a great variability depending on the strain, species, and genus (90, 92, 93, 110, 111). These results suggest that they have the capacity to adhere to the intestinal mucus, which could assist the pathogens in the invasion into the human intestinal mucosa. It has been cautioned that the *in vitro* studies should always be corroborated by studies in animal model and human trial (83, 92, 95).

The ability to inhibit the adhesion of pathogen appears to depend on both the probiotic and the pathogen strains, with a high specificity (92, 95, 96, 110, 111, 120, 144). These results, in accordance with previous reports (90, 110, 120), indicate that the displacement was not related to the adhesion ability of the strains. In addition, the displacement and competition profiles are very different from those observed for the inhibition of pathogens. These results appear to confirm that different mechanisms are implied in the processes of pathogen inhibition.

5.2.7 Probiotics in Combination

Since the beneficial effects of probiotics are strain dependent, it has been suggested that combinations of different probiotic strains may be more effective than monostrain probiotics (143). However, the impact can be counteractive, if strains are not selected in a scientifically sound manner. Recently, it has been shown that a combination of probiotic strains may complement or improve health benefits given by individual strains (82, 86, 105–107, 115, 117, 123). The adhesion properties of single bacteria to the intestinal mucus could be improved when a combination of probiotics is used. Furthermore, adhesion *in vivo* could be influenced by both the normal microbiota and the specific probiotics included in each food preparation. In *in vitro* trials, the probiotic properties have mainly been tested alone or in combination with yogurt bacteria such as *L. delbrueckii* and *L. acidophilus* (127) but rarely combined with other probiotics (93, 127). Figure 5.3a and b demonstrates the influence of other probiotic strains in the adhesion of probiotics. This influence may be positive (enhance the adhesion) or negative (decrease).

In addition, a recent report (138) describes how the presence of exopolysaccharides produced by probiotic bacteria can modify the bacterial adhesion to intestinal mucus.

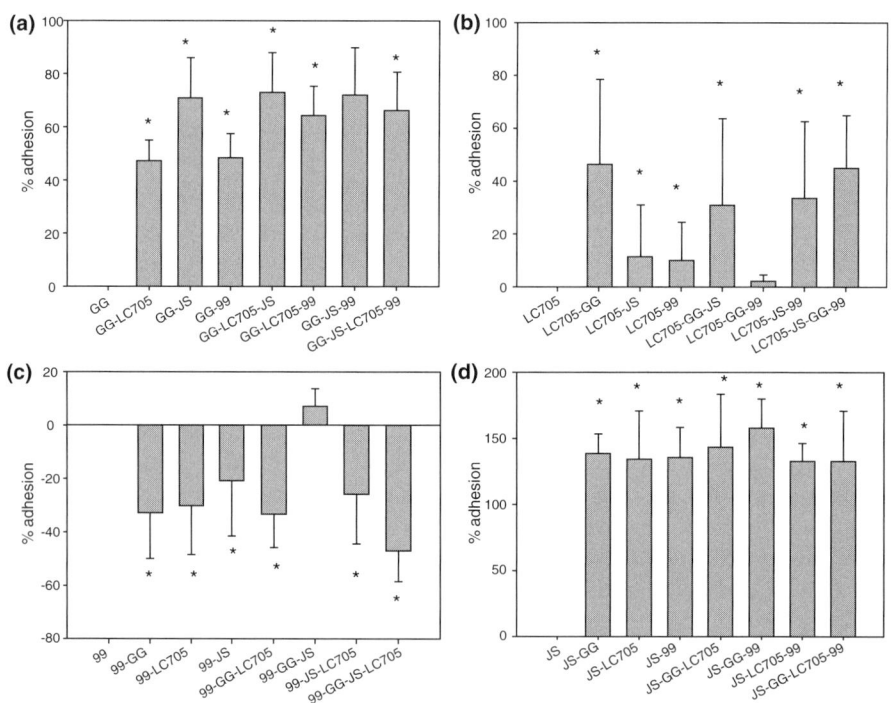

FIGURE 5.3 Changes in adhesion percentages of (a) *L. rhamnosus* GG (b) *L. rhamnosus* LC705 (c) *B. breve* 99, and (d) *P. freudenrenchii* JS in double, triple and quadruple combination (adapted from Ref. 93).

Thus, it is necessary to further research on the influence of coculture and metabolic productions in the adhesion of probiotic strains in order to find the best combinations to be included in functional food products (92, 93).

5.2.8 Conclusion

The studies so far demonstrate that it is possible to characterize specific probiotic combinations that increase the beneficial health effects beyond single strains. Such demonstrations include the inhibition of pathogen adhesion properties and their influence in their colonization. We suggest that combinations of different probiotics strains may be more effective *in vivo* than single probiotic strains especially in adult and elderly populations and in subjects with complicated microbiota aberrancies. Several reports also exist to support this hypothesis (92, 105, 123, 131, 143, 150). Taken together, the results show a very high specificity in the inhibition of the adhesion and displacement of enteropathogens by different probiotic strain combinations, indicating the need of a case-by-case characterization of these combinations. It must also be considered that at the end *in vivo* studies are always necessary to confirm the potential health effects prior to introducing such combinations to clinical intervention studies.

The presented new strategy and methodology allow the development of new specific probiotic combinations for the treatment or prevention of specific diseases caused by known pathogens or microbiota deviations. It can therefore be hypothesized that a specific combination of these microorganisms may have the potential to increase the efficacy of future probiotic preparations. As a basis, careful characterization and assessment of microbiota aberrancies are required. Thereafter, adhering organisms and their suitability to counteract the microbiota aberrancy in question has to be assessed. In the best cases, the selected strains may complement each other's probiotic effect and even work synergistically and such combinations can then be further characterized in human intervention studies in target populations.

5.3 PRODUCTION OF ANTIMICROBIAL SUBSTANCES

SATU VESTERLUND

Functional Foods Forum, University of Turku, Turku, Finland

Although health aspects of fermented milks appeared in the early 1900s because of Metchnikoff observations (174), lactic acid bacteria (LAB) have been used in the preservation of milk and vegetables for centuries (179). Today, a wide variety of strains including LAB, propionic acid bacteria (PAB), and bifidobacteria are industrially used as starter cultures, cocultures, or bioprotective cultures to improve preservation, flavors, and texture of milk, vegetable, meat, and cereal products. The scientific basis of improved preservation is based on fermentation in which the pH value is lowered, the amount of available carbohydrates is reduced, and many antimicrobial compounds are produced by fermenting bacteria. The widespread ability of LAB, PAB, and bifidobacteria to produce antimicrobial substances such as organic acids, hydrogen peroxide, and bacteriocins implies an important biological role for bacteria and has been one of the key elements in the selection of these bacteria for probiotic use (164). The aim of this chapter is to discuss the production, activity, and mode of action of different antimicrobial substances produced by probiotics.

5.3.1 Organic Acids

Organic acids are produced by all LAB upon fermentation of hexose. The levels and types of organic acids produced depend on the species and growth conditions of the bacteria. Homofermentative LAB use the glycolytic pathway to produce two pyruvates, which are further converted to lactate. Theoretically, the yield is 2 mol of lactic acid and net gain of 2 ATP per mole of glucose. When there is a lack of sugar or oxygen, heterofermentative LAB such as leuconostocs and group III lactobacilli as well as homofermentative LAB use the pentose phosphate pathway to produce 1 mol each of lactic acid, acetic acid/ethanol, CO_2, and ATP per mole of glucose consumed. The undissociated, more hydrophobic form of the acid diffuses over the cell membrane and dissociates inside the cell, releasing H^+ ions that acidify the cytoplasm (166). Acetic acid and propionic acid have higher pK_a values than lactic acid (4.87, 4.75, and 3.08, respectively) and therefore have a higher proportion of undissociated acid at a certain pH (165). Acetic acid is more inhibitory than lactic acid and can inhibit yeasts, molds, and bacteria (156). In addition to the pH effect, the undissociated acid collapses the electrochemical proton gradient, causing bacteriostasis and finally death of susceptible bacteria (166). Moreover, lactic acid has been shown to permeabilize the outer membrane of Gram-negative bacteria by liberation of its lipopolysaccharides, leading to loss of viability (151). Thus by permeabilization, the action of lactic acid may facilitate the action of other antimicrobial factors. The effect of organic acids against bacteria, yeasts, and molds is well documented (156, 157).

5.3.2 Hydrogen Peroxide

In the presence of flavoprotein oxidases and oxygen, bacteria are able to produce hydrogen peroxide (H_2O_2). The antimicrobial effect of hydrogen peroxide is attributed to a strong oxidizing effect on the bacterial cell. Under certain natural conditions as in milk, the effect of hydrogen peroxide may be enhanced by the lactoperoxidase system resulting in the production of hypothiocyanate ($OSCN^-$) or higher oxyacids (O_2SCN^- and O_3SCN^-) (161). The exposure of bacteria to $OSCN^-$ has been shown to damage metabolic enzymes of bacteria by oxidizing sulfhydryl groups of enzymes such as glyceraldehyde-3-phosphate dehydrogenase (158) or disrupt bacterial membranes (170).

The production of hydrogen peroxide by lactobacilli is supposed to be a nonspecific antimicrobial defense mechanism of the normal vaginal ecosystem (175). This is seen in a number of H_2O_2-generating lactobacilli when they are present in the vagina of most normal women but are absent from most women with bacterial vaginosis (167, 172). In addition, *L. crispatus* and *L. jensenii*, the most common lactobacilli in the female genital tract, have been shown to inhibit gonococci by producing H_2O_2 (177). However, it is argued that H_2O_2 can accumulate and be inhibitory to some microorganisms as in the presence of catalase produced by other bacteria, H_2O_2 breaks down (177).

5.3.3 Carbon Dioxide

Carbon dioxide (CO_2) is mainly formed from heterolactic fermentation. CO_2 can directly create an anaerobic environment and is toxic to some aerobic food microorganisms through its action on cell membranes. The antimicrobial activity of the carbon dioxide molecule is thought to be due to the inhibition of enzymatic decarboxylation and accumulation of carbon dioxide in the membrane lipid bilayers causing dysfunction in membrane permeability (173). As CO_2 can effectively inhibit the growth of many food spoilage microorganisms, especially Gram-negative psychrotrophic bacteria (162), the effect has been commercially applied to many refrigerated products as modified atmosphere packaging.

5.3.4 Bacteriocins

Bacteriocins are defined as proteinaceous, antibacterial substances synthesized by many different bacterial species in order to kill or inhibit growth of other bacteria. They are heterogeneous group of proteins that vary in spectrum of activity, mode of action, molecular weight, genetic origin, and biochemical properties. Bacteriocins are often confused with antibiotics in the literature, but they differ from antibiotics in several ways: (a) they are ribosomally synthesized, (b) host cells are immune to them due to the production of immunity proteins, (c) their mode of action is different, and (d) they can have narrow killing spectrum (160). Typically, bacteriocins are produced by Gram-negative bacteria and they have wide spectrum of activity as compared to those produced by Gram-positive bacteria. Bacteriocins produced by Gram-positive bacteria, usually LAB, inhibit strains of the same or closely related species (168).

Bacteriocins produced by LAB are classified into three major classes: (I) lantibiotics, (II) small heat-stable peptides, and (III) large heat-labile proteins (171). A fourth class of complex bacteriocins with lipid or carbohydrate moieties has been suggested, but the information related to this class is very limited and is supposed to include partially purified bacteriocins (171). Class I comprises bacteriocins, which are posttranslationally modified to contain unusual amino acids not normally found in the nature (e.g., lanthionine and β-methyllanthionine). The group is further divided into two subgroups: elongated, amphipathic, pore-forming type A lantibiotics and rigid, globular peptides of the type B category (154). Class II includes unmodified bacteriocins that are subdivided into three subclasses, namely, class IIa (*Listeria*-active or pediocin-like bacteriocins), class IIb (two-peptide bacteriocins), and IIc (other bacteriocins) (168). Class III includes large bacteriocins whose mode of action is poorly described.

The best characterized bacteriocin is nisin, which belongs to class I, and is produced by several *Lactococcus lactis* strains. Nisin has a broad spectrum of activity against

Gram-positive bacteria including most LAB, *Staphylococcus aureus*, *Listeria monocytogenes*, vegetative cells of *Bacillus* ssp., and *Clostridium* spp. as well as preventing outgrowth of spores in *Bacillus* and *Clostridium* species (171). The cytoplasmic membrane of susceptible bacteria is the target for nisin activity as nisin has been shown to bind the membrane-bound cell wall precursor lipid II (154, 168). Binding to lipid II results in two modes of action, that is, pore formation and inhibition of cell wall synthesis. The latter is, however, a comparatively slow process and pore formation, which results in the rapid efflux of small cytoplasmic compounds such as amino acids, potassium, inorganic phosphate, and ATP, is still considered the primary mode of action (154).

Bacteriocins of LAB have attracted interest for use as natural preservatives in food or in probiotic applications. For these purposes, it is an advantage that the bacteriocins have a broad spectrum of activity, that is, inhibit many pathogens. However, many bacteriocins of LAB have a narrow spectrum of activity and they are not effective against Gram-negative bacteria. Thus, the combination of bacteriocins with other preservation mechanisms has been proposed to extend its inhibitory activity to Gram-negative species (152). Also interpretation of spectra of inhibitory activity can sometimes be difficult as many of the producer strains can release more than one bacteriocin. In probiotic applications, bacteriocins should be active in the intestinal conditions and this can be difficult due to the proteinaceous nature of bacteriocins. For example, current evidence has shown that nisin is degraded or inactivated in intestinal conditions (155).

5.3.5 Low Molecular Weight Antimicrobial Compounds

Reuterin is the most intensively studied low molecular mass inhibitory compound of probiotics. It is produced by *Lactobacillus reuteri*, an inhabitant of the gastrointestinal tract of humans and animals. Reuterin is produced during stationary phase by anaerobic growth on a mixture of glucose and glycerol or glyceraldehydes (153). During log growth phase, reuterin is not produced due to the reducing power of glucose metabolism. When the cells enter into the stationary phase of growth, reuterin starts to accumulate. Moreover, reuterin accumulates when contacted with the target cells (159). Reuterin has a wide antimicrobial spectrum affecting both Gram-positive and Gram-negative bacteria as well as yeasts, fungi, and protozoa (153, 159, 178). The broad spectrum of activity is explained by its mechanism of action as it has been shown to inhibit ribonucleotide reductase, thereby interfering DNA synthesis (163).

5.3.6 Other Antimicrobial Agents

Probiotic bacteria can also produce antimicrobial substances such as diacetyl, acetaldehyde, and ethanol. Diacetyl is a product of citrate metabolism and is responsible for the aroma and flavor of butter. Gram-negative bacteria, yeasts, and molds are more sensitive to diacetyl than Gram-positive bacteria and its mode of action is believed to be due to interference with the utilization of arginine (169). The contribution of acetaldehyde to biopreservation is minor since the flavor threshold is much lower than the levels that are considered necessary to achieve inhibition of microorganisms (157). In addition, the levels of diacetyl, acetaldehyde, and ethanol produced are so low that the contribution to antimicrobial effect is minimal.

5.4 IMMUNE EFFECTS OF PROBIOTIC BACTERIA

MIMI TANG

Department of Allergy and Immunology, Royal Children's Hospital, Australia

5.4.1 The Neonatal Intestinal Microbiota

The largest microbial exposure in early life occurs through the acquisition of an intestinal microbiota. Newborns have their first exposure to bacteria in the birthing process. Prior to this, the intestinal tract is sterile. The process of colonization and the types of organisms that become established are influenced by many factors including mode of delivery (vaginal or caesarean section), mother's diet including use of probiotics, and whether the infant is bottle or breast-fed. The newborn microbiota changes rapidly in the first weeks of life, but stabilizes within months and by 2 years reflects adult patterns. Bifidobacteria are the dominant bacterial group in the infant intestine by 1 week of age. Studies using molecular methods show that they compose up to 60–90% of the intestinal microbiota in breast-fed infants. Lower counts of bifidobacteria are found in formula-fed infants (188) and adults; bacteroides are the most prevalent organisms with lesser number of bifidobacteria, lactobacilli, staphylococci, enterobacteria, streptococci, and clostridia species.

5.4.2 The Importance of the Intestinal Microbiota in Immune Development

Intestinal microbes have important effects on the intestinal mucosal barrier function and on intestinal maturation and are necessary for full development of the body's largest collection of lymphoid tissue, the gut-associated lymphoid tissue. It has been shown in mouse models that the first bacteria to colonize the newborn intestine ("pioneer bacteria") can modulate gene expression in host intestinal epithelial cells (192). This results in an altered intestinal microenvironment that influences the nature of subsequent intestinal colonization. The initial neonatal colonization is also important in stimulating normal immune development. Absence or an inadequacy of early microbial stimuli has been shown to cause defects in intestinal barrier function, reduced inflammatory responses, defective IgA responses, and deficient oral tolerance induction. Mice raised in a germ-free environment fail to develop oral tolerance and have a persistent Th2-dependent antibody response (214). This immune deviation can be corrected by reconstitution of intestinal microbiota, but only if this occurs during the neonatal period (214).

These findings suggest that early microbial exposure and a healthy intestinal microflora are crucial in providing immune stimulation for normal immune development (182).

5.4.3 Interaction of Commensal and Pathogenic Bacteria with the Intestinal Immune System

There is a constant interaction between bacteria in the intestinal lumen and the epithelial and immune cells within the gut, and this continuous interaction is central to the maintenance of immune homeostasis. Bacterial and food antigens are continuously sampled by specialized epithelial cells overlying the dome area of the follicles and by dendritic cells (DCs) that send dendrite processes into the gut lumen between the epithelial cell tight junctions. These cells represent the first cells of the mucosal immune system to encounter commensal and pathogenic bacteria. The immune system here has important opposing roles of responding to pathogens while maintaining suppressed responses against commensal microbial antigens.

To sense microbes within the gastrointestinal ecosystem, the gastrointestinal epithelium and dendritic cells in the gut-associated lymphatic tissue are equipped with pattern recognition receptors (PRRs) that recognize specific conserved molecular patterns on pathogens (215). Interestingly, these molecular patterns are not unique to bacterial pathogens, but are shared by entire classes of bacteria, both commensal and pathogen, and also viruses. There are a number of families of PRRs, including the TLRs that are usually expressed on the cell surface and the nucleotide binding oligomerization domain-like receptor (NLR) family, which are expressed in the cytosol (186, 215).

Signaling through TLR or NLR molecules typically leads to proinflammatory gene expression (183). Interaction of bacterial components with PRR on DCs leads to upregulation of cell surface costimulatory molecules, such as CD80 and CD86, and DC migration to lymph nodes where they activate and influence the differentiation of naïve T cells toward T regulatory or T helper (h) cell pathways. DCs that produce interleukin (IL)-12 can promote Th1 responses, while production of IL-4 preferably promotes Th2 responses, and IL-10 or TGF-β promotes induction of T regulatory cells. Therefore, DCs play a key role in guiding the regulation of immune responsiveness or tolerance. Interactions between commensal bacteria and DCs result in anti-inflammatory or tolerogenic immune responses, while pathogenic bacteria induce active immune responses.

5.4.4 Probiotic Effects on Immune Responses

Specific probiotics can assist in maintaining gut immune homeostasis by directly modulating immune responses, enhancing epithelial barrier function, and inhibiting pathogen growth. Probiotics interact with the mucosal immune system via the same pathways as commensal bacteria, specifically via interaction with epithelial cells and DCs, to influence both innate and adaptive immune responses. Probiotic effects on immune responses appear to be immune regulating rather than immune activating. *In vitro* and *in vivo* studies in mice and humans have shown that probiotics may predominantly modulate DC and T regulatory cell activity rather than T helper responses per se (203).

5.4.5 Probiotic Effects on Epithelial Cells

The close association between gut barrier integrity and immune response has been demonstrated in a suckling rat model, where *Lactobacillus rhamnosus* GG (LGG) reversed the increased intestinal permeability induced by cow's milk (195). Probiotics may modulate epithelial barrier function through interactions with TLR2 (197). Probiotic bacteria have been shown to modulate epithelial cell signal transduction pathways and cytokine production, thereby suppressing systemic inflammatory responses (200, 206, 210). Other nonpathogenic enteric bacteria are known to have an immunosuppressive effect on intestinal epithelial cells by inhibiting the transcription factor NF-κB pathway (205)

5.4.6 Probiotic Effects on DCs

Given the central role of DCs in directing the T cell response toward T regulatory, Th1, or Th2 pathways, probiotics are likely to modulate immune responses via these, in a species- and strain-dependent manner (189). Individual bacterial strains were shown to differentially modulate IL-10 and IL-12 production and the expression of costimulatory and maturation markers by DCs. The bifidobacteria strains showed markedly enhanced IL-10 production by both myeloid DCs and plasmacytoid DCs. In that study, *Lactobacillus* strains downregulated or had no effect upon IL-10 production. The bifidobacteria strains also showed downregulation of costimulatory molecules CD80 and CD40, and in some cases reduced IFN-γ production that was IL-10 dependent. Nevertheless, specific *Lactobacillus* strains were shown to induce T regulatory cells and tolerance by generating semimature DCs with upregulation of costimulatory molecules but low production of proinflammatory cytokines (189).

5.4.7 Probiotic Effects on Adaptive Immune Responses: T Helper Cells and T Regulatory Cells

As for DC effects, the effects of probiotics on T helper and T regulatory responses are species specific. Some *Lactobacillus* strains have been shown to stimulate Th1 cytokine production while others have increased Th2 responses or induced a mixed Th1/Th2 response. The cytokine patterns induced by intestinal bifidobacteria have also been shown to be strain specific and different *Bifidobacterium* strains may induce distinct and even opposing responses (190). In infants with cow's milk allergy and eczema, treatment with LGG was reported to increase PBMC production of IFN-γ, while a mixture of four probiotics that included LGG had no effect on IFN-γ and increased IL-4 production (208). In animal models of autoimmune arthritis, some probiotic bacteria have been reported to inhibit Th1 responses providing beneficial effects while others have been demonstrated to aggravate disease by inducing Th1 cytokine responses (201).

Probiotic bacteria have been shown to enhance IgA immune responses to oral and parenteral vaccines (191, 194, 204). A mixture of probiotic bacteria given to infants for the first 6 months of life was also shown to increase IgG responses to parenteral HIB vaccine (199).

Probiotic bacteria have also been demonstrated to induce regulatory cytokine production and T regulatory cells *in vitro* in animal models of disease and in human clinical trials. Several *Lactobacillus* strains have been shown to inhibit T cell proliferation, induce IL-10 and TGF-β production, and modify Th1 and Th2 cytokine production *in vitro* in various models of autoimmune inflammatory disease (201, 207, 212, 213, 217). In some animal models and human clinical trials, beneficial clinical effects associated with probiotic treatment have been attributed to increased IL-10 and/or TGF-β production and increased $CD4^+$ T regulatory cells (185, 209). Of note, some probiotic bacteria, for example, *B. lactis*, have been shown to inhibit TGF-β production *in vivo* (193). These findings together demonstrate that probiotic bacteria can have varied effects on both innate and adaptive immune responses. Interactions with DCs in particular may be important for probiotic effects on immune responses. DCs may be modulated by probiotic bacteria to induce T regulatory/tolerogenic responses or T helper immune responses. Selection of probiotic bacteria for clinical applications should take into account the specific effects on immune responses, underlining the importance of preclinical testing in specific target functions and populations.

5.4.8 Delivery of Probiotic Bacteria

Beneficial effects of probiotic bacteria are dependent upon ingestion in the live form. Bacteria must survive the host's digestive process, colonize the intestine, and attach to and adhere to gut epithelium to mediate their effects *in vivo*. Moreover, heat-inactivated probiotic bacteria have been reported to cause adverse gastrointestinal symptoms and diarrhea requiring discontinuation of therapy (198). Probiotics produce bacteriocins, hydrogen peroxide, and biosurfactants to aid their survival in the intestinal tract. They may also upregulate mucin encoding genes that stimulate the production of mucus to form a protective barrier. An effective probiotic will adapt to healthy flora and not displace native bacteria.

The best studied probiotic is LGG. Orally administered LGG survives passage in the gastrointestinal tract and adheres to intestinal mucus and epithelial cells. It can persist in the intestine for a week after oral administration ends when fecal recovery is used as a marker of colonization (187). LGG persists for longer than this in infants and in infants with rotavirus diarrhea (196, 211). Studies using colonic biopsies in adults show that LGG colonization continues for longer than that indicated by fecal recovery studies (180).

Lactic acid-producing organisms can lower intestinal pH, which favors the growth of more beneficial organisms. LGG has been shown to alter the intestinal microecology by creating an environment that promotes bifidobacterial growth (181) and has been shown to increase total levels of anaerobic bacteria, particularly bifidobacteria, bacteroides, and clostridia. LGG does not displace other species of lactobacilli or increase total counts of lactobacilli. Colonization is transient and LGG is no longer cultured from the stool by 7–10 days after discontinuation of its use.

5.4.9 The Specificity of Probiotic Effects

It is clear that different probiotic species and strains can have very different effects both *in vivo* and *in vitro*. The clinical or laboratory effects of one probiotic cannot be assumed for another probiotic species, or even for different strains of the same species. For example, in a double-blind placebo-controlled trial comparing LGG and a mixture of four probiotic strains (LGG, *L. ramnosus* LC705, *Bifidobacterium breve* Bbi99, *Propionibacterium* JS) for the treatment of infant eczema, beneficial effects were observed only for LGG and not the probiotic mix (216). Furthermore, LGG has been shown to enhance IgA responses against rotavirus, which are not found with different strains of the same species (202). *In vitro*, *Lactobacillus* species vary in their capacity to induce IL-12 production by murine dendritic cells. For example, a strain of *Lactobacillus reuteri* was found to specifically inhibit *Lactobacillus casei*-induced IL-12, IL-6, and TNF-α production by murine dendritic cells and to inhibit *Lactobacillus casei*-induced upregulation of dendritic cell costimulatory markers (184).

5.4.10 Summary

As a class of bacteria, probiotics appear to have predominantly immune modulating effects on DC and T regulatory activity rather than directly influencing Th1 or Th2 activity. Their potential application in the clinical setting holds great promise and selection of specific probiotic species and strains will be dictated by the spectrum of species-specific effects.

5.5 ALTERATION OF MICROECOLOGY IN HUMAN INTESTINE

5.5.1 Impact on Human Health: in Infants and the Elderly

SEPPO SALMINEN AND REETTA SATOKARI

Functional Foods Forum, University of Turku, Finland

The indigenous microbiota of an infant gastrointestinal tract is created through contact with the microbiota of the parents and the infant's immediate environment. Nature-induced initial colonization is optimally enhanced by oligosaccharides in breast milk and the microbiota of the mother. This process directs the later microbiota succession and health of the infant throughout the rest of his/her life (238). It also has an impact on the type of microbiota the individual will harbor at the time of aging. Thus, understanding and positive guidance of the process through dietary means is an important target when facilitating the different stages of the colonization starting from the mother to infant microbiota transfer during birth, breast feeding and weaning, and maintenance of healthy microbiota during the first years of life and throughout the lifetime and especially old age.

5.5.1.1 Stepwise Establishment of Microbiota

1. *Source of Microbiota.* The basis of the healthy infant gut microbiota is created by the mother during pregnancy and microbiota transfer at birth. The microbiota of a newborn develops rapidly after birth and it is initially dependent on the mother's microbiota, mode of delivery, gestational age, and birth environment (218, 238). The microbiota of the mother is determined by genetic and environmental factors. Recently, it has been demonstrated that stress and dietary habits during later stage of pregnancy and prior to birth may have a significant impact on the microbiota at the time of delivery and thus influence the quality and quantity of the first colonizers of the newborn. It has been shown that a probiotic strain can be transferred from mother to child during birth and that maternal consumption of probiotics influences the transfer and establishment of neonatal microbiota (224, 251). Subsequently, feeding practices including formula feeding and breast feeding and the nursing environment of the infant influence the microbiota, at the level of both species composition and number of bacteria.

   The stepwise establishment of the gut microbiota is usually characterized by early colonization at birth and during the first 1 or 2 weeks of life with facultative anaerobes such as the enterobacteria, coliforms, lactobacilli, and streptococci, followed by anaerobic genera such as *Bifidobacterium, Bacteroides, Clostridium*, and *Eubacterium* (238, 243).

   The classical view on the infant intestinal microbiota is that breast-fed infants are more frequently colonized with bifidobacteria and have higher counts of fecal bifidobacteria than formula-fed infants. This has also been supported by new molecular methods indicating that bifidobacteria can range from 60% up to 90% of the total fecal microbiota in breast-fed infants while lactic acid-producing bacteria may account for less than 1% of the total microbiota. However, many papers report equally high fecal counts of bifidobacteria in breast-fed and formula-fed infants (235, 245). In general, the microbiota of formula-fed infants seems more complex and more often colonized also by other bacteria (224, 239). The differences between the breast-fed and formula-fed infants at bacterial groups' level decreased along with the development and improved composition of infant formulas (244, 247). The most commonly found species of bifidobacteria in infants are *Bifidobacterium bifidum, B. breve, B. infantis*, and *B. longum*. Recent molecular studies indicate that there might be a significant difference in the species composition of bifidobacteria between breast-fed and formula-fed infants (220), between infants of allergic, atopic mothers, and nonallergic mothers (221), and between allergic and nonallergic children (241). The relevance of specific species to infant health is still unclear and warrants for further investigations. The most common lactobacilli in infant feces are lactobacilli belonging to the *Lactobacillus acidophilus* group (*L. acidophilus, L. gasseri*, and *L. johnsonii*) irrespective of the type of feeding (240, 249).

   The practice of breast feeding for 4–6 months has been considered to assist in the development of healthy gut microbiota. Major changes in the microbiota composition that occur during breast feeding are related to breast milk components, especially oligosaccharides that act as a selective agent for the infant microbiota and stimulate the growth of bifidobacteria. Recent pilot study by Rinne et al. (247) on 6-month-old infants showed that infants receiving formula supplemented with prebiotic oligosaccharides developed *Bifidobacterium* microbiota resembling that of breast-fed in terms of both bacterial counts and species diversity. Further, breast milk contains bacteria, notably also bifidobacteria, and thus serves as a constant inoculum for the infant (221, 224, 237). However, the data concerning the breast milk microbiota are scarce and this form of exposure needs to be further assessed. Breast feeding also provides an optimal environment for exchange of microbes between mother and infant via the skin contact and exposure to microbiota present in the immediate environment. As a result, every individual has a unique characteristic microbiota during later phases of breast feeding. Also formula-fed infants acquire their individually specific microbiota through various contacts with the environment. Thus, the intestinal microbiota as a defined entity does not exist as this microbial population rather comprises a dynamic mixture of microbes typical to each individual. Weaning will break the contact and constant supply of oligosaccharides and microbes from the mother and introduction of solid foods as well as possible antimicrobial drug treatment period further influences the microbiota development. By the age of 1 year, the infant microbiota starts to resemble that of an adult.

2. *Basis for Healthy Gut.* The desired outcome of the infant gut colonization process is a complex microbial community that provides the barrier against foreign microbes and some harmful components in the diet. In addition, the colonization process creates the basis for the establishment of a "noninflammatory" status of the gut. Such an environment in infants is hallmarked with a large Gram-positive bacterial population with a significant number of bifidobacteria in a species composition typical to the healthy infant (mainly *B. Bifidum*, *B. longum*, *B. breve*, and *B. infantis*). Lactic acid bacteria may have a role in providing the right conditions for bifidobacteria to dominate. The collective composition of the colonizing strains in infancy also provides the basis for healthy gut microbiota later in life as the development of the disease-free state of the gut lies in the host–microbe interaction in infancy. As the disturbed succession during early infancy has been linked to the risk of developing infectious, inflammatory, and allergic diseases later in life, it is still of great interest to further characterize both composition and succession of microbiota during infancy. Some deviations of the infant microbiota related to diseases are compiled in Table 5.9.

5.5.1.2 Methodological Improvements in Microbiota Assessment Our understanding of the intestinal microbiota has improved stage by stage along with the methodological improvements in the microbiota characterization. While early studies relied on cultivation, and only organisms that grew on laboratory media could be studied, today molecular biological techniques allow us to study the microbiota more

TABLE 5.9 Characteristics of Infant Microbiota: Deviations Associated with Diseases or Disease Risk

Subject Group	Microbiota at 0–12 Months	Microbiota at 24 Months	References
Autistic children	Higher number of clostridia	Higher number of clostridia	(252)
Wheezing infants	High *Clostridium*	Less diverse microbiota	(245, 256)
Infants at risk of diarrhea	Low bifidobacteria, high *Clostridium*, less diverse microbiota	Less diverse microbiota	(231)
Allergic infants	At 6 months lower bifidobacteria and higher clostridia	Less lactobacilli, high number of aerobic bacteria, high coliforms, higher *Staphylococcus aureus* counts	(233, 234, 245, 254)
Infants later developing allergic disease	Early microbiota (already at 2–3 weeks or 1 month), less bifidobacteria, and different species composition, often higher number of *Bifidobacterium adolescentis*, higher clostridia	Differences similar but not so pronounced, but still present even at 5 years of age (including the higher number of *B. adolescentis* in allergic infants)	(232, 245, 254)
Normal infants	At 6 months high bifidobacteria, low clostridia (especially in breast-fed infants)	High number of aerobic bacteria, high diversity, number of unculturable bacteria increase	(222, 232, 245)

comprehensively also in a culture-independent way (249). In the first phase, molecular techniques such as amplified ribosomal DNA restriction analysis (ARDRA), randomly amplified polymorphic DNA (RAPD), pulsed field gel electrophoresis (PFGE), and ribotyping were used to characterize cultivated colonies. Later, community profiling techniques such as PCR coupled to temperature and denaturing gradient gel electrophoresis (PCR–TGGE and PCR–DGGE, respectively) and sequencing of amplified small subunit ribosomal RNA (SSU rRNA) gene products were introduced to perform culture-independent analysis. Specific primers and probes in qPCR and fluorescent *in situ* hybridization (FISH), respectively, are widely used for the quantitative analysis of specific species/groups of bacteria. The major recent methodological improvement is the development of high-density microarrays for the analysis of human microbiota (243, 246). The arrays consist of thousands of SSU rRNA gene-targeted oligonucleotide probes designed to have selected specificity to different taxonomic groups or species. The methods of array design and analysis are imperfect and still evolving, but it is already evident that they provide us with a powerful high-throughput tool. It is anticipated that in future we will be able to analyze the microbiota composition in great detail at species or even strain level by using arrays. Large-scale studies with detailed microbiota descriptions will become possible and this will tremendously increase our knowledge about the human intestinal microbiota.

5.5.1.3 Microbiota After Infancy Following the first 6 months of life, the microbiota succession diverts toward a more diverse community after weaning (238, 243, 255). The differences between breast-fed and formula-fed infants disappear gradually due to the increase in the number of enterobacteria and enterococci and the succession of *Bacteroides*, *Clostridium*, and anaerobic cocci including peptococci, peptostreptococci in the former group (253). Increases in *E. coli* and enterococci have been reported after weaning. The levels of bacteroides and anaerobic Gram-positive cocci also appear to increase gradually during and following weaning, whereas enterobacteria decrease (236, 255).

A small study reports the microbiota follow-up for two infants for a period of 2 years using molecular methods. At 6 months, the T-RFLP profiles were dominated by *Bacteroides* and *Clostridium*. Between 6 and 12 months, more members of clostridia (*C. coccoides–Eub. rectale* group [*Clostridium* cluster XIV]) appeared in the feces of the infants. At 1 year, there was a new shift in the microbiota and it became more diverse with *Bacteroides*, *Vellionella*, and *Fusobacterium prausnitzii* increasing. Contrary to that Guerin-Danan and coworkers (223) report that in 10–18-month infants bifidobacteria predominates, followed by *Bacteroides*, enterobacteria, and enterococci. In the comprehensive study of Palmer et al. (243), the microbiota development of 14 healthy infants was followed throughout the first year of life by using 16S-based microarray combined with sequencing. In consistence with earlier studies, the colonization started with aerobic organisms followed by strict anaerobes. Strikingly, the observed *Bifidobacterium* levels were very low in this study and constituted only a minor part of the infant microbiota. Thus, there seem to be also geographical and demographic differences in the prevalence of bifidobacteria and in microbiota composition in general. Several abrupt shifts occurred in the population structure, but these could not be linked to any specific age or event, although the transition to an adult-like profile often followed the introduction of solid foods (243). At the age of 1–2 years, microbiota resembles that of adults (236, 243). However, it has been reported that children (16 months to 7 years) still harbor higher levels of bifidobacteria and enterobacteria than adults (229). Along with the maturation of gut and normal mucus production also mucin-degrading bacteria colonize the gut. *Akkermansia muciniphila*-like bacteria start to colonize infants at the age of few months and by the age of 1 year reach a level close to that observed in adults (219).

5.5.1.4 Host–Microbe Cross Talk Components of the human intestinal microbiota or organisms entering the intestine may have harmful or beneficial effects on human health and a complex community is required for the individual balance. Abundant evidence exists to document that specific strains of the healthy gut microbiota exhibit powerful antipathogenic and anti-inflammatory capabilities and are consequently involved with enhanced colonization resistance in the intestine.

Following weaning, the healthy microbiota is gradually created. In the gastrointestinal tract, there is a constant challenge by diverse antigens such as microbial antigens, foods, and allergens. Such priming of gut-associated lymphoid tissue is important for two opposing functions: mounting a response to pathogens and maintaining hyporesponsiveness to innocuous antigens. An important question is how the inflammation is kept under control during weaning and how the microbiota is altered during the adaptive process. The strains of the healthy gut microbiota are likely to provide the host with an anti-inflammatory stimulus directing the host–microbe interaction toward a healthy gut. The host–microbe cross talk during and after breast feeding seems important in this respect. At this stage, the bifidobacteria-dominated environment may provide the child a more anti-inflammatory stimulus than bacteria from adults, which have been shown to be more proinflammatory (226, 227, 242).

5.5.1.5 Microbiota in the Elderly Earlier culture-based methods assessing the human intestinal microbiota demonstrated changes in composition in the old age (239). Culture-based studies reported the decline of bifidobacteria in concentration in the elderly (218) and some support for the conception is offered by recent studies using molecular methods (230). Bifidobacteria are regarded as key members of healthy and protective intestinal microbiota of life (248) and the decline in their number and species diversity may contribute to the often decreased gut health in elderly. Further, *Bacteroides* decrease, while facultative anaerobes, fusobacteria, clostridia, and eubacteria increase (257). Also significant intestinal microbiota alterations have been reported among the elderly with potentially important clinical consequences related to bowel regularity and intestinal infections (228–230). Table 5.10 lists some characteristics of intestinal microbiota among elderly. The defining factors in microbiota composition and fluctuation in old age are largely unknown. Decreased secretion of mucus (a major source of nutrients for gut microbes) is a very probable factor causing changes in the microbiota. This is supported by the recent finding that mucin-degrading bacterium *Akkermansia muciniphila* or *A. muciniphila*-like bacteria decrease significantly in elderly (219).

TABLE 5.10 Characteristics of Intestinal Microbiota Among Elderly Subject in Different Intestinal Conditions

Subjects	Potential Microbiota Deviations	General Microbiota Characteristics
Normal elderly	Lower number of bifidobacteria, higher number of staphylococci	Less diversity, decreased barrier
Elderly with constipation	Lower number of lactic acid bacteria and bifidobacteria	Lowered diversity, reduced acid production, altered bifidobacterial species composition
Elderly with diarrhea	Higher number of clostridia, different clostridial species composition	Less diversity, altered metabolic activity

5.5.1.6   Maintenance of Healthy Microbiota   Creating a healthy gut microbiota during early life must be followed by proper maintenance and enhancement of the individual balance. During times of disease or following detectable deviations in initial microbiota development, the gut microbiota can be directed into healthy balance by dietary means, for instance, by using probiotics or prebiotics. Probiotics are members of the healthy gut microbiota and assist in mimicking the healthy microbiota of both infants and adults. Each probiotic strain has its specific effects, which have to be evaluated prior to application. A recent brief overview of the effects of probiotics on microbial diseases in humans was given by Woodmansey (257).

Prebiotics act through promotion of specific microbes with potential to maintain health. The prerequisite of prebiotic activity is that bacteria to be stimulated are already present in the gut. Most prebiotic components have been shown to enhance the *Bifidobacterium* microbiota, but different prebiotic oligosaccharides have different microbiota modifying properties, and the prebiotic effects should be defined more clearly. First, the bifidogenic change alone cannot be considered as a prebiotic effect. It has been reported that some fructooligosaccharides also enhance the levels of unknown microbes in human gut, thus potentially facilitating undesirable effects. Second, the desired *Bifidobacterium* strains should be present in the gut for the prebiotic effect. Finally, a clinical benefit has to be documented before a prebiotic effect can be verified.

There is a need for the development of novel specific prebiotics with defined microbiota and health effects. The mother-to-infant transfer of both microbes and oligosaccharides could be considered as a model for the development of future functional foods. Carefully designed combinations of probiotics and prebiotics would offer optimal means for creating and maintaining a healthy microbiota.

5.5.1.7   Conclusion   The healthy human microbiota is metabolically active and acts as a defense mechanism for our body. Deviations in its composition are related to multiple disease states not only within the intestine but also beyond the gastrointestinal tract. Infant microbiota and the first colonization steps are thought to have crucial role in the health later. Especially, bifidobacteria seem to have a key role in this process. The mother–infant contact has an important impact on initial development. The mother provides the first inoculum at birth, promotes the bifidogenic environment through (prebiotic) oligosaccharides in breast milk, and introduces environmental bacteria through her skin and other contact with the infant. Taken together, this guides the development of individually optimized microbiota under the existing environmental conditions for each infant. The process may also influence the microbiota diversity and development during old age.

The future target is to further clarify both the composition and the succession of microbial communities from early infancy to weaning and during the first years of life in order to understand how the specific composition and characteristics of the microbiota of each individual develop. In a similar manner, we need to characterize more carefully the microbiota changes, temporal alterations, and decreasing diversity in old age. Such knowledge would greatly assist in finding out the ways to support the healthy microbiota development and maintenance from birth to old age and understand the role of specific probiotics and prebiotics in the dietary management especially in the case of health-related microbiota deviations.

5.5.2 Impact on Animal Health: Designer Probiotics for the Management of Intestinal Health and Colibacillosis in Weaner Pigs

JAMES J.C. CHIN AND TONI A. CHAPMAN
Immunology and Molecular Diagnostic Research Unit (IMDRU), Elizabeth Macarthur Agricultural Institute, NSW Department of Primary Industries, Australia

Escherichia coli is a naturally ubiquitous group of Gram-negative facultative anaerobic rod-shaped and sometimes flagellated bacteria capable of fermenting lactose, which are normal enteric residents of the gastrointestinal tract (GIT) of vertebrates and even some invertebrates. Most *E. coli* are fairly resilient to stomach acidity and generally survive to colonize the duodenum, jejunum, ileum, and colon, with the greatest number of commensals preferentially located in the small intestine. *E. coli* are normally shed in the feces and this association has been used as an environmental estimate of fecal coliform contamination (260), particularly for evaluation of sanitary quality of water or as a general indicator of sanitary conditions in the food processing environment.

5.5.2.1 The Farrowing Environment The GIT of newborn mammals is sterile. Within a short period of time, uptake of microorganisms from the mother through skin contact and breast feeding (263) very quickly results in the introduction of bacteria capable of colonizing the microenvironment of both the small and large intestines (283). With the current intensive farming practices, farrowing sows are confined to crates (308) that limit movement so that neonate suckers may access teats more readily. Since all animals are grounded in the crates, there is a very high incidence of fecal contamination of the lower belly and teats. Consequently, newborn piglets become rapidly colonized with maternal fecal microbiota (295). Although the dynamics of fecal colonization of the neonatal GIT is under investigation in our laboratory, it is nevertheless intuitively evident that pathogenic strains, if present, would be competing on an equal footing with commensal biotypes. The ultimate balance of successful colonizers in different intestinal compartments would establish "niche-adapted" diverse communities (276) with varying abilities to function as competitive excluders.

5.5.2.2 The Weaning Environment In most intensive pig production systems, piglets are weaned usually between 21 and 28 days of age. Weaning is stressful because young piglets are separated from their mothers, mixed with piglets from other litters, and experience change in environment and diets (267). Weaning stress has been found to activate the enteric neuronal network in pigs that is accompanied by coactivation of the prostanoid pathway. This is associated with increased expression of intestinal corticotropin releasing factor receptors that trigger mucosal dysfunctions such as a reduction in transepithelial electrical resistance (TER) and increased intestinal permeability (289). Nutritional stress is a major factor in predisposing weaners to enteric infection because piglets have been accustomed to sow milk at hourly intervals in relatively small proportions. At weaning, there is a drastic change to more granular and particulate formulations such as whey, yeast, fish meal, soybean meal, fat, and oil. If provided *ad libitum*, overzealous eating can cause indigestion and malabsorbtion during the early days of weaning, predisposing animals to enteric infection. Much of this is attributed to transient and long-lasting modifications of the piglet intestine (266) and the impact of stress and feeding behavior on the hypothalamic–pituitary–adrenal (HAP) axis (268).

5.5.2.3 Colibacillosis in Pigs Pathogenic *E. coli* is mainly responsible for colibacillosis diarrhea in newborn, preweaning, and postweaning pigs. These pathogenic strains are grouped as enteropathogenic *E. coli* (EPECs) but display a number of unique phenotypes such as the production of toxins that qualify them as members of two subgroups—enterotoxigenic *E. coli* (ETECs) and verotoxigenic *E. coli* or (VTECs). Diarrheagenic porcine ETECs elaborate a high molecular weight heat-labile enterotoxin (LT) and smaller sized heat-stable enterotoxins (STa/STb), while VTECs produce verotoxin (VTe) that causes edema disease. VTe is synonymous with Shiga toxin 2e (Stx2e) and VTEC strains are also known as Shiga toxigenic *E. coli* (STECs). Although rare, some strains of *E. coli* responsible for postweaning diarrhea (PWD) secrete both entero- and verotoxins and are referred to as ETECs and VTECs, respectively (290). Traditionally, porcine ETECs and VTECs are identified by serological typing based on specific antibodies directed against the O antigen, determined by the chemistry of the polysugar side chain bound to the core of the lipopolysaccharide in the outer membrane (181 recognized types), K or acid polysaccharide capsular antigen (60 recognized types), and H or flagellar antigen (53 recognized types). As most of these antibody reagents were produced and cross-absorbed against pathogenic isolates, many commensal strains remain untypable and these generally are designated ONT (O-antigen nontypable). Antibody-based methods of serotyping (284) have been made more efficient through miniaturized antibody arrays (259), but even these may soon be replaced by molecular serotyping technology (261).

Throughout the world, there are only a number of specific serogroups associated with colibacillosis in pigs. These are O8, O64, O138, O139, O141, O147, O149, and O157 (265). Apart from the toxin repertoire, a third feature associated with identification of porcine ETEC/STEC strains is the presence of fimbrial adhesins. Fimbriae are colonization factors that facilitate attachment of *E. coli* to the lumenal surface of enterocytes. There are five major fimbrial adhesin types—F4 (K88, *fae*), F5 (K99, *fanA*), F6 (P987, *fasA*), F18 (*fedA*), and F41 (*fimF$_{41a}$*). The ownership of genes encoding enterotoxins, verotoxins, and fimbrial types can now be determined by PCR (265). While the restricted number of virulence genes (VGs) associated with diarrheagenic *E. coli* is an advantage for genotyping clonal groupings in outbreaks, they do not provide sufficient signal to noise discrimination to distinguish more subtle differences in gene carriage that can describe the signatures of individual pathogenic and commensal strains with greater precision and accuracy.

5.5.2.4 Control of Colibacillosis Unlike free-range farming, intensive animal industries for growing poultry, pigs, and feedlot cattle create conditions conducive to increase in ill-thrift and reduced growth performance. From therapeutic antibiotic treatment of sick animals in the 1950s, production systems have incorporated antibiotics such as tylosin, avilamycin, tetracycline, and penicillin into stock feed. These supplements were initially used prophylactically for individual healthy animals believed to be at risk of infection. Some of these antibiotics are used presently as growth promotants to increase growth rates and improve efficiency of feed conversion. Using NAHMS swine data, antibiotic growth promotants (AGPs) improve average daily gain by 0.5% and feed conversion ratio by 1.1% and reduce mortality rates by 0.22%. In the United States, these productivity improvements translate into a profitability gain of US$ 0.59 per pig marketed or an improvement of 9% in net profits associated with growth promotion antibiotics (288). Unfortunately, indiscriminate use of AGPs has caused significant development of antimicrobial resistance in bacteria associated with farm animals and increasing the risk of these resistance genes to be acquired by human pathogens via the food chain (277). So great is the perception of this risk that the European Union has banned all AGPs from being used in animal production systems from 2006 (http://www.ucsusa.org/food_and_environment/antibiotics_and_food/european-union-ban.html). Alternative strategies to control production from disease have focused on dietary strategies such as acidification, immunization (258), and the use of probiotics (287).

5.5.2.5 Mechanism of Action Parker's (292) and Fuller's (278) definitions of probiotics (see Section 1.1) have unknowingly incorporated a basic concept underwriting one of the key aspects of probiotic functionality and mechanism of action (298), that is, the ability to alter microbial balance. Upon closer scrutiny, balance is probably an inappropriate choice to describe microbial dynamics in the GIT because even minor changes in diet and environmental/community/social stressors can modify balance (293, 294). Since balance of good versus bad bacteria is too simple a concept, it may be more realistic to interpret balance as a continuum of shifting community dynamics of microflora in various microniches of each gastrointestinal compartment. Inherent in this notion of shifting balances in microbial community dynamics is our understanding of the role of probiotics in competitive exclusion (CE). Because community dynamics is changing, it is impossible to establish a persistent autochthonous commensal community, coexisting in static equilibrium with a minority of orally administered probiotic bacteria in any one intestinal microniche. How then can a snapshot be taken of shifting balances in the microbial community if a significant component of that community is not culturable and therefore impossible to enumerate to demonstrate an impact of probiotics on that balance? To address this issue, we have explored the use of culture-independent gene signatures of pathogenic and commensal *E. coli* strains on a clonal and population basis to generate snapshot profiles of the impact of probiotic administration on microbial community dynamics. Indeed, while it has been demonstrated very clearly *in vitro* that probiotic LAB strains can modulate Toll receptors and influence immune responses (271), it may be pertinent to consider that changing microbial community dynamics, particularly in the small intestine, can also play an important role *in vivo* in modifying both active immunity and tolerance in the host animal (272).

5.5.2.6 Pathogenic and Commensal E. coli—the Concept of Gene Signatures

Competitive exclusion is an important mechanism of action of probiotic bacteria but it is by no means an exclusive phenomenon. Microbial communities in the lumen that are in peristaltic transit, those in the mucus substratum or attached to receptors on the epithelial face of enterocytes in different compartments of the GIT, are continuously engaged in competition with each other. The role of *E. coli* in the pig gut provides an ideal model to evaluate the dynamics of such clonal interactions within a single species community and the possible impact of nutrition and probiotic supplements as well as host immunity (innate and acquired) in changing the composition of this population.

At least six *E. coli* strains and their genomes have been fully sequenced. This information base has been a motivating force in both phylogenetic and pathogenomic studies of *E. coli*. *E. coli* are phylogenetically grouped as A, B1, B2, and D phylotypes based on multilocus enzyme electrophoresis (MLEE) and multilocus sequence typing (MLST). Commensals belong to phylotype A and B1 while most pathogens belong to B2 and D, respectively. Functionally, pathogenic *E. coli* are grouped as diarrheagenic (DEC) or extraintestinal pathogenic *E. coli* (ExPEC). Diarrheagenic members have been subclassified as enteropathogenic (EPEC), EHEC (enterohemorrhagic), and atypical EPEC or ATEC. EPECs are a major cause of infant diarrhea in underdeveloped countries while EHECs cause bloody diarrhea and hemolytic uremic syndrome (HUS). The EPEC attaching and effacing gene (*eae*) is located within a specific chromosomal locus known as the locus for enterocytes effacement (LEE) pathogenicity island and encodes an attachment factor intimin that is shared between almost all EPEC and EHEC strains. EHECs are distinguished from EPECs by the presence of Shiga toxin genes (*stx*). A distinguishing feature of EPECs is the presence of an EPEC adherence factor plasmid (EAF) that also encodes a type IV bundle-forming pilus (Bfp) gene (*bfp*) and plasmid-encoded regulator gene *per*. Colonies expressing Bfp not only adhere to each other, but also form compact microcolonies that locally adhere (LA) to HEp-2 epithelial cells as part of the pathology of colonization and invasion (LEE$^+$). ATECs are EPECs that are LEE$^+$, *stx*$^-$, EAF$^-$, and *bfp*$^-$. Although strains that are LA$^+$ show a close association with ownership of Bfp, a number of *bfp*$^-$ EPEC strains are still capable of local adherence. Recent studies to examine this property with ATEC O26:H11 have revealed that LA is mediated via a novel afimbrial adhesin located in a new locus designated as the locus of diffuse adherence (*lda*) with homologies to the plasmid-encoded porcine K88 *fae* and ETEC CS31A *clp* fimbrial operons (296). Additional polymorphisms associated with 11 *bfp* and 20 *per* alleles have been used to subtype 12 EAF plasmids associated with various EPEC clones (286). Together with preferred insertions of the LEE, loci at the *sel*C or *phe*U site demonstrate the versatility of horizontal gene transfer acquisitions and loss as part of the evolving phylogeny of DECs.

ExPECs are primarily responsible for extraintestinal disease and classified pathotypically as UPECs (uropathogenic *E. coli* that cause urinary tract infections (UTIs), pyelonephritis and cystitis, sepsis/bacteremia, and neonatal meningitis (NMEC). Many of the VGs of UPECs have been derived from studies with strains CFT073, J96, and 536. UPEC and EPEC pathotypes appear to maintain a high degree of synthetically preserved and vertically evolved core genes. Punctuating this backbone are many pathogenicity islands acquired by different horizontal transfer events, but UPECs do not contain genes for the type III secretion system. These islands contain strain-specific VGs including fimbrial adhesins (*pap*), hemolysin (*hlyBACD*), secreted autotransporters (*sat, vat*), and phase-switch recombinases. Strains vary in other virulence genes: J96 has *cnf-1* (cytotoxic necrotizing factor), *hra* (heat resistant agglutinin), and no aerobactin genes while CFT073 only has the former (305).

5.5.2.7 Mosaicism and Genome Plasticity in Porcine E. coli (Clone Gene Signatures)

Genome plasticity and diversity (264) in pathogenic and commensal *E. coli* have arisen because of horizontal gene transfers and mutations into a "core" gene pool responsible for essential cellular functions and a "flexible" gene pool made up of ectochromosomal DNA (ECDNA) that encode adaptive functionalities such as pathogenicity, antibiotic/antiseptic/metal resistances, and fitness traits. Core and flex genes are generally located downstream of tDNA loci (DNA encoding transfer RNA). Contrary to expectations, only a limited number of tDNA loci remain hotspots for ECDNA insertions and members of different phylogenetic groups display different preferential patterns of tDNA loci polymorphisms (279). In view of the mosaic nature of the *E. coli* genome and assuming that gene acquisitions are not confined solely to only pathogenic strains and could also occur in commensals, we assembled a 58 virulence gene panel selected from virulence factors known to be associated with DEC(diarrheagenic) and ExPEC/UPEC/NMEC pathotypes (270). With the aid of uni- and multiplex PCR assays, the VG distribution was determined for 23 commensal strains (275) and 52 clinical isolates (20 neonatals and 32 weaners) (270). The primary intention of this investigation was to test the hypothesis that virulence gene ownership is not specific and exclusive to different *E. coli* pathotypes and commensal strains. The second hypothesis was that the presence as well as the absence of combinations of VGs can be applied to genotype both pathogenic and commensal isolates. If this was the case, then combinatorial sorting of various virulence gene combinations could be used to assign groupings for *E. coli* isolates from healthy pigs as well as diarrheic swine. As shown in Fig. 5.4, gene signature combinations for each clone when plotted after multiple regression to two principal coordinates (PCOs) successfully displayed the spatial distribution of clinical and commensal isolates. Commensal clones were more tightly clustered but the separation was not absolute as a small number of commensal isolates were rastered left toward the clinical zone. However, a proportion of clinical clones were also positioned in the commensal zone. Sporadic introgression of occasional clinical and healthy *E. coli* clone genotypes into pathogenic and commensal signature zones reflects the dynamic nature of gene acquisitions and loss in this group of enterobacteriaceae and supports the contention that pathogenicity ultimately resides in a clone not just being able to acquire virulence genes, but to be

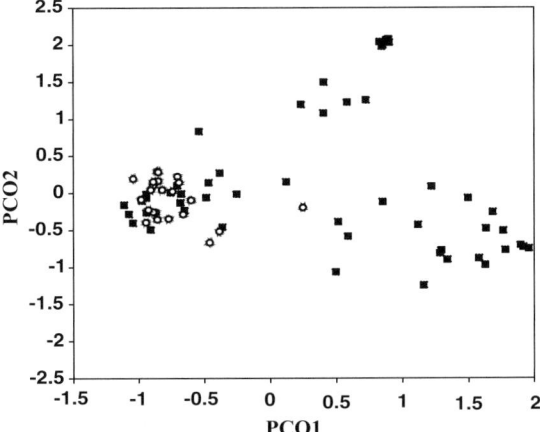

FIGURE 5.4 Clone gene signatures of commensal and clinical *E. coli*.— Principal coordinate analysis of single clones isolated from individual healthy pigs and diarrheic swine. Twenty-three commensal clones (from healthy pigs) and 52 clinical clones (from diarrheic swine) are represented by open circles and solid squares, respectively. The data matrix scoring system of 1 and 0 for presence and absence of 58 virulence genes was converted by regression to two dimensions using multiple linear equations (270). This was then plotted to display the spatial relationships between clones. Some of the clones depicted below may overlap if they possess identical virulence gene profiles, and such overlapping clones have been marginally displaced for visualization purposes by digital jittering.

able to express the right combination at the most optimal time in the best suited environmental niche.

The significance of this discovery led us to question whether commensals and ETECs also carried ExPEC virulence genes. We expanded and rescreened 158 porcine intestinal isolates and 47 ETECs for 36 extraintestinal virulence genes and discovered in these the presence of *fimH* (D-mannose-specific adhesin, type 1 fimbriae), *traT* (ColV plasmid gene encoding serum resistance), *fyuA* (*Yersinia* siderophore receptor), *hlyA* (alpha-hemolysin), *kpsMt*II (group 2 polysaccharide synthesis), *kpsMt k5* (non-K1 and non-K2 group II *kpsMii*), *iha* (nonhemagglutinin adhesin), and *ompT* (outer membrane protein T). One or more of these ExPEC genes were present at various frequencies in all commensal isolates (from 0.6% in the case of *hlyA* to 84.8% for *fimH*). Amongst ETECs, *fimH* was completely absent in all O149 serogroups but always present in O141 (307). These results justify the use of combinatorial gene signatures to track the clinical and commensal status of *E. coli* isolated from pigs. At the same time when this was published, others had reported that virulence gene-carrying *E. coli* strains are a normal part of intestinal bacterial populations and high number of these that carry hemolysin cannot always be correlated with disease (297). This conclusion was reached with only a restricted panel of porcine ETEC-specific

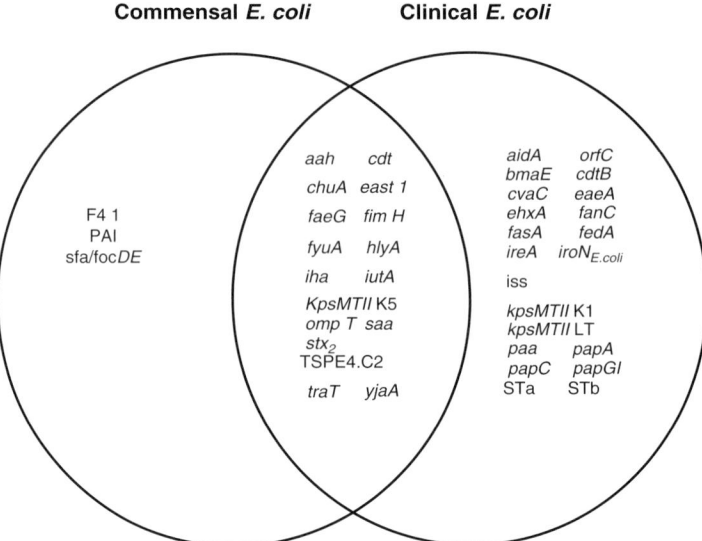

FIGURE 5.5 Venn diagram depicting virulence clone gene signatures identified in commensal and clinical *E. coli* clones isolated from pigs. VGs located in the overlapping zone of both circles represent virulence genes that have been found in clone isolates from both commensals and ETECs.

genes and does not reflect the diverse repertoire of virulence genes that can be present in both commensals and pathogens. It is important to ascribe biological relevance of VGs because presence or absence of individual virulence genes only facilitates genotyping, while the presence or absence of specific gene combinations defines the pathotype and virulence. Commensal and ETEC clones not only possess a number of VGs unique to their phenotype but also share some others in common (Fig. 5.5). It is the number of genes that are shared by different pathogenic clones that is responsible for attaining either a tight cluster in the clinical zone or shifting each pathogenic clone toward the commensal signature cluster.

5.5.2.8 Population Gene Signatures in Epidemiological Study

If virulence gene signatures of clones permit each commensal and clinical isolate to be spatially and separately clustered, then it should also be possible to group healthy and diarrheic swine based on their ownership of prevalent/dominant commensal or clinical clones within a population of *E. coli* from each respective host. In short, clinical pigs with diarrhea will tend to shed more ETECs than commensals while *E. coli* found in the feces of healthy animals would be dominated by commensal clones. The approach undertaken to test this concept was to deploy a rectal swab sampling technique followed by dispersal and immobilization of bacteria on a hydrophobic grid membrane filter (HGMF) (Fig. 5.6). *E. coli* colonies on the filter were then enriched by transfer of filter grids onto MacConkey agar medium. With appropriate dilution it was possible to

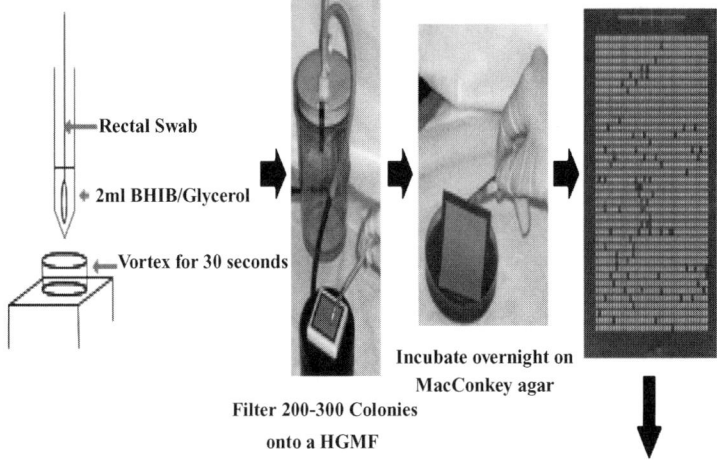
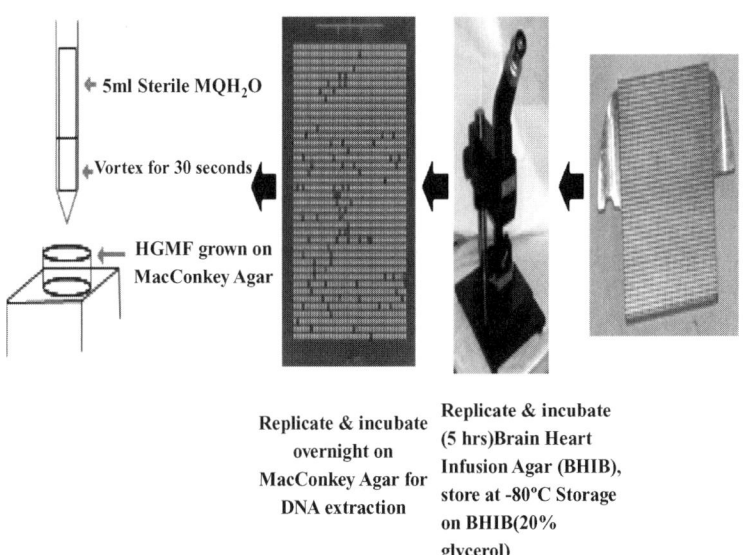

FIGURE 5.6 Processing of rectal swab samples for enrichment of *E. coli* on hydrophobic grid membrane filters before enumeration, differentiation, replication, and harvesting for assessment of population gene signatures.

display 200–300 colonies on each filter, which can then be immortalized by freezing or replica plating. Otherwise, the entire population of 200–300 *E. coli* colonies from a representative filter can be harvested and total DNA can be extracted. Each of these samples would then represent an entire *E. coli* population of the respective host and the composite DNA serve as a template for profiling population VG signatures.

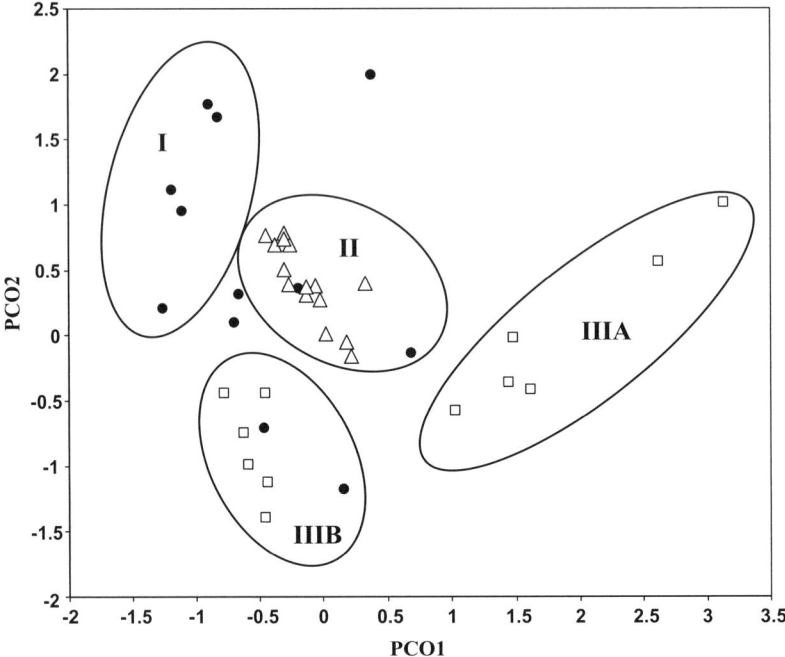

FIGURE 5.7 Population gene signatures of individual animals based on virulence gene profiling of DNA extracted from enriched *E. coli* clones (200–300 cfu) per HGMF. Healthy (nondiarrheic) pigs ($n = 12$) are represented by open squares; subclinical animals ($n = 15$) exposed to diarrheic swine but not displaying any clinical signs of colibacillosis at time of sampling are represented by open triangles; clinical pigs ($n = 12$) with at least three episodes of watery diarrhea per day are represented by closed circles. In the population gene signature analysis, only a subset of 35 out of a total of 58 genes nominated in Figure 5.4 was assayed. The abbreviations and names of all of these genes are listed in Table 5.11.

Figure 5.7 shows results of VG signatures of composite DNA from 12 healthy, 15 subclinical, and 12 diarrheic swine. Healthy animals were pigs that did not show any clinical signs of colibacillosis (open squares), while diarrheic swine represented animals that had produced more than two to three episodes of watery stools per day (solid circles). Subclinical animals were pigs that had been mingled with ETEC secretors but had, at the time of sampling, not displayed any signs of diarrhea (open triangles). Three primary clusters can be discerned even with such a small subset of animals analyzed. Pigs with diarrhea were grouped mainly in zone I while healthy animals were separated into two subclusters—zones IIIA and IIIB. The two major clusters, zones 1 and 2, were separated because of differences in virulence gene composition. Zone III animals were identifiable primarily by the absence of combinations of VGs in contrast to animals represented in zone 1. The most interesting observation was that subclinical animals were tightly bunched in zone II and these were marked by VG combinations intermediate between healthy and subclinical pigs.

There were two "clinical" pigs (solid circles) located in zone II and another two in zone IIIB. These cases probably represent differences in levels of gene expression and variations in the immune status of the pigs. Smaller and weaker animals in sibling cohorts are usually more susceptible to lower doses of ETECs with earlier onset of diarrhea. The clustering of subclinical pigs provides a very useful epidemiological tool to rapidly assess the clinical status of animals on farm before making decisions about preemptive treatments to prevent diarrhea.

Composite DNA extracted from an enriched population of *E. coli* is a more statistically reliable sample for epidemiological analysis than that obtained by separate clonal replicates from each animal even though the population signature approach is limited because it provides a once-off snapshot of the complexity of interactive microbial dynamics. Nonetheless, VG signatures of populations provide a rapid method for establishing the bacteriological status of the animal without the time demands of more traditional single clone isolation, characterization, and phenotypic/genotypic analysis (282). Indeed, population signatures from individual animals can also be pooled to represent herds from different production units in pig farming systems. Pooling not only reduces the cost of analysis but also facilitates temporal monitoring of subclinical disease and timely therapeutic intervention before an outbreak (unpublished results). Population signatures of different gene sets such as antimicrobial genes can also be used to monitor on-farm use of antibiotics (269, 299), paving the way to novel epidemiological surveillance applications with population gene signatures.

5.5.2.9 Designer Lactic Acid Bacteria as Probiotics The sustained use of AGPs in intensive animal industries has established a background of antimicrobial resistance in the microbial community of the GIT of pigs in piggeries throughout the world. In the United States, and also in Australia (281, 269), more than 75% of farming operations deploy antibiotics prophylactically to control infections due to postweaning diarrhea and edema disease (303). With increasing public scrutiny and concerns over the risk of antibiotic resistance genes transferring to human pathogens via the food chain, zero tolerance may be mandated despite several internationally prescribed principles of self-regulation (307). In the course of evolution of social behavior within the *E. coli* community (274), survival is dependent to a large extent on the ability of a proportion of the established clonal community to ward off invaders through the production of colicins (275). Stahl (301) showed that purified ColE1 and ColN were effective in suppressing the growth of fimbriated porcine ETECs (F4 and F18). Lactic acid bacteria are phylogenetically diverse group of Gram-positive, nonsporulating cocci- or rod-shaped lactic acid-producing microorganisms. LAB produce three classes of bacteriocins: (I) lantibiotics that are greatly posttranslationally modified, for example, nisin; (II) nonlantibiotic heat-stable single and bipeptide bacteriocins similar to colicin V that have not been extensively modified posttranslationally; and (III) nonlantibiotic heat-labile proteins (304).

Instead of utilizing probiotic formulation off the shelf (285, 302), we investigated the practicalities of producing designer LAB with specific targeted antagonism against predominantly Australian ETEC isolates O141, O149, and O8G7 by screening more than 100 internationally and commercially available GRAS (generally regarded as safe) LAB strains. The screening strategy employed a microtiter protocol known as kinetic inhibition microtiter assay (KIMA), where the ETEC bacteria are embedded in an agarose–brain heart infusion broth (BHIB) medium, subsequently overlaid with culture supernatant harvested from end of log-phase LAB strains grown in MRS broth. The reason for exploring LAB rather than *E. coli* was because bacteriocin production of LAB is predominantly constitutive while colicin production occurs in only a proportion of the clonal community in response to SOS-induced stress that can be simulated artificially *in vitro* through the use of growth modulating agents such as mitomycin C. The growth of the ETEC was determined temporally (hence a kinetic assay) against control MRS (bacteriocin-free) and BHIB (colicin-free) supernatants. Figure 5.8 shows the growth inhibition kinetics of four of the GRAS strains with the

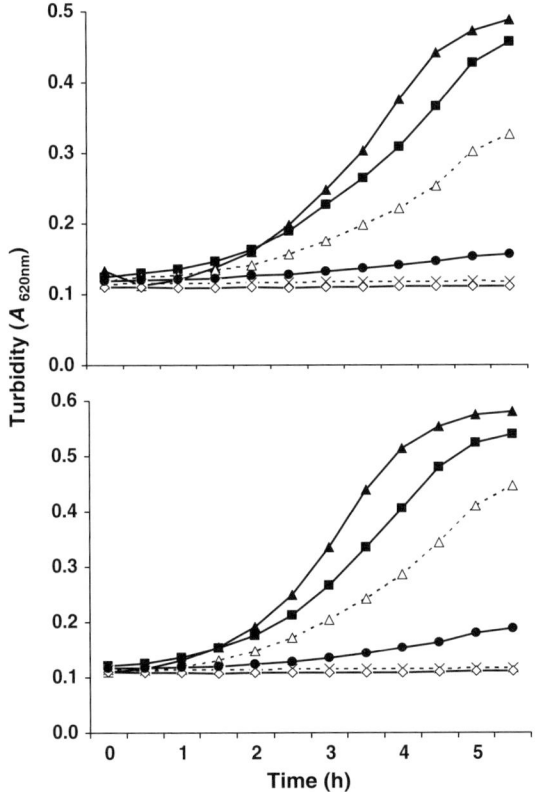

FIGURE 5.8 Kinetic inhibition microtiter assay—time course assessment of growth kinetics of targeted ETECs (O141 upper and O149 lower panels) grown in nutrient agarose and overlaid with culture supernatant (CS). From top to bottom, each symbol set designates the growth curve of targeted bacteria in the presence of BHIB (solid triangles), MRS broth (solid squares), and CS from stationary phase cultures of *L. acidophilus* (open triangles), *L. plantarum* (closed circles), *L. casei* (crosses), and *L. salivarus* (open diamonds) grown in MRS for 36–48 h.

most to least inhibitory activity displayed by *L. salivarus*, *L. casei*, *L. plantarum*, and *L. acidophilus*, respectively. These four strains were selected as a prototype combinatorial formulation (Top4) for further evaluation in pig trials. Top4 has been commercially registered as ColiGuard®.

5.5.2.10 Population Gene Signatures as a Measure of Probiotic Bioefficacy

To optimize conditions for trial of probiotics, rectal swabs ($n = 50$) were collected from a large number of pig farms ($n = 9$) in New South Wales and the population gene signature for subclinical endemic ETEC (according to Fig. 5.7) determined. Based on these results, one test farm with a high incidence of subclinical pigs was selected for evaluation of Top4 as a probiotic. A subset of 98 pigs on this farm was weaned at 21 days and 48 of these were immediately fed weaner mash supplemented with ColiGuard® (10^9 cfu/g each pig) daily for 10 days. The remaining 50 pigs were not supplemented and served as controls. All pigs consumed on the average approximately 300 g of feed per day. Rectal swab samples were collected from all animals after Top4 supplementation and processed on HGMF as shown in Fig. 5.6. Population signatures were collated and the average change in frequency of occurrence for each of 35 virulence genes (Table 5.11) was assessed.

The underlying assumption behind any justification for a gene signature shift is the null hypothesis; that is, if Top4 were ineffective against ETECs, then it would not act antagonistically against them. Under these conditions, there would be little or no change in the frequency of each VG. However, if the *E. coli* community were to be perturbed to the disadvantage of ETECs, then they would be excluded and purged in the feces. These changes in E. coli population structure would be reflected in specific gene shifts as shown in Fig. 5.9. Insignificant shifts were registered for 16 out of 35 VGs. Two significant gene shift patterns can be identified for the remaining 19 genes. Eleven of the significantly shifted genes were excreted at higher frequencies in ColiGuard®-treated animals compared to controls and the remaining eight VGs were found at reduced frequencies in the feces of these animals (Table 5.12 and Fig. 5.9).

Closer examination of the VGs in each of the two significantly shifted population signatures shows a partitioning of *bmaE*, *cvaC*, *eaeA*, F18, *hlyA*, *iha*, *kpsMTII*, *paa*, *saa*, *stx$_2$*, and *traT* from *cnf1*, *east1*, *fimH*, *fyuA*, *iron$_{E.coli}$*, *iss*, K5, and STb. When population gene signatures are compared to clone gene signatures on a single gene basis, it is clear that VGs that contribute to significant shifts in population signatures cannot be used to exactly match VG combinations that characterize clone signatures of commensals and ETECs (compare with Fig. 5.5). The deployment of singly shifted VGs in population signatures is not intended to provide a rationale for how ETEC clones within the community microflora of subclinical animals are disengaged from their intended role of causing disease. The shifts in individual gene frequencies seen in population signatures simply reflect perturbation of the population structure. Although Top4 (ColiGuard®) may most likely exert its impact directly against ETECs, one should consider another alternative mechanism. Instead of direct antagonism of ColiGuard® against the *E. coli* community, the presence of bioactive LAB in the GIT can provide SOS (e.g., stress-induced by competition and nutrient deprivation) (280) signals to perturb *E. coli* community dynamics and increase the production

TABLE 5.11 List of 35 Virulence Genes Used for Establishing Population Gene Signatures

Virulence Genes	Description/Function
Adhesins	
aah	Autotransporter adhesin heptosyltransferase encoding the AAH protein that modifies the AIDA-I adhesin
aidA	Adhesin involved in diffuse adherence, consists of AIDA-I (orf B) and AIDA^c (orf B^c)
AIDA^c (orfC)	AIDA-I adhesion
bmaE	M-agglutinin subunit
eaeA	Intimin
faeG	F4 fimbrial adhesin
fedA	F18 fimbrial adhesin
fimH	D-Mannose-specific adhesin, type 1 fimbriae
focG	Pilus tip molecule, F1C fimbriae (sialic acid specific)
iha	Novel nonhemagglutinin adhesin (from O157:H7 and FT073)
paa	Porcine A/E-associated gene
saa	STEC autoagglutinating adhesion
Toxins	
cdt	Cytolethal distending toxin
cnf1	Cytotoxic necrotizing factor 1
cvaC	Colicin V; conjugative plasmids (traT, iss, and antimicrobial resistance)
east1	EaggEC heat-stable enterotoxin
ehxA	Enterohemolysin
hlyA	α-Hemolysin
LT	Heat-labile toxin
STa	Heat-stable enterotoxin a
STb	Heat-stable enterotoxin b
stx$_2$	Shiga toxin II
univcnf	Universal primer for cytotoxic necrotizing factor 1
Capsule synthesis	
kpsMT II	Group II capsular polysaccharide synthesis (e.g., K1, K5, and K12)
kpsMT "K5"	Specific for non-K1 and non-K2 group II kpsMT
Siderophores	
fyuA	Yersinia siderophore receptor (ferric yersiniabactin uptake)
ireA	Iron-regulated element, a siderophore receptor
iroN$_{E.coli}$	Novel catecholate siderophore
iutA	Ferric aerobactin receptor (iron uptake: transport)
Additional virulence genes	
chuA	Gene required for heme transport in EHEC O157:H7
iss	Serum survival gene
ompT	Outer membrane protein A and T (protease)
traT	Surface exclusion, serum survival
TspE4.C2	Anonymous DNA fragment
yjaA	Identified in E. coli K12, function currently unknown

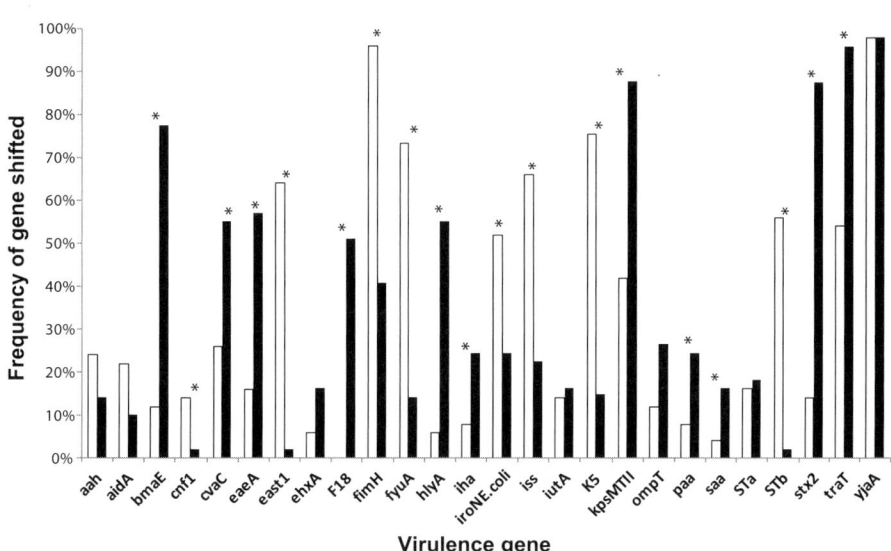

FIGURE 5.9 Demonstration of shifts in the population gene signatures of virulence genes profiled from enriched *E. coli* clones (200–300 per pig per HGMF grid). Data set represents profiles obtained from 48 pigs supplemented with lactic acid bacteria formulation (Top4) for a period of 10 days after weaning (solid bars) compared to 50 control animals that were not administered with LAB (open bars). LAB-treated and control pigs were housed in different pens in the same shed. Significant differences in the frequency of each VG between treatment groups are asterisked ($p < 0.05$). Only 19 out of 35 VGs were significantly shifted (all shown in the figure) while only 7 out of 17 nonsignificantly shifted genes are represented.

of *E. coli*-specific colicins that would also purge pathogenic ETECs in subclinical animals. Irrespective of these two proposed mechanisms, the shift in gene signature of fecal E. coli demonstrates a modulating effect of ColiGuard® on the enteric community of *E. coli*.

A major benefit of *E. coli* population signatures for the pig industry is to provide a facile tool for monitoring treatment programs intended to control or eradicate colibacillosis. Administration of ColiGuard® to weaners on a farm characterized by population gene signatures of subclinical pigs with endemic colibacillosis was able to significantly shift individual VGs. It can be concluded that Top4 not only reshaped the enteric *E. coli* population gene signatures but also confirmed the effectiveness of this treatment by increasing the growth weight of probiotic supplemented weaners.

5.5.2.11 Creation of Enteric Microbial Communities for Sustainable Intestinal Health (Probiosis) One of the proposed alternatives to AGPs for the control of ND and PWD is maternal vaccination with autogenous ETEC strains. While maternal antibodies can undoubtedly provide transient protection for the neonate, the lack of fimbriae and O antigen-specific antibodies in bacterin vaccines opens up the real possibility that common antigens such as membrane porins and heat-shock proteins in

TABLE 5.12 List of Virulence Genes that Were Either Affected or Remained Unaffected in the Feces of Weaners Following Probiotic (Top4) Supplementation

Insignificant Gene Shifts	Significant Genes Shifts	
	Increased Fecal Frequency	Decreased Fecal Frequency
aah	bmaE	cnf1
aidA	cvaC	east1
ehxA	eaeA	fimH
iutA	fedA (F18)	fyuA
ompT	hlyA	$iroN_{E.\ coli}$
STa	iha	iss
yjaA	KpsMTII	KpsMTII K5
$AIDA^c$(orfC)	paa	STb
faeG (F4)	saa	
focG	stx_2	
cdt	traT	
unicnf		
LT		
ireA		
chuA		
TspE4.C2		

the vaccine strain can elicit maternal antibodies with the ability to interfere with effective colonization by beneficial commensals and niche-specific E. coli. The implication behind maternal antibodies reshaping community dynamics of E. coli in the newborn is only being unraveled with the use of gene signatures and comparative diversity and typing analysis of colonizing commensal E. coli strains (e.g., SNPs, MLST, subtraction PCR libraries) (unpublished data). In future, VG profiling will be further enhanced by other sequences obtained by subtractive hybridization (300) to identify housekeeping and fitness genes that are unique to commensal populations. These genes will be added to the binary data set used currently for VG profiling and hence further increase the resolving power of gene signatures. Indeed, it appears that E. coli strains from different intestinal compartments of healthy grower pigs possess unique colicin and microcin gene signatures distinct from ETECs (291). These results support the notion that early neonatal intervention with a cocktail of niche-adapted E. coli can provide a sound basis for the use of E. coli strains as probiotics for promoting intestinal health (probiosis). Complementation of this strategy with LAB delivery to weaners will probably form the basis of new management strategies for AGP-free rearing of pigs in intensive production systems (267).

ACKNOWLEDGMENTS

The authors acknowledge the cumulative contributions (2001–2007) of graduate students at IMDRU, including S. Dixit, X.-Y. Wu, J. Patterson, and P. Njuguna. The expertise and input of Bernadette Turner merits special mention for well-performed technical work. This program was funded substantively by a Commonwealth Research and Development Start Grant awarded to International Animal Health (C. Lawlor and K. Healey) in conjunction with CSIRO and NSW DPI. ColiGuard® probiotic strains were provided by International Animal Health, Sydney, Australia.

REFERENCES

1. Adlerberth I, Ahrne S, Johansson ML, Molin G, Hanson LA, and Wold AE. A mannose-specific adherence mechanism in *Lactobacillus plantarum* conferring binding to the human colonic cell line HT-29. *Appl. Environ. Microbiol.* 1996; 62: 2244–2251.
2. Alander M, Korpela R, Saxelin M, Vilpponen-Salmela T, Mattila-Sandholm T, and von Wright A. Recovery of *Lactobacillus rhamnosus* GG from human colonic biopsies. *Lett. Appl. Microbiol.* 1997; 24: 61–64.
3. Alander M, Satokari R, Korpela R, Saxelin M, Vilpponen-Salmela T, Mattila-Sandholm T, and von Wright A. Persistence of colonization of human colonic mucosa by a probiotic strain, *Lactobacillus rhamnosus* GG, after oral consumption. *Appl. Environ. Microbiol.* 1999; 65: 351–354.
4. Bergonzelli GE, Granato D, Pridmore RD, Marvin-Guy LF, Donnicola D, and Corthesy-Theulaz IE. GroEL of *Lactobacillus johnsonii* La1 (NCC 533) is cell surface associated: potential role in interactions with the host and the gastric pathogen *Helicobacter pylori*. *Infect. Immun.* 2006; 74: 425–434.
5. Bernet MF, Brassart D, Neeser JR, and Servin AL. Adhesion of human bifidobacterial strains to cultured human intestinal epithelial cells and inhibition of enteropathogen–cell interactions. *Appl. Environ. Microbiol.* 1993; 59: 4121–4128.
6. Bernet MF, Brassart D, Neeser JR, and Servin AL. *Lactobacillus acidophilus* LA 1 binds to cultured human intestinal cell lines and inhibits cell attachment and cell invasion by enterovirulent bacteria. *Gut* 1994; 35: 483–489.
7. Bezkorovainy A. Probiotics: determinants of survival and growth in the gut. *Am. J. Clin. Nutr.* 2001; 73: 399S–405S.
8. Blum S, Reniero R, Schiffrin EJ, Crittenden R, Mattila-Sandholm T, von Wright A, Saarela M, Saxelin M, Collins K, and Morelli L. Adhesion studies for probiotics: need for validation and refinement. *Trends Food Sci. Technol.* 1999; 10: 405–410.
9. Buck BL, Altermann E, Svingerud T, and Klaenhammer TR. Functional analysis of putative adhesion factors in *Lactobacillus acidophilus* NCFM. *Appl. Environ. Microbiol.* 2005; 71: 8344–8351.
10. Cesena C, Morelli L, Alander M, Siljander T, Tuomola E, Salminen S, Mattila-Sandholm T, Vilpponen-Salmela T, and von Wright A. *Lactobacillus crispatus* and its nonaggregating mutant in human colonization trials. *J. Dairy Sci.* 2001; 84: 1001–1010.
11. Chauviere G, Coconnier MH, Kerneis S, Darfeuille-Michaud A, Joly B, and Servin AL. Competitive exclusion of diarrheagenic *Escherichia coli* (ETEC) from human enterocyte-like Caco-2 cells by heat-killed *Lactobacillus*. *FEMS Microbiol. Lett.* 1992; 70: 213–217.
12. Chauviere G, Coconnier MH, Kerneis S, Fourniat J, and Servin AL. Adhesion of human *Lactobacillus acidophilus* strain LB to human enterocyte-like Caco-2 cells. *J. Gen. Microbiol.* 1992; 138(Part 8): 1689–1696.

13. Coconnier MH, Bernet MF, Chauviere G, and Servin AL. Adhering heat-killed human *Lactobacillus acidophilus*, strain LB, inhibits the process of pathogenicity of diarrhoeagenic bacteria in cultured human intestinal cells. *J. Diarrhoeal Dis. Res.* 1993; 11: 235–242.
14. Coconnier MH, Klaenhammer TR, Kerneis S, Bernet MF, and Servin AL. Protein-mediated adhesion of *Lactobacillus acidophilus* BG2FO4 on human enterocyte and mucus-secreting cell lines in culture. *Appl. Environ. Microbiol.* 1992; 58: 2034–2039.
15. Collado MC, Gueimonde M, Sanz Y, and Salminen S. Adhesion properties and competitive pathogen exclusion ability of bifidobacteria with acquired acid resistance. *J. Food Prot.* 2006; 69: 1675–1679.
16. Crociani J, Grill JP, Huppert M, and Ballongue J. Adhesion of different bifidobacteria strains to human enterocyte-like Caco-2 cells and comparison with *in vivo* study. *Lett. Appl. Microbiol.* 1995; 21: 146–148.
17. Del Re B, Sgorbati B, Miglioli M, and Palenzona D. Adhesion, autoaggregation and hydrophobicity of 13 strains of *Bifidobacterium longum*. *Lett. Appl. Microbiol.* 2000; 31: 438–442.
18. Elo S, Saxelin M, and Salminen S. Attachment of *Lactobacillus casei* strain GG to human colon carcinoma cell line Caco-2: comparison with other dairy strains. *Lett. Appl. Microbiol.* 1991; 13: 154–156.
19. Frece J, Kos B, Svetec IK, Zgaga Z, Mrsa V, and Suskovic J. Importance of S-layer proteins in probiotic activity of *Lactobacillus acidophilus* M92. *J. Appl. Microbiol.* 2005; 98: 285–292.
20. Fuller R and Brooker BE. Lactobacilli which attach to the crop epithelium of the fowl. *Am. J. Clin. Nutr.* 1974; 27: 1305–1312.
21. Gopal PK, Prasad J, Smart J, and Gill HS. *In vitro* adherence properties of *Lactobacillus rhamnosus* DR20 and *Bifidobacterium lactis* DR10 strains and their antagonistic activity against an enterotoxigenic *Escherichia coli*. *Int. J. Food Microbiol.* 2001; 67: 207–216.
22. Granato D, Bergonzelli GE, Pridmore RD, Marvin L, Rouvet M, and Corthesy-Theulaz IE. Cell surface-associated elongation factor Tu mediates the attachment of *Lactobacillus johnsonii* NCC533 (La1) to human intestinal cells and mucins. *Infect. Immun.* 2004; 72: 2160–2169.
23. Granato D, Perotti F, Masserey I, Rouvet M, Golliard M, Servin A, and Brassart D. Cell surface-associated lipoteichoic acid acts as an adhesion factor for attachment of *Lactobacillus johnsonii* La1 to human enterocyte-like Caco-2 cells. *Appl. Environ. Microbiol.* 1999; 65: 1071–1077.
24. Greene JD and Klaenhammer TR. Factors involved in adherence of lactobacilli to human Caco-2 cells. *Appl. Environ. Microbiol.* 1994; 60: 4487–4494.
25. Gueimonde M, Noriega L, Margolles A, de los Reyes-Gavilan CG, and Salminen S. Ability of *Bifidobacterium* strains with acquired resistance to bile to adhere to human intestinal mucus. *Int. J. Food Microbiol.* 2005; 101: 341–346.
26. Gusils C, Cuozzo S, Sesma F, and Gonzalez S. Examination of adhesive determinants in three species of *Lactobacillus* isolated from chicken. *Can. J. Microbiol.* 2002; 48: 34–42.
27. He F, Isolauri E, Morita H, Hosoda M, Hashimoto H, Fuse T, Mizumachi K, Kurisaki J, Benno Y, Salminen S, and Ouwehand AC. Bifidobacteria isolated from allergic and healthy infants: differences in taxonomy, mucus adhesion and immune modulatory effects. *Microecol. Ther.* 2002; 29: 103–108.

28. He F, Ouwehand AC, Hashimoto H, Isolauri E, Benno Y, and Salminen S. Adhesion of *Bifidobacterium* spp. to human intestinal mucus. *Microbiol. Immunol.* 2001; 45: 59–262.
29. He F, Ouwehand AC, Isolauri E, Hashimoto H, Benno Y, and Salminen S. Comparison of mucosal adhesion and species identification of bifidobacteria isolated from healthy and allergic infants. *FEMS Immunol. Med. Microbiol.* 2001; 30: 43–47.
30. He F, Ouwehand AC, Isolauri E, Hosoda M, Benno Y, and Salminen S. Differences in composition and mucosal adhesion of bifidobacteria isolated from healthy adults and healthy seniors. *Curr. Microbiol.* 2001; 43: 351–354.
31. Henriksson A, Szewzyk R, and Conway PL. Characteristics of the adhesive determinants of *Lactobacillus fermentum* 104. *Appl. Environ. Microbiol.* 1991; 57: 499–502.
32. Hood SK and Zottola EA. Effect of low pH on the ability of *Lactobacillus acidophilus* to survive and adhere to human intestinal cells. *J. Food Sci.* 1988; 53: 1514–1516.
33. Hynönen U, Westerlund-Wikström B, Palva A, and Korhonen TK. Identification by flagellum display of an epithelial cell- and fibronectin-binding function in the SlpA surface protein of *Lactobacillus brevis*. *J. Bacteriol.* 2002; 184: 3360–3367.
34. Ibnou-Zekri N, Blum S, Schiffrin EJ, and von der Weid T. Divergent patterns of colonization and immune response elicited from two intestinal *Lactobacillus* strains that display similar properties *in vitro*. *Infect. Immun.* 2003; 71: 428–436.
35. Jacobsen CN, Rosenfeldt Nielsen V, Hayford AE, Moller PL, Michaelsen KF, Paerregaard A, Sandstrom B, Tvede M, and Jakobsen M. Screening of probiotic activities of forty-seven strains of *Lactobacillus* spp. by *in vitro* techniques and evaluation of the colonization ability of five selected strains in humans. *Appl. Environ. Microbiol.* 1999; 65: 4949–4956.
36. Jakava-Viljanen M and Palva A. Isolation of surface (S) layer protein carrying *Lactobacillus* species from porcine intestine and faeces and characterization of their adhesion properties to different host tissues.*Vet. Microbiol.* 2007, doi: 10.1016/j.vetmic.2007.04.029.
37. Johansson ML Molin G, Jeppsson B, Nobaek S, Ahrne S, and Bengmark S. Administration of different *Lactobacillus* strains in fermented oatmeal soup: *in vivo* colonization of human intestinal mucosa and effect on the indigenous flora. *Appl. Environ. Microbiol.* 1993; 59: 15–20.
38. Juntunen M, Kirjavainen PV, Ouwehand AC, Salminen SJ, and Isolauri E. Adherence of probiotic bacteria to human intestinal mucus in healthy infants and during rotavirus infection. *Clin. Diagn. Lab. Immunol.* 2001; 8: 293–296.
39. Kankaanpää P, Tuomola E, El-Nezami H, Ahokas J, and Salminen SJ. Binding of aflatoxin B_1 alters the adhesion properties of *Lactobacillus rhamnosus* strain GG in a Caco-2 model. *J. Food Prot.* 2000; 63: 412–414.
40. Kapczynski DR, Meinersmann RJ, and Lee MD. Adherence of *Lactobacillus* to intestinal 407 cells in culture correlates with fibronectin binding. *Curr. Microbiol.* 2000; 41: 136–141.
41. Kirjavainen PV, Ouwehand AC, Isolauri E, and Salminen SJ. The ability of probiotic bacteria to bind to human intestinal mucus. *FEMS Microbiol. Lett.* 1998; 167: 185–189.
42. Klarin B, Johansson ML, Molin G, Larsson A, and Jeppsson B. Adhesion of the probiotic bacterium *Lactobacillus plantarum* 299v onto the gut mucosa in critically ill patients: a randomised open trial. *Crit. Care* 2005; 9: R285–R293.

43. Lahtinen SJ, Saarinen NM, Ämmälä J, Mäkelä SI, Salminen S, and Ouwehand AC. Interactions between lignans and probiotics. *Microb. Ecol. Health Dis.* 2002; 14: 106–109.
44. Larsen N, Nissen P, and Willats WG. The effect of calcium ions on adhesion and competitive exclusion of *Lactobacillus* ssp. and *E. coli* O138. *Int. J. Food Microbiol.* 2007; 114: 113–119.
45. Lehto EM and Salminen S. Adhesion of twelve different *Lactobacillus* strains to Caco-2 cell cultures. *Nutr. Today* 1996; 31(Suppl.): 49S–50S.
46. Matsumoto M, Tani H, Ono H, Ohishi H, and Benno Y. Adhesive property of *Bifidobacterium lactis* LKM512 and predominant bacteria of intestinal microflora to human intestinal mucin. *Curr. Microbiol.* 2002; 44: 212–215.
47. Miyoshi Y, Okada S, Uchimura T, and Satoh E. A mucus adhesion promoting protein, MapA, mediates the adhesion of *Lactobacillus reuteri* to Caco-2 human intestinal epithelial cells. *Biosci. Biotechnol. Biochem.* 2006; 70: 1622–1628.
48. Morelli L, Garbagna N, Rizzello F, Zonenschain D, and Grossi E. *In vivo* association to human colon of *Lactobacillus paracasei* B21060: map from biopsies. *Digest. Liver Dis.* 2006; 38: 894–898.
49. Morita H, He F, Fuse T, Ouwehand AC, Hashimoto H, Hosoda M, Mizumachi K, and Kurisaki J. Adhesion of lactic acid bacteria to Caco-2 cells and their effect on cytokine secretion. *Microbiol. Immunol.* 2002; 46: 293–297.
50. Op den Camp HJ, Oosterhof A, and Veerkamp JH. Interaction of bifidobacterial lipoteichoic acid with human intestinal epithelial cells. *Infect. Immun.* 1985; 47: 332–334.
51. Ouwehand A, Parhiala R, Salminen S, Rantala A, Huhtinen H, Sarparanta H, and Salminen E. Influence of the endogenous mucosal microbiota on the adhesion of probiotic bacteria *in vitro*. *Microb. Ecol. Health Dis.* 2004; 16: 202–204.
52. Ouwehand A, Tölkkö S, and Salminen S. The effect of digestive enzymes on the adhesion of probiotic bacteria *in vitro*. *J. Food Sci.* 2001; 66: 856–859.
53. Ouwehand AC, Isolauri E, Kirjavainen PV, and Salminen SJ. Adhesion of four *Bifidobacterium* strains to human intestinal mucus from subjects in different age groups. *FEMS Microbiol. Lett.* 1999; 172: 61–64.
54. Ouwehand AC, Isolauri E, Kirjavainen PV, Tolkko S, and Salminen SJ. The mucus binding of *Bifidobacterium lactis* Bb12 is enhanced in the presence of *Lactobacillus* GG and *Lact. delbrueckii* subsp. *bulgaricus*. *Lett. Appl. Microbiol.* 2000; 30: 10–13.
55. Ouwehand AC, Kirjavainen PV, Grönlund M-M, Isolauri E, and Salminen SJ. Adhesion of probiotic micro-organisms to intestinal mucus. *Int. Dairy J.* 1999; 9: 623–630.
56. Ouwehand AC, Niemi P, and Salminen SJ. The normal faecal microflora does not affect the adhesion of probiotic bacteria *in vitro*. *FEMS Microbiol. Lett.* 1999; 177: 35–38.
57. Ouwehand AC, Salminen S, Roberts PJ, Ovaska J, and Salminen E. Disease-dependent adhesion of lactic acid bacteria to the human intestinal mucosa. *Clin. Diagn. Lab. Immunol.* 2003; 10: 643–646.
58. Ouwehand AC, Salminen S, Tölkkö S, Roberts P, Ovaska J, and Salminen E. Resected human colonic tissue: new model for characterizing adhesion of lactic acid bacteria. *Clin. Diagn. Lab. Immunol.* 2002; 9: 184–186.
59. Ouwehand AC, Salminen S, Tölkkö S, Roberts PJ, Ovaska J, and Salminen E. Dependent adhesion of lactic acid bacteria to colonic tissue *in vitro*. *Microecol. Ther.* 2002; 29: 95–102.

60. Ouwehand AC, Tuomola EM, Tölkkö S, and Salminen S. Assessment of adhesion properties of novel probiotic strains to human intestinal mucus. *Int. J. Food Microbiol.* 2001; 64: 119–126.
61. Ouwehand AC, Tölkkö S, Kulmala J, Salminen S, and Salminen E. Adhesion of inactivated probiotic strains to intestinal mucus. *Lett. Appl Microbiol.* 2000; 31: 82–86.
62. Pretzer G, Snel J, Molenaar D, Wiersma A, Bron PA, Lambert J, de Vos WM, van der Meer R, Smits MA, and Kleerebezem M. Biodiversity-based identification and functional characterization of the mannose-specific adhesin of *Lactobacillus plantarum*. *J. Bacteriol.* 2005; 187: 6128–6136.
63. Resta-Lenert S, and Barrett KE. Live probiotics protect intestinal epithelial cells from the effects of infection with enteroinvasive *Escherichia coli* (EIEC). *Gut* 2003; 52: 988–997.
64. Riedel CU, Foata F, Goldstein DR, Blum S, and Eikmanns BJ. Interaction of bifidobacteria with Caco-2 cells—adhesion and impact on expression profiles. *Int. J. Food Microbiol.* 2006; 110: 62–68.
65. Rinkinen M, Westermarck E, Salminen S, and Ouwehand AC. Absence of host specificity for *in vitro* adhesion of probiotic lactic acid bacteria to intestinal mucus. *Vet. Microbiol.* 2003; 97: 55–61.
66. Rojas M, Ascencio F, and Conway PL. Purification and characterization of a surface protein from *Lactobacillus fermentum* 104R that binds to porcine small intestinal mucus and gastric mucin. *Appl. Environ. Microbiol.* 2002; 68: 2330–2336.
67. Roos S and Jonsson H. A high-molecular-mass cell-surface protein from *Lactobacillus reuteri* 1063 adheres to mucus components. *Microbiology* 2002; 148: 433–442.
68. Savage DC. Growth phase, cellular hydrophobicity, and adhesion *in vitro* of lactobacilli colonizing the keratinizing gastric epithelium in the mouse. *Appl. Environ. Microbiol.* 1992; 58: 1992–1995.
69. Sherman LA and Savage DC. Lipoteichoic acids in *Lactobacillus* strains that colonize the mouse gastric epithelium. *Appl. Environ. Microbiol.* 1986; 52: 302–304.
70. Tallon R, Arias S, Bressollier P, and Urdaci MC. Strain- and matrix-dependent adhesion of *Lactobacillus plantarum* is mediated by proteinaceous bacterial compounds. *J. Appl. Microbiol.* 2007; 102: 442–451.
71. Tuomola EM, Ouwehand AC, and Salminen SJ. Chemical, physical and enzymatic pretreatments of probiotic lactobacilli alter their adhesion to human intestinal mucus glycoproteins. *Int. J. Food Microbiol.* 2000; 60: 75–81.
72. Tuomola EM and Salminen SJ. Adhesion of some probiotic and dairy *Lactobacillus* strains to Caco-2 cell cultures. *Int. J. Food Microbiol.* 1998; 41: 45–51.
73. Walter J, Chagnaud P, Tannock GW, Loach DM, Dal Bello F, Jenkinson HF, Hammes WP, and Hertel C. A high-molecular-mass surface protein (Lsp) and methionine sulfoxide reductase B (MsrB) contribute to the ecological performance of *Lactobacillus reuteri* in the murine gut. *Appl. Environ. Microbiol.* 2005; 71: 979–986.
74. Vesterlund S, Paltta J, Karp M, and Ouwehand AC. Measurement of bacterial adhesion—*in vitro* evaluation of different methods. *J. Microbiol. Methods* 2005; 60: 225–233.
75. von Wright A, Vilpponen-Salmela T, Llopis MP, Collins K, Kiely B, Shanahan F, and Dunne C. The survival and colonic adhesion of *Bifidobacterium infantis* in patients with ulcerative colitis. *Int. Dairy J.* 2002; 12: 197–200.

76. Zarate G, Morata de Ambrosini V, Perez Chaia A, and Gonzalez S. Some factors affecting the adherence of probiotic *Propionibacterium acidipropionici* CRL 1198 to intestinal epithelial cells. *Can. J. Microbiol.* 2002; 48: 449–457.
77. Aissi EA, Lecocq M, Brassart C, and Buoquelet S. Adhesion of some Bifidobacteria strains to human enterocyte-like cells and binding to mucosal glycoproteins. *Microb. Ecol. Health Dis.* 2001; 13: 32–39.
78. Alander M, Satokari R, Korpela R, Saxelin M, Vilpponen-Salmela T, Mattila-Sandholm T, and von Wright A. Persistence of colonization of human colonic mucosa by a probiotic strain, *Lactobacillus rhamnosus* GG, after oral consumption. *Appl. Environ. Microbiol.* 1999; 65: 351–354.
79. Bachrach G, Ianculovici C, Naor R, and Weiss EI. Fluorescence based measurements of *Fusobacterium nucleatum* co-aggregation and of fusobacterial attachment to mammalian cells. *FEMS Microbiol. Lett.* 2005; 248(2): 235–240.
80. Beachey EH. Bacterial adherence: adhesin–receptor interactions mediating the attachment of bacteria to mucosal surfaces. *J. Infect. Dis.* 1981; 143: 325–345.
81. Bernet MF, Brassart D, Nesser JR, and Servin AL. Adhesion of human bifidobacterial strains to cultured human intestinal epithelial cells and inhibition of enteropathogen–cell interactions. *Appl. Environ. Microbiol.* 1993; 59: 4121–4128.
82. Bibiloni R, Fedorak RN, Tannock GW, Madsen KL, Gionchetti P, Campieri M, De Simone C, and Sartor RB. VSL#3 probiotic-mixture induces remission in patients with active ulcerative colitis. *Am. J. Gastroenterol.* 2005; 100(7): 1539–1546.
83. Bibiloni R, Perez PF, and de Antoni GL. Will a high adhering capacity in a probiotic strain guarantee exclusion of pathogens from intestinal epithelia? *Anaerobe* 1999; 5: 519–524.
84. Bibiloni R, Perez PF, Garrote GL, Disalvo EA, and De Antoni GL. Surface characterization and adhesive properties of bifidobacteria. *Methods Enzymol.* 2001; 336: 411–427.
85. Boris S, Suarez JE, and Barbes C. Characterization of the aggregation promoting factor from *Lactobacillus gasseri*, a vaginal isolate. *J. Appl. Microbiol.* 1997; 83(4): 413–420.
86. Camilleri M. Probiotics and irritable bowel syndrome: rationale, putative mechanisms, and evidence of clinical efficacy. *J. Clin. Gastroenterol.* 2006; 40(3): 264–269.
87. Canzi E, Guglielmetti S, Mora D, Tamagnini I, and Parini C. Conditions affecting cell surface properties of human intestinal bifidobacteria. *Antonie van Leeuwenhoek* 2005; 88 (3–4): 207–219.
88. Cesena C, Morelli L, Alander M, Siljander T, Tuomola E, Salminen S, Mattila-Sandholm T, Vilpponen-Salmela T, and Von Wright A. *Lactobacillus crispatus* and its nonaggregating mutant in human colonization trials. *J. Dairy Sci.* 2001; 84: 1001–1010.
89. Chouraqui JP, Van Egroo LD, and Fichot MC. Acidified milk formula supplemented with *Bifidobacterium lactis*: impact on infant diarrhea in residential care settings. *J. Pediatr. Gastroenterol. Nutr.* 2004; 38(3): 288–292.
90. Collado MC, Gueimonde M, Hernández M, Sanz Y, and Salminen S. Adhesion of selected *Bifidobacterium* strains to human intestinal mucus and its role in enteropathogen exclusion. *J. Food Prot.* 2005; 68(12): 2672–2678.
91. Collado MC, Hernández M, and Sanz Y. Production of bacteriocin-like inhibitory compounds by potentially probiotic bifidobacteria. *J. Food Prot.* 2005; 68(5): 1034–1040.

92. Collado MC, Meriluoto J, and Salminen S. Development of new probiotics by strains combination: is it possible to improve the adhesion to intestinal mucus? *J. Dairy Sci.* 2007; 90: 2710–2716.

93. Collado MC, Meriluoto J, and Salminen S. *In vitro* analysis of probiotic strains combinations to inhibit pathogen adhesion to human intestinal mucus. *Food Res. Int.* 2007; 40(5): 629–636.

94. Collado MC, Meriluoto J, and Salminen S. Interactions between pathogens and lactic acid bacteria: aggregation and coaggregation abilities. *Eur. J. Food Res. Technol.* 2008; 226(5): 1065–1073.

95. Collado MC, Meriluoto J, and Salminen S. Measurement of aggregation properties of probiotic strains with pathogens: *in vitro* evaluation of different methods. *J. Microbiol. Methods* 2007; 71(1): 71–74.

96. Collado MC, Meriluoto J, and Salminen S. Role of commercial probiotic strains against human pathogen adhesion to intestinal mucus. *Lett. Appl. Microbiol.* 2007; 45(4): 454–460.

97. Del Re B, Sgorbati B, Miglioli M, and Palenzona D. Adhesion, autoaggregation and hydrophobicity of 13 strains of *Bifidobacterium longum*. *Lett. Appl. Microbiol.* 2000; 31: 438–442.

98. Femia AP, Luceri C, Dolara P, Giannini A, Biggeri A, Salvadori M, Clune Y, Collins KJ, Paglierani M, and Caderni G. Antitumorigenic activity of the prebiotic inulin enriched with oligofructose in combination with the probiotics *Lactobacillus rhamnosus* and *Bifidobacterium lactis* on azoxymethane-induced colon carcinogenesis in rats. *Carcinogenesis* 2002; 23(11): 1953–1960.

99. Fooks LJ, Fuller R, and Gibson GR. Prebiotics, probiotics and human gut microecology. *Int. Dairy J.* 1999; 9: 53–61.

100. Forestier C, De Champs C, Vatoux C, and Joly B. Probiotic activities of *Lactobacillus casei rhamnosus*: *in vitro* adherence to intestinal cells and antimicrobial properties. *Res. Microbiol.* 2001; 152: 167–173.

101. Freter M, Factors affecting the microecology of the gut. In: Fuller R editor. Probiotics. The Scientific Basis. Chapman and Hall, Glasgow, 1992, pp. 111–145.

102. Fujiwara S, Hashiba H, Hirota T, and Forstner JF. Inhibition of the binding of enterotoxigenic *Escherichia coli* Pb176 to human intestinal epithelial cell line HCT-8 by an extracellular protein fraction containing BIF of *Bifidobacterium longum* SBT2928: suggestive evidence of blocking of the binding receptor gangliotetraosylceramide on the cell surface. *Int. J. Food Microbiol.* 2001; 67: 97–106.

103. Gagnon M, Kheadr EE, Le Blay G, and Fliss I. *In vitro* inhibition of *Escherichia coli* O157: H7 by bifidobacterial strains of human origin. *Int. J. Food Microbiol.* 2004; 92: 69–78.

104. Gaudier E, Michel C, Segain JP, Cherbut C, and Hoebler C. The VSL# 3 probiotic mixture modifies microflora but does not heal chronic dextran–sodium sulfate-induced colitis or reinforce the mucus barrier in mice. *J. Nutr.* 2005; 135(12): 2753–2761.

105. Gionchetti P, Amadini C, Rizzello F, Venturi A, and Campieri M. Review article: treatment of mild to moderate ulcerative colitis and pouchitis. *Aliment. Pharmacol. Ther.* 2002; 16(4): 13–19.

106. Gionchetti P, Lammers KM, Rizzello F, and Campieri M. VSL#3: an analysis of basic and clinical contributions in probiotic therapeutics. *Gastroenterol. Clin. North Am.* 2005; 34(3): 499–513.

107. Gomez Zavaglia A, Kociubinski G, Perez P, Disalvo E, and De Antoni G. Effect of bile on the lipid composition and surface properties of bifidobacteria. *J. Appl. Microbiol.* 2002; 93(5): 794–799.
108. Green JD and Klaenhammer TR. Factors involved in adherence of lactobacilli to human Caco-2 cells. *Appl. Environ. Microbiol.* 1994; 60: 4487–4494.
109. Guandalini S. Probiotics for children: use in diarrhea. *J. Clin. Gastroenterol.* 2006; 40(3): 244–248.
110. Gueimonde M, Jalonen L, He F, Hiramatsu M, and Salminen S. Adhesion and competitive inhibition and displacement of human enteropathogens by selected lactobacilli. *Food Res. Int.* 2006; 39(4): 467–471.
111. Gueimonde M, Noriega L, Margolles A, De los Reyes-Gavilan CG, and Salminen S. Ability of *Bifidobacterium* strains with acquired resistance to bile to adhere to human intestinal mucus. *Int. J. Food Microbiol.* 2005; 101: 341–346.
112. Handley PS, Harty DW, Wyatt JE, Brown CR, Doran JP, and Gibbs AC. A comparison of the adhesion, coaggregation and cell-surface hydrophobicity properties of fibrillar and fimbriate strains of *Streptococcus salivarius*. *J. Gen. Microbiol.* 1987; 133(11): 3207–3217.
113. Jankovic I, Ventura M, Meylan V, Rouvet M, Elli M, and Zink R. Contribution of aggregation-promoting factor to maintenance of cell shape in *Lactobacillus gasseri* 4B2. *J. Bacteriol.* 2003; 185: 3288–3296.
114. Juntunen M, Kirjavainen PV, Ouwehand AC, Salminen SJ, and Isolauri E. Adherence of probiotic bacteria to human intestinal mucus in healthy infants and during rotavirus infection. *Clin. Diagn. Lab. Immunol.* 2001; 8(2): 293–296.
115. Kajander K, Hatakka K, Poussa T, Farkkila M, and Korpela R. A probiotic mixture alleviates symptoms in irritable bowel syndrome patients: a controlled 6-month intervention. *Aliment. Pharmacol. Ther.* 2005; 22(5): 387–394.
116. Katz JA. Probiotics for the prevention of antibiotic-associated diarrhea and *Clostridium difficile* diarrhea. *J. Clin. Gastroenterol.* 2006; 40(3): 249–255.
117. Kim HJ, Vazquez Roque MI, Camilleri M, Stephens D, Burton DD, Baxter K, Thomforde G, and Zinsmeister AR. A randomized controlled trial of a probiotic combination VSL#3 and placebo in irritable bowel syndrome with bloating. *Neurogastroenterol. Motil.* 2005; 17(5): 687–696.
118. Kirjavainen PE, Ouwehand AC, Isolauri E, and Salminen S. The ability of probiotic bacteria to bind to human intestinal mucus. *FEMS Microbiol. Lett.* 1998; 167: 185–189.
119. Kos B, Suskovic J, Vukovic S, Simpraga M, Frece J, and Matosic S. Adhesion and aggregation ability of probiotic strain *Lactobacillus acidophilus* M92. *J. Appl. Microbiol.* 2003; 94: 981–987.
120. Lee Y-J, Yu W-K, and Heo T-R. Identification and screening for antimicrobial activity against *Clostridium difficile* of *Bifidobacterium* and *Lactobacillus* species isolated from healthy infant faeces. *Int. J. Antimicrob. Agents* 2003; 21: 340–346.
121. Li J and Ellen RP. Coaggregation of *Porphyromonas* (*Bacteroides*) *gingivalis*, other species of bacteroides, and *Actinomyces viscosus*. Methodological evaluation. *J. Microbiol. Methods* 1990; 12: 91–96.
122. Malik A, Sakamoto M, Ono T, and Kakii K. Coaggregation between *Acinetobacter johnsonii* S35 and *Microbacterium esteraromaticum* strains isolated from sewage activated sludge. *J. Biosci. Bioeng.* 2003; 96: 10–15.

123. Myllyluoma E, Veijola L, Ahlroos T, Tynkkynen S, Kankuri E, Vapaatalo H, Rautelin H, and Korpela R. Probiotic supplementation improves tolerance to *Helicobacter pylori* eradication therapy—a placebo-controlled, double-blind randomized pilot study. *Aliment. Pharmacol. Ther.* 2005; 21(10): 1263–1272.

124. Nagayama M, Sato M, Yamaguchi R, Tokuda C, and Takeuchi H. Evaluation of co-aggregation among *Streptococcus mitis*. Fusobacterium nucleatum and Porphyromonas gingivalis. *Lett. Appl. Microbiol.* 2001; 33(2): 122–125.

125. Olivares M, Díaz-Ropero MP, Gómez N, Lara-Villoslada F, Sierra S, Maldonado JA, Martín R, Rodríguez JM, and Xaus J. The consumption of two new probiotic strains, *Lactobacillus gasseri* CECT 5714 and Lactobacillus coryniformis CECT 5711, boosts the immune system of healthy humans. *Int. Microbiol.* 2006; 9(1): 47–52.

126. Ouwehand AC and Salminen S. *In vitro* adhesion assays for probiotics and their *in vivo* relevance: a review. *Microbial Ecol. Health Dis.* 2003; 15: 175–184.

127. Ouwehand AC, Isolauri E, Kirjavainen PV, and Salminen SJ. Adhesion of four *Bifidobacterium* strains to human intestinal mucus from subjects in different age groups. *FEMS Microbiol. Lett.* 1999; 172: 61–64.

128. Ouwehand AC, Isolauri E, Kirjavainen PV, Tolkko S, and Salminen SJ. The mucus binding of *Bifidobacterium lactis* Bb12 is enhanced in the presence of *Lactobacillus* GG and *Lact. delbrueckii* subsp. *bulgaricus*. *Lett. Appl. Microbiol.* 2000; 30(1): 10–13.

129. Ouwehand AC, Salminen S, and Isolauri E. Probiotics: an overview of beneficial effects. *Antonie van Leeuwenhoek* 2002; 82: 279–289.

130. Pelletier C, Bouley C, Cayuela C, Bouttier S, Bourlioux P, and Bellon-Fontaine MN. Cell surface characteristics of *Lactobacillus casei* subsp. *casei*, *Lactobacillus paracasei* subsp. *paracasei*, and *Lactobacillus rhamnosus* strains. *Appl. Environ. Microbiol.* 1997; 63: 1725–1731.

131. Perdigón G, Maldonado Galdeano C, Valdez JC, and Medici M. Interaction of lactic acid bacteria with the gut immune system. *Eur. J. Clin. Nutr.* 2002; 56(4): S21–S26.

132. Peréz PF, Minnaard Y, Disalvo EA, and De Antoni GL. Surfaceproperties of bifidobacterial strains of human origin. *Appl. Environ. Microbiol.* 1998; 64: 21–26.

133. Reid G, McGroarty JA, Angotti R, and Cook RL. *Lactobacillus* inhibitor production against *Escherichia coli* and coaggregation ability with uropathogens. *Can. J. Microbiol.* 1988; 34: 344–351.

134. Rickard AH, Gilbert P, High NJ, Kolenbrander PE, and Handley PS. Bacterial coaggregation: an integral process in the development of multi-species biofilms. *Trends Microbiol.* 2003; 11: 94–100.

135. Rinkinen M, Westermarck E, Salminen S, and Ouwehand AC. Absence of host specificity for *in vitro* adhesion of probiotic lactic acid bacteria to intestinal mucus. *Vet. Microbiol.* 2003; 97: 55–61.

136. Rojas M and Conway PL. Colonization by lactobacilli of piglet small intestinal mucus. *J. Appl. Bacteriol.* 1996; 81: 474–480.

137. Roselli M, Finamore A, Britti MS, and Mengheri E. Probiotic bacteria *Bifidobacterium animalis* MB5 and *Lactobacillus rhamnosus* GG protect intestinal Caco-2 cells from the inflammation-associated response induced by enterotoxigenic *Escherichia coli* K88. *Br. J. Nutr.* 2006; 95(6): 1177–1184.

138. Ruas-Madiedo P, Gueimonde M, Margolles A, de los Reyes-Gavilan CG, and Salminen S. Exopolysaccharides produced by probiotic strains modify the adhesion of probiotics and enteropathogens to human intestinal mucus. *J. Food Prot.* 2006; 69(8): 2011–2015.

139. Schachtsiek M, Hammes WP, and Hertel C. Characterization of *Lactobacillus coryniformis* DSM 20001T surface protein Cpf mediating coaggregation with and aggregation among pathogens. *Appl. Environ. Microbiol.* 2004; 70(12): 7078–7085.

140. Schellenberg J, Smoragiewicz W, and Karska-Wysocki B. A rapid method combining immunofluorescence and flow cytometry for improved understanding of competitive interactions between lactic acid bacteria (LAB) and methicillin-resistant *S. aureus* (MRSA) in mixed culture. *J. Microbiol. Methods* 2006; 65(1): 1–9.

141. Servin AL. Antagonistic activities of lactobacilli and bifidobacteria against microbial pathogens. *FEMS Microbiol. Rev.* 2004; 28: 405–440.

142. Spencer RJ and Chesson A. The effect of *Lactobacillus* spp. on the attachment of enterotoxigenic *Escherichia coli* to isolated porcine enterocytes. *J. Appl. Bacteriol.* 1994; 77: 215–220.

143. Timmerman HM, Koning CJ, Mulder L, Rombouts FM, and Beynen AC. Monostrain, multistrain and multispecies probiotics—a comparison of functionality and efficacy. *Int. J. Food Microbiol.* 2004; 96: 219–233.

144. Toure R, Kheardr E, Lacroix C, Moroni O, and Fliss I. Production of antibacterial substances by bifidobacterial isolates from infant stool active against *Listeria monocytogenes*. *J. Appl. Microbiol.* 2003; 95: 1058–1069.

145. Tuomola EM, Ouwehand AC, and Salminen S. Chemical, physical and enzymatic pretreatments of probiotic lactobacilli alter their adhesion to human intestinal mucus glycoproteins. *Int. J. Food Microbiol.* 2000; 60: 75–81.

146. Vandevoorde L, Christiaens H, and Verstraete W. Prevalence of coaggregation among chicken lactobacilli. *J. Appl. Bacteriol.* 1992; 72: 214–219.

147. Vesterlund S, Paltta J, Karp M, and Ouwehand AC. Measurement of bacterial adhesion—in vitro evaluation of different methods. *J. Microbiol. Methods* 2005; 60: 225–233.

148. Vinderola CG, Medici M, and Perdigón G. Relationship between interaction sites in the gut hydrophobicity, mucosal immunomodulating capacities and cell wall protein profiles in indigenous and exogenous bacteria. *J. Appl. Bacteriol.* 2004; 96: 230–243.

149. Wadström T, Anderson K, Sydow M, Axelsson L, Lindgren S, and Gullmar B. Surface properties of lactobacilli isolated from the small intestine of pigs. *J. Appl. Bacteriol.* 1997; 62: 513–520.

150. Zoppi G, Cinquetti M, Benini A, Bonamini E, and Minelli EB. Modulation of the intestinal ecosystem by probiotics and lactulose in children during treatment with ceftriaxone. *Curr. Ther. Res.* 2001; 62: 418–435.

151. Alakomi HL, Skyttä E, Saarela M, Mattila-Sandholm T, Latva-Kala K, and Helander IM. Lactic acid permeabilizes Gram-negative bacteria by disrupting the outer membrane. *Appl. Environ. Microbiol.* 2000; 66: 2001–2005.

152. Arques JL, Fernandez J, Gaya P, Nunez M, Rodriguez E, and Medina M. Antimicrobial activity of reuterin in combination with nisin against food-borne pathogens. *Int. J. Food Microbiol.* 2004; 95: 225–229.

153. Axelsson LT, Chung TC, Dobrogosz WJ, and Lindgren SE. Production of a broad spectrum antimicrobial substance by *Lactobacillus reuteri*. *Microb. Ecol. Health Dis.* 1989; 2: 131–136.

154. Bauer R and Dicks LM. Mode of action of lipid II-targeting lantibiotics. *Int. J. Food Microbiol.* 2005; 101: 201–216.

155. Bernbom N, Licht TR, Brogren CH, Jelle B, Johansen AH, Badiola I, Vogensen FK, and Nørrung B. Effects of *Lactococcus lactis* on composition of intestinal microbiota: role of nisin. *Appl. Environ. Microbiol.* 2006; 72: 239–244.

156. Blom H and Mortvedt C. Anti-microbial substances produced by food associated microorganisms. *Biochem. Soc. Trans.* 1991; 19: 694–698.

157. Caplice E and Fitzgerald GF. Food fermentations: role of microorganisms in food production and preservation. *Int. J. Food Microbiol.* 1999; 50: 131–149.

158. Carlsson J, Iwami Y, and Yamada T. Hydrogen peroxide excretion by oral streptococci and effect of lactoperoxidase thiocyanate–hydrogen peroxide. *Infect. Immun.* 1983; 40: 70–80.

159. Chung TC, Axelsson LT, Lindgren SE, and Dobrogosz WJ. *In vitro* studies on reuterin synthesis by *Lactobacillus reuteri*. *Microb. Ecol. Health Dis.* 1989; 2: 137–144.

160. Cleveland J, Montville TJ, Nes IF, and Chikindas ML. Bacteriocins: safe, natural antimicrobials for food preservation. *Int. J. Food Microbiol.* 2001; 71: 1–20.

161. Condon S. Responses of lactic acid bacteria to oxygen. *FEMS Microbiol. Rev.* 1987; 46: 269–280.

162. Daniels JA, Krishnamurthi R, and SSH R. A review of the effects of carbon dioxide on microbial growth and food quality. *J. Food Prot.* 1985; 6: 532–537.

163. Dobrogosz WJ, Casas IA, Pagano GA, Talarico TL, Sjöberg B-M, and Karlsson M. *Lactobacillus reuteri* and the enteric microbiota. In: Gruff R, Midtvedt T, and Norin E, editors. *The Regulatory and Protective Role of the Normal Microflora*. Stockton Press, New York, 1989, pp. 283–292.

164. Dunne C, O'Mahony L, Murphy L, Thornton G, Morrissey D, O'Halloran S, Feeney M, Flynn S, Fitzgerald G, Daly C, Kiely B, O'Sullivan GC, Shanahan F, and Collins JK. *In vitro* selection criteria for probiotic bacteria of human origin: correlation with *in vivo* findings. *Am. J. Clin. Nutr.* 2001; 73: 386S–392S.

165. Eklund T. The antimicrobial effect of dissociated and undissociated sorbic acid at different pH levels. *J. Appl. Bacteriol.* 1983; 54: 383–389.

166. Eklund T. Organic acids, and esters. In: Gould GW, editor. *Mechanisms of Action of Food Preservation Procedures*. Elsevier, New York, 1989, pp. 161–200.

167. Eschenbach DA, Davick PR, Williams BL, Klebanoff SJ, Young-Smith K, Critchlow CM, and Holmes KK. Prevalence of hydrogen peroxide-producing *Lactobacillus* species in normal women and women with bacterial vaginosis. *J. Clin. Microbiol.* 1989; 27: 251–256.

168. Héchard Y and Sahl HG. Mode of action of modified and unmodified bacteriocins from Gram-positive bacteria. *Biochimie* 2002; 84: 545–557.

169. Jay JM. Antimicrobial properties of diacetyl. *Appl. Environ. Microbiol.* 1982; 44: 525–532.

170. Kamau DN, Doores S, and Pruitt K. Enhanced thermal destruction of *Listeria monocytogenes* and *Staphylococcus aureus* by the lactoperoxidase system. *Appl. Environ. Microbiol.* 1990; 56: 2711–2716.

171. Klaenhammer TR. Genetics of bacteriocins produced by lactic acid bacteria. *FEMS Microbiol. Rev.* 1993; 12: 39–85.

172. Klebanoff SJ, Hillier SL, Eschenbach DA, and Waltersdorph AM. Control of the microbial flora of the vagina by H_2O_2-generating lactobacilli. *J. Infect. Dis.* 1991; 164: 94–100.

173. Lindgren SE and Dobrogosz WJ. Antagonistic activities of lactic acid bacteria in food and feed fermentations. *FEMS Microbiol. Rev.* 1990; 87: 149–163.
174. Metchnikoff E. The Prolongation of Life. Optimistic Studies. Butterworth-Heinemann, London, 1907.
175. Reid G and Burton J. Use of *Lactobacillus* to prevent infection by pathogenic bacteria. *Microbes Infect.* 2002; 4: 319–324.
176. Ryan CS and Kleinberg I. Bacteria in human mouths involved in the production and utilization of hydrogen peroxide. *Arch. Oral Biol.* 1995; 40: 753–763.
177. St, Amant DC, Valentin-Bon IE, and Jerse AE. Inhibition of *Neisseria gonorrhoeae* by *Lactobacillus* species that are commonly isolated from the female genital tract. *Infect. Immun.* 2002; 70: 7169–7171.
178. Talarico TL and Dobrogosz WJ. Chemical characterization of an antimicrobial substance produced by *Lactobacillus reuteri*. *Antimicrob. Agents Chemother.* 1989; 33: 674–679.
179. Tamime AY. Fermented milks: a historical food with modern applications—a review. *Eur. J. Clin. Nutr.* 2002; 56(Suppl. 4): S2–S15.
180. Alander M, Satokari R, Korpela R, et al. Persistence of colonization of human colonic mucosa by *Lactobacillus rhamnosus* GG, after oral consumption. *Appl. Environ. Microbiol.* 1999; 65: 351–354.
181. Benno Y, He F, and Hosoda M. Effects of *Lactobacillus* GG yoghurt on human intestinal microecology in Japanese subjects. *Nutr. Today* 1996; 31: 9S–11S.
182. Bjorksten B, Sepp E, Julge K, Voor T, and Mikelsaar M. Allergy development and the intestinal microflora during the first year of life. *J. Allergy Clin. Immunol.* 2001; 108(4): 516–520.
183. Carmody RJ and Chen YH. Nuclear factor-kappaB: activation and regulation during toll-like receptor signaling. *Cell. Mol. Immunol.* 2007; 4(1): 31–41.
184. Christensen HR, Frokiaer H, and Pestka JJ. Lactobacilli differentially modulate expression of cytokines and maturation surface markers in murine dendritic cells. *J. Immunol.* 2002; 168(1): 171–178.
185. Di Giacinto C, Marinaro M, Sanchez M, Strober W, and Boirivant M. Probiotics ameliorate recurrent Th1-mediated murine colitis by inducing IL-10 and IL-10-dependent TGF-beta-bearing regulatory cells. *J. Immunol.* 2005; 174(6): 3237–3246.
186. Franchi L, McDonald C, Kanneganti TD, Amer A, and Nunez G. Nucleotide-binding oligomerization domain-like receptors: intracellular pattern recognition molecules for pathogen detection and host defense. *J. Immunol.* 2006; 177(6): 3507–3513.
187. Goldin BR, Gorbach SL, Saxelin M, Barakat S, Gualtieri L, and Salminen S. Survival of *Lactobacillus* species (strain GG) in human gastrointestinal tract. *Digest. Dis. Sci.* 1992; 37(1): 121–128.
188. Harmsen HJ, Wildeboer-Veloo AC, Raangs GC, Wagendorp AA, Klijn N, Bindels JG, et al. Analysis of intestinal flora development in breast-fed and formula-fed infants by using molecular identification and detection methods. *J. Pediatr. Gastroenterol. Nutr.* 2000; 30(1): 61–67.
189. Hart AL, et al. Modulation of human dendritic cell phenotype and function by probiotic bacteria. *Gut* 2004; 53(11): 1602–1609.
190. He F, Morita H, Hashimoto H, et al. Intestinal *Bifidobacterium* species induce varying cytokine production. *J. Allergy Clin. Immunol.* 2002; 109: 1035–1036.

191. He F, Tuomola E, Arvilommi H, and Salminen S. Modulation of humoral immune response through probiotic intake. *FEMS Immunol. Med. Microbiol.* 2000; 29: 47–52.
192. Hooper LV, Wong MH, Thelin A, Hansson L, Falk PG, and Gordon JI. Molecular analysis of commensal host–microbial relationships in the intestine. *Science* 2001; 291 (5505): 881–884.
193. Isolauri E, Arvola T, Sutas Y, Moilanen E, and Salminen S. Probiotics in the management of atopic eczema. *Clin. Exp. Allergy* 2000; 30(11): 1604–1610.
194. Isolauri E, Joensuu J, Suomalainen H, Luomala M, and Vesikari T. Improved immunogenicity of oral D × RRV reassortant rotavirus vaccine by *Lactobacillus casei* GG. *Vaccine* 1995; 13: 310–312.
195. Isolauri E, Majamaa H, Arvola T, Rantala I, Virtanen E, and Arvilommi H. *Lactobacillus casei* strain GG reverses increased intestinal permeability induced by cow milk in suckling rats. *Gastroenterology* 1993; 105: 1643–1650.
196. Kaila M, Isolauri E, Salminen S, Mikelsaar M, and Sepp E. Colonisation of infants with *Lactobacillus* GG during acute rotavirus diarrhea. *Biosci. Microflora* 1998; 17: 149–151.
197. Kalliomäki MA and Walker WA. Physiologic and pathologic interactions of bacteria with gastrointestinal epithelium. *Gastroenterol. Clin. North Am.* 2005; 34: 383–399.
198. Kirjavainen PV, Salminen SJ, and Isolauri E. Probiotic bacteria in the management of atopic disease: underscoring the importance of viability. *J. Pediatr. Gastroenterol. Nutr.* 2003; 36: 223–227.
199. Kukkonen K, Nieminen T, Poussa T, Savilahti E, and Kuitunen M. Effect of probiotics on vaccine antibody responses in infancy—a randomized placebo-controlled double-blind trial. *Pediatr. Allergy Immunol.* 2006; 17(6): 416–421.
200. Lammers KM, et al. Effect of probiotic strains on interleukin 8 production by HT29/19A cells. *Am. J Gastroenterol.* 2002; 97(5): 1182–1186.
201. Maassen CB, et al. Orally administered Lactobacillus strains differentially affect the direction and efficacy of the immune response. *Vet. Q.* 1998; 20(Suppl. 3): S81–S83.
202. Majamaa H, Isolauri E, Saxelin M, and Vesikari T. Lactic acid bacteria in the treatment of acute rotavirus gastroenteritis. *J. Pediatr. Gastroenterol. Nutr.* 1995; 20 (3): 333–338.
203. Matsuzaki T, et al. Prevention of onset in an insulin-dependent diabetes mellitus model, NOD mice, by oral feeding of *Lactobacillus casei*. *Acta Pathol. Microbiol. Immunol. Scand.* 1997; 105(8): 643–649.
204. Mullie C, Yazourh A, Thibault H, et al. Increased poliovirus-specific intestinal antibody response coincides with promotion of *Bifidobacterium longum-infantis* and *Bifidobacterium breve* in infants: a randomized, double-blind, placebo-controlled trial. *Pediatr. Res.* 2004; 56(5): 791–795.
205. Neish AS, Gewirtz AT, Zeng H, et al. Prokaryotic regulation of epithelial responses by inhibition of IkappaB-alpha ubiquitination. *Science* 2000; 289(5484): 1560–1563.
206. Neish AS. Bacterial inhibition of eukaryotic pro-inflammatory pathways. *Immunol. Res.* 2004; 29(1–3): 175–186.
207. Pessi T, Sutas Y, Hurme M, and Isolauri E. Interleukin-10 generation in atopic children following oral *Lactobacillus rhamnosus* GG. *Clin. Exp. Allergy* 2000; 30(12): 1804–1808.
208. Pohjavuori E, et al. *Lactobacillus* GG effect in increasing IFN-gamma production in infants with cow's milk allergy. *J. Allergy Clin. Immunol.* 2004; 114(1): 131–136.

209. Rautava S, Kalliomaki M, and Isolauri E. Probiotics during pregnancy and breast-feeding might confer immunomodulatory protection against atopic disease in the infant. *J. Allergy Clin. Immunol.* 2002; 109: 119–121.
210. Ruiz PA, et al. Innate mechanisms for *Bifidobacterium lactis* to activate transient pro-inflammatory host responses in intestinal epithelial cells after the colonization of germ-free rats. *Immunology* 2005; 115(4): 441–450.
211. Sepp E, Mikelsaar M, and Salminen S. Effect of *Lactobacillus casei* strain GG administration on the gastrointestinal microbiota of newborns. *Microb. Ecol. Health Dis.* 1993; 6: 309–314.
212. Sheil B, McCarthy J, O'Mahony L, et al. Is the mucosal route of administration essential for probiotic function? Subcutaneous administration is associated with attenuation of murine colitis and arthritis. *Gut* 2004; 53(5): 694–700.
213. Sturm A, Rilling K, Baumgart DC, et al. *Escherichia coli* Nissle 1917 distinctively modulates T-cell cycling and expansion via toll-like receptor 2 signaling. *Infect. Immun.* 2005; 73(3): 1452–1465.
214. Sudo N, Sawamura S, et al. The requirement of intestinal bacterial flora for the development of an IgE production system susceptible to oral tolerance induction. *J. Immunol.* 1997; 159: 1739–1745.
215. Takeda K and Akira S. Toll-like receptors in innate immunity. *Int. Immunol.* 2005; 17: 1–14.
216. Viljanen M, Savilahti E, Haahtela T, Juntunen-Backman K, Korpela R, Poussa T, Tuure T, and Kuitunen M. Probiotics in the treatment of atopic eczema/dermatitis syndrome in infants: a double blind placebo-controlled trial. *Allergy* 2004, doi:10.1111/j.1398-9995.
217. von der Weid T, Bulliard C, and Schiffrin EJ. Induction by a lactic acid bacterium of a population of CD4($^+$) T cells with low proliferative capacity that produce transforming growth factor beta and interleukin-10. *Clin. Diagn. Lab. Immunol.* 2001; 8(4): 695–701.
218. Benno Y and Mitsuoka T. Development of intestinal microflora in humans and animals. *Bifidobacteria Microfi.* 1986; 5: 13–25.
219. Collado MC, Derrien M, Isolauri E, de Vos WM, and Salminen S. Intestinal integrity and *Akkermansia muciniphila*: a mucin-degrading member of the intestinal microbiota present in infants, adults and the elderly. *Appl. Environ. Microbiol.* 2007, doi:10.1128/AEM.01477-07.
220. Gore C, Munro K, Lay C, Bibiloni R, Morris J, Woodstock A, Custovic A, and Tannock GW, *Bifidobacterium pseudocatenulatum* is associated with atopic eczema: a nested case-control study investigating the fecal microbiota of infants. *J. Allergy Clin. Immunol.*, 2008; 121: 135–140.
221. Grönlund MM, Gueimonde M, Laitinen K, Kociubinski G, Salminen S, and Isolauri E. Maternal breast-milk and intestinal bifidobacteria guide the compositional development of the *Bifidobacterium* microbiota in infants at risk of allergic disease. *Clin. Exp. Allergy* 2007, doi:10.1111/j.1365-2222.2007.02849.x.
222. Guarner F and Malagelada JR. Gut flora in health and disease. *Lancet* 2003; 361: 512–519.
223. Guerin-Danan C, Meslin JC, Chambard A, Charpilienne A, Relano P, Bouley C, Cohen J, and Andrieux C. Food supplementation with milk fermented by *Lactobacills casei* DN-114 001 protects suckling rats from rotavirus-associated diarrhea. *J. Nutr.* 2001; 131: 111–117.

224. Gueimonde M, Sakato S, Kalliomäki M, Isolauri E, Benno Y, and Salminen S. Effect of maternal consumption of *Lactobacillus* GG on transfer and establishment of fecal bifidobacterial microbiota in neonates. *J. Pediatr. Gastroenterol. Nutr.* 2006; 42: 166–170.
225. Harmsen HJ, Wildeboer Veloo AC, Raangs GC, Wagendorp AA, Klijn N, Bindels JG, and Welling GW. Analysis of intestinal flora development in breast-fed and formula-fed infants by using molecular identification and detection methods. *J. Pediatr. Gastroenterol. Nutr.* 2000; 30: 61–67.
226. He F, Morita H, Hashimoto H, Hosoda M, Kurisaki J, Ouwehand AC, Isolauri E, Benno Y, and Salminen S. Intestinal *Bifidobacterium* species induce varying cytokine production. *J. Allergy Clin. Immunol.* 2002; 109: 1035–1036.
227. He F, Morita H, Kubota A, Ouwehand AC, Hosoda M, Hiramatsu M, and Kurisaki J. Effect of orally administered non-viable *Lactobacillus* cells on murine humoral immune responses. *Microbiol. Immunol.* 2005; 49: 993–997.
228. Hebuterne X. Gut changes attributed to ageing: effects on intestinal microflora. *Curr. Opin. Clin. Nutr. Metab. Care* 2003; 6: 49–54.
229. Hopkins MJ, Sharp R, and Macfarlane GT. Age and disease related changes in intestinal bacterial populations assessed by cell culture, 16S rRNA abundance, and community cellular fatty acid profiles. *Gut* 2001; 48: 198–205.
230. Hopkins MJ and Macfarlane GT. Changes in predominant bacterial populations in human faeces with age and with *Clostridium difficile* infection. *J. Med. Microbiol.* 2002; 51: 448–454.
231. Juntunen M, Kirjavainen P, Ouwehand AC, Salminen SJ, and Isolauri E. Gut microflora changes and probiotics in children in day care centers. *Biosci. Microflora* 2003; 22: 99–107.
232. Kalliomäki M, Kirjavainen P, Eerola E, et al. Distinct patterns of neonatal gut microflora in infants in whom atopy was and was not developing. *J. Allergy Clin. Immunol.* 2001; 107: 129–134.
233. Kirjavainen PV, Apostolou E, Arvola T, Salminen SJ, Gibson GR, and Isolauri E. Characterizing the composition of intestinal microflora as a prospective treatment target in infant allergic disease. *FEMS Immunol. Med. Microbiol.* 2001; 32: 1–7.
234. Kirjavainen PV, Arvola T, Salminen SJ, and Isolauri E. Aberrant composition of gut microbiota of allergic infants: a target of bifidobacterial therapy at weaning? *Gut* 2002; 51: 51–55.
235. Kleessen B, Bunke H, Tovar K, Noack J, and Sawatzki G. Influence of two infant formulas and human milk on the development of the faecal flora in newborn infants. *Acta Pediatr.* 1995; 84: 1347–1356.
236. Mackie RI, Sghir A, and Gaskins HR. Developmental microbial ecology of the neonatal gastrointestinal tract. *Am. J. Clin. Nutr.* 1999; 69(Suppl.): 1035S–1045S.
237. Martin R, Heilig HGHJ, Zoetendal EG, Jimenez E, Fernandez L, Smidt H, and Rodriguez JM. Cultivation-independent assessment of the bacterial diversity of breast milk among healthy women. *Res. Microbiol.* 2007; 158: 31–37.
238. Mikelsaar M, Mändar R, Sepp E, and Annuk H, Human lactic acid microflora and its role in the welfare of host. In: Salminen S, von Wright A, Ouwehand A, editors. Lactic Acid Bacteria. Marcel Dekker Inc., New York, 2004.

239. Mitsuoka T and Hayakawa K. Die faecalflora bei Menschen I. Mitteilung: die zusammensetzung der faecalflora der verschiedenen altersgruppen. *ZBL Bakt. Hyg.* 1972; 233: 333–42.

240. Morelli LC, Cesena C, de Haën, and Gozzini L. Taxonomic Lactobacillus composition of feces from human newborns during the first few days. *Microb. Ecol.* 1998; 35: 205–212.

241. Ouwehand AC, Isolauri E, He F, Hashimoto H, Benno Y, and Salminen S. Differences in *Bifidobacterium* flora composition in allergic and healthy infants. *J. Allergy Clin. Immunol.* 2001; 108: 144–145.

242. Ouwehand A, Isolauri E, and Salminen S. The role of the intestinal microflora for the development of the immune system in early childhood. *Eur. J. Nutr.* 2002; 41(Suppl. 1): I32–I37.

243. Palmer C, Bik EM, DiGiulio DB, Relman DA, and Brown PO. Development of the human infant intestinal microbiota. *PLoS Biol.* 2007; 5(7): e177.

244. Penders J, Thijs C, Vink C, Stelma FF, Snijders B, Kummeling I, van den Brandt PA, and Stobberingh EE. Factors influencing the composition of the intestinal microbiota in early infancy. *Pediatrics* 2006; 118: 511–521.

245. Penders J, Thijs C, van den Brandt PA, Kummeling I, Snijders B, Stelma F, Adams H, van Ree R, and Stobberingh EE. Gut microbiota composition and development of atopic manifestations in infancy: the KOALA Birth Cohort Study. *Gut* 2007; 56: 661–667.

246. Rajilic-Stojanovic M, Smidt H, and de Vos WM. Diversity of the human gastrointestinal tract microbiota revisited. *Environ. Microbiol.* 2007; 9: 2125–2136.

247. Rinne M, Gueimonde M, Kalliomäki M, Hoppu U, Salminen S, and Isolauri E. Similar bifidogenic effects of prebiotic-supplemented partially hydrolyzed infant formula and breastfeeding on infant gut microbiota. *FEMS Immunol. Med. Microbiol.* 2005; 43: 59–65.

248. Salminen S, Bouley C, Boutron-Ruault MC, et al. Gastrointestinal physiology and function—targets for functional food development. *Br. J. Nutr.* 1998; 80(Suppl.): 147–171.

249. Satokari RM, Vaughan EE, Favier CF, Dore J, Edwards C, and de Vos WM. Diversity of *Bifidobacterium* and *Lactobacillus* spp. in breast-fed and formula-fed infants as assessed by 16S rDNA sequence differences. *Microb. Ecol. Health Dis.* 2002; 14: 97–105.

250. Satokari RM, Vaughan EE, Smidt H, Saarela M, Mättö J, and de Vos WM. Molecular approaches for the detection and identification of bifidobacteria and lactobacilli in the human gastrointestinal tract. *Syst. Appl. Microbiol.* 2003; 26: 572–584.

251. Schultz M, Gottl C, Young RJ, Iwen P, and Vanderhoof JA. Administration of oral probiotic bacteria to pregnant women causes temporary infantile colonization. *J. Pediatr. Gastroenterol. Nutr.* 2004; 38: 293–297.

252. Song Y, Liu C, and Finegold SM. Real-time PCR quantitation of clostridia in feces of autistic children. *Appl. Environ. Microbiol.* 2004; 70: 6459–6465.

253. Stark PL and Lee A. The microbial ecology of the large bowel of breast-fed and formula-fed infants during the first year of life. *J. Med. Microbiol.* 1982 189–203.

254. Stsepetova J, Sepp E, Julge K, Vaughan E, Mikelsaar M, and de Vos WM. Molecularly assessed shifts of *Bifidobacterium* ssp. and less diverse microbial communities are

characteristic of 5-year-old allergic children. *FEMS Immunol. Med. Microbiol.* 2007; 51: 260–269.

255. Wang M, Ahrne S, Antonsson M, and Molin G. T-RFLP combined with principal components analysis and 16S rRNA sequencing: an effective strategy for comparison of fecal microbiota in infants of different ages. *J. Microbiol. Methods* 2004; 59: 53–69.

256. Woodcock A, Moradi M, Smillie FI, Murray CS, Burnie JP, and Custovic A. *Clostridium difficile*, atopy and wheeze during the first year of life. *Pediatr. Allergy Immunol.* 2002; 13 (5): 357–360.

257. Woodmansey EJ. Intestinal bacteria and ageing. *J. Appl. Microbiol.* 2007; 102: 1178–1186.

258. Alexa P, Hamrik J, Stouracova K, and Salajka E. Oral immunization against enterotoxigenic colibacillosis in weaned piglets by non-pathogenic *E. coli* strain with K88 (F4) colonizing factors. *Vet. Med. Czech.* 2005; 50: 315–320.

259. Anjum M, Tucker J, Sprigings K, Woodward M, and Ehricht R. Use of miniaturized protein arrays for *E. coli* O serotyping. *Clin. Vaccine Immunol.* 2006; 13: 561–567.

260. America Public Health Association (APHA), *Standard Methods for the Examination of Water and Wastewater*, 20th ed. APHA, Washington, DC, 1998.

261. Ballmer K, Korczak B, Kuhnert P, Slickers P, Ehricht R, and Hachler H. Fast DNA serotyping of *E. coli* by use of an oligonucleotide microarray. *J. Clin. Microbiol.* 2007; 45: 370–379.

262. Beale L, Njuguna P, Chapman T, Al-Jassim R, Trott D, and Chin JC. Fitness genes of commensal and enterotoxigenic *E. coli*. In: Paterson JE, editor. *Manipulating Pig Production XI*. Australasian Pig Science Association Book Publication, Australia, 2007, pp. 193.

263. Bertschinger HW, Eng V, and Wegmann P. Relationship between coliform contamination of floor and teats and the incidence of puerperal mastitis in two types of farrowing accommodations. In: Eskesbo I, editor. *Proceedings of the Sixth International Congress on Animal Hygiene*. Swedish University of Agricultural Science, Skara, Sweden, 1988, pp. 86–88.

264. Bielaszewska M, Dobrindt U, Gartner J, Gallitz I, Hacker J, Karch H, Muller D, Schubert S, Schmidt M, Sorsa L, and Zdziarski J. Aspects of genome plasticity in pathogenic *E. coli*. *Int. J. Med. Microbiol.* 2007, .

265. Blanco M, Lazo L, Blanco J, Dahbi G, Mora A, Lopez C, Gonzalez E, and Blanco J. Serotypes, virulence genes, and PFGE patterns of enteropathogenic *E. coli* isolated from Cuban pigs with diarrhea. *Int. Microbiol.* 2006; 9: 53–60.

266. Boudry G, Peron V, Le Huerou-Luron I, Lalles J, and Seve B. Weaning induces both transient and long-lasting modifications of absorptive, secretory and barrier properties of piglet intestine. *J. Nutr.* 2004; 134: 2256–2262.

267. Bruininx E, Binnendijk G, Van der Peet-Schwering C, Schrama J, Den Hartog L, Everts H, and Beynen A. Effect of creep feed consumption on individual feed intake characteristics and performance of group-housed weanling pigs. *J. Anim. Sci.* 2002; 80: 1413–1418.

268. Carroll J, Touchette K, Matteri B, Dyer C, and Allee G. Effect of spray-dried plasma and lipopolysaccharide exposure on weaned pigs. II. Effects on the hypothalamic–pituitary–adrenal axis of weaned pigs. *J. Anim. Sci.* 2002; 80: 502–509.

269. Chapman T, Smith AM, Daniels P, Jordan D, Trott D, and Chin JC. Phenotyping *Escherichia coli* antimicrobial resistances in pigs. In: Paterson JE, editor. *Manipulating Pig Production XI*. Australasian Pig Science Association, Book Publication, Australia, 2007, pp. 194.
270. Chapman T, Wu XY, Barchia I, Bettelheim K, Driesen S, Trott D, Wilson M, and Chin J. Comparison of virulence gene profiles of *E. coli* strains isolated from healthy and diarrheic swine. *Appl. Environ. Microbiol.* 2006; 72: 4782–4795.
271. Chin JC, Chapman TA. Turner B, and Wu ZY. A re-appraisal of early signaling events and gene activation in a human intestinal epithelial cell line *in vitro* by probiotics and enteric pathogenic bacteria species. *Int. J. Probiotics Prebiotics*, in press.
272. Chin JC and Mullbacher A, Immune activation versus hyporesponsiveness and tolerance in the gut: is there a role for probiotics in shaping an unbalanced response against commensals and pathogens? In: Fuller DR and Perdigon G, editors. *Gut Flora, Nutrition and Health*. Blackwell Science Ltd., 2002, pp. 178–200.
273. Correa MG and Marin J. O-serogroups, eae gene and EAF plasmid in *E. coli* isolates from cases of bovine mastitis in Brazil. *Vet. Microbiol.* 2002; 85: 125–132.
274. Crespi B. The evolution of social behavior in microorganisms. *Trends Ecol. Evol.* 2001; 16: 178–181.
275. Diez-Gonzalez F, Applications of bacteriocins in livestock. *Curr. Iss. Intest. Microbiol.* 2005; 8: 15–24.
276. Dixit S, Gordon D, Wu XY, Chapman T, Kailasapathy K, and Chin J. Diversity analysis of commensal porcine *E. coli* – associations between genotypes and habitat in the porcine gastrointestinal tract. *Microbiology* 2004; 150: 1735–1740.
277. Dixon B. Antibiotics as growth promoters: risks and alternatives. *ASM News* 2000; 66(No. 5)
278. Fuller R. Probiotics in man and animals. *J. Appl. Bacteriol.* 1989; 66: 365–378.
279. Germon P, Roche D, Melo S, Mignon-Grasteau S, Dobrindt U, Hacker J, Schouler C, and Moulin-Schouleur M. tENA locus polymorphism and ecto-chromosomal DNA insertion hotspots are related to the phylogenetic group of *E. coli* strains. *Microbiology* 2007; 153: 826–837.
280. Janion C, Sikora (nee Wójcik) A, Nowosielska A, and Grzesiuk E. SOS response in starved *E. coli*. *Environ. Mol. Mutagen.* 2002; 40: 129–133.
281. Jordan D, Chin J, Trott D, Antimicrobial use in the Australian pig industry: which drugs, how often and why. In: Paterson JE, editor. *Manipulating Pig Production XI*. Australasian Pig Science Association Book Publication, Australia, 2007, pp. 122.
282. Katouli M, Lund A, Wallgren P, Kuhn I, Soderlind O, and Mollby R. Phenotypic characterization of intestinal *E. coli* of pigs during suckling, postweaning and fattening periods. *Appl. Environ. Microbiol.* 1995; 61: 778–783.
283. Kuhn L, Tullus K, and Molby R. Colonization and persistence of *E. coli* phenotypes in intestines of children aged 0–18 months. *Infection* 1986; 14: 7–12.
284. Kunin C and Beard M. Serological studies of O antigens of *E. coli* by means of the hemagglutination test. *J. Bacteriol.* 1963; 85: 541–548.
285. Kyriakis S, Tsiloyiannis V, Vlemmas J, Sarris K, Tsinas A, Alexopoulos C, and Jansegers L. The effect of probiotic LSP 122 on the control of post-weaning diarrhea syndrome of piglets. *Res. Vet. Sci.* 1999; 67: 223–228.

286. Lacher D, Strinsland H, Blank T, Donnenberg M, and Whittam T. Molecular evolution of typical EPEC: clonal analysis by multilocus sequence typing and virulence gene allelic profiling. *J. Bacteriol.* 2007; 189: 342–350.

287. Manner K and Spieler A. Probiotics in piglets—an alternative to traditional growth promoters. *Microecol. Ther.* 1997; 26: 243–256.

288. Miller G, Algozin K, McNamara P, and Bush E. Productivity and economic effects of antibiotics used for growth promotion in U.S. pork production. *J. Agric. Appl. Econ.* 2003; 35: 469–482.

289. Moeser A, Klok C, Ryan K, Wooten J, Little D, Cook V, and Blikslager A. Stress signaling pathways activated by weaning mediate intestinal dysfunction in the pig. *Am. J. Physiol. Gastrointest. Liver Physiol.* 2007; 292: 173–181.

290. Naomani B, Fairbrother J, and Gyles C. Virulence genes of O149 enterotoxigenic *E. coli* from outbreaks of postweaning diarrhea in pigs. *Vet. Microiobiol.* 2003; 97: 87–101.

291. Njuguna P. Studies on the manipulation of gastrointestinal tract bacteria. MSc. thesis, 2005, http://www.library.uow.edu.au/adt-NWU/public/adt-NWU20060719.125402/index.html.

292. Parker RB. Probiotics: the other half of the antibiotic story. *Anim. Nutr. Health* 1974; 29: 4–8.

293. Patterson J, Chapman T, Hegedus E, Barchia I, and Chin J. Selected culturable enteric bacterial populations are modified by diet acidification and the growth promotant tylosin. *Letts. Appl. Microbiol.* 2005; 41: 119–124.

294. Patterson J, Chandrasekara A, Bao S, Hegedus E, Caterson I, Denyer G, Turner B, Mullbacher A, Chin JC, Prebiotic supplementation with resistant starch alters health indices associated with metabolic changes and intestinal dysbiosis. *Int. J. Probiotics Prebiotics*, in press.

295. Sansom BF and Gleed PT. The ingestion of sow's faeces by suckling piglets. *Br. J. Nutr.* 1981; 46: 451–456.

296. Scaletsky I, Michalski J, Torres A, Dulguer M, and Kaper J. Identification and characterization of the locus for diffuse adherence, which encodes a novel afimbrial adhesin found in atypical EPEC. *Infect. Immun.* 2005; 73: 4753–4765.

297. Schierack P, Steinruck H, Kleta S, and Vahjen W. Virulence factor gene profiles of *E. coli* isolates from clinically healthy pigs. *Appl. Environ. Microbiol.* 2006; 72: 6680–6686.

298. Simon O, Jadamus A, and Vahjen W. Probiotic feed additives – Effectiveness and expected modes of action. *J. Anim. Feed Sci.* 2001; 10: 51–67.

299. Smith MG, Chapman TA, Jordan D, Chin JC, and Trott D. Resistant phenotypes of enteric microbial communities from Australian pig farms. In: Paterson JE, editor. *Manipulating Pig Production XI*. Australasian Pig Science Association Book Publication, 2007, pp. 121.

300. Srinivasan U, Zhang L, France A, Ghosh D, Shalaby W, Xie J, Marrs C, and Foxman B. Probe hybridization array typing: a binary typing method for *E. coli*. *J. Clin. Microbiol.* 2007; 45: 206–214.

301. Stahl CH, Callaway TR, Lincoln L, Lonergan S, and Genovese K. Inhibitory activities of colicins against *E. coli* strains responsible for postweaning diarrhea and edema disease in swine. *Antimicrob. Agents Chemother.* 2004; 48: 3119–3121.

302. Taras D, Vahjen W, Macha M, and Simon O. Performance, diarrhea incidence, and occurrence of *E. coli* virulence genes during long-term administration of a probiotic *Enterococcus faecium* strain to sows and piglets. *J. Anim. Sci.* 2006; 84: 608–607.

303. U.S. Department of Agriculture. Reference of swine health and management in the United States. 2002. N338.0801. National Animal Health Monitoring System, Fort Collins, CO, 2001.
304. Van Belkum M and Stiles ME. Nonlantibiotic antibacterial peptides from lactic acid bacteria. *Nat. Prod. Rep.* 2000; 17: 323–335.
305. Welch R, Burland V, PlunkettIIIG, Redford P, Roesch P, Rasko D, Buckles E, Liou S, Boutin A, Hackett J, Stroud D, Mayhew G, Rose D, Zhou S, Schwartz D, Perna N, Mobley H, Donnenberg M, and Blattner F. Extensive mosaic structure revealed by the complete genome sequence of uropathogenic *E. coli. Proc. Natl. Acad. Sci. USA* 2002; 99: 17020–17024.
306. World Health Organization. Global principles for the containment of antimicrobial resistance in animals intended for food, 2000, http://www.who.int/cmc-documents/zoonoses/docs/whocdscsraph2004.pdf.
307. Wu XY, Chapman TA, Trott DJ, Bettelheim KA, Do TN, Driesen S, Walker MJ, and Chin JC. Comparative analysis of virulence genes, genetic diversity and phylogeny between commensal and enterotoxigenic *Escherichia coli* from weaned pigs. *Appl. Environ. Microbiol.* 2007; 73: 83–91.
308. www.thepigsite.com/pighealth/article/227/farrowing-house-design.

6

COMMERCIALLY AVAILABLE HUMAN PROBIOTIC MICROORGANISMS

6.1 *Lactobacillus acidophilus*, LA-5®

CAMILLA HOPPE AND CHARLOTTE NEXMANN LARSEN

Health & Nutrition Division, Chr. Hansen A/S, Denmark

Lactobacillus acidophilus, LA-5 originally selected by Chr. Hansen for the production of probiotic dairy products, has been used in dietary supplements and fermented milk products worldwide. LA-5 is clinically well documented. LA-5 has no adverse effects on the taste, appearance, or palatability of the product. Furthermore, it is able to survive in the product until consumption.

LA-5 has many probiotic features. It is able to survive passage through the stomach and upper small intestine due to its tolerance of stomach and bile acid and resistance to digestive enzymes (1–5). It is able to adhere to the intestinal mucosa (6, 7), and the survivability of LA-5 in the intestines has been demonstrated showing a good recovery in feces following oral administration (8).

6.1.1 Gastrointestinal Effects

6.1.1.1 Intestinal Microbial Balance   *Lactobacillus acidophilus* belongs to the group of Gram-positive non-sporulating facultative or anaerobic rods, and it is a natural inhabitant of the gut. The main end products of glucose fermentation by Lactobacilli are lactic acid, acetic acid, and H_2O_2. These metabolites make

---

Handbook of Probiotics and Prebiotics, Second Edition   Edited by Yuan Kun Lee and Seppo Salminen
Copyright © 2009 John Wiley & Sons, Inc.

the environment less favorable for the growth of potentially pathogenic microorganisms (9). Lactobacilli play a significant role in controlling intestinal pH through the production of acids, which decrease the intestinal pH thereby restricting the growth of many potentially pathogenic and putrefactive bacteria. *In vitro*, LA-5 caused an increase in acetate and propionate production by colonic bacteria when a stable microbiota had not been established, suggesting that supplementation with LA-5 modifies colonic fermentation (10).

LA-5 also produces bacteriocin CH5 that is characterized not only by a particularly wide antibacterial range (11), but also by inhibitory action against certain yeasts and moulds (12). More than 60% inhibition of *Salmonella typhimurium* was demonstrated by LA-5 (13), and the growth of *Campylobacter jejuni* decreased as the concentration of cell-free extract fermented with LA-5 increased *in vitro* (14).

In a number of intestinal inflammatory diseases, disturbances of the normal intestinal microbiota seem to play a pivotal role. An increase of bifidobacteria and lactobacilli and a decrease in the number of stools per day (15) as well as an improvement of gastrointestinal disorders (16) were observed in patients with ileal-pouch-anal anastomosis, receiving LA-5 and *Bifidobacterium animalis* ssp. *lactis* (BB-12®). In patients with collagenous colitis and diarrhea, days per week with liquid feces were reduced from 6 to 1 in patients receiving LA-5 and BB-12, whereas they increased from 4.5 to 6 in the placebo group ($p = 0.052$) (17).

6.1.1.2 Diarrhea

Diarrhea is the most common health problem encountered by visitors traveling to regions with poor hygiene. The condition is characterized by acute diarrhea caused mainly by intake of food or water infected with *Escherichia coli*, *Salmonella*, *Shigella*, or *Campylobacter*. Several studies have been carried out addressing the issue of probiotic prophylaxis of traveler's diarrhea (18). In travelers going to Egypt, the intake of LA-5 together with BB-12, *Streptococcus thermophilus* (STY-31™), and *Lactobacillus delbrueckii* spp. *bulgaricus* (LBY-27™) significantly decreased the occurrence of diarrhea in the test group (43%) compared to the placebo group (71%) (19).

Diarrhea is a potential consequence of a disturbed intestinal flora during an oral therapy with antibiotics, possibly leading to an unfortunate amplification of potentially pathogenic bacteria. The triple treatment for *Helicobacter pylori* is particularly harsh resulting in many adverse effects. *H. pylori* infected patients experienced less vomiting, constipation, and diarrhea during antibiotic treatment when supplemented with LA-5 and BB-12 containing yogurt compared to placebo yogurt (20, 21). In another study, administration of yogurt with LA-5 and BB-12 significantly reduced the content of *H. pylori* and gastritis activity in the antrum of the stomach after 6 weeks of therapy (22). Other studies have shown that volunteers treated with ampicillin or clindamycin and those supplemented with LA-5 and BB-12 experienced a faster normalization of the intestinal flora than those receiving placebo (23, 24).

6.1.1.3 Other Gastrointestinal Effects

Constipation is very common in developed countries, and supplementation with lactobacillus might improve natural bowel transit time. Elderly patients suffering from chronic constipation were randomized to receive either unfermented milk or fermented milk with LA-5 and BB-12. The probiotic group experienced a significant improvement in frequency of bowel movement, and reported no negative side effects (25).

Some probiotic bacteria have been shown to improve lactose digestion by release of β-galactosidase. In lactose intolerant subjects symptoms of lactose intolerance, such as stomachache, flatulence, diarrhea, and constipation have been found to be improved (26) or unchanged (27) by LA-5, BB-12, STY-31, and LBY-27. However, supplementing LA-5 in milk significantly decreased breath hydrogen values in lactose intolerant subjects when comparing with milk alone, indicating improved lactose degradation (28). Also *in vitro* studies demonstrated that LA-5 enhances lactose fermentation, as lactose concentration decreased and β-galactosidase activity increased in media supplemented with LA-5 (29, 30).

6.1.2 Immunomodulatory Effects

The gut-associated lymphoid tissue (GALT) is the largest organ of the immune system, and numerous probiotic health effects can be related to the modulation of immune functions. Administration of LA-5 has been associated with nonspecific stimulatory effects on the production of cytokines and phagocytic activity as well as more specific immune reactions such as antibody production.

6.1.2.1 Nonspecific Immune Responses In an *in vitro* study with a murine macrophage-like cell line that was cultured in the presence of cell-free extracts of LA-5, LA-5 enhanced phagocytosis of inert particles or viable Salmonella (31).

The induction of signaling compounds for immune reactions, cytokines, can also be affected by LA-5. *In vitro*, 10 lactic acid bacteria strains were compared as to their ability to induce the release of three cytokines: tumor necrosis factor alpha (TNF-α), interleukin-6 (IL-6), and IL-10. LA-5 was among the four best inducers of TNF-α release inducing amounts significantly greater than those obtained with lipopolysaccharide as a control (32).

6.1.2.2 Specific Immune Responses In a study, mice fed conventional yogurt, yogurt plus LA-5, and BB-12 or unfermented milk without bacteria (control) were orally immunized with cholera toxin to mimic enteral infections. Higher concentrations of cholera toxin-specific antibodies were found in the feces and serum of LA-5, and BB-12 fed mice compared with the yogurt and milk group (33).

6.1.3 Other Health Effects

Health effects of LA-5 in humans seem to include inhibited growth of cancer breast cells (34). In patients with acute leukemia, administration of LA-5 and BB-12 during and after chemotherapy significantly delayed the occurrence of fever, on average from day 8 in the placebo group to day 12 in the probiotic group (35).

Also, an improvement in lipid metabolism by reducing cholesterol has been shown both in humans (36) and rodents (37, 38) as well as *in vitro* (39).

6.1.4 Safety

In relation to the use of microorganisms as human probiotics, safety is clearly a major concern. Being a part of the normal human intestinal microbiota, Lactobacilli are generally considered safe, and millions of humans have consumed LA-5 without any reported negative side effects.

6.2 *Lactobacillus acidophilus* NCDO 1748

RANGNE FONDÉN[1] AND ULLA SVENSSON[2]

[1]*Finnboda Kajväg 15, Sweden*
[2]*Arla Foods, Sweden*

Since the beginning of the former century *L. acidophilus* has been the original probiotic bacteria used in health studies (44, 48, 69, 74, 79). These early studies were probably been done with species belonging to the acidophilus group but not restricted to strains belonging to the present definition of *L. acidophilus*. The original type strain was lost and later a new strain was isolated and given the status of type strain (55, 63). The strain *L. acidophilus* NCDO 1748 (1748) is identical to ATCC 4356T, DSM 20079, JCM 5342 and is present in several strain collections given different numbers. 1748 was selected as the strain to be used in food products during 1970 due to its growth in milk and its survival through the GI (64, 70, 71). Human studies have shown that 1748 influences the gut microbiota and gut function in a favorable way (57). Similarly 1748 alone or as a strain in a mixture favorably balances bacterial flora during treatment with antibiotics (56, 58, 65, 67, 76). 1748 binds and immobilizes food mutagens both *in vitro* and *in vivo* and might thereby reduce their level in internal organs like kidney and liver (59, 60, 66, 68). 1748 was used for 25 years on its own or in

TABLE 6.1 Identification and Safety of *L. acidophilus* NCDO 1748

Area	Data
Identification	Type strain isolated from human pharynx (55)
	Available at NCIMB, UK as strain 8690
	(originally named NCDO1748, later NCFB 1748)
Safety Confirmation	
Antibiotic resistance	Inherited antibiotic resistance of NCDO 1748 is in accordance with *L. acidophilus* group (46, 47).
Virulence	The virulence in a rabbit infective endocarditis model was negligible (43)
Hospitalized persons	Minor presence of *L. acidophilus* in isolates from hospitalized persons (45)
Bacteraemia	1748 not found in 44 persons suffering from bacteraemia caused by lactobacillus group including 4 infected with *L. acidophilus* (77)
Production of lactate isomer	No D-lactate was produced when milk was fermented (41)

combinations with other probiotics in fermented milk products produced by a patented process from Arla Foods and in natural remedies from Semper (52, 53).

6.2.1 Origin and Safety

The ATCC strain 4356T was isolated from human pharynx (Table 6.1). The species *L. acidophilus* was seldom present in the clinical isolates, and the strain 1748 was not present in such isolates from Sweden during the time it was used in probiotic products (45, 77).

6.2.2 *In Vitro* and Animal Studies

1748 survives when exposed to low pH and/or bile acids as well as *in vivo* in humans (Tables 6.2 and 6.3) (51, 70). It binds several food mutagens and

TABLE 6.2 *In Vitro* and Animal Studies on 1748

Area	Data
Survival, resistance to low pH and bile acids	Good survival when exposed to pH 2.0 or to a mixture of product/gastric juice at pH 1.4. 45% survived in a medium containing 0.3% bile for 3 h (51, 70). Survival confirmed *in vivo* (Table 6.3).
Adherance	Binding to surface layer protein of rat colonic mucin (62)
Micellaneous	
Antioxidative capacity	The strain had the highest radical scavenging activity of 25 strains tested. Intact cells reduced the cytotoxicity of 4-nitroquinoline-*N*-oxide by half (61, 80)
Binding of food mutagens	Almost complete binding of Trp-P-1 and Trp-P-2, less effective binding of PhIP, MeIQx, and IQ (66, 81)
Uptake and distribution of food mutagen Trp-P-2 in mice	Uptake and distribution to kidney, liver, lung, and thymus in mice was reduced[a] (68)
Protect intestinal cells alone or in combination with *S. thermophilus*	Limits adhesion, invasion, and physiological dysfunction induced by enteroinvasive *E. coli* and reversal of epithelial damage produced by cytokines (72, 73)
Production of bacteriocins	Formation of Acidocin J1132 and Acidocin D20079 (50, 51, 78).
Production of antifungal metabolites	Showing antifungal activity proven by agar-diffusion methods (49)
Technological aspects/ performance in foods	Ferments milk enriched with yeast extract Good survival in fermented milks and in dried products (52, 53)

[a]Results with 1748 + *Bifidobacterium longum*.

TABLE 6.3 Human Studies on 1748 Alone or Together with Other Probiotic Strains

Area	Data
Tolerance	NCDO 1748 has been well tolerated in all groups included in investigations, for example, healthy adults[a–c] (57, 60, 64), elderly[a] (42), healthy adults given antibiotics[a–c] (58, 65, 67, 76), colon cancer patients[a] (59), lactose intolerants[a] (40)
Survival in the gut	The survival has been analyzed as part of compliance in several of the human studies conducted. Immediate after consumption good survival, not isolated thereafter[a,c] (56, 64, 75, 76)
Influence on gut microbiota and balance	Number of 1748 increased[a] (57). Number of Lactobacillus generally increased[a] (57, 71) Sustained/favorable balanced bacterial flora during antibiotic treatment[a–c] (56, 58, 65, 67, 76)
Influence on gut function and well being	Stabilizes stool frequency in elderly[a] (42, 54) Increased tolerance to lactose[a] (40). Decreased uptake of mutagens[a] (60)

[a]Results with 1748.
[b]Results with 1748 + B. longum.
[c]Results with 1748 + F19 + BB-12.

decreases their levels in internal organs and has a high antioxidative capacity (60, 61, 66, 68, 80, 81). Further it protects intestinal cells from deleterious effects of cytokines on epithelial function and protects the cells from negative effects of enteroinvasive E. coli (72, 73).

Technologically 1748 could, as a single strain, be used to ferment milk enriched with a minute amount of yeast extract. It survives well and grows to the level of up to $2 \times 10^9 \, g^{-1}$ of fermented milk (52, 53). However, the product tastes a bit acrid, but on flavoring it is acceptable to most people.

6.2.3 Human Studies

1748 was well tolerated in all the investigated groups, and no side effects were reported during the 25 years of use even when daily intake was well above 10^{11} cfu. Further human studies proved that 1748 survived in the gut. The number of lactobacilli increased during the intake of 1748 and antibiotics (58). When taken together with bifidobacteria or with two complementary probiotics, it could counteract the negative impact of antibiotics on the normal GI flora (56, 65, 76). The ability of 1748 to bind mutagens *in vitro* and in animal studies was further proven in the studies of human volunteers (60). On ingestion of 1748 the mutagens formed during frying of meat were found be reduced in the urine and feces.

6.3 *Lactobacillus acidophilus* NCFM®

ARTHUR OUWEHAND AND SAMPO LAHTINEN

Health & Nutrition, Danisco, Finland

Lactobacillus acidophilus NCFM was isolated from a human fecal sample in the early 1970s. The strain designation NCFM is derived from "North Carolina Food Microbiology," the research laboratory at North Carolina State University (NCSU) where isolation took place (105). The strain is marketed by Danisco (Danisco Cultures Division, Paris). *L. acidophilus* NCFM is to date the only strain of this species for which the genome has been sequenced, annotated, and published (82). Genomic work has identified several regions important in probiotic functionality and supporting the role of *L. acidophilus* NCFM in maintaining or restoring gastrointestinal well being. This includes the genes involved in bacteriocin production, sugar and prebiotic metabolism (86), adherence to human cells, lactose metabolism, and tolerance to physiologically relevant stresses including acid and bile (83). Genomic reconstruction of metabolic pathways reflects the adaptation of *L. acidophilus* NCFM to the gastrointestinal environment. Furthermore, analysis of the *L. acidophilus* NCFM genome sequence has confirmed the absence of known transferable genetic elements related to antibiotic resistance (82).

6.3.1 *L. acidophilus* NCFM Basic Properties

In vitro studies have shown that *L. acidophilus* NCFM has the necessary prerequisites to survive gastrointestinal transit; resistance to bile, low pH, and digestive enzymes (89). Furthermore, *L. acidophilus* NCFM has been shown to adhere to human epithelial cell lines (96, 99) and human intestinal mucus (87). This adhesion property could be further improved by the addition of Ca^{2+}. By adhering to intestinal mucus and by coaggregation, *L. acidophilus* NCFM may also prevent the adhesion of certain pathogens (87, 88). Genome analysis of *L. acidophilus* NCFM has indicated the presence of several genes of potential importance to the adherence process, including mucus-binding proteins, fibrinonectin proteins, and others. Exopolysaccharides encoded by clusters of genes may also contribute to adherence capabilities (82). Oral adhesion and survival of *L. acidophilus* NCFM has also been investigated (97). Finally, *L. acidophilus* NCFM has been indicated to exert antimicrobial activity (84, 85). The ability of *L. acidophilus* NCFM to produce antimicrobial substances was confirmed by the genome sequence of the strain, where 12 putative genes were identified and implicated in the production and processing of antimicrobial substances (82, 90).

6.3.2 Survival of Intestinal Transit and Change in Intestinal Microbiota Composition and Activity

Consumption of 10^{10} cfu of *L. acidophilus* NCFM by healthy adult volunteers caused an increase in the number of fecal lactobacilli during the intervention and in particular *L. acidophilus,* which were not detectable prior to the intervention, became dominant (108), though no change in the levels of bifidobacteria and sulphide producing bacteria was observed (109). Recently, the survival of *L. acidophilus* NCFM through the gastrointestinal tract of humans was confirmed in a clinical trial, in which the strain in combination with lactitol also increased the levels of fecal *Bifidobacterium* levels (102). In a study with immune compromised mice experimentally infected with *Candida albicans,* prophylactic feeding with *L. acidophilus* NCFM significantly reduced the incidence of systemic candidiasis and prolonged the survival of the mice (111, 112). Moreover, prophylactic administration of *L. acidophilus* NCFM has been shown to have a protective effect on colonic hyperplasia caused by *Citrobacter rodentium* in mice (110). In combination with four other probiotic strains, *L. acidophilus* NCFM has been shown to stabilize the intestinal microbiota in humans following antibiotic therapies (92), and in combination with soluble fiber and glutamine, it has been shown to improve diarrhea associated with antiretroviral therapies (98).

In addition to changing composition of the intestinal microbiota, changes in the intestinal metabolic activity have been observed. Three human intervention trials showed a reduction in azoreductase and β-glucuronidase (93–95), and one showed a reduction in nitroreductase (94) when subjects consumed 10^9–10^{10} cfu *L. acidophilus* NCFM a day, added as a concentrate to milk. In animal studies, these enzyme activities have been related to intestinal tumor; the relevance for humans is however less clear. Also, in small bowel bacterial over growth, *L. acidophilus* NCFM has been shown to beneficially alter the metabolism of the intestinal microbiota by reducing the levels of dimethylamine and nitrosodimethylamine (91, 107).

6.3.3 Lactose Intolerance

Lactose maldigestion is a common state for most of adults of non-North-West European descent. An objective measure for lactose intolerance is the excretion of H_2 in breath after consumption of lactose. Although *L. acidophilus* NCFM does not always show a reduction in breath H_2, it has been shown to reduce lactose intolerance symptoms in children (100). Other mechanisms may therefore underlie this beneficial effect, such as relief of intestinal pain, as discussed below.

6.3.4 Relief of Intestinal Pain

One of the most recent findings on *L. acidophilus* NCFM is its ability to induce μ-opioid and canabinoid receptors in the intestine of experimental animals (104). The relief of intestinal pain may have great implications in the area of, for example, irritable bowel syndrome (IBS) and infant colic. In animal model of IBS, *L. acidophilus* NCFM has been shown to be effective (104), but this requires further confirmation in humans.

6.3.5 Prevention of Common Respiratory Infections and Effects on Immunity

In a recent large-scale randomized controlled trial *L. acidophilus* NCFM, administered either as a single strain or in combination with *Bifidobacterium lactis* Bi-07, has been shown to reduce the incidence of symptoms of common respiratory tract infections such as fever, cough, and runny nose in children attending child-care centers (101). Symptom duration, absenteeism due to disease, and the use of antibiotics were also significantly reduced during the 6-month administration of probiotics.

L. acidophilus NCFM is capable of beneficially modulating immune responses. In a combination with lactitol, *L. acidophilus* NCFM was shown to increase the levels of fecal PGE_2 and spermidine levels in healthy elderly volunteers, suggesting reduced inflammation and enhanced cytoprotection in the intestine (102). Modulation of antibody responses to *C. albicans* has been demonstrated using a murine model (113). Recently, immunomodulatory effects of *L. acidophilus* NCFM on serum immunoglobulins were observed in healthy adults following oral vaccination (103). The observed improvement of immune responses in healthy volunteers may partly explain the efficacy of *L. acidophilus* NCFM in the prevention of common respiratory infections (101).

6.3.6 Application

L. acidophilus NCFM has been commercialized in supplements, fermented, and nonfermented foods, and has been shown to have excellent stability (106) and provides good flavor to fermented products. Furthermore, *L. acidophilus* NCFM is capable of utilizing fructooligosaccharides (86) and other prebiotic substances, providing opportunities for formulating synbiotic functional foods.

6.3.7 Conclusion

From the studies performed, it is clear that *L. acidophilus* NCFM's best-documented health benefits mainly lie in the area of intestinal health and well being. Claims in the area of intestinal health have been submitted for the so-called Article 13 process in the European Union.

6.4 *Lactobacillus casei* SHIROTA

Mayumi Kiwaki and Koji Nomoto
Yakult Central Institute for Microbiology Research, Tokyo, Japan

Reported characters of *Lactobacillus casei* Shirota associated with probiotic effects are summarized in the following tables.

6.4.1 Effects on Intestinal Environment

Human Studies

Reported Effects/Type of Study	References
Change in fecal microbiota by continuous ingestion of fermented milk	(199)
Modulation of the composition and metabolic activity of the microbiota/randomized placebo-controlled trial	(187)
Decrease of the urinary mutagenicity, possible relation to the change in microbiota/comparative study	(133)
Suppression of the toxic fermentation metabolites/randomized placebo-controlled crossover study	(123)
Favorable effect on nitrogen–protein metabolism/randomized placebo-controlled crossover study	(125)
Decrease of the intestinal bacterial enzyme activity in feces/randomized crossover study	(124)

Animal Experiments

Reported Effects/Administration Route/Animal	References
Enhancement of the radical scavenging activity of milk/feeding of fermented milk/hamster	(167)
Effect on cecal fermentation pattern depending on indigenous microbiota/oral administration/pig	(173)
Mild effect on the metabolic activity in the intestinal mucosa/oral administration/rat	(148)

6.4.2 Adhesive Property

Human Studies

Reported Effects	Reference
Survival in transit through the gastrointestinal tract after ingestion of the fermented milk	(203)

Animal Experiments

Reported Effects/Administration Route/Animal	References
Low colonization potential by long average doubling times/oral administration/mouse	(151)

In Vitro Experiments

Reported Effects/Target	References
Changes in adhesive abilities by different inactivation methods/human mucus	(175)
Low adhesion, no change by pretreatments of the LcS/human intestinal mucus glycoproteins	(196)
Low adherence to mucus prepared from fecal samples/human mucus	(139)
Low number but high affinity adhesion, multiple binding mechanisms/human mucus, Caco-2 cells	(152)
Estimation of two types of adhesins on the surface of LcS/human cells	(153)
Isolation of biosurfactants containing several collagen-binding proteins/collagen	(138)
Strongly positive binding of bovine fibrinogen and porcine fibronectin/animal ECM	(187)
Promoting effects of PUFA on adhesion ability/Caco-2 cell	(145)

6.4.3 Intestinal Physiology

Human Studies

Reported Effects/Type of Study	References
Improvement of the intestinal function in a short bowel syndrome (SBS) patient by synbiotic therapy/Case Reports	(141)
Beneficial effect on gastrointestinal symptoms of patients with chronic constipation/double-blind, placebo-controlled, randomized study	(150)

Animal Experiments

Reported Effects/Administration Route/Animal	References
Activation of the large intestine motility response to feeding/feed of fermented milk/pig	(171)
Increase of the number and the diversity of indigenous Lactobacilli/feed of fermented milk/pig	(170)
Transition of LcS with maximum level at cecum 6 h after dosing/feed of fermented milk/pig	(174)
Effect on gene expression in the ileal epithelial cells/oral administration/mouse	(184)

6.4.4 Immunomodulation

Human Studies

Reported Effects/Type of Study	References
No effects on the immune system by ingestion of fermented milk/randomized placebo-controlled trial	(185)
Increase and maintenance of elevated activity by ingestion of fermented milk	(165)
IL-12 may be responsible for enhancement of NK cell activity/placebo-controlled, crossover trial	(192)
Positive effect on NK-cell activity by daily intake of LcS/controlled clinical trial	(191)

Animal Experiments

Reported Effects/Administration Route/Animal	References
Induction of cytokines and prolonging survival of Meth A-bearing mice/i.pl. administration/mouse	(158)
Effective inhibition of IgE production in serum/oral feeding of heat-killed LcS/mouse	(163)
Induction of TNF-α playing an important role in the antitumor effect/i.pl. injection/mouse	(202)
Raising NK activity of blood mononuclear cells and splenocytes/feed on a diet/mouse	(137)
Enhanced antigen-specific humoral and cellular Th1 cell-associated immunity/oral administration/rat	(128)
Suppression of IgE responses and systemic anaphylaxis in a food allergy model/i.pl. injection/mouse	(183)
Suppression of the humoral and cellular immune responses to C II and reduction of the development of CIA/oral administration/mouse	(146)
Immunological modulations and prolongation of the lifespan of autoimmune disease model mouse/feeding and i.p. injection/mouse	(164)
Regulation of the host immune response in the IDDM model, NOD mice/oral feeding/mouse	(160)
Modification of the host immune responses in NIDDM model mice/oral administration/mouse	(162)
Activation of systemic and local cellular immunity against influenza virus/oral administration/mouse	(136)
Varying effects on the Th1 responses among the different model/oral administration/rodent	(117)
Divergent immune effects induced by timing of administration/oral administration/rodent	(129)

In Vitro Experiments

Reported Effects/Target	References
LTAs from LcS elicit strong TNF-α-inducing activities in macrophages/mouse cells	(156)
Patterns of cytokine induction of LcS and Gram-negative probiotic strain/mouse cell line	(122)
Induction of cytokines and augmentation of NK cell activity/human PBMNC	(182)
Inhibition of antigen-induced IgE production through induction of IL-12/mouse splenocytes	(181)
Activation of X-irr-Spl followed by IL-12 secretion and stimulation of IFN-γ/mouse splenocytes	(147)
Stimulation of macrophages to induce IL-12 by rigid cell wall structure/mouse peritoneal macrophages	(180)

6.4.5 Effects on Cancer

Human Studies

Reported Effects/Type of Study	References
Safe and effective prevention of recurrence of superficial bladder cancer by oral administration of BLP/randomized, controlled double-blind trial	(116)
Habitual intake of LcS reduces the risk of bladder cancer/case-control study	(172)

Animal Experiments

Reported Effects/Administration Route/Animal	References
Potent and safe therapeutic effects in the bladder tumors implantation model/intravesical instillation/mouse	(190)
Inhibition of MC-induced tumorigenesis, modulation of the host immune response, enhancement of NK cell cytotoxicity/feed on a diet/mouse	(188, 189)
Preventive effect against chemically induced colon carcinogenesis/oral administration/rat	(200)
Suppressive effects on the incidence of thymic lymphoma/i.p. administration/mouse	(198)
Induction of TNF-α playing an important role in the antitumor effect/i.pl. injection/mouse	(202)

In Vitro Experiments

Reported Effects/Target	Reference
Induction of cytotoxic effects in bladder cancer cells primarily via necrosis/human cells	(178)

6.4.6 Prevention of Infectious Diseases

Animal Experiments

Reported Effects/Administration Route/Animal	References
Protective activity on a novel mouse model of abdominal sepsis/i.p. pretreatment/mouse	(194)
Enhancement of the local immune responses to STEC cells and Stxs leading to elimination of STEC from intestine/feeding of fermented milk/rabbit	(168)
A heat-killed preparation of LcS exerted significant antimicrobial effects in a UTI model/intraurethrally administration/mouse	(114)
Enhancement of host resistance against oral *Listeria monocytogenes* infection, increased cell-mediated immunity, enhancement of cell-mediated immunological memory/oral administration/rodent	(126, 127)
Enhancement of cellular immunity in the respiratory tract and protection against IFV infection/intranasal administration/mouse	(135)
Activation of systemic and local cellular immunity, or immature immune system, amelioration of or protection against IFV infection/oral administration/mouse	(136, 201)

***In Vitro* Experiments**

Reported Effects/Target	References
Competition with, exclusion and displacement of pathogenic GI bacteria/human intestinal mucus glycoprotein and Caco-2 cell	(154)
Bactericidal effect depending on its lactic acid production and pH reductive effect/STEC O157:H7	(169)
Inhibition of *S. typhimurium* adhesion to human intestinal mucus/human intestinal glycoproteins	(195)
Decrease in the viability of Salmonella due to non-lactic acid molecule(s) in the CFCS/*Salmonella enterica* serovar *typhimurium*	(130)
Antibacterial activity towards Salmonella solely due to the production of lactic acid/*Salmonella enterica* serovar *typhimurium*	(155)

6.4.7 Prevention of Life Style Diseases

Human Studies

Reported Effects/Type of Study	Reference
Lex lowered the blood pressure and showed beneficial effects on glucose and lipid metabolism in patients with hypertension/double-blind, placebo-controlled study	(166)

Animal Experiments

Reported Effects/Administration Route/Animal	References
Blood pressure-lowering effect of heat extract from LcS/oral administration/rat	(132)
Purification of SG-1, the most effective antihypertensive compounds in Lex/oral administration/rat	(177)
Antihypertensive effect of SG-1 resulted from an enhancement of PGI_2 biosynthesis and decrease in PR/oral administration/mouse	(131)
Effective reduction of the onset of AXN-induced diabetes/feeding or oral administration/mouse	(159)
Effective inhibition of the onset of diabetes in an IDDM model/oral feeding of heat-killed cell/mouse	(160)
Significant decrease in plasma glucose level in an NIDDM model/feeding or oral administration/mouse	(162)
Decrease in serum cholesterol levels by feeding of the cell wall components/oral administration/rat	(121)
Suppression of the plasma triglyceride level without lowering the plasma cholesterol concentration/feeding of fermented milk/hamster	(149)

6.4.8 Clinical Application

Human Studies

Reported Effects/Type of Study	References
Reduction of postoperative infections in biliary cancer patients by synbiotics administration/randomized controlled trial	(144)
Beneficial effect on gastrointestinal symptoms of patients with chronic constipation/double-blind, placebo-controlled, randomized study	(150)
Improvement of intestinal flora and increase of SCFA in short bowel patients with refractory enterocolitis/synbiotic therapy	(143)
Non-significant trend towards a suppressive effect on *H. pylori* by frequent consumption of a fermented milk drink/randomized controlled intervention study	(120)
Increase in a frequency of eradication of *H. pylori* by combination with traditional triple treatment/clinical comparative randomized study	(176)
Improvement of spasticity and urinary symptoms in patients with HTLV-1 associated of HAM/TSP by oral administration/uncontrolled preliminary trial	(161)
Delay on the occurrence of allergic symptoms induced by JCP/randomized double-blind, placebo-controlled study	(193)
Reconstruction of microflora by synbiotic therapy in a infant with MASA enterocolitis after vancomycin treatment/Case Reports	(140)
Beneficial effect on intestinal function of a pediatric patient with LTEC by synbiotic therapy/Case Reports	(142)
Improvement of the intestinal function in a girl with SBS by synbiotics therapy/Case Reports	(141)
Prompt and dramatic improvement in sodium balance in a child with SBS/Case Reports	(119)

Animal Experiments

Reported Effects/Administration Route/Animal	References
Reduction of *H. pylori* colonization, a significant decline in gastric mucosal inflammation/administration in the water/mouse	(179)
Improvement in murine chronic IBD, reduction of IL-6 synthesis by LI-LPMC/feed on a diet/mouse	(157)
No prevention of the ulcerative colitis induction by DSS, improvement of the clinical parameters/feeding via intragastric tube/mouse	(134)

In Vitro Experiments

Reported Effects/Target	References
Reduction of the number of bacteria and yeasts in oropharyngeal biofilms on voice prostheses after perfusion of fermented milk or suspension of LcS/biofilm bacteria	(118, 197)

6.4.9 Safety Assessment

Human Studies

Reported Effects/Type of Study	Reference
Clinical safety assessment by bacteriologic surveillance suggesting that the use of LcS as a probiotic in enterally fed CIC is safe/randomized controlled trial	(186)

Animal Experiments

Reported Effects/Administration Route/Animal	References
No infectivity in a rabbit IE model/intravenous injection/rabbit	(115)
Aggravation of EAE observed in Lewis rats/oral administration/rodent	(117)
Significant increase in the duration of clinical symptoms of EAE and exacerbated lung inflammation by early administration during lactation/oral administration/rodent	(129)

Abbreviations

AXN	Alloxan
BCG	Bacillus Calmette-Guérin
BLP	*Lactobacillus casei* preparation
C II	Type II collagen
CFCS	Cell-free culture supernatants
CIA	Type II collagen-induced arthritis
CIC	Critically ill children
DSS	Dextran sodium sulphate

EAE	Experimental autoimmune encephalomyelitis
ECM	Extracellular matrix
GI	Gastrointestinal
HAM/TSP	HTLV-1-associated myelopathy/tropical spastic paraparesis
HTLV-1	Human T-cell lymphotropic virus type-1
i.p.	intraperioneal
i.pl.	Intrapleural
IBD	Inflammatory bowel disease
IDDM	Insulin-dependent diabetes mellitus
IE	Infective endocarditis
IFN	Interferon
IFV	Influenza virus
Ig	Immunoglobulin
IL	Interleukin
JCP	Japanese cedar pollen
LcS	*Lactobacillus casei* strain Shirota
Lex	Extract of autologous *Lactobacillus casei* cell lysate
LI-LPMC	Large intestinal lamina propria mononuclear cells
LLNA	Local Lymph Node Assay
LTAs	Lipoteichoic acids
LTEC	laryngotracheo-esophageal cleft
MASA	methicillin-resistant *Staphylococcus aureus*
MC	3-methylcholanthrene
NIDDM	Non-insulin-dependent diabetes mellitus
NK	Natural killer
NOD	Nonobese diabetic
PBMNC	Peripheral blood mononuclear cells
PR	Peripheral vascular resistance
PUFA	Polyunsaturated fatty acids
SG-1	Polysaccharide-glycopeptide complex from LcS
SCFA	Short chain fatty acids
STEC	Shiga toxin-producing *Escherichia coli*
Stxs	Shiga toxins
TNF	Tumor necrosis factor
UTI	Urinary tract infection
X-irr-Spl	X-ray-irradiated splenocytes

6.5 *Lactobacillus gasseri* OLL2716 (LG21)

KATSUNORI KIMURA

Food Science Institute, Division of Research and Development, Meiji Dairies Corporation, Japan

The evidence of a relationship between *Helicobacter pylori* and gastric diseases has gradually become clear since Warren and Marshall succeeded in cultivating this bacterium in 1982. Since the infection rate of *H. pylori* is very high, about 50% in Japanese, it is difficult to eradicate *H. pylori* in all the infected people. The eradication therapy using antibiotics is often accompanied by side effects and is not always successful because of antibiotic-resistant strains. Moreover, the number of antibiotic-resistant strains has recently been increasing (208, 210).

Probiotics have been proposed as safe and simple alternatives to antibiotic therapy for *H. pylori* infection (204, 207). In order to select a strain that exhibited excellent anti-*H. pylori* activities, we screened 203 *Latobacillus* strains in *in vitro* assays and animal studies and selected *Lactobacillus gasseri* OLL2716 (LG21) as the most suitable strain for use as a probiotic for *H. pylori* infection. We examined the efficacy of LG21 against *H. pylori* infection in humans.

6.5.1 *Helicobacter pylori*

H. pylori is a spiral Gram-negative bacterium with some unipolar flagella. This bacterium displays strong urease activity, and thus can catalyze urea in stomach to ammonia and carbon dioxide. Because of this ammonia production, *H. pylori* can survive strong acidic conditions.

H. pylori inhabits the human stomach and can cause chronic inflammation and ulcers in the stomach and duodenum. *H. pylori* has also been identified as a risk factor for stomach cancer (213).

6.5.2 Selection of a Probiotic for *H. pylori* Infection

We selected *Lactobacillus gasseri* OLL2716 (LG21), *Lactobacillus gasseri* No. 6 and *Lactobacillus salivarius* WB1004 by screening 203 *Lactobacillus* strains for *in vitro* anti-*H. pylori* abilities such as resistance to gastric juice, proliferation under acidic conditions, adherence to cultured gastric epithelial cells, and suppression of *H. pylori* during cofermentation.

Following these *in vitro* experiments, the effects of administration of the selected three *Lactobacillus* strains on *H. pylori*-infected mice were examined. In a study using *H. pylori*-infected mice, orally administered LG21 and *L. salivarius* WB1004 eradicated *H. pylori* in mice, and the anti-*H. pylori* IgG levels of the LG21 administered mice were found to be the lowest (209, Fig. 6.1). Therefore LG21 was selected as the most suitable strain for use as a probiotic for *H. pylori* infection. LG21 is a Gram-positive rod, which was isolated from a healthy human (Fig. 6.2).

LG21 was also found to act as a probiotic for clarithromycin-resistant *H. pylori* infection in mice (214).

6.5.3 Effects of LG21 on *H. pylori* Infection in Humans

The effects of yogurt containing LG21 on *H. pylori* infection in humans were examined (211). Thirty-one subjects infected with *H. pylori* ingested 90 g of normal yogurt containing no LG21 twice daily for 8 weeks. This 8-week period was followed

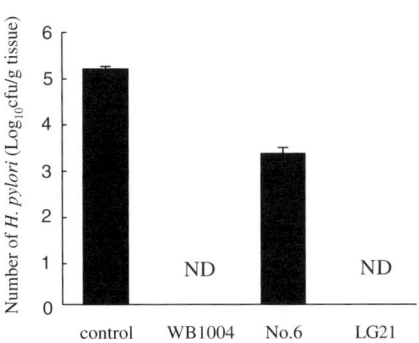 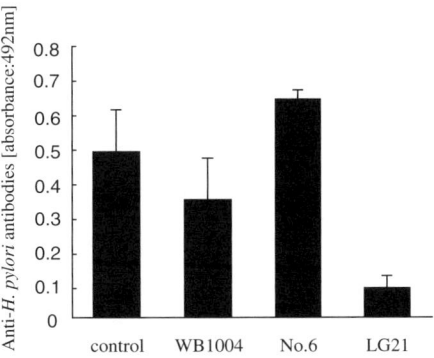

FIGURE 6.1 Effects of administration of LG21 on *H. pylori* infected mice 8 weeks, $n = 5$.

by a 1-week period without yogurt ingestion. This in turn was followed by an 8-week period during which the subjects ingested 90 g of yogurt containing 10^9 cfu of LG21 twice daily. The urea breath test and assays of serum pepsinogen I/II ratios were performed to measure the population of *H. pylori* and to evaluate the degree of mucosal inflammation in the stomach at three points, respectively: prior to ingestion of yogurt, after the first 8-week period, and after the second 8-week period. In addition gastric biopsy specimens were taken from six subjects for quantitative culture both before and after ingestion of LG21 yogurt. The $\Delta^{13}C$ value for the urea breath test did not significantly change after ingestion of normal yogurt, but did significantly decrease following ingestion of LG21 yogurt (Fig. 6.3). The serum pepsinogen I/II ratio did not significantly change after ingestion of normal yogurt, but did significantly increase following ingestion of LG21 yogurt (Fig. 6.4). The number of *H. pylori* in the gastric

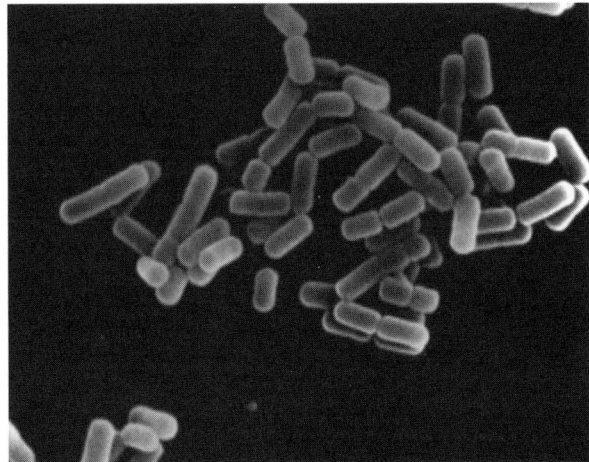

FIGURE 6.2 *Lactobacillus gasseri* OLL2716 (LG21).

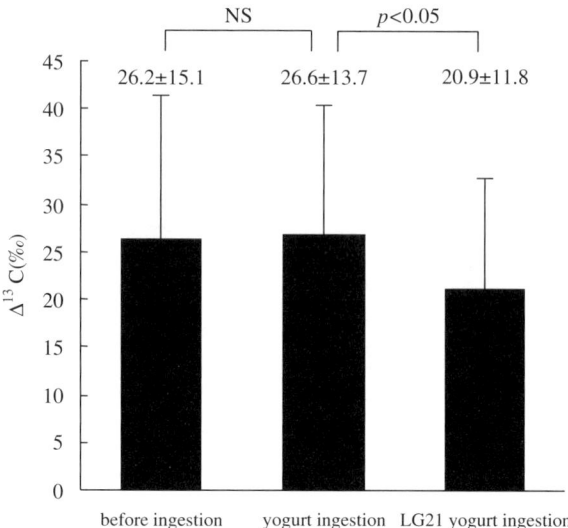

FIGURE 6.3 Effect of yogurt containing LG21 on $\Delta^{13}C$ value of subjects infected with *H. pylori* ($n = 29$).

biopsy specimens from all six volunteers decreased after ingestion of LG21 yogurt (Fig. 6.5).

Similarly, the long-term (12 months) ingestion of LG21 yogurt significantly decreased the $\Delta^{13}C$ value in the urea breath test and significantly increased the serum pepsinogen I/II ratio (209).

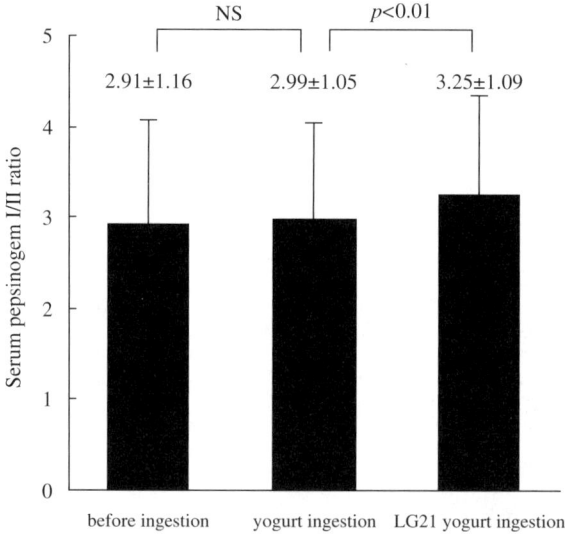

FIGURE 6.4 Effect of yogurt containing LG21 on serum pepsinogen I/II ratio of subjects infected with *H. pylori* ($n = 30$).

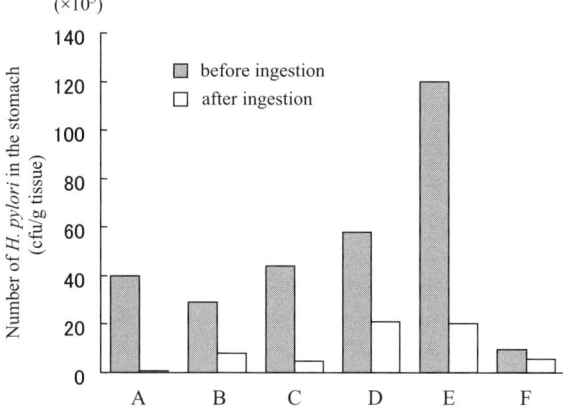

FIGURE 6.5 The number of *H. pylori* before and after ingestion of yogurt containing LG21.

Accumulating evidence indicates that IL-8 plays a major role in the mucosal inflammation caused by *H. pylori* infection. Since LG21 suppressed *H. pylori*-induced IL-8 production *in vitro* (212, 213), the effect of LG21 on IL-8 production in the gastric mucosa of *H. pylori* infected subjects was examined in a study with a randomized double-blind placebo-controlled design. Twenty-five subjects infected with *H. pylori* ingested 120 g of yogurt containing 10^9 cfu of LG21 or placebo yogurt twice daily for 8 weeks. Biopsy specimens were taken both before and after ingestion of each yogurt to measure IL-8 levels in the gastric mucosa of subjects. IL-8 concentrations in the gastric mucosa of LG21 treated group significantly decreased after ingestion of LG21 yogurt. In contrast, IL-8 concentrations in the gastric mucosa of the placebo treated group did not significantly change after ingestion of placebo yogurt (212).

These results suggest that the ingestion of LG21 yogurt decreases the number of *H. pylori* and alleviates the mucosal inflammation in the stomach.

6.5.4 Mechanisms of Therapeutic Effects of LG21 on *H. pylori* Infection

LG21 displays high resistance to gastric juice, grows well under acidic conditions, and can be detected in the stomach of subjects who have ingested LG21 (unpublished data). LG21 also produced large amounts of lactic acid and inhibited the growth of *H. pylori* in *in vitro* experiments. Lactic acid exerted a stronger inhibitory effect on the growth of *H. pylori* than hydrochloric acid at the same pH (204). In addition, LG21 suppressed *H. pylori*-induced IL-8 production in both gastric cell line and gastric mucosa of *H. pylori* infected subjects (212, 213). Considering these results, a decrease in *H. pylori* by the antimicrobial activity of lactic acid and suppression of *H. pylori*-induced IL-8 production by LG21 appear to be involved in the therapeutic effects of LG21 on *H. pylori* infection.

6.5.5 Conclusion

Patients who are *H. pylori*-positive and have peptic ulcers should receive eradication therapy. However, for many otherwise healthy people infected with *H. pylori*, probiotic therapy may be a safe and simple alternative to antibiotic therapy. The density of *H. pylori* colonization in the stomach is of importance in the pathogenesis of infection associated with this bacterium. It was reported that there was a correlation between *H. pylori* density and gastric inflammation or duodenal ulceration (205, 206). Since the ingestion of LG21 yogurt suppresses *H. pylori* in addition to alleviating inflammation in the stomach, LG21 may reduce the risk of the *H. pylori*-induced gastrointestinal diseases.

6.6 *Lactobacillus paracasei* ssp. *paracasei*, F19®

ULLA SVENSSON[1] AND RANGNE FONDÉN[2]

[1]*Arla Foods, Sweden*
[2]*Finnboda Kajväg 15, Sweden*

Lactobacillus paracasei ssp. *paracasei* F19 (F19) was originally isolated from the healthy human colonic mucosa (226, 227). It was selected due to its robustness in gut environment, in foods, and as a freeze-dried culture (226, 227, 232, 235) and also due to its favorable food properties (232) and promising probotic character (220). *Lactobacillus paracasei* occurs naturally in foods and humans and F19 has been isolated from cheese and several humans (216, 220, 222). Human studies show that F19 influences favorably the gut microbiota, and gut function of the healthy adults, children, and elderly (220, 239, 240). F19 given to children during weaning reduces the number of secondary infections noticed as a decreased need for antibiotic treatment (245) and when given to women it reduces symptoms of vaginosis (221). The influence of F19 alone or in combination with other probiotic bacteria on gut microbiota, gut well being and infections are shown by the stabilization of a healthy gut flora (223, 240, 241) and stool frequency (220), relieve from IBS symptoms (237) and a lower rate of infections in immunocomprimised patients and children (234, 238, 245). Global gene expression in mice given F19 showed influence on a group of genes involved in insulin and fat metabolism (231, 233). F19 is used in fermented milks, e.g. Cultura® in Sweden and Denmark, in cheese, food products for children, and in health remedies.

6.6.1 Identification and Safety

F19 has been identified to the species level, both pheno- and genotypically (216, 219) (Table 6.4). It is patented (244) in many countries and deposited at BCCM/LMG Gent (LMG P-17806). The safety documentation includes a study on sepsis patients in Sweden (242) and its influence on gene expression (231) in mice. Both studies confirm F19 to be safe.

TABLE 6.4 Identification and Safety *L. paracasei* ssp. *paracasei* F19

Area	Data
Identification	Identified as *L. paracasei* ssp *paracasei* by phenotypic typing (219), FTIR and Riboprinter (216)
	Identity confirmed upon deposition for patent purposes LMG, Gent (244)
Safety Confirmation:	
General aspects	Overview (222)
Antibiotic resistance	Inherited antibiotic resistance of F19 is in accordance with the *L. casei* group and no transmissible antibiotic resistance found (217)
	Slight diminished presence of antibiotic resistant strains in persons receiving F19 together with antibiotics (241)
Translocation in immunosuppressed rats and mice	Decreased translocation in rats[a] (236), and of enterococci in mice in the presence of F19 (222)
Immunosuppressed persons	No adverse effects in immunosuppressed or seriously ill persons[a] (234, 236, 238)
Sepsis	Not found in persons suffering from sepsis caused by lactobacillus (243)
Production of lactate	Only L-lactate and no D-lactate were produced in the Shime gut model (215)
Influence on gene expression	No unwanted major effect on global gene expression in mice (compared to Bacterioides) (231)
Genetic stability	Pheno- and geno-typically stable, for example, plasmid content, during 6 years of industrial production (229)

[a]Result with F19 + 3 other probiotics and 4 fibers.

6.6.2 *In Vitro* Studies

F19 has good survival in *in vitro* models of the gut (Table 6.5) (215). It has a hydrophobic surface character and moderate adhesion property to epithelial cells, mucus, and surface proteins (224–227). Its antioxidative property has been compared to that of vitamin C (226) and bacteriocin production has been described as well as the sole production of the L-lactate isomer (215, 226). The technological properties of F19 allow the production of concentrated cultures (235), fermented milks, and fruit drinks with good taste and high survival of the bacteria (232, 235).

6.6.3 Global Gene Expression

Global gene expression in the small intestine of mice given F19 shows an influence on a number of genes involved in insulin metabolism, satiety, and fat metabolism

TABLE 6.5 *In Vitro* Studies and Technological Properties of *L. paracasei* ssp. *paracasei* F19

Area	Data
Survival	
Resistance to low pH and bile acids	Good survival at low pH and together with bile (218, 226, 227)
	Enhanced survival in presence of milk proteins (215)
	Survival confirmed in Shime gut model (215)
Adherence	
Adhesion to epithelial cells	Moderate adherence to gut epithelial cells (222, 230)
Adhesion to surface proteins/mucus	Binding to collagen I, III, fibronectin, fibrinogen, feutin, vitronectin, and heparin (226, 227)
	Low to moderate adhesion to mucus (224, 225)
Interaction with immune system	Induction of Il 8 and Il 10 (226)
	Low induction of TNF-α, Il 6, Il 10 (228)
Micellaneous	
Antioxidative capacity	Antioxidative effect *in vitro*, $1 \times 10E7$ comparable to 100 µg vitamin C (226)
Deconjugation of bile acids	No deconjugation of bile acids (222, 244)
Antibacterial properties	Bacteriocin against *L. monocytogenes* produced (226)
Technological aspects/ performance in foods	A robust strain easy to produce and freeze dry (235)
	Good survival and influence on taste and texture in food products, for example, milk and fruit based products, cheese, powders (infant formula) and porridge (232)

(231). In the same investigation, a significant reduction in food intake occurred and in a second study on mice given a high fat diet and F19 a lower storage of abdominal fat was noticed (233) (Table 6.6).

6.6.4 Human Studies

Human studies confirmed that F19 was well tolerated in all groups investigated and that the bacterium had good survival in the gut, as analyzed from biopsies and in

TABLE 6.6 Global Gene Expression in Mice in the Presence of *L. paracasei* ssp. *paracasei* F19

Area	Data
Global gene expression in gnotobiotic and normal flora Swiss Webster mice	Influence on genes of the immune system, host defense, metabolism, and miscellaneous (231)
Influence on food intake, weight, and fat distribution in normal flora Bl6 mice	Diminished food intake, weight gain, and fat deposit in mice (233)

fecal samples (220, 230, 237, 239–242). The number of F19 in feces increased during consumption (220, 239, 241) and a favorable balance of the gut flora could be detected when F19 was given together with two other probiotic bacteria, in the Cultura, during antibiotic treatment (241). Stool frequency was stabilized in two studies on elderly and children (220, 238) and a large proportion of IBS patients reported sufficient leaning after one week consumption of the Cultura containing F19 (237). Effect of F19 on infection had been studied on children during weaning (245) and on women suffering from vaginite (221). The effect of F19 in combination with three other probiotics (Synbiotic 2000® Medipharm, Kågeröd, Sweden) was studied on liver transplant patients or other seriously ill patients (234, 238). All four studies show the potential for F19 or a product containing F19 to reduce the risk of infections (Table 6.7).

TABLE 6.7 Human Studies on F19 Alone or Together with Other Probiotic Strains

Area	Data
Tolerance	F19 was well tolerated in all groups included in investigations, for example, healthy adults, children, elderly, lactose intolerants, IBS patients, liver transplanted patients, so on (220, 230, 235, 239–242)
Survival in the gut	Survival has been analyzed as part of compliance in several human studies (220, 230, 237, 239–241)
	Immediately after consumption very good survival (220, 230, 237, 239–241)
	2 weeks after consumption present in a few patients (220)
	8 weeks after consumption isolated in two patients (220)
Influence on gut microbiota and balance	Number of F19 increased[a, b] (220, 237, 239, 240)
	Number of Lactobacillus generally increased and sustained in elderly[b] (220, 239, 240)
	Favorable balanced bacterial flora during antibiotic treatment[b] (223, 241)
Influence on gut function and well being	Stabilizes stool frequency/texture in elderly and children[a] (220, 239, 240)
	Increased tolerance to lactose[a] (220)
	Rapid leaning in IBS patients[b] (237)
Influence on infection	Lower rate of antibiotic treatment during weaning in small children[a] (245)
	Less symptoms from vaginosis[a] (221)
	Lower rate of infections after liver transplantation[c] (234)
	Reduced rate of infections in intensive care units[c] (238)

[a]Results with F19.
[b]Result with F19 + L.a + BB-12.
[c]Result with F19 + 3 other probiotics and 4 fibers.

6.7 *Lactobacillus paracasei* ssp *paracasei, L. casei* 431®

DORTE ESKESEN AND CHARLOTTE NEXMANN LARSEN
Chr. Hansen A/S, Health & Nutrition Division, Denmark

Lactobacillus paracasei ssp. *paracasei, L. casei* 431 is a strain originally isolated from feces of a healthy child. It has been used for more than 10 years by the dairy industry showing good survival properties in fermented milk products. *L. casei* 431 has also been used as a freeze-dried product in dietary supplements. *L. casei* 431 has shown to beneficially influence gastrointestinal health as well as the immune system. In the literature, *L. casei* 431 is also referred to as *L. paracasei* ssp. *paracasei* CRL-431 or *Lactobacillus casei* CRL-431. The term *L. casei* 431 will be used throughout this text.

6.7.1 Adhesion and Survival Through the GI Tract

In an *in vitro* study investigating adhesion and survival properties of several *Lactobacillus* sp. strains, *L. casei* 431 (in this study named CHCC 3264) showed moderate adhesion, tolerance to bile, but no survival after 4 h at pH levels of 2.5 (246). In human studies, survival of the strain after oral intake has been investigated with contradictory results from fecal recovery (247–249).

6.7.2 Gastrointestinal Effects

6.7.2.1 Intestinal Microbial Balance The balance of the intestinal microflora is of significant importance for gastrointestinal health. In a human study with healthy children ingesting fermented milk containing *L. casei* 431 and *L. acidophilus*, an inverse relation between the level of Lactobacilli in feces and the risk of developing diarrhea was observed (250). In various studies in mice, administration of *L. casei* 431 either alone or together with *L. acidophilus* has been shown to be effective in protection against infections with *S. sonnei, E. coli, Listeria monocytogenes* and *S. typhimurium* (251–255), the latter with 100% survival rate compared to 20% in the control group (254). The ability of *L. casei* 431 alone or in combination with *L. acidophilus* to inhibit growth of E. coli and *S. sonnei* has also been shown *in vitro* (256, 257).

6.7.2.2 Diarrhea Diarrhea is a common disorder among infants and young children. As severe and persistent diarrhea can have a major impact on infant and child development and is a common cause of infant morbidity and mortality, prevention of this condition is of vital importance. In a double-blind placebo-controlled study, 93 children (6–24 months) with persistent diarrhea were randomized to three groups, receiving either pasteurized cow's milk with either *L. casei* 431 and *L. acidophilus* or *S. boulardii* (10^{10} cfu/g) or placebo (pasteurized cow's milk). The dietary treatment of 175 g was given twice daily for 5 days. Compared to the control, both the treatment groups significantly reduced the number of stools per day and number of days with diarrhea. After the 5-day treatment period, 90% in the Lactobacilli group and 83% in the *S. boulardii* group had recovered completely compared to only 10% in the control group (248).

In another double-blind placebo-controlled trial, 49 healthy children (5–29 months) were randomized to receive either 240 mL of skimmed milk plus 15 mL of fermented milk containing a combination of *L. casei* 431 and *L. acidophilus* or 240 mL skimmed milk plus 15 mL of unfermented milk and studied over a period of 3 months. The incidence of diarrhea among children receiving the milk fermented with Lactobacilli was much lower (17%) compared to the control group (59%), and levels of lactobacilli in feces ($>10^6$) correlated inversely with the risk of diarrhea. Furthermore, the children receiving the fermented milk gained more weight compared to the control group (250).

The effect of using Lactobacilli for children hospitalized with post-gastroenteritis syndrome was examined in a small study. Thirteen children (6–24 months), who had suffered from diarrhea between 27 and 180 days, received fermented milk containing a mixture of *L. casei* 431 and *L. acidophilus*. The milk was given at 6 h intervals for 7 consecutive days. The children served as their own control, as they were all resistant to conventional treatment. Use of fermented milk for a maximum of 4 days was effective in eliminating diarrhea symptoms in 11 of the children (258).

Twenty-four persons suffering from bacterial overgrowth related diarrhea were randomized to daily receive two capsules containing either a mixture of *L. casei* 431 and *L. acidophilus* or placebo for 21 days. After 2 weeks, a significant reduction in the number of stools per day was observed in the probiotic group compared to the control group (Fig. 6.6). This effect was sustained up to 2 weeks after the supplementation period (247).

L. casei 431 has also shown effects against antibiotic-induced diarrhea. In mice treated with ampicilin, *L. casei* 431 given together with ampicilin or shortly after introduction of the antibiotic therapy improved the intestinal microflora and eliminated diarrhea (259).

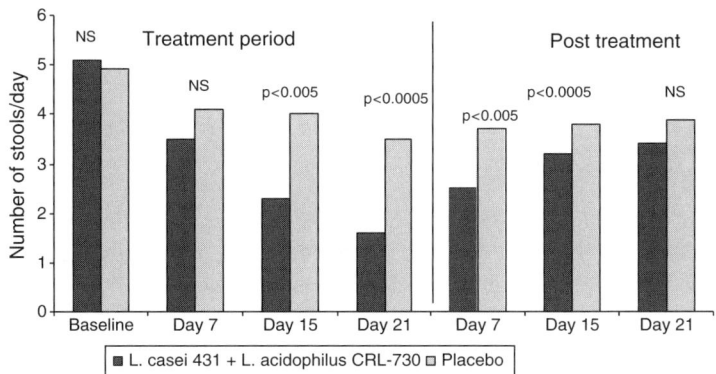

FIGURE 6.6 *L. casei* 431 and *L. acidophilus* versus placebo. Treatment period and post treatment. Number of stools per day during 7 days of each period. Slightly modified with permission from Medicina (247).

6.7.3 Immunomodulatory Effects

L. casei 431 has been shown to influence the immune system by modulating functions such as phagocytosis, production of antibodies, and cytokines. Stimulation of phagocytic activity in healthy mice has been shown using L. casei 431 in fermented (260) as well as in nonfermented milk (261), and a phagocytic effect has also been demonstrated in immunosuppressed mice (262). Increased phagocytosis has also been seen in mice challenged with a pathogen often causing pneumonia in susceptible people. In addition, enhanced clearance of the pathogen from the lungs was observed (263). Studies with specific cell wall components of L. casei 431 on phagocytosis support these findings (264, 265).

Immunoglobulins are part of the specific immune system response, with IgA mainly being associated with intestinal mucosal immunity, functioning as a local defense against bacteria, viruses, toxins, and other food allergens, and IgM and IgG being active in the systemic immune response. As a model for measuring the systemic immune response to enteroviruses, the oral poliovirus neutralization test can be used. This was done in a randomized, placebo-controlled, and double-blind human study with the purpose of investigating the immune response of L. casei 431 and Lactobacillus rhamnosus GG (LGG®). Sixty-six healthy young men were randomized to receive over 5 weeks 100 g/day of acidified milk either as placebo with no probiotics or with probiotic bacteria either L. casei 431 or LGG. Polio vaccination, using three serotypes, was given at day 8. L. casei 431 increased poliovirus neutralizing antibody titers and induced a significant increase in specific IgM (serotype 2) compared to the control (266).

The effect of L. casei 431 on specific immunity has also been demonstrated in several studies in mice, where feeding with L. casei 431 was shown to stimulate production of IgA-secreting cells (267–269) and of $CD4^+$ cells (269). In mice, pretreated with L. casei 431 and subsequently challenged with S. typhimurium, a significant increase was shown in specific antibodies (254, 270), in IgA producing cells (271), and in levels of IgA (270, 271). Also, in ampicilin-treated mice, the administration of L. casei 431 significantly increased IgA producing cells (259).

The ability to modulate cytokine production has also been investigated in mice, showing increases in TNF-α, interferon-gamma (INF-γ, IL-10, IL-4, and IL-12 (267, 272).

6.7.4 Other Health Effects

Antitumor effects of L. casei 431 have been studied in mice showing effects on tumor growth, which could be mediated via the immune system, as increased levels of TNF-α and IgA producing cells were observed in the same study (273). A cytolytic effect against tumor cells in L. casei 431 treated mice has also been demonstrated (274).

6.7.5 Safety

The species L. paracasei ssp. paracasei (or L. casei) is a common member of the normal gut microflora and due to its long history of use L. casei 431 is regarded as safe.

6.8 *Lactobacillus rhamnosus* GG, LGG®

MAIJA SAXELIN, AND KAJSA KAJANDER
Valio Ltd, R&D, Finland

Lactobacillus rhamnosus GG (ATCC 53103; LGG) is one of the best known probiotic strains. It is also commonly known by the name *Lactobacillus* GG (275). The strain is undisputedly classified as a strain of the *L. rhamnosus* species, although the phenotypic characterization is not completely typical for the species (276, 277). The first commercial probiotic products with LGG, under the Gefilus® brand, were launched in Finland in 1990 (278), and presently the annual local per capita consumption is the highest in the world (279). *Lactobacillus* GG is in use in various product applications, including several dairy-based products, such as yogurt, fermented milk, daily-dose mini-bottles, pasteurized (uncultured) milk, semihard cheese, and a few milk-free products are also in the market, such as juice drinks and food supplements (capsules, tablets, sachets).

6.8.1 Storage Stability

There are at least two reasons for the success of *Lactobacillus* GG: (1) the abundant scientific research published on it and (2) the robust character and good survival of the strain. The storage stability of refrigerated *Lactobacillus* GG in yogurt is excellent, without any decline in viability in a month follow-up (280, 281). Storage stability is also good in other fresh dairy products as well as in food supplements (data not shown). In addition to cow milk fermentation, *Lactobacillus* GG has been used successfully for soy fermentation (282).

6.8.2 Gastrointestinal Persistence and Colonization

Good survival in artificial gastric juice and good adhesion properties on intestinal epithelial cells were the basic criteria to isolate *Lactobacillus* GG (275). The strain showed good adhesion properties on intestinal mucus and various cell culture models (283), and also on tissue samples from different parts of the human intestine (284, 285). The strain was also recovered in biopsy samples taken from colon mucosa during administration and for at least 1 week after finishing the oral ingestion (286). Milk-based products are especially good matrixes for the administration of probiotics; however, other product forms are found to be able to deliver the strain to the target site in the intestine, and a good fecal recovery during the treatment was also observed (279).

6.8.3 Health Benefits

Lactobacillus GG has a balancing effect on the intestinal ecosystem, that is, increases the level of lactobacilli and bifidobacteria, formation of SCFAs, lowering the activity of procarcinogenic enzymes, and improving as well as normalizing the mucosal barrier [for review, see (276, 287)].

TABLE 6.8 Health Effects of *Lactobacillus* GG as Seen in Human Intervention Studies

Health Effect	References
Reduction of antibiotic-associated symptoms	(293–295)
Shortening the duration of acute diarrhea	(288)
Alleviation of food allergy/atopic dermatitis in children (conflicting results)	(296–299)
Alleviation of IgE-associated atopic dermatitis in infants	(300–303)
Reduction of atopic dermatitis in infants at high risk	(304–309)
Reduction of infections in children	(291)
Relief of pulmonary exacerbation and intestinal inflammation in cystic fibrosis	(310, 311)
Enhancement of antibody formation and immune response	(312–320)
Modulation of gene expression in gut mucosa	(321)
Eradicate harmful microbes in gastrointestinal tract	(322, 323)
Relief of arthritis rheumatoid symptoms	(324)
Prolongation of relapses in pouchitis (conflicting results)	(325, 326)
Maintaining remission of ulcerative colitis	(327)
No effect on relapses in Crohn's disease	(328)

Clinical trials confirmed the following efficacies of LGG: prevention and treatment of antibiotic associated and traveler's diarrhea, as well as acute nonspecific and rotavirus-induced diarrhea in children (288, 289). The suggested mechanisms involve the enhancement of immune response, balancing of intestinal microbiota, and restoration of the mucosal barrier (290). Furthermore, LGG-enriched milk reduced the risk of respiratory infections and the development of tooth decay in children (291, 292). In addition to these, several other health benefits have been studied with promising results (276, 287) and are presented in Table 6.8.

6.8.4 Source of LGG®

LGG is commercially available through Valio Ltd, Finland, and its qualified suppliers.
Contact information: kalle.leporanta@valio.fi
LGG® is a registered trade mark owned by Valio Ltd.

6.9 *Lactobacillus rhamnosus*, GR-1® AND *Lactobacillus reuteri* RC-14®

Gregor Reid and Andrew W. Bruce

Canadian Research and Development Centre for Probiotics, Lawson Health Research Institute, Canada

Lactobacillus rhamnosus GR-1 and *Lactobacillus reuteri* RC-14 are the world's most documented probiotic strains for women's health. They were selected for their ability to interfere with infectivity of a range of bacteria and yeast in the vagina, as well as confer benefits to the intestine and reduce the risk of bladder infections.

6.9.1 The Strains

Lactobacillus rhamnosus GR-1 was originally isolated in 1980 from the distal urethra of a healthy woman. It was first classified as *Lactobacillus* species (329), then *L. casei* (330), then *L. casei* subsp. *rhamnosus* GR-1 (331, 332).

L. reuteri RC-14 was originally isolated in 1985 from the vagina of a healthy woman. It was first classified as *L. acidophilus* RC-14 (332), then *L. fermentum* RC-14 (333), before being grouped with the *L. reuteri* species (334).

6.9.2 *In Vitro* Properties

Lactobacillus rhamnosus GR-1 has been shown to produce a thin hydrophilic capsule (330, 335–337), with little or no hemagglutination activity (338). It adheres to uroepithelial, vaginal, intestinal cells, and biomaterials (329, 332, 339–341), displaces and prevents adhesion by intestinal and urogenital pathogens (332, 342–346), inhibits the growth of intestinal and urogenital pathogens likely via a bacteriocin-like mechanism (347–349) as well as via lactic acid production (350), and interferes with yeast viability and biofilm formation (351, 352). It is innately resistant to vancomycin and spermicidal nonoxynol-9 (337). The strain induces an anti-inflammatory effect via a G-CSF non-IL-10 pathway (353) and modulates factors involved in the preterm labor pathway, namely through reduction in TNF-α and Cox-2, and possible increase in PDGH production (354).

L. reuteri RC-14 is hydrophobic (336, 337) with little or no hemagglutination activity (338). It adheres to uroepithelial, vaginal, intestinal cells, and biomaterials (332, 355, 356), displaces and prevents adhesion by intestinal and urogenital pathogens (334, 344, 345, 352, 355, 357–363), inhibits the growth of intestinal and urogenital pathogens via lactic acid production and other means but not via reuterin which it does not produce (344, 350, 350a). It produces hydrogen peroxide and is sensitive to spermicidal nonoxynol-9 ((338), unpublished data). It produces signaling factors that downregulate toxin production and release by *Staphylococcus aureus* (334). It also possesses an immune modulatory activity.

6.9.3 Animal Safety, Toxicity, and Effectiveness Studies

- *Lactobacillus* GR-1 intravesical instillation in rats and mice strain did not colonize, adhere, or persist and had no adverse effects on bladder or renal tissues, but did reduce infectivity of uropathogenic *E. coli* (331, 367).
- *Lactobacillus* GR-1 and RC-14 oral use in salmonella model—reduced pathogen translocation, disease, and deaths (344).
- *Lactobacillus* GR-1 and RC-14 oral use at excessive dose in rats—no significant adverse effects on blood parameters (365).
- *Lactobacillus* GR-1 and RC-14 oral use in pregnancy in rats—reduced death rate and improved weights for newborns (366).

6.9.4 Clinical Evidence

There is extensive clinical documentation on these two probiotics. This includes early studies performed with *L. rhamnosus* GR-1 alone and in combination with *L. fermentum* B-54.

6.9.4.1 Safety, Effectiveness, and Efficacy With more than 20 million doses administered worldwide, the safety record of the two strains is excellent. A summary of safety includes:

- *Lactobacillus* GR-1 intravesical instillation in humans with neurogenic bladder disease—strain did not colonize, adhere, or persist in bladder (337).
- *Lactobacillus* GR-1 intravaginal instillation in humans led to repopulation of the vagina for several weeks (368, 370) and longer time to next bladder infection in small pilot study (371).
- *Lactobacillus* GR-1 and B-54 intravaginal instillation in humans led to repopulation of the vagina and reduced recurrence of urinary tract infection (372) after antibiotic therapy (373).
- *Lactobacillus* GR-1 and B-54 intravaginal instillation in humans led to repopulation of the vagina and reduced recurrence of urinary tract infection (374).
- *Lactobacillus* GR-1 and RC-14 intravaginal instillation in humans led to repopulation of the vagina and cure of bacterial vaginosis (BV) in randomized, placebo-controlled, double blind study (375).
- *Lactobacillus* GR-1 and RC-14—orally twice daily for 14 days ($>6 \times 10^9$ cfu) with milk—did not induce adverse immunological reactions (376).
- *Lactobacillus* GR-1 and RC-14—orally once daily in capsules for 14 days showed ability to populate the vagina from this administration, but no adverse persistence in intestine or vagina at 21 days in this independent study (377).
- *Lactobacillus* GR-1 and RC-14—orally once daily with milk showed repopulation of the vagina by lactobacilli (378).
- *Lactobacillus* GR-1 and RC-14—orally once or twice daily at different doses showed retention of a normal lactobacilli-dominated vaginal flora, and 50% cure of BV (379).
- *Lactobacillus* GR-1 and RC-14—orally showed ability of the lactobacilli to displace BV pathogens (345).
- *Lactobacillus* GR-1 and RC-14—randomized, placebo-controlled double blind study of 64 women given lactobacilli orally once daily for 2 months showed significant reduction for transfer of yeast and pathogenic bacteria from rectum to vagina and retention of a normal lactobacilli-dominated vaginal flora (380).
- *Lactobacillus* GR-1 and RC-14—randomized, placebo-controlled double blind study of 24 patients given antibiotics for 10 days plus probiotics for 30 days,

showed significant reduction in onset of BV, and retention of a normal lactobacilli-dominated vaginal flora (381).

- *Lactobacillus* GR-1 and RC-14—randomized, placebo-controlled double blind study of 106 women treated with antimicrobial plus oral probiotics to treat BV. Significantly improved cure of BV compared to metronidazole plus placebo (382).
- *Lactobacillus* GR-1 and RC-14—orally for 30 days to 15 Crohn's disease and 5 ulcerative colitis patients in food—no adverse effects, no bacteremia, no adverse immunological outcomes, but significant reduction in inflammation in subset of patients (383).
- *Lactobacillus* GR-1 and RC-14—orally to HIV positive subjects leading to resolution of diarrhea and increase in CD4 count, with no cases of bacteremia or major side effects (384).

6.9.5 Summary

A combination of *Lactobacillus* GR-1 and RC-14 has been created over a 25-year period to provide a safe and natural restoration and maintenance of urogenital health in women. These well-characterized strains have been produced in food and capsule form and are now commercialized by Chr. Hansen, Denmark, formulated to survive at room temperature for up to 2 years. They are currently sold under different brand names in Malaysia, India, Austria, Poland, France, USA, Croatia, Russia, Canada, Italy and New Zealand.

6.10 *Lactobacillus rhamnosus* HN001 AND *Bifidobacterium lactis* HN019

ARTHUR OUWEHAND, SAMPO LAHTINEN, AND PÄIVI NURMINEN

Health & Nutrition, Danisco, Sokeritehtaantie 20, 02460 Kantvik, Finland

Lactobacillus rhamnosus HN001 and *Bifidobacterium animalis* ssp. *lactis* HN019 were selected from the culture collection of the New Zealand Dairy Research Institute on the basis of their immune modulating properties and their acid and bile resistance (385). The strains are marketed by Danisco as HOWARU Rhamnosus and HOWARU Bifido (Danisco Cultures Division, Paris), and by Fonterra (New Zealand) as *L. rhamnosus* DR20 and *B. lactis* DR10 (386).

6.10.1 Basic Properties of *L. rhamnosus* HN001 and *B. lactis* HN019

In vitro studies have shown that *L. rhamnosus* HN001 and *B. lactis* HN019 have the necessary prerequisites to survive gastrointestinal transit; they are resistant to bile, low pH, and digestive enzymes (385). Furthermore, the strains have been shown to adhere to human epithelial cell lines (387).

The strains have thoroughly been investigated for their safety and have not been found to contain clinically significant transmissible antibiotic resistance genes (388). While the strains are intrinsically vancomycin resistant, as many other probiotic strains including all strains of *L. rhamnosus* (389), they are sensitive to a number of clinically effective antibiotics. Toxicity tests have not indicated any negative effects. The LD50 of the strains for mice is more than 50 g/kg/day, and their acceptable daily intake value is 35 g dry bacteria per day for a 70-kg person (390). The strains do not degrade intestinal mucus and therefore will not affect the intestinal mucosal barrier (391), as is indicated by their inability to translocate (390). The safety of the strains has also been shown in other toxicity tests (392, 393). Furthermore, the absence of negative effects during human studies and the history of commercial use indicate the safety of the strains.

6.10.2 Survival During the Intestinal Transit and Modulation of the Intestinal Microbiota

After consumption of 1.6×10^9 cfu of *L. rhamnosus* HN001 daily for 6 months, the strain could be detected in the feces of all volunteers (394). Furthermore, the consumption caused a transient increase in the fecal *Lactobacillus* and *Enterococcus* populations. *B. lactis* HN019 has been shown to increase the level of fecal bifidobacteria and lactobacilli in humans (395, 396). No other changes in the fecal microbiota or metabolic activity were observed upon consumption of the strains.

6.10.3 Modulation of the Immune System

The main selection criterion for both *L. rhamnosus* HN001 and *B. lactis* HN019 has been their ability to modulate the immune system. This has indeed been shown in numerous animal (397–399) and human studies (Tables 6.9 and 6.10). The studies have been performed with middle-aged to elderly subjects as a model for people with reduced immune function. In these people, in particular, the nonspecific (natural) immune responses were found to be improved. This type of immune response does not need to be induced and is always "ready for action", providing the first line of defense against infections. Consumption of *L. rhamnosus* HN001 and *B. lactis* HN019 has been demonstrated to increase the cytotoxic activity of natural killer cells and the phagocytic activity of peripheral blood mononucleocytes. This improvement is particularly clear in those subjects who initially have poorer than average immune response. However, as usual with probiotics, the improvement in nonspecific immune responses disappears after consumption of the strain(s) is ceased, although in the case of *B. lactis* HN019, enhanced phagocytic activity has been observed 6 weeks after the consumption. Although *L. rhamnosus* HN001 and *B. lactis* HN019 increase the nonspecific immune response, this does not happen indiscriminately. Animal studies have shown that there is no risk for an inflammatory response (400).

TABLE 6.9 Human Clinical Studies with *Lactobacillus rhamnosus* HN001 in Healthy Volunteers

Matrix	Dose	Duration of Consumption	Volunteers, n	Volunteers, Age	Effect	References
Reconstituted milk	1.6×10^9 cfu	6 months	10	25–55 years	Increase in fecal lactobacilli and enterococci	(394)
Reconstituted low fat milk	2.5×10^{10} cfu, $2 \times$ daily	3 weeks	13	62–77 years	Increased leucocyte phagocytic activity	(410)
Reconstituted low fat milk or reconstituted lactose-hydrolyzed low fat milk	2.5×10^{10} cfu, $2 \times$ daily	3 weeks	52	44–80 years	Increased PMN cell phagocytic activity and NK cell activity	(411)
Reconstituted lactose-hydrolyzed low fat milk	2.5×10^{10} cfu, $2 \times$ daily	3 weeks	13	60–84 years	Enhanced NK cell activity	(412)
Reconstituted low lactose/low fat milk	2.5×10^9 cfu, $2 \times$ daily	3 weeks	13	65–85 years	Increased leucocyte phagocytic activity. Enhanced NK cell activity	(413)
Supplement	1×10^{10} cfu[a]	12 weeks	59	1–11 years	Improved atopic dermatitis in the group of food-sensitized subjects	(414)

[a] In combination with 1×10^{10} cfu *Bifidobacterium lactis* HN019.

TABLE 6.10 Human Clinical Studies with *Bifidobacterium lactis* HN019 in Healthy Volunteers

Matrix	Dose	Duration of Consumption	Volunteers, n	Volunteers, Age	Effect	References
Reconstituted milk	3×10^{10} cfu	4 weeks	30	20–60 years	Increased fecal bifidobacteria and lactobacilli	(395)
Reconstituted skim milk	5×10^9, 1×10^9, or 6.5×10^7 cfu	4 weeks	80	60–87 years	Increased fecal bifidobacteria, lactobacilli and enterococci	(396)
Reconstituted low fat milk	1.5×10^{11} cfu	12 weeks	25	60–83 years	Increased phagocytic activity	(415)
Reconstituted low fat milk	2.5×10^{10} cfu	9 weeks	50	41–18 years	Increased phagocytic activity and increased natural killer cell activity	(416)
Reconstituted low fat milk	5×10^9 or, 5×10^{10} cfu	9 weeks	30	63–84 years	Increased phagocytic activity and increased natural killer cell activity	(417)
Reconstituted low fat milk	2.5×10^{10} cfu, 2 × daily	3 weeks	13	60–84 years	Enhanced NK cell activity	(412)
Reconstituted milk	10^7–10^8 cfu[a]	6 months	634	1–3 years	Reduced disease incidence, improved iron status	(406)
Supplement	1×10^{10} cfu[b]	12 weeks	59	1–11 years	Improved atopic dermatitis in the group of food-sensitized subjects	(414)

[a] In combination with galactooligosaccharides.
[b] In combination with 1×10^{10} cfu *L. rhamnosus* HN001.

6.10.4 Reduction of Disease Risk

As humans can obviously not be exposed to pathogens on purpose, animal studies are commonly used to investigate the efficacy of, for example, a probiotic in protecting or curing a disease. This approach has also been followed for *L. rhamnosus* HN001 and *B. lactis* HN019. A group of mice was fed *B. lactis* HN019 daily for a week, while a control group was not. The mice were subsequently challenged with *Salmonella typhimurium*. Three weeks after the challenge, only 7% of the control mice had survived, while 80% of the *B. lactis* HN019-fed mice were still alive (401). Similarly, 90% of the mice fed for *L. rhamnosus* HN001 survived *Salmonella typhimurium* challenge, while only 4% of the control mice survived (402). Other studies have confirmed the activity of *L. rhamnosus* HN001 and *B. lactis* HN019 against gastrointestinal diseases (398, 403–405).

A double-blind, placebo-controlled study involving more than 600 young children (1–3 years) has indicated that consumption of *B. lactis* HN019 for a year, in combination with galactooligosaccharides, reduces both the incidence and prevalence of dysentery. In addition, the administration reduced the number of days of severe illness or fever and the incidence of middle ear infection. Iron status and growth were also improved compared to the control group (406).

6.10.5 Application

In addition to their health benefits, *B. lactis* HN019 and *L. rhamnosus* HN001 have been found to have excellent stability in supplements, dairy, and nondairy products (386) and provide good flavor to fermented products. Furthermore, the strains are capable of utilizing galactooligosaccharides, providing options for formulating synbiotic functional foods (407).

6.10.6 Conclusions

From the studies performed, it is clear that most of the extensive health documentation of *L. rhamnosus* HN001 and *B. lactis* HN019 lie in the area of immune health (408, 409). Claims in the area of immune health and natural defenses have been submitted for the so-called Article 13 process in the European Union.

6.11 LGG®EXTRA, A MULTISPECIES PROBIOTIC COMBINATION

MAIJA SAXELIN, KAJSA KAJANDER, AND RIITTA KORPELA
Valio Ltd, R&D, Finland

6.11.1 Strain Selection for the Combination

Multispecies probiotics may in some conditions be more efficient than a monostrain probiotic due to, for instance, enhanced adhesion and a greater variety of antimicrobial compounds. It may also be beneficial to combine strains that are able to induce proinflammatory and anti-inflammatory immune response in immune cells. The general criteria for all industrially interesting probiotics are safety, adhesion capacity, and survival in the gastrointestinal tract, technological feasibility, and documented physiological effects shown in clinical trials. With this in mind, a new combination, LGG®Extra, was developed on the basis LGG. Other strains selected for the combination were *L. rhamnosus* Lc705 (DSM 7061; Lc705), *Propionibacterium freudenreichii* subsp. *shermanii* JS (DSM 7067; PJS), and a *Bifidobacterium*, preferably *B. animalis* subsp. *lactis* BB-12 (DSM 15954; Chr. Hansen, Denmark). All the strains have been used in food applications in Europe since 1995 or earlier.

Lc705 together with PJS significantly inhibited the growth of yeasts and moulds in food and feed applications (418) and thus, also have the potential to suppress the growth of harmful microbes in the oral cavity and intestine. Lc705 alone produces an antimicrobial agent (419). Combination of Lc705 and PJS has a tendency to alleviate constipation (420), and it also yielded promising results in toxin binding and metabolic activity enhancement (421). *B. animalis* subsp. *lactis* BB-12 is the most studied individual *Bifidobacterium* strain (422). Of the strains in combination, LGG has the highest adherence to intestinal mucus (20–39%, depending on the study and the origin of mucus), followed by BB-12 (24%), whereas Lc705 and PJS are poorly adhesive on mucus (0.3–1.3%) (423–427). The adhesion on Caco-2 cell culture is at a similar level (10–12%) for LGG, PJS, and Lc705 (428, 429).

6.11.2 Adhesion and Gastrointestinal Survival

Regarding synergistic effects, especially adhesion can be notably increased or decreased in probiotic combinations: the presence of LGG significantly enhances the mucus adhesion of BB-12 strain (430), and the presence of Lc705 slightly decreases the adhesion of PJS on Caco-2 cell line (428). However, Collado et al. (431) showed that each individual strain has stronger adhesion on mucus when the other strains are present. Coaggregation does not explain the enhanced adhesion in combinations. Intestinal survival of the strain combination in a long-term intervention in capsule form was good for LGG and bifidobacteria: recovered in 89–95% and 79–89%, respectively, of fecal samples during a 6-month intervention, whereas the recovery was lower for the other two strains: 50% for Lc705 and 63–84% for PJS (432, 433).

6.11.3 Health Benefits

When used as a single strain probiotic, LGG has great potential to enhance immune response, and thus reduces the risk and helps in the treatment of infections. LGG Extra was developed to relieve gastrointestinal symptoms, other than infections, especially in the adult population. Among the most common disturbances is IBS, for which the current medical treatments are regarded unsatisfactory. Thus, effective dietary therapy is the most welcome mechanism to relieve it. Multispecies combination of LGG®Extra is really effective in reducing gastrointestinal discomfort by reducing the total symptom score by 42%, versus 6% in

TABLE 6.11 Health Effects of LGG®Extra as Shown in Human Intervention Studies

Health Effect	References
Alleviation of IBS^a symptoms	(434)
Alleviation of IBS symptoms, improvement in quality of life, and balancing microbiota	(435)
Minor effect in the microbiota and enzyme activities of IBS patients	(432)
Prevention of antibiotic-associated symptoms, decrease in serum gastrin-17, enhancement of the eradication of *H. pylori*, and slightly balancing the composition of microbiota	(436–438)
No effect in the treatment of atopic eczema/dermatitis syndrome in infants, but a tendency to increase fecal IgA and plasma IL-10 content in infants with atopic eczema/dermatitis syndrome, and increase in the secretion of IL-4 by peripheral blood mononuclear cells isolated from infants with cow milk allergy	(433, 439–441)
Relief of atopic eczema in high-risk infants (in combination with GOS^b)	(442)
Enhancement of antibody response during vaccination	(443)
Suppression of the growth of oral Candida	(444)
Decline in the occurrence of recurrent respiratory infection but not the occurrence of acute otitis media	(445)

^{a}IBS: irritable bowel syndrome.
^{b}GOS galactooligosaccharide.

placebo in a well-designed, 6-month intervention study (434). In the second clinical trial, the positive results were confirmed, with a positive response also in the quality of life questionnaire (435). Beneficial effects of LGG®Extra were confirmed by the acceptance of the health claim that "LGG®Extra helps to reduce IBS symptoms or helps to relieve lower gastrointestinal discomfort" by the Swedish Nutrition Foundation.

Carriage of gastritis inducing *H. pylori* is very common in many parts of the world. Standard therapy for eradication causes commonly severe gastrointestinal symptoms, yielding treatment failure. Also, treatment is not always effective enough, and a person can remain a carrier after the treatment. In a recent study, LGG®Extra enhances the eradication effect of *H. pylori* when used together with the standard therapy (436).

LGG Extra orms is a good match for LGG by giving a more precise target group in adult population. A summary of human intervention studies is presented in Table 6.11.

6.11.4 Technological Characteristics

The combination of the strains is suitable for fermented dairy products, and all the strains showed excellent storage stability in a low fat yogurt (Table 6.12).

TABLE 6.12   Typical Storage Stability of the LGG®Extra Strains in Low Fat Yogurt

Strain	Colony Forming Units (±SD)/g, $n=4$		
	Fresh	1 Week	2 Weeks
LGG	$2.55\ (0.74) \times 10^7$	$2.35\ (0.90) \times 10^7$	$2.33\ (0.25) \times 10^7$
Lc705	$1.39\ (0.31) \times 10^7$	$1.65\ (0.55) \times 10^7$	$1.67\ (0.41) \times 10^7$
PJS	$4.18\ (0.38) \times 10^7$	$3.73\ (0.37) \times 10^7$	$3.40\ (0.58) \times 10^7$
BB-12	$7.80\ (1.96) \times 10^7$	$7.28\ (1.75) \times 10^7$	$6.93\ (1.38) \times 10^7$

Pasteurized, lactose-hydrolyzed milk (11% total solid, 0.5% fat) was fermented with standard yogurt culture together with the LGGPlus strains, and mixed with 18% fruit preparation after fermentation, and stored refrigerated.

6.11.5   Source of LGG®Extra

LGG®Extra combination is available for licensing from Valio Ltd only. Contact information: kalle.leporanta@valio.fi.

LGG® is a registered trade mark owned by Valio Ltd.

6.12   *Bifidobacterium animalis* ssp. *lactis*, BB-12

CAMILLA HOPPE AND CHARLOTTE NEXMANN LARSEN

Chr. Hansen A/S, Health & Nutrition Division, Denmark

Bifidobacterium animalis ssp. *lactis*, BB-12, was specially selected by Chr. Hansen for the production of probiotic dairy products. BB-12 has been used in infant formula, dietary supplements, and fermented milk products worldwide and is clinically very well documented. Technologically, BB-12 is expressing good stability and high acid and bile tolerance, also as freeze-dried products in dietary supplements. Furthermore, BB-12 has no adverse effects on taste, appearance, or palatability of the product, and is also able to survive in the product until consumption.

6.12.1   Adhesion and Survival Through the GI Tract

As BB-12 is tolerant of stomach and bile acid, is resistant to digestive enzymes (446), and expresses relatively high oxygen tolerance (447); it is able to survive the passage through the stomach and upper small intestine (448). This has been demonstrated in numerous recovery studies (449–453). Several studies (454–457) also demonstrated that BB-12 has excellent adhesion properties.

6.12.2   Gastrointestinal Effects

6.12.2.1   Intestinal Microbial Balance   Bifidobacteria are Gram-positive, non-motile, anaerobic bacteria, and take a variety of shapes. These organisms are natural inhabitants in the gut of humans and animals, and play a significant role in reducing intestinal pH by producing lactic and acetic acids, thereby restricting the growth of many potentially pathogenic bacteria.

When analyzing the changes in fecal microflora, recovery of BB-12, adhesion abilities, and the production of acids and antibacterial substances (449–452, 458–462), it is shown that BB-12 increases the number of bifidobacteria and lactobacilli, and decreases that of the clostridia and other pathogenic bacteria (463). The protection against pathogenic bacteria is supported by an animal study by Silva and colleagues. Here 80% of the mice receiving BB-12 were still alive 4 weeks after oral infection with *Salmonella typhimurium*, while only 20% of the control group survived (464).

In a number of intestinal inflammatory diseases, disturbances of the normal intestinal microbiota seem to play a pivotal role. In patients with an ileal-pouch-anal anastomosis, a significant increase in bifidobacteria and lactobacilli was observed in those receiving BB-12 and *Lactobacillus acidophilus* (LA-5), and the number of stools per day decreased with no changes recorded in the placebo group (465). In patients with collagenous colitis and diarrhea, reduction in bowel frequency of 50% occurred in 6 of 21 patients receiving BB-12 and LA-5, and in 1 of 8 receiving placebo (466).

Administration of BB-12 not only modifies the composition of the intestinal flora but also its metabolic activity. An increased concentration of short chain organic acids, a decrease of the pH-value, and a lower concentration of ammonia, indoles and other putrefactive substances have been observed in feces (449).

6.12.2.2 Diarrheas

Diarrhea is the most common health problem encountered by visitors traveling to regions with poor hygiene. The condition is characterized by acute diarrhea caused mainly by intake of food or water infected with *Escherichia coli*, *Salmonella*, *Shigella*, or *Campylobacter*. Several studies have been carried out addressing the issue of probiotic prophylaxis of traveler's diarrhea (467). In travelers to Egypt, the intake of BB-12 together with LA-5, *Streptococcus thermophilus* (STY-31) and *Lactobacillus delbrueckii* ssp. *bulgaricus* (LBY-27) significantly decreased the frequency of diarrhea in the supplemented group (43%) compared to the placebo group (71%) (468).

Diarrhea is a potential consequence of oral therapy with antibiotics, causing a disturbed balance between beneficial and harmful bacteria in the intestinal tract, potentially leading to an amplification of pathogenic bacteria. The triple antibiotic treatment for *H. pylori* is particularly harsh. *H. pylori* infected patients experienced less vomiting, constipation, and diarrhea during antibiotic treatment when supplemented with BB-12 and LA-5 containing yogurt compared to a placebo yogurt (469, 470). Other studies have shown that volunteers prescribed to ampicillin or clindamycin treatment and supplemented with BB-12 and LA-5 experience a faster normalization of the intestinal flora than those receiving antibiotics and placebo (471, 472).

Acute diarrhea is a severe cause of child morbidity and infant mortality. Even though diarrhea attacks are usually self-limiting with a duration of 2–5 days, symptoms can persist for months or years, a recent evidence shows that 10–17% may develop post-infectious IBS (473). In a placebo-controlled study, supplementation of an infant formula with BB-12 and TH-4™ significantly reduced

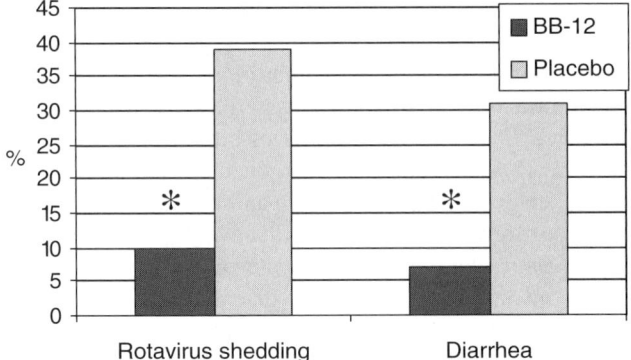

FIGURE 6.7 In the placebo group, 31% developed diarrhea and 39% shed rotavirus, while in the probiotic group, only 7% developed diarrhea and 10% shed rotavirus. Reprinted with permission from Elsevier (474).

the incidence of diarrhea and shedding of rotavirus in hospitalized children (Fig. 6.7) (474).

6.12.2.3 Gastrointestinal Health of Infants It has been shown that the intestinal flora of formula-fed infants differs from that of breast-fed infants (475). Temporary intestinal colonization of infants with bifidobacteria by feeding BB-12 has been demonstrated (476). Additionally, in infants fed a formula with BB-12, the colonization with bifidobacteria was similar to that of breast-fed infants and significantly higher than that in infants fed standard formula (451). Formula with BB-12 decreased the number of days with acute diarrhea in 4-month-old infants compared with conventional formula (477). Also, formulas with BB-12 and TH-4 were associated with a significantly lower frequency of reported colic and irritability, and antibiotic use in 3–24-month-old infants, when compared to unsupplemented formula. There were no significant differences between the probiotic and the control group in growth, health care attention seeking, daycare absenteeism, or other health variables (478). Better growth, however, was observed in children receiving BB-12 alone or with TH-4 in their formula compared to children receiving unsupplemented formula in a double-blind study with 148 infants and toddlers (479).

In very low birth weight neonates who were randomized to receive either a probiotic mixture (ABC Dophilus) of BB-12, *Streptococcus thermophilus* (TH-3), and *B. infantis* (BB-02), or no supplements, and the incidence and severity of necrotizing enterocolitis (NEC) was significantly reduced in the probiotic group (480).

6.12.2.4 Other Gastrointestinal Effects Constipation is very common in developed countries, and supplementation with bifidobacteria might improve natural bowel function. The elderly patients suffering from chronic constipation were randomized to receive either unfermented milk or fermented milk with BB-12 and LA-5. The probiotic group experienced a significant improvement in the frequency of bowel movement and did not report any negative side effects (481). In a recent study, it was possible to normalize bowel movements in frail nursing home residents, as the group receiving a fermented oat drink with BB-12 had more frequent bowel movements than the group receiving fermented oat drink alone did (482).

Some probiotic bacteria have been shown to improve lactose digestion by release of β-galactosidase. In lactose intolerant subjects, symptoms of lactose intolerance, such as stomachache, flatulence, diarrhea, and constipation have been found to be improved (483) or unchanged (448) by BB-12, LA-5, STY-31, and LBY-27.

6.12.3 Immunomodulatory Effects

The GALT is the largest organ of the immune system, and numerous probiotic health effects are related to the modulation of immune functions. Administration of BB-12 has been associated with nonspecific stimulatory effects on the production of cytokines and phagocytic activity, as well as more specific immune reactions such as antibody production.

6.12.3.1 Nonspecific Immune Responses Consumption of a fermented milk containing BB-12 resulted in increased phagocytic activity of granulocytes and monocytes in healthy subjects (461). The impact of BB-12 on the macrophage function was also examined *in vitro* with a murine macrophage-like cell line that was cultured in the presence of cell-free extracts of BB-12. BB-12 enhanced phagocytosis of inert particles or viable Salmonella (484).

Cytokines are a group of proteins and peptides that are used in organisms as signaling compounds. Due to their central role in the immune system, cytokines are involved in a variety of immunological, inflammatory, and infectious diseases, and thus play an important role in interactions between probiotic bacteria and the immune system.

In an *in vitro* study, 10 lactic acid bacteria strains were compared with regard to their ability to induce, reduce, or inhibit the release of three cytokines; TNF-α, IL-6, and IL-10. BB-12 was among the four best inducers of TNF-α release, and also among the three best inducers of IL-6 release. BB-12 induced the production of TNF-α (485), and this amounts significantly greater than those obtained with lipopolysaccharide, a widely used antigen, as a stimulant (486). BB-12 also inhibited IL-2 and IL-4 production in an *in vitro* study investigating the influence of nonviable BB-12 on cytokine production (487).

6.12.3.2 Specific Immune Responses Noncellular immunity results in the production of immunoglobulins (Igs) and antigen-specific antibodies. Administration of a formula containing BB-12 significantly increased fecal concentrations of total IgA in healthy Japanese children and anti-polio-specific IgA as a response to polio vaccination (450). Infants receiving formula with BB-12 and *L. rhamnosus* GG (LGG) had significantly higher numbers of cow's milk specific IgA secreting cells compared to infants receiving placebo (488).

In another study, healthy adults consumed fermented milk with BB-12 and Nestlés LA-1, and an attenuated *S. typhi* was administered to mimic an enteropathogenic infection. Serum total and *S. typhi*-specific IgA were significantly higher in the BB-12 supplemented group compared with the control group, and IgG remained high over a longer period in the BB-12 group (452). Furthermore, in mice, orally immunized with cholera toxin, significantly higher concentrations of cholera toxin-specific antibodies were found in the feces and serum of BB-12 and LA-5 fed mice (489).

6.12.3.3 Other Immunomodulatory Effects Immunomodulatory effects of BB-12 have also been seen in subjects with allergy, respiratory infections, or with a compromised immune system.

Over the last decades, the incidence of allergic diseases has increased in industrialized countries, and consequently there has been much interest in exploring the ability of certain bacteria not only to stimulate the defense mechanisms but also to regulate it, for example, in cases related to food allergies or atopic diseases. Both in infants (458, 490) and in adults (491), supplementation with BB-12 seems to improve allergic conditions when compared to placebo groups.

Infectious disease is the most important cause of morbidity in infants. A randomized double-blind placebo controlled study was conducted to test the safety and efficacy of BB-12 and LGG in reducing the risk of infections in infants. Supplementation significantly reduced the incidence of acute otitis media, the prescription of antibiotics, and fewer recurrent respiratory infections compared to the placebo group (492).

The advantage of immunomodulatory stimulation by probiotics can also be demonstrated in subjects with a compromised immune defense. In patients with acute leukemia, BB-12 and LA-5 during and after chemotherapy significantly postponed the occurrence of fever (493). BB-12 also prolonged survival, decreased systemic dissemination, inhibited *Candida albicans* in the alimentary tract, stimulated antibody and cell-mediated immunity in sterile, severely immunodeficient bg/bg-nu/nu mice, which lack the thymus (494, 495).

6.12.4 Other Health Effects

BB-12 seems to have an inhibitory impact on tumorigenesis as well as on lipid metabolism.

The effect of BB-12 on several aspects of tumorigenesis has been studied in a number of *in vitro* studies (496–504, 51–59), including mice (499, 505) and rat studies (506–508) and human studies (509, 510). BB-12 seems to produce substances during fermentation, which inhibit the growth and division of cancer cells (496).

The ability of BB-12 to reduce the reabsorption of cholesterol and other sterols (bile acids) from the intestine has been examined in noninsulin dependent diabetics, consuming fermented milk with BB-12, LA-5, and STY-31 showing a 20% decrease in total cholesterol (511). This has also been studied in mice (512) and in rats (513). An *in vitro* study demonstrated that BB-12 precipitated cholesterol and bile salts, and a decrease in the cholesterol content during the fermentation was detected. Removal of cholesterol from culture medium was not due to cholesterol uptake by the strains, but as a result of bacterial bile salt deconjugation (514).

6.12.5 Safety

In relation to the use of microorganisms as human probiotics, safety is a major concern. BB-12 is considered safe. It is not pathogenic or toxic for humans (478, 515), and millions of humans have consumed BB-12 without reporting any adverse effect. BB-12 has been notified Generally Recognized As Safe (GRAS Notice No. GRN 000049), together with TH-4 as ingredients in milk-based infant formula intended for consumption by infants 4 months and older.

6.13 *Bifidobacterium breve* STRAIN YAKULT

HIROKAZU TSUJI AND KOJI NOMOTO

Yakult Central Institute for Microbiology ResearchTokyo, Japan

Reported characters of *Bifidobacterium breve* strain Yakult associated with probiotic effects are summarized in the following tables.

6.13.1 Effects on Intestinal Environment

Human Studies

Reported Effects/Type of Study	Reference
Favorable effect on nitrogen–protein metabolism/randomized placebo-controlled crossover study	(519)

6.13.2 Intestinal Physiology

Animal Experiments

Reported Effects/Administration Route/Animal	Reference
Effect on gene expression in the ileal and colonic epithelial cells/oral administration/mouse	(535)

***In Vitro* Experiments**

Reported Effects/Target	Reference
Absence of either deoxycholic acid or 7-keto-deoxycholic acid/bacterial culture	(537)

6.13.3 Effects on Cancer

Animal Experiments

Reported Effects/Administration Route/Animal	References
Inhibition of the PhIP-induced mammary carcinogenesis in rats by feeding of FSM/feed on a diet/rat	(533)
Suppressive effects on the number of ACF by colonization of *B. breve* in GB rat/oral administration/rat	(534)

6.13.4 Prevention of Infectious Diseases

Human Studies

Reported Effects/Type of Study	Reference
Eradicating *Campylobacter* and restoring the normal intestinal flora in *Campylobacter enteritis*/randomized trial	(538)

Animal Experiments

Reported Effects/Administration Route/Animal	References
Anti-infectious activity of Bifidobacteria in combination with TOS against *Salmonella enterica* serovar *typhimurium* LT-2 in an antibiotic-induced murine infection model/oral administration/mouse	(516)
Protective effect of oral administration of fermented milk against the endogenous *E. coli* infection/oral administration/mouse	(518)
Anti-infectious activity of Bifidobacteria against STEC O157:H7 in a fatal mouse STEC infection model/oral administration/mouse	(517)

6.13.5 Prevention of Life Style Diseases

Animal Experiments

Reported Effects/Administration route/Animal	References
Reduction of plasma total cholesterol levels by feeding of FSM in a ovariectomized model/oral feeding/mouse	(528)
Hypolipidemic effect by feeding of FSM in a ovariectomized model/oral feeding hamster	(530)
Decrease in liver and plasma cholesterol levels, and fecal steroids by feeding of FSM/oral feeding/hamster	(529)

6.13.6 Clinical Application

Human Studies

Reported Effects/Type of Study	References
Reduction of postoperative infections in biliary cancer patients by synbiotics administration/randomized controlled trial	(526)
Improvement of intestinal flora and increase of SCFA in short bowel patients with refractory enterocolitis/synbiotic therapy	(525)
Effect for preventing recurrence in patients with ulcerative colitis by drinking of the BFM product/randomized controlled intervention trial	(521)
Effect for endoscopic activity index and histological score by administration of BFM on active ulcerative colitis/randomized placebo-controlled trial	(527)
Improving the intestinal flora and systemic immunonutritional status of patients with SBS/synbiotic therapy	(539)
Effect for body weight gain in VLBW infants/randomized controlled trial	(531)
Enhancement of immune responses, attenuation of systemic postoperative inflammatory responses, and improvement of intestinal microbial environment by preoperative oral administration of synbiotics/randomized controlled trial	(536)
Reconstruction of microflora by synbiotic therapy in an infant with MASA enterocolitis after vancomycin treatment/case reports	(522)
Beneficial effect on intestinal function of a pediatric patient with LTEC by synbiotic therapy/case reports	(524)
Improvement of the intestinal function in a girl with SBS by synbiotics therapy/case reports	(523)
Improvement of the intestinal microflora in infants with intractable diarrhea by administration of *Bifidobacterium*/uncontrolled trial	(520)

Animal Experiments

Reported Effects/Administration Route/Animal	Reference
Improvement murine chronic IBD in SAMP1/Yit strain mice by feeding of fermented milk/oral feeding/mouse	(532)

Abbreviations

ACF	Aberrant crypt foci
BFM	Bifidobacteria-fermented milk
FSM	Bifidobacterium-fermented soya milk
GB	Gnotobiotic
IBD	Inflammatory bowel disease
LTEC	Laryngotracheo-esophageal cleft
MASA	Methicillin-resistant *Staphylococcus aureus*
PhIP	2-Amino-1-methyl-6-phenylimidazo[4,5-*b*]pyridine
SBS	Short bowel syndrome
SCFA	Short chain fatty acid

STEC   Shiga toxin-producing *Escherichia coli*
TOS   Transgalactosylated oligosaccharides
VLBW   Very low birth weight

6.14 *Bifidobacterium longum* BB536

JIN-ZHONG XIAO

Food Science and Technology Institute, Morinaga Milk Industry Co. Ltd, Japan

Bifidobacteria are the major components of intestinal microflora in humans. As probiotic agents, bifidobacteria have been studied for their efficacy in the prevention and treatment of a broad spectrum of animal and/or human disorders, such as constipation, colonic transit disorders, intestinal infections, colonic cancer, and allergic diseases.

Bifidobacterium longum BB536 was originally isolated from a healthy infant in 1969. BB536 was first commercially available in Japan in 1977, with the launch of Morinaga Bifidus Milk. At present, a large number of products ranging from dairy products to supplements have been marketed in Japan. Presently, BB536 is also broadly available in the European, USA, and Asian marketplaces. BB536 is characteristic for its high survivability in food applications and its high accessibility to the gastrointestinal tract. Lines of evidence including *in vitro*, *in vivo*, and clinical studies and consumption history have supported the safety of BB536. Accumulated data have also shown the health benefits for BB536 in various hosts (Fig. 6.8).

6.14.1 Evaluation of Safety of BB536

Despite the general safe use of bifidobacteria, some side effects in susceptible individuals are theoretically possible. In consideration of the potential adverse effects,

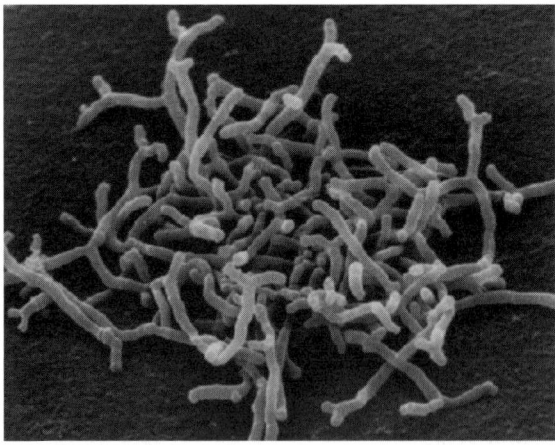

FIGURE 6.8 Bifidobacterium longum BB536.

FAO/WHO guidelines for the evaluation of microbes for probiotic use in foods have recommended testing for several parameters, including antibiotic resistance, metabolic activities (e.g., D-lactate production, bile salt deconjugation), toxin production, hemolytic activity, infectivity, side effects during human studies, and adverse incidents in consumers (FAO/WHO 2002, ftp://ftp.fao.org/es/esn/food/wgreport2.pdf).

These parameters have been examined for *B. longum* BB536. Data from *in vitro* studies, acute, chronic, and repeat dose animal studies, clinical studies involving healthy and unhealthy adults, or children, as well as a long historical consumption of almost 30 years have provided information that supports the safety of the use of *B. longum* BB536. For example (1) an evaluation on antibiotics, which demonstrated that BB536 is not an antibiotic-resistant strain, and which also reported that resistant gene was not found there; (2) BB536 was found to produce L-lactic acid predominantly, while the production of D-lactic acid was negligible; (3) *B. longum* BB536 was observed to possess a conjugated bile salt hydrolase that was able to deconjugate 80–95% of the selected bile salts seen in concurrent bacterial growth, with the only compounds produced being the deconjugated bile salts; (4) genomic analysis of BB536 failed to find any high homological sequences with amino acid sequences of known bacterial toxins that are listed in the GeneBank database; and (5) tests of hemolytic activity of BB536 by using BL agar plates supplemented with horse blood indicated that *B. longum* BB536 does not have any hemolytic activity.

On the basis of these safety investigations, clinical observations, and the long use experiences within the food category, BB536 has been accepted as a GRAS strain for its intended use in the USA.

6.14.2 Physiological Effects of BB536

6.14.2.1 Improvement of Intestinal Environment Probiotics are defined as a live microbial food supplement, which beneficially affects the host animal by improving its intestinal microbial balance. Improvements of the intestinal environment are considered to the pedestal for its diverse physiological effects. Several studies have been carried out to examine the effect of BB536 on the intestinal environment in healthy adults with a constipation tendency. With the consumption of milk or yogurt supplemented with BB536, there were increased defecation frequencies, increased cell numbers and relative percentages of bifidobacteria, and decreased concentrations of ammonia and β-glucuronidase in fecal samples (543, 550).

On the contrary, BB536 intake has been demonstrated to suppress antibiotic-induced gastrointestinal disorders. In a double-blind placebo-controlled study in adults given erythromycin, intake of yogurt supplemented with BB536 suppressed the erythromycin-induced increase of the stool weight, the stool number, and the abdominal discomfort, as compared to placebo yogurt (540). Other studies have reported a trend in decreased proliferation of harmful bacteria that are typically seen during the administration of anticancer therapies or immunosuppressive drugs, or increases in *Bifidobacterium* and *L. acidophilus* levels in the feces. These results suggest that there is a bio-modulatory effect of BB536 in maintaining a normal, healthy intestinal environment.

6.14.2.2 Effects on Immunity and Cancer

The effects of BB536 on an antiintestinal infection, the immune stimulation, and the suppression of cancer were assessed in several studies. Sekine et al. (545) examined the effect of BB536 administered in milk over the course of a year on the chemiluminescence (CL) reaction of peripheral leukocytes and mean corpuscular volume of red blood cells in 18 children with leukemia. CL reactions were enhanced during the treatment. It was speculated that this effect was due to the activation of the monocyte–macrophage cell populations.

BB536 was found to protect germ-free mice from infection by pathogenic *E. coli*. BB536-inoculated mice were also observed to have lower concentrations of endotoxin and E. coli in organs and feces as compared to germ-free mice (541, 551). Other studies have been conducted to examine the production of cytokines in mouse peritoneal cells (546). Following an intraperitoneal injection of *B. longum* BB536 in male BALB/c mice, inductions of inflammatory cytokine expressions, including IL-1β, IL-6 and TNF-α, and IL-10 were observed. The incidence of food-mutagen-induced tumors was significantly lower among mice that received *B. longum* BB536 (544). In a preliminary double-blind, placebo-controlled trial, administration of BB536 to the elderly volunteers was found to suppress the occurrence of influenza virus infection and fever (541a). NK activity and the bactericidal activity of neutrophils were enhanced by BB536 intake (541a).

6.14.2.3 Antiallergic Activity

The prevalence of allergic diseases has rapidly increased worldwide over the past decades, especially in industrialized countries. Japanese cedar pollinosis (JCPsis) is an immunoglobulin E-mediated type I allergy caused by exposure to Japanese cedar pollen. It represents a public health issue affecting more than 16% of the Japanese population, with an increasing prevalence over the past decades.

The effect of BB536 in the treatment of JCPsis was investigated in randomized, double-blind, placebo-controlled human studies. It was found that subjective symptoms were alleviated following the intake of a yogurt supplemented with BB536, while decreased IFN-γ levels were particularly suppressed at the early stages of the pollen season, as compared to that seen with a placebo yogurt (547). Furthermore, fecal microflora significantly fluctuated during the pollen season and BB536 intake was found to suppress the changes of the microflora, such as the increases in *Bacteroides fragilis* during the pollen season (542). In a trial that employed BB536 lyophilized powder, BB536 intake was associated with a significant reduction in the number of subjects prematurely terminated due to severe symptoms and pollinosis medication (548). In addition, there was a significant repression in subjective symptom scores and a significant suppression of the Th2-skewed immune response that occurred along with the pollen dispersion (Fig. 6.9). Similar to the trial that used BB536 yogurt, fluctuations of fecal microbiota were observed and BB536 intake tended to suppress such fluctuations. Using a double-two-way crossover design, the efficacy of BB536 on the output of symptoms in JCPsis patients exposed to Japanese cedar pollen (JCP) was also studied in an environmental exposure unit (EEU) outside the normal JCP season (549). BB536 intake significantly reduced the ocular symptom scores during the JCP

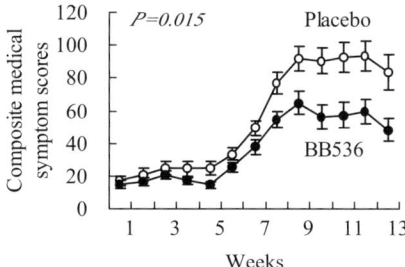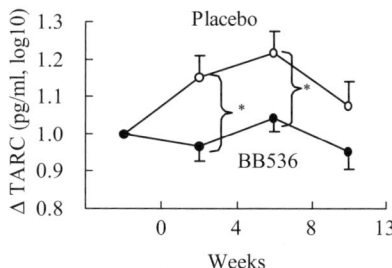

FIGURE 6.9 Effects of intake of *Bifidobacterium longum* BB536 on subjective symptoms and blood markers in patients of Japanese cedar pollen allergy (547).

exposures, alleviated some of the delayed symptoms, and reduced the prevalence of medication uses after exposures, when compared with placebo.

These results suggest the efficacy of BB536 in relieving JCPsis symptoms, which were probably due to the modulation of the Th2-skewed immune response.

In conclusion, BB536 has been validated for its pronounced effects on improving intestinal environments, stimulating immunity, suppressing harmful bacterial infection, reducing cancer risk, and preventing and treating allergic disorders. Evidence has also been presented on lowering of serum cholesterol and enhancement of bone strength for BB536.

6.14.3 Technologies in BB536 Applications

As an anaerobic bacterium, BB536 is known to be relatively tolerant to oxygen and a low pH. BB536 is characteristic for its high stability in various food applications and has been successfully applied in chocolate, lozenges, gum, nutrition powders, milk, and yogurt (Fig. 6.10).

In summary, *B. longum* BB536 is a human-originated probiotic that has been documented to be safe in humans, with further evidence that supports healthy benefits associated with its use, and technologically, it is a highly applicable probiotic. It is expected in the future that BB536 will be used worldwide in various kinds of foods, such as yogurt, milk, infant formula, powdered foods, chocolate, tablets, capsules, so on.

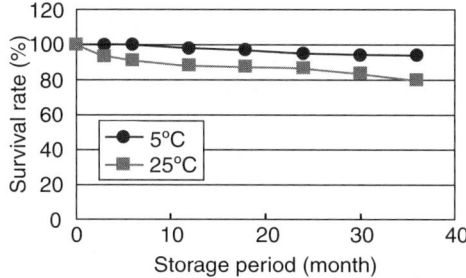

FIGURE 6.10 Stability of *Bifidobacterium longum* BB536 (lyophilized powder).

6.15 *Bifidobacterium longum* STRAINS BL46 AND BL2C—PROBIOTICS FOR ADULTS AND AGEING CONSUMERS

SAMPO LAHTINEN AND SEPPO SALMINEN

Functional Foods Forum, University of Turku, Finland

To date, most probiotics have been designed for the common healthy population or the young infants and children. Also, there is clearly a need for other target groups, especially the ageing population considering their intestinal microbiota deviations and aberrancies that may have detrimental effects on their health. For this purpose, some 10 years ago a search was initiated first by characterizing the microbiota of healthy adults, healthy elderly, and extremely old subjects in Finland and Japan. Based on these studies it became apparent that the important factor in the microbiota composition was found in the genus *Bifidobacterium*. The microbiota of the elderly subjects has been reported to become less diverse and less stable against outside factors such as pathogens or harmful substances in the diet. Moreover, the mean bifidobacterial populations are believed to be decreased in the elderly subjects. Thus, a need for stabilizing microbes was recognized. For this purpose, bifidobacterial strains from very old healthy subjects were isolated and characterized for probiotic use. The selection criteria included adhesion to human intestinal mucus, stability against acid, and ability to bind toxins, notably mycotoxins. Aflatoxin B_1 binding was used as the model for toxin binding. Based on screening eight strains were originally characterized. Two of these strains, *Bifidobacterium longum* 46 (BL46; DSM 14583) and *B. longum* 2C (BL2C; DSM 14579), were identified for studies including technical properties, *in vitro* studies, and clinical intervention studies (563, 568). The objective was to provide a bifidobacterial supplement that would balance the intestinal microbiota and enhance the healthy microbial barrier function in the intestine. Studies conducted on the chosen strains and characterization of their properties in *in vitro*, *in vivo*, and in clinical intervention are reported below.

6.15.1 Safety of BL2C and BL46

Safety of the probiotics is of utmost importance. The current safety record of probiotics is excellent. Nevertheless, the safety of new probiotic strains should always be assessed in a clinical setting before the strains are used in the commercial products. *Bifidobacterium longum* is a common member of the healthy intestinal tract and has been assessed as safe by the European Food Safety Authority system of Qualified Presumption of Safety (552). Bifidobacteria have never been reported in the cases of bacteremia in healthy subjects and this also attests to their safety. The safety of BL2C and BL46 has been confirmed in a clinical safety, tolerance, and side effect assessment trial (564). In the trial, the product containing BL2C and BL46 was well tolerated and did not cause adverse effects or symptoms in healthy adult volunteers. Furthermore, BL2C and BL46 had been used by over 300 healthy elderly subjects in two clinical intervention trials (560, 567). Again, the probiotic strains were well tolerated and had no adverse effects on the volunteers. In addition, *B. longum* is a common member of healthy breast-fed infant intestinal microbiota, adult microbiota, and the microbiota of healthy elderly subjects.

6.15.2 The Health Effects of BL2C and BL46

The health effects of BL2C and BL46 have been assessed in a number of scientific studies. Here, the current scientific evidence is summarized.

6.15.2.1 BL2C and BL46 Stabilize the Gut Function in the Elderly The ability of BL2C and BL46 to stabilize the gut function has recently been demonstrated. In a clinical study by Pitkälä and coworkers (567), the consumption of BL2C and BL46 resulted in more frequent bowel movements in elderly subjects consuming the probiotic strains compared to the placebo group. The study showed that it is possible to normalize bowel movements in frail nursing home residents by including BL46 and BL2C in the diet.

6.15.2.2 Modulation of Gut Microbiota by BL2C and BL46 BL46 and BL2C may induce positive changes in the gut microbiota of the elderly subjects. Two recent clinical intervention trials have shown that consumption of BL2C and BL46 led to synergistic effects on the *Bifidobacterium* species naturally present in the gut microbiota (560, 566). These effects correlated with changes in the immune function (566).

6.15.2.3 BL46 is Effective Against Harmful Bacteria Recently published studies have shown that BL46 has antimicrobial activities against potentially harmful bacteria. Lahtinen and colleagues (559) showed that BL46 was able to produce compounds that were effective against *Staphylococcus aureus*. Moreover, BL46 had been shown to possess antimicrobial activity against *Helicobacter pylori* and *Escherichia coli* (554).

6.15.2.4 Effects of BL2C and BL46 on the Immune System and Infections The combination of BL2C and BL46 may be used to modulate the immune system. The immunomodulatory effects of the two strains on the elderly subjects have been assessed during a clinical intervention trial in Helsinki, Finland (567). The results of the study indicated that the consumption of BL2C and BL46 positively influenced the cytokine levels of the serum (566). The net effect of the modulation of the immune system by BL2C and BL46 was shown to be anti-inflammatory rather than proinflammatory.

6.15.2.5 BL2C and BL46 Can Bind Toxic Compounds Laboratory studies showed that both BL2C and BL46 were capable of binding aflatoxin B_1, a potent carcinogenic contaminant present in certain foods around the world (563). BL2C and BL46 are also able to bind heavy metals. Halttunen and colleagues showed that both BL2C and BL46 were capable of effectively binding lead and cadmium (553). Of all the probiotic strains tested, BL46 was the most effective in binding these heavy metals. Binding of microcystin-LR by BL46 had also been demonstrated (565).

Binding of toxins and heavy metals by probiotic bacteria may reduce the exposure of these toxins to humans by removing them from the gastrointestinal tract via feces providing new possibilities for modulating the exposure to harmful components in the human diet.

6.15.3 Technical Properties and Sensory Qualities of BL2C and BL46

Fermentations carried out with BL2C and BL46 require knowledge of their technical and metabolic properties. The strains require anaerobic growth conditions (563) and should be protected from light (555). During storage in food products, the strains survive by entering a dormant state, in which the cells maintain their critical functions but do not undergo cell division (556–558, 561). The culturability of the two strains can be improved by cell encapsulation (562). The two strains lack the overly acidic and bitter flavors associated with many other probiotic and starter strains, and their fruity and smooth flavors compose a genuinely unique probiotic combination for fermentation processes.

6.15.4 Conclusions

Bifidobacterium longum 2C and *B. longum* 46 are specific probiotic bacteria targeted at adult and elderly and ageing consumers. The two strains have been subjected to a number of scientific studies. They possess scientifically demonstrated health effects such as modulation of gut function, bowel regularity, and immune function. The sensory qualities of the products fermented with BL2C and BL46 are exceptional, making the two strains a unique combination for fermentation processes. BL2C and BL46 represent a new type of probiotic strains, characterized for the needs of adult and elderly consumers.

REFERENCES

1. Noh DO and Gilliland SE. Influence of bile on cellular integrity and β-galactosidase activity of *Lactobacillus acidophilus*. *J. Dairy Sci.* 1993; 76: 1253–1259.
2. Plockova M, Chumchalova J, and Pluharova B. The pH tolerance, bile resistance and production of antimicrobial compounds by lactobacilli. *Potrav. Vedy* 1996; 14(3): 165–174.
3. Holcomb JE and Frank JF. Viability of *Lactobacillus acidophilus* and *Bifidobacterium bifidum* in soft-serve frozen yogurt. *Cult. Dairy Prod. J.* 1991; August: 4–5.
4. Hoier E and Hier E. Saaure and gallentoleranz von *Lactobacillus acidophilus* und bifidobackterien. *Lebensmittelindustrie Milchwirtschaft* 1992; 26: 769–772.
5. Sanders ME, Walker D, Walker K, Aoyama K, and Klaenhammer T. Performance of commercial cultures in fluid milk applications. *J. Dairy Sci.* 1996; 79: 943–955.
6. Juntunen M, Kirjavainen PV, Ouwehand AC, Salminen SJ, and Isolauri E. Adherence of probiotic bacteria to human intestinal mucus in healthy infants and during rotavirus infection. *Clin. Diag. Lab. Immunol.* 2001; 8: 293–296.

7. Apostolou E, Kirjavainen PV, Saxelin M, Rautelin H, Valtonen V, Salminen S, and Ouwehand AC. Good adhesion properties of probiotics: a potential risk for bacteremia? *FEMS Immunol. Med. Microbiol.* 2001; 31: 35–39.

8. Saarela M, Maukonen J, von Wright A, Vilpponen-Salmela T, Patterson AJ, Scott K, Hämynen H, and Mätto J. Tetracycline susceptibility of the ingested *Lactobacillus acidophilus* LaCH-5 and *Bifidobacterium animalis* subsp. *lactis* BB-12 strains during antibiotic/probiotic intervention. *Int. J. Antimicrob. Agents* 2007; 29: 271–280.

9. Gorbach SL. Lactic acid bacteria and human health. Intestinal Microflora and Health. *Ann. Med.* 1990; 22: 37–41.

10. Jiang T and Savaiano DA. Lactobacilli supplementation and short chain fatty acid production by human fecal bacteria grown in continuous culture. *Suppl. Gastroenterol.* 1995; 108(4): A293

11. Chumchalova J, Josephsen J, and Plockova M. Characterization of acidocin CH5, a saccharolytic sensitive bacteriocin of *Lactobacillus acidophilus* CH5. *Chem. Mikrobiol. Techhnol. Lebensm.* 1995; 17(5/6): 145–150.

12. Plockova M, Tomanova J, and Chumchalova J. Inhibition of mould growth and spore production by *Lactobacillus acidophilus* CH5 metabolites. *Bull. Food Res.* 1997; 36: 237–247.

13. Salminen S, Laine M, von Wright A, Vuopio-Varkila J, Korhonen T, and Mattila-Sandholm T. Development of selection criteria for probiotic strains to assess their potential in functional foods: a Nordic and European approach. *Biosci. Microflora* 1996; 15(2): 61–67.

14. Ding W, Wang H, and Griffiths MW. Probiotics down-regulate flaA sigma28 promotor in *Campylobacter jejuni*. *J. Food Prot.* 2005; 68(11): 2295–2300.

15. Laake KO, Bjorneklett A, Bakka A, Midtvedt T, Norin KE, Eide TJ, Jacobsen MB, Lingaas E, Axelsen AK, Lotveit T, and Vatn MH. Influence of fermented milk on clinical state, fecal bacterial counts and biochemical characteristics in patients with ileal-pouch-anal-anastomosis. *Microb. Ecol. Health Dis.* 1999; 11: 211–217.

16. Laake KO, Line PD, Aabakken L, Løtveit T, Bakka A, Eide J, Røseth A, Grzyb K, Bjørneklett A, and Vatn MH. Assessment of mucosal inflammation and circulation in response to probiotics in patients operated with ileal pouch anal anastomosis for ulcerative colitis. *Scand. J. Gastroenterol.* 2003; 4: 409–414.

17. Wildt S, Munck LK, Vinter-Jensen L, Hansen BF, Nordgaard-Lassen I, Christensen S, Avnstroem S, Rasmussen SN, and Rumessen J. Probiotic treatment of collagenous colitis: a randomized, double-blind, placebo-controlled trial with *L. acidophilus* and *Bifidobacterium animalis* subsp. *lactis*. *Inflamm. Bowel. Dis.* 2006; 12(5): 395–401.

18. McFarland LV. Meta-analysis of probiotics for the prevention of traveler's diarrhea. *Travel Med. Infect. Dis.* 2007; 5: 97–105.

19. Black FT, Anderson PL, Orskov J, Orskov F, Gaarslev K, and Laulund S. Prophylactic efficacy of lactobacilli on traveler's diarrhea. *Travel Med.* 1989; 333–335.

20. Sheu BS, Wu J, Lo CY, Wu HW, Chen JH, Lin YS, and Lin MD. Impact of supplement with Lactobacillus- and Bifidobacterium-containing yoghurt on triple therapy for *Helicobacter pylori* eradication. *Aliment. Pharmacol. Ther.* 2002; 16: 1669–1675.

21. Sheu BY, Cheng HC, Kao AW, Wang ST, Yang YJ, Yang HB, and Wu JJ. Pretreatment with Lactobacillus- and Bifidobacterium-containing yogurt can improve the efficacy of

quadruple therapy in eradicating residual *Helicobacter pylori* infection after failed triple therapy. *Am. J. Clin. Nutr.* 2006; 83: 864–869.

22. Wang KY, Li SN, Liu CS, Perng DS, Su YC, Wu DC, Jan CM, Lai CH, Wang T, and Wang WM. Effects of ingesting Lactobacillus- and Bifidobacterium-containing yoghurt in subjects with colonized *Helicobacter pylori*. *Am. J. Clin. Nutr.* 2004; 80: 737–741.

23. Black FT, Einarsson K, Lidbeck A, Orrhage K, and Nord CE. Effect of lactic acid producing bacteria on the human intestinal microflora during ampicillin treatment. *Scand. J. Infect. Dis.* 1991; 23: 247–254.

24. Nord CE, Lidbeck A, Orrhange K, and Sjostedt S. Oral supplementation with lactic acid bacteria during intake of clindamycin. *Clin. Microbiol. Infect.* 1997; 3(1): 124–132.

25. Alm L, Ryd-Kjellen E, Setterberg G, Blomquist L. Effect of a new fermented milk product "CULTURA" on constipation in geriatric patients. *1st Lactic Acid Bacteria Computer Conference Proceedings*. Horizon Scientific Press, Norfolk, England, 1993.

26. Virta P, Otterström K, Niemi L, Wieser-Aho MT, Lähteenmäki AL, Leppänen T. Effect of a preparation containing freeze-dried lactic acid bacteria on lactose intolerance. *External report* 1993.

27. Hove H, Nordgaard-Andersen I, and Mortensen PB. Effect of lactic acid bacteria on the intestinal production of lactate and short-chain fatty acids, and the absorption of lactose. *Am. J. Clin. Nutr.* 1994; 59: 74–79.

28. Lin MY, Savaiano D, and Harlander S. Influence of nonfermented dairy products containing bacterial starter cultures on lactose maldigestion in humans. *J. Dairy Sci.* 1991; 74: 87–95.

29. Jiang TA, Savaiano DA. Impact of Lactobacilli supplements on colonic fermentation of lactose *in vitro*. Food Science and Nutrition, University of Minnesota, 1993.

30. Jiang T and Savaiano DA. In vitro lactose fermentation by human bacteria is modified by *Lactobacillus acidophilus* supplementation. *J. Nutr.* 1997; 27: 1489–1495.

31. Hatcher G and Lambrecht R. Augmentation of macrophage phagocytic activity by cell-free extracts of selected lactic acid-producing bacteria. *J. Dairy Sci.* 1993; 76: 2485–2492.

32. Miettinen M, Vuopio-Varkila J, and Varkila K. Production of human tumor necrosis factor alpha, interleukin-6 and interleukin 10 is induced by lactic acid bacteria. *Infect. Immun.* 1996; 64(12): 5403–5405.

33. Tejada-Simon MV, Lee JH, Ustunol Z, and Pestka JJ. Ingestion of yogurt containing *Lactobacillus acidophilus* and Bifidobacterium to potentiate Immunoglobulin A responses to cholera toxin in mice. *J. Dairy Sci.* 1999; 82: 649–660.

34. Biffi A, Coradini D, Larsen R, Riva L, and DiFronzo G. Antiproliferative effect of fermented milk on the growth of a human breast cancer cell line. *Nutr. Cancer* 1997; 28(1): 93–99.

35. Ellegaard J, Peterslund NA, Black FT. Infection prophylaxis in neutropenic patients by oral administration of Lactobacilli. *The Seventh International Symposium on Infections in the Immunocompromised Host*, Boulder, CO, June 21–24, 1992

36. Obradovic D, Curic M, Ivanovic M, Trbojevic B, Djordjevic M. Probiotic function of the fermented milk Jogurt Plus. *FEMS Conference (Fifth Symposium on Lactic Acid Bacteria)*, Holland, September 8–12, 1996.

37. Cardona M.E, Vanay VV de, Midtvedt T, and Norin KE. Probiotics in gnotobiotic mice. Conversion of cholesterol to coprostanol *in vitro* and *in vivo* and bile acid deconjugation *in vitro*. *Microb. Ecol. Health Dis.* 2000; 12: 219–224.
38. El-Gawad IA, El-Sayed EM, Hafez SA, El-Zeini HM, and Saleh FA. The hypo cholesterolaemic effect of milk yoghurt and soy–yoghurt containing bifidobacteria in rats fed on a cholesterol-enriched diet. *Int. Dairy J.* 2005; 15: 37–44.
39. Klaver F, van der Meer R. The assumed assimilation of cholesterol by Lactobacilli and *Bifidobacterium bifidum* is due to their bile salt-deconjugating activity. *Appl. Environ. Microb.* 1993; 59(4): 1120–1124.
40. Alm L. Effect of fermentation on lactose, glucose, and galactose content in milk and suitability of fermented milk products for lactose intolerant individuals. *J. Dairy Sci.* 1982a; 65: 346–352.
41. Alm L. Effect of fermentation on l(+) and d(−) lactic acid in milk. *J. Dairy Sci.* 1982b; 65: 515–520.
42. Alm L, Humble E, Ryd-Kjellen E, Setterberg G. The effect of acidophilus milk in the treatment of constipation in hospitalised geriatric patients.In: Hallgren B editor. Nutrition and the Intestinal Microflora, Almqvist & Wiksell International, Stockholm, Sweden, 1983, pp. 131–138.
43. Asahara T, Takahashi M, Nomoto K, Takayama H, Onoue M, Morotomi M, Tanaka R, Yokokura T, and Yamashita N. Assessment of safety of lactobacillus strains based on resistance to host innate defense mechanisms. *Clin. Diagn. Lab. Immunol.* 2003; 10: 169–173.
44. Calland M. Clinical results of the administration of a lyophilized and antibiotic-resistant "*Lactobacillus acidophilus*" powder during antibiotic treatment of gastroenteritis in premature infants. *Sem. Hop. Ther. Paris* 1962; 38: 431–432.
45. Cannon JP, Lee TA, Bolanos JT, and Danziger LH. Pathogenic relevance of Lactobacillus: a retrospective review of over 200 cases. *Eur. J. Clin. Microbiol. Infect. Dis.* 2005; 24(1): 31–40.
46. Charteris WP, Kelly PM, Morelli L, and Collins JK. Antibiotic susceptibility of potentially probiotic Lactobacillus species. *J. Food Prot.* 1998; 61: 1636–1643.
47. Charteris WP, Kelly PM, Morelli L, and Collins JK. Gradient diffusion antibiotic susceptibility testing of potentially probiotic lactobacilli. *J. Food Prot.* 2001; 64: 2007–2014.
48. Cheplin HA and Rettger LF. Studies on the transformation of the intestinal flora, with special reference to the implantation of *Bacillus Acidophilus*: II. Feeding experiments on man. *Proc. Natl. Acad. Sci. U.S.A.* 1920; 6: 704–705.
49. DeMuynck C, Leroy AI, DeMaeseneire S, Arnaut F, Soetaert W, and Vandamme EJ. Potential of selected lactic acid bacteria to produce food compatible antifungal metabolites. *Microbiol. Res.* 2004; 159: 339–346.
50. Deraz SF, Karlsson EN, Hedstrom M, Andersson MM, and Mattiasson B. Purifucation and characterization of acidocin D20079, a bacteriocin produced by *Lactobacillus acidophilus* DSM 20079. *J. Biotechnol.* 2005; 117: 343–354.
51. Deraz SF, Karlsson EN, Khalil AA, and Mattiasson B. Mode of action of acidocin D20079, a bacteriocin produced by the potential probiotic strain, *Lactobacillus acidophilus* DSM 20079. *J. Ind. Microbiol. Biotechnol.* 2007; 34: 373–379.

52. Fondén R, and Holgersson S, Method of cultivating, in milk, organisms having a slow growth capacity, and organisms produced by the method, and milk products containing such organisms. European Patent 0154614B1, 1985.
53. Fondén R. Lactobacillus acidophilus In: Les laits fermentés. Actualité de la Recherche, John Libbey Eurotext, London, Paris, 1989; pp. 35–40.
54. Graf W. Studies on the therapeutic properties of acidophilus milk. In: Hallgren B editor. Nutrition and the Intestinal Microflora, Ed. Almqvist & Wiksell International, Stockholm, Sweden, 1983, pp. 119–121.
55. Hansen PA and Mocquot G. *Lactbacillus acidophilus* (Moro) comb. Nov. *Int. J. Syst. Bacteriol.* 1970; 20: 325–327.
56. Jernberg C, Sullivan A, Edlund C, and Jansson J. Alterations in the human intestinal microflora and detection of probiotic strains by use of terminal restriction fragment length polymorphism. *Appl. Environ. Microbiol.* 2005; 71: 501–506.
57. Lidbeck A, Gustafsson JA, and Nord CE. Impact of L*actobacillus acidophilus* supplements on human oropharyngeal and intestinal microflora. *Scand. J. Infect. Dis.* 1987; 19: 531–537.
58. Lidbeck A, Edlund C, Gustafsson JA, Kager L, and Nord CE. Impact of *Lactobacillus acidophilus* supplements administration on the intestinal microflora after clindamycin treatment. *J. Chemother.* 1989; (4 Suppl.): 630–632.
59. Lidbeck A, Geltner Allinger U, Orrhage K, Ottova L, Brismar B, Gustafsson JA, Rafter J, and Nord CE. Impact of *Lactobacillus acidophilus* supplements on the faecal microflora and soluble faecal bile acids in colon cancer patients. *Microb. Ecol. Health Dis.* 1991; 4: 81–88.
60. Lidbeck A, Overvik J, Rafter J, and Nord CE. Effect of *Lactobacillus acidophilus* supplements on mutagen excretion in faeces and urine in humans. *Microb. Ecol. Health Dis.* 1992; 5: 59–67.
61. Lin MY and Chang FJ. Antioxidative effect of intestinal bacteria *Bifidobacterium longum* ATCC 15708 and *Lactobacillus acidophilus* ATCC 4356. *Dig. Dis. Sci.* 2000; 45: 1617–1622.
62. Matsamura A, Saito T, Arakuni M, Kitazawa H, Kawai Y, and Itoh T. New binding assay and preparative trial of cell-surface lectin from *Lactobacillus acidophilus* group lactic acid bacteria. *J. Dairy Sci.* 1999; 82: 2525–2529.
63. Moro E. Über den *Bacillus acidophilus* n.sp. *Jahrb. Kinderheilk.* 1900; 52: 38–55.
64. Mättö J, Fondén R, Tolvanen T, von Wright A, Vilppon en-Salmela T, Satokari R, and Saarela M. Intestinal survival and persistence of probiotic Lactobacillus and Bifidobacterium strains administered in triple-strain yoghurt. *Int. Dairy J.* 2006; 16: 1174–1180.
65. Orrhage K, Brismar B, and Nord CE. Effect of supplements with *Bifidobacterium longum* and *Lactobacillus acidophilus* on the intestinal microbiota during administration of clindamycin. *Microb. Ecol. Health Dis.* 1994a; 7: 17–25.
66. Orrhage K, Sillerstrom E, Gustafsson JA, Nord CE, and Rafter J. Binding of mutagenic heterocyclic amines by intestinal and lactic acid bacteria. *Mutat. Res.* 1994b; 311: 239–248.
67. Orrhage K, Sjostedt S, CE Nord. Effect of supplements with lactic acid bacteria and oligofructose on the intestinal microflora during administration of cefpodoxime proxetil. *J. Antimicrob. Chemother.* 2000; 46: 603–612.

68. Orrhage K, Annas A, Nord CE, Brittebo EB, and Rafter J. Effects of lactic acid bacteria on the uptake and distribution of the food mutagen Trp-P-2 in mice. *Scand. J. Gastroenterol.* 2002; 37: 215–222.
69. Otto W. Purified acidophilus culture in treatment of infected superficial wounds. *Dtsch. Med. Wochenschr.* 1952; 77: 1511–1513.
70. Pettersson L, Graf W, Alm L, Lindwall S, Strömberg A. Survival of *Lactobacillus acidophilus* NCDO 1748 in the human gastrointestinal tract 1. Incubation in gastric juice. In: Hallgren B, editor. *Nutrition and the Intestinal Microflora*, Almqvist & Wiksell International, Stockholm, Sweden, 1983a, pp. 123–126.
71. Pettersson L, Graf W, Sewelin U. Survival of *Lactobacillus acidophilus* NCDO 1748 in the human gastrointestinal tract 2. Ability to pass the stomach and intestine *in vivo*. In: Hallgren B, editor. *Nutrition and the Intestinal Microflora*, Almqvist & Wiksell International, Stockholm, Sweden, 1983b, pp. 127–13.
72. Resta-Lenert S, and Barett KE. Live probiotics protect intestinal epithelial cells from the effects of infection with enteroinvasive *Escherichia coli* (EIEC). *Gut* 2003; 52: 988–997.
73. Resta-Lenert S and Barett KE. Probiotics and commensals reverse TNF-alpha- and IFN-gamma-induced dysfunction in human intestinal epithelial cells. *Gastroenterology* 2006; 130: 731–746.
74. Robinson EL and Thompson WL. Effect on weight gain of the addition of *Lactobacillus acidophilus* to the formula of newborn infants. *J. Pediatr.* 1952; 41: 395–398.
75. Sarra PG and Dellaglio F. Colonization of a human intestine by four different genotypes of *Lactobacillus acidophilus*. *Microbiologica* 1984; 7: 331–339.
76. Sullivan A, Barkholt L, and Nord CE. *Lactobacillus acidophilus*. *Bifidobacterium lactis* and *Lactobacillus* F19 prevent antibiotic-associated ecological disturbances of *Bacteroides fragilis* in the intestine. *J. Antimicrob. Chemother.* 2003; 52: 308–311.
77. Sullivan A and Nord CE. Probiotic lactobacilli and bacteraemia in Stockholm. *Scand. J. Infect. Dis.* 2006; 38: 327–331.
78. Tahara T, Oshimura M, and Umezawa K. Isolation, partial characterization, and mode of action of Acidocin J1132, a two-component bacteriocin produced by *Lactobacillus acidophilus* JCM 1132. *Appl. Environ. Microbiol.* 1966; 62: 892–897.
79. Wenstein L, Weiss JE, and Gillespie RW. The influence of diet on the *L. acidophilus* content and H-ion concentration of the intestine. *J. Bacteriol.* 1938; 35: 515–525.
80. Virtanen T, Pihlanto A, Akkanen S, and Korhonen H. Development of antioxidant activity in milk whey during fermentation with lactic acid bacteria. *J. Appl. Microbiol.* 2007; 102: 106–115.
81. Zhang XB and Otha Y. *In vitro* binding of mutagenic pyrolyzates to lactic acid bacterial cells in human gastric juice. *J. Dairy Sci.* 1991; 74: 752–757.
82. Altermann E, Russell WM, Azcarate-Peril MA, Barrangou R, Buck BL, McAuliffe O, et al. Complete genome sequence of the probiotic lactic acid bacterium *Lactobacillus acidophilus* NCFM. *Proc. Natl. Acad. Sci. U.S.A.* 2005a; 102(11): 3906–3912.
83. Azcarate-Peril MA, Altermann E, Hoover-Fitzula RL, Cano RJ, and Klaenhammer TR. Identification and inactivation of genetic loci involved with *Lactobacillus acidophilus* acid tolerance. *Appl. Environ. Microbiol.* 2004; 70(9): 5315–5322.
84. Barefoot SF and Klaenhammer TR. Detection and activity of Lactacin B, a bacteriocin produced by *Lactobacillus acidophilus*. *Appl. Environ. Microbiol.* 1983; 45(6): 1808–1815.

85. Barefoot SF and Klaenhammer TR. Purification and characterization of the *Lactobacillus acidophilus* bacteriocin Lactacin B. *Antimicrob. Agents Chemother.* 1984; 26(3): 328–334.
86. Barrangou R, Altermann E, Hutkins R, Cano R, and Klaenhammer TR. Functional and comparative genomic analyses of an operon involved in fructooligosaccharide utilization by *Lactobacillus acidophilus*. *Proc. Natl. Acad. Sci.* 2003; 100(15): 8957–8962.
87. Collado MC, Meriluoto J, and Salminen S. Role of commercial probiotic strains against human pathogen adhesion to intestinal mucus. *Lett. Appl. Microbiol.* 2007; 45(4): 454–460.
88. Collado MC, Meriluoto J, and Salminen S. Adhesion and aggregation properties of probiotic and pathogen strains. *Eur. Food Res. Technol.* 2008; 226(5): 1065–1073.
89. Daniel C, Poiret S, Goudercourt D, Dennin V, Leyer G, and Pot B. Selecting lactic acid bacteria for their safety and functionality by use of a mouse colitis model. *Appl. Environ. Microbiol.* 2006; 72(9): 5799–5805.
90. Dobson AE, Sanozky-Dawes RB, and Klaenhammer TR. Identification of an operon and inducing peptide involved in the production of lactacin B by *Lactobacillus acidophilus*. *J. Appl. Microbiol.* 2007; 103: 1766–1778.
91. Dunn SR, Simenhoff ML, Ahmed KE, Gaughan WJ, Eltayeb BO, Fitzpatrick M-ED, et al. Effect of oral administration of freeze-dried *Lactobacillus acidophilus* on small bowel bacterial overgrowth in patients with end stage kidney disease. Reducing uremic toxins and improving nutrition. *Int. Dairy J.* 1998; 8: 545–553.
92. Engelbrektson AL, Korzenik JR, Sanders ME, Clement BG, Leyer G, Klaenhammer TR, and Kitts CL. Analysis of treatment effects on the microbiological ecology of the human intestine. *FEMS Microb. Ecol.* 2006; 57: 239–250.
93. Goldin BR and Gorbach SL. The effect of milk and *Lactobacillus* feeding on human intestinal bacterial enzyme activity. *Am. J. Clin. Nutr.* 1984; 39: 756–761.
94. Goldin BR and Gorbach SL. The effect of oral administration on *Lactobacillus* and antibiotics on intestinal bacterial activity and chemical induction of large bowel tumors. *Dev. Indust. Microbiol.* 1984; 25: 139–150.
95. Goldin BR, Swenson L, Dwyer J, Sexton M, and Gorbach SL. Effect of diet and *Lactobacillus acidophilus* supplements on human fecal bacterial enzymes. *J. Natl. Cancer Inst.* 1980; 64(2): 255–261.
96. Greene JD and Klaenhammer TR. Factors involved in adherence of lactobacilli to human Caco-2 cells. *Appl. Environ. Microbiol.* 1994; 60(12): 4487–4494.
97. Haukioja A, Yli-Knuuttila H, Loimaranta V, Kari K, Ouwehand AC, Meurman JH, and Tenovuo J. Oral adhesion and survival of probiotic and other lactobacilli and bifidobacteria *in vitro*. *Oral. Microbiol. Immunol.* 2006; 21: 326–332.
98. Heiser CR, Ernst JA, Barrett JT, French N, Schultz M, and Dube MB. Probiotics, soluble fiber, and L-glutamine (GLN) reduce nelfinavir (NFV)- or lopinavir/ritonavir (LPV/r)-related diarrhea. *J. Int. Assoc. Physicians AIDS Care* 2004; 3(4): 121–129.
99. Kleeman EG and Klaenhammer TR. Adherence of *Lactobacillus* Species to human fetal intestinal cells. *J. Dairy Sci.* 1982; 65(11): 2063–2069.
100. Montes RG, Bayless TM, Saavedra JM, and Perman JA. Effect of milks inoculated with *Lactobacillus acidophilus* or a yogurt starter culture in lactose-maldigesting children. *J. Dairy Sci.* 1995; 78: 1657–1664.
101. Ouwehand AC, Carcano D, Li S, and Leyer G. Probiotics reduce incidence and duration of respiratory tract infection symptoms in 3–5-year-old children. *Cibus* 2007; 3(2): 60

102. Ouwehand AC, Tiihonen K, Saarinen M, Putaala H, and Rautonen N. Influence of a combination of *Lactobacillus acidophilus* NCFM and lactitol on healthy elderly: intestinal and immune parameters. *Br. J. Nutr.* 2008; .

103. Paineau D, Carcano D, Leyer G, Darquy S, Alyanakian MA, Simoneau G, Bergmann JF, Brassart D, Bornet F, and Ouwehand AC. Effects of seven potential probiotic strains on specific immune responses in healthy adults: a double-blind, randomized, controlled trial. *FEMS Immunol. Med. Microbiol.* 2008; 53(1): 107–113.

104. Rousseaux C, Thuru X, Gelot A, Barnich N, Neut C, Dubuquoy L, et al. *Lactobacillus acidophilus* modulates intestinal pain and induces opioid and cannabinoid receptors. *Nat. Med.* 2007; 13(1): 35–37.

105. Sanders ME and Klaenhammer TR. The scientific basis of *Lactobacillus acidphilus* NCFM functionality as a probiotic. *J. Dairy Sci.* 2001; 84: 319–331.

106. Sanders ME, Walker DC, Walker KM, Aoyama K, and Klaenhammer TR. Performance of commercial cultures in fluid milk applications. *J. Dairy Sci.* 1996; 79: 943–955.

107. Simenhoff ML, Dunn SR, Zollner GP, Fitzpatrick ME, Emery SM, Sandine WE, et al. Biomodulation of the toxic and nutritional effects of small bowel bacterial overgrowth in end-stage kidney disease using freeze-dried *Lactobacillus acidophilus*. *Miner. Electrolyte Metab.* 1996; 22(1–3): 92–96.

108. Sui J, Leighton S, Busta F, and Brady L. 16S ribosomal DNA analysis of the faecal lactobacilli composition of human subjects consuming a probiotic strain *Lactobacillus acidophilus* NCFM®. *J. Appl. Microbiol.* 2002; 93: 907–912.

109. Varcoe J, Zook C, Sui J, Leighton S, Busta F, and Brady L. Variable response to exogenous *Lactobacillus acidophilus* NCFM® consumed in different delivery vehicles. *J. Appl. Microbiol.* 2002; 93: 900–906.

110. Varcoe JJ, Krejcarek G, Busta F, and Brady L. Prophylactic feeding of *Lactobacillus acidophilus* NCFM to mice attenuates overt colonic hyperplasia. *J. Food Prot.* 2003; 66(3): 457–465.

111. Wagner RD, Pierson C, Warner T, Dohnalek M, Farmer J, Roberts L, et al. Biotherapeutic effects of probiotic bacteria on candidiasis in immunodeficient mice. *Infect. Immun.* 1997; 65(10): 4165–4172.

112. Wagner RD, Warner T, Roberts L, Farmer J, Dohnalek M, Hilty M, and Balish E. Variable biotherapeutic effects of *Lactobacillus acidophilus* isolates on orogastric and systemic candidiasis in immunodeficient mice. *Rev. Iberoam Micol.* 1998; 15: 271–276.

113. Wagner RD, Dohnalek M, Hilty M, Vazquez-Torres A, and Balish E. Effects of probiotic bacteria on humoral immunity to *Candida albicans* in immunodeficient bg/bg−nu/nu and bg/bg−nu/+ mice. *Rev. Iberoam Micol.* 2000; 17: 55–59.

114. Asahara T, Nomoto K, Watanuki M, and Yokokura T. Antimicrobial activity of intra-urethrally administered probiotic *Lactobacillus casei* in a murine model of *Escherichia coli* urinary tract infection. *Antimicrob. Agents Chemother.* 2001; 45: 1751–1760.

115. Asahara T, Takahashi M, Nomoto K, Takayama H, Onoue M, Morotomi M, Tanaka R, Yokokura T, and Yamashita N. Assessment of safety of lactobacillus strains based on resistance to host innate defense mechanisms. *Clin. Diagn. Lab. Immunol.* 2003; 10: 169–173.

116. Aso Y, Akaza H, Kotake T, Tsukamoto T, Imai K, and Naito S. Preventive effect of a *Lactobacillus casei* preparation on the recurrence of superficial bladder cancer in a double-blind trial The BLP Study Group. *Eur. Urol.* 1995; 27: 104–109.

117. Baken KA, Ezendam J, Gremmer ER, de Klerk A, Pennings JL, Matthee B, Peijnenburg AA, van Loveren H. Evaluation of immunomodulation by *Lactobacillus casei* Shirota: immune function, autoimmunity and gene expression. *Int. J. Food Microbiol.* 2006; 112: 8–18.
118. Busscher HJ, Free RH, Van Weissenbruch R, Albers FW, Van Der Mei HC. Preliminary observations on influence of dairy products on biofilm removal from silicone rubber voice prostheses *in vitro*. *J. Dairy Sci.* 2000; 83: 641–647.
119. Candy DC, Densham L, Lamont LS, Greig M, Lewis J, Bennett H, and Griffiths M. Effect of administration of *Lactobacillus casei* shirota on sodium balance in an infant with short bowel syndrome. *J. Pediatr. Gastroenterol. Nutr.* 2001; 32: 506–508.
120. Cats A, Kuipers EJ, Bosschaert MA, Pot RG, Vandenbroucke-Grauls CM, and Kusters JG. Effect of frequent consumption of a *Lactobacillus casei*-containing milk drink in *Helicobacter pylori*-colonized subjects. *Aliment. Pharmacol. Ther.* 2003; 17: 429–435.
121. Chonan O, Makino K, Ishikawa H, Iwabuti A, and Watanuki M. Effects of cell wall components from *Lactobacillus casei* on serum chorestrol levels in chorestrol-fed rats. *Biosci. Microflora* 1997; 16: 19–21.
122. Cross ML, Ganner A, Teilab D, and Fray LM. Patterns of cytokine induction by gram-positive and gram-negative probiotic bacteria. *FEMS Immunol. Med. Microbiol.* 2004; 42: 173–180.
123. De Preter V, Geboes K, Verbrugghe K, De Vuyst L, Vanhoutte T, Huys G, Swings J, Pot B, and Verbeke K. The *in vivo* use of the stable isotope-labelled biomarkers lactose-[15N] ureide and [2H4]tyrosine to assess the effects of pro- and prebiotics on the intestinal flora of healthy human volunteers. *Br. J. Nutr.* 2004; 92: 439–446.
124. De Preter V, Raemen H, Cloetens L, Houben E, Rutgeerts P, and Verbeke K. Effect of dietary intervention with different pre- and probiotics on intestinal bacterial enzyme activities. *Eur. J. Clin. Nutr.* 2008; 62: 225–231.
125. De Preter V, Vanhoutte T, Huys G, Swings J, De Vuyst L, Rutgeerts P, and Verbeke K. Effects of *Lactobacillus casei* Shirota, *Bifidobacterium breve*, and oligofructose-enriched inulin on colonic nitrogen-protein metabolism in healthy humans. *Am. J. Physiol. Gastrointest. Liver Physiol.* 2007; 292: G358–G368
126. de Waard R, Claassen E, Bokken GC, Buiting B, Garssen J, and Vos JG. Enhanced immunological memory responses to *Listeria monocytogenes* in rodents, as measured by delayed-type hypersensitivity (DTH), adoptive transfer of DTH, and protective immunity, following *Lactobacillus casei* Shirota ingestion. *Clin. Diagn. Lab. Immunol.* 2003; 10: 59–65.
127. de Waard R, Garssen J, Bokken GC, and Vos JG. Antagonistic activity of *Lactobacillus casei* strain shirota against gastrointestinal *Listeria monocytogenes* infection in rats. *Int. J. Food Microbiol.* 2002; 73: 93–100.
128. de Waard R, Garssen J, Snel J, Bokken GC, Sako T, Veld JH, and Vos JG. Enhanced antigen-specific delayed-type hypersensitivity and immunoglobulin G2b responses after oral administration of viable *Lactobacillus casei* YIT9029 in Wistar and Brown Norway rats. *Clin. Diagn. Lab. Immunol.* 2001; 8: 762–767.
129. Ezendam J, van Loveren H. *Lactobacillus casei* Shirota administered during lactation increases the duration of autoimmunity in rats and enhances lung inflammation in mice. *Br. J. Nutr.* 2008; 99: 83–90.

130. Fayol-Messaoud D, Berger CN, Coconnier-Polter MH, Liévin-Le Moal V, and Servin AL. pH-, Lactic acid-, and non-lactic acid-dependent activities of probiotic Lactobacilli against *Salmonella enterica* Serovar *typhimurium*. *Appl. Environ. Microbiol.* 2005; 71: 6008–6013.

131. Furushiro M, Hashimoto S, Hamura M, and Yokokura T. Mechanism for the antihypertensive effect of a polysaccharide-glycopeptide complex from *Lactobacillus casei* in spontaneously hypertensive rats (SHR). *Biosci. Biotechnol. Biochem.* 1993; 57: 978–981.

132. Furushiro M, Sawada H, Hirai K, Motoike M, Sansawa H, Kobayashi S, Watanuki M, and Yokokura T. Blood pressure-lowering effect of extract from *Lactobadllus casei* in spontaneously hypertensive rats (SHR). *Agric. Biol. Chem.* 1990; 54: 2193–2198.

133. Hayatsu H and Hayatsu T. Suppressing effect of *Lactobacillus casei* administration on the urinary mutagenicity arising from ingestion of fried ground beef in the human. *Cancer Lett.* 1993; 73: 173–179.

134. Herías MV, Koninkx JF, Vos JG, Huis in't Veld JH, and van Dijk JE. Probiotic effects of *Lactobacillus casei* on DSS-induced ulcerative colitis in mice. *Int. J. Food Microbiol.* 2005; 103: 143–155.

135. Hori T, Kiyoshima J, Shida K, and Yasui H. Effect of intranasal administration of *Lactobacillus casei* Shirota on influenza virus infection of upper respiratory tract in mice. *Clin. Diagn. Lab. Immunol.* 2001; 8: 593–597.

136. Hori T, Kiyoshima J, Shida K, and Yasui H. Augmentation of cellular immunity and reduction of influenza virus titer in aged mice fed *Lactobacillus casei* strain Shirota. *Clin. Diagn. Lab. Immunol.* 2002; 9: 105–108.

137. Hori T, Kiyoshima J, and Yasui H. Effect of an oral administration of *Lactobacillus casei* strain Shirota on the natural killer activity of blood mononuclear cells in aged mice. *Biosci. Biotechnol. Biochem.* 2003; 67: 420–422.

138. Howard JC, Heinemann C, Thatcher BJ, Martin B, Gan BS, and Reid G. Identification of collagen-binding proteins in *Lactobacillus* spp. with surface-enhanced laser desorption/ionization-time of flight ProteinChip technology. *Appl. Environ. Microbiol.* 2000; 66: 4396–4400.

139. Juntunen M, Kirjavainen PV, Ouwehand AC, Salminen SJ, and Isolauri E. Adherence of probiotic bacteria to human intestinal mucus in healthy infants and during rotavirus infection. *Clin. Diagn. Lab. Immunol.* 2001; 8: 293–296.

140. Kanamori Y, Hashizume K, Kitano Y, Tanaka Y, Morotomi M, Yuki N, and Tanaka R. Anaerobic dominant flora was reconstructed by synbiotics in an infant with MRSA enteritis. *Pediatr. Int.* 2003; 45: 359–362.

141. Kanamori Y, Hashizume K, Sugiyama M, Morotomi M, and Yuki N. Combination therapy with *Bifidobacterium breve*, *Lactobacillus casei*, and galactooligosaccharides dramatically improved the intestinal function in a girl with short bowel syndrome: a novel synbiotics therapy for intestinal failure. *Dig. Dis. Sci.* 2001; 46: 2010–2016.

142. Kanamori Y, Hashizume K, Sugiyama M, Mortomi M, Yuki N, and Tanaka R. A novel synbiotic therapy dramatically improved the intestinal function of a pediatric patient with laryngotracheo-esophageal cleft (LTEC) in the intensive care unit. *Clin. Nutr.* 2002; 21: 527–530.

143. Kanamori Y, Sugiyama M, Hashizume K, Yuki N, Morotomi M, and Tanaka R. Experience of long-term synbiotic therapy in seven short bowel patients with refractory enterocolitis. *J. Pediatr. Surg.* 2004; 39: 1686–1692.

144. Kanazawa H, Nagino M, Kamiya S, Komatsu S, Mayumi T, Takagi K, Asahara T, Nomoto K, Tanaka R, and Nimura Y. Synbiotics reduce postoperative infectious complications: a randomized controlled trial in biliary cancer patients undergoing hepatectomy. *Langenbecks Arch. Surg.* 2005; 390: 104–113.

145. Kankaanpää PE, Salminen SJ, Isolauri E, and Lee YK. The influence of polyunsaturated fatty acids on probiotic growth and adhesion. *FEMS Microbiol. Lett.* 2001; 194: 149–153.

146. Kato I, Endo-Tanaka K, and Yokokura T. Suppressive effects of the oral administration of *Lactobacillus casei* on type II collagen-induced arthritis in DBA/1 mice. *Life Sci.* 1998; 63: 635–644.

147. Kato I, Tanaka K, and Yokokura T. Lactic acid bacterium potently induces the production of interleukin-12 and interferon-gamma by mouse splenocytes. *Int. J. Immunopharmacol.* 1999; 21: 121–131.

148. Kato R, Yuasa H, Inoue K, Iwao T, Tanaka K, Ooi K, and Hayashi Y. Effect of *Lactobacillus casei* on the absorption of nifedipine from rat small intestine. *Drug. Metab. Pharmacokinet.* 2007; 22: 96–102.

149. Kikuchi-Hayakawa H, Shibahara-Sone H, Osada K, Onodera-Masuoka N, Ishikawa F, and Watanuki M. Lower plasma triglyceride level in Syrian hamsters fed on skim milk fermented with *Lactobacillus casei* strain Shirota. *Biosci. Biotechnol. Biochem.* 2000; 64: 466–475.

150. Koebnick C, Wagner I, Leitzmann P, Stern U, and Zunft HJ. Probiotic beverage containing *Lactobacillus casei* Shirota improves gastrointestinal symptoms in patients with chronic constipation. *Can. J. Gastroenterol.* 2003; 17: 655–659.

151. Lee YK, Ho PS, Low CS, Arvilommi H, and Salminen S. Permanent colonization by *Lactobacillus casei* is hindered by the low rate of cell division in mouse gut. *Appl. Environ. Microbiol.* 2004; 70: 670–674.

152. Lee YK, Lim CY, Teng WL, Ouwehand AC, Tuomola EM, and Salminen S. Quantitative approach in the study of adhesion of lactic acid bacteria to intestinal cells and their competition with enterobacteria. *Appl. Environ. Microbiol.* 2000; 66: 3692–3697.

153. Lee YK and Puong KY. Competition for adhesion between probiotics and human gastrointestinal pathogens in the presence of carbohydrate. *Br. J. Nutr.* 2002; 88(Suppl. 1): S101–S108.

154. Lee YK, Puong KY, Ouwehand AC, and Salminen S. Displacement of bacterial pathogens from mucus and Caco-2 cell surface by lactobacilli. *J. Med. Microbiol.* 2003; 52: 925–930.

155. Makras L, Triantafyllou V, Fayol-Messaoudi D, Adriany T, Zoumpopoulou G, Tsakalidou E, Servin A, De Vuyst L. Kinetic analysis of the antibacterial activity of probiotic lactobacilli towards *Salmonella enterica* serovar Typhimurium reveals a role for lactic acid and other inhibitory compounds. *Res. Microbiol.* 2006; 157: 241–247.

156. Matsuguchi T, Takagi A, Matsuzaki T, Nagaoka M, Ishikawa K, Yokokura T, and Yoshikai Y. Lipoteichoic acids from *Lactobacillus* strains elicit strong tumor necrosis factor alpha-inducing activities in macrophages through Toll-like receptor 2. *Clin. Diagn. Lab. Immunol.* 2003; 10: 259–266.

157. Matsumoto S, Hara T, Hori T, Mitsuyama K, Nagaoka M, Tomiyasu N, Suzuki A, and Sata M. Probiotic *Lactobacillus*-induced improvement in murine chronic inflammatory bowel disease is associated with the down-regulation of pro-inflammatory cytokines in lamina propria mononuclear cells. *Clin. Exp. Immunol.* 2005; 140: 417–426.

158. Matsuzaki T, Hashimoto S, and Yokokura T. Effects on antitumor activity and cytokine production in the thoracic cavity by intrapleural administration of *Lactobacillus casei* in tumor-bearing mice. *Med. Microbiol. Immunol.* 1996; 185: 157–161.
159. Matsuzaki T, Nagata Y, Kado S, Uchida K, Hashimoto S, Yokokura T, *Effect of oral administration of Lactobacillus casei on alloxan-induced diabetes in mice. APMIS* 1997; 105: 637–642.
160. Matsuzaki T, Nagata Y, Kado S, Uchida K, Kato I, Hashimoto S, and Yokokura T. Prevention of onset in an insulin-dependent diabetes mellitus model, NOD mice, by oral feeding of *Lactobacillus casei*. *APMIS* 1997; 105: 643–649.
161. Matsuzaki T, Saito M, Usuku K, Nose H, Izumo S, Arimura K, and Osame M. A prospective uncontrolled trial of fermented milk drink containing viable *Lactobacillus casei* strain Shirota in the treatment of HTLV-1 associated myelopathy/tropical spastic paraparesis. *J. Neurol. Sci.* 2005; 237: 75–81.
162. Matsuzaki T, Yamazaki R, Hashimoto S, and Yokokura T. Antidiabetic effects of an oral administration of *Lactobacillus casei* in a non-insulin-dependent diabetes mellitus (NIDDM) model using KK-Ay mice. *Endocr. J.* 1997; 44: 357–365.
163. Matsuzaki T, Yamazaki R, Hashimoto S, and Yokokura T. The effect of oral feeding of *Lactobacillus casei* strain Shirota on immunoglobulin E production in mice. *J. Dairy Sci.* 1998; 81: 48–53.
164. Mike A, Nagaoka N, Tagami Y, Miyashita M, Shimada S, Uchida K, Nanno M, and Ohwaki M. Prevention of B220+ T cell expansion and prolongation of lifespan induced by *Lactobacillus casei* in MRL/lpr mice. *Clin. Exp. Immunol.* 1999; 117: 368–375.
165. Nagao F, Nakayama M, Muto T, and Okumura K. Effects of a fermented milk drink containing *Lactobacillus casei* strain Shirota on the immune system in healthy human subjects. *Biosci. Biotechnol. Biochem.* 2000; 64: 2706–2708.
166. Nakajima K, Hata Y, Osono Y, Hamura M, Kobayashi S, and Watanuki M. Antihypertensive effect of extracts *of Lactobacillus casei* in patients with hypertension. *J. Clin. Biochem. Nutr.* 1995; 18: 181–187.
167. Nishino T, Shibahara-Sone H, Kikuchi-Hayakawa H, and Ishikawa F. Transit of radical scavenging activity of milk products prepared by Maillard reaction and *Lactobacillus casei* strain Shirota fermentation through the hamster intestine. *J. Dairy Sci.* 2000; 83: 915–922.
168. Ogawa M, Shimizu K, Nomoto K, Takahashi M, Watanuki M, Tanaka R, Tanaka T, Hamabata T, Yamasaki S, and Takeda Y. Protective effect of *Lactobacillus casei* strain Shirota on Shiga toxin-producing *Escherichia coli* O157:H7 infection in infant rabbits. *Infect. Immun.* 2001; 69: 1101–1108.
169. Ogawa M, Shimizu K, Nomoto K, Tanaka R, Hamabata T, Yamasak S, Takeda T, and Takeda Y. Inhibition of *in vitro* growth of Shiga toxin-producing *Escherichia coli* O157:H7 by probiotic *Lactobacillus* strains due to production of lactic acid. *Int. J. Food Microbiol.* 2001; 68: 135–140.
170. Ohashi Y, Inoue R, Tanaka K, Matsuki T, Umesaki Y, and Ushida K. *Lactobacillus casei* strain Shirota-fermented milk stimulates indigenous *Lactobacilli* in the pig intestine. *J. Nutr. Sci. Vitaminol. (Tokyo)* 2001; 47: 172–176.
171. Ohashi Y, Inoue R, Tanaka K, Umesaki Y, and Ushida K. Strain gauge force transducer and its application in a pig model to evaluate the effect of probiotic on colonic motility. *J. Nutr. Sci. Vitaminol. (Tokyo)* 2001; 47: 351–356.

172. Ohashi Y, Nakai S, Tsukamoto T, Masumori N, Akaza H, Miyanaga N, Kitamura T, Kawabe K, Kotake T, Kuroda M, Naito S, Koga H, Saito Y, Nomata K, Kitagawa M, and Aso Y. Habitual intake of lactic acid bacteria and risk reduction of bladder cancer. *Urol. Int.* 2002; 68: 273–280.

173. Ohashi Y, Tokunaga M, and Ushida K. The effect of *Lactobacillus casei* strain Shirota on the cecal fermentation pattern depends on the individual cecal microflora in pigs. *J. Nutr. Sci. Vitaminol. (Tokyo)* 2004; 50: 399–403.

174. Ohashi Y, Umesaki Y, and Ushida K. Transition of the probiotic bacteria, *Lactobacillus casei* strain Shirota, in the gastrointestinal tract of a pig. *Int. J. Food Microbiol.* 2004; 96: 61–66.

175. Ouwehand AC, Tölkkö S, Kulmala J, Salminen S, and Salminen E. Adhesion of inactivated probiotic strains to intestinal mucus. *Lett. Appl. Microbiol.* 2000; 31: 82–86.

176. Sahagun-Flores JE, Lopez-Pena LS, de la Cruz-Ramirez Jaimes J, Garcia-Bravo MS, Peregrina-Gomez R, and de Alba-Garcia JE. Eradication of *Helicobacter pylori*: triple treatment scheme plus *Lactobacillus* vs. triple treatment alone. *Cir. Cir.* 2007; 74: 333–336.

177. Sawada H, Furushiro M, Hirai K, Motoike M, Watanabe T, and Yokokura T. Purification and characterization of an antihypertensive compound from *Lactobacillus casei*. *Agric. Biol. Chem.* 1990; 54: 3211–3219.

178. Seow SW, Rahmat JN, Mohamed AA, Mahendran R, Lee YK, and Bay BH. *Lactobacillus* species is more cytotoxic to human bladder cancer cells than Mycobacterium Bovis (bacillus Calmette-Guerin). *J. Urol.* 2002; 168: 2236–2239.

179. Sgouras D, Maragkoudaki P, Petraki K, Martinez-Gonzalez B, Eriotou E, Michopoulos S, Kalantzopoulos G, Tsakalidou E, and Mentis A. *In vitro* and *in vivo* inhibition of *Helicobacter pylori* by *Lactobacillus casei* strain Shirota. *Appl. Environ. Microbiol.* 2004; 70: 518–526.

180. Shida K, Kiyoshima-Shibata J, Nagaoka M, Watanabe K, and Nanno M. Induction of interleukin-12 by *lactobacillus* strains having a rigid cell wall resistant to intracellular digestion. *J. Dairy Sci.* 2006; 89: 3306–3317.

181. Shida K, Makino K, Morishita A, Takamizawa K, Hachimura S, Ametani A, Sato T, Kumagai Y, Habu S, and Kaminogawa S. *Lactobacillus casei* inhibits antigen-induced IgE secretion through regulation of cytokine production in murine splenocyte cultures. *Int. Arch. Allergy Immunol.* 1998; 115: 278–287.

182. Shida K, Suzuki T, Kiyoshima-Shibata J, Shimada S, and Nanno M. Essential roles of monocytes in stimulating human peripheral blood mononuclear cells with *Lactobacillus casei* to produce cytokines and augment natural killer cell activity. *Clin. Vaccine Immunol.* 2006; 13: 997–1003.

183. Shida K, Takahashi R, Iwadate E, Takamizawa K, Yasui H, Sato T, Habu S, Hachimura S, and Kaminogawa S. *Lactobacillus casei* strain Shirota suppresses serum immunoglobulin E and immunoglobulin G1 responses and systemic anaphylaxis in a food allergy model. *Clin. Exp. Allergy* 2002; 32: 563–570.

184. Shima T, Fukushima K, Setoyama H, Imaoka A, Matsumoto S, Hara T, Suda K, and Umesaki Y. Differential effects of two probiotic strains with different bacteriological properties on intestinal gene expression, with special reference to indigenous bacteria. *FEMS Immunol. Med. Microbiol.* 2008; 52: 69–77.

185. Spanhaak S, Havenaar R, and Schaafsma G. The effect of consumption of milk fermented by *Lactobacillus casei* strain Shirota on the intestinal microflora and immune parameters in humans. *Eur. J. Clin. Nutr.* 1998; 52: 899–907.
186. Srinivasan R, Meyer R, Padmanabhan R, and Britto J. Clinical safety of *Lactobacillus casei* shirota as a probiotic in critically ill children. *J. Pediatr. Gastroenterol. Nutr.* 2006; 42: 171–173.
187. Styriak I, Nemcová R, Chang YH, and Ljungh A. Binding of extracellular matrix molecules by probiotic bacteria. *Lett. Appl. Microbiol.* 2003; 37: 329–333.
188. Takagi A, Matsuzaki T, Sato M, Nomoto K, Morotom M, and Yokokura T. Inhibitory effect of oral administration of *Lactobacillus casei* on 3-methylcholanthrene-induced carcinogenesis in mice. *Med. Microbiol. Immunol.* 1999; 188: 111–116.
189. Takagi A, Matsuzaki T, Sato M, Nomoto K, Morotomi M, and Yokokura T. Enhancement of natural killer cytotoxicity delayed murine carcinogenesis by a probiotic microorganism. *Carcinogenesis* 2001; 22: 599–605.
190. Takahashi T, Kushiro A, Nomoto K, Uchida K, Morotomi M, Yokokura T, and Akaza H. Antitumor effects of the intravesical instillation of heat killed cells of the *Lactobacillus casei* strain Shirota on the murine orthotopic bladder tumor MBT-2. *J. Urol.* 2001; 166: 2506–2511.
191. Takeda K and Okumura K. Effects of a fermented milk drink containing *Lactobacillus casei* strain Shirota on the human NK-cell activity. *J. Nutr.* 2007; 137: 791S–793S.
192. Takeda K, Suzuki T, Shimada SI, Shida K, Nanno M, and Okumura K. Interleukin-12 is involved in the enhancement of human natural killer cell activity by *Lactobacillus casei* Shirota. *Clin. Exp. Immunol.* 2006; 146: 109–115.
193. Tamura M, Shikina T, Morihana T, Hayama M, Kajimoto O, Sakamoto A, Kajimoto Y, Watanabe O, Nonaka C, Shida K, and Nanno M. Effects of probiotics on allergic rhinitis induced by Japanese cedar pollen: randomized double-blind, placebo-controlled clinical trial. *Int. Arch. Allergy Immunol.* 2007; 143: 75–82.
194. Tsunoda A, Shibusawa M, Tsunoda Y, Watanabe M, Nomoto K, and Kusano M. Effect of *Lactobacillus casei* on a novel murine model of abdominal sepsis. *J. Surg. Res.* 2002; 107: 37–43.
195. Tuomola EM, Ouwehand AC, and Salminen SJ. The effect of probiotic bacteria on the adhesion of pathogens to human intestinal mucus. *FEMS Immunol. Med. Microbiol.* 1999; 26: 137–142.
196. Tuomola EM, Ouwehand AC, and Salminen SJ. Chemical, physical and enzymatic pretreatments of probiotic lactobacilli alter their adhesion to human intestinal mucus glycoproteins. *Int. J. Food Microbiol.* 2000; 60: 75–81.
197. van der Mei HC, Free RH, Elving GJ, Van Weissenbruch R, Albers FW, and Busscher HJ. Effect of probiotic bacteria on prevalence of yeasts in oropharyngeal biofilms on silicone rubber voice prostheses *in vitro*. *J. Med. Microbiol.* 2000; 49: 713–718.
198. Watanabe T. Suppressive effects of *Lactobacillus casei* cells, a bacterial immunostimulant, on the incidence of spontaneous thymic lymphoma in AKR mice. *Cancer Immunol. Immunother.* 1996; 42: 285–290.
199. Yamagishi T, Serikawa T, Morita R, Takahashi K, and Nishida S. Effect of a lactobacillus product administration on the anaerobic intestinal flora of aged adults. *Jpn. J. Microbiol.* 1974; 18: 211–216.

200. Yamazaki K, Tsunoda A, Sibusawa M, Tsunoda Y, Kusano M, Fukuchi K, Yamanaka M, Kushima M, Nomoto K, and Morotomi M. The effect of an oral administration of *Lactobacillus casei* strain shirota on azoxymethane-induced colonic aberrant crypt foci and colon cancer in the rat. *Oncol Rep.* 2000; 7: 977–982.

201. Yasui H, Kiyoshima J, and Hori T. Reduction of influenza virus titer and protection against influenza virus infection in infant mice fed *Lactobacillus casei* Shirota. *Clin. Diagn. Lab. Immunol.* 2004; 11: 675–679.

202. Yasutake N, Matsuzaki T, Kimura K, Hashimoto S, Yokokura T, and Yoshikai Y. The role of tumor necrosis factor (TNF)-alpha in the antitumor effect of intrapleural injection of *Lactobacillus casei* strain Shirota in mice. *Med. Microbiol. Immunol.* 1999; 188: 9–14.

203. Yuki N, Watanabe K, Mike A, Tagami Y, Tanaka R, Ohwaki M, and Morotom M. Survival of a probiotic, *Lactobacillus casei* strain Shirota, in the gastrointestinal tract: selective isolation from faeces and identification using monoclonal antibodies. *Int. J. Food Microbiol.* 1999; 48: 51–57.

204. Aiba Y, Suzuki N, Kabir AMA, et al. Lactic acid-mediated suppression of *Helicobacter pylori* by the oral administration of *Lactobacillus salivarius* as a probiotic in a gnotobiotic murine model. *Am. J. Gastroenterol.* 1998; 93: 2097–2101.

205. Atherton JC, Tham KT, Peek RM, et al. Density of *Helicobacter pylori* infection *in vivo* as assessed by quantitative culture and histology. *J. Infect. Dis.* 1996; 174: 552–556.

206. Furuta T, Kaneko E, Baba S, et al. Percentage changes in serum pepsinogens are useful as indices of eradication of *Helicobacter pylori. Am. J. Gastroenterol.* 1997; 92: 84–88.

207. Kabir AMA, Aiba Y, Takagi, et al. Prevention of *Helicobacter pylori* infection by lactobacilli in a gnotobiotic murine model. *Gut* 1997; 41: 49–55.

208. Kato S, Konno M, Maisawa S, et al. Results of triple eradication therapy in Japanese children: a retrospective multicenter study. *J. Gastroenterol.* 2004; 9: 838–843.

209. Kimura K. Health benefits of probiotics: Probiotics for *Helicobacter pylori* infection. *Food Sci. Technol. Res.* 2004; 10: 1–5.

210. Masuda H, Hiyama T, Yoshihara M, et al. Characteristics and trends of clarithromycin-resistant *Helicobacter pylori* isolates in Japan over a decade. *Pathobiology* 2004; 71: 159–163.

211. Sakamoto I, Igarashi M, Kimura K, et al. Suppressive effect of *Lactobacillus gasseri* OLL2716 (LG21) on *Helicobacter pylori* infection in humans. *J. Antimicrob. Chemother.* 2001; 47: 709–710.

212. Tamura A, Kumai H, Nakamichi N, et al. Suppression of *Helicobacter pylori*-induced interleukin-8 production *in vitro* and within the gastric mucosa by a live *Lactobacillus* strain. *J. Gastroenterol. Hepatol.* 2006; 21: 1399–1406.

213. Uemura N, Okamoto S, Yamamoto S, et al. *Helicobacter pylori* infection and the development of gastric cancer. *N. Engl. J. Med.* 2001; 345: 784–789.

214. Ushiyama A, Tanaka K, Aiba Y, et al. *Lactobacillus gasseri* OLL2716 as a probiotic in clarithromycin-resistant *Helicobacter pylori* infection. *J. Gastroenterol. Hepatol.* 2003; 18: 986–991.

215. Alander M, De Smet I, Nollet L, Verstraete W, von Wright, and Mattila-Sandholm T. The effect of probiotic strains on the microbiota of the Simulator of the Human Intestinal Microbial Ecosystem (SHIME). *Int. J. Food. Microbiol.* 1999; 46: 71–79.

216. Björneholm S, Eklöw A, Saarela M, and Mättö J. Enumeration and identification of *Lactobacillus paracasei* ssp. *paracasei* F19. *Microb. Ecol. Health Dis. Suppl. (Lactobacillus F19—Closing the broken circle)* 2002; 3: 7–13.
217. Charteris W, Kelly P, Morelli L, and Collins K. Antibiotic susceptibility of potentially probiotic Lactobacillus species. *J. Food Prot.* 1998; 61: 1636–1643.
218. Charteris W, Kelly P, Morelli L, and Collins K. Development and application of an *in vitro* methodology to determine the transit tolerance of potentially probiotic Lactobacillus and Bifidobacterium species in the upper gastrointestinal tract. *J. Appl. Microbiol.* 1998; 84: 759–768.
219. Charteris W, Kelly P, Morelli L, and Collins K. Quality control Lactobacillus strains for use with API 50CH and API ZYM systems at 37 degrees C. *J. Basic Microbiol.* 2001; 41(5): 241–251.
220. Crittenden R, Saarela M, Mättö J, Ouwehand AC, Salminen S, Pelto L, Vaughan EE, de Vos WM, von Wright A, Fondén R, and Mattila-Sandholm T. *Lactobacillus paracasei* ssp. *paracasei* F19: survival, ecology and safety in the human intestinal tract—a survey of feeding studies within the PROBDEMO project. *Microb. Ecol. Health Dis. Suppl. (Lactobacillus F19—Closing the broken circle)* 2002; 3: 22–26.
221. Delia A, Morgante G, Rago G, Musacchio MC, Petraglia F, De Leo V. Effectiveness of Oral administration of *Lactobacillus paracasei* ssp. *paracasei* F19 in association with vaginal suppositories of *Lactobacillus acidophilus* in the treatment of vaginosis and in the prevention of recurrent vaginitis. *Minerva Ginecol.* 2006; 58(3): 227–231.
222. Fondén R, Björneholm S, Ohlson K, Lactobacillus F19-safety considerations in practice. In: Fermented Milk. IDF, Brussels 2003, pp. 159–167.
223. Jernberg C, Sullivan Å, Edlund C, and Jansson J. Alterations in the human intestinal microflora and detection of probiotic strains by use of terminal restriction fragment length polymorphism. *Appl. Environ. Microbiol.* 2005; 71: 501–506.
224. Juntunen M, Kirjavainen PV, Ouwehand AC, Salminen SJ, and Isolauri E. Adherence of probiotic bacteria to human intestinal mucus in healthy infants and during rotavirus infection. *Clin. Diagn. Lab. Immunol.* 2001; 8: 293–296.
225. Kirjavainen P, Ouwehand AC, Isolauri E, and Salminen SJ. The ability of probiotic bacteria to bind to human intestinal mucus. *FEMS Microbiol. Lett.* 1998; 167: 185–189.
226. Kruszenska D, Lan J, Lorca G, Yanagisawa N, Marklinder I, and Ljung Å. Selection of lactic acid bacteria as probiotic strains by *in vitro* tests. *Microecol. Therapy* 2002; 29: 37–49.
227. Ljungh Å, Lan J, and Yanagisawa N. Isolation, selection and characteristics of *Lactobacillus paracasei* subsp. *paracasei* F19. *Microb. Ecol. Health Dis. Suppl. (Lactobacillus F19—Closing the broken circle)* 2002; 3: 4–6.
228. Miettinen M, Vuopio-Varkila J, and Varkila K. Production of human tumour necrosis factor alpha, interleukin-6 and lactic acid bacteria induce interleukin-10. *Infect. Immun.* 1996; 64: 5403–5404.
229. Morelli L and Campominosi E. Genetic stability of *Lactobacillus paracasei* subsp. *paracasei* F19. *Microb. Ecol. Health Dis. Suppl. (Lactobacillus F19—Closing the broken circle)* 2002; 3: 14–16.
230. Mättö J, Fondén R, Tolvanen T, von Wright A, Vilpponen-Salmela T, Satokari R, and Saarela M. Intestinal survival and persistence of probiotic *Lactobacillus* and *Bifidobacterium* strains administered in triple-strain yoghurt. *Int. Dairy J.* 2006; 16: 1174–1180.

231. Nerstedt A, Nilsson EC, Ohlson K, Håkansson J, Svensson LT, Löwenadler B, Svensson UK, and Mahlapuu M. Administration of *Lactobacillus* evokes coordinated changes in the intestinal expression profile of genes regulating energy homeostasis and immune phenotype in mice. *Br. J. Nutr.* 2007; 16: 1–11.
232. Ohlson K, Björneholm S, Fondén R, and Svensson U. *Lactobacillus* F19—a probiotic strain suitable for consumer products. *Microb. Ecol. Health Dis. Suppl. (Lactobacillus F19—Closing the broken circle)* 2002; 3: 27–32.
233. Ohlson K, Mahlapuu M, and Svensson U, Probiotics to influence fat metabolism and obesity. Swedish Patent 529185 2007.
234. Rayes N, Seehofer D, Theruvath T, Schiller RA, Langrehr JM, Jonas S, Bengmark S, Neuhaus P. Supply of pre- and probiotics reduces bacterial infection rates after liver transplantation—a randomized, double blind trial. *Am. J. Transplant.* 2005; 5: 125–130.
235. Saxelin M, Grenov B, Svensson U, Fondén R, Reniero R, and Mattila-Sandholm T. The technology of probiotics. *Trends Food Sci. Technol.* 1999; 10: 387–392.
236. Seehöfer D, Rayes N, Schiller. Probiotics partly reverses increased bacterial translocation after simultaneous liver resection and colonic anastomosis in rats. *J. Surg. Res.* 2004; 117: 262–271.
237. Simrén M, Lindh A, Samuelsson L, Olsson J, Posserud I, Strid H, and Abrahamsson H. Effect of yoghurt containing three probiotic bacteria in patients with irritable bowel syndrome (IBS)—a randomized, double-blind, controlled trial. *Gastroenterology* 2007; 132(4) Suppl. 2: S1269.
238. Spindler-Vesel A, Bengmark S, Vovk I, and Kompan L. Synbiotics, prebiotics, glutamine, or peptide in early enteral nutrition: a randomized study in traumatic patients. *J. Parent. Enteral Nutr.* 2007; 31: 1–8.
239. Sullivan Å, Palmgren AC, and Nord CE. Effect of *Lactobacillus paracasei* on intestinal colonisation of Lactobacilli, Bifidobacteria and *Clostridium difficile* in elderly persons. *Anerobe* 2001; 07: 67–70.
240. Sullivan Å, Bennet R, Viitanen M, Palmgren AC, and Nord CE. Influence of *Lactobacillus* F19 on intestinal microflora in children and elderly persons and impact on *Helicobacter pylori* infections. *Microb. Ecol. Health Dis. Suppl. (Lactobacillus F19—Closing the broken circle)* 2002; 3: 17–21.
241. Sullivan Å, Barkholt L, and Nord CE. *Lactobacillus acidophilus, Bifidobacterium lactis* and *Lactobacillus* F19 prevent antibiotic-associated ecological disturbances of Bacteroides fragilis in the intestine. *J. Antimicrob. Chemother.* 2003; 52: 308–311.
242. Sullivan Å, Johansson A, Svennungsson B, and Nord CE. Effect of *Lactobacillus* F19 on the emergence of antibiotic-resistant microorganisms in the intestinal microflora. *J. Antimicrob. Chemother.* 2004; 54: 791–797.
243. Sullivan Å and Nord CE. Probiotic lactobacilli and bacteraemia in Stockholm. *Scand. J. Inf. Dis.* 2006; 38: 327–331.
244. Wadström T, Aleljung P, Svensson U, Fondén R, Strain of bacteria of the species *Lactobacillus paracasei* subsp. *paracasei* compositions thereof for use in food and product containing said strain. European Patent 1036160B1 2006.
245. West CE, Gothefors L, Granström M, Käyhty H, Hammarström M-L, Hernell O, Effects of feeding probiotics during weaning on infections and antibody responses to diphteria, tetanus and Hib vaccines *Pediatr. Allergy Immunol.* 2008; 19: 53–60.

246. Jacobsen CN, Rosenfeldt Nielsen V, Hayford AE, Møller PL, Michaelsen KF, Pærregaard A, Sandström B, Tvede M, and Jakobsen M. Screening of probiotic activities of forty-seven strains of Lactobacillus spp. by *in vitro* techniques and evaluation of the colonization ability of five selected strains in humans. *Appl. Environ. Microbiol.* 1999; 65: 4949–4956.
247. Gaon D, Garmendia C, Murrielo NO, de Cucco Games A, Cerchio A, Quintas R, Gonzalez SN, and Oliver G. Effect of Lactobacillus strains (*L. casei* and *L. acidophilus* Cerela) on bacterial overgrowth-related chronic diarrhea. *Medicina* 2002; 62: 159–163.
248. Gaon D, Garcia H, Winter L, Rodriguez N, Quintas R, Gonzalez SN, and Oliver G. Effect of Lactobacillus strains and *Saccharomyces boulardii* on persistent diarrhea in children. *Medicina* 2003; 63: 293–298.
249. Larsen CN, Nielsen S, Kæstel P, Brockmann E, Bennedsen M, Christensen HR, Eskesen DC, Jacobsen BL, and Michaelsen KF. Dose-response study of probiotic bacteria *Bifidobacterium animalis* subsp. *lactis* BB-12 and *Lactobacillus paracasei* subsp. *paracasei* CRL-431 in healthy young adults. *Eur. J. Clin. Nutr.* 2006; 60(11): 1284–1293.
250. Gonzalez S, Albarracin G, Locascio de Ruiz Pesce M, Male M, Apella MC, Pesce de Ruiz Holgado A, and Oliver G. Prevention of infantile diarrhea by fermented milk. *Microbiol. Aliment. Nutr.* 1990; 8: 349–354.
251. Macias M, Apella M, Romero N, Gonzalez S, and Oliver G. Inhibition of *Shigella sonnei* by *Lactobacillus casei* and *Lactobacillus acidophilus*. *J. Appl. Bacteriol.* 1992; 73: 407–411.
252. Nader de Macias ME, Romero C, Apella MC, Gonzalez SN, and Oliver G. Prevention of infection produced by *Escherichia coli* and *Listeria monocytogenes* by feeding milk fermented with Lactobacilli. *J. Food Protect.* 1993; 56(5): 401–405.
253. Perdigon G, Alvarez S, Pesce de Ruiz Holgado A. Immunoadjuvant activity of oral *Lactobacillus casei*: influence of dose on the secretory immune response and protective capacity in intestinal infections. *J. Dairy Res.* 1991; 58: 485–496.
254. Perdigon G, Nader de Macia M, Alvarez S, Oliver G, Pesce de Ruiz Holgado A. Prevention of gastrointestinal infection using immunobiological methods with milk fermented with *Lactobacillus casei* and *Lactobacillus acidophilus J. Dairy Res.* 1990; 57: 255–264.
255. Perdigon G, Alvarez S, Medici M, Pesce de Ruiz Holgado A. Influence of the use of *Lactobacillus casei* as an oral adjuvant on the levels of secretory IgA during an infection with *Salmonella typhimurium*. *Food Agric. Immunol.* 1993; 5: 27–37.
256. Gonzalez S, Apella M, Romero N, Macias M, and Oliver G. Inhibition of enteropathogens by lactobacilli strains used in fermented milk. *J. Food Protect.* 1993; 56(9): 773–776.
257. Apella MC, Gonzalez SN, Nader de Macias ME, Romero N, and Oliver G. In vitro studies on the inhibition of the growth of *Shigella sonnei* by *Lactobacillus casei* and *Lactobacillus acidophilus*. *J. Appl. Bacteriol.* 1992; 73: 480–483.
258. Gonzalez S, Cardozo R, Apella M, and Oliver G. Biotherapeutic role of fermented milk. *Biotherapy* 1995; 8: 129–134.
259. de Petrino S, Eugenia M, de Jorrat B, de Budeguer M, and Perdigon G. Influence of the oral administration of different lactic acid bacteria on intestinal microflora and IgA-secreting cells in mice treated with ampicillin. *Food Agric. Immunol.* 1997; 9: 265–275.

260. Perdigon G, Nader de Macias ME, Alvarez S, Oliver G, Pesce de Ruiz Holgado A. Systemic augmentation of the immune response in mice by feeding fermented milk with *Lactobacillus casei* and *Lactobacillus acidophilus*. *Immunology* 1988; 63: 17–23.

261. Perdigon G, Nader de Macias ME, Alvarez S, Oliver G, Pesce de Ruiz Holgado A. Effect of perorally administered lactobacilli on macrophage activation in mice. *Infect. Immun.* 1986; 53(2): 404–410.

262. Petrino S, Jorrat M, and Perdigon G. Effect of different lactic acid bacteria on immune response in corticoid-immunosuppressed mice. *Microbiol. Aliments. Nutr.* 1996; 14: 227–236.

263. Alvarez S, Herrero C, Bru E, and Perdigón G. Effect of *Lactobacillus casei* and yoghurt administration on prevention of *Pseudomonas aeruginosa* infection in young mice. *J. Food Protect.* 2001; 64: 1768–1774.

264. Ambrosini V, Gonzalez S, Perdigon G, Holgado A, and Oliver G. Chemical composition of the cell wall of lactic acid bacteria and related species. *C.P. Bull.* 1996; 44(12): 2263–2267.

265. Ambrosini V, Gonzalez S, Perdigon G, Holgado A, and Oliver G. Immunostimulation activity of cell walls from lactic acid bacteria and related species. *Food Agric. Immunol.* 1998; 10: 183–191.

266. de Vrese M, Rautenberg P, Laue C, Koopmans M, Herremans T, and Schrezenmeir J. Probiotic bacteria stimulate virus-specific neutralizing antibodies following a booster polio vaccination. *Eur. J. Nutr.* 2005; 44: 406–413.

267. Maldonado Galdeano C, and Perdigon G. Role of viability of probiotic strains in their persistence in the gut and in mucosal immune stimulation. *J. Appl. Microbiol.* 2004; 97: 673–681.

268. Vitiñi E, Alvarez S, Medina M, Medici M, de Buduguer MV, and Perdigón G. Gut mucosal immunostimulation by lactic acid bacteria. *Biocell* 2000; 24(3): 223–232.

269. Perdigon G, Vintini E, Alvarez S, Medina M, and Medici M. Study of the possible mechanisms involved in the mucosal immune system activation by lactic acid bacteria. *J. Dairy Sci.* 1999; 82: 1108–1114.

270. Perdigon G, Alvarez S, Nader de Macias M, Roux M, Pesce de Ruiz Holgado A. The oral administration of lactic acid bacteria increase the mucosal intestinal immunity in response to enteropathogens. *J. Food Protect.* 1990; 53(5): 404–410.

271. Perdigon G, Alvarez S, Gobbato N, de Budeguer MV, de Ruiz Holgado AAP. Comparative effect of the adjuvant capacity of *Lactobacillus casei* and lipopolysaccharide on the intestinal secretory antibody response and resistance to Salmonella infection in mice. *Food Agric. Immunol.* 1995; 7: 283–294.

272. Perdigon G, Maldonado Galdeano C, Valdez JC, and Medici M. Interaction of lactic acid bacteria with the gut immune system. *Eur. J. Clin. Nutr.* 2002; 56(Suppl. 4): S21–S26.

273. Bonet MEB, de Petrino SF, Meson O, and Perdigon G. Antitumour effect of *Lactobacillus casei* CRL 431 on different experimental tumours. *Food Agric. Immunol.* 2005; 16(1–4): 181–191.

274. Perdigon G, Eugenia de Jorrat M, Valdez JC, de Budeguer M, and Oliver G. Cytolytic effect of the serum of mice fed with *Lactobacillus casei* on tumor cells. *Microbiol. Aliment. Nutr.* 1995; 13: 15–24.

275. Gorbach SL. The discovery of Lactobacillus GG. *Nutr. Today* 1996; 31: 2S–4S.

276. Saxelin M. *Lactobacillus* GG—a human probiotic strain with thorough clinical documentation. *Food Rev. Int.* 1997; 13: 293–313.
277. Tynkkynen S, Satokari R, Saarela M, Mattila-Sandholm T, and Saxelin M. Comparison of ribotyping, randomly amplified polymorphic DNA analysis, and pulsed-field gel electrophoresis in typing of *Lactobacillus rhamnosus* and *L. casei* strains. *Appl. Environ. Microbiol.* 1999; 65: 3908–3914.
278. Salminen S, Salminen K, and Gorbach S. *Lactobacillus* GG fermented whey drink and yoghurt—new clinically tested dairy product to promote health. *Scand. Dairy Inform.* 1991; 3: 66–67.
279. Saxelin M, Formulations, applications, current marketplace, changes in the marketplace—A European perspective *Clin. Infect. Dis.* 2007; 46, Suppl. 2: 576–9.
280. Nighswonger BD, Brashears MM, and Gilliland SE. Viability of *Lactobacillus acidophilus* and *Lactobacillus casei* in fermented milk products during refrigerated storage. *J. Dairy Sci.* 1996; 79: 212–219.
281. Donkor ON, Tsangalis D, and Shah P. Viability of probiotic bacteria and concentrations of organic acids in commercial yoghurts during refrigerated storage. *Food Aust.* 2007; 59: 121–126.
282. Farnworth ER, Mainville I, Desjardins MP, Gardner N, Fliss I, and Champagne C. Growth of probiotic bacteria and bifidobacteria in a soy yogurt formulation. *Int. J. Food Microbiol.* 2007; 116: 174–181.
283. Tuomola EM, Ouwehand AC, and Salminen SJ. The effect of probiotic bacteria on the adhesion of pathogens to human intestinal mucus. *FEMS Immunol. Med. Microbiol.* 1999; 26: 137–142.
284. Ouwehand AC, Salminen S, Roberts PJ, Ovaska J, and Salminen E. Disease-dependent adhesion of lactic acid bacteria to the human intestinal mucosa. *Clin. Diagn. Lab. Immunol.* 2003; 10: 643–646.
285. Vesterlund S, Paltta J, Karp M, and Ouwehand AC. Adhesion of bacteria to resected human colonic tissue: quantitative analysis of bacterial adhesion and viability. *Res Microbiol.* 2005; 156: 238–244.
286. Alander M, Satokari R, Korpela R, Saxelin M, Vilpponen-Salmela T, Mattila-Sandholm T, von Wright A. Persistence of colonization of human colonic mucosa by a probiotic strain, *Lactobacillus rhamnosus* GG, after oral consumption. *Appl. Environ. Microbiol.* 1999; 65: 351–354.
287. Doron S, Snydman DR, and Gorbach SL. *Lactobacillus* GG: bacteriology and clinical applications. *Gastroenterol. Clin. North Am.* 2005; 34: 483–498.
288. Szajewska H, Skorka A, Ruszczynski M, and Gieruszczak-Bialek D. Meta-analysis: *Lactobacillus* GG for treating acute diarrhea in children. *Aliment. Pharmacol. Ther.* 2007; 25: 871–881.
289. Hawrelak JA, Whitten DL, and Myers SP. Is *Lactobacillus rhamnosus* GG effective in preventing the onset of antibiotic-associated diarrhea: a systematic review. *Digestion* 2005; 72: 51–56.
290. Saxelin M, Tynkkynen S, Mattila-Sandholm T, de Vos WM. Probiotic and other functional microbes: from markets to mechanisms. *Curr. Opin. Biotechnol.* 2005; 16: 204–211.
291. Hatakka K, Savilahti E, Ponka A, Meurman JH, Poussa T, Nase L, Saxelin M, and Korpela R. Effect of long term consumption of probiotic milk on infections in children attending day care centres: double blind, randomised trial. *BMJ* 2001; 322: 1327–1329.

292. Nase L, Hatakka K, Savilahti E, Saxelin M, Ponka A, Poussa T, Korpela R, and Meurman JH. Efect of long-term consumption of a probiotic bacterium, *Lactobacillus rhamnosus* GG, in milk on dental caries and caries risk in children. *Caries Res.* 2001; 35: 412–420.

293. D'Souza AL, Rajkumar C, Cooke J, and Bulpitt CJ, Probiotics in prevention of antibiotic associated diarrhoea: meta-analysis. *BMJ* 2002; 324: 1361

294. McFarland LV. Meta-analysis of probiotics for the prevention of antibiotic associated diarrhea and the treatment of *Clostridium difficile* disease. *Am. J. Gastroenterol.* 2006; 101: 812822.

295. Wenus C, Goll R, Loken EB, Biong AS, Halvorsen DS, and Florholmen J. Prevention of antibiotic-associated diarrhoea by a fermented probiotic milk drink. *Eur. J Clin. Nutr.* Mar 14: epub. 2007.

296. Brouwer ML, Wolt-Plompen SA, Dubois AE, van der Heide S, Jansen DF, Hoijer MA, Kaufman HF, and Duiverman EJ. No effects of probiotics on atopic dermatitis in infancy: a randomized placebo-controlled trial. *Clin. Exp. Allergy* 2006; 6: 899–906.

297. Folster-Holst R, Muller F, Schnopp N, Abeck D, Kreiselmaier I, Lenz T, von Rüden U, Schrezenmeir J, Christophers E, and Weichenthal M. Prospective, randomized controlled trial on *Lactobacillus rhamnosus* in infants with moderate to severe atopic dermatitis. *Br. J. Dermatol.* 2006; 155: 1256–1261.

298. Isolauri E, Arvola T, Sutas Y, Moilanen E, and Salminen S. Probiotics in the management of atopic eczema. *Clin. Exp. Allergy* 2000; 30: 1604–1610.

299. Majamaa H and Isolauri E. Probiotics: a novel approach in the management of food allergy. *J. Allergy Clin. Immunol.* 1997; 99: 179–185.

300. Pohjavuori E, Viljanen M, Korpela R, Kuitunen M, Tiittanen M, Vaarala O, and Savilahti E. *Lactobacillus* GG effect in increasing IFN-gamma production in infants with cow's milk allergy. *J. Allergy Clin. Immunol.* 2004; 114: 131–136.

301. Viljanen M, Kuitunen M, Haahtela T, Juntunen-Backman K, Korpela R, and Savilahti E. Probiotic effects on faecal inflammatory markers and on faecal IgA in food allergic atopic eczema/dermatitis syndrome infants. *Pediatr. Allergy Immunol.* 2005; 16: 65–71.

302. Viljanen M, Pohjavuori E, Haahtela T, Korpela R, Kuitunen M, Sarnesto A, Vaarala O, and Savilahti E. Induction of inflammation as a possible mechanism of probiotic effect in atopic eczema-dermatitis syndrome. *J. Allergy Clin. Immunol.* 2005; 115: 1254–1259.

303. Viljanen M, Savilahti E, Haahtela T, Juntunen-Backman K, Korpela R, Poussa T, Tuure T, and Kuitunen M. Probiotics in the treatment of atopic eczema/dermatitis syndrome in infants: a double-blind placebo-controlled trial. *Allergy* 2005; 60: 494–500.

304. Kalliomäki M, Kirjavainen P, Eerola E, Kero P, Salminen S, and Isolauri E. Distinct pattern of neonatal gut microflora in infants in whom atopy was and was not developing. *J. Allergy Clin. Immunol.* 2001; 107: 129–134.

305. Kalliomaki M, Salminen S, Arvilommi H, Kero P, Koskinen P, and Isolauri E. Probiotics in primary prevention of atopic disease: a randomised placebo-controlled trial. *Lancet* 2001; 357: 1076–1079.

306. Kalliomaki M, Salminen S, Poussa T, Arvilommi H, and Isolauri E. Probiotics and prevention of atopic disease: 4-year follow-up of a randomised placebo-controlled trial. *Lancet* 2003; 361: 1869–1871.

307. Kalliomaki M, Salminen S, Poussa T, and Isolauri E. Probiotics during the first 7 years of life: a cumulative risk reduction of eczema in a randomized, placebo-controlled trial. *J. Allergy Clin. Immunol.* 2007; 119: 1019–1021.

308. Laitinen K, Kalliomaki M, Poussa T, Lagstrom H, and Isolauri E. Evaluation of diet and growth in children with and without atopic eczema: follow-up study from birth to 4 years. *Br. J. Nutr.* 2005; 94: 565–574.

309. Rautava S, Kalliomaki M, and Isolauri E. Probiotics during pregnancy and breast-feeding might confer immunomodulatory protection against atopic disease in the infant. *J. Allergy Clin. Immunol.* 2002; 109: 119–121.

310. Bruzzese E, Raia V, Gaudiello G, et al. Intestinal inflammation is a frequent feature of cystic fibrosis and is reduced by probiotic administration. *Aliment. Pharmacol. Ther.* 2004; 20: 813–819.

311. Bruzzese E, Raia V, Spagnuolo MI, Polito G, Buccigrossi V, Formicola V, and Guarino A. Effect of *Lactobacillus* GG supplementation on pulmonary exacerbations in patients with cystic fibrosis: a pilot study. *Clin. Nutr.* 2007; 26: 322–328.

312. de Vrese M, Rautenberg P, Laue C, Koopmans M, Herremans T, and Schrezenmeir J. Probiotic bacteria stimulate virus-specific neutralizing antibodies following a booster polio vaccination. *Eur. J. Nutr.* 2005; 44: 406–413.

313. He F, Tuomola E, Arvilommi H, and Salminen S. Modulation of humoral immune response through probiotic intake. *FEMS Immunol. Med. Microbiol.* 2000; 29: 47–52.

314. Isolauri E, Joensuu J, Suomalainen H, Luomala M, and Vesikari T. Improved immunogenicity of oral DxRRV reassortant rotavirus vaccine by *Lactobacillus casei* GG. *Vaccine* 1995; 13: 310–312.

315. Kaila M, Isolauri E, Soppi E, Virtanen E, Laine S, and Arvilommi H. Enhancement of the circulating antibody secreting cell response in human diarrhea by a human *Lactobacillus* strain. *Pediatr. Res.* 1992; 32: 141–144.

316. Majamaa H, Isolauri E, Saxelin M, and Vesikari T. Lactic acid bacteria in the treatment of acute rotavirus gastroenteritis. *J. Pediatr. Gastroenterol. Nutr.* 1995; 20: 333–338.

317. Rinne M, Kalliomaki M, Arvilommi H, Salminen S, and Isolauri E. Effect of probiotics and breastfeeding on the bifidobacterium and lactobacillus/enterococcus microbiota and humoral immune responses. *J. Pediatr.* 2005; 147: 186–191.

318. Pelto L, Isolauri E, Lilius EM, Nuutila J, and Salminen S. Probiotic bacteria down-regulate the milk-induced inflammatory response in milk-hypersensitive subjects but have an immunostimulatory effect in healthy subjects. *Clin. Exp. Allergy* 1998; 28: 1474–1479.

319. Malin M, Suomalainen H, Saxelin M, and Isolauri E. Promotion of IgA immune response in patients with Crohn's disease by oral bacteriotherapy with *Lactobacillus* GG. *Ann. Nutr. Metab.* 1996; 40: 137–145.

320. Schultz M, Linde HJ, Lehn N, et al. Immunomodulatory consequences of oral administration of *Lactobacillus rhamnosus* strain GG in healthy volunteers. *J. Dairy Res.* 2003; 70: 165–173.

321. Di Caro S, Tao H, Grillo A, et al. Effects of *Lactobacillus* GG on genes expression pattern in small bowel mucosa. *Dig. Liver Dis.* 2005; 37: 320–329.

322. Manley KJ, Fraenkel MB, Mayall BC, and Power DA. Probiotic treatment of vancomycin-resistant enterococci: a randomised controlled trial. *Med. J. Aust.* 2007; 186: 454–457.

323. Manzoni P, Mostert M, Leonessa ML, Priolo C, Farina D, Monetti C, Latino MA, and Gomirato G. Oral supplementation with *Lactobacillus casei* subspecies *rhamnosus* prevents enteric colonization by *Candida* species in preterm neonates: a randomized study. *Clin. Infect. Dis.* 2006; 42: 1735–1742.

324. Hatakka K, Martio J, Korpela M, Herranen M, Poussa T, Laasanen T, Saxelin M, Vapaatalo H, Moilanen E, and Korpela R. Effects of probiotic therapy on the activity and activation of mild rheumatoid arthritis—a pilot study. *Scand. J. Rheumatol.* 2003; 32: 211–215.

325. Gosselink MP, Schouten WR, Van Lieshout LM, Hop WC, Laman JD, Ruseler-Van Embden JG. Delay of the first onset of pouchitis by oral intake of the probiotic strain *Lactobacillus rhamnosus* GG. *Dis. Colon Rectum* 2004; 47: 876–184.

326. Kuisma J, Mentula S, Jarvinen H, Kahri A, Saxelin M, and Farkkila M. Effect of *Lactobacillus rhamnosus* GG on ileal pouch inflammation and microbial flora. *Aliment. Pharmacol Ther.* 2003; 17: 509–515.

327. Zocco MA, dal Verme LZ, Cremonini F, Piscaglia AC, Nista EC, Candelli M, Novi M, Rigante D, Cazzato IA, Ojetti V, Armuzzi A, Gasbarrini G, and Gasbarrini A. Efficacy of *Lactobacillus* GG in maintaining remission of ulcerative colitis. *Aliment. Pharmacol. Ther.* 2006; 3: 1567–1574.

328. Rolfe VE, Fortun PJ, Hawkey CJ, and Bath-Hextall F. Probiotics for maintenance of remission in Crohn's disease. *Cochrane Database Syst. Rev.* 2006; CD004826

329. Chan RCY, Reid G, Irvin RT, Bruce AW, and Costerton JW. Competitive exclusion of uropathogens from uroepithelial cells by *Lactobacillus* whole cells and cell wall fragments. *Infect. Immun.* 1985; 47: 84–89.

330. Cook RL, Harris RJ, and Reid G. Effect of culture media and growth phase on the morphology of lactobacilli and on their ability to adhere to epithelial cells. *Curr. Microbiol.* 1988; 17(3): 159–166.

331. Reid G, Chan RCY, Bruce AW, and Costerton JW. Prevention of urinary tract infection in rats with an indigenous *Lactobacillus casei* strain. *Infect. Immun.* 1985; 49(2): 320–324.

332. Reid G, Cook RL, and Bruce AW. Examination of strains of lactobacilli for properties which may influence bacterial interference in the urinary tract. *J. Urol.* 1987; 138: 330–335.

333. Zhong W, Millsap K, Bialkowska-Hobrzanska H, and Reid G. Differentiation of *Lactobacillus* species by molecular typing. *Appl. Environ. Microbiol.* 1998; 64: 2418–2423.

334. Laughton J, Devillard E, Heinrichs D, Reid G, and McCormick J. Inhibition of expression of a staphylococcal superantigen-like protein by a secreted signaling factor from *Lactobacillus reuteri*. *Microbiology* 2006; 152: 1155–1167.

335. Reid G, Cook RL, Harris RJ, Rousseau JD, and Lawford H. Development of a freeze sustitution technique to examine the structure of *Lactobacillus casei* GR-1 grown in agar and under batch and chemostat culture conditions. *Curr. Microbiol.* 1988; 17(3): 151–158.

336. Reid G, Cuperus PL, Bruce AW, Tomeczek L, van der Mei HC, Khoury AH, and Busscher HJ. Comparison of contact angles and adhesion to hexadecane of urogenital, dairy and poultry lactobacilli: effect of serial culture passages. *Appl. Environ. Microbiol.* 1992; 58(5): 1549–1553.

337. Tomeczek L, Reid G, Cuperus PL, McGroarty JA, van der Mei H, Bruce AW, Khoury AH, and Busscher HJ. Correlation between hydrophobicity and resistance to nonoxynol-9 and vancomycin for urogenital isolates of lactobacilli. *FEMS Microbiol. Lett.* 1992; 94: 101–104.

338. Reid G, Zalai C, and Gardiner G. Urogenital lactobacilli probiotics, reliability, and regulatory issues. *J. Dairy Sci.* 2001; 84(E Suppl.): E164–E169.

339. Reid G, Lam D, Bruce AW, van der Mei HC, and Busscher HJ. Adhesion of lactobacilli to urinary catheters and diapers: effect of surface properties. *J. Biomed. Mater. Res.* 1994; 28: 731–734.

340. Reid G, Hawthorn LA, Mandatori R, Cook RL, and Beg HS. Adhesion of lactobacillus to polymer surfaces *in vivo* and *in vitro*. *Microb. Ecol.* 1988; 16(3): 241–251.

341. Reid G. *In vitro* analysis of a dairy strain of *Lactobacillus acidophilus* NCFM™ as a possible probiotic for the urogenital tract. *Int. Dairy J.* 2000; 10: 415–419.

342. Reid G, Tieszer C, and Lam D. Influence of lactobacilli on the adhesion of *Staphylococcus aureus* and *Candida albicans* to diapers. *J. Ind. Microbiol.* 1995; 15: 248–253.

343. Reid G and Tieszer C. Use of lactobacilli to reduce the adhesion of *Staphylococcus aureus* to catheters. *Int. Biodeter. Biodegrad.* 1995; 34: 73–83.

344. Reid G, Charbonneau D, Gonzalez S, Gardiner G, Erb J, and Bruce AW. Ability of *Lactobacillus* GR-1 and RC-14 to stimulate host defences and reduce gut translocation and infectivity of *Salmonella typhimurium*. *Nutraceut. Food* 2002; 7: 168–173.

345. Burton JP, Cadieux P, and Reid G. Improved understanding of the bacterial vaginal microbiota of women before and after probiotic instillation. *Appl. Environ. Microbiol.* 2003; 69: 97–101.

346. Reid G, Kim SO, and Kohler G. Selection, testing and understanding probiotic microbes. *FEMS Immunol. Med. Microbiol.* 2006; 46: 149–157.

347. Reid G, McGroarty JA, Angotti R, and Cook RL. Lactobacillus inhibitor production against *E. coli* and coaggregation ability with uropathogens. *Can. J. Microbiol.* 1988; 34: 344–351.

348. McGroarty JA and Reid G. Detection of a lactobacillus substance which inhibits *Escherichia coli*. *Can. J. Microbiol.* 1988; 34: 974–978.

349. McGroarty JA and Reid G. Inhibition of enterococci by *Lactobacillus* species *in vitro*. *Microb. Ecol. Health Dis.* 1988; 1: 215–219.

350. Cadieux P, Identification of anti-infective signals from lactobacilli. Ph. D. Thesis, University of Western Ontario, Canada 2006.

350a. Cadieux P., A. Wind, P. Sommer, L. Schaefer, K. Crowley, R.A. Britton, and G. Reid. 2008. Evaluation of reuterin production in urogenital probiotic *Lactobacillus reuteri* RC-14. *Appl. Environ. Microbiol.* 74(15): 4645–9.

351. Koehler G, Reid G, Mechanisms of probiotic interference with Candida albicans. ASM Conference on Candida and Candidiasis, 2006.

352. Saunders SG, Bocking A, Challis J, and Reid G. Disruption of *Gardnerella vaginalis* biofilms by *Lactobacillus*. *Coll. Surf. B: Biointerf.* 2007; 55(2): 138–142.

353. Kim SO, Sheik HI, Ha SD, Martins A, and Reid G. G-CSF mediated inhibition of JNK is a key mechanism for *Lactobacillus rhamnosus*-induced anti-inflammatory effects in macrophages. *Cell Microbiol.* 2006, online Aug 2, Nov 6; 8(12): 1958–1971.

354. Yeganegi M, Watson M, Kim C, Reid G, Challis J, and Bocking A, Lactobacilli supernatant inhibits TNF-α production and COX_2 expression in LPS-activated placental trophoblasts. ISAPP Open Forum, London, UK, June 27, oral presentation, 2007.

355. Reid G, Servin A, Bruce AW, and Busscher HJ. Adhesion of three *Lactobacillus* strains to human urinary and intestinal epithelial cells. *Microbios* 1993; 75: 57–65.

356. Velraeds MC, van der Mei HC, Reid G, and Busscher HJ. Inhibition of initial adhesion of uropathogenic *Enterococcus faecalis* by biosurfactants from *Lactobacillus* isolates. *Appl. Environ. Microbiol.* 1996; 62: 1958–1963.

357. Velraeds MC, van der Mei HC, Reid G, and Busscher HJ. Physicochemical and biochemical characterization of biosurfactants released from *Lactobacillus* strains. *Coll. Surf. B: Biointerf.* 1996; 8: 51–61.

358. Velraeds M, van der Mei HC, Reid G, and Busscher HJ. Inhibition of initial adhesion of uropathogenic *Enterococcus faecalis* to solid substrate by an adsorbed biosurfactant layer from *Lactobacillus acidophilus* strains. *Urology* 1997; 49: 790–794.

359. Velraeds MC, van der Belt B, van der Mei HC, Reid G, and Busscher HJ. Interference in initial adhesion of uropathogenic bacteria and yeasts silicone rubber by a *Lactobacillus acidophilus* biosurfactant. *J. Med. Microbiol.* 1998; 49: 790–794.

360. Heinemann C, Van Hylckama Vlieg JET, Janssen DB, Busscher HJ, van der Mei HC, and Reid G. Purification and characterization of a surface-binding protein from *Lactobacillus fermentum* RC-14 inhibiting *Enterococcus faecalis* 1131 adhesion. *FEMS Microbiol. Lett.* 2000; 190: 177–180.

361. Howard J, Heinemann C, Thatcher BJ, Martin B, Gan BS, and Reid G. Identification of collagen-binding proteins in *Lactobacillus* spp. With surface-enhanced laser desorption/ionization-time of flight ProteinChip technology. *Appl. Environ. Microbiol.* 2000; 66: 4396–4400.

362. Gan BS, Kim J, Reid G, Cadieux P, and Howard JC. *Lactobacillus fermentum* RC-14 inhibits *Staphylococcus aureus* infection of surgical implants in rats. *J. Infect. Dis.* 2002; 185: 1369–1372.

363. Gan BS, Kim J, Reid G, Cadieux P, and Howard JC. Probiotic *Lactobacillus* RC-14 and its biosurfactant prevent *Staph. aureus* infection: a new approach in the prevention of implant infection? *Plastic Surg. Forum* 2003; 25: 164–167.

364. Reid G, Cook RL, Hagberg L, Bruce AW. Lactobacilli as competitive colonizers of the urinary tract. In: Kass EH and Svanborg Eden C, editors. *Host–Parasite Interactions in Urinary Tract Infections*. University of Chicago Press, 1989, pp. 390–396.

365. Anukam KC, Osazawa EO, and Reid G. Feeding probiotic strains *Lactobacillus rhamnosus* GR-1 and *Lactobacillus fermentum* RC-14 does not significantly alter hematological parameters of Sprague–Dawley rats *HAEMA (J. Helenic Soc. Haematol.)* 2004; 7(4): 497–501.

366. Anukam KC, Osazuwa EO, and Reid G. Improved appetite of pregnant rats and increased birth weight of newborns following feeding with probiotic *Lactobacillus rhamnosus* GR-1 and *L. rueteri* RC-14. *J. Appl. Res.* 2005; 5: 46–52.

367. Hagberg L, Bruce AW, Reid G, Svanborg Eden C, Lincoln K, Lidin-Janson G, Colonization of the urinary tract with live bacteria from the normal fecal and urethral flora in patients with recurrent symptomatic urinary tract infections. In: Kass EH and Svanborg Eden C, editors. *Host–Parasite Interactions in Urinary Tract Infections*. University of Chicago Press 1989, pp. 194–97.

368. Reid G, Millsap K, and Bruce AW. Implantation of *Lactobacillus casei* var *rhamnosus* into the vagina. *Lancet* 1994; 344: 1229

369. Cadieux P, Burton J, Kang CY, Gardiner G, Braunstein I, Bruce AW, Reid G, *Lactobacillus* strains and vaginal ecology. *JAMA* 2002; 287: 1940–1941.

370. Gardiner G, Heinemann C, Beuerman D, Bruce AW, and Reid G. Persistence of *Lactobacillus fermentum* RC-14 and *L. rhamnosus* GR-1, but not *L. rhamnosus* GG in the human vagina as demonstrated by randomly amplified polymorphic DNA (RAPD). *Clin. Diag. Lab. Immunol.* 2002; 9: 92–96.

371. Bruce AW and Reid G. Intravaginal instillation of lactobacilli for prevention of recurrent urinary tract infections. *Can. J. Microbiol.* 1988; 34: 339–343.
372. Bruce AW, Reid G, McGroarty JA, Taylor M, and Preston C. Preliminary study on the prevention of recurrent urinary tract infections in ten adult women using intravaginal lactobacilli. *Int. Urogynecol. J.* 1992; 3: 22–25.
373. Reid G, Bruce AW, and Taylor M. Influence of three day antimicrobial therapy and lactobacillus suppositories on recurrence of urinary tract infection. *Clin. Therap.* 1992; 14(1): 11–16.
374. Reid G, Bruce AW, and Taylor M. Instillation of Lactobacillus and stimulation of indigenous organisms to prevent recurrence of urinary tract infections. *Microecol. Ther.* 1995; 23: 32–45.
375. Anukam KC, Osazuwa E, Osemene GI, Ehigiagbe F, Bruce AW, and Reid G. Clinical study comparing probiotic *Lactobacillus* GR-1 and RC-14 with metronidazole vaginal gel to treat symptomatic bacterial vaginosis. *Microb. Infect.* 2006; 8(12–13): 2772–2776.
376. Gardiner G, Heinemann C, Baroja ML, Bruce AW, Beuerman D, Madrenas J, and Reid G. Oral administration of the probiotic combination *Lactobacillus rhamnosus* GR-1 and *L. fermentum* RC-14 for human intestinal applications. *Int. Dairy J.* 2002; 12(2–3): 191–196.
377. Morelli L, Zonenenschain D, Del Piano M, and Cognein P. Utilization of the intestinal tract as a delivery system for urogenital probiotics. *J. Clin. Gastroenterol.* 2004; (6 Suppl.): S107–S110.
378. Reid G, Bruce AW, Fraser N, Heinemann C, Owen J, and Henning B. Oral probiotics can resolve urogenital infections. *FEMS Immunol. Med. Microbiol.* 2001; 30: 49–52.
379. Reid G, Beuerman D, Heinemann C, and Bruce AW. Probiotic *Lactobacillus* dose required to restore and maintain a normal vaginal flora. *FEMS Immunol. Med. Microbiol.* 2001; 32: 37–41.
380. Reid G, Charbonneau D, Erb J, Kochanowski B, Beuerman D, Poehner R, and Bruce AW. Oral use of *Lactobacillus rhamnosus* GR-1 and *L. fermentum* RC-14 significantly alters vaginal flora: randomized, placebo-controlled trial in 64 healthy women. *FEMS Immunol. Med. Microbiol.* 2003; 35: 131–134.
381. Reid G, Hammond JA, and Bruce AW. Effect of lactobacilli oral supplement on the vaginal microflora of antibiotic treated patients: randomized, placebo-controlled study. *Nutraceut. Food* 2003; 8: 145–148.
382. Anukam K, Osazuwa E, Ahonkhai I, Ngwu M, Osemene G, Bruce AW, and Reid G. Augmentation of antimicrobial metronidazole therapy of bacterial vaginosis with oral probiotic *Lactobacillus rhamnosus* GR-1 and *Lactobacillus reuteri* RC-14: randomized, double-blind, placebo controlled trial. *Microb. Infect.* 2006; 8(6): 1450–1454.
383. Baroja ML, Kirjavainen PV, Hekmat S, Reid G, Anti-inflammatory effects of probiotic-yogurt in inflammatory bowel disease patients *Clin. Exp. Immunol.* 2007; 149: 470–479.
384. Anukam KC, Osazuwa EO, Osadolor BE, Bruce AW, Reid G, Yogurt containing probiotic *Lactobacillus rhamnosus* GR-1 and *L. reuteri* RC-14 helps resolve moderate diarrhea and increases CD4 count in HIV/AIDS patients *J. Clin. Gastroenterol.* 2008; 42(3): 239–43.
385. Prasad J, Gill H, Smart J, and Gopal PK. Selection and characterisation of *Lactobacillus* and *Bifidobacterium* strains for use as probiotics. *Int. Dairy J.* 1998; 8: 993–1002.

386. Gopal P, Dekker J, Prasad Y, Pillidge CJ, Delabre M-L, and Collett M. Development and commercialisation of Fonterra's probiotic strains. *Aust. J. Dairy Technol.* 2005; 60: 173–182.

387. Gopal PK, Prasad J, Smart J, and Gill HS. *In vitro* adherence properties of *Lactoacillus rhamnosus* DR20 and *Bifidobacterium lactis* DR10 strains and their antagonistic activity against an enterotoxigenic *Escherichia coli. Int. J. Food Microbiol.* 2001; 67: 207–216.

388. Zhou JS, Pillidge CJ, Gopal PK, and Gill HS. Antibiotic susceptibility profiles of new probiotic *Lactobacillus* and *Bifidobacterium* strains. *Int. J. Food Microbiol.* 2005; 98: 211–217.

389. Delgado S, Belén Flórez A, and Mayo B. Antibiotic susceptibility of *Lactobacillus* and *Bifidobacterium* species from the human gastrointestinal tract. *Curr. Microbiol.* 2005; 50: 1–6.

390. Zhou JS, Shu Q, Rutherfurd KJ, Prasad J, Gopal PK, and Gill H. Acute oral toxicity and bacterial translocation studies on potentially probiotic strains of lactic acid bacteria. *Food Chem. Toxicol.* 2000; 38: 153–161.

391. Shu Q, Zhou JS, Rutherfurd KJ, Britles MJ, Prasad J, Gopal PK, et al. Probiotic lactic acid bacteria (*Lactobacillus acidophilus* HN017, *Lactobacillus rhamnosus* HN001 and *Bifidobacterium lactis* HN019) have no adverse effects on the health of mice. *Int. Dairy J.* 1999; 9: 831–836.

392. Zhou JS, Shu Q, Rutherfurd KJ, Prasad J, Britles MJ, Gopal PK, et al. Safety assessment of potential probiotic lactic acid bacteria strains *Lactobacillus rhamnosus* HN001, *Lb. acidophilus* HN017, and *Bifidobacterium lactis* HN019 in BALB/c mice. *Int. J. Food Microbiol.* 2000; 56: 87–96.

393. Zhou JS, Rutherfurd KJ, and Gill HS. Inability of probiotic bacterial strains *Lactobacillus rhamnosus* HN001 and *Bifidobacterium lactis* HN019 to induce human platelet aggregation *in vitro*. *J. Food Prot.* 2005; 68: 2459–2464.

394. Tannock GW, Munro K, Harmsen HJM, Welling GW, Smart J, and Gopal PK. Analysis of the fecal microflora of human subjects consuming a probiotic product containing *Lactobacillus rhamnosus* DR20. *Appl. Environ. Microbiol.* 2000; 66: 2578–2588.

395. Gopal P, Prasad J, and Gill HS. Effects of the consumption of *Bifidobacterium lactis* HN019 (DR10™) and galacto-oligosaccharides on the microflora of the gastrointestinal tract in human subjects. *Nutr. Res.* 2003; 23: 1313–1328.

396. Ahmed M, Prasad J, Gill H, Stevenson L, and Gopal P. Impact of consumption of different levels of Bifidobacterium lactis HN019 on the intestinal microflora of elderly human subjects. *J. Nutr. Health Aging* 2007; 11: 26–31.

397. Gill HS, Rutherfurd KJ, Prasad J, and Gopal PK. Enhancement of natural and acquired immunity by *Lactobacillus rhamnosus* (HN001), *Lactobacillus acidophilus* (HN017) and *Bifidobacterium lactis* (HN019). *Br. J. Nutr.* 2000; 83: 167–176.

398. Shu Q, Qu F, Lin K, Rutherfurd KJ, Zhou J, Gill HS. Bifidobacterium lactis HN019 enhances host immunity and resistance to gastrointestinal pathogens. In: Tuijtelaars ACJ, Samson RA, Rombouts FM, and Notermans S, editors. *Food Microbiology and Food Safety into the Next Millenium.* Foundation Food Micro '99, Wageningen, The Netherlands, 1999, pp. 858–861.

399. Gill HS and Rutherfurd KJ. Viability and dose-response studies on the effects of the immunoenhancing lactic acid bacterium *Lactobacillus rhamnosus* in mice. *Br. J. Nutr.* 2001; 86: 285–289.

400. Zhou JS and Gill HS. Immunostimulatory probiotic *Lactobacillus rhamnosus* HN001 and *Bifidobacterium lactis* HN019 do not induce pathological inflammation in mouse model of experimental autoimmune thyroiditis. *Int. J. Food Microbiol.* 2005; 103: 97–104.
401. Shu Q, Lin H, Rutherfurd KJ, Fenwick SG, Prasad J, Gopal PK, et al. Dietary *Bifidobacterium lactis* (HN019) enhances resistance to oral *Salmonella typhimurium* infection in mice. *Microbiol. Immunol.* 2000; 44: 213–22.
402. Gill HS, Shu Q, Lin H, Rutherfurd KJ, and Cross ML. Protection against translocating *Salmonella typhimurium* infection in mice by feeding the immuno-enhancing probiotic *Lactobacillus rhamnosus* strain HN001. *Med. Microbiol. Immunol.* 2001; 190: 97–104.
403. Shu Q and Gill HS. A dietary probiotic (*Bifidobacterium lactis* 019) reduces the severity of *Escherichia coli* 0157:H7 infection in mice. *Med. Microbiol. Immunol.* 2001; 189: 147–152.
404. Shu Q, Qu F, and Gill HS. Probiotic treatment using *Bifidobacterium lactis* HN019 reduces weanling diarrhea associated with rotavirus and *Escherichia coli* infection in a piglet model. *J. Pediatr. Gastroenterol Nutr.* 2001; 33: 171–177.
405. Shu Q and Gill HS. Immune protection mediated by the probiotic *Lactobacillus rhamnosus* HN001 (DR20™) against *Escherichia coli* 0157:H7 infection in mice. *FEMS Immunol. Med. Microbiol.* 2002; 34: 59–64.
406. Sazawal S, Dhingra U, Sarkar A, Dhingra P, Deb S, Marwah D, et al. Efficacy of milk fortified with a probiotic *Bifidobacterium lactis* (DR-10™) and prebiotic galacto-oligosaccharides in prevention of morbidity and on nutritional status. *Asia Pac. J. Clin. Nutr.* 2004; 13: S28.
407. Gopal PK, Sullivan PA, and Smart JB. Utilisation of galacto-oligosaccharides as selective substrates for growth by lactic acid bacteria including *Bifidobacterium lactis* DR10 and *Lactobacillus rhmanosus* DR20. *Int. Dairy J.* 2001; 11: 19–25.
408. Ouwehand AC and Philipp S. *Bifidobacterium lactis* HN019; the good taste of health. *Agrofood Ind. Hi-Tech.* 2004; 15: 10–12.
409. Sanders ME. Summary of probiotic activities of Bifidobacterium lactis HN019. *J. Clin. Gastroenterol.* 2006; 40: 776–783.
410. Gill HS and Rutherfurd KJ. Probiotic supplementation to enhance natural immunity in the elderly: effects of a newly characterized immunostimulatory strain *Lactobacillus rhamnosus* HN001 (DR20™) on leucocyte phagocytosis. *Nutr. Res.* 2001; 1: 183–189.
411. Sheih Y-H, Chiang B-L, Wang L-H, Liao C-K, and Gill HS. Systemic immunity-enhancing effects in healthy subjects following dietary consumption of the lactic acid bacterium *Lactobacillus rhamnosus* HN001. *J. Am. Coll. Nutr.* 2001; 20: 149–156.
412. Gill H, Rutherfurd KJ, and Cross ML. Dietary probiotic supplementation enhances natural killer cell activity in the elderly: an investigation of age-related immunological changes. *J. Clin. Immunol.* 2001; 21: 264–271.
413. Gill HS, Cross ML, Rutherfurd KJ, and Gopal PK. Dietary probiotic supplementation to enhance cellular immunity in the elderly. *Br. J. Biomed. Sci.* 2001; 58: 94–96.
414. Sistek D, Kelly R, Wickens K, Stanley T, Fitzharris P, and Crane J. Is the effect of probiotics on atopic dermatitis confined to food sensitized children? *Clin. Exp. Allergy* 2006; 36: 629–633.
415. Arunachalam K, Gill HS, and Chandra RK. Enhancement of natural immune function by dietary consumption of *Bifidobacterium lactis* (HN019). *Eur. J. Clin. Nutr.* 2000; 54: 263–267.

416. Chiang BL, Sheih YH, Wang LH, Liao CK, and Gill HS. Enhancing immunity by dietary consumption of a probiotic lactic acid bacterium (*Bifidobacterium lactis* HN019): optimization and definition of cellular immune responses. *Eur. J. Clin. Nutr.* 2000; 54: 849–855.

417. Gill H, Rutherfurd KJ, Cross ML, and Gopal PK. Enhancement of immunity in the elderly by dietary supplementation with the probiotic *Bifidobacterium lactis* HN019. *Am. J. Clin. Nutr.* 2001; 74: 833–839.

418. Suomalainen TH and Mäyrä-Mäkinen AM. Probionic acid bacteria as protective cultures in fermented milk and breads. *Lait* 1999; 79: 165–174.

419. Yang Z, Suomalainen T, Mäyrä-Mäkinen A, and Huttunen E. Antimicrobial activity of 2-pyrrolidone-5-carboxylic acid produced by lactic acid bacteria. *J. Food Prot.* 1997; 60: 786–790.

420. Ouwehand AC, Lagstrom H, Suomalainen T, and Salminen S. Effect of probiotics on constipation, fecal azoreductase activity and fecal mucin content in the elderly. *Ann. Nutr. Metab.* 2002; 46: 159–162.

421. El-Nezami HS, Polychronaki N, Ma J, Zhu H, Ling W, Salminen EK, Juvonen RO, Salminen SJ, Poussa T, and Mykkanen H. Probiotic supplemention reduces a biomarker for increased risk of liver cancer in young men from Southern China. *Am. J. Clin. Nutr.* 2006; 83: 1199–1203.

422. Salminen S, Benno Y, De Vos W. Intestinal colonisation, microbiota and future probiotics? *Asia Pac. J. Clin. Nutr.* 2006; 15: 558–562.

423. Tuomola EM, Ouwehand AC, and Salminen SJ. Chemical, physical and enzymatic pretreatments of probiotic lactobacilli alter their adhesion to human intestinal mucus glycoproteins. *Int. J. Food Microbiol.* 2000; 60: 75–81.

424. Tuomola EM, Ouwehand AC, and Salminen SJ. The effect of probiotic bacteria on the adhesion of pathogens to human intestinal mucus. *FEMS Immun. Med. Microbiol.* 1999; 26: 137–142.

425. Ouwehand AC, Suomalainen T, Tölkkö S, and Salminen S. *In vitro* adhesion of probiotic acid bacteria to human intestinal mucus. *Lait* 2002; 82: 123–130.

426. Ouwehand AC, Tuomola EM, Tolkko S, and Salminen S. Assessment of adhesion properties of novel probiotic strains to human intestinal mucus. *Int. J. Food Microbiol.* 2001; 64: 119–126.

427. Collado MC, Meriluoto J, and Salminen S. *In vitro* analysis of probiotic strain combinations to inhibit pathogen adhesion to human intestinal mucus. *Food Res. Int.* 2007; 40: 629–636.

428. Lehto EM and Salminen S. Adhesion of two *Lactobacillus* strains, one *Lactococus* and one *Propionibacterium* strain to cultured human intestinal Caco-2 cell line. *Biosci. Microflora* 1997; 16: 13–17.

429. Tuomola EM and Salminen SJ. Adhesion of some probiotic and dairy *Lactobacillus* strains to Caco-2 cell cultures. *Int. J. Food Microbiol.* 1998; 41: 45–51.

430. Ouwehand AC, Isolauri E, Kirjavainen PV, Tolkko S, and Salminen SJ. The mucus binding of *Bifidobacterium lactis* Bb12 is enhanced in the presence of *Lactobacillus* GG and *Lact. delbrueckii* subsp. *bulgaricus*. *Lett. Appl. Microbiol.* 2000; 30: 10–13.

431. Collado MC, Meriluoto J, and Salminen S. Development of new probiotics by strain combinations: is it possible to improve the adhesion to intestinal mucus? *J. Dairy Sci.* 2007; 90: 2710–2716.

432. Kajander K, Krogius-Kurikka L, Rinttilä T, Karjalainen H, Palva A, and Korpela R. Effects of a multispecies probiotic combination on intestinal microbiota and markers of microbial activity in IBS patients. *Aliment. Pharmacol. Ther.* 2007; 26: 463–473.
433. Viljanen M, Savilahti E, Haahtela T, Juntunen-Backman K, Korpela R, Poussa T, Tuure T, and Kuitunen M. Probiotics in the treatment of atopic eczema/dermatitis syndrome in infants: a double-blind placebo-controlled trial. *Allergy* 2005; 60: 494–500.
434. Kajander K, Hatakka K, Poussa T, Färkkilä M, and Korpela R. A probiotic mixture alleviates symptoms in irritable bowel syndrome patients: a controlled 6-month intervention. *Aliment. Pharmacol. Ther.* 2005; 22: 387–394.
435. Kajander K, Myllyluoma E, Rajilic-Stojanovic M, Kyronpalo S, Rasmussen M, Jarvenpaa S, Zoetenda EG, De Vos WM, Vapaatalo H, and Korpela R. Clinical trial: multispecies probiotic supplementation alleviates the symptoms of irritable bowel syndrome and stabilizes intestinal microbiota. *Aliment. Pharmacol. Ther.* 2008; 27: 48–57.
436. Myllyluoma E, Veijola L, Ahlroos T, Tynkkynen S, Kankuri E, Vapaatalo H, Rautelin H, and Korpela R. Probiotic supplementation improves tolerance to *Helicobacter pylori* eradication therapy—a placebo-controlled, double-blind randomized pilot study. *Aliment. Pharmacol. Ther.* 2005; 21: 1263–1272.
437. Myllyluoma E, Ahlroos T, Veijola L, Rautelin H, Tynkkynen S, and Korpela R. Effects of anti-*Helicobacter pylori* treatment and probiotic supplementation on intestinal microbiota. *Int. J. Antimicrob. Agents* 2007; 29: 66–72.
438. Myllyluoma E, Kajander K, Mikkola H, Kyronpalo S, Rasmussen M, Kankuri E, Sipponen P, Vapaatalo H, and Korpela R. Probiotic intervention decreases serum gastrin-17 in *Helicobacter pylori* infection. *Dig. Liver Dis.* 2007; 39: 516–523.
439. Viljanen M, Kuitunen M, Haahtela T, Juntunen-Backman K, Korpela R, and Savilahti E. Probiotic effects on faecal inflammatory markers and on faecal IgA in food allergic atopic eczema/dermatitis syndrome infants. *Pediatr. Allergy Immunol.* 2005; 16: 65–71.
440. Viljanen M, Pohjavuori E, Haahtela T, Korpela R, Kuitunen M, Sarnesto A, Vaarala O, and Savilahti E. Induction of inflammation as a possible mechanism of probiotic effect in atopic eczema-dermatitis syndrome. *J. Allergy Clin. Immunol.* 2005; 115: 1254–1259.
441. Pohjavuori E, Viljanen M, Korpela R, Kuitunen M, Tiittanen M, Vaarala O, and Savilahti E. *Lactobacillus* GG effect in increasing IFN-gamma production in infants with cow's milk allergy. *J. Allergy Clin. Immunol.* 2004; 114: 131–136.
442. Kukkonen K, Savilahti E, Haahtela T, Juntunen-Backman K, Korpela R, Poussa T, Tuure T, and Kuitunen M. Probiotics and prebiotic galacto-oligosaccharides in the prevention of allergic diseases: a randomized, double-blind, placebo-controlled trial. *J. Allergy Clin. Immunol.* 2007; 119: 192–198.
443. Kukkonen K, Nieminen T, Poussa T, Savilahti E, and Kuitunen M. Effect of probiotics on vaccine antibody responses in infancy—a randomized placebo-controlled double-blind trial. *Pediatr. Allergy Immunol.* 2006; 17: 416–421.
444. Hatakka K, Ahola AJ, Yli-Knuuttila H, Richardson M, Poussa T, Meurman JH, and Korpela R. Probiotics reduce the prevalence of oral candida in the elderly—a randomized controlled trial. *J. Dent. Res.* 2007; 86: 125–30.
445. Hatakka K, Blomgren K, Pohjavuori S, Kaijalainen T, Poussa T, Leinonen M, Korpela R, and Pitkäranta A. Treatment of acute otitis media with probiotics in otitis-prone children—a double-blind, placebo-controlled randomised study. *Clin. Nutr.* 2007; 26: 314–321.

446. Hoier E and Hier E. Saaure and gallentoleranz von *Lactobacillus acidophilus* und bifidobackterien. Lebensmittelindustrie und Milchwirtschaft 1992; 26: 769–772.
447. Meile L, Ludwig W, Rueger U, Gut C, Kaufmann P, Dasen G, Wenger S, and Teuber M. *Bifidobacterium lactis* sp. nov., a moderately oxygen tolerant species isolated from fermented milk. *System. Appl. Microbiol.* 1997; 20: 57–64.
448. Hove H, Nordgaard-Andersen I, and Mortensen PB. Effect of lactic acid bacteria on the intestinal production of lactate and short-chain fatty acids, and the absorption of lactose. *Am. J. Clin. Nutr.* 1994; 59: 74–79.
449. Fukushima Y, Li S-T, Hara H, Terada A, and Mitsuoka T. Effect of follow-up formula containing bifidobacteria (NAN BF) on fecal flora and fecal metabolites in healthy children. *Biosci. Microflora* 1997; 16(2): 65–72.
450. Fukushima Y, Kawata Y, Hara H, Terada A, and Mitsuoka T. Effect of a probiotic formula on intestinal immunoglobulin A production in healthy children. *Int. J. Food Microbiol.* 1998; 42: 39–44.
451. Langhendries JP, Detry J, Van Hees J, Lamboray JM, Darimont J, Mozin MJ, Secretin MC, and Senterre J. Effect of a fermented infant formula containing viable bifidobacteria on the fecal flora composition and pH of healthy full-term infants. *J. Pediatr. Gastroenterol. Nutr.* 1995; 21: 177–181.
452. Link-Amster H, Rochat F, Saudan KY, Mignot O, and Aeschlimann JM. Modulation of a specific humoral immune response and changes in intestinal flora mediated through fermented milk intake. *FEMS Immunol. Med. Microbiol.* 1994; 10: 55–63.
453. Larsen CN, Nielsen S, Kæstel P, Brockmann E, Bennedsen M, Christensen HR, Eskesen DC, Jacobsen BL, and Michaelsen KF. Dose-response study of probiotic bacteria *Bifidobacterium animalis* subsp. *lactis* BB-12 and *Lactobacillus paracasei* subsp. *paracasei* CRL-431 in healthy young adults. *Eur. J. Clin. Nutr.* 2006; 60: 1284–1293.
454. Salminen S, Laine M, von Wright A, Vuopio-Varkila J, Korhonen T, and Mattila-Sandholm T. Development of selection criteria for probiotic strains to assess their potential in functional foods: a Nordic and European approach. *Biosci. Microflora* 1996; 15(2): 61–67.
455. Juntunen M, Kirjavainen PV, Ouwehand AC, Salminen SJ, and Isolauri E. Adherence of probiotic bacteria to human intestinal mucus in healthy infants and during rotavirus infection. *Clin. Diag. Lab. Immunol.* 2001; 8: 293–296.
456. Ouwehand AC, Isolauri E, Kirjavainen PV, Tolkko S, and Salminen SJ. The mucus binding of *Bifidobacterium lactis* BB12 is enhanced in the presence of *Lactobacillus* GG and *Lact. delbrueckii* subsp. *bulgaricus*. *Lett. Appl. Microbiol.* 2000; 30: 10–13.
457. Kirjavainen P, Ouwehand A, Isolauri E, and Salminen S. The ability of probiotic bacteria to bind to human intestinal mucus. *FEMS Microbiol. Lett.* 1998; 167: 185–189.
458. Kirjavainen PV, Arvola T, Salminen SJ, and Isolauri E. Aberrant composition of gut microbiota of allergic infants: a target of bifidobacterial therapy at weaning? *Gut* 2002; 51: 51–55.
459. Alander M, Määtö J, Kneifel W, Johansson M, Kögler B, Crittenden R, Mattila-Sandholm T, and Saarela M. Effect of galacto-oligosaccharide supplementation on human faecal microflora and on survival and persistence of *Bifidobacterium lactis* BB-12 in the gastrointestinal tract. *Int. Dairy J.* 2001; 11: 817–825.

460. Fukushima Y, Kawata Y, Mizumachi K, Kurisaki J, and Mitsuoka T. Effect of bifidobacteria feeding on fecal flora and production of immunoglobulins in lactating mouse. *Int. J. Food Microbiol.* 1999; 46: 193–197.
461. Schiffrin EJ, Rochat F, Link-Amster H, Aeschlimann JM, and Donnet-Hughes A. Immunomodulation of human blood cells following the ingestion of lactic acid bacteria. *J. Dairy Sci.* 1995; 78: 491–497.
462. Miettinen M, Alander M, von Wright A, Buopio-Varkila J, Marteau P, Veld J, and Mattila-Sandholm T. The survival of and cytokine induction by lactic acid bacteria after passage through a gastrointestinal model. *Microb. Ecol. Health Dis.* 1998; 10: 141–147.
463. Collado MC, Jalonen L, Meriluoto J, and Salminen S. Protection Mechanism of probiotic combination against human pathogens: *in vitro* adhesion to human intestinal mucus. *Asia Pac. J. Clin. Nutr.* 2006; 15(4): 570–575.
464. Silva AM, Bambirra EA, Oliveira AL, Souza PP, Gomes DA, Vieira EC, and Nicoli JR. Protective effect of bifidus milk on the experimental infection with *Salmonella enteritidis* subsp. *typhimurium* in conventional and gnotobiotic mice. *J. Appl. Microbiol.* 1999; 86: 331–336.
465. Laake KO, Bjorneklett A, Bakka A, Midtvedt T, Norin KE, Eide TJ, Jacobsen MB, Lingaas E, Axelsen AK, Lotveit T, and Vatn MH. Influence of fermented milk on clinical state, fecal bacterial counts and biochemical characteristics in patients with ileal-pouch-anal-anastomosis. *Microb. Ecol. Health Dis.* 1999; 11: 211–217.
466. Wildt S, Munck LK, Vinter-Jensen L, Hansen BF, Nordgaard-Lassen I, Christensen S, Avnstroem S, Rasmussen SN, and Rumessen J. Probiotic treatment of collagenous colitis: a randomized, double-blind, placebo-controlled trial with *L. acidophilus* and *Bifidobacterium animalis* subsp. *lactis. Inflamm. Bowel Dis.* 2006; 12(5): 395–401.
467. McFarland LV. Meta-analysis of probiotics for the prevention of traveler's diarrhea. *Travel Med. Infect. Dis.* 2007; 5: 97–105.
468. Black FT, Anderson PL, Orskov J, Orskov F, Gaarslev K, and Laulund S. Prophylactic efficacy of lactobacilli on traveler's diarrhea. *Travel Med.* 1989; 333–335.
469. Sheu BS, Wu J, Lo CY, Wu HW, Chen JH, Lin YS, and Lin MD. Impact of supplement with Lactobacillus- and Bifidobacterium-containing yoghurt on triple therapy for *Helicobacter pylori* eradication. *Aliment. Pharmacol. Ther.* 2002; 16: 1669–1675.
470. Sheu BY, Cheng H-C, Kao A-W, Wang S-T, Yang Y-J, Yang H-B, and Wu J-J. Pretreatment with Lactobacillus- and Bifidobacterium-containing yogurt can improve the efficacy of Quadruple therapy in eradicating residual Helicobacter pylori infection after failed triple therapy. *Am. J. Clin. Nutr.* 2006; 83: 864–869.
471. Black FT, Einarsson K, Lidbeck A, Orrhage K, and Nord CE. Effect of lactic acid producing bacteria on the human intestinal microflora during ampicillin treatment. *Scand. J. Infect. Dis.* 1991; 23: 247–254.
472. Nord CE, Lidbeck A, Orrhange K, and Sjostedt S. Oral supplementation with lactic acid bacteria during intake of clindamycin. *Clin. Microbiol. Infect.* 1997; 3(1): 124–132.
473. Frieling T. Functional and inflammatory bowel disorders. *Med. Klin. (Munich)* 2006; 101 (Suppl. 1): 139–142.
474. Saavedra JM, Bauman NA, Oung I, Perman JA, and Yolken RH. Feeding of *Bifidobacterium bifidum* and *Streptococcus thermophilus* to infants in hospital for prevention of diarrhea and shedding of rotavirus. *Lancet* 1994; 344: 1046–1049.

475. Harmsen HJ, Wildeboer-Veloo AC, Raangs GC, Wagendorp AA, Klijn N, Bindels JG, and Welling GW. Analysis of intestinal flora development in breast-fed and formula-fed infants by using molecular identification and detection methods. *J. Pediatr. Gastroenterol. Nutr.* 2000; 30: 61–67.

476. Haschke F, Wang W, Ping G, Varavithya W, Podhipak A, Rochat F, Link-Amster H, Pfeifer A, Diallo-Ginstl E, and Steenhout P. Clinical trials prove the safety and efficacy of the probiotic strain Bifidobacterium BB12 in follow-up formula and growing-up milks. *Monatsschr. Kinderheilkd.* 1998; 146(Suppl. 1): S26–S30.

477. Chouraqui JP, Van Egroo LD, and Fichot MC. Acidified milk formula supplemented with *Bifidobacterium lactis*: impact on infant diarrhea in residential care settings. *J. Pediatr. Gastroenterol. Nutr.* 2004; 38: 288–292.

478. Saavedra JM, Abi-Hanna A, Moore N, and Yolken RH. Long-term consumption of infant formulas containing live probiotic bacteria: tolerance and safety. *Am. J. Clin. Nutr.* 2004; 79: 261–267.

479. Nopchinda S, Varavithya W, Phuapradit P, Sangchai R, Suthutvoravut U, Chantraruksa V, and Haschke F. Effect of *Bifidobacterium* BB12 with or without *Streptococcus thermophilus* supplemented formula on nutritional status. *J. Med. Assoc. Thai.* 2002; 85(Suppl. 4): 1225–1231.

480. Bin-Nun A, Bromiker R, Wilschanski M, Kaplan M, Rudensky B, Caplan M, and Hammerman C. Oral probiotics prevent necrotizing enterocolitis in very low birth weight neonates. *J. Pediatr.* 2005; 147: 192–196.

481. Alm L, Ryd-Kjellen E, Setterberg G, Blomquist L, Effect of a new fermented milk product "CULTURA" on constipation in geriatric patients. In: *1st Lactic Acid Bacteria Computer Conference Proceedings*. Horizon Scientific Press, Norfolk, England, 1993.

482. Pitkälä KH, Strandberg TE, Finne Soveri UH, Ouwehand AC, Poussa T, and Salminen S. Fermented cereal with specific bifidobacteria normalizes bowel movements in elderly nursing home residents. A randomized, controlled trial. *J. Nutr. Health Aging.* 2007; 11(4): 305–311.

483. Virta P, Otterström K, Niemi L, Wieser-Aho M-T, Lähteenmäki A-L, and Leppänen T, The effect of a preparation containing freeze-dried lactic acid bacteria on lactose intolerance, 1993, External report.

484. Hatcher G and Lambrecht R. Augmentation of macrophage phagocytic activity by cell-free extracts of selected lactic acid-producing bacteria. *J. Dairy Sci.* 1993; 76: 2485–2492.

485. Matsumoto M, Hara K, and Benno Y. The influence of the immunostimulation by bacterial cell components derived from altered large intestinal microbiota on probiotic anti-inflammatory benefits. *FEMS Immunol. Med. Microbiol.* 2007a; 49: 387–390.

486. Miettinen M, Vuopio-Varkila J, and Varkila K. Production of human tumor necrosis factor alpha, interleukin-6 and interleukin 10 is induced by lactic acid bacteria. *Infect. Immun.* 1996; 64(12): 5403–5405.

487. Kankaanpaa P, Sütas Y, Salminen S, and Isolauri E. Homogenates derived from probiotic bacteria provide down-regulatory signals for peripheral blood mononuclear cells. *Food Chem.* 2003; 83: 269–277.

488. Rautava S, Arvilommi H, and Isolauri E. Specific probiotics in enhancing maturation of IgA responses in formula-fed infants. *Pediatr. Res.* 2006; 60(2): 222–225.

489. Tejada-Simon MV, Lee JH, Ustunol Z, and Pestka JJ. Ingestion of yogurt containing *Lactobacillus acidophilus* and *Bifidobacterium* to potentiate Immunoglobulin A responses to cholera toxin in mice. *J. Dairy Sci.* 1999; 82: 649–660.
490. Isolauri E, Arvola T, Sutas Y, Moilanen E, and Salminen S. Probiotics in the management of atopic eczema. *Clin. Exp. Allergy* 2000; 30: 1604–1610.
491. Matsumoto M, Aranami A, Ishige A, Watanabe K, and Benno Y. LKM512 yogurt consumption improves the intestinal environment and induces the T-helper type 1 cytokine in adult patients with intractable atopic dermatitis. *Clin. Exp. Allergy* 2007b; 37(3): 358–370.
492. Rautava S, Probiotics in maturing the immune system and reducing the risk of disease in infancy. Ph. D. Thesis, University of Turku, 2005; ISBN 951-29-2923-6.
493. Ellegaard J, Peterslund NA, Black FT, Infection prophylaxis in neutropenic patients by oral administration of Lactobacilli. *The Seventh International Symposium on Infections in the Immunocompromised Host*, 1992, June 21–24, Boulder, CO.
494. Wagner RD, Pierson C, Warner T, Dohnalek M, Farmer J, Roberts L, Hilty M, and Balish E. Biotherapeutic effects of probiotic bacteria on candidiasis in immunodeficient mice. *Infect. Immun.* 1997; 65(10): 4165–4172.
495. Wagner RD, Warner T, Pierson C, Roberts L, Farmer J, Dohnalek M, Hilty M, and Balish E. Biotherapeutic effects of *Bifidobacterium* spp. on orogastric and systemic candidiasis in immunodeficient mice. *Rev. Iberoam. Micol.* 1998; 15: 265–270.
496. Biffi A, Coradini D, Larsen R, Riva L, and DiFronzo G. Antiproliferative effect of fermented milk on the growth of a human breast cancer cell line. *Nutr. Cancer* 1997; 28(1): 93–99.
497. Lo P-R, Yu R-C, Chou C-C, and Tsai Y-H. Antimutagenic activity of several probiotic bacteria against benzoapyrene. *J. Biosci. Bioeng.* 2002; 94(2): 148–153.
498. Lo P-R, Yu R-C, Chou C-C, and Huang E-C. Determinations of the antimutagenic activities of several probiotic bifidobacterias under acidic and bile conditions against benzo[a]pyrene by a modified Ames test. *Int. J. Food Microbiol.* 2004; 93: 249–257.
499. El-Gawad IAA, El-Sayed EM, Hafez SA, El-Zeini HM, and Saleh FA. Inhibitory effect of yoghurt and soya yoghurt containing bifidobacteria on the proliferation of Ehrlich ascites tumour cells *in vitro* and *in vivo* in a mouse tumour model. *Br. J. Nutr.* 2004; 92: 81–86.
500. Peltonen KD, El-Nezami HS, Salminen SJ, and Ahokas JT. Binding of aflatoxin B1 by probiotic bacteria. *J. Sci. Food Agric.* 2000; 80: 1942–1945.
501. Burns AJ and Rowland IR. Antigenotoxicity of probiotics and prebiotics on feacal water-induced DNA damage in human colon adenocarcinoma cells. *Mut. Res.* 2004; 551: 233–243.
502. Commane DM, Shortt CT, Silvi S, Cresci A, Hughes RM, and Rowland IR. Effects of fermentation products of pro- and prebiotics on trans-epithelial electrical resistance in an *in vitro* model of the colon. *Nutr. Cancer* 2005; 51(1): 102–109.
503. Meriluoto J, Gueimonde M, Haskard CA, Spoof L, Sjöwall O, and Salminen S. Removal of the cyanobacterial toxin microcystin-LR by human probiotics. *Toxicon* 2005; 46: 111–114.
504. Raipulis J, Toma MM, and Semjonovs P. The effect of probiotics on the genotoxicity of furazolidone. *Int. J. Food Microbiol.* 2005; 102: 343–347.
505. Kim H-Y, Bae H-S, and Baek Y-J. In vivo antitumor effects of lactic acid bacteria on sarcoma 180 and mouse lewis lung carcinoma. *J. Korean Cancer* 1991; 23: 188–195.

506. Femia AP, Luceri C, Dolara P, Giannini A, Biggeri A, Salvadori M, Clune Y, Collins KJ, Paglierani M, and Caderni G. Antitumorigenic activity of the prebiotic inulin enriched with oligofructose in combination with the probiotics *Lactobacillus rhamnosus* and *Bifidobacterium lactis* on azoxymethane-induced colon carcinogenesis in rats. *Carcinogenesis* 2002; 23: 1953–1960.

507. Klinder A, Förster A, Caderni G, Femia AP, and Pool-Zobel BL. Fecal water genotoxicity is predictive of tumor preventive activities by inulin-like oligofructoses, probiotics (*Lactobacillus rhamnosus* and *Bifidobacterium lactis*), and their synbiotic combination. *Nutr. Cancer* 2004; 49(2): 144–155.

508. Roller M, Femia AP, Caderni G, Rechkemmer G, and Watzl B. Intestinal immunity of rats with colon cancer is modulated by oligofructose-enriched inulin conbined with *Lactobacillus rhamnosus* and *Bifidobacterium lactis*. *Br. J. Nutr.* 2004; 92: 931–938.

509. Matsumoto M and Benno Y. Consumption of *Bifidobacterium lactis* LKM512 yogurt reduces gut mutagenicity by increasing gut polyamine contents in healthy adult subjects. *Mut. Res.* 2004; 568: 147–153.

510. Collins JK, Clune Y, Meaney K, o'Donoghue M, Klinder A, Roller M, Karlsson P, Bennett M, O'Riordan M, Dunne C, O'Sullivan G, Rafter J, Watzl B, Rechkemmer G, and Pool-Zobel BL. The influence of synbiotic consumption on cancer risk biomarkers in previously resected colon cancer subjects. *J. Biotech.* 2005; 118S1: S1–S1189

511. Obradovic D, Curic M, Ivanovic M, Trbojevic B, Djordjevic M. Probiotic function of the fermented milk Jogurt Plus. *FEMS Conference (Fifth Symposium on Lactic Acid Bacteria)*, Holland, September 8–12, 1996.

512. Cardona ME, Vanay VV de, Midtvedt T, and Norin KE. Probiotics in gnotobiotic mice. Conversion of cholesterol to coprostanol *in vitro* and *in vivo* and bile acid deconjugation *in vitro*. *Microb. Ecol. Health Dis.* 2000; 12: 219–24.

513. El-Gawad IA, El-Sayed EM, Hafez SA, El-Zeini HM, and Saleh FA. The hypo cholesterolaemic effect of milk yoghurt and soy–yoghurt containing bifidobacteria in rats fed on a cholesterol-enriched diet. *Int. Dairy J.* 2005; 15: 37–44.

514. Klaver F, van der Meer R. The assumed assimilation of cholesterol by Lactobacilli and *Bifidobacterium bifidum* is due to their bile salt-deconjugating activity. *Appl. Environ. Microb.* 1993; 59(4): 1120–1124.

515. Abi-Hanna A, Moore N, Yolken R, and Saavedra J. Long term consumption of infant formulas with live probiotic bacteria: safety and tolerance. *J Pediatr Gastroenterol Nutr.* 1998; 27(4): 484

516. Asahara T, Nomoto K, Shimizu K, Watanuki M, and Tanaka R. Increased resistance of mice to *Salmonella enterica* serovar typhimurium infection by synbiotic administration of Bifidobacteria and transgalactosylated oligosaccharides. *J. Appl. Microbiol.* 2001; 91: 985–996.

517. Asahara T, Shimizu K, Nomoto K, Hamabata T, Ozawa A, and Takeda Y. Probiotic bifidobacteria protect mice from lethal infection with Shiga toxin-producing *Escherichia coli* O157:H7. *Infect. Immun.* 2004; 72: 2240–2247.

518. Asahara T, Shimizu K, Nomoto K, Watanuki M, and Tanaka R. Antibacterial effect of fermented milk containing *Bifidobacterium breve*, *Bifidobacterium bifidum* and *Lactobacillus acidophilus* against indigenous *Escherichia coli* infection in mice. *Microb. Ecol. Health Dis.* 2001; 13: 1–24.

519. De Preter V, Vanhoutte T, Huys G, Swings J, De Vuyst L, Rutgeerts P, and Verbeke K. Effects of *Lactobacillus casei* Shirota, *Bifidobacterium breve*, and oligofructose-enriched inulin on colonic nitrogen–protein metabolism in healthy humans. *Am. J. Physiol. Gastrointest. Liver Physiol.* 2007; 292: G358–G368
520. Hotta M, Sato Y, Iwata S, Yamashita N, Sunakawa K, Oikawa T, Tanaka R, Watanabe K, Takayama H, Yajima M, et al. Clinical effects of *Bifidobacterium* preparations on pediatric intractable diarrhea. *Keio J. Med.* 1987; 36: 298–314.
521. Ishikawa H, Akedo I, Umesaki Y, Tanaka R, Imaoka A, and Otani T. Randomized controlled trial of the effect of bifidobacteria-fermented milk on ulcerative colitis. *J. Am. Coll. Nutr.* 2003; 22: 56–63.
522. Kanamori Y, Hashizume K, Kitano Y, Tanaka Y, Morotomi M, Yuki N, and Tanaka R. Anaerobic dominant flora was reconstructed by synbiotics in an infant with MRSA enteritis. *Pediatr. Int.* 2003; 45: 359–362.
523. Kanamori Y, Hashizume K, Sugiyama M, Morotomi M, and Yuki N. Combination therapy with *Bifidobacterium breve*, *Lactobacillus casei*, and galactooligosaccharides dramatically improved the intestinal function in a girl with short bowel syndrome: a novel synbiotics therapy for intestinal failure. *Dig. Dis. Sci.* 2001; 46: 2010–2016.
524. Kanamori Y, Hashizume K, Sugiyama M, Mortomi M, Yuki N, and Tanaka R. A novel synbiotic therapy dramatically improved the intestinal function of a pediatric patient with laryngotracheo-esophageal cleft (LTEC) in the intensive care unit. *Clin. Nutr.* 2002; 21: 527–530.
525. Kanamori Y, Sugiyama M, Hashizume K, Yuki N, Morotomi M, and Tanaka R. Experience of long-term synbiotic therapy in seven short bowel patients with refractory enterocolitis. *J. Pediatr. Surg.* 2004; 39: 1686–1692.
526. Kanazawa H, Nagino M, Kamiya S, Komatsu S, Mayumi T, Takagi K, Asahara T, Nomoto K, Tanaka R, and Nimura Y. Synbiotics reduce postoperative infectious complications: a randomized controlled trial in biliary cancer patients undergoing hepatectomy. *Langenbecks Arch. Surg.* 2005; 390: 104–113.
527. Kato K, Mizuno S, Umesaki Y, Ishii Y, Sugitani M, Imaoka A, Otsuka M, Hasunuma O, Kurihara R, Iwasaki A, and Arakawa Y. Randomized placebo-controlled trial assessing the effect of bifidobacteria-fermented milk on active ulcerative colitis. *Aliment. Pharmacol. Ther.* 2004; 20: 1133–1141.
528. Kikuchi-Hayakawa H, Onodera N, Matsubara S, Yasuda E, Chonan O, Takahashi R, and Ishikawa F. Effects of soy milk and *Bifidobacterium* fermented soy milk on lipid metabolism in aged ovariectomized rats. *Biosci. Biotechnol. Biochem.* 1998; 62: 1688–1692.
529. Kikuchi-Hayakawa H, Onodera N, Matsubara S, Yasuda E, Shimakawa Y, and Ishikawa F. Effects of soya milk and *Bifidobacterium*-fermented soya milk on plasma and liver lipids, and faecal steroids in hamsters fed on a cholesterol-free or cholesterol-enriched diet. *Br. J. Nutr.* 1998; 79: 97–105.
530. Kikuchi-Hayakawa H, Onodera-Masuoka N, Kano M, Matsubara S, Yasuda E, and Ishikawa F. Effect of soy milk and *Bifidobacterium*-fermented soy milk on plasma and liver lipids in ovariectomized Syrian hamsters. *J. Nutr. Sci. Vitaminol. (Tokyo)* 2000; 46: 105–108.
531. Kitajima H, Sumida Y, Tanaka R, Yuki N, Takayama H, and Fujimura M. Early administration of *Bifidobacterium breve* to preterm infants: randomised controlled trial. *Arch. Dis. Child. Fetal. Neonatal. Ed.* 1997; 76: F101–F107.

532. Matsumoto S, Watanabe N, Imaoka A, and Okabe Y. Preventive effects of *Bifidobacterium*- and *Lactobacillus*-fermented milk on the development of inflammatory bowel disease in senescence-accelerated mouse P1/Yit strain mice. *Digestion* 2001; 64: 92–99.

533. Ohta T, Nakatsugi S, Watanabe K, Kawamori T, Ishikawa F, Morotomi M, Sugie S, Toda T, Sugimura T, and Wakabayashi K. Inhibitory effects of *Bifidobacterium*-fermented soy milk on 2-amino-1-methyl-6-phenylimidazo[4, 5-b]pyridine-induced rat mammary carcinogenesis, with a partial contribution of its component isoflavones. *Carcinogenesis* 2000; 21: 937–941.

534. Onoue M, Kado S, Sakaitani Y, Uchida K, and Morotomi M. Specific species of intestinal bacteria influence the induction of aberrant crypt foci by 1, 2-dimethylhydrazine in rats. *Cancer Lett.* 1997; 113: 179–186.

535. Shima T, Fukushima K, Setoyama H, Imaoka A, Matsumoto S, Hara T, Suda K, and Umesaki Y. Differential effects of two probiotic strains with different bacteriological properties on intestinal gene expression, with special reference to indigenous bacteria. *FEMS Immunol. Med. Microbiol.* 2008; 52: 69–77.

536. Sugawara G, Nagino M, Nishio H, Ebata T, Takagi K, Asahara T, Nomoto K, and Nimura Y. Perioperative synbiotic treatment to prevent postoperative infectious complications in biliary cancer surgery: a randomized controlled trial. *Ann. Surg.* 2006; 244: 706–714.

537. Takahashi T and Morotomi M. Absence of cholic acid 7 alpha-dehydroxylase activity in the strains of *Lactobacillus* and *Bifidobacterium*. *J. Dairy Sci.* 1994; 77: 3275–3286.

538. Tojo M, Oikawa T, Morikawa Y, Yamashita N, Iwata S, Satoh Y, Hanada J, and Tanaka R. The effects of *Bifidobacterium breve* administration on campylobacter enteritis. *Acta Paediatr. Jpn.* 1987; 29: 160–167.

539. Uchida K, Takahashi T, Inoue M, Morotomi M, Otake K, Nakazawa M, Tsukamoto Y, Miki C, and Kusunoki M. Immunonutritional effects during synbiotics therapy in pediatric patients with short bowel syndrome. *Pediatr. Surg. Int.* 2007; 23: 243–248.

540. Colombel JF, et al. Yogurt with *Bifidobacterium longum* reduces erythromycin-induced gastrointestinal effects. *Lancet* 1987; 2(2549): 43

541. Namba K, et al. Inhibitory effects of *Bifidobacterium longum* on enterohemorrhagic *Escherichia coli* O157:H7. *Biosci. Microflora* 2003; 22: 85–91.

541a. Namba K, et al. Effects of ingestion of *Bifidobacterium longum* BB536 on the phylaxis to influenza virus infections in the elderly. *J. Intest. Microbiol* (Jpn). 2006; 20: 133.

542. Odamaki T, et al. Fluctuation of fecal microbiota in individuals with Japanese cedar pollinosis during the pollen season and influence of probiotic intake. *J. Invest. Allergol. Clin. Immunol.* 2007; 17: 92–100.

543. Ogata T, et al. Effect of *Bifidobacterium longum* BB536 administration on the intestinal environment, defecation frequency and fecal characteristics of human volunteers. *Biosci. Microflora* 1997; 16: 53–58.

544. Reddy BS and Rivenson A. Inhibitory effects of *Bifidobacterium longum* on colon, mammary, and liver carcinogenesis induced by 2-amino-3-methylimidazo[4, 5-f]quinoline, a food mutagen. *Cancer Res.* 1993; 53: 3914–3918.

545. Sekine I, et al. Effects of *Bifidobacterium* containing milk on chemiluminescence reaction of peripheral leukocytes and mean corpuscular volume of red blood cells: a possible role

of *Bifidobacterium* on activation of macrophages. *Biomed. Therapeut. (Jpn.)* 1985; 14: 691–695.

546. Sekine K, et al. Comparison of the TNF-α levels induced by human-derived *Bifidobacterium longum* and rat-derived *Bifidobacterium animalis* in mouse peritoneal cells. *Bifidobacteria Microflora* 1994; 13: 79–89.

547. Xiao JZ, et al. Effect of Probiotic *Bifidobacterium longum* BB536 in relieving clinical symptoms and modulating plasma cytokine levels of Japanese cedar pollinosis during the pollen season, a randomized, double-blind, placebo-controlled trial. *J. Invest. Allergol. Clin. Immunol.* 2006a; 16: 86–93.

548. Xiao JZ, et al. Probiotics in the treatment of Japanese cedar pollinosis: a double-blind placebo-controlled trial. *Clin. Exp. Allergy* 2006b; 36: 1425–1435.

549. Xiao JZ, et al. Clinical efficacy of probiotic *Bifidobacterium longum* for the treatment of symptoms of Japanese cedar pollen allergy in subjects evaluated in an environmental exposure unit. *Allergol. Int.* 2007; 56: 67–75.

550. Yaeshima T, et al. Effect of yogurt containing *Bifidobacterium longum* BB536 on the intestinal environment, fecal characteristics and defecation frequency: a comparison with standard yogurt. *Biosci. Microflora* 1997; 16: 73–77.

551. Yamazaki S, et al. Protective effect of Bifidobacterium-monoassociation against lethal activity of *Escherichia coli*. *Bifidobacteria Microflora* 1982; 1: 55–59.

552. European Food Safety Authority Scientific, Committee. EFSA public consultation on the Qualified Presumption of Safety (QPS) approach for the safety assessment of microorganisms deliberately added to food and, feed. Annex 3: assessment of Gram positive non-sporulating bacteria with respect to a qualified presumption of safety, 2007. http://www.efsa.europa.eu/en/science/sc_commitee/sc_consultations/sc_consultation_qps.html.

553. Halttunen T, Salminen S, and Tahvonen R. Rapid removal of lead and cadmium from water by specific lactic acid bacteria. *Int. J. Food Microbiol.* 2007; 114: 30–35.

554. Hutt P, Shchepetova J, Loivukene K, Kullisaar T, and Mikelsaar M. Antagonistic activity of probiotic lactobacilli and bifidobacteria against entero- and uropathogens. *J. Appl. Microbiol.* 2006; 100: 1324–1332.

555. Kiviharju K, Leisola M, von Weymarn N. Light sensitivity of *Bifidobacterium longum* in bioreactor cultivations. *Biotechnol. Lett.* 2004; 26: 539–542.

556. Lahtinen SJ. New insights into the viability of probiotic bacteria. Ph. D. thesis, University of Turku, Finland 2007.

557. Lahtinen SJ, Gueimonde M, Ouwehand A, Reinikainen JP, and Salminen S. Comparison of four methods to enumerate probiotic bifidobacteria in a fermented food product. *Food Microbiol.* 2006; 23: 571–577.

558. Lahtinen SJ, Gueimonde M, Ouwehand AC, Reinikainen JP, and Salminen SJ. Probiotic bacteria may become dormant during storage. *Appl. Environ. Microbiol.* 2005; 71: 1662–1663.

559. Lahtinen SJ, Jalonen L, Ouwehand AC, and Salminen SJ. Specific *Bifidobacterium* strains isolated from elderly subjects inhibit growth of *Staphylococcus aureus*. *Int. J. Food Microbiol.* 2007; 117: 125–128.

560. Lahtinen SJ, Tammela L, Korpela J, Parhiala R, Ahokoski H, Mykkänen H, and Salminen S. Probiotics modulate the *Bifidobacterium* microbiota of elderly nursing home residents Age. Accepted 2008.

561. Lahtinen SJ, Ouwehand AC, Reinikainen JP, Korpela JM, Sandholm J, and Salminen SJ. Intrinsic properties of so-called dormant probiotic bacteria, determined by flow cytometric viability assays. *Appl. Environ. Microbiol.* 2006; 72: 5132–5134.
562. Lahtinen SJ, Ouwehand AC, Salminen SJ, Forssell P, and Myllarinen P. Effect of starch- and lipid-based encapsulation on the culturability of two *Bifidobacterium longum* strains. *Lett. Appl. Microbiol.* 2007; 44: 500–505.
563. Laine R, Salminen S, Benno Y, and Ouwehand AC. Performance of bifidobacteria in oat-based media. *Int. J. Food Microbiol.* 2003; 83: 105–109.
564. Mäkeläinen H, Tahvonen R, Salminen S, and Ouwehand AC. *In vivo* safety assessment of two *Bifidobacterium longum* strains. *Microbiol. Immunol.* 2003; 47: 911–914.
565. Nybom SM, Salminen SJ, and Meriluoto JA. Removal of microcystin-LR by strains of metabolically active probiotic bacteria. *FEMS Microbiol. Lett.* 2007; 270: 27–33.
566. Ouwehand AC, Bergsma N, Parhiala R, Gueimonde M, Lahtinen SJ, Pitkälä KH, and Salminen S. *Bifidobacterium* microbiota and parameters of immune function in elderly subjects. *FEMS Immunol. Med. Microbiol.* 2008; 53: 18–25.
567. Pitkälä KH, Strandberg TE, Finne-Soveri UH, Ouwehand AC, Poussa T, and Salminen S. Fermented cereal with specific bifidobacteria normalizes bowel movements in elderly nursing home residents. A randomized, controlled trial. *J. Nutr. Health Aging* 2007; 11: 305–311.
568. Salminen S, Ouwehand A, Salminen E, and Isolauri E. Method for screening probiotic strains of the genus *Bifidobacterium*. Patent WO 02/38798, 2002.

PART II

PREBIOTICS

7

PREBIOTICS

Ross Crittenden[1] and Martin J. Playne[2]
[1]*Food Science Australia, Australia*
[2]*Melbourne Biotechnology, Australia*

Our intestinal tract is colonized by a complex ecosystem of microorganisms that increase in numbers from 10^2 to 10^4 per gram of contents in the stomach, to 10^6–10^8 per gram in the small intestine, and 10^{10}–10^{12} per gram in the colon (1). It has also become increasingly clear that these bacteria are not merely commensals, but have coevolved with us in a truly symbiotic relationship. Our intestinal microbes provide us with a barrier to infection by intestinal pathogens (2), provide much of the metabolic fuel for our colonic epithelial cells (3), and contribute to normal immune development and function (4, 5). Members of the intestinal microbiota can also be involved in acute and chronic diseases such as antibiotic-associated diarrhea (6) and inflammatory bowel disease (IBD) (7). Undesirable metabolic activity of the intestinal microbiota may play a role in the development of colorectal cancer (8, 9). Hence, it is reasonable to hypothesize that modifying the intestinal microbiota to develop, restore or maintain a beneficial balance of microorganisms and microbial activities may improve health.

7.1 THE PREBIOTIC CONCEPT

A number of different strategies can be applied to modify microbial intestinal populations. Antibiotics can be effective in eliminating pathogenic organisms within the intestinal microbiota. However, they carry the risk of side effects and cannot be routinely used for longer periods or prophylactically.

Handbook of Probiotics and Prebiotics, Second Edition   Edited by Yuan Kun Lee and Seppo Salminen
Copyright © 2009 John Wiley & Sons, Inc.

The consumption of probiotics aims to directly supplement the intestinal microbiota with live beneficial organisms. Lactobacilli and bifidobacteria are numerically common members of the human intestinal microbiota, and are nonpathogenic, nonputrefactive, nontoxigenic, saccharolytic organisms that appear from available knowledge to provide little opportunity for deleterious activity in the intestinal tract (10, 11). As such, they are reasonable candidates to target in terms of restoring a favorable balance of intestinal species.

Prebiotics represent a third strategy to manipulate the intestinal microbiota. Rather than supplying an exogenous source of live bacteria, prebiotics are nondigestible food ingredients that selectively stimulate the proliferation and/or activity of desirable bacterial populations *already resident* in the consumer's intestinal tract. Most prebiotics identified so far are nondigestible, fermentable carbohydrates. Intestinal populations of bifidobacteria, in particular, are stimulated to proliferate upon consumption of a range of prebiotics, increasing in numbers by as much as 10–100-fold in feces (12, 13).

There is an obvious potential to use prebiotics and probiotics together in a complementary and synergistic manner. Therefore, foods containing both probiotic and prebiotic ingredients have been termed synbiotics (14).

7.2 A BRIEF HISTORY OF PREBIOTICS

TABLE 7.1 Seminal Events in the Development of Prebiotics

Year	Event	References
1900	Tissier described bifidobacteria (*Bacillus bifidus*) in the feces of babies.	(15)
1906	Tissier proposed the oral feeding of bifidobacteria to prevent infant diarrhea.	(16)
1954	Gyorgy reported that components of human milk (*N*-acetyl-glucosamine) promoted the growth of a *Bifidobacterium* strain.	(17)
1957	Petuely recognized lactulose as a "bifidus" factor.	(18)
1970s–1980s	Japanese researchers discovered that a number of different nondigestible oligosaccharides were "bifidus" factors.	(19, 20)
1995	The term "prebiotic" coined by Gibson and Roberfroid to link the concepts of prebiotics and probiotics for promoting beneficial populations of intestinal bacteria.	(14)

7.3 ADVANTAGES AND DISADVANTAGES OF THE PREBIOTIC STRATEGY

The prebiotic strategy offers a number of advantages over modifying the intestinal microbiota using probiotics or antibiotics.

Advantages over probiotics

- Stable in long shelf life foods and beverages;
- Heat and pH stable and can be used in a wide range of processed foods and beverages;
- Have physicochemical properties useful to food taste and texture;
- Resistant to acid, protease, and bile during intestinal passage;
- Stimulate organisms already resident in the host, and so avoid host/strain compatibilities, and the need to compete with an already established microbiota;
- Stimulate fermentative activity of the microbiota and health benefits from SCFA (short chain fatty acids);
- Lower intestinal pH and provide osmotic water retention in the gut.

Advantages over antibiotics

- Safe for long-term consumption and prophylactic approaches;
- Do not stimulate side effects such as antibiotic-associated diarrhea, sensitivity to UV radiation, or liver damage;
- Do not stimulate antimicrobial resistance genes;
- Not allergenic.

Disadvantages of prebiotics

- Unlike probiotics, overdose can cause intestinal bloating, pain, flatulence, or diarrhea.
- Not as potent as antibiotics in eliminating specific pathogens.
- May exacerbate side effects of simple sugar malabsorption during active diarrhea.

A consumed probiotic strain must compete with an already established microbiota, and in most cases they persist only transiently in the intestine (21–26). Individuals also harbor their own specific combination of species and unique strains within their intestinal bacteria (27–34) suggesting that certain host–microbiota compatibilities exist. By targeting those strains that are already resident in the intestinal tract of an individual, the prebiotic strategy overcomes the need for probiotic bacteria to compete with intestinal bacteria that are well established in their niche.

7.4 TYPES OF PREBIOTICS

Most identified prebiotics are carbohydrates. Within these, there is a wide diversity of molecular structures. However, these carbohydrates share a number of physiological traits important to their beneficial effects. They are

- nondigestible (or only partially digested);
- nonabsorbable in the small intestine;
- poorly fermented by bacteria in the mouth;
- well fermented by purportedly beneficial bacteria in the gut;
- poorly fermented by potentially pathogenic bacteria in the gut.

To date, the largest number of reported studies and the most consistent evidence accumulated for prebiotic effects have been for several nondigestible oligosaccharides (NDOs) (Table 7.2). These include fructooligosaccharides (FOS), and the polyfructan inulin, galactooligosaccharides (GOS) and lactulose. A number of other NDOs, to which less rigorous study has been so far applied, have at least indications of prebiotic potential. These include lactosucrose, xylo- (XOS), isomalto- (IMO), and soybean (SOS) oligosaccharides. Indeed, it appears that a wide range of NDOs can stimulate the growth of bifidobacteria and new potential prebiotics continue to emerge. *In vitro* and animal feeding trial data showing potential bifidogenic effects have been reported for gluco- and galactomannan oligosaccharides (35, 36), alpha-glucooligosaccharides (37),

TABLE 7.2 Known Prebiotics and the Major Emerging Prebiotics

Prebiotics	Effects on the Intestinal Microbiota	References
Fructans (FOS and inulin)	Prebiotics	
	Good evidence of prebiotic effects in numerous randomized, double blind, placebo-controlled human feeding studies	(13, 41–48)
	Doses generally 7–15 g/day, though doses as low as 2.5 g/day FOS and 5 g/day inulin are bifidogenic	(49, 50)
	Both luminal and mucosal populations of bifidobacteria (and lactobacilli) respond in proximal and distal colon	(51)
	Prebiotic potency is inversely related to molecular weight of fructans	(41)
	Inulin fermented more distally in the colon than FOS	(52)
GOS	Prebiotic	
	Good evidence of prebiotic effects in human infants, but results in adults have been mixed	(13, 53–55)
	Effective doses usually between 10 and 20 g/day	
	Stimulates the proliferation of lactobacilli as well as bifidobacteria	
Lactulose	Prebiotic	
	Good evidence of prebiotic effects in a number of randomized, double-blind, placebo-controlled human feeding studies	(56–59)
	Effective doses usually in the order of 8–20 g/day	(60–62)

TABLE 7.2 (*Continued*)

Prebiotics	Effects on the Intestinal Microbiota	References
	Stimulates the proliferation of lactobacilli as well as bifidobacteria	
Lactitol	Probable prebiotic	
	Similar effects to lactulose, though slightly less potent	(56, 63, 64)
Xylooligosaccharides	Potential prebiotics	
	In vitro studies show XOS are efficiently and relatively selectively fermented by bifidobacteria with the intestinal microbiota	(65–71)
	When compared to other NDOs including FOS in two independent trials in rodents, XOS induced the largest increase in fecal bifidobacterial numbers	(72, 73)
	Dearth of human studies	
Isomaltooligosaccharides	Potential prebiotics	
	Considerable digestion of IMO occurs during intestinal transit, but some arrives in the colon	(74, 75)
	Doses 10–20 g/day shown bifidogenic effects in small, uncontrolled feeding trials	(76–79)
	Doses <10 g/day failed to produce a bifidogenic effect in a double blind, placebo-controlled human feeding trial	(58)
Soybean oligosaccharides (raffinose and stachyose)	Probable prebiotics	
	Evidence of prebiotic effects in double-blind placebo-controlled human feeding studies	(58, 80)
	Earlier evidence of prebiotic effects in a number of noncontrolled human feeding studies	(81–84)
	Effective doses as low as 2–4 g/day	(58, 80, 82, 84)
Lactosucrose	Possible prebiotic	
	Preliminary evidence of prebiotic effects in a number of noncontrolled human feeding studies	(85–88)
Resistant starch	Possible prebiotic	
	Preliminary evidence of prebiotic effects from numerous animal studies; often both bifidobacteria and lactobacilli are stimulated	(89–97)
Cereal fibers	Possible prebiotic effects	
	Arabinoxylans are relatively selectively fermented by some bifidobacteria	(68–70)
	Bifidogenic effect in mice and a small human trial	(98, 99)

pectic-oligosaccharides (38), gentiooligosaccharides, and oligosaccharides from agarose (39) among others.

The evidence that some polysaccharide dietary fibers, such as resistant starches, arabinoxylan, and plant gums (40) have prebiotic potential is accumulating, but to date remains limited largely to *in vitro* and animal studies.

7.5 PRODUCTION OF PREBIOTICS

Some prebiotics can be extracted from plant sources, but most are synthesized commercially using enzymatic or chemical methods. Overall, prebiotics are manufactured by four major routes (Table 7.3).

A typical production process for nondigestible oligosaccharides is shown in Fig. 7.1, and major prebiotic producers and products are listed in Table 7.4. Food-grade oligosaccharides are not pure products, but are mixtures containing oligosaccharides of different degrees of polymerization (dp), the parent polysaccharide or disaccharide, and monomer sugars. An example of a typical product mixture produced by transfructosylation of sucrose is shown in Fig. 7.2. Oligosaccharide products are sold at this level of purity, often as syrups. Chromatographic purification processes are used to remove contaminating mono- and disaccharides to produce

TABLE 7.3 Main Approaches Used for the Production of Prebiotic Carbohydrates

Approach	Process	Prebiotic Examples
Direct extraction	Extraction from raw plant materials	Soybean oligosaccharides from soybean whey
		Inulin from chicory
		Resistant starch from maize
Controlled hydrolysis	Controlled enzymatic hydrolysis of polysaccharides; may be followed chromatography to purify the prebiotics	Fructooligosaccharides from inulin
		Xylooligosaccharides from arabinoxylan
Transglycosylation	Enzymatic process to build up oligosaccharides from disaccharides; may be followed by chromatography to purify the prebiotics	Galactooligosaccharides from lactose
		Fructooligosaccharides from sucrose
		Lactosucrose from lactose + sucrose
Chemical processes	Catalytic conversion of carbohydrates	Lactulose from alkaline isomerization of lactose
		Lactitol from hydrogenation of lactose

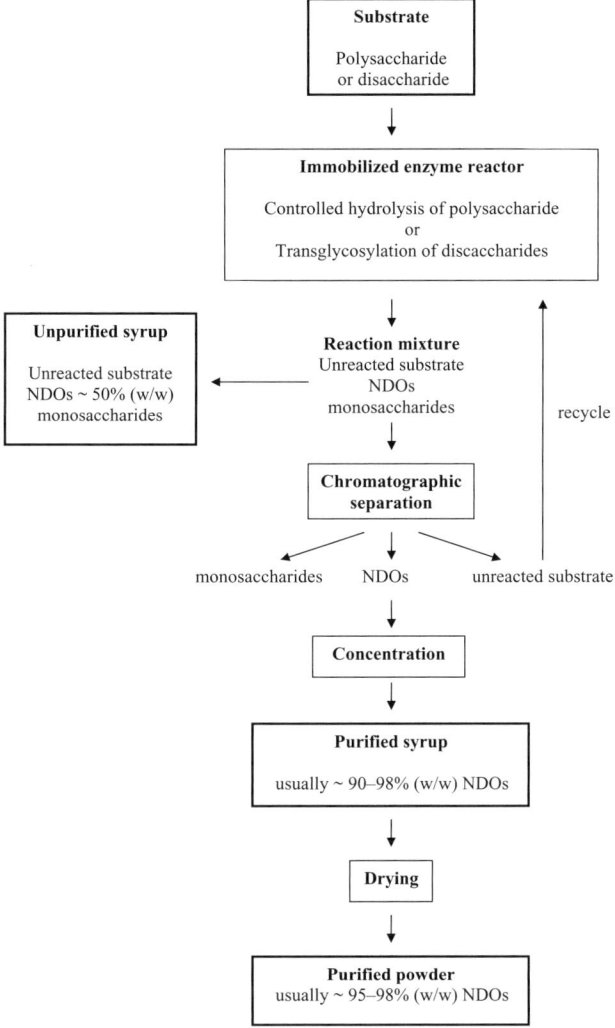

FIGURE 7.1 A typical production flowchart for the manufacture of prebiotic oligosaccharides.

higher purity oligosaccharide products containing between 85 and 99% oligosaccharides, which are often dried to powders.

Different manufacturing processes also produce slightly different oligosaccharide mixtures. For example, FOS mixtures produced by transfructosylation of sucrose contain oligosaccharides between three and five monomer units, with the proportion of each oligosaccharide decreasing with increasing molecular size. These oligosaccharides contain a terminal glucose with β-1 → 2 linked fructose moieties. In comparison,

TABLE 7.4 Major Manufacturers and Their Prebiotic Products

Company Names and Addresses	Product Names	Product Description
Galactooligosaccharides		
Yakult Honsha Co. Ltd, Japan; www.yakult.co.jp	Oligomate 55	Syrup containing 75% (w/v) solids Oligosaccharides >55% of solids
	Oligomate 55P	Powder, >55% oligosaccharides
	TOS-100	Powder, >99% oligosaccharides
Nissin Sugar Mfg. Co. Ltd; Japan; www.nissin-sugar.co.jp	Cup-Oligo H-70	Syrup containing 75% (w/v) solids Oligosaccharides ≈ 70% of solids
	Cup-Oligo P	Powder, 70% oligosaccharides
Friesland Foods Domo; The Netherlands; www.borculodomo.com	Vivinal GOS	Syrup containing 75% (w/v) solids Oligosaccharides ≈ 60% of solids
Lactulose		
Morinaga Milk Industry Co., Japan; www.morinagamilk.co.jp	MLS-50	Syrup containing 70% solids Lactulose ≈ 70% of solids
	MLC-A	Powder (anhydride) 98% lactulose
Solvay, Germany; www.solvay.com	Duphalac	Syrup containing >70% solids Lactulose >74% of solids
	Cephulac	Powder, >95% lactulose
Lactosucrose		
Ensuiko Sugar Refining Co., Japan; www.ensuiko.co.jp	Nyuka-Origo LS-40L	Syrup containing 72% (w/v) solids Lactosucrose ≈ 42% of solids
	Nyuka-Origo LS-55L	Syrup containing 75% (w/v) solids Lactosucrose ≈ 55% of solids
	Nyuka-Origo LS-55P	Powder, 55% lactosucrose
	Pet-Oligo L55	Syrup containing 75% (w/v) solids Lactosucrose ≈ 57% of solids
	Pet-Oligo P55	Powder, 45% lactosucrose
Hayashibara Shoji Inc., Japan; www.hayashibara.co.jp	Newka-Oligo LS-35	Syrup containing 72% (w/v) solids Lactosucrose ≥35% of solids
	Newka-Oligo LS-55L	Syrup containing 75% (w/v) solids Lactosucrose ≥55% of solids
	Newka-Oligo LS-55P	Powder, ≥55% lactosucrose
Lactitol		
PURAC, The Netherlands; www.purac.com	LACTY	Lactitol powder 96%
Danisco; www.danisco.com	Lactitol	Milled and crystalline powders
	Finlac DC	Compressible form for tablets

TABLE 7.4 (*Continued*)

Company Names and Addresses	Product Names	Product Description
Transfructosylation Fructooligosaccharides		
Meiji Seika Kaisha, Japan; www.meiji.co.jp	Meioligo P	Powder, 95% oligosaccharides
	Meioligo G	Syrup containing 75% (w/v) solids Oligosaccharides ≈ 55% of solids
Beghin-Meiji Industries, France; www.beghin-say.fr	Actilight P	Powder, 95% oligosaccharides
	Actilight G	Syrup containing 75% (w/v) solids
	Actisucre	Oligosaccharides ≈ 55% of solids
Golden Technologies Co., GTC Nutrition, USA; www.nutraflora.com	NutraFlora scFOS	Powder, 95% oligosaccharides
Cheil Foods and Chemicals, C.P.O Box 1155, Seoul, Korea	Oligo-sugar (syrup)	Syrup containing 70% (w/v) solids Oligosaccharides ≈ 60% of solids
	Oligo-sugar (powder)	Powder, 23% oligosaccharides (animal feed)
Inulin Fructooligosaccharides and Inulin		
Beneo-ORAFTI, Belgium; www.beneo.orafti.com	Orafti L60	Syrup containing 75% (w/v) solids Oligosaccharides >60% of solids
	Orafti L85	Syrup containing 75% (w/v) solids Oligosaccharides >85% of solids
	Orafti L95	Syrup containing 75% (w/v) solids Oligosaccharides >95% of solids
	Orafti P95	Powder, >95% oligosaccharides
	Orafti Synergy1	Powder, mix of oligofructose and inulin
	Orafti ST	Powder inulin >90%
	Orafti GR	Granulated powder – inulin >90%
	Orafti HP	Powder–inulin >99.5%
	Orafti HP-Gel	Instant powder–inulin >99.5%
Cosucra SA, Belgium; www.cosucra.com	Fibrulose F90 and F97	Fructooligosaccharides
	Fibruline standard Fibruline instant Fibruline long chain	Inulins
Sensus C.V., The Netherlands; www.sensus.nl	Frutafit –inulin	White powder from chicory plants
Isomaltooligosaccharides		
Showa Sangyo Co., Japan; www.showa-sangyo.co.jp	Isomalto-500	Syrup containing 75% (w/v) solids Oligosaccharides > 50% of solids
	Isomalto-900	Syrup containing 75% (w/v) syrup Oligosaccharides >85% of solids
	Isomalto-900P	Powder, >85% oligosaccharides

(*continued*)

TABLE 7.4 (*Continued*)

Company Names and Addresses	Product Names	Product Description
Nihon Shokuhin Kako Co., (Japan Maize Products), Japan; www.nisshoku.co.jp	Biotose #50	Syrup containing 75% (w/v) solids Oligosaccharides > 50% of solids
	Panorich	Syrup containing 75% (w/v) solids Oligosaccharides ≥ 50% of solids including ≥25% panose
Hayashibara Shoji Inc., Japan; www.hayashibara.co.jp	Panorup	Syrup containing ≥74% (w/v) solids Oligosaccharides ≥ 50% of solids including ≥25% panose
Soybean-oligosaccharides		
The Calpis Food Industry, Co. Japan; www.calpis.net	Soya-oligo	Syrup containing 75% (w/v) solids Soybean oligosaccharides 35% of solids
Xylooligosaccharides		
Suntory Ltd, Japan; www.suntory.com	Xylo-oligo 70	Syrup containing 75% (w/v) solids Oligosaccharides ≈ 70% of solids
	Xylo-oligo 20P	Powder, ≈ 20% oligosaccharides
	Xylo-oligo 35P	Powder, ≈ 35% oligosaccharides
	Xylo-oligo 95P	Powder, ≈ 95% oligosaccharides
Resistant Starches		
Penford Australia Ltd, Australia; www.penford.com	Hi-Maize	High amylose maize starches
	Culture-Pro	Modified resistant starches with different prebiotic properties also available
National Chemical & Starch Company, USA; www.nationalstarch.com	Hylon VII	High amylase (70%) corn starch (a RS2 type granular starch)
	Novelose 330	From maize; a RS3 type (nongranular, retrograded starch)
	Novelose 240	Higher dietary fiber type of resistant starch

FOS produced by the controlled hydrolysis of inulin contain a wider range of β-1 → 2 fructooligosaccharide sizes (dp 2–9), relatively few of which possess a terminal glucose residue. Even different β-galactosidases used in the production of GOS will produce oligosaccharide mixtures with different proportions of β-1 → 4 and β-1 → 6 linkages. Hence, there can be some diversity between the structures of oligosaccharides produced by different manufacturers. The precise impact of these differences in their health effects remains to be determined. Some differences have been observed in the ability of lactobacilli and bifidobacteria to use oligosaccharides of different degrees of polymerization (100, 101), which has also translated to different bifidogenic potencies *in vivo* (41, 51).

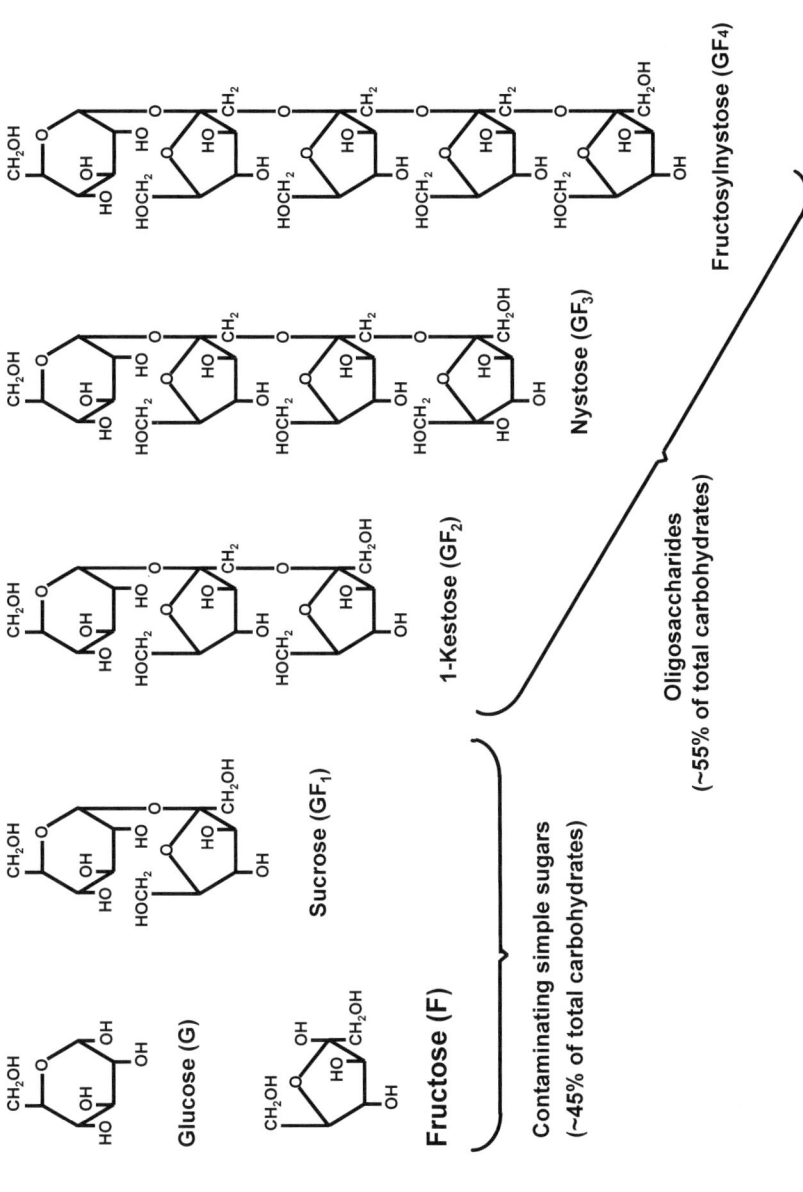

FIGURE 7.2 Commercial fructooligosaccharide production via transfructosylation of sucrose. Following the enzymatic reaction, the reaction mixture contains 50–60% oligosaccharides. Chromatographic processes are used to remove monosaccharides and unreacted sucrose to produce higher purity oligosaccharides.

7.6 PREBIOTIC MECHANISMS

The molecular structure of the prebiotic can be expected to determine physiological effects and which microbial species are able to utilize it as a carbon and energy source in the bowel. However, despite the diversity in molecular sizes, sugar compositions, and structural linkages within the range of prebiotic carbohydrates, it is the bifidobacteria that are almost universally observed to respond. How prebiotics promote the relatively specific proliferation of bifidobacteria within the intestinal microbiota remains speculative, but the following points provide some insights.

- Bifidobacteria possess the metabolic machinery to utilize a wide variety of oligosaccharides and complex carbohydrates as carbon and energy sources (102).
- Bifidobacteria grow well on many NDOs in comparison to putrefactive or potentially pathogenic bacterial species in the gut (76, 103–105).
- Although other intestinal genera (e.g. lactobacilli (100, 106–108), bacteriodes (109), and eubacteria (110)) can grow *in vitro* using prebiotic carbohydrates, bifidobacteria appear to grow more efficiently.
- Bifidobacteria are tolerant to the resulting production of short chain fatty acids (SCFA) and acidification of the intestinal environment.
- Bifidobacteria generally do not hydrolyze NDOs extracellularly, but possess permeases to internalize these substrates before hydrolyzing and metabolizing them (111, 112). This presumably minimizes the availability of released simple sugars for cross-feeding by other intestinal bacteria.

7.7 MODULATING THE INTESTINAL MICROBIOTA IN INFANTS

7.7.1 Breast Milk

The composition of the human intestinal microbiota changes naturally with age, and in early infancy the microbiota is believed to be particularly important in correct functioning of the gut and maturation of the immune system. Bifidobacteria colonize the human intestinal tract during or soon after birth and in breast-fed infants they eventually dominate the microbiota (113). The numerical dominance of bifidobacteria is induced by bifidogenic components in breast milk, including oligosaccharides (113, 114). Indeed, human milk oligosaccharides (HMOs) are the original prebiotics. The concentration of HMOs in breast milk (5–12 g/L) is about 100 times than that found in cow's milk (0.03–0.06 g/L) (115, 116) representing the third largest solid component behind lactose and fat (117). The fact that mothers direct so much energy toward components of breast milk that are not directly metabolized by the infant, but rather selectively modulate the composition of the microbiota, shows evolutionarily just how important the infant gut microbiota is to well-being.

7.7.2 Infant Milk Formulas

In contrast to breast-fed infants, infants fed traditional cow's milk-based formulas develop a more mixed intestinal microbiota, with lower counts of bifidobacteria and higher counts of clostridia and enterococci (118). Formula-fed infants have also been observed to have higher levels of fecal ammonia and other potentially harmful bacterial products (119, 120). The bifidogenic effect of HMO can be emulated using oligosaccharides such as GOS and FOS. GOS share some structural similarities with the backbone of HMOs, although HMOs are considerably more complex. Over 200 different HMO structures have been identified. They are based on the combinations of glucose, galactose, sialic acid, fucose and/or N-acetylglucosamine, with varied sizes and linkages accounting for the considerable variety (115). In recent years, a number of studies have investigated the effect of infant formula containing 8 g/L of a mixture of 9:1 GOS:FOS designed to mimic the molecular size distribution of HMOs (121). Table 7.5 shows formulae supplemented with the GOS:FOS can mimic many of the physiological impacts of HMOs.

Although the bifidogenic effect of HMOs can be emulated using FOS and GOS, there is increasing evidence for roles of HMOs outside their bifidogenic impact in the gut. These include blocking adhesion of pathogens to the intestinal mucosa (131–133) and roles in developing cognition (134). Hence, N-acetylneuraminic acid derivatives or sialyl-lactose are also commonly added to infant milk formulae. The complexity of HMOs has thwarted attempts to synthesize their full range of structures commercially, although specific oligosaccharides have been synthesized using chemical and biotechnological approaches (135–138). There is a ready market in infant milk formulas

TABLE 7.5 Infant Formulae Containing Prebiotics Mimic Many of the Impacts of Breast Milk on the Infant Intestinal Microbiota

Traditional cow's milk-based infant formula	Cow's milk-based infant formula supplemented with 8 g/L of a 9:1 mixture of GOS/FOS
Contains little or no nondigestible oligosaccharides (115)	Contains nondigestible oligosaccharides in similar quantity to human breast milk (116)
Stimulates a diverse, adult-like intestinal microbiota (118)	Stimulates an intestinal microbiota dominated by bifidobacteria (121, 122)
Adult-like *Bifidobacterium* and *Lactobacillus* species composition (123, 124)	*Bifidobacterium* and *Lactobacillus* species composition similar to breast-fed infants (123, 124)
Adult-like composition of SCFA in stools high in butyrate and propionate and higher pH (121, 125–127)	Breast-fed infant-like composition of SCFA high in acetate and more acidic stools (121, 125–127)
Reduced stool frequency and firmer stool consistency (125, 128–130)	Breast-fed infant-like, higher stool frequency and softer consistency (125, 128–130)
Lower levels of protective secretory IgA in the gut (127)	Increased levels of protective secretory IgA in the gut (127)

for oligosaccharides that more closely replicate all of the properties of HMOs, and research to synthesize them will no doubt continue.

7.8 MODULATING THE INTESTINAL MICROBIOTA IN ADULTS

Following weaning, the composition of the colonic microbiota becomes increasingly complex. In adulthood, the microbiota consists of more than 500 different species, though is dominated by 30–40 of them, and its composition becomes quite stable (1). Bifidobacteria remain numerically important, though their proportion of total microbes declines from 60 to 80% in breast-fed infants to 1–5% in adults. In the feces of healthy adults, bifidobacteria are found in numbers generally in the order of 10^8–10^{10} cells/g (31, 34, 139–141). While these figures represent the typical *Bifidobacterium* cell density, a proportion of healthy adults harbor considerably lower numbers of *Bifidobacteria* in their gut (by several orders of magnitude) without any discernable adverse effects (31, 142–144). It is yet to be determined how the total number of bifidobacteria within a stable microbiota influences the long-term health of the human host.

7.8.1 Effects at the Genus Level

Feeding prebiotics to adults typically induces 10–100-fold increases in the size of the intestinal *Bifidobacterium* population during the feeding period (12). However, a range of factors may influence the magnitude of any increase in *Bifidobacterium* numbers, the most important being the initial size of the population within the intestinal tract. In individuals colonized with an already large population of bifidobacteria (in the order of 10^8 cfu/g in feces) prebiotic consumption appears not to increase *Bifidobacterium* numbers further (58, 145). Lactobacilli have often been noted to proliferate upon feeding various prebiotics, particularly in feeding studies with lactulose and galactooligosaccharides.

The rise in beneficial bacterial populations during prebiotic feeding has often been shown to be accompanied by concomitant reductions in the numbers of putrefactive organisms such as clostridia and *Bacteriodes* spp. and *Enterobacteriaceae* (14, 52, 56, 57, 61, 83). The changes in population dynamics appear to persist only while the prebiotics are being consumed. The bacterial populations at the genus level tend to revert to prefeeding levels over a period of days once prebiotic consumption ceases (49, 60, 73).

7.8.2 Effects at the Species Level

B. adolescentis, *B. catenulatum/pseudocatenulatum*, *B. bifidum*, and *B. longum* are the most frequently reported *Bifidobacterium* species in the intestines of adults, with considerable variation between individuals (31, 142, 146–148). To date, no clear rationale for promoting one species of *Bifidobacterium* over others has emerged. Indeed, it may be quite difficult to achieve major shifts within the population dynamic of bifidobacteria at the species level even if this was desirable. Studies that monitored the species composition of bifidobacteria during and after prebiotic feeding using GOS (21, 149, 150), inulin (48), and raffinose (80) have shown that the prebiotics did not substantially alter the relative proportions of *Bifidobacterium* species present. The species composition within intestinal bifidobacteria has been shown to remain fairly stable over many months in adults (27, 28, 30, 31, 139) suggesting that day-to-day fluctuations in normal diets have little impact on the species dynamic.

7.8.3 Altering the Physiology of the Microbiota

Physiological changes to bacterial metabolism rather than phylogenic changes to microbiota composition may be more important for many health effects. Examples include increased production of SCFA or vitamins that benefit the health of the colonic epithelium, or synthesis of antagonistic metabolites that augment colonization resistance against pathogens. Tannock et al. (29) used molecular techniques to show that feeding low levels of GOS or FOS could induce changes in relative gene expression with no discernable changes in population dynamics. The fermentation of different prebiotics by the consortia of gut bacteria can yield different profiles of SCFA. Generally, small, rapidly fermented prebiotics yield more acetate and lactate, while more slowly fermentable prebiotics tend to yield butyrate as a major end product (101, 151–155).

7.9 MODIFYING THE INTESTINAL MICROBIOTA IN THE ELDERLY

Early studies using culture methods showed that *Bifidobacterium* levels substantially decreased as a proportion of the total fecal microbiota in the elderly Japanese, while the numbers of putrefactive bacteria such as clostridia increased (155). This suggests that prebiotics may benefit the elderly by restoring the *Bifidobacterium* populations while reducing the levels of deleterious bacteria. More recent studies using molecular techniques have confirmed that some elderly subjects do indeed have reduced levels of bifidobacteria, but this is not generally the case (156–158). It remains to be seen if increasing *Bifidobacterium* levels in elderly subjects can lead to improved health.

7.10 HEALTH EFFECTS AND APPLICATIONS OF PREBIOTICS

Prebiotics have physicochemical and organoleptic properties that make them useful food ingredients. For example, NDOs are sweet and are used as low-cariogenic and low-calorific sugar substitutes, while polysaccharides such as inulin are used as fat replacers. Their indigestibility and subsequent reduced impact on glucose and insulin responses also make them suitable for diabetics. In terms of food technology, NDOs supply a number of valuable physicochemical functionalities. They can be used to increase viscosity, reduce Malliard reactions, alter water retention, depress freezing points, and suppress crystal formation (159). Hence, they are used commercially in a wide variety of foods and beverages.

A number of largely prophylactic health targets have been proposed for prebiotics. As might be expected, these overlap considerably with the targets of probiotic interventions. Some effects have therapeutic value for specific disorders while others are potentially beneficial to the population at large. Hence, prebiotics have found applications both as pharmaceuticals and as functional food ingredients.

7.10.1 Laxatives

Lactulose is widely used as a pharmaceutical to treat constipation. It has proven efficacy in a number of placebo-controlled trials at doses between 10 and 20 g/day (160, 161) even in patients with chronic constipation. The effect is not caused by modifications to the composition of the intestinal microbiota.

Rather, since it is a relatively small molecule that is not digested or absorbed, lactulose has an osmotic effect, trapping fluid, accelerating transit in the small bowel, and increasing ileocecal flow. Its rapid fermentation to SCFA and hydrogen also contributes to this effect and induces peristalsis (162). A number of other NDOs, such as inulin has been shown to mildly improve stool frequency and consistency in adults (163–167) although their applications are targeted towards functional foods rather than pharmaceutical applications.

7.10.2 Hepatic Encephalopathy

Lactulose (and lactitol) are also front-line therapeutic agents for the treatment of hepatic encephalopathy (HE). This neuropsychiatric condition results from liver dysfunction caused by cirrhosis or hepatitis. It includes a spectrum of symptoms ranging from subtle changes in cognition and personality to lethargy, stupor, and coma (168). A dysfunctional liver is unable to clear ammonia from the blood stream, which then accumulates to levels toxic to the central nervous system. The ammonia is produced by the intestinal microbiota as an end product of protein metabolism, and ammonia readily crosses the intestinal epithelium to enter circulation. Lactulose and lactitol act by limiting both ammonia production by the microbiota and the absorption of ammonia from the intestinal lumen (168, 169). Acidification of the colonic lumen resulting from SCFA production inhibits urease positive and deaminating bacteria (implicated in intestinal ammonia production) and importantly leads to the protonation of ammonia to ammonium ions in the intestinal lumen. Ammonium ions cannot readily cross the intestinal epithelium, and so the drop in pH effectively traps ammonia in the lumen. Lactulose and lactitol have similar efficacy, although lactitol is more palatable and produces more rapid results with fewer side effects (168). However, to be effective lactulose and lactitol are delivered in high doses (30–60 g/day). This of course has a large laxative effect, causing significant discomfort. Therefore, there is an interest in using larger NDOs and soluble fibers that ferment rapidly, but with less osmotic effect in the gut.

7.10.3 Primary Prevention of Allergy in Infants

There is accumulating evidence that early colonization of the intestinal tract by an appropriate intestinal microbiota is important for the healthy maturation of the immune system, including appropriate programming of oral tolerance to dietary antigens (170, 171). Additionally, differences in certain bacterial populations within the intestinal tract have been noted between allergic and nonallergic infants (172). These include differences within the genus *Bifidobacterium*, which is found in lower numbers in the feces of allergic infants (173–179) and with a more adult-like species composition dominated by *B. adolescentis* (180, 181) rather than the usual species associated with the infant intestine such as *B. bifidum*, *B. breve*, and *B. longum* (= *B. infantis*) (142, 182). Hence, dietary interventions to augment the intestinal microbiota of infants have been explored as a means of decreasing allergic disease.

In allergies, the immune response at the T cell level is skewed towards T helper type 2 cells (Th2) that signal the production of allergenic antibodies of the immunoglobulin E (IgE) class. It is believed that allergies can be prevented by inducing a counterbalancing T helper type 1 (Th1) cell response, and/or by the induction of regulatory T cells (T reg) that produce cytokines such as interleukin-10 (IL-10) and transforming growth factor-beta (TGF-β) that are involved in tolerance and suppress inflammation. Studies in animals demonstrated that feeding probiotics could redress Th1/Th2 imbalances, induce regulatory T cell activity, and reduce the development of allergy (183–185). In humans, a number of randomized controlled trials have shown a preventative effect of probiotic feeding on the development and severity of atopic dermatitis in infants (177, 186–190) although not all probiotic strains and trials have proven beneficial (191–195).

Positive effects from using probiotics stimulated interest in determining if similar effects could be achieved with prebiotics. A number of animal feeding studies with NDOs have shown that prebiotics can induce Th1 immune responses, re-balance Th2-biased immune responses and suppress allergenic IgE synthesis (Table 7.6). Primary prevention of atopic dermatitis in infants genetically at risk of atopic disease has now been reported in randomized controlled trials using a GOS:FOS mixture and a synbiotic combination including GOS (Table 1.6). These results indicate that prebiotics may be able to replicate the benefits seen for probiotics in allergy prevention.

7.10.4 Amelioration of Inflammatory Bowel Disease

Inflammatory bowel disease (IBD) describes a group of chronic, severe and relapsing inflammatory conditions of the gut that includes Crohn's disease, ulcerative colitis, and pouchitis. The precise etiology of IBD remains unknown. However, there is accumulating evidence that a genetic predisposition to develop an overzealous inflammatory immune response to components of the intestinal microbiota is responsible (203). The drugs that have been developed to manage IBD have mostly targeted

TABLE 7.6 Prebiotic Immunomodulation on Preventing the Development of Allergies

Prebiotics	Models	Effects	References
Raffinose	Sensitized mice	Stimulation of Th1 cells, reduced Th2 cell activity, suppressed IgE production to allergen	(196)
Alginate oligosaccharides	Sensitized mice	Stimulation of Th1 cells, reduced Th2 cell activity, suppressed IgE production to allergen	(197)
Isomaltooligosaccharides	Mice	Stimulation of Th1 immunity	(198)
Fructooligosaccharides	Sensitized mice	Trend to reduce Th2 cell numbers in the gut; reduced mast cell numbers and reduced allergy symptoms	(199)
Pectin-oligosaccharide and GOS:FOS mixture	Mice vaccination model	Enhanced Th1 response; reduced Th2 response	(200)
GOS;FOS mixture	Human infants at risk of atopy ($n = 206$)	Significant reduction in atopic dermatitis	(201)
Synbiotic including four probiotic strains and GOS	Human infants at risk of atopy ($n = 925$)	Significant reduction in atopic dermatitis	(202)

suppression of the host immune response and all are associated with significant side effects when used for long periods (204).

Since the microbiota is the likely source of the inflammatory stimulus, modification of the intestinal microbiota using antibiotics, probiotics, and prebiotics have all been proposed and trialed as approaches to treat IBD. However, as no specific microbial groups or antigens have been confirmed as causatives, these approaches have been empirical rather than targeted. Elimination of specific bacterial antigens, immunomodulation, and trophic effects of SCFA on the intestinal epithelium have all been proposed as mechanisms by which prebiotics could alleviate IBD. The size of the intestinal *Bifidobacterium* population has been shown to be relatively small in subjects afflicted with IBD (205–207). However, cause and effect links between disease and a diminished intestinal *Bifidobacterium* population remain to be established.

Prebiotic and synbiotic interventions have ameliorated colitis in different rodent models of IBD, although not all studies have demonstrated benefit (Table 7.7). Studies in both animal models and human subjects have shown that prebiotic-induced stimulation of *Bifidobacterium* numbers has been associated with downregulation of inflammatory markers in intestinal mucosa (208–210) and evidence of increased immune regulation (208, 211). However, direct evidence that a change in the microbial

TABLE 7.7 Recent Prebiotic Feeding Studies Using Animal Models of IBD

Prebiotics	Models	Effects	References
FOS	TNBS[1] rats	Amelioration of colitis Increased SCFA and LAB in the intestine	(212)
	TNBS rats	Amelioration of colitis Reduction in inflammatory markers in intestinal mucosa Increased numbers of lactobacilli and bifidobacteria and SCFA production	(210)
	DSS[2] rats	No decrease in colitis score	(213)
	DSS rats	No significant reduction in colitis Resistant starch treatment was effective in the same study	(214)
	DSS rats	Significant reduction in colitis and faster mucosal recovery	(215)
Inulin in synbiotic combination	HLA-B27-β_2-microglobulin transgenic rats	Significant amelioration of colitis Significant increase in numbers of bifidobacteria	(203)
FOS + inulin (Synergy 1)	HLA-B27-β_2-microglobulin transgenic rats	Amelioration of colitis Reduction in inflammatory markers in intestinal mucosa Increase in markers of immune regulation Increased numbers of lactobacilli and bifidobacteria	(208)
Lactulose	DSS rats	Dose-dependant amelioration of colitis	(216)
	TNBS rats	Decreased colonic inflammation and inflammatory cytokines Increased numbers of bifidobacteria and lactobacilli	(217)
GOS	TNBS rats	Increased colonic levels of bifidobacteria, but no attenuation of inflammation	(218)
Resistant starch	DSS rats	Significant amelioration of colitis Increased luminal SCFA concentrations FOS treatment was ineffective in the same study	(214)

[1]TNBS-Trinitrobenzene sulfuric acid
[2]DSS-Dextran sodium sulfate

TABLE 7.8 Human Studies Using Prebiotic or Synbiotic Treatments to Ameliorate IBD

	Study Design	Result	References
Pouchitis	Small, open label study $n = 10$ with active disease Synbiotic combination of *Lactobacillus rhamnosus* GG and FOS	Complete clinical and endoscopic remission	(219)
	Small, randomized, double-blind, crossover study $n = 20$ with active disease Inulin	Reduction in mucosal inflammation, accompanied by a reduced concentration of *Bacteroides fragilis*, increased butyrate concentrations and lowered pH, but no effect on the numbers of lactobacilli or bifidobacteria.	(220)
Ulcerative colitis	Randomized controlled trial $n = 18$ with active disease Synbiotic combination of Synergy 1 (FOS + inulin)	Trend toward reduced inflammation Improved regeneration of epithelial tissue. A decrease in mucosal proinflammatory cytokines was associated with an increase in the numbers of mucosal-associated bifidobacteria	(207, 209)
	Open labeled, controlled trial $n = 18$ with active disease Fermented barley feed (rich in arabinoxylan)	Significant reduction in clinical activity and mucosal inflammation Increased numbers of fecal bifidobacteria and increased fecal butyrate concentration	(221, 222)
Crohn's disease	Open label trial $n = 10$ with moderately active disease FOS	Significant reduction in clinical disease score Increased number of both fecal and mucosa-associated bifidobacteria in patients entering remission Enhanced immunoregulatory response (increased proportion of IL-10-positive mucosal dendritic cells)	(211)

antigen population within the microbiota leads to reduced inflammation with increased immune regulation remains to be established.

Only a few pilot human studies have been reported using prebiotics or synbiotics to treat IBD (Table 7.8). The results of these trials suggest that prebiotics have at least some potential to induce remission in patients with active IBD. The use of prebiotics to extend or maintain remission in IBD has been scantly explored, but studies reporting promising results using probiotics (see review by Hedin et al. (204)) indicate that the maintenance of remission would be a worthy target for prebiotics. Overall, there are enough encouraging indications that prebiotics could be beneficial in the treatment of IBD to warrant further exploration of efficacy and mechanisms in large, randomized, controlled trials.

7.10.5 Prevention of Infections

Studies in which animals have been challenged with enteric pathogens have indicated that prebiotics have some potential in preventing intestinal infections (Table 7.9). In addition to direct antagonism of pathogens, a number of animal studies have indicated that prebiotics can enhance protective immune responses (223). These include stimulation of secretory immunoglobulin A (sIgA) production in the gut (198), improved humoral antibody responses (224) and enhanced Th1-dependent immune responses (198, 200).

Prebiotic oligosaccharides may provide protection against enteric infections through competitive inhibition of pathogen adherence to the mucosa. Many intestinal pathogens, such as *Escherichia coli*, *Salmonellae* and *Campylobacters* utilize oligosaccharide receptor sites in the gut for attachment (225). NDOs can act as structural mimics of the pathogen binding sites and act as soluble decoys. Human milk oligosaccharides act in this way to block the initial binding of a range of pathogens to inhibit colonization (225–227). *In vitro* experiments using epithelial cell culture models (227, 228) have shown that GOS and lactulose have the ability to interfere with the adhesion of enteropathogenic *E. coli*.

The efficacy of prebiotics in preventing infections in human studies has been mixed. Lactulose consumption at high dose (up to 60 g/day) is effective in eliminating salmonella from the intestinal tract of chronic human carriers and it is used as a pharmaceutical for this purpose in some countries (232). Several small human trials

TABLE 7.9 Animal Pathogen Challenge Studies Showing Benefits from Prebiotic Ingestion

Prebiotics	Models	Results	References
Lactulose	Rats with obstructive jaundice	Significantly reduced cecal overgrowth and translocation of *Escherichia coli*	(229)
Lactulose	Rats challenged with the invasive pathogen *Salmonella enteriditis*	Significantly reduced translocation of the pathogen	(230)
FOS and inulin	Puppies challenged with *Salmonella typhimurium*	Significant attenuation of infection symptoms including loss of appetite and enterocyte sloughing	(231)

have indicated that prebiotics or synbiotics can prevent infections in some specific circumstances (Table 7.10). Prebiotics have so far generally failed to demonstrate anti-infective effects in trials of healthy subjects likely to be exposed to a diverse range of intestinal pathogens (e.g. traveler's diarrhea) or for antibiotic-associated diarrhea.

Overall, the evidence to date suggests that prebiotics have the capacity to prevent intestinal infections in some specific cases, particularly for acid-sensitive pathogens that infect the large bowel. However, prebiotics do not represent "silver bullets" that are effective against all intestinal pathogens. Further research is required to determine which prebiotics are effective against which pathogens, at what doses, for which populations, and by what mechanisms.

7.10.6 Mineral Absorption

As for nondigestible, fermentable carbohydrates in general, a number of prebiotics have been shown to increase mineral absorption. The precise mechanisms of prebiotic-mediated improvements in mineral uptake remain unclear, but fermentative activities of the microbiota are believed to be involved (239, 240). Bifidogenic effects, although evident, are probably not relevant.

Calcium and magnesium are the main minerals for which uptake is improved. In rats, increased calcium uptake has led to improved bone mineralization for animals fed GOS (241), FOS (242–245), lactulose (246), and resistant starch (247, 248). A number of small, randomized, and controlled human studies have confirmed the beneficial impact of prebiotic consumption on mineral absorption, particularly for calcium. Positive results have been reported for fructans (recently reviewed by Frank (249)), other NDOs (239), and resistant starch (154). Increased calcium absorption stimulated by prebiotics has further been demonstrated to improve markers of bone health in humans (250, 251).

Under normal circumstances dietary calcium is predominately absorbed in the small intestine with little calcium absorbed in the colon (252). However, prebiotic fermentation is believed to extend calcium uptake into the colon (43, 249). Sustainable fermentation in the gut appears to be important. Animal and human studies comparing prebiotics of differing chain lengths, fermentation patterns, and doses have shown that higher doses and more persistent fermentation profiles are more effective (42, 242, 253–256).

The encouraging results seen in animal and human studies suggest that there is good potential for prebiotics to improve calcium and magnesium absorption from the gut and to improve bone health. Further research is warranted to investigate links between long-term prebiotic consumption and improved bone density in humans at risk of developing osteoporosis.

7.10.7 Prevention of Colorectal Cancer

Due to the difficulties in conducting long-term preventative studies in humans, there is little epidemiological data available for the use of pro- or prebiotics in cancer prevention. Evidence of benefits is largely limited to *in vitro* and animal studies,

TABLE 7.10 Anti-Infective Outcomes in Human Studies with Prebiotics or Synbiotics

Prebiotics	Study Designs	Results	References
FOS/inulin	Small ($n = 20$) randomized, controlled trial of young children in day care	Significant reduction in the number of infectious diseases requiring antibiotic treatment, and the incidence of diarrhea, vomiting, and fever	(233)
	Randomized, controlled trial of elderly subjects ($n = 435$) administered broad-spectrum antibiotics	Failed to reduce the incidence of diarrhea, *Clostridium difficile* infection, or hospital stays	(234)
	Randomized controlled trial of 244 healthy volunteers travelling to destinations with medium to high risk of developing infective diarrhea	Nonsignificant ($p = 0.08$) trend to reduce episodes of traveler's diarrhea	(235)
	Two randomized controlled trials of Peruvian infants living in low socio-economic areas, $n = 282$ and $n = 289$	No significant reduction in the prevalence of diarrhea	(236)
Lactitol	Placebo-controlled study $n = 60$ patients with chronic viral hepatitis prone to endotoxemia	Significantly reduced levels of plasma endotoxin Significantly increased numbers of fecal bifidobacteria and lactobacilli and reduced numbers of *Clostridium perfringens*	(64)
Synbiotic 2000 (synbiotic combination of several probiotics and prebiotics)	Randomized controlled trial. $n = 65$ critically ill trauma patients	Significantly reduced rates of infections, systemic inflammatory response syndrome (SIRS), severe sepsis and mortality	(237)
	Randomized controlled trial $n = 62$ patients with severe acute pancreatitis	Significantly lower incidence of SIRS, organ failure and number of patients recovering with complications	(238)

TABLE 7.11 Impacts of Prebiotics on Improving Risk-Factors of Colorectal Cancer

Beneficial Effects of Prebiotics	References
Fermentation of prebiotics to SCFA (particularly butyrate) Induces colonocyte differentiation and apoptosis	(257–259)
Butyrate stimulated by larger fermentable carbohydrates such as inulin and resistant starch that may also provide SCFA production at the requited site of action in the distal colon	(153, 154, 260, 261)
Reduce mutagenic enzyme activities (e.g. β-glucuronidase and azoreductase) and bacterial metabolites (e.g. secondary bile acids, phenols, and indoles) that are purportedly associated with colon cancer risk	
Examples include studies with:	
• Lactulose	(56, 59, 61)
• GOS	(262)
• Resistant starch	(92, 263–266)
• Lactosucrose	(87)
• Inulin	(267)
• Not all prebiotic feeding studies have shown improvements in these parameters.	(57, 58, 89, 268)

which have shown that prebiotics have the capacity to reduce purported risk factors for colon cancer in the colonic environment (Table 7.11). There are also numerous studies reporting protection by prebiotics against the development of preneoplastic lesions and/or tumors in rodent models of colon carcinogenesis (Table 7.12). Overall, the

TABLE 7.12 Protective Effects of Prebiotics in Carcinogen Challenge Animal Studies

Beneficial Effects of Prebiotics	References
Protection against the development of preneoplastic lesions and/or tumors in rodent models of colon carcinogenesis	
Examples include:	
• FOS and inulin	(269)
• Lactulose	(270, 271)
• XOS	(272)
• Resistant starch	(273, 274)
Stimulation of anticancer immune activity	(275, 276)
Studies with different molecular sized fructan prebiotics reported increased protection with the larger, more slowly fermented prebiotics	(46, 278)
A number of studies comparing interventions with probiotics, prebiotics, and synbiotics have shown the synbiotic combination to be the most effective. These studies have used varying combination of pro- and prebiotics, and the exact nature of the synergies that lead to improved cancer markers remain speculative	(96, 271, 276, 278–280)

capacity of prebiotics to significantly contribute to a reduced incidence of colorectal cancer in humans remains unproven. However, the results of preliminary human and animal experiments provide sufficient encouragement to maintain the impetus for continued research into the protective effects of prebiotics.

7.10.8 Reduction in Serum Lipid Concentrations

Numerous animal and human studies have focused on the effects of prebiotics on serum concentrations of cholesterol and triacylglycerides. While convincing beneficial results have often been obtained in animal studies, the results of trials in humans have been somewhat mixed, although no deleterious effects have been reported. Overall, the most convincing evidence is for inulin fructans, which demonstrate mainly a reduction in triglyceridaemia and only a relatively slight decrease in cholesterolaemia, and mostly in hypertriglyceridaemic conditions (42). Positive results have been observed in feeding studies using lactulose or lactitol (281–283) and resistance starch (reviewed by Nugent (154)). The mechanisms have been speculated to be regulation of host *de novo* lipogenesis via SCFA (particularly acetate) absorbed from the gut (284), or reduced intestinal fat absorption (282). Changes to the bacterial composition of the intestinal tract do not appear to be important.

7.10.9 Use in Weight Management and Improving Insulin Sensitivity

Since NDOs are sweet and not digested, they have a low calorific value and are used as low energy, low glycaemic index sweeteners that are also suitable for individuals with diabetes. Preliminary data also suggests that SCFA production resulting from prebiotic fermentation (and in particular acetate) could improve insulin sensitivity (281, 285, 286). The potential of prebiotics, and in particular small, rapidly fermented NDOs in the management of metabolic disorders linked to insulin resistance warrants further study.

7.11 FUNCTIONAL FOODS FOR ANIMALS

Prebiotics have been trialed for use in both farm animal feeds and for companion animals. The advantage of feeding oligosaccharides in dogs and cats is the reduction of odor and improvement in volume and consistency of feces (287). With respect to farm animals, prebiotics have been studied for their potential to replace antibiotics in maintaining high feed conversion efficiencies, particularly in poultry and pork, and also to suppress methane production with ruminants (288–291). Alternatives to antibiotic use in animal feeds are urgently sought, and there has been considerable interest in the use of both prebiotics and probiotics to aid production. Although they have shown some promise in preliminary studies (292–294), further research is needed into their application within an overall management strategy in order to match the performance of antibiotics.

TABLE 7.13 Safety of Prebiotics

	References
• Nondigestible carbohydrates are consumed as part of the normal daily diet, as they are natural components of most plants	(296)
• Lactulose, short-chain oligosaccharides, inulin, resistant starch, and dietary fiber are not toxic, even at high doses	(295)
• Unsupplemented intakes of FOS and inulin are between 1 and 10 g/day from normal diets in Europe and the United States	(43)
• Recommended effective doses of prebiotic oligosaccharides in adults usually range from 10 to 15 g/day	(43, 297–299)
• With the shorter chain oligosaccharides, intakes exceeding 15 g/day in adults can lead to flatulence, abdominal discomfort, and osmotic diarrhea at excessive doses	(43, 297–299)
• Excessive consumption of rapidly fermented, acidogenic prebiotics may cause mucosal irritation	(300–302)

7.12 SAFETY OF PREBIOTICS

Most prebiotics can be found naturally in various foods. For example, most NDOs are natural components of many common foods including honey, milk, and various fruits and vegetables (295, 296) although usually in low concentrations. It is well established that lactulose, short-chain oligosaccharides, inulin, resistant starch, and dietary fiber are not toxic, even at high doses (Table 7.13).

7.13 REGULATION OF PREBIOTICS

The regulatory regimes for nondigestible carbohydrates have been under active review in many countries in recent years. A marked development since the year 2000 has been the inclusion of most NDOs in the category "dietary fiber." This has allowed recognition of these products as having some health benefits on product labels. Prior recognition in a regulatory sense had been restricted to Japan, who for many years has regulated the functional food sector with the FOSHU (Foods for Specified Health Use) system. The use of GOS and FOS as ingredients in infant milk formulas has been the subject of intensive regulatory inquiry and its acceptance varies among countries. Revised Codex standards were released in November 2006 (Codex ALINORM 07/30/26). Readers are referred to the Codex website for current standards (CODEX STAN 72–1981; and 156–1987). The website address is www.codexalimentarius.net.

7.14 CONCLUSION

- It is clear from the volume of the accumulated evidences from human and animal studies using both culture and culture-independent methods that the consumption of prebiotics can modulate the composition of the intestinal microbiota.
- Bifidobacteria remain the population that is most often stimulated, although a number of prebiotics have also been shown to promote the proliferation of lactobacilli too.
- The fructans, FOS and inulin, and the lactose derivatives, GOS and lactulose, have the largest number of studies showing prebiotic effects. However, a range of other NDOs and dietary fibers have emerging evidence of prebiotic action.
- The success of prebiotics in replicating the bifidogenic effect of human milk oligosaccharides is a good example of where prebiotics can be applied with a clearly targeted objective for modulation of the composition and activity of the intestinal microbiota.
- In healthy adults, an imperative to maintain a numerically large population of bifidobacteria and lactobacilli has not been clearly demonstrated.
- Diminished or altered populations of intestinal bifidobacteria have been observed in IBD and atopy, but cause/effect links remain to be demonstrated.
- Prebiotics exert beneficial effects not only by manipulating the composition of the microbiota, but also by modulating its metabolic activity. It is clear that the production of SCFA and lowering luminal pH are important mechanisms for a number of potential health benefits.
- The most convincing evidence accumulated so far is for laxative effects, control of hepatic encephalopathy, and increased mineral absorption from the gut.
- There is preliminary evidence from randomized controlled trials for primary prevention of atopic eczema in infants; amelioration of IBD inflammation; prevention of infections; and control of serum lipids.
- The profile of SCFA produced varies with different prebiotics and the intestinal site and persistence of fermentation will clearly be important in determining which prebiotics are the most effect for specific health targets.
- We are just at the beginning of understanding how fermentable dietary fibers affect the composition and activity of the intestinal microbiota. In future, the distinction between current prebiotics and other fermentable fibers may prove somewhat arbitrary and artificial. What is likely to prove important is the choice of appropriate fermentable substrate(s) for the maintenance of health in specific populations. This may be highly bifidogenic NDOs in infants, larger fermentable fibers in adults, or indeed combinations of these as is already being explored.
- We are still grappling with the question "what is normal?" in terms of the composition and activities of the intestinal microbiota in different age groups and populations and in health and disease. Fundamental research into the

composition and activities of the intestinal microbiota to determine which bacterial groups are important in health and disease remains an important prerequisite to developing targeted strategies to manipulate the intestinal microbiota using prebiotics to benefit the host.
- While acknowledging that the science of manipulating the intestinal microbiota to achieve improved health is still very much in its infancy, progress is being made, and strategies that may lead to tangible health benefits in specific populations are emerging.

REFERENCES

1. McCartney AL and Gibson GR. The normal microbiota of the human gastrointestinal tract: history of analysis, succession, and dietary influences In: Ouwehand AC and Vaughan EE, editors. *Gastrointestinal Microbiology*. Taylor and Francis, New York, 2006, pp. 51–73.
2. Bourlioux P, Koletzko B, Guarner F, and Braesco V. The intestine and its microflora are partners for the protection of the host: report on the Danone Symposium "The Intelligent Intestine" held in Paris, June 14, 2002. *Am. J. Clin. Nutr.* 2003; 78: 675–683.
3. Topping DL and Clifton PM. Short-chain fatty acids and human colonic function: roles of resistant starch and nonstarch polysaccharides. *Physiol. Rev.* 2001; 81: 1031–1064.
4. Blum S and Schiffrin EJ. Intestinal microflora and homeostasis of the mucosal immune response: implications for probiotic bacteria?. *Curr Iss. Intest. Microbiol.* 2003; 4: 53–60.
5. Tlaskalova-Hogenova H, Stepankova R, Hudcovic T, Tuckova L, Cukrowska B, Lodinova-Zadnkova R, Kozakova H, Rossmann P, Bartova J, Sokol D, Funda DP, Borovska D, Rehakova Z, Sinkora J, Hofman J, Drastich P, and Kokesova A. Commensal bacteria (normal microflora), mucosal immunity and chronic inflammatory and autoimmune diseases. *Immunol. Lett.* 2004; 93: 97–108.
6. Cummings JH, Macfarlane GT, and Macfarlane S. Intestinal bacteria and ulcerative colitis. *Curr. Iss. Intesti. Microbiol.* 2003; 4: 9–20.
7. Marteau P, Seksik P, and Shanahan F. Manipulation of the bacterial flora in inflammatory bowel disease. *Best Pract. Res.: Clin. Gastroenterol.* 2003; 17: 47–61.
8. Saunier K and Doré J. Gastrointestinal tract and the elderly: functional foods, gut microflora and healthy ageing. *Digest. Liver Dis.* 2002; 34(Suppl.) S19–S24.
9. Guarner F and Malagelada JR. Gut flora in health and disease. *Lancet* 2003; 361: 512–519.
10. Crittenden R. An Update on Probiotic Bifidobacteria. In: Salminen S, von Wright A and Ouwerhand A. editors. *Lactic Acid Bacteria: Microbiological and Functional Aspects*. Marcel Dekker, New York, 2004, pp. 125–157.
11. Salminen S, Gorbach S, Lee YK, and Benno Y. Human Studies on Probiotics: What is Scientifically Proven Today? In: Salminen S, von Wright A and Ouwerhand A, editors *Lactic Acid Bacteria: Microbiological and Functional Aspects*. Marcel Dekker, New York, 2004, pp. 515–530.
12. Crittenden RG, Prebiotics. In: Tannock GW,edtitor. Probiotics: *A Critical review.* Horizon Scientific Press, Wymondham, United Kingdom, 1999, pp. 141–156.

13. Boehm G and Stahl B. Oligosaccharides. In: Mattila-Sandholm and Saarela M, editors. *Functional Dairy Products*, Woodhead Publishing, CRC Press, England, 2003, pp. 203–243.
14. Gibson GR and Roberfroid MB. Dietary modulation of the human colonic microbiota - introducing the concept of prebiotics. *J. Nutr.* 1995; 125: 1401–1412.
15. Tissier H. Recherches sur la flore intestinale des nourrissons (état normal et pathologique). Thesis. Paris, France: University of Paris; 1900 [In French].
16. Tissier H. Traitement des infections intestinales par la méthode de la flore bactérienne de l'intestin. *Crit. Rev. Soc. Biol.* 1906; 60: 359–361 [In French].
17. Gyorgy P. Norris RF, and Rose CS. Bifidus factor I. A variant of *Lactobacillus bifidus* requiring a special growth factor. *Arch. Biochem. Biophys.* 1954; 48: 193–201.
18. Petuely F. Bifidusflora bei flaschenkindern durch bifidogene substanzen (Bifidusfaktor). *Z. Kinderheilkunde* 1957; 79: 174–179 [In German].
19. Yazawa K, Imai K, and Tamura Z. Oligosaccharides and polysaccharides specifically utilizable by bifidobacteria. *Chem. Pharm. Bull. (Tokyo)* 1978; 26: 3306–3311.
20. Yazawa K and Tamura Z. Search for sugar sources for selective increase of bifidobacteria. *Bifidobact. Microflora* 1982; 1: 34–44.
21. Satokari RM, Vaughan EE, Akkermans ADL, Saarela M, and de Vos WM. Polymerase chain reaction and denaturing gradient gel electrophoresis monitoring of fecal *Bifidobacterium* populations in a prebiotic and probiotic feeding trial. *Syst. Appl. Microbiol.* 2001; 24: 227–231.
22. Mattila-Sandholm T, Blum S, Collins JK, Crittenden R, de Vos W, Dunne C, Fondén R, Grenov B, Isolauri E, Kiely B, Marteau P, Morelli L, Ouwehand A, Reniero R, Saarela M, Salminen S, Saxelin M, Schiffrin E, Shanahan F, Vaughan E, and von Wright A. Probiotics: towards demonstrating efficacy. *Trends Food Sci. Technol.* 1999; 10: 393–399.
23. von Wright A, Vilpponen-Salmela T, Llopis MP, Collins K, Kiely B, Shanahan F, and Dunne C. The survival and colonic adhesion of *Bifidobacterium infantis* in patients with ulcerative colitis. *Int. Dairy J.* 2002; 12: 197–200.
24. Shimakawa Y, Matsubara S, Yuki N, Ikeda M, and Ishikawa F. Evaluation of *Bifidobacterium breve* strain Yakult-fermented soymilk as a probiotic food. *Int. J. Food Microbiol.* 2003; 81: 131–136.
25. Fujiwara S, Seto Y, Kimura A, and Hashiba H. Intestinal transit of an orally administered streptomycin-rifampicin-resistant variant of *Bifidobacterium longum* SBT2928: its long-term survival and effect on the intestinal microflora and metabolism. *J. Appl. Micrbiol.* 2001; 90: 43–52.
26. Brigidi P, Swennen E, Vitali B, Rossi M, and Matteuzzi D. PCR detection of *Bifidobacterium* strains and *Streptococcus thermophilus* in feces of human subjects after oral bacteriotherapy and yogurt consumption. *Int. J. Food Microbiol.* 2003; 81: 203–209.
27. Zoetendal EG, Akkermans AD, and de Vos WM. Temperature gradient gel electrophoresis analysis of 16S rRNA from human fecal samples reveals stable and host-specific communities of active bacteria. *Appl. Environ. Microbiol.* 1998; 64: 3854–3859.
28. Tannock GW. A fresh look at faeces. *Microbiol. Aust.* 2003; 24: 34–35.
29. Tannock GW. Munro K, Bibiloni R, Simon MA, Hargreaves P, Gopal P, Harmsen H, and Welling G. Impact of consumption of oligosaccharide-containing biscuits on the fecal microbiota of humans. *Appl. Environ. Microbiol.* 2004; 70: 2129–2136.
30. Requena T, Burton J, Matsuki T, Munro K, Simon MA, Tanaka R, Watanabe K, and Tannock GW. Identification, detection, and enumeration of human *Bifidobacterium*

31. Satokari RM, Vaughan EE, Akkermans ADL, Saarela M, and de Vos WM. Bifidobacterial diversity in human feces detected by genus-specific PCR and denaturing gradient gel electrophoresis. *Appl. Environ. Microbiol.* 2001; 67: 504–513.

species by PCR targeting the transaldolase gene. *Appl. Environ. Microbiol.* 2002; 68: 2420–2427.

32. McCartney AL, Wenzhi W, and Tannock GW. Molecular analysis of the composition of the bifidobacterial and lactobacillus microflora of humans. *Appl. Environ. Microbiol.* 1996; 62: 4608–4613.

33. Kimura K, McCartney AL, McConnell MA, and Tannock GW. Analysis of fecal populations of bifidobacteria and lactobacilli and investigation of the immunological responses of their human hosts to the predominant strains. *Appl. Environ. Microbiol.* 1997; 63: 3394–3398.

34. Mangin I, Bouhnik Y, Bisetti N, and Decaris B. Molecular monitoring of human intestinal *Bifidobacterium* strain diversity. *Res. Microbiol.* 1999; 150: 343–350.

35. Chen H-L, Fan Y-H, Chen M-E, and Chan Y. Unhydrolyzed and hydrolyzed konjac glucomannans modulated cecal and fecal microflora in Balb/c mice. *Nutrition* 2005; 21: 1059–1064.

36. Vulevic J, Rastall RA, and Gibson GR. Developing a quantitative approach for determining the *in vitro* prebiotic potential of dietary oligosaccharides. *FEMS Microbiol. Lett.* 2004; 236: 153–159.

37. Chung C-H, and Day DF. Glucooligosaccharides from *Leuconostoc mesenteroides* B-742 (ATCC 13146): A potential prebiotic. *J. Ind. Microbiol. Biotechnol.* 2002; 29: 196–199.

38. Mandalari G, Nueno Palop C, Tuohy K, Gibson GR, Bennett RN, Waldron KW, Bisignano G, Narbad A, and Faulds CB. In vitro evaluation of the prebiotic activity of a pectic oligosaccharide-rich extract enzymatically derived from bergamot peel. *Appl. Microbiol. Biotechnol.* 2007; 73: 1173–1179.

39. Hu B, Gong Q, Wang Y, Ma Y, Li J, and Yu W. Prebiotic effects of neoagarooligosaccharides prepared by enzymatic hydrolysis of agarose. *Anaerobe* 2006; 12: 260–266.

40. Okubo T, Ishihara N, Takahashi H, Fujisawa T, Kim M, Yamamoto T, and Mitsuoka T. Effects of partially hydrolyzed guar gum intake on human intestinal microflora and its metabolism. *Biosci. Biotechnol. Biochem.* 1994; 58: 1364–1369.

41. Van Loo J. Inulin-type fructans as prebiotics. In: Gibson GR and Rastall RA, editors. *Prebiotics: Development and Application.* John Wiley and Sons, Chichester, UK, 2006, pp. 57–100.

42. Roberfroid MB. Introducing inulin-type fructans. *Br. J. Nutr.* 2005; 93 (Suppl. 1): S13–S25.

43. van Loo J, Cummings J, Delzenne N, Englyst H, Franck A, Hopkins M, Kok N, Macfarlane G, Newton D, Quigley M, Roberfroid M, van Vliet T, and van den Heuvel E. Functional food properties of nondigestible oligosaccharides: a consensus report from the ENDO project (DGXII AIRII-CT94-1095). *Br J. Nutr.* 1999; 81: 121–132.

44. Kolida S, Tuohy K, and Gibson GR. Prebiotic effects of inulin and oligofructose. *Br. J. Nutr.* 2002; 87(Suppl. 2): S193–S197.

45. Taper HS and Roberfroid MB. Inulin/oligofructose and anticancer therapy. *Br. J. Nutr.* 2002; 87(Suppl. 2): S283–S286.

46. Pool-Zobel B, van Loo J, Rowland I, and Roberfroid MB. Experimental evidences on the potential of prebiotic fructans to reduce the risk of colon cancer. *Br. J. Nutr.* 2002; 87 (Suppl. 2): S273–S281.
47. Buddington KK, Donahoo JB, and Buddington RK. Dietary oligofructose and inulin protect mice from enteric and systemic pathogens and tumor inducers. *J. Nutr.* 2002; 132: 472–477.
48. Harmsen HJM, Raangs GC, Franks AH, Wildeboer-Veloo ACM, and Welling GW. The effect of the prebiotic inulin and the probiotic *Bifidobacterium longum* on the fecal microflora of healthy volunteers measured by FISH and DGGE. *Microb. Ecol. Health Dis.* 2002; 14: 211–219.
49. Bouhnik Y, Raskine L, Simoneau G, Paineau D, and Bornet F. The capacity of short-chain fructo-oligosaccharides to stimulate faecal bifidobacteria: a dose-response relationship study in healthy humans. *J. Nutr.* 2006; 5: 8.
50. Langlands SJ, Hopkins MJ, Coleman N, and Cummings JH. Prebiotic carbohydrates modify the mucosa associated microflora of the human large bowel. *Gut* 2004; 53: 1610–1616.
51. Suzuki N, Aiba Y, Takeda H, Fukumori Y, and Koga Y. Superiority of 1-kestose, the smallest fructo-oligosaccharide, to a synthetic mixture of fructo-oligosaccharides in the selective stimulating activity on Bifidobacteria. *Biosci. Microflora* 2006; 25: 109–116.
52. Roberfroid MB, Van Loo JAE, and Gibson GR. The bifidogenic nature of Chicory inulin and its hydrolysis products. *J. Nutr.* 1998; 128: 11–19.
53. Ben X-M, Zhou X-Y, Zhao W-H, Yu W-L, Pan W, Zhang W-L, Wu S-M, Beusekom CM, and van Schaafsma A. Growth and development of term infants fed with milk with long-chain polyunsaturated fatty acid supplementation. *Chin. Med. J.* 2004; 117: 1268–1270.
54. Boehm G, Lidestri M, Casetta P, Jelinek J, Negretti F, Stahl B, and Marini A. Supplementation of a bovine milk formula with an oligosaccharide mixture increases counts of faecal bifidobacteria in preterm infants. *Arch. Dis. Childhood* 2002; 86: F178–F181.
55. Rastall RA. Galacto-oligosaccharides as prebiotics. In: Gibson GR and Rastall RA, editors. *Prebiotics: Development and Application*. John Wiley & Sons, Chichester, England, 2006, pp. 101–110.
56. Ballongue J, Schumann C, and Quignon P. Effects of lactulose and lactitol on colonic microflora and enzymatic activity. *Scand. J. Gastroenterol.* 1997; 32 (Suppl. 222): 41–44.
57. Tuohy KM, Ziemer CJ, Klinder A, Knobel Y, Pool-Zobel BL, and Gibson GR. A human volunteer study to determine the prebiotic effects of lactulose powder on human colonic microbiota. *Microbial. Ecol. Health Dis.* 2002; 14: 165–173.
58. Bouhnik Y, Attar A, Joly FA, Riottot M, Dyard F, and Flourie B. Lactulose ingestion increases faecal bifidobacterial counts: a randomised double-blind study in healthy humans. *Eur. J. Clin. Nutr.* 2004; 58: 462–466.
59. De Preter V, Vanhoutte T, Huys G, Swings J, Rutgeerts P, and Verbeke K. Effect of lactulose and *Saccharomyces* boulardii administration on the colonic urea-nitrogen metabolism and the bifidobacteria concentration in healthy human subjects. *Aliment. Pharmacol. Ther.* 2006; 23: 963–974.
60. Vanhoutte T, De Preter V, De Brandt E, Verbeke K, Swings J, and Huys G. Molecular monitoring of the fecal microbiota of healthy human subjects during administration of lactulose and *Saccharomyces* boulardii. *Appl. Environ. Microbiol.* 2006; 72: 5990–5997.

61. Terada A, Hara H, Katoaka M, and and Mitsuoka T. Effect of lactulose on the composition and metabolic activity of human faecal flora. *Microbial. Ecol. Health Dis.* 1992; 5: 43–50.
62. Bouhnik Y, Raskine L, Simoneau G, Vicaut E, Neut C, Flourié B, Brouns F, and Bornet FR. The capacity of nondigestible carbohydrates to stimulate fecal bifidobacteria in healthy humans: a double-blind, randomized, placebo-controlled, parallel-group, dose-response relation study. *Am. J. Clin. Nutr.* 2004; 80: 1658–1664.
63. Kummel KF and Brokx S. Lactitol as a functional prebiotic. *Cer. Foods World* 2001; 46: 424–429.
64. Chen C, Li L, Wu Z, Chen H, and Fu S. Effects of lactitol on intestinal microflora and plasma endotoxin in patients with chronic viral hepatitis. *J. Infect.* 2007; 54: 98–102.
65. Okazaki M, Fujikawa S, and Matsumomo N. Effect of xylooligosaccharide on the growth of bifidobacteria. *Bifidobact. Microflora* 1990; 9: 77–86.
66. Yamada H, Itoh K, Morishita Y, and Taniguchi H. Structure and properties of oligosaccharides from wheat bran. *Cer. Foods World* 1993; 38: 490–492.
67. Jaskari J, Kontula P, Siitonen A, Jousimies-Somer H, Mattila-Sandholm T, and Poutanen K, Oat β-glucan and xylan hydrolysates as selective substrates for *Bifidobacterium* and *Lactobacillus* strains. *Appl. Microbiol. Biotechnol.* 1998; 49: 175–181.
68. van Laere KM, Hartemink R, Bosveld M, Schols HA, and Voragen AG. Fermentation of plant cell wall derived polysaccharides and their corresponding oligosaccharides by intestinal bacteria. *J. Agric. Food Chem.* 2000; 48: 1644–1652.
69. Rycroft CE, Jones MR, Gibson GR, and Rastall RA. A comparative *in vitro* evaluation of the fermentation properties of prebiotic oligosaccharides. *J. Appl. Microbiol.* 2001; 91: 878–887.
70. Crittenden R, Karppinen S, Ojanen S, Tenkanen M, Fagerström R, Mättö J, Saarela M, Mattila-Sandholm T, and Poutanen K. *In vitro* fermentation of cereal dietary fibre carbohydrates by probiotic and intestinal bacteria. *J. Sci. Food Agric.* 2002; 82: 1–9.
71. Al-Tamimi MAHM, Palframan RJ, Cooper JM, Gibson GR, and Rastall RA. *In vitro* fermentation of sugar beet arabinan and arabino-oligosaccharides by the human gut microflora. *J. Appl. Microbiol.* 2006; 100: 407–414.
72. Campbell JM, Fahey GC Jr, and Wolf BW. Selected indigestible oligosaccharides affect large bowel mass, cecal and fecal short-chain fatty acids, pH and microflora in rats. *J. Nutr.* 1997; 127: 130–136.
73. Santos A, San Mauro M, and Marquina Diaz D. Prebiotics and their long-term influence on the microbial populations of the mouse bowel. *Food Microbiol.* 2006; 23: 498–503.
74. Oku T and Nakamura S. Comparison of digestibility and breath hydrogen gas excretion of fructo-oligosaccharide, galactosyl-sucrose, and isomalto-oligosaccharide in healthy human subjects. *Eur. J. Clin. Nutr.* 2003; 57: 1150–1156.
75. Kohmoto T, Tsuji K, Kaneko T, Shiota M, Fukui F, Takaku H, Nakagawa Y, Ichikawa T, and Kobayashi S. Metabolism of ^{13}C-isomaltooligosaccharides in healthy men. *Biosci. Biotechnol. Biochem.* 1992; 56: 937–940.
76. Kohmoto T, Fukui F, Takaku H, Machida Y, Arai M, and Mitsuoka T. Effect of isomalto-oligosaccharides on human facal flora. *Bifidobact. Microflora* 1998; 7: 61–69.
77. Kohmoto T, Fukui F, Takaku H, and Mitsuoka T. Dose-response test of isomaltooligosaccharides for increasing fecal bifidobacteria. *Agric. Biol. Chem.* 1991; 55: 2157–2159.

78. Kaneko T, Kohmoto T, Kikuchi H, Shiota M, Yatake T, Iino H, and Tsuji K. Effects of isomaltoologosaccharides intake on defecation and intestinal environment in healthy volunteers. *J. Home Econ. Japan* 1993; 44: 245–254 [In Japanese].

79. Kaneko T, Kohmoto T, Kikuchi H, Shiota M, Iino H, and Mitsuoka T. Effects of isomaltooligosaccharides with different degrees of polymerisation on human fecal bifidobacteria. *Biosci. Biotechnol. Biochem.* 1994; 58: 2288–2290.

80. Dinoto A, Marques TM, Sakamoto K, Fukiya S, Watanabe J, Ito S, and Yokota A. Population dynamics of *Bifidobacterium* species in human feces during raffinose administration monitored by fluorescence in situ hybridization-flow cytometry. *Appl. Environ. Microbiol.* 2006; 72: 7739–7747.

81. Hayakawa K, Mizutani J, Wada K, Masai T, Yoshihara I, and Mitsuoka T. Effects of soybean oligosaccharides on human faecal microflora. *Microbial. Ecol. Health Dis.* 1990; 3: 293–303.

82. Wada K, Watabe J, Mizutani J, Tomoda M, Suzuki H, and Saitoh Y. Effects of soybean oligosaccharides in a beverage on human fecal flora and metabolites. *J. Agric. Chem. Soc. Japan* 1992; 66: 127–135.

83. Benno Y, Endo K, Shiragami N, Sayama K, and Mitsuoka T. Effects of raffinose intake on human fecal microflora. *Bifido'a.* 1987; 6: 59–63.

84. Hara T, Ikeda N, Hatsumi K, Watabe J, Iino H, and Mitsuoka T. Effects of small amount ingestion of soybean oligosaccharides on bowel habits and fecal flora of volunteers. *Japanese J. Nutr.* 1997; 55: 79–84.

85. Fujita K, Hara K, Sakai S, Miyake T, Yamashita M, Tsunstomi Y, and Mitsuoka T. Effects of 4-β-D-galactosylsucrose (lactosucrose) on intestinal flora and its digestibility in humans. *J. Japanese Soc. Starch Sci.* 1991; 38: 249–255.

86. Yoneyama M, Mandai T, Aga H, Fujii K, Sakai S, and Katayama Y. Effects of 4-β-D-galactosylsucrose (lactosucrose) intake on intestinal flora in healthy humans. *J. Japanese Soc. Nutr. Food Sci.* 1992; 45: 101–107.

87. Hara H, Li S-T, Sasaki M, Maruyama T, Terada A, Ogata Y, Fujita K, Ishigami H, Hara K, Fujimori I, and Mitsuoka T. Effective dose of lactosucrose on fecal flora and fecal metabolites of humans. *Bifidobact Microflora* 1994; 13: 51–63.

88. Ohkusa T, Ozaki Y, Sato C, Mikuni K, and Ikeda H. Long-term ingestion of lactosucrose increases *Bifidobacterium* sp. in human fecal flora. *Digestion* 1994; 56: 415–420.

89. Kleessen B, Stoof G, Proll J, Schmiedl D, Noack J, and Blaut M. Feeding resistant starch affects fecal and cecal microflora and short-chain fatty acids in rats. *J. Anim. Sci.* 1997; 75: 2453–2462.

90. Brown IL, Wang X, Topping DL, Playne MJ, and Conway PL. High amylose maize starch as a versatile prebiotic for use with probiotic bacteria. *Food Aust.* 1998; 50: 603–610.

91. Wang X, Brown IL, Evans AJ, and Conway PL. The protective effects of high amylose maize (amylomaize) starch granules on the survival of *Bifidobacterium* spp. in the mouse intestinal tract. *J. Appl. Microbiol.* 1999; 87: 631–639.

92. Silvi S, Rumney CJ, Cresci A, and Rowland IR. Resistant starch modifies gut microflora and microbial metabolism in human flora-associated rats inoculated with faeces from Italian and UK donors. *J. Appl. Microbiol.* 1999; 86: 521–530.

93. Bielecka M, Biedrzycka E, Majkowska A, Juskiewicz J, and Wroblewska M. Effect of non-digestible oligosaccharides on gut microecosystem in rats. *Food Res. Int.* 2002; 35: 139–144.

94. Wang X, Brown IL, Khaled D, Mahoney MC, Evans AJ, and Conway PL. Manipulation of colonic bacteria and volatile fatty acid production by dietary high amylose maize (amylomaize) starch granules. *J. Appl. Microbiol.* 2002; 93: 390–397.
95. Le Blay G, Michel C, Blottiere H, and Cherbut C. Raw potato starch and short-chain fructo-oligosaccharides affect the composition and metabolic activity of rat intestinal microbiota differently depending on the caecocolonic segment involved. *J. Appl. Microbiol.* 2003; 94: 312–320.
96. Le Leu RK, Brown IL, Hu Y, Bird AR, Jackson M, Esterman A, and Young GP. A Synbiotic combination of resistant starch and *Bifidobacterium lactis* facilitates apoptotic deletion of carcinogen-damaged cells in rat colon. *J. Nutr.* 2005; 135: 996–1001.
97. Bird AR, Vuaran M, Brown I, and Topping DL. Two high-amylose maize starches with different amounts of resistant starch vary in their effects on fermentation, tissue and digesta mass accretion, and bacterial populations in the large bowel of pigs. *Br. J. Nutr.* 2007; 97: 134–144.
98. Oikarinen S, Heinonen S, Karppinen S, Mättö J, Adlecreutz H, Poutanen K, and Mutanen M. Plasma enterolactone or intestinal *Bifidobacterium* levels do not explain adenoma formation in multiple intestinal neoplasia (Min) mice fed with two different types of rye-bran fractions. *Br. J. Nutr.* 2003; 90: 119–125.
99. Kanauchi O, Fujiyama Y, Mitsuyama K, Araki Y, Ishii T, Nakamura T, Hitomi Y, Agata K, Saiki T, Andoh A, Toyonaga A, and Bamba T. Increased growth of *Bifidobacterium* and *Eubacterium* by germinated barley foodstuff, accompanied by enhanced butyrate production in healthy volunteers. *Int. J. Mol. Med.* 1999; 3: 175–179.
100. Saulnier DMA, Molenaar D, de Vos WM, Gibson GR, and Kolida S. Identification of prebiotic fructooligosaccharide metabolism in *Lactobacillus plantarum* WCFS1 through microarrays. *Appl. Environ. Microbiol.* 2007; 73: 1753–1765.
101. Van de Wiele T, Boon N, Possemiers S, Jacobs H, and Verstraete W. Inulin-type fructans of longer degree of polymerization exert more pronounced in vitro prebiotic effects. *J. Appl. Microbiol.* 2007; 102: 452–460.
102. Schell MA, Karmirantzou M, Snel B, Vilanova D, Berger B, Pessi G, Zwahlen MC, Desiere F, Bork P, Delley M, Pridmore RD, and Arigoni F. The genome sequence of *Bifidobacterium longum* reflects its adaptation to the human gastrointestinal tract. *Proc. Nat. Acad. Sci. USA* 2002; 99: 14422–14427.
103. Okazaki M, Fujikawa S, and Matsumoto N. Effect of xylo-oligosaccharide on the growth of bifidobacteria. *Bifidobact. Microflora* 1990; 9: 77–86.
104. Gibson GR and Wang X. Bifidogenic properties of different types of fructo-oligosacharides. *Food Microbiol.* 1994; 11: 491–498.
105. Wang X and Gibson GR. Effects of the in vitro fermentation of oligofructose and inulin by bacteria growing in the human large intestine. *J. Appl. Bacteriol.* 1993; 75: 373–380.
106. Goh Y-J, Zhang C, Benson AK, Schlegel V, Lee J-H, and Hutkins RW. Identification of a putative operon involved in fructo-oligosaccharide utilization by *Lactobacillus paracasei*. *Appl. Environ. Microbiol.* 2006; 72: 7518–7530.
107. Barrangou R, Altermann E, Hutkins R, Cano R, and Klaenhammer TR. Functional and comparative genomic analyses of an operon involved in Fructooligosaccharide utilization by *Lactobacillus acidophilus*. *Proc. Nat. Acad. Sci. USA* 2003; 100: 8957–8962.
108. Kaplan H and Hutkins RW. Metabolism of fructooligosaccharides by *Lactobacillus paracasei* 1195. *Appl. Environ. Microbiol.* 2003; 69: 2217–2222.

109. Van der Meulen R, Makras L, Verbrugghe K, Adriany T, and De Vuyst L. *In vitro* kinetic analysis of oligofructose consumption by bacteroides and *Bifidobacterium* spp. indicates different degradation mechanisms. *Appl. Environ. Microbiol.* 2006; 72: 1006–1012.
110. Manderson K, Pinart M, Tuohy KM, Grace WE, Hotchkiss AT, Widmer W, Yadhav MP, Gibson GR, and Rastall RA. *In vitro* determination of prebiotic properties of oligosaccharides derived from an orange juice manufacturing by-product stream. *Appl. Environ. Microbiol.* 2005; 71: 8383–8389.
111. Ryan SM, Fitzgerald GF, and van Sinderen Douwe. Transcriptional regulation and characterization of a novel beta-fructofuranosidase-encoding gene from *Bifidobacterium breve* UCC2003. *Appl. Environ. Microbiol.* 2005; 71: 3475–3482.
112. Parche S, Amon J, Jankovic I, Rezzonico E, Beleut M, Barutcu H, Schendel I, Eddy MP, Burkovski A, Arigoni F, and Titgemeyer F. Sugar transport systems of *Bifidobacterium longum* NCC2705. *J. Mol. Microbiol. Biotechnol.* 2007; 12: 9–19.
113. Harmsen HJM, Wildeboer-Veloo ACM, Raangs CG, Wagendorp AA, Klijn N, Bindels JG, and Welling GW. Analysis of intestinal flora development in breast-fed and formula-fed infants by using molecular identification and detection methods. *J. Pediatr. Gastroenterol. Nutr.* 2000; 30: 61–67.
114. Mountzouris KC, McCartney AL, and Gibson GR. Intestinal microflora of human infants and current trends for its nutritional modulation. *Br. J. Nutr.* 2002; 87: 405–420.
115. Kunz C, Rudloff S, Baier W, Klein N, and Strobel S. Oligosaccharides in human milk: structural, functional, and metabolic aspects. *Annu. Rev. Nutr.* 2000; 20: 699–722.
116. Boehm G and Stahl B. Oligosaccharides from milk. *J. Nutr.* 137 (Suppl.) 2007; 847S–849S.
117. Prentice AM and Prentice A. Evolutionary and environmental influences on human lactation. *Proc. Nutr. Soc.* 1995; 54: 391–400.
118. Adlerberth I. Establishment of a normal intestinal microflora in the newborn infant. In: Probiotics, other nutritional factors and intestinal microflora. Eds. Hansen LA and Yolken RH. Nestle Nutrition Workshop Series, Vol 42, Nestec Ltd, Vevey. Lippincott-Raven, Philadelphia. pp. 63–78, 1999
119. Heavey PM, Savage SA, Parrett A, Cecchini C, Edwards CA, and Rowland IR. Protein-degradation products and bacterial enzyme activities in faeces of breast-fed and formula-fed infants. *Br. J. Nutr.* 2003; 89: 509–515.
120. Edwards CA and Parrett AM. Intestinal flora during the first months of life: new perspectives. *Br. J. Nutr.* 2002; 88 (Suppl. 1): S11–S18.
121. Knol J, Boehm G, Lidestri M, Negretti F, Jelinek J, Agosti M, Stahl B, Marini A, and Mosca F. Increase of faecal bifidobacteria due to dietary oligosaccharides induces a reduction of clinically relevant pathogen germs in the faeces of formula-fed preterm infants. *Acta Paediatrica* 2005; 94: 31–33.
122. Rinne MM, Gueimonde M, Kalliomäki M, Hoppu U, Salminen SJ, and Isolauri E. Similar bifidogenic effects of prebiotic-supplemented partially hydrolyzed infant formula and breastfeeding on infant gut microbiota. *FEMS Immunol. Med. Microbiol.* 2005; 43: 59–65.
123. Haarman M and Knol J. Quantitative real-time PCR assays to identify and quantify fecal *Bifidobacterium* species in infants receiving a prebiotic infant formula. *Appl. Environ. Microbiol.* 2005; 71: 2318–2324.
124. Haarman M and Knol J. Quantitative real-time PCR analysis of fecal *Lactobacillus* species in infants receiving a prebiotic infant formula. *Appl. Environ. Microbiol.* 2006; 72: 2359–2365.

125. Boehm G, Stahl B, Jelinek J, Knol J, Miniello V, and Moro GE. Prebiotic carbohydrates in human milk and formulas. *Acta Paediatr.* 2005; 94 (Suppl.): 18–21.
126. Fanaro S, Boehm G, Garssen J, Knol J, Mosca F, Stahl B, and Vigi V. Galacto-oligosaccharides and long-chain fructo-oligosaccharides as prebiotics in infant formulas: a review. *Acta Paediatr.* 2005; 94: 22–26.
127. Bakker-Zierikzee AM, Alles MS, Knol J, Kok FJ, Tolboom JJM, and Bindels JG. Effects of infant formula containing a mixture of galacto- and fructo-oligosaccharides or viable *Bifidobacterium animalis* on the intestinal microflora during the first 4 months of life. *Br. J. Nutr.* 2005; 94: 783–790.
128. Bruzzese E, Volpicelli M, Squaglia M, Tartaglione A, and Guarino A. Impact of prebiotics on human health. *Digest. Liver Dis.* 2006; 38 (Suppl. 2): S283–S287.
129. Ziegler E, Vanderhoof JA, Petschow B, Mitmesser SH, Stolz SI, Harris CL, and Berseth CL. Term infants fed formula supplemented with selected blends of prebiotics grow normally and have soft stools similar to those reported for breast-fed infants. *J. Pediatr. Gastroenterol. Nutr.* 2007; 44: 359–364.
130. Colome G, Sierra C, Blasco J, Garcia MV, Valverde E, and Sanchez E. Intestinal permeability in different feedings in infancy. *Acta Paediatric.* 2007; 96: 69–72.
131. Newburg DS. Human milk glycoconjugates that inhibit pathogens. *Curr. Medicinal Chem.* 1999; 6: 117–127.
132. Martin-Sosa S, Martin MJ, and Hueso P. The sialylated fraction of milk oligosaccharides is partially responsible for binding to enterotoxigenic and uropathogenic *Escherichia coli* human strains. *J. Nutr.* 2002; 132: 3067–3072.
133. Morrow AL, Ruiz-Palacios GM, Altaye M, Jiang X, Guerrero ML, Meinzen-Derr JK, Farkas T, Chaturvedi P, Pickering LK, and Newburg DS. Human milk oligosaccharides are associated with protection against diarrhea in breast-fed infants. *J. Pediatr.* 2004; 145: 297–303.
134. Wang B and Brand-Miller J. The role and potential of sialic acid in human nutrition. *Eur. J. Clin. Nutr.* 2003; 57: 1351–1369.
135. Vincent SJ, Faber EJ, Neeser J-R, Stingele F, and Kamerling JP. Structure and properties of the exopolysaccharide produced by *Streptococcus macedonicus* Sc136. *Glycobiology* 2001; 11: 131–139.
136. Rencurosi A, Poletti L, Guerrini M, Russo G, and Lay L. Human milk oligosaccharides: an enzymatic protection step simplifies the synthesis of 3'- and 6'-O-sialyllactose and their analogues. *Carbohydrate Res.* 2002; 337: 473–483.
137. La Ferla B, Prosperi D, Lay L, Russo G, and Panza L. Synthesis of building blocks of human milk oligosaccharides. Fucosylated derivatives of the lacto- and neolacto-series. *Carbohydrate Res.* 2002; 337: 1333–1342.
138. Dumon C, Samain E, and Priem B. Assessment of the two *Helicobacter pylori* alpha-1,3-fucosyltransferase ortholog genes for the large-scale synthesis of LewisX human milk oligosaccharides by metabolically engineered *Escherichia coli*. *Biotechnol. Progr.* 2004; 20: 412–419.
139. Vaughan EE, de Vries MC, Zoetendal EG, Ben-Amor K, Akkermans ADL, and de Vos WM. The intestinal LABs. Antonie van Leeuwenhoek. *Int. J. Gen. Mol. Microbiol.* 2002; 82: 341–352.
140. Tannock GW. Analysis of the intestinal microflora using molecular methods. *Eur. J. Clin. Nutr.* 2002; 56(Suppl. 4): S44–S49.

141. Tannock GW. The bifidobacterial and lactobacillus microflora of humans. *Clin. Rev. Allergy Immunol.* 2002; 22: 231–253.

142. Matsuki T, Watanabe K, Tanaka R, Fukuda M, and Oyaizu H. Distribution of bifidobacterial species in human intestinal microflora examined with 16S rRNA-gene-targeted species-specific primers. *Appl. Environ. Microbiol.* 1999; 65: 4506–4512.

143. Matsuki T, Watanabe K, Fujimoto J, Miyamoto Y, Takada T, Matsumoto K, Oyaizu H, and Tanaka R. Development of 16S rRNA-gene-targeted group-specific primers for the detection and identification of predominant bacteria in human feces. *Appl. Environ. Micrbiol.* 2002; 68: 5445–5451.

144. Hayashi H, Sakamoto M, and Benno Y. Phylogenetic analysis of the human gut microbiota using 16S rDNA clone libraries and strictly anaerobic culture-based methods. *Microbiol. Immunol.* 2002; 46: 535–548.

145. Rao AV. Dose-response effects of inulin and oligofructose on intestinal bifidogenesis effects. *J. Nutr.* 1999; 129(Suppl. 7): 1442–1445.

146. Requena T, Burton J, Matsuki T, Munro K, Simon MA, Tanaka R, Watanabe K, and Tannock GW. Identification, detection, and enumeration of human *Bifidobacterium* species by PCR targeting the transaldolase gene. *Appl. Environ. Microbiol* 2002; 68: 2420–2427.

147. Saito T, Kato S, Maeda T, Suzuki S, Iijima S, and Kobayashi T. Overproduction of thermostable β-galactosidase in *Escherichia coli*, its purification and molecular structure. *J. Ferment. Bioeng.* 1992; 74: 12–16.

148. Ventura M, Elli M, Reniero R, and Zink R. Molecular microbial analysis of *Bifidobacterium* isolates from different environments by the species-specific amplified ribosomal DNA restriction analysis (ARDRA). *FEMS Microbiol. Ecol.* 2001; 36: 113–121.

149. Alander M, Mättö J, Kneifel W, Johansson M, Kogler B, Crittenden R, Mattila-Sandholm T, Saarela M. Effect of galacto-oligosaccharide supplementation on human faecal microflora and on survival and persistence of *Bifidobacterium lactis* Bb-12 in the gastrointestinal tract. *Int. Dairy J.* 2001; 11: 817–825.

150. Malinen E, Mättö J, Salmitie M, Alander M, Saarela M, and Palva A. PCR-ELISA - II: Analysis of *Bifidobacterium* populations in human faecal samples from a consumption trial with *Bifidobacterium lactis* Bb-12 and a galacto-oligosaccharide preparation. *Syst. Appl. Microbiol.* 2002; 25: 249–258.

151. Loh G, Eberhard M, Brunner RM, Hennig U, Kuhla S, Kleessen B, and Metges CC. Inulin alters the intestinal microbiota and short-chain fatty acid concentrations in growing pigs regardless of their basal diet. *J. Nutr.* 2006; 136: 1198–1202.

152. Mountzouris KC, Balaskas CFF, Tuohy KM, Gibson GR, and Fegeros K. Profiling of composition and metabolic activities of the colonic microflora of growing pigs fed diets supplemented with prebiotic oligosaccharides. *Anaerobe* 2006; 12(4): 178–185.

153. Rossi M, Corradini C, Amaretti A, Nicolini M, Pompei A, Zanoni S, and Matteuzzi D. Fermentation of fructooligosaccharides and inulin by bifidobacteria: a comparative study of pure and fecal cultures. *Appl. Environ. Microbiol.* 2005; 71: 6150–6158.

154. Nugent AP. Health properties of resistant starch. *Nutr. Bull.* 2005; 30: 27–54.

155. Mitsuoka T. Recent trends in research on intestinal flora. *Bifidobact. Microflora* 1982; 1: 3–24.

156. Hopkins MJ, Sharp R, and Macfarlane GT. Age and disease related changes in intestinal bacterial populations assessed by cell culture, 16S rRNA abundance, and community cellular fatty acid profiles. *Gut* 2001; 48: 198–205.

157. Canzi E, Casiraghi MC, Zanchi R, Gandolfi R, Ferrari A, Brighenti F, Bosia R, Crippa A, Maestri P, Vesely R, and Salvadori BB. Yogurt in the diet of the elderly: a preliminary investigation into its effect on the gut ecosystem and lipid metabolism. *Lait* 2002; 82: 713–723.

158. Silvi S, Verdenelli MC, Orpianesi C, and Cresci A. EU project Crownalife: functional foods, gut microflora and healthy ageing - Isolation and identification of *Lactobacillus* and *Bifidobacterium* strains from faecal samples of elderly subjects for a possible probiotic use in functional foods. *J. Food Eng.* 2003; 56: 195–200.

159. Crittenden RG and Playne MJ. Production, properties and applications of food-grade oligosaccharides. *Trends Food Sci. Technol.* 1996; 7: 353–361.

160. Fernández-Bañares F. Nutritional care of the patient with constipation. *Clin. Gastroenterol.* 2006; 20: 575–587.

161. Quah HM, Ooi BS, Seow-Choen F, Sng KK, and Ho KS. Prospective randomized crossover trial comparing fibre with lactulose in the treatment of idiopathic chronic constipation. *Tech. Coloproctol* 2006; 10: 111–114.

162. Jouët P, Sabate J-M, Cuillerier E, Coffin B, Lemann M, Jian R, and Flourie B. Low-dose lactulose produces a tonic contraction in the human colon. *Neurogastroenterol. Motil.* 2006; 18: 45–52.

163. Gibson GR, Beatty ER, Wang X, and Cummings JH. Selective stimulation of bifidobacteria in the human colon by oligofructose and inulin. *Gastroenteroogy* 1995; 108: 975–982.

164. Castiglia-Delavaud C, Verdier E, Besle JM, Vernet J, Boirie Y, Beaufrère B, de Baynast R, and Vermorel M. Net energy value of non-starch polysaccharide isolates (sugarbeet fibre and commercial inulin) and their impact on nutrient digestive utilization in healthy human subjects. *Br. J. Nutr.* 1998; 80: 343–352.

165. Chen H-L, Lu Y-H, Lin J-J, Ko L-Y. Effects of fructooligosaccharide on bowel function and indicators of nutritional status in constipated elderly men. *Nutr. Res.* 2000; 20: 1725–1733.

166. Chen H-L, Lu Y-H, Lin J-J, Ko L-Y. Effects of isomalto-oligosaccharides on bowel functions and indicators of nutritional status in constipated elderly men. *J. Am. Coll. Nutr.* 2001; 20: 44–49.

167. Tateyama I, Hashii K, Johno I, Iino T, Hirai K, Suwa Y, and Kiso Y. Effect of xylooligosaccharide intake on severe constipation in pregnant women. *J. Nutr. Sci. Vitaminol.* 2005; 51: 445–448.

168. Dbouk N and McGuire BM. Hepatic encephalopathy: a review of its pathophysiology and treatment. *Curr. Treatment Opt. Gastroenterol.* 2006; 9: 464–474.

169. Bongaerts G, Severijnen R, and Timmerman H. Effect of antibiotics, prebiotics and probiotics in treatment for hepatic encephalopathy. *Med. Hypotheses* 2005; 64: 64–68.

170. Björkstén B. Effects of intestinal microflora and the environment on the development of asthma and allergy. *Springer Semin. Immunopathol.* 2004; 25: 257–270.

171. Sudo N, Aiba Y, Oyama N, Yu XN, Matsunaga M, Koga Y, and Kubo C. Dietary nucleic acid and intestinal microbiota synergistically promote a shift in the Th1/Th2 balance toward Th1-skewed immunity. *Int. Arch. Allergy Immunol.* 2004; 135: 132–135.

172. Ouwehand AC. Antiallergic effects of probiotics. *J. Nutr.* 2007; 137 (Suppl.): 7S.
173. Björkstén B, Naaber P, Sepp E, and Mikelsaar M. The intestinal microflora in allergic Estonian and Swedish 2-year old children. *Clin. Exp. Allergy* 1999; 29: 342–346.
174. Björkstén B, Sepp E, Julge K, Voor T, and Mikelsaar M. Allergy development and the intestinal microflora during the first year of life. *J. Allergy Clin. Immunol.* 2001; 108: 516–520.
175. Böttcher M, Sandin A, Norin E, Midtvedt T, and Björkstén B. Microflora associated characteristics in faeces from allergic and non-allergic children. *Clin. Exp. Allergy* 2000; 30: 590–596.
176. Grönlund MM, Arvilommi H, Kero P, Lehtonen OP, and Isolauri E. Importance of intestinal colonisation in the maturation of humoral immunity in early infancy: a prospective follow up study of healthy infants aged 0–6 months. *Arch. Dis. Child* 2000; 83: F186–F192.
177. Kalliomäki M, Kirjavainen P, Eerola E, Kero P, Salminen S, and Isolauri E. Distinct patterns of neonatal gut microflora in infants in whom atopy was and was not developing. *J. Allergy Clin. Immunol.* 2001; 107: 129–134.
178. Kirjavainen PV, Apostolou E, Arvola T, Salminen SJ, Gibson GR, and Isolauri E. Characterizing the composition of intestinal microflora as a prospective treatment target in infant allergic disease. *FEMS Immunol. Med. Microbiol.* 2001; 32: 1–7.
179. Kirjavainen PV, Arvola T, Salminen SJ, and Isolauri E. Aberrant composition of gut microbiota of allergic infants: a target of bifidobacterial therapy at weaning?. *Gut* 2002; 51: 51–55.
180. He F, Ouwehand AC, Isolauri E, Hashimoto H, Benno Y, and Salminen S. Comparison of mucosal adhesion and species identification of bifidobacteria isolated from healthy and allergic infants. *FEMS Immunol. Med. Microbiol.* 2001; 30: 43–47.
181. Ouwehand AC, Isolauri E, He F, Hashimoto H, Benno Y, and Salminen S. Differences in *Bifidobacterium* flora composition in allergic and healthy infants. *J. Allergy Clin. Immunol.* 2001; 108: 144–145.
182. Mackie RI, Sghir A, and Gaskins HR. Developmental microbial ecology of the neonatal gastrointestinal tract. *Am. J. Clin. Nutr.* 1999; 69: 1035S–1045S.
183. Sashihara T, Sueki N, and Ikegami S. An analysis of the effectiveness of heat-killed lactic acid bacteria in alleviating allergic diseases. *J. Dairy Sci.* 2006; 89: 2846–2855.
184. Takahashi N, Kitazawa H, Shimosato T, Iwabuchi N, Xiao JZ, Iwatsuki K, Kokubo S, and Saito T. An immunostimulatory DNA sequence from a probiotic strain of *Bifidobacterium longum* inhibits IgE production *in vitro*. *FEMS Immunol. Med. Microbiol.* 2006; 46: 461–469.
185. Sawada J, Morita H, Tanaka A, Salminen S, He F, and Matsuda H. Ingestion of heat-treated *Lactobacillus rhamnosus* GG prevents development of atopic dermatitis in NC/Nga mice. *Clin. Exp. Allergy* 2007; 37: 296–303.
186. Isolauri E, Arvola T, Sutas Y, Moilanen E, and Salminen S. Probiotics in the management of atopic eczema. *Clin. Exp. Allergy* 2000; 30: 1604–1610.
187. Kalliomaki M, Salminen S, Arvilommi H, Kero P, and Isolauri E. Probiotics in primary prevention of atopic disease: a randomised placebo-controlled trial. *Lancet* 2001; 357: 1076–1079.

188. Kalliomaki M, Salminen S, Poussa T, Arvilommi H, and Isolauri E. Probiotics and prevention of atopic disease: 4-year follow-up of a randomised placebo-controlled trial. *Lancet* 2003; 361: 1869–1871.

189. Viljanen M, Savilahti E, Haahtela T, Juntunen-Backman K, Korpela R, Poussa T, Tuure T, and Kuitunen M. Probiotics in the treatment of atopic eczema/dermatitis syndrome in infants: a double-blind placebo-controlled trial. *Allergy* 2005; 60: 494–500.

190. Weston S, Halbert A, Richmond P, and Prescott SL. Effects of probiotics on atopic dermatitis: a randomised controlled trial. *Arch. Dis. Child* 2005; 90: 892–897.

191. Taylor AL, Hale J, Wiltschut J, Lehmann H, Dunstan JA, and Prescott SL. Effects of probiotic supplementation for the first 6 months of life on allergen- and vaccine-specific immune responses. *Clin. Exp. Allergy* 2006; 36: 1227–1235.

192. Taylor AL, Dunstan JA, and Prescott SL. Probiotic supplementation for the first 6 months of life fails to reduce the risk of atopic dermatitis and increases the risk of allergen sensitization in high-risk children: a randomized controlled trial. *J. Allergy Clin. Immunol.* 2007; 119: 184–191.

193. Brouwer ML, Wolt-Plompen SA, Dubois AE, van der HS, Jansen DF, Hoijer MA, Kauffman HF, and Duiverman EJ. No effects of probiotics on atopic dermatitis in infancy: a randomized placebo-controlled trial. *Clin. Exp. Allergy* 2006; 36: 899–906.

194. Fölster-Holst R, Müller F, Schnopp N, Abeck D, Kreiselmaier I, Lenz T, von Rüden U, Schrezenmeir J, Christophers E, and Weichenthal M. Prospective, randomized controlled trial on *Lactobacillus rhamnosus* in infants with moderate to severe atopic dermatitis. *Br. J. Dermatol.* 2006; 155: 1256–1261.

195. Moreira A, Kekkonen R, Korpela R, Delgado L, and Haahtela T. Allergy in marathon runners and effect of *Lactobacillus* GG supplementation on allergic inflammatory markers. *Respir. Med.* 2007; 101: 1123–1131.

196. Nagura T, Hachimura S, Hashiguchi M, Ueda Y, Kanno T, Kikuchi H, Sayama K, and Kaminogawa s. Suppressive effect of dietary raffinose on T-helper 2 cell-mediated immunity. *Br. J. Nutr.* 2002; 88: 421–427.

197. Yoshida T, Hirano A, Wada H, Takahashi K, and Hattori M. Alginic acid oligosaccharide suppresses Th2 development and IgE production by inducing IL-12 production. *Int. Arch. Allergy Immunol.* 2004; 133: 239–247.

198. Mizubuchi H, Yajima T, Aoi N, Tomita T, and Yoshikai Y. Isomalto-oligosaccharides polarize Th1-like responses in intestinal and systemic immunity in mice. *J. Nutr.* 2005; 135: 2857–2861.

199. Fujitani S, Ueno K, Kamiya T, Tsukahara T, Ishihara K, Kitabayashi T, and Itabashi K. Increased number of CCR4-positive cells in the duodenum of ovalbumin-induced food allergy model NC/jic mice and antiallergic activity of fructooligosaccharides. *Allergol. Int.* 2007; 56: 131–138.

200. Vos AP, Haarman M, van Ginkel J-WH, Knol J, Garssen J, Stahl B, Boehm G, and M'Rabet L. Dietary supplementation of neutral and acidic oligosaccharides enhances Th1-dependent vaccination responses in mice. *Pediatr. Allergy Immunol.* 2007; 18: 304–312.

201. Moro G, Arslanoglu S, Stahl B, Jelinek J, Wahn U, and Boehm G. A mixture of prebiotic oligosaccharides reduces the incidence of atopic dermatitis during the first six months of age. *Arch. Dis. Child* 2006; 91: 814–819.

202. Kukkonen K, Savilahti E, Haahtela T, Juntunen-Backman K, Korpela R, Poussa T, Tuure T, and Kuitunen M. Probiotics and prebiotic galacto-oligosaccharides in the

prevention of allergic diseases: a randomized, double-blind, placebo-controlled trial. *J. Allergy Clin. Immunol.* 2007; 119: 192–198.

203. Schultz M, Timmer A, Herfarth H, Sartor RB, Vanderhoof JA, and Rath HC. Lactobacillus GG in inducing and maintaining remission of Crohn's disease. *BMC Gastroenterol.* 2004; 15: 5.

204. Hedin C, Whelan K, and Lindsay JO. Evidence for the use of probiotics and prebiotics in inflammatory bowel disease: a review of clinical trials. *Proc. Nutr. Soc.* 2007; 66: 307–315.

205. Linskens RK, Huijsdens XW, Savelkoul PHM, Vandenbroucke-Grauls CMJE, Meuwissen SMG. The bacterial flora in inflammatory bowel disease: Current insights in pathogenesis and the influence of antibiotics and probiotics. *Scand. J. Gastroenterol.* 2001; 36: S29–S40.

206. Favier C, Neut C, Mizon C, Cortot A, Colombel JF, and Mizon J. Fecal beta-D-galactosidase production and bifidobacteria are decreased in Crohn's disease. *Digest. Dis. Sci.* 1997; 42: 817–822.

207. Macfarlane S, Furrie E, Kennedy A, Cummings JH, and Macfarlane GT. Mucosal bacteria in ulcerative colitis. *Br. J. Nutr.* 2005; 93(Suppl. 1): S67–S72.

208. Hoentjen F, Welling GW, Harmsen HJM, Zhang X, Snart J, Tannock GW, Lien K, Churchill TA, Lupicki M, and Dieleman LA. Reduction of colitis by prebiotics in HLA-B27 transgenic rats is associated with microflora changes and immunomodulation. *Inflamm. Bowel Dis.* 2005; 11: 977–985.

209. Furrie E, Macfarlane S, Kennedy A, Cummings JH, Walsh SV, O'Neil DA, Macfarlane GT. Synbiotic therapy (*Bifidobacterium longum*/Synergy 1) initiates resolution of inflammation in patients with active ulcerative colitis: a randomised controlled pilot trial. *Gut* 2005; 54: 242–249.

210. Lara-Villoslada F, de Haro O, Camuesco D, Comalada M, Velasco J, Zarzuelo A, Xaus J, and Galvez J. Short-chain fructooligosaccharides, in spite of being fermented in the upper part of the large intestine, have anti-inflammatory activity in the TNBS model of colitis. *Eur. J. Nutr.* 2006; 45: 418–425.

211. Lindsay JO, Whelan K, Stagg AJ, Gobin P, Al-Hassi HO, Rayment N, Kamm MA, Knight SC, and Forbes A. Clinical, microbiological, and immunological effects of fructo-oligosaccharide in patients with Crohn's disease. *Gut* 2006; 55: 348–355.

212. Cherbut C, Michel C, and Lecannu G. The prebiotic characteristics of fructooligosaccharides are necessary for reduction of TNBS-induced colitis in rats. *J. Nutr.* 2003; 133: 21–27.

213. Geier MS, Butler RN, Giffard PM, and Howarth GS. Prebiotic and synbiotic fructooligosaccharide administration fails to reduce the severity of experimental colitis in rats. *Dis. Colon Rectum* 2007; 50: 1061–1069.

214. Moreau NM, Martin LJ, Toquet CS, Laboisse CL, Nguyen PG, Siliart BS, Dumon HJ, and Champ MMJ. Restoration of the integrity of rat caeco-colonic mucosa by resistant starch, but not by fructo-oligosaccharides, in dextran sulfate sodium-induced experimental colitis. *Br. J. Nutr.* 2003; 90: 75–85.

215. Winkler J, Butler R, and Symonds E. Fructo-oligosaccharide reduces inflammation in a dextran sodium sulphate mouse model of colitis. *Digest. Dis. Sci.* 2007; 52: 52–58.

216. Rumi G, Tsubouchi R, Okayama M, Kato S, Mózsik G, and Takeuchi K. Protective effect of lactulose on dextran sulfate sodium-induced colonic inflammation in rats. *Digest. Dis. Sci.* 2004; 49: 1466–1472.

217. Camuesco D, Peran L, Comalada M, Nieto A, Di Stasi LC, Rodriguez-Cabezas ME, Concha A, Zarzuelo A, and Galvez J. Preventative effects of lactulose in the trinitrobenzenesulphonic acid model of rat colitis. *Inflamm. Bowel Dis.* 2005; 11: 265–271.
218. Holma R, Juvonen P, Asmawi MZ, Vapaatalo H, and Korpela R. Galacto-oligosaccharides stimulate the growth of bifidobacteria but fail to attenuate inflammation in experimental colitis in rats. *Scand. J. Gastroenterol.* 2002; 37: 1042–1047.
219. Friedman G and George J. Treatment of refractory 'pouchitis' with prebiotic and probiotic therapy. *Gastroenterology* 2000; 118: G4167.
220. Welters CFM, Heineman E, Thunnissen BJM, Van den Bogaard AEJM, Soeters PB, and Baeten CGMI. Effect of dietary inulin supplementation on inflammation of pouch mucosa in patients with an ileal pouch-anal anastomosis. *Dis. Colon Rectum* 2002; 45: 621–627.
221. Kanauchi O, Suga T, Tochihara M, Hibi T, Naganuma M, Homma T, Asakura H, Nakano H, Takahama K, Fujiyama Y, Andoh A, Shimoyama T, Hida N, Haruma K, Koga H, Mitsuyama K, Sata M, Fukuda M, Kojima A, and Bamba T. Treatment of ulcerative colitis by feeding with germinated barley foodstuff: first report of a multicenter open control trial. *J. Gastroenterol.* 2002; 37(Suppl. 14): 67–72.
222. Bamba T, Kanauchi O, Andoh A, and Fujiyama Y. A new prebiotic from germinated barley for nutraceutical treatment of ulcerative colitis. *J. Gastroenterol. Hepatol.* 2002; 17: 818–824.
223. Fukasawa T, Murashima K, Matsumoto I, Hosono A, Ohara H, Nojiri C, Koga J, Kubota H, Kanegae M, Kaminogawa S, Abe K, and Kono T. Identification of marker genes for intestinal immunomodulating effect of a fructooligosaccharide by DNA microarray analysis. *J. Agric. Food Chem.* 2007; 55: 3174–3179.
224. Adogony V, Respondek F, Biourge V, Rudeaux F, Delaval J, Bind J-L, Salmon H. Effects of dietary scFOS on immunoglobulins in colostrums and milk of bitches. *J. Anim. Physiol. Anim. Nutr.* 2007; 91: 169–174.
225. Gibson GR, McCartney AL, and Rastall RA. Prebiotics and resistance to gastrointestinal infections. *Br. J. Nutr.* 2005; 93: 31–34.
226. Kunz C and Rudloff S. Biological functions of oligosaccharides in human milk. *Acta Paediatr.* 1993; 82: 903–912.
227. Shoaf K, Mulvey GL, Armstrong GD, and Hutkins RW. Prebiotic galactooligosaccharides reduce adherence of enteropathogenic *Escherichia coli* to tissue culture cells. *Infect. Immun.* 2006; 74: 6920–6928.
228. Tzortzis G, Goulas AK, and Gibson GR. Synthesis of prebiotic galactooligosaccharides using whole cells of a novel strain, *Bifidobacterium bifidum* NCIMB 41171. *Appl. Microbiol. Biotechnol.* 2005; 68: 412–416.
229. Özaslan C, Türkçapar AG, Kesenci M, Karayalçin K, Yerdel MA, Bengisun S, and Törüner A. Effect of lactulose on bacterial translocation. *Eur. J. Surg.* 1997; 163: 463–467.
230. Bovee-Oudenhoven IMJ, Termont DMSL, Heidt PJ, and Van der Meer R. Increasing the intestinal resistance of rats to the invasive pathogen *Salmonella enteritidis*: additive effects of dietary lactulose and calcium. *Gut* 1997; 40: 497–504.
231. Apanavicius CJ, Powell KL, Vester BM, Karr-Lilienthal LK, Pope LL, Fastinger ND, Wallig MA, Tappenden KA, and Swanson K. Fructan supplementation and infection affect food intake, fever, and epithelial sloughing from salmonella challenge in weanling puppies. *J. Nutr.* 2007; 137: 1923–1930.

232. Schumann C. Medical, nutritional and technological properties of lactulose. An update. *Eur. J. Nutr.* 2002; 41(Suppl. 1): 17–25.
233. Waligora-Dupriet AJ, Campeotto F, Nicolis I, Bonet A, Soulaines P, Dupont C, and Butel MJ. Effect of oligofructose supplementation on gut microflora and well-being in young children attending a day care. *Int. J. Food Microbiol.* 2007; 113: 108–113.
234. Lewis S, Burmeister S, Cohen S, Brazier J, and Awasthi A. Failure of dietary oligofructose to prevent antibiotic-associated diarrhoea. *Aliment. Pharmacol. Ther.* 2005; 21: 469–477.
235. Cummings JH, Christie S, and Cole TJ. A study of fructo oligosaccharides in the prevention of travellers' diarrhea. *Aliment. Pharmacol. Ther.* 2001; 15: 1139–1145.
236. Duggan C, Penny ME, Hibberd P, Gil A, Huapaya A, Cooper A, Coletta F, Emenhiser C, and Kleinman RE. Oligofructose-supplemented infant cereal: 2 randomized, blinded, community-based trials in Peruvian infants. *Am. J. Clin. Nutr.* 2003; 77: 937–942.
237. Kotzampassi K, Giamarellos-Bourboulis EJ, Voudouris A, Kazamias P, and Eleftheriadis E. Benefits of a synbiotic formula (Synbiotic 2000Forte) in critically Ill trauma patients: early results of a randomized controlled trial. *World J. Surg.* 2006; 30: 1848–1855.
238. Oláh A, Belágyi T, Pótó L, Romics L, and Bengmark S. Synbiotic control of inflammation and infection in severe acute pancreatitis: a prospective, randomized, double blind study. *Hepato-gastroenterology* 2007; 54: 590–594.
239. Scholz-Ahrens KE, Ade P, Marten B, Weber P, Timm W, Açil Y, Glüer C-C, Schrezenmeir J. Prebiotics, probiotics, and synbiotics affect mineral absorption, bone mineral content, and bone structure. *J. Nutr.* 2007; 137: 838S–846S.
240. Morohashi T. The effect on bone of stimulated intestinal mineral absorption following fructooligosaccharide consumption in rats. *Biosci. Microflora* 2002; 21: 21–25.
241. Chonan O, Matsumoto K, and Watanuki M. Effect of galacto-oligosaccharides on calcium absorption and preventing bone loss in ovariectomized rats. *Biosci. Biotechnol. Biochem.* 1995; 59: 236–239.
242. Coudray C, Demigne C, and Rayssiguier Y. Effects of dietary fibers on magnesium absorption in animals and humans. *J. Nutr.* 2003; 133: 1–4.
243. Taguchi A, Ohta A, Abe M, Baba S, Ohtsuki M, Takizawa T, Yuda Y, and Adachi T. The influence of fructo-oligosaccharides on the bone of model rats with ovariectomized osteoporosis. *Sci. Rep. Meiji Seika Kaisha* 1995; 33: 37–44.
244. Ohta A, Osakabe N, Yamada K, Saito Y, and Hidaka H. The influence of fructooligosaccharides and various other oligosaccharides on the absorption of Ca, Mg and P in rats. *J. Japanese Soc. Nutr. Food Sci.* 1993; 46: 123–129.
245. Devareddy L, Khalil DA, Korlagunta K, Hooshmand S, Bellmer DD, and Arjmandi BH. The effects of fructo-oligosaccharides in combination with soy protein on bone in osteopenic ovariectomized rats. *Menopause* 2006; 13: 692–699.
246. Mizota T. Lactulose as a growth promoting factor for *Bifidobacterium* and its physiological aspects. *Bull. Int. Dairy Fed.* 1996; 313: 43–48.
247. Lopez HW, Levrat-Verny M-A, Coudray C, Besson C, Krespine V, Messager A, Demigné C, and Rémésy C. Class 2 Resistant starches lower plasma and liver lipids and improve mineral retention in rats. *J. Nutr.* 2001; 131: 1283–1289.
248. Younes H, Demigné C, and Rémésy C. Acidic fermentation in the caecum increases absorption of calcium and magnesium in the large intestine of the rat. *Br. J. Nutr.* 1996; 75: 301–314.

249. Frank A. Oligofructose-enriched inulin stimulates calcium absorption and bone mineralisation. *Br. Nutr. Foundation Nutr. Bull.* 2006; 31: 341–345.
250. Holloway L, Moynihan S, Abrams SA, Kent K, Hsu AR, and Friedlander AL. Effects of oligofructose-enriched inulin on intestinal absorption of calcium and magnesium and bone turnover markers in postmenopausal women. *Br. J. Nutr.* 2007; 97: 365–372.
251. Abrams SA, Griffin IJ, Hawthorne KM, Liang L, Gunn SK, Darlington G, and Ellis KJ. A combination of prebiotic short- and long-chain inulin-type fructans enhances calcium absorption and bone mineralization in young adolescents. *Am. J. Clin. Nutr.* 2005; 82: 471–476.
252. Hillman LS, Tack E, Covell DG, Vieira NE, and Yergey AL. Measurement of true calcium absorption in premature infants using intravenous ^{46}Ca and oral ^{44}Ca. *Pediatr. Res.* 1988; 23: 589–594.
253. Van den Heuvel EG, Muijs T, van Dokkum W, and Schaafsma G. Lactulose stimulates calcium absorption in postmenopausal women. *J. Bone Mineral Res.* 1999; 14: 1211–1216.
254. Younes H, Coudray C, Bellanger J, Demigne C, Rayssiguier Y, and Remesy C. Effects of two fermentable carbohydrates (inulin and resistant starch) and their combination on calcium and magnesium balance in rats. *Br. J. Nutr.* 2001; 86: 479–485.
255. Kruger MC, Brown KE, Collett G, Layton L, and Schollum LM. The effect of fructooligosaccharides with various degrees of polymerization on calcium bioavailability in the growing rat. *Exp. Biol. Med.* 2003; 228: 683–688.
256. Griffin IJ, Davila PM, and Abrams SA. Non-digestible oligosaccharides and calcium absorption in girls with adequate calcium intakes. *Br. J. Nutr.* 2002; 87(Suppl. 2): S187–S191.
257. Orchel A, Dzierewicz Z, Parfiniewicz B, Weglarz L, and Wilczok T. Butyrate-induced differentiation of colon cancer cells is PKC and JNK dependent. *J. Digest. Dis. Sci.* 2005; 50: 490–498.
258. Daly K, Shirazi-Beechey SP. Microarray analysis of butyrate regulated genes in colonic epithelial cells. *DNA Cell. Biol.* 2006; 25: 49–62.
259. Lupton Microbial degradation products influence colon cancer risk: the butyrate controvosy. *J. Nutr.* 2004; 134: 479–482.
260. Duncan SH, Louis P, and Flint HJ. Lactate-utilizing bacteria, isolated from human feces, that produce butyrate as a major fermentation product. *Appl. Environ. Microbiol.* 2004; 70: 5810–5817.
261. Morrison DJ, Mackay WG, Edwards CA, Preston T, Dodson B, and Weaver LT. Butyrate production from oligofructose fermentation by the human faecal flora: what is the contribution of extracellular acetate and lactate?. *Br. J. Nutr.* 2006; 96: 570–577.
262. Rowland IR and Tanaka R. The effects of transgalactosylated oligosaccharides on gut flora metabolism in rats associated with a human faecal microflora. *J. Appl. Bacteriol.* 1993; 74: 667–674.
263. Phillips J, Muir JG, Birkett A, Lu ZX, Jones GP, O'Dea K, Young GP. Effect of resistant starch on fecal bulk and fermentation-dependent events in humans. *Am. J. Clin. Nutr.* 1995; 62: 121–130.
264. Hylla S, Gostner A, Dusel G, Anger H, Bartram HP, Christl SU, Kasper H, and Scheppach W. Effects of resistant starch on the colon in healthy volunteers: possible implications for cancer prevention. *Am. J. Clin. Nutr.* 1998; 67: 136–142.

265. Grubben MJ, van den Braak CC, Essenberg M, Olthof M, Tangerman A, Katan MB, and Nagengast FM. Effect of resistant starch on potential biomarkers for colonic cancer risk in patients with colonic adenomas: a controlled trial. *Digest. Dis. Sci.* 2001; 46: 750–756.
266. Jacobasch G, Dongowski G, Schmiedl D, Müller-Schmehl K. Hydrothermal treatment of Novelose 330 results in high yield of resistant starch type 3 with beneficial prebiotic properties and decreased secondary bile acid formation in rats. *Br. J. Nutr.* 2006; 95: 1063–1074.
267. Bouhnik Y, Raskine L, Champion K, Andrieux C, Penven S, Jacobs H, and Simoneau G. Prolonged administration of low-dose inulin stimulates the growth of bifidobacteria in humans. *Nutr. Res.* 2007; 27: 187–193.
268. Bouhnik Y, Flourié B, Ouarne F, Riottot M, Bissetti N, Bornet F, and Rambaud JC. Effects of prolonged ingestion of fructo-oligosaccharides (FOS) on colonic bifidobacteria, faecal enzymes and bile acids in humans. *Gastroenterology* 1994; 106: A598.
269. Pool-Zobel BL. Inulin-type fructans and reduction in colon cancer risk: review of experimental and human data. *Br. J. Nutr.* 2005; 93(Suppl. 1): S73–S90.
270. Rowland IR, Bearne CA, Fischer R, Pool-Zobel BL. The effect of lactulose on DNA damage induced by DMH in the colon of human flora-associated rats. *Nutr. Cancer* 1996; 26: 37–47.
271. Challa A, Rao DR, Chawan CB, and Shackelford L. *Bifidobacterium longum* and lactulose suppress azoxymethane-induced colonic aberrant crypt foci in rats. *Carcinogenesis* 1997; 18: 517–521.
272. Hsu C-K, Liao J-W, Chung Y-C, Hsieh C-P, Chan Y-C. Xylooligosaccharides and fructooligosaccharides affect the intestinal microbiota and precancerous colonic lesion development in rats. *J. Nutr.* 2004; 134: 1523–1528.
273. Perrin P, Pierre F, Patry Y, Champ M, Berreur M, Pradal G, Bornet F, Meflah K, and Menanteau J. Only fibers promoting a stable butyrate producing colonic ecosystem decrease the rate of aberrant crypt foci in rats. *Gut* 2001; 48: 53–61.
274. Nakanishi S, Kataoka K, Kuwahara T, and Ohnishi Y. Effects of high amylose maize starch and *Clostridium butyricum* on metabolism in colonic microbiota and formation of azoxymethane-induced aberrant crypt foci in the rat colon. *Microbiol. Immunol.* 2003; 47: 951–958.
275. Watzl B, Girrbach S, and Roller M. Inulin, oligofructose and immunomodulation. *Br. J. Nutr.* 2005; 93(Suppl. 1): S49–S55.
276. Roller M, Pietro Femia A, Caderni G, Rechkemmer G, and Watzl B. Intestinal immunity of rats with colon cancer is modulated by oligofructose-enriched inulin combined with *Lactobacillus rhamnosus* and *Bifidobacterium lactis*. *Br. J. Nutr.* 2004; 92: 931–938.
277. Verghese M, Walker LT, Shackelford L, and Chawan CB. Inhibitory effects of nondigestible carbohydrates of different chain lengths on azoxymethane-induced aberrant crypt foci in Fisher 344 rats. *Nutr. Res.* 2005; 25: 859–868.
278. Rowland IR, Rumney CJ, Coutts JT, and Lievense LC. Effect of *Bifidobacterium longum* and inulin on gut bacterial metabolism and carcinogen-induced aberrant crypt foci in rats. *Carcinogenesis* 1998; 19: 281–285.
279. Gallaher DD and Khil j. The effect of synbiotics on colon carcinogenesis in rats. *J. Nutr.* 1999; 129: S1483–S1487.
280. Femia AP, Luceri C, Dolara P, Giannini A, Biggeri A, Salvadori M, Clune Y, Collins JK, Paglierani M, and Caderni G. Antitumorigenic activity of the prebiotic inulin enriched

with oligofructose in combination with the probiotics *Lactobacillus rhamnosus* and *Bifidobacterium lactis* on azoxymethane-induced colon carcinogenesis in rats. *Carcinogenesis* 2002; 23: 1953–1960.

281. Ferchaud-Roucher V, Pouteau E, Piloquet H, Zaïr Y, and Krempf M. Colonic fermentation from lactulose inhibits lipolysis in overweight subjects. *Am. J. Physiol.* 2005; 289: E716–E720.

282. Shimomura Y, Maeda K, Nagasaki M, Matsuo Y, Murakami T, Bajotto G, Sato J, Seino T, Kamiwaki T, and Suzuki M. Attenuated response of the serum triglyceride concentration to ingestion of a chocolate containing polydextrose and lactitol in place of sugar. *Biosci. Biotechnol. Biochem.* 2005; 69: 1819–1823.

283. Vogt JA, Ishii-Schrade KB, Pencharz PB, Jones PJH, and Wolever TMS. L-rhamnose and lactulose decrease serum triacylglycerols and their rates of synthesis, but do not affect serum cholesterol concentrations in men. *J. Nutr.* 2006; 136: 2160–2166.

284. Williams CM and Jackson KG. Inulin and oligofructose: effects on lipid metabolism from human studies. *Br. J. Nutr.* 2002; 87(Suppl. 2): S261–S264.

285. Brighenti F, Benini L, Del Rio D, Casiraghi C, Pellegrini N, Scazzina F, Jenkins DJA, and Vantini I. Colonic fermentation of indigestible carbohydrates contributes to the second-meal effect. *Am. J. Clin. Nutr.* 2006; 83: 817–822.

286. Juskiewicz J, Klewicki R, and Zdunczyk Z. Consumption of galactosyl derivatives of polyols beneficially affects cecal fermentation and serum parameters in rats. *Nutr. Res.* 2006; 26: 531–536.

287. Swanson KS and Fahey GC Jr. Prebiotic impacts on companion animals. In: Gibson GR and Rastall RA, editors. *Prebiotics: Development and Application.* John Wiley & Sons, Chichester, England, 2006, pp. 213–236.

288. Mwenya B, Santoso B, Sar C, Gamo Y, Kobayashi T, Arai I, and Takahashi J. Effects of including beta1-4 galacto-oligosaccharides, lactic acid bacteria or yeast culture on methanogenesis as well as energy and nitrogen metabolism in sheep. *Animal Feed Sci. Technol.* 2004; 115: 313–326.

289. Mwenya B, Zhou X, Santoso B, Sar C, Gamo Y, Kobayashi T, and Takahashi J. Effects of probiotic-vitacogen and beta1-4 galacto-oligosaccharides supplementation on methanogenesis and energy and nitrogen utilization in dairy cows. *Asian-Australasian J. Anim. Sci.* 2004; 17: 349–354.

290. Sar C, Santoso B, Gamo Y, Kobayashi T, Shiozaki S, Kimura K, Mizukoshi H, Arai I, and Takahashi J. Effects of combination of nitrate with beta1-4 galacto-oligosaccharides and yeast (*Candida kefyr*) on methane emission from sheep. *Asian-Australasian J. Anim. Sci.* 2004; 17: 73–79.

291. Santoso B, Mwenya B, Sar C, Gamo Y, Kobayashi T, Morikawa R, Kimura K, Mizukoshi H, and Takahashi J. Effects of supplementing galacto-oligosaccharides, *Yucca schidigera* or nisin on rumen methanogenesis, nitrogen and energy metabolism in sheep. *Livestock Production Sci.* 2004; 91: 209–217.

292. Patterson JA and Burkholder KM. Application of prebiotics and probiotics in poultry production. *Poultry Sci.* 2003; 82: 627–631.

293. Flickinger EA, Van Loo J, and Fahey GC Jr. Nutritional responses to the presence of inulin and oligofructose in the diets of domesticated animals: A review. *Crit. Rev. Food Sci. Nutr.* 2003; 43: 19–60.

294. Shim SB, Verstegen MWA, Kim IH, Kwon OS, and Verdonk JMAJ. Effects of feeding antibiotic-free creep feed supplemented with oligofructose, probiotics or synbiotics to suckling piglets increases the preweaning weight gain and composition of intestinal microbiota. *Arch. Animal Nutr.* 2005; 59: 419–427.

295. Playne MJ and Crittenden R. Commercially available oligosaccharides. *Bull. Int. Dairy Fed.* 1996; 313: 10–22.

296. Moshfegh AJ, Friday JE, Goldman JP, and Chug Ahuja JK. Presence of inulin and oligofructose in the diets of Americans. *J. Nutr.* 1999; 129(Suppl. 7): S1407–S1411.

297. Smith PB. Safety of short-chain fructo-oligosaccharides and GRAS affirmation by the US FDA. *Biosci. Microflora* 2002; 21: 27–29.

298. Deguchi Y, Matsumoto K, Ito A, and Watanuki M. Effects of β 1-4 galacto-oligosaccharides administration on defaecation of healthy volunteers with a tendency to constipation. *Japanese J. Nutr.* 1997; 55: 13–22.

299. Saavedra JM and Tschernia A. Human studies with probiotics and prebiotics: clinical implications. *Br. J. Nutr.* 2002; 87(Suppl. 2): S241–S246.

300. Bovee-Oudenhoven IMJ, Ten Bruggencate SJM, Lettink-Wissink MLG, and van der Meer R. Dietary fructo-oligosaccharides and lactulose inhibit intestinal colonisation but stimulate translocation of *Salmonella* in rats. *Gut* 2003; 52: 1572–1578.

301. Ten Bruggencate SJM, Bovee-Oudenhoven IMJ, Lettink-Wissink MLG, and van der Meer R. Dietary fructo-oligosaccharides dose-dependently increase translocation of *Salmonella* in rats. *J. Nutr.* 2003; 133: 2313–2318.

302. Ten Bruggencate SJM, Bovee-Oudenhoven IMJ, Lettink-Wissink MLG, Katan MB, and van der Meer R. Dietary fructooligosaccharides affect intestinal barrier function in healthy men *J. Nutr.* 2006; 136: 70–74.

AUTHOR INDEX

Abelardo Margolles, 4, 6, 190
Allan Lim, 123, 321
Andrew W. Bruce, 30, 472
Arthur C. Ouwehand, 21, 23, 106, 377, 379, 382–384, 386, 389, 390, 447, 451, 469, 477, 551

Baltasar Mayo, 4, 37

Camilla Hoppe, 441, 480
Charlotte Nexmann Larsen, 38, 41, 441, 466, 480
Clara de los Reyes-Gavilán, 6, 25

Diana Donohue, 75, 83
Dorte Eskesen, 38, 41, 466, 480

Fred H Degnan, 111

Gregor Reid, 4, 20, 79, 83, 85, 90, 386, 388, 470–472

Hai–Meng Tan, 123, 321
Hirokazu Tsuji, 220, 539, 485

James J.C. Chin, 406, 408, 409
Jin–zhong Xiao, 306, 488, 490, 491
Jose M Saavedra, 111, 482, 485, 560
J.M. Laparra, 259, 260, 262

Kajsa Kajander, 258, 285, 385, 390, 469, 477–479
Katsunori Kimura, 14, 15, 21, 457
Koji Nomoto, 449, 485

M. Carmen Collado, 1, 6, 20, 23, 24, 38, 41, 257, 263, 382, 384–391, 397, 403, 404, 447, 478, 481
Maija Saxelin, 63, 67, 70, 92, 182, 462–465, 469, 470
Martin J. Playne, 62, 80, 560
Mayumi Kiwaki, 217, 226, 449
Miguel Gueimonde, 20, 21, 23–25, 36, 38, 41, 79, 94, 262, 382, 389, 400, 401
Mimi Tang, 395

Paivi Nurminen, 30, 473
Patricia Ruas–Madiedo, 4, 190, 390

Handbook of Probiotics and Prebiotics, Second Edition   Edited by Yuan Kun Lee and Seppo Salminen
Copyright © 2009 John Wiley & Sons, Inc.

Rafael Frias, 79, 94, 263
Rangne Fondén, 444–446, 462–464, 473
Reetta Satokari, 7, 14, 16, 33, 400, 402, 548, 549
Riitta Korpela, 270, 271, 477
Ross Crittenden, 20, 58, 62–66, 68, 69, 462, 465, 535, 536, 539, 548, 550

Sampo Lahtinen, 41, 42, 106, 377, 382, 447, 473, 492–494
Satu Vesterlund, 81, 377, 391, 469

Seppo Salminen, 3, 23, 35, 36, 59, 64, 81, 83, 84, 85, 90, 92, 257, 267, 384, 399, 469, 478, 480, 492, 536

Toni A Chapman, 43, 406, 410, 411, 415

Ulla Svensson, 63, 67, 70, 219, 444, 462–465

William Hung Chang Tien, 95, 205, 212, 216

Yuan Kun Lee, 3, 52, 54, 57, 177, 180, 181, 267, 321, 322, 450, 451, 454

SUBJECT INDEX

7-keto-deoxycholic acid, 485
16s sequencing, 9, 10

Aberrant crypt foci (ACF), 486, 487
Absenteeism, reduction of, 449, 482
Acid, 6, 15, 29, 39, 40, 64, 66, 91, 98, 99, 180, 190, 191, 267, 268, 331, 341, 392, 398, 400, 415, 418, 441, 447, 454, 461, 471, 480, 485, 487, 489
  effect on adhesion, 382, 480
  tolerance, 64, 73, 180, 197, 480
Acquired immunity, 76, 409
Adhesion, 19–21, 23, 24, 56, 81, 181, 190, 222, 262, 263, 322, 348, 377–391, 418, 445, 447, 451, 454, 463, 464, 466, 469, 471, 477, 478, 480, 481, 492, 547
  factors affecting, 382–383
  to epithelial cells, 23, 377–378, 382–383, 386, 389
  to human intestinal mucus, 23, 377–379, 384, 385, 389, 390, 454, 469, 478, 492
  to oral cavity, 21, 388, 478
  to fibronectin, 381
  prevention of pathogen adhesion, 19–21, 24, 377, 383–386, 389, 391, 492
AFLP, 11, 13, 18
Aggregation, 384–386, 388
Agricultural Compounds and Veterinary Medicines Regulation (ACVM), 135
Allergic infants, 384, 402, 551
Allergy, 89, 91, 105, 302, 305, 307, 309, 310, 397, 452, 470, 479, 484, 490, 491, 551, 552
Amplified Ribosomal DNA Restriction Analysis (ARDRA), 571
Animal model, 5, 263–267, 397, 398, 448, 552, 553
Antagonistic, 22, 23, 385, 549
Antibiotic, 25, 38, 41, 54, 78, 81–86, 90, 91, 124, 181, 258, 276–279, 292, 321–323, 331, 385, 393, 408, 410, 415, 442, 446, 448, 449, 458, 462, 463, 467, 470, 472, 474, 481, 482, 489, 535–537, 552, 559
  associated diarrhea, 54, 55, 57, 91, 258, 276, 279, 535, 537
  growth promoter (AGP), 321, 322, 408, 415, 419, 420

Antibiotic (*Continued*)
  resistance, 78, 83, 85, 86, 124, 181, 190, 415, 444, 463, 489
  therapies, 86, 258, 385, 448, 458, 461
Antibody responses, 449, 479, 555
Antigenotoxic, 21, 24
Antimicrobial, 19–22, 24, 66, 84, 88, 181, 257, 263, 385, 391–395, 401, 408, 415, 418, 447, 454, 461, 473, 493, 537
  against *Escherichia coli*, 27, 76, 493
  against *Helicobacter pylor*, 76, 461, 493
  against *Salmonella*, 24
  against *Staphylococcus aureus*, 76, 493
  resistant, 408, 415, 418, 537
Antioxidant, 66, 67, 124, 262, 445, 446, 463, 464
Antitumor, 93, 310, 468
Arthritis rheumatoides, 470
Aspergillus niger, 124, 126, 136
Aspergillus oryzae, 329, 337, 339, 341–343
Association of American Feed Control Authority (AAFCO), 123
Atopy, 35, 302, 561
Autistic children, 402
Australia, 125, 136, 138, 415, 544
Australian Pesticides and Veterinary Medicines Authority (APVMA), 125, 135
Autochthonous commensal, 408
Azoreductase, 448, 558

β-glucuronidase, 200, 448, 489, 558
Bacillus badius, 49, 126, 136
Bacillus cereus, 76, 200, 336, 338, 346
  mutagenesis, 200
  transposon, 200
Bacillus clausii, 277, 293, 295
Bacillus coagulans, 105, 124, 325
Bacillus lentus, 124, 126, 136
Bacillus licheniformis, 124, 126, 136, 336, 337
Bacillus pumilus, 124, 126, 136, 336
Bacillus subtilis, 124, 127, 136, 221, 322–324, 327, 328, 330, 336, 341, 342
  vector, 212, 214
Bacillus toyoi, 127, 136, 337
Bacterial community profiling, 43, 44, 47, 48, 51, 402s

Bacteriocin, 190, 199, 226, 388, 392–394, 398, 415, 416, 442, 445
  CH-5, 442
Bacteriodes amylophilus, 124, 127, 136
Bacteriodes capillosus, 124, 127, 136
Bacteriodes ruminocola, 124, 127, 136
Bacteriodes suis, 124, 127, 136
Bacteroides vulgatus, 29, 31, 32, 36, 388, 389
Bifidobacterium, 4–10, 13, 14, 19, 20, 22, 23, 25, 29–33, 35–41, 49, 50, 60, 62, 64, 70, 72, 75, 86, 123, 124, 127, 136, 200, 221, 262, 263, 281, 284, 293, 294, 303, 306, 350, 380, 387, 397, 399–401, 403, 405, 442, 448, 478, 485, 488, 489, 491–493, 536, 547–549, 551, 552
  characterization, 4–6, 19, 24
  identification, 4–18, 24
  typing of, 5, 7, 9–12, 14, 24
Bifidobacterium adolescentis, 31, 101, 124, 127, 136, 212, 221, 402
  M101-4, 268
  vector, 212–221
Bifidobacterium angulatum, 7, 262
Bifidobacterium animalis, 7, 72, 86, 87, 108, 200, 221, 262, 284, 442, 473, 480
  BB12, 41, 86, 90, 108, 442, 478, 480
  DN-173010, 284
  mutagenesis, 199
  subsp. *animalis*, 7, 13, 14, 62–64, 66, 70–74, 86, 478
  subsp. *lactis*, 7, 13, 14, 62–64, 66, 70–74, 86, 478
  vector, 221
Bifidobacterium asteroids, 7, 8, 202, 221
  plasmid, 202, 221
Bifidobacterium bifidum, 7, 39, 62, 63, 71, 72, 80, 88, 101, 124, 127, 136, 200, 202, 262, 268, 301, 329, 381, 400, 401, 548, 551
  A234-4, 268
  MF20/5, 301
  mutagenesis, 199
  plasmid, 200, 202
  vector, 212–221
Bifidobacterium boum, 7, 8
Bifidobacterium breve, 7, 31, 58, 62, 63, 66, 71, 101, 200, 202, 221, 262, 267, 285, 286, 293, 306, 307, 399, 400, 401, 482, 486

Bb99, 286, 293, 307, 320
I-53–8, 268
mutagenesis, 199
plasmid, 200, 202, 221
vector, 212–221
Yakult, 296, 485
Bifidobacterium catenulatum, 7, 31, 33, 202, 221, 548
  plasmid, 202, 221, 222
Bifidobacterium choerinum, 7
Bifidobacterium coryneforme, 7
Bifidobacterium cuniculi, 7
Bifidobacterium dentium, 7, 9, 33, 221
  vector, 221
Bifidobacteria fermented milk (BFM), 40, 487
Bifidobacteria fermented soya milk (FMS), 487
Bifidobacterium gallicum, 7
Bifidobacterium gallinarum, 7
Bifidobacterium indicum, 7
Bifidobacterium infantis, 31, 54, 62, 63, 66, 71, 88, 101, 124, 127, 136, 200, 221, 262, 267, 268, 273, 277, 280, 290, 291, 384, 400, 401, 551
  I-10-5, 268
  mutagenesis, 199
  vector, 212–221
Bifidobacterium lactis, 20, 85, 90, 114, 277, 278, 298, 303, 310, 348, 379, 385, 398, 473, 474, 477
  Bb-12, 90, 298, 303, 310, 348, 379, 385
  Bi-07, 449
  HN019 (DR10), 473–477
Bifidobacterium longum, 7, 8, 31, 62, 63, 66, 71, 72, 74, 103, 200, 202, 221, 262, 267, 268, 277, 280, 290, 291, 294, 301, 306, 314, 316, 378, 400, 401, 488–492, 494, 548, 551
  913, 314
  antiallergic activity, 490
  BB536, 488, 489–491
  BL2C, 491, 492, 494
  BL46, 492, 494
  effects on immunity and cancer, 490
  improvement of intestinal environment, 489
  M101-2, 268
  mutagenesis, 199

plasmids, 200, 202, 221
safety, 103, 267, 488, 489, 492
SP07/0, 301
subsp. *infantis*, 7, 62, 63, 66, 71, 401, 551
subsp. *longum*, 7, 62, 63, 66, 71, 401, 551
subsp. *suis*, 7
vectors, 221
Bifidobacterium magnum, 7, 221
  vector, 221
Bifidobacterium merycicum, 7
Bifidobacterium minimum, 7
Bifidobacterium pseudocatenulatum, 7, 33, 221
  vector, 221
Bifidobacterium pseudolongum, 7, 33, 221
  subsp. *globosum*, 7, 133, 202
  subsp. *pseudolongum*, 60, 262
  vector, 221
Bifidobacterium pseudomonas subsp. *globosum*, 202
  plasmid, 202
Bifidobacterium psychraerophilum, 7
Bifidobacterium pullorum, 7
Bifidobacterium ruminantium, 7
Bifidobacterium saeculare, 7
Bifidobacterium scardovii, 7
Bifidobacterium subtile, 7
Bifidobacterium thermacidophilum, 7
  subsp. *thermoacidophilum*, 7
  subsp. *porcinum*, 7
Bifidobacterium thermophilum, 7, 49, 185, 336, 342
Biliary cancer, 296, 455, 487
Bile, 6, 19–22, 24, 25, 37, 38, 39, 52, 98, 180, 181, 184, 190, 191, 193, 200, 211, 262, 263, 268, 296, 313, 357, 382, 441, 445, 464, 466, 473, 480, 484, 489, 537
  effect on adhesion, 382
  tolerance to, 6, 20, 180, 183, 184, 190, 446, 465, 466
Biopsies, 6, 25, 290, 379, 380, 384, 398, 459–461, 464, 469
Bladder cancer, 182, 311, 313, 453
Blood mononuclear cells, 308, 452, 457, 479
Blood pressure, 454, 455
Bowel function, 108, 182

Caco-2, 23, 260–263, 378, 380–383, 451, 454, 478

Calves, 126–128, 131, 134, 321, 331, 340–343
Campylobacter enteritis, 486
Campylobacter jejuni, 280, 322, 323, 348, 442
Cancer, 84, 105, 118, 182, 260, 292, 296, 310–313, 443, 446, 453, 455, 486, 487, 48, 490, 556, 558
Candida albicans, 87, 118, 448, 449, 484
Candida glabrata, 127
Candida pinolepessi, 127, 136
Candida utilis, 101, 127
Candidiasis, 448
Cannabinoid receptors, 448
Carbon dioxide, 393, 458
Cell line model, 259, 263
Cellular fatty acids, 29, 37
Cellular immunity, 89, 452, 454
Characterization, probiotics, 4–6, 19, 24
China, 100, 125, 136, 138
Cheese, 15, 21, 67, 70, 71, 95, 464, 469
Children, improvement of growth, 108
Cholesterol, 22, 104, 105, 221, 313–317, 323–327, 329, 334, 345, 455, 484–486, 559
  liver, 104, 313, 317, 324, 326, 486
  plasma, 313, 345, 455, 486
Claims, 76–79, 89, 90, 94, 99, 100, 103, 104, 106–111, 113, 115–122, 310, 350, 449, 477, 479
  content, 116
  benefit or efficacy, 116
  health, 76, 79, 89, 94, 99, 106–110, 117–119, 121, 479
  qualified health, 117, 119
  structure or function, 117–121
Classification, probiotics, 4, 5, 7, 13–16
Clostridium butyricum, 127, 137, 277
  MIYAIRI, 277
Clostridium cluster XIV, 403
Clostridium coccoides-Eubacterium rectale group, 32, 35
Clostridium perfringens, 321, 322, 348, 389, 557
Clostridium sporogenes phage, 127
Closridium tyrobutyricum phage, 127
Coaggregation, 35, 195, 200, 385, 386, 388, 447, 478
Coccidiostats, 125

Colic, infant, 448
Colicins, 415, 416, 418–420
Colibacillosis, 406–408, 414, 419
Collagen, 81, 381, 451, 456, 464
Colon carcinogenesis, 453, 558
Colonic epithelial cell, 261, 485, 535
Colonic hyperplasia, 448
Colonization, 15, 20, 21, 23, 87, 98, 181, 184, 262, 270, 298, 309, 310, 345, 377–380, 384–386, 388, 389, 391, 395, 398–401, 403–407, 409, 420, 450, 456, 462, 469, 482, 486, 549, 551, 553
Competition, 24, 54–56, 385, 389, 409, 417, 454
Competitive exclusion, 23, 321, 322, 408, 409
Confectionary, 74
Conjugated bile salt hydrolase, 200, 489
Cough, 449
Critically ill children (CIC), 456
Crohn's disease, 36, 76, 93, 287, 288, 470, 473, 551, 554
Cystic fibrosis, 76, 470
Cytokines, 88, 89, 93, 259, 308, 378, 397, 398, 443, 445, 446, 452, 453, 468, 483, 490, 493, 551, 553, 554
Cytoprotection, 449
Cytotoxic activity, 474

Digestive enzymes, 52, 98, 102, 382, 441, 447, 473, 480
  effect on adhesion, 382
  tolerance to, 447
Dimethylamine, 448
Deconjugate bile salts, 21, 24, 489
Decrease cholesterol, 24
Denaturing Gradient Gel Electrophoresis (DGGE), 11–13, 16, 27, 30, 32–34, 42, 43, 51, 290, 402
Denaturing /Thermal Gradient Gel Electrophoresis (DGGE/TGGE), 33, 43
Denaturing High Performance Liquid Chromatography, 33
Dendritic cells, 89, 396, 399, 554
Dental caries, 270
Deoxycholic acid, 485
Department of Livestock Development, Thailand, 125
Desserts, 72
Designer lactic acid bacteria, 415

SUBJECT INDEX 589

Diabetes, 76, 84, 88, 259, 455, 457, 484, 550, 559
Diarrhea, 52, 53, 54, 90, 258, 281, 466, 481
 antibiotic-associated, 54, 55, 181, 258, 276
 Colibacillosis, 406, 407, 414
 Clostridium difficile-associated, 279
 infants at risk, 402, 434, 552
 radiation-induced, 279, 280, 361
 Rotavirus, 53, 258, 272, 385, 398
 tube-fed patients, 281, 362
Dietary Supplement Health and Education Act, USA (DSHEA), 115
Digestibility enhancer, 125
Digestive juices, tolerance, 180–181
Displacement, 321, 389, 391
Dysentery, 477

Ear infections, 477
Effective dosage, 52, 53, 55
 acute (Rotavirus) diarrhea, 53, 272
 antibiotic-associated diarrhea, 54. 55, 57, 82 181, 258, 276, 385
 Helicobacter pyroli, 58
Egg quality, 323
Elderly, 23, 32, 34, 36, 90, 91, 152, 492, 493
 normal, 60, 77, 82, 86, 87, 89, 103, 556
 with constipation, 284, 404
 with diarrhea, 279, 404, 414, 467
ELISA, 13, 124
Encapsulation, 65, 68, 69
Endogenous microbiota, 382
Enteric Microbial Community Profiling (EMCoP), 43, 47, 51
Enterobacterial Repetitive Intergenic Consensus Sequence-PCR (ERIC-PCR), 13
Enterobacter sakazakii, 24
Enterococcus cremoris, 127
Enterococcus diacetylactis, 127
Enterococcus durans, 221
 vector, 212, 214, 215, 216
Enterococcus faecalis, 18, 75
 homologous recombination, 198, 199, 222, 224
 plasmid insertion, 226
 transposable element insertion, 226
 vector, 212, 214, 215, 216
Enterococcus faecium, 80, 322, 350
 M-74, 314

mutangenesis, 190–201
plasmid, 38, 83, 206–211
SF68, 277, 329, 337, 347
vector, 212, 214, 215, 216
Enterococcus mundtii, 129
Enteropathogen, 23, 391, 407, 409
Enteropathogenic *Escherichia coli* (EPEC), 407, 409, 484
Epithelial cells, 451, 458, 463, 469, 535
Epithelia model, 259–267
Escherichia coli, 24, 33, 36, 48, 76, 389, 406
 extraintestinal pathogenic, 409
 H10407, 262
 JPN15, 262
 O6:K5:H1, 289
 serogroup, 407, 411
Eubiosis, 321
European Food Safety Authority (EFSA), 5, 125
European Parlement, 106, 123
European Union (EU), 91, 123, 135, 449
Exopolysaccharides ERIC-PCR, 390, 447
Experimental autoimmune encephalomyelitis (EAE), 88
Extracellular matrix (ECM), 457

Fecal microbiota, 31, 33, 34, 35, 36, 406
Feed, 14, 15, 50
 additives, 114, 123
 conversion ratio, 321, 408
Fermented foods, 14, 63, 69 78, 83
Fermented milk, 38, 58, 60, 64, 69, 92, 99, 263, 269, 450
Fever, 82, 83, 443, 449
Fibrinogen, 81
Fibronectin, 228, 451, 464
Fimbriae, 407, 411, 418, 419
Fimbrial adhesions, 407, 409
Fluorescence *in situ* hybridization (FISH), 30, 35
Food and Agricultural Materials Inspection Centre (FAMIC), Japan, 125
Food and Drug Administration (FDA), USA, 97, 101, 103, 111, 118, 135
Food Safety Authority of Ireland, 110
Food Standards Agency (UK), 110
Freeze-dried probiotics, 65, 66
Fructose-6-phosphate phosphoketolase (F6PPK) 9, 201

Fructooligosaccharides (FOS) 108, 405, 449
Fusobacterium prausnitzii, 31, 403

Galactooligosaccharides (GOS), 296, 307, 320
Gastrointestinal, 296, 307, 320
   diseases, 4, 7, 31, 52, 76, 82, 88
   environnent, 447
   model, 5, 23, 47, 89, 93, 259, 260, 261, 263, 264
   survival during, 59, 64, 179
   symptom, 83, 88, 90, 94, 96, 257, 258, 259, 266
   transit, 4, 9, 19, 60, 403
Gene expression, in gut, 470
Gene signatures, 51, 408, 411, 412, 413
Generally recognized as safe (GRAS), 112, 135, 485
Genome, 13, 18, 19, 34, 356, 42
   plasticity, 410
Genotoxicity, 124
Generally Recognized as Safe (GRAS), 112, 135, 485
Glucose metabolism, 394
Gut, 23, 30, 31, 33, 35, 36, 77, 87, 123, 259
   associated lymphoid tissue, 23, 395, 404, 443
   flora stabilizers, 125
   immune system, 7, 23, 24, 75, 87, 90, 94
   microbiota, 4, 6, 25, 27, 28, 29, 30, 31, 32, 33, 34, 35
   permeability, 27, 35, 260
   transit, 4, 9, 19, 60, 63
Health benefits, 59, 60, 69, 75, 76
Healthy microbiota, 399, 404, 405
Heavy metals, binding of, 493
Helicobacter pylori, 7, 76, 258, 292
Histomonostats, 125
Host immunity, 409
HOWARU, 473
   Rhamnosus, 18, 54, 62, 84, 92
   Bifido, 4, 5, 6, 7, 29, 31, 32, 33, 35, 63
HT29, 260, 261, 262
HT-29-MTX, 378
HTB-37, 23, 260
Human T-cell lymphotropic virus type-1 (HTLV-1), 457
Hybridization probes, 31
Hydrogen peroxide, 231, 392, 398

Hydrophobic grid membrane filtration (HGMF), 412, 413
Hydrophobicity, 386, 388

Ice cream, 60, 67, 72
Identification, 4, 5, 6, 9, 13, 14
   Bifidobacterium, 5, 6, 7, 8, 9, 13, 31, 492
   Lactobacillus, 4, 5, 14, 15, 16
Infectious diseases, 454, 483
Immunoglobulin, 259, 449, 490
   A (IgA), 468, 483
   E (IgE), 551
   G (IgG), 319–320, 458, 468
   M (IgM), 319, 468
Immunological memory, 454
Immunomodulation, 15, 59, 89, 552
Indigenous lactobacilli, 451
Indonesia, 104
Infant formula, 60, 90, 114, 485
Inflammation, 287, 292, 298, 554
Inflammatory bowel disease (IBD), 287, 535
Inflammatory diseases, 442, 481
Influenza virus (IFV), 490
Inhibition, 15, 24, 339
Innate immunity, 434
Insulin-dependent diabetes mellitus (IDDM), 452, 455, 457
Interferon-gamma (INF-γ), 468
Interleukin, 396, 443
   4 (IL-4), 396
   6 (IL-6), 443
   8 (IL-8), 508
   12 (IL-12), 396
Intestinal, 3, 4, 6, 23, 25, 33, 42
   bacterial enzyme activity, 450
   barrier, 19, 65, 66
   epithelial cell, 23, 56, 377
   metabolic activity, 42, 258, 383, 474
   microbiota, 4, 6, 29
   mucus, 19, 23, 24, 81, 378
Intergenic spacer (IGS), 12, 44
Internal transcribed spacer (ITS), 44
Irritable bowel syndrome (IBS), 258, 448
Isolation, 6, 25, 43, 82
   bifidobacteria, 5, 6, 9, 13, 33, 36
   enterococci, 25, 52, 84, 403
   lactobacilli, 5, 6, 14, 18, 64, 66
   streptococci, 270, 395
   yeast, 4, 52, 95, 394, 407

Japan, 96, 99, 105, 125, 136, 488, 560
Joint Health Claims Initiative (UK), 106
Juice, 19, 73, 180

Korea, 102–103, 125, 136
Kinetic Inhibition Microtiter Assay (KIMA), 416
Kluyveromyces marxianus, 129
Kluyveromyces marxianus-fragilis, 129
Knock-outs, 265

Labeling, 8, 94, 98, 109, 116, 118
Lactina, 322, 329
Lactic acid, 14, 79, 91, 191, 195, 392
Lactitol, 448, 449, 550, 557
Lactobacillus, 4, 7, 10, 20–22, 33, 61, 81–92, 339–340, 386, 397–398, 547
   characterization, 4–5
   identification, 4–6, 10–11, 18, 444, 462–463
   mutagenesis, 190–200
   plasmid, 203–206
   typings, 5, 8, 10–11, 18
Lactobacillus acidophilus, 62, 66, 80, 101, 129, 137, 203, 293, 441, 444, 447 145, 314
   antibiotic-associate diarrhea, 276
   CH 5, 86, 180–181
   DDS Plus, 316
   DNA fragment replacement, 224
   homologous recombination, 198–199, 222, 224
   L-1, 316
   LA-1, 316
   LAVRI-A1, 309
   La5, 222, 298, 343
   LB, 293, 383
   NCDO 1748, 444, 446
   NCFM, 447–449
   NCIMB 8690, 180–181
   plasmid, 38, 83, 202–210
   SBT2062, 268
   vector, 212–221
Lactobacillus agilis, 381
Lactobacillus amyloliticus, 129
Lactobacillus amylovorans, 129
Lactobacillus amylovorus, 142, 201
   mutagenesis, 190–200
Lactobacillus bifidus, 137

Lactobacillus brevis, 124, 325, 381
   DNA fragment replacement, 224
   homologous recombination, 198–199, 222, 224
   plasmid, 203
   vector, 212–221
Lactobacillus bucheri, 124, 129, 137
Lactobacillus bulgaricus, 54, 101, 113
Lactobacillus casei, 82, 101, 203, 449–456
   DN-114001, 301
   GG, 390–391
   homologous recombination, 198–199, 222–224
   NCIMB 8822, 179–181
   NCIMB 11970, 179–181
   plasmid, 203
   recombination, 222–228
   Shirota, 41, 88, 89, 179–181, 283, 296, 311, 345, 449
   subsp. *alactosus*, 221
   transposable element, 255
   vector, 212–221
Lactobacillus cellobiosus, 129, 137
Lactobacillus collinoides, 129
Lactobacillus crispatus, 50, 352, 380
   vector, 212–221
Lactobacillus curvatus, 129, 137
   vectors, 212–221
Lactobacillus delbrueckii, 101, 203, 442
   100-18, 268
   100-21, 268
   mutagenesis, 190–200
   NCIMB 11778, 181
   plasmid, 203
   subsp. *bulgaricus* (also *Lactobacillus bulgaricus*), 39, 70, 101, 181, 203, 345
   subsp. *lactis*, 221
   vectors, 212–221
Lactobacillus farciminis, 130
Lactobacillus fareiminis, 124, 130
Lactobacillus fermentum, 204, 221
   100-20, 268
   104-R, 381, 383
   DNA fragment replacement, 224
   KLD, 280
   homologous recombination, 198–199, 222–224
   PCC, 306, 317
   plasmid, 204

Lactobacillus fermentum (Continued)
   recombination, 222–228
   vector, 212–221
   VRI-033 PCC, 306
Lactobacillus fructosus, 221
   Vector, 212–221
Lactobacillus gasseri, 221, 457–459
   DNA fragment replacement, 224
   F71, 263
   homologous recombination, 198–199, 222–224
   L1, 206
   LG21, 457–462
   mutagenesis, 190–200
   PA16/8, 301
   plasmid, 202–211
   vector, 212–221
Lactobacillus helveticus, 204, 221
   DNA fragment replacement, 224
   homologous recombination, 198–199, 222–224
   plasmid, 204
   transposable element insertion, 226
   vector, 212–221
Lactobacillus hilgardii, 221
   plasmid, 204
Lactobacillus jensenii, 221
   vector, 212–221
Lactobacillus johnsonii, 221
   DNA fragment replacement, 224
   homologous recombination, 198–199, 222–224
   La1, 41, 381
   plasmid, 204
   vector, 212–221
Lactobacillus lactis, 114, 201
   Bb-12, 480–485
   DNA fragment replacement, 224
   homologous recombination, 198–199, 222–224
   mutagenesis, 190–200
   plasmid, 206
   subsp. *cremoris*, 207
   subsp. *lactis*, 200
   subsp. *lactis* biovar *diacetylactis*, 132, 208
   vector, 212–221
Lactobacillus mucosa, 49
Lactobacillus murinus,
Lactobacillus paracasei, 67–68, 462, 466

   BA3, 263
   F76, 263
   NCIMB 8001, 180–181
   NCIMB 9709, 180–181
   NCIMB 9713, 180–181
   plasmid, 204
   subsp. *paracasei* F19, 462–465
   vector, 212–221
Lactobacillus paraplantarum, 221
   plasmid, 205
   vector, 212–221
Lactobacillus pentosus, 221
Lactobacillus plantarum, 205
   299v, 41, 283, 298, 384
   DNA fragment replacement, 224
   homologous recombination, 198–199, 222, 224
   mutagenesis, 190–200
   plasmid, 205
   recombination, 222–228
   transposable element, 226
   vector, 212–221
Lactobacillus reuteri, 49, 350, 394, 399, 470
   100-23, 206, 268, 381
   DSM12246, 305
   plasmid, 206
   RC-14, 470–471
   vector, 212–221
Lactobacillus rhamnosus, 81, 101, 397, 469–475
   19070-2, 305
   GR-1, 470–471
   antibiotic-associated diarrhea, 54, 82, 276
   GG, 60, 397, 469
   HN001 (DR20), 473–475
   LC705, 293–294, 390, 478, 480
   mutagenesis, 190–200
   NCIMB 6375, 180–181
   plasmid, 190–200
   vector, 212–221
Lactobacillus sake, 223–226
   homologous recombination, 198–199, 222, 224
   plasmid, 190–200
   transposable element insertion, 226
Lactobacillus sakei, 206
   DNA fragment replacement, 224

homologous recombination, 198–199, 222–224
mutagenesis, 190–200
plasmid, 206
vector, 212–221
Lactobacillus salivarius, 33, 206
  subsp. *salivarius*, 206
  plasmid, 206
  vector, 212–221
Lactobacillus sanfranciscensis, 221
  vectors, 212–221
Lactose, 194, 211, 222–228, 258, 268–270, 344, 443, 446–448
  Intolerance, 443, 448
  Maldigestion, 268–269
Lantibiotics, 393, 415
laryngotracheo-esophageal cleft (LTEC), 457, 487
Legal Status, in
  Asia, 95–105
  China, 100, 125, 136–138
  Japan, 488, 542–544
  Korean, 103Malaysia, 103, 139
  Taiwan, 103, 104
  Thailand, 104, 125, 136–138
Legislation, 106, 108, 110
Leuconostoc citreum, 209
  plasmids, 209
  vectors, 212–221
Leuconostoc mesenteroides, 124, 132, 137, 209
  mutagenesis, 190–200
  plasmid, 209
  subsp. *mesenteroides*, 209
  vector, 212–221
Leuconostoc oeno, 132
Leuconostoc pseudo-mesenteroides, 132, 137
Lignans, 382
Lipoteichoic acids (LTAs), 381, 457
Listeria monocytogenes, 394, 454
  LS174T, 260

Macrophage, 182, 443, 483, 490
Maintenance, 75, 177–186, 405
  bifidobacteria, 5, 9, 13, 14
  enterococci, 177–178
  lactobacilli, 177–178
  streptococci, 177–178
  yeast, 177–178

Malaysia, 103, 139
Mammary carcinogenesis, 486
Manufacturing criteria 179
Market size, 96
Maternal fecal microbiota, 406
Meat, 62, 179, 321, 446
Metabolism, 18, 64, 67, 72, 261, 266, 331, 381, 549
  carbohydrates, 64, 331
  lactose, 268, 448
  prebiotics, 405, 536, 550,
  metagenomic, 36
Metaproteomic, 30
Methicillin-resistant *Staphylococcus aureus* (MRSA), 430, 457
Microarray, 12, 14
Microbial, 4, 43, 47, 55, 265, 395, 409, 466
  analysis, 9, 14, 16, 33, 48
  diversity, 6, 47
Microbiota, 42, 258, 331, 383, 384, 395, 399, 400, 402, 403, 405, 492, 552, 562
  assessment, 25–33, 37
  elderly, 34, 36, 404, 492, 493, 549
  infancy, 87, 401–405, 546, 547
Microcystin-LR, 493
Microecology, 310
Microencapsulation, 68, 69
Minerals, 67, 107, 124
  effect on adhesion, 382
  improvement of nutritional status, 107, 124
Ministry of Agriculture, People's Republic of China, 176
Ministry of Health (Belgium, The Netherlands), 110
Monostrain, 390
Mosaicism, 410
MUC5AC mucin gene, 261
Mucosa, 23
Mucus, 377, 378, 385
  glycoprotein, 450–455
Multilocus enzyme electrophoresis (MLEE), 409
Multilocus sequence syping (MLST), 13
Multiple strains, 10, 75
Multispecies, 88, 91, 477
Mutagen binding, 310, 445, 446
Mutagenicity, 124

Mycotoxins, 382
 binding of, 382
 effect on adhesion, 382

Natural killer (NK)-cell, 474–476
Necrosis, 91
New Dietary Ingredient (NDI) USA, 114
New Zealand, 69, 135
New Zealand Food authority, 69
Nisin, 393
Nitroreductase, 448
Nitrosodimethylamine, 448
Non-insulin-dependent diabetes mellitus (NIDDM), 457
Nonfermented foods, 69
Nonlantibiotics, 393, 418
 heat-stable, 393, 418
 heat-labile, 393, 418
Nonobese diabetic (NOD), 457
Nonviable probiotics, 380, 383

Oligonucleotide, 34
Operational Taxonomic Units, 16, 49
(μ-)Opioid receptors, 448
Oral infection, 270
Organic acid, 64
Oropharyngeal biofilms, 456
Overage, 59

Pediocuceus, 132, 138
 plasmid, 132–138
 vector, 132–138
Pediococcus cerevisiae, 124, 133
Pediococcus damnosu, 208–211
 Plasmid, 208–211
Pediococcus parvulus, 209
 plasmid, 209
Pediococcus pentosaceus, 209
 plasmid, 209
 vector, 209
Peripheral blood mononuclear cells (PBMNC), 308
Pets, 339
PCR, 13, 30, 31, 35, 38
PCR-DGGE, 33
PCR-TGGE, 32
PFGE, 14
Phagocytic activity, 468
Phenotyping, 10

Philippines, 104, 139
Pigs, 47, 323, 407
Phylogenetic diversity, 44
Phylogenetic typing, 18, 44, 410, 415
Phylogeny, 409
Polydextrose, 108
Polyunsaturated fatty acids (PUFA), 457
Population virulence gene signatures, 412, 414, 417, 418
Postoperative infection, 299
Pouchitis, 287
Prebiotics, 3–123, 536–560
Principal Coordinate Analysis (PCO), 411
Probiosis, 419
Probiotics, 3–123, 536–560
 antagonistic properties, 19–24
 bioefficacy, 417
 definition, 408
 mechanism of action, 408, 409
 persistence, 469, 561
 product classification, 111
 tracking of, 16
Proprionibacterium freudenreichii, 285
 subsp. *shemanii* JS, 285
Proprionibacterium globosum, 133, 138
Proprionibacterium shermanii, 133, 138, 285, 306, 307, 320
Protozoa, 331
Pulsed Field Gel Electrophoresis (PFGE), 13, 29

qPCR, 402
Qualified Presumption of Safety, 5, 492

Rabbits, 339, 345, 346
Radical scavenging, 445, 450
Random Amplification of Polymorphic DNA (RAPD), 13
Real-Time PCR, 35, 36
Registration process, 97, 98
Regulations, 79, 94, 95, 99, 100, 103, 112, 119, 182
REP-PCR, 11–13
Resected colonic tissue, 377, 384
Respiratory tract infections, 299, 449
Reuterin, 394, 471
Reverse transcriptase quantitative PCR, 36
Rhodospeudomonas palustris, 133, 138
Ribosomal database, 17, 44

rRNA-stable isotope probing, 37
Ribotyping, 11, 14, 18, 402
Rotavirus, 53–54, 267, 272–274, 319, 384, 398, 399, 482,
  digestion, 268, 443
  microbiota, 258, 470
  rumen, 331
Ruminal acidosis, 331
Ruminants, 331, 339, 559

Saccharomyces boulardii, 54, 57, 275, 278–279, 281, 293
Saccharomyces cerevisiae, 101
Safety, 4–5, 18, 75–83, 85–98, 103–105, 110–111, 113–115, 119–120, 122–125, 182–183, 444–445, 456, 462–463, 468, 472, 474, 485, 488, 492, 560
Salmonella enterica serovar *typhimurium*, 454, 486
  SL1344, 262
SAMP1/Yit SAMP1/Yit, 487
Screening, probiotics, 4–20, 60
Selection, probiotics, 4–20, 178–184, 399, 458
Serum cholesterol, 313, 324–325, 329, 334, 455, 491
Serum pepsinogen, 459, 460
Shelf life, 63, 74, 98, 104, 179, 183
Shiga toxin-producing *Escherichia coli* (STEC), 407, 457, 488
Shiga toxins (Stxs), 407, 409, 457

Shigella, 281, 385, 442, 481
Short bowel syndrome (SBS), 487
Short chain fatty acid (SCFA), 313, 457, 487, 537
Single stranded site conformational polymorphism (SSCP), 43
Source of microbiota, 400–401
Soybean. 102, 324, 539, 540, 544
Spray-dried probiotics, 60
SSU rRNA, 402
Standing Committee on the Food and Animal Health, 125
Streptococcus, 221
  vector, 212–221
Staphylococcus aureus, 76, 221, 394, 402, 471, 493
Streptococcus gordonii, 201, 221

mutagenesis, 190–200
  vector, 212–221
Streptococcus intermedius, 134, 138
Streptococcus mutans, 221
  vector, 212–221
Streptococcus pneumoniae, 322–321
Streptococcus salivarus, 138, 209
  antibiotic-associated diarrhea, 54–58, 276
  DNA fragment replacement, 224–228
  homologous recombination, 198–199, 222, 224
  mutagenesis, 190–200
  NCIMB 10387, 180–181
  plasmid, 209–210
  subsp. *thermophilus* (also *S. thermophilus*), 138, 180vectors, 212–221
Streptococcus thermophilus, 54–55, 101, 210, 383
Stress proteins, 63
Survival, probiotics, 6, 19–22, 58–75, 336, 445–452, 463–466, 474, 478, 480
  effect of food ingredients,
  effect of freeze-thawing,
  effect of oxygen,
  effect of pH,
  effect of sheer,
  effect of temperature,
  effect of water activity,
  inter-species differences,
  strain-to-strain differences,
SW-116, 260
SW-480, 260
Swedish Nutrition Foundation, 106, 110, 479

Swine, 323, 408, 410–414

Synbiotics, 67, 322, 339, 455, 487, 536, 557–558,

T helper cells, 88, 397–398, 551
  (Th1), 88–89, 396–399
  (Th2), 89, 395–399, 551
T regulatory cells, 396–398
TAP-PCR, 12, 18
Taxonomy, probiotics, 7–8
Terminal Restriction Fragment Length Polymorphism (T-RFLP), 27, 30, 33–34, 43–48, 51–52

Temperature Gradient Gel Electrophoresis (TGGE), 11–13, 27, 32–33, 42–43
Terminal Labeled Restriction Fragments (TRF), 43–51
Terminal Restriction Fragment Length Polymorphism (TRFLP)
Tetragenococcus halophila, 211
  plasmid, 211
Thailand, 104, 125, 136, 138, 302
Thymic lymphoma, 453
Toxins, 124, 407, 418, 468, 492, 494
Transepithelial Electrical Resistance (TER), 261, 406
Transgalactosylated oligosaccharides (TOS), 486, 488, 542
Transgenic animal, 265
Translocation, 82, 298, 463, 555
Traveler's diarrhea, 280–281, 385, 442, 470, 481, 557
Triglyceride, plasma, 314–317, 455
Tumor necrosis factor-alpha (TGF-α), 443

Ulcerative colitis, 288–291, 470
United States, 95, 111–124, 408, 415
Urease, 192–194

Urinary tract infection (UTI), 457, 472
Urogenital infection, 258

Vaccination, 419, 468, 552
Vaginal infection, 98
Vellionella, 403
Very low birth weight (VLBW), 488
Viability, probiotics, 38, 52, 59, 60–67, 69, 71–74, 98, 101, 179
Vietnam, 104, 139
Virulence, 83, 84, 407, 411, 412, 414, 418–420, 444
  factors, 81, 24, 410
  genes, 83–84, 410, 411, 412, 418–420
Voedings centrum, 106

Wheezing infants, 402

X-ray-irradiated splenocytes (X-irr-Spl), 457
Yersinia pseudotuberculosis, 262
  YPIII pYV, 262
Yogurt, 64, 70, 80

Zootechnical additives, 123, 125